PROGRESS IN FILTRATION AND SEPARATION

PROGRESS IN FILTRATION AND SEPARATION

Edited by

STEVE TARLETON
Department of Chemical Engineering
Loughborough University
Loughborough, UK

ELSEVIER

Amsterdam • Boston • Heidelberg • London
New York • Oxford • Paris • San Diego
San Francisco • Singapore • Sydney • Tokyo

Academic Press is an imprint of Elsevier

Academic Press is an imprint of Elsevier
32 Jamestown Road, London NW1 7BY, UK
525 B Street, Suite 1800, San Diego, CA 92101-4495, USA
225 Wyman Street, Waltham, MA 02451, USA
The Boulevard, Langford Lane, Kidlington, Oxford OX5 1GB, UK

ISBN: 978-0-12-384746-1

British Library Cataloguing-in-Publication Data
A catalogue record for this book is available from the British Library

Library of Congress Cataloging-in-Publication Data
A catalog record for this book is available from the Library of Congress

For information on all Academic Press publications
visit our website at http://store.elsevier.com/

CONTENTS

Section 2 Force field assisted separators

LIST OF CONTRIBUTORS

Chris Aldrich
Department of Mining Engineering and Metallurgical Engineering, Western Australian School of Mines, Curtin University of Technology, Perth, WA, Australia

John Bratby
Brown and Caldwell, Golden, CO, USA

Luzheng Chen
State Key Laboratory of Complex Nonferrous Metal Resources Clean Utilization, Kunming University of Science and Technology, Kunming, Yunnan, China

Morgane Citeau
CREOL, Pessac, France

Claudia Cobzaru
Technical University "Gh Asachi" Isai, Faculty of Chemical Engineering and Environmental Protection, Romania

Bart Van der Bruggen
Department of Chemical Engineering, Laboratory of Applied Physical Chemistry and Environmental Technology, KU Leuven, Leuven, Belgium

Luhui Ding
Technological University of Compiegne, Compiegne Cedex, France

Gjergj Dodbiba
Department of Systems Innovation, The University of Tokyo, Tokyo, Japan

John Dodds
Professeur honoraire, Ecole des Mines d'Albi, France

B. Ekberg
Outotec Oyj, Finland, Turku, Finland

Mohamed Essalhi
Department of Applied Physics I, Faculty of Physics, University Complutense of Madrid, Madrid, Spain

Toyohisa Fujita
Department of Systems Innovation, The University of Tokyo, Tokyo, Japan

Aloke Kumar Ghoshal
Department of Chemical Engineering, Indian Institute of Technology Guwahati, Guwahati, India

N.F. Gray
Water Technology Research Group, Trinity Centre for the Environment, Trinity College, University of Dublin, Dublin, Ireland

Thomas Harlacher
Aachener Verfahrenstechnik, Chemical Process Engineering, RWTH Aachen University, Aachen, Germany

Steffen Heidenreich
Pall Filtersystems GmbH, Werk Schumacher Crailsheim, Crailsheim, Germany

Vassilis Inglezakis
Chemical Engineer (MSc, PhD) Environmental Consulting & Project Management Romania

Michel Y. Jaffrin
Technological University of Compiegne, Compiegne Cedex, France

Mohamed Khayet
Department of Applied Physics I, Faculty of Physics, University Complutense of Madrid, Madrid, Spain; Madrid Institute for Advanced Studies of Water (IMDEA Water Institute), Alcalá de Henares, Madrid, Spain

Olivier Larue
Solvay, Center for Research & Technology of Shanghai, Xin Zhuang Industrial Zone, Shanghai, China

Patricia Luis
Department of Chemical Engineering, Laboratory of Applied Physical Chemistry and Environmental Technology, KU Leuven, Leuven, Belgium

Jianquan Luo
Technological University of Compiegne, Compiegne Cedex, France

Pentti Pirkonen
Technical Research Centre of Finland (VTT), Jyväskylä, Finland

Majid Naderi
Surface Measurement Systems, Ltd., Alperton, London, UK

Marie-Noëlle Pons
Laboratoire Réactions et Génie des Procédés, Université de Lorraine, Nancy cedex, France

Prabirkumar Saha
Department of Chemical Engineering, Indian Institute of Technology Guwahati, Guwahati, India

Yoshinobu Tanaka
IEM Research, Ushiku-Shi, Ibaraki, Japan

Eugene Vorobiev
Département de Génie des Procédés Industriels, Laboratoire Transformations Intégrées de la Matière Renouvelable, Université de Technologie de Compiègne, Centre de Recherches de Royallieu, Compiègne, France

Matthias Wessling
Aachener Verfahrenstechnik, Chemical Process Engineering, RWTH Aachen University, Aachen, Germany

Dahe Xiong
SLon Magnetic Separator Co., Ltd., Ganzhou, Jiangxi, China

CHAPTER ONE

Hydrocyclones

Chris Aldrich
Department of Mining Engineering and Metallurgical Engineering, Western Australian School of Mines, Curtin University of Technology, Perth, WA, Australia

Contents

Progress in Filtration and Separation
http://dx.doi.org/10.1016/B978-0-12-384746-1.00001-X

GLOSSARY

Cut Size diameter of particle with equal probability to report to the overflow and underflow of the cyclone; often a measure of the performance of the classifier.

Fish hook effect the phenomenon where partition curves in hydrocyclones do not have a sigmoidal shape, but a shape resembling that of a fish hook.

Partition curve a graphical representation of the recovery of each particle size in the hydrocyclone underflow in relation to its availability in the feed; also known as a Tromp curve.

Spigot the outlet part or apex at the lower conical end of a hydrocyclone.

Vortex finder a short, removable top mounted overflow pipe in a hydrocyclone that extends a short distance into the cylindrical body part of the cyclone to prevent short-circuiting of feed directly into the overflow.

NOMENCLATURE

Symbol	Meaning	SI Units
α	Coefficient representing sharpness of separation	—
θ	Cone angle of hydrocyclone	°
μ	Viscosity of liquid	Pa s
ρ_f	Density of fluid	kg/m^3
ρ_s	Density of solids	kg/m^3
A_0	Cross-sectional area of particle	m^2
C_D	Drag coefficient	—
C_v	Volumetric concentration of solids in the feed	—
D_c	Diameter of hydrocyclone cylinder	m
D_i	Diameter of inlet	m
D_o	Diameter of overflow	m
D_u	Diameter of underflow	m
d	Particle diameter	m
d_{25}	Diameter of particle with 25% probability to report to the underflow of the cyclone	m
d_{50}	Cut size; diameter of particle with equal probability to report to the overflow and underflow of the cyclone	m
$d_{50(c)}$	Corrected cut size	
d_{75}	Diameter of particle with 75% probability to report to the underflow of the cyclone	m
Eu	Euler number, $Eu = \frac{2\Delta P}{\rho v^2}$	—
F_C	Centrifugal force	N
F_D	Drag force	N
F_i	Material parameter, $i = 1, 2, 3$	—
h	Distance between vortex finder and apex of cyclone	m
I	Inefficiency of separation	—

—cont'd

Symbol	Meaning	SI Units
k	Constant	
L	Length of cyclone	m
l	Length of vortex finder	m
P	Pressure	Pa
Q_f	Throughput	m^3/s
R	Fraction of liquid in the feed recovered in the underflow stream	—
Re	Reynolds number, $Re = \frac{\rho v d}{\mu}$	—
r	Radial position of particle inside hydrocyclone	m
Stk_{50}	Stokes number, $Stk_{50} = \frac{(\rho_s - \rho) v d_{50}^2}{18 \mu D_c}$	—
u_r	Radial velocity of particle in hydrocyclone	m/s
v_r	Radial velocity of fluid in hydrocyclone	m/s
v_θ	Tangential velocity of fluid in hydrocyclone	m/s
V_p	Volume of particle	m^3
v	Velocity	m/s
y	Uncorrected efficiency	—
y'	Reduced efficiency	—

1. BACKGROUND

Hydrocyclones have been in use in industry since the 1940s, although the first patent can be traced back to the nineteenth century (Bretnai, 1891). Owing to their simple design, low cost, easy operation, and low maintenance, they have assumed an important role in the separation of solids and liquids. Although hydrocyclones are widely used at present, such as in closed circuit grinding (Casali et al., 1998), desliming (Yalamanchili and Miller, 1995), liquid clarification (Puprasert et al., 2004), degritting (Murray, 1980), and thickening operations (Woodfield and Bickert, 2004; Yang et al., 2004), the phenomena leading to separation are still not fully understood yet.

In a hydrocyclone, a slurry enters through a tangential inlet, giving rise to a vortex in the stationary body. The particles and fluid are accelerated centrifugally and separation occurs in the radial direction. Denser materials migrate to the outer wall of the hydrocyclone chamber, while less dense materials move toward the inner axis. Flow in the cyclone cylinder is characterized by two vortices that flow in opposite directions. Denser flow tends to travel along the primary vortex to the underflow and less dense material travels along the secondary vortex in the opposite direction to the overflow, as indicated in Figure 1.

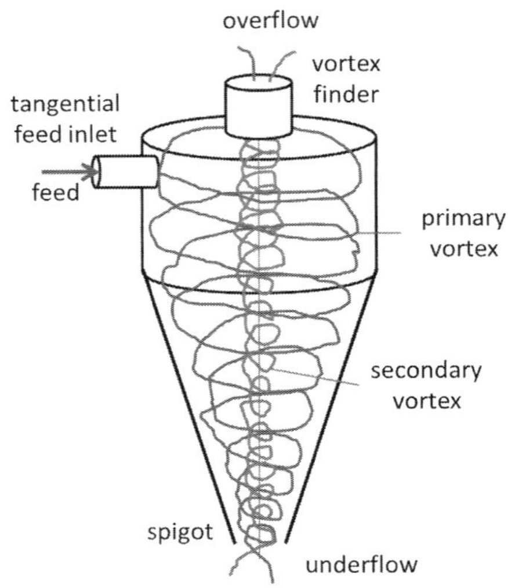

Figure 1 Features of a hydrocyclone in operation.

According to classical theory, particles in the cyclone are subject to two opposing forces, i.e., an outwardly acting centrifugal force and an inwardly acting drag force, as indicated in Figure 2. Particles are separated by the accelerating centrifugal force based on size, shape, and density, while the drag force moves slower settling particles to the low pressure zone along the axis of the cyclone, where they are carried upward through the vortex finder to the overflow. Figure 2 also shows the equations representing the classical force balance acting on a spherical particle with diameter *d*, at a position *r* in the cyclone, i.e., at equilibrium, the centrifugal force in the cyclone is equal to the drag force and the buoyancy forces acting on the particle, or $F_C - F_B - F_D = 0$.

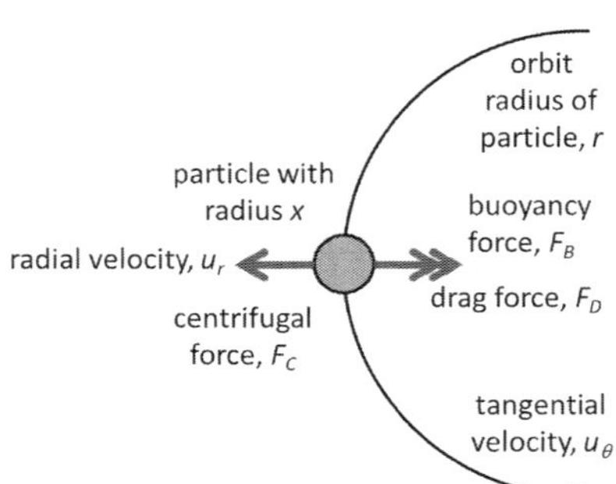

$$F_D = 3\pi d \mu u_r$$

$$F_C = \frac{\pi d^3}{6} \rho_s \frac{u_\theta^2}{r}$$

$$F_B = \frac{\pi d^3}{6} \rho_f \frac{u_\theta^2}{r}$$

$$F_C - F_B - F_D = 0$$

Figure 2 Force balance on a particle in a hydrocyclone.

2. BASIC DESIGN

2.1 Geometry

Various designs of hydrocyclones have been proposed in order to exploit the use of inertial and gravitational forces to separate particulate matter from fluids. The main parameter of a hydrocyclone is its diameter, i.e., the inside diameter of the cylindrical feed chamber. This diameter can range from 10 mm to 2.5 m in commercial hydrocyclones, capable of separating particles of sizes ranging from 1.5 to 300 μm (and densities of approximately 2700 kg/m^3). This is followed by the area of the inlet nozzle at the point of entry into the feed chamber. The inlet nozzle is usually a rectangular orifice, with the larger dimension parallel to the axis of the cyclone. Typically, the area of the inlet nozzle is approximately 5% of that of the square of the diameter of the cyclone.

The primary function of the vortex finder is to control both the separation and the flow leaving the cyclone. The size of the vortex finder is approximately 35% of the diameter of the cyclone. The upper cylindrical section of the cyclone is located between the feed chamber and the conical section, and it has the same diameter as the feed chamber. It serves to extend the length of the cyclone in order to increase the retention time of particles. Its length is typically equal to the cyclone diameter.

The lower conical section further adds to the retention time of particles and has an included angle normally between 10° and 20°. The conical section terminates in the apex orifice and the critical dimension of the orifice is its inside diameter at the discharge point. It has to be of sufficient size to permit solids in the underflow to exit without plugging. The normal minimum orifice size is typically from 10% to 35% of the cyclone diameter. Finally, a splash skirt below the apex often helps to contain the underflow slurry.

2.2 Inlet Design

The feed inlet of the cyclone can be circular or rectangular, while the inlet opening can also vary in size. Different inlet configurations are shown in Figure 3. Modification of the inlet section of hydrocyclones is seen as a simple approach to control particle cut size and to improve the classification performance of hydrocyclones (Nenu and Yoshida, 2009). Modifications include the use of movable circular guide plates (Yoshida et al., 2006) or rotational top blades (Norimoto et al., 2004).

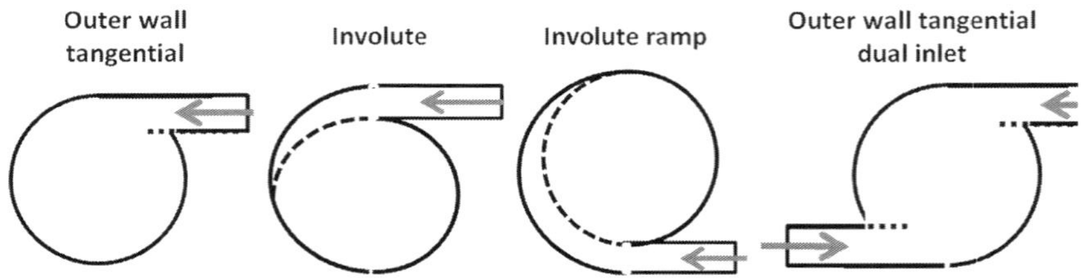

Figure 3 Feed inlet configurations in hydrocyclones.

Likewise, Nenu and Yoshida (2009) have compared one-inlet with two-inlet cyclones and have concluded that the particle collection efficiency of the two-inlet cyclone was better than that of the one-inlet cyclone with the same total flow rate, owing to enhancement of the tangential velocity profile of the particles in the former. The particle collection efficiency in the two-inlet cyclone was also found to be marginally better than that in the one-inlet cyclone under the same pressure drop. Similar results were observed by other authors for gas cyclones (Lim et al., 2003; Zhao et al., 2004) and hydrocyclones (Yoshida et al., 2006).

Wet size classification of particulate materials with a hydrocyclone with an axial inlet was investigated by Yalcin et al. (2003) as an alternative to tangential inlet cyclones traditionally used in the mineral processing industry. Experimental work was done with copper–nickel mill tailings having a particle size of 91% passing 300 μm at different inlet pressures, feed pulp densities, and vortex finder lengths. Higher throughputs, coarser cut sizes in relatively dilute pulps, and greater flexibility and control over the cyclone separation process were observed in comparison with tangential inlet cyclones.

2.3 Materials of Construction

The construction of hydrocyclones varies widely, depending on the application, but the majority of the designs include metal housings with replaceable liners. The most common liner material is natural gum rubber, owing to its comparatively low cost, high resistance to wear, and ease of handling. These liners are not suitable in conditions where temperatures may exceed approximately 60 °C or in aggressive chemical environments, such as where the slurry may contain oil or large amounts of other hydrocarbons. Under these circumstances, other elastomers, such as neoprene may perform better. In addition, urethane may be used in areas exposed to relatively fine solids.

In the presence of highly abrasive slurries, ceramic materials such as silicon carbide can be used for the apex orifice, as well as other areas which may require protection, such as the lower cone or vortex finder (Madge et al., 2004). Nihard, a nickel-based steel alloy with a martensitic microstructure, has also proven to be an acceptable wear material, especially for vortex finders and other areas which require strength as well as abrasion resistance.

3. CHARACTERIZATION OF PERFORMANCE

The practical range of particle sizes that can be classified by hydrocyclones is from 40 μm to 400 μm, with some specialized applications separating fines in the submicron range (Endres et al., 2012) or as coarse as 1000 μm. Operating pressures range from 50 kPa for large units to 1 MPa for smaller ones (Cilliers, 2000).

3.1 Partition Curves

The commonest approach to describe the efficiency of a cyclone is through its partition or performance curve in which the mass fraction of each particle size in the feed that reports to the underflow is related to the particle size itself, as indicated in Figure 3. The sharpness of the cut is indicated by the slope of the central portion of the graph, and the larger this slope (the closer to the vertical), the more efficient the cyclone.

Correction of the classification curve may be necessary to account for solids of all sizes that are entrained in the underflow in direct proportion to the fraction of feed water reporting to the underflow. The corrected partition curve (y') can be obtained from the uncorrected curve (y) by using Eqn (1). In this equation, y is the actual mass fraction of a given particle size reporting to the underflow, y' is the corrected mass fraction of a given particle size reporting to the underflow, and R is the fraction of liquid in the feed recovered in the underflow stream. This is indicated by the blue (uncorrected) curve and black (corrected) curve in Figure 4.

$$y' = \frac{y - R}{1 - R} \quad (1)$$

This efficiency or imperfection (I) can be expressed by using the points at which 75% and 25% of the feed particles report to the underflow (d_{75} and d_{25}, respectively) in addition to the cut size d_{50}, i.e., the particle size having an equal probability of reporting to the overflow or underflow, as follows:

$$I = \frac{d_{75} - d_{25}}{2d_{50}} \quad (2)$$

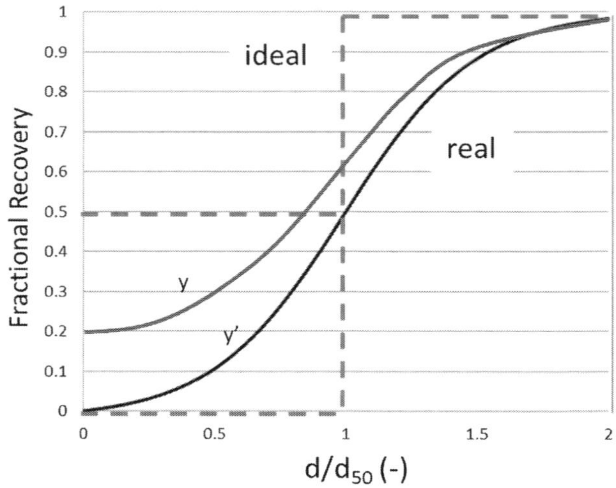

Figure 4 Typical reduced efficiency curve for a hydrocyclone showing recovery as a function of normalized particle size.

The so-called fish hook effect may occur in the partition curve when progressively higher partition numbers are observed for particle sizes finer than that at the minimum partition value (Kraipech et al., 2002; Neesse et al., 2004a).

The effect is more prevalent in smaller hydrocyclones and may be attributed to turbulent dispersion of the particles. Moreover, under these circumstances water recovery may be considerably lower than the lowest observed partition value and correction of such curves would be meaningless. Fish hook partition curves can be modelled by summation of a corrected partition curve (such as represented by Eqn (3)) and the product of an inverted partition curve and a bypass fraction (Cilliers, 2000).

$$y' = \frac{e^{\alpha X} - 1}{e^{\alpha X} + e^{\alpha} - 2} \tag{3}$$

3.2 Factors Affecting the Performance of Hydrocyclones

In general, the following trends or heuristics have been accumulated from experience. Classification efficiency depends on the cyclone size (diameter) and geometry, generally decreases with an increase in feed solids concentration and increase in viscosity. Efficiency is increased by limiting water to the underflow.

Table 1 Effect of cyclone design and operating variables on performance

Cyclone	D_i/D_c	D_o/D_c	L/D_c	l/D_c	θ
Bradley	0.133–0.143	0.200	6.850	0.333	9°
Krebs	0.267	0.159	5.874	-	12.7°
Rietema	0.280	0.340	5	0.400	15–20°

Table 2 Geometric proportions of hydrocyclones (Castilho and Medronho, 2000, Silva et al. (2012))

Factor	Throughput (Q)	Cut size (d_{50})	Sharpness of classification
Cyclone diameter	+	+	+
Feed inlet	+	−	−
Vortex finder diameter	+	+	+
Spigot diameter	+	−	−
Cone angle	N/A	+	+
Volumetric feed solids concentration	+	+	−
Pressure drop	+	−	+ Or -
Free vortex height	+	+	+

Asomah and Napier-Munn (1997) have found that angles of inclination exceeding 45° with the vertical axis of the hydrocyclone can play a significant role in its operation.

The effects of cyclone geometry (summarized in Table 2) and operational variables on the throughput, cut size, and sharpness of classification in hydrocyclones are summarized in Table 1, with "+" indicating an increase and "–" indicating a decrease in the performance criterion with an increase in the factor (Table 2).

4. HYDROCYCLONE MODELS

A large variety of hydrocyclone models have been proposed to estimate separation efficiencies of solid particles and pressure drops in these devices. These include empirical models encapsulating experimental data by use of fitted formulae or equations, as well as semiempirical models based on equilibrium orbit theory, residence time, and turbulent flow theory.

4.1 Fundamental and Empirical Models

These models are typically based on heuristics or some theoretical description of the phenomena in a hydrocyclone, as follows.

4.1.1 Equilibrium Orbit Theory

According to the equilibrium orbit theory (Bradley and Pulling, 1959), first proposed by Driessen (1951), a particle attains an equilibrium radial position in the cyclone, where its terminal settling velocity is equal to the radial velocity of the liquid (Coelho and Medronho, 2001). At this point, the centrifugal force on the particle is equal to the drag force in the opposite direction. If the liquid at that point flows inward towards the central axis of the hydrocyclone, the particles will move with the liquid to the overflow, and conversely, if the liquid flows outward, away from the central axis of the hydrocyclone, the particles will move to the wall and become separated through the underflow.

Conditions defining an equilibrium orbit can be defined as follows:

$$F_D = \frac{1}{2} C_D (v_r - u_r)^2 \rho_f A_0 \tag{4}$$

Here, F_D is the drag force experienced by the particle, C_D is the drag coefficient, v_r is the radial velocity of the fluid, u_r is the radial velocity of the particle, ρ_f is the density of the fluid, and A_0 is the cross-sectional area of the particle.

$$F_C = \frac{v_\theta^2}{r} V_p (\rho_s - \rho_f) \tag{5}$$

The centrifugal force F_C is represented by Eqn (5) as a function of the tangential component of the particle velocity vector, v_θ, the volume of the particle, V_p, and the difference between the fluid (ρ_f) and particle densities (ρ_s). When the drag force and the centrifugal forces are equal, the particle finds itself in an equilibrium orbit.

4.1.2 Crowding Theory

Crowding theory (Bloor et al., 1980; White, 1991) is based on the observation that at higher feed concentrations, separation size is chiefly determined by the discharge capacity of the spigot and the feed size distribution (e.g., Slechta and Firth, 1984). Theoretically, any cut size within the feed size distribution can be obtained by controlling the outlet dimensions of the hydrocyclone.

4.1.3 Residence Time Theory

Residence time theory models (Rietema, 1961; Dwari et al., 2004) are not concerned with particular flow phenomena, other than that a particle can be considered separated, if it can traverse the distance to the cyclone wall region within the residence time of the particle in the hydrocyclone.

4.1.4 Turbulent Two-Phase Flow

In turbulent two-phase models (Neesse et al., 2004a). separation arises owing to the turbulent cross flow moving perpendicular to the direction of the force field.

4.1.5 Empirical Modelling

Empirical models are not based on any specific theory of hydrocyclone operation and are derived by fitting models to experimental data. A variety of these were constructed in the previous century to predict the performance of industrial equipment, e.g., Shepherd and Lapple, (1939), Lapple, (1951), Leith and Licht (1972). More recently, models have been proposed by among other Nageswararo et al. (2004) and Kraipech et al. (2006).

One of the most widely used empirical models for hydrocyclones is that of Plitt (1976) and Lynch and Rao (1975). In the Plitt model, which has also been incorporated in a number of commercial steady simulation packages (Cilliers and Hinde, 1991), such as DYNAFRAG (Flament et al., 1993), the corrected cut size in micron is expressed as

$$d_{50(c)} = \frac{F_1 39.7 D_c^{0.46} D_i^{0.6} D_o^{1.21} \mu^{0.5} \exp(0.063 C_v)}{D_u^{0.71} h^{0.38} Q_f^{0.45} \left(\frac{\rho_s - 1}{1.6}\right)^k} \tag{6}$$

In this model, D_c, D_i, D_o, and D_u represent the inside diameters in cm of the hydrocyclone, inlet, vortex finder, and apex, respectively, C_v is the volumetric concentration of the solids in the feed, μ is the liquid viscosity (cP), ρ_s is the density of the solids (g/cm^3), F_1 is a material specific constant that needs to be determined from tests with the feed material, Q_f is the feed low rate (L/min), h is the distance between the vortex finder and the apex (cm), and k is a hydrodynamic exponent that needs to be estimated experimentally (default value of $k = 0.5$ for laminar flow conditions).

Likewise, the volumetric flow rate of the slurry to the hydrocyclone can be expressed by

$$Q_f = \frac{F_2 P^{0.56} D_c^{0.21} D_i^{0.53} h^{0.16} \left(D_u^2 + D_o^2\right)^{0.49}}{\exp(0.0031 C_v)} \tag{7}$$

Here, like F_1, F_2 is a material-specific constant that needs to be determined from tests with the feed material and P is the pressure drop across the cyclone (kPa). Similar equations for the efficiency of the hydrocyclone, as well as the water split between the underflow and overflow are reported by Plitt (1976).

Implicit models based on artificial neural networks (Stange, 1993; Van der Walt et al., 1993) and genofuzzy systems (Karr et al., 2000) have been proposed more recently with the growing use of machine-learning algorithms in process engineering. These models are theoretically more powerful than explicit equations, but are also more complex to use and have not found general acceptance in industry yet.

4.2 Numerical Models

With advances in computer hardware and software, numerical modelling of hydrocyclones is becoming more popular (Chen et al., 2000; Brennan, 2006; Delgadillo and Rajamani, 2007). Numerical methods are used to predict the multiphase flow fields in cyclones by solution of the basic equations governing flow, including the effects of turbulence and liquid–solid interactions. Numerical models can be categorized according to their characteristics, i.e., as two-dimensional or three-dimensional models, or as Eulerian (Dueck et al., 2000) or Lagrangian models. In the latter, the observer follows an individual fluid element moving through space and time, while in the former, the focus is one specific location in space through which fluid flows as time passes. In the Eulerian approach, or mixture model, the particle phases are treated as a pseudo-continuum and the transport equations for phase concentrations and phase momentum are solved. State-of-the-art approaches currently tend to focus on three-dimensional Lagrangian models (Dyakowski et al., 1999; Nowakowski et al., 2004; Schuetz et al., 2004).

Although the particle and fluid equations can be treated independently in dilute slurries, this is not the case as the particle size concentrations increase, since this affects the density of the mixtures as well as their effective viscosities (Pericleous and Rhodes, 1986).

The air core in the cyclone is treated similar to the particles. Density differences between the air and the liquid give rise to centrifugal forces acting on the air, which leads to a slip velocity of the particles, directed toward the axis of the cyclone. Over time, air accumulates along the cyclone axis to form the air core.

4.2.1 Modelling Turbulence

Fluid motion is usually represented by steady state, incompressible Navier–Stokes flow and continuity equations. Turbulence is represented by the effective pulp viscosity, which is variant under local flow conditions. These include simple mixing length models (Pericleous and Rhodes, 1986) or higher order turbulence models, such as the k-ε model, requiring equations to be solved for the kinetic energy (k), as well as the dissipation rate of the energy (ε). The standard k–ε model may not be able to account for the high degree of swirl found in hydrocyclones. As a result of this, more advanced representations of turbulence phenomena have been developed over the last couple of decades.

The three main approaches to the modelling of turbulence are RANS model, LES model, and DNS model. RANS is the oldest approach to model turbulence in cyclones, and in unsteady RANS, an ensemble averaged representation of the governing equations, including transient terms, are solved (Utikar et al., 2010). Turbulence closure can be accomplished by use of the so-called Boussinesq hypothesis. This involves the use of an algebraic equation for the Reynolds stresses or by solving the transport equations for the Reynolds stresses. Early examples of the use of RANS models are those of Hsieh (1988) and Hsieh and Rajamani (1988) that solved the Reynolds averaged Navier–Stokes (RANS) equations to simulate single phase flow through a hydrocyclone. These results could be compared with the velocity measurements obtained from a 75 mm diameter hydrocyclone.

In the LES approach, larger carrying eddies are simulated and smaller eddies are filtered and modelled by means of a sub-grid scale model. DNS models are very expensive, since fully resolved Navier–Stokes equations are solved and all relevant scales of turbulent motion are simulated. As a consequence, these models are not yet practical for solving large-scale industrial problems. Although LES models are more costly than unsteady RANS models, they can be used for modelling the performance of full-scale equipment. For example, Delgadillo and Rajamani (2005) have compared the renormalized group k–ε model, the Reynolds stress model and LES model to predict the dimensions of the air core and velocity profiles in a cyclone. The LES model was better able to predict the experimental data.

Several other examples of the use of RANS and LES models to study the impact of various geometrical and operational variables on the performance of hydrocyclones can be found in the literature (Cullivan et al., 2004; Nowakowski et al., 2004; Schuetz et al., 2004; Mangadoddy et al., 2005).

4.2.2 Model Validation

Earlier studies were conducted with pitot tubes (Yoshioka and Hotta, 1955) and kinematic photography (Knowles et al., 1973). Later on, Das and Miller (1996) have used X-ray tomography to analyse the swirl flow characteristics in an air-sparged hydrocyclone flotation system. Apart from gaining crucial insights into flow phenomena in hydrocyclones, experimental studies are increasingly used in conjunction with numerical modelling for this purpose.

The data required to validate numerical models can be obtained through use of among other, particle image velocimetry (PIV) and laser Doppler velocimetry (LDV) (Lim et al., 2010; Marins et al., 2010). With PIV photographic frames containing a number of particles are cross-correlated. In phase Doppler anemometry, the velocity distribution and size of particles are acquired by detecting laser light scattered by the particle at two angles (e.g., Monredon et al., 1992; Fisher and Flack, 2002; Marins et al., 2010).

Another efficient method of measuring flow processes in hydrocyclones is by use of radioisotope tracers. Stegowski and Leclerc (2002) have used ^{64}Cu with a half-life of 12.7 h to investigate the classification of copper ore slurries.

Methods such as PIV and LDV track the average velocity distributions of particles, as opposed to Lagrangian tracking, where individual particles of a dispersed phase can be tracked in space and time. By examining the particle trajectory, it can be seen how a particle interacts with the flow around it under the specific conditions and any local instabilities can be observed in detail.

Similarly, Bamrungsri et al. (2008) have used a high-speed camera to record the motion of a dyed oil drop in water in a hydrocyclone. Wang et al. (2008) have also used a high-speed camera to record the motion of a vegetable seed with density of 1140 kg/m^3 in a water hydrocyclone. Two-dimensional trajectories could be plotted by recording the positions in consecutive frames. The two-dimensional particle paths obtained by Wang et al. (2008) show that local or instantaneous instabilities in the flow field can have a major effect on the particle trajectories, and hence on the separation performance of the hydrocyclone. These insights cannot be gained from studies of average velocity distributions. Although an additional camera can be set up in an orthogonal direction to obtain three-dimensional trajectories of particles, the spatial resolution can become an issue.

Lagrangian tracking also includes the use of positron emission particle tracking (PEPT), which is a more recent development in process engineering. PEPT is closely related to the medical method referred to as positron emission tomography (PET). The difference is that while PET is based on detection of spatially distributed radioactive sources (positron emitters) and image reconstruction, and therefore provides the time-dependent concentrations of radioactively labelled substances, PEPT locates a point-like positron emitter by cross-triangulation. The spatiotemporal resolution of PEPT is considerably higher than that for PET. For example, Chang et al. (2011) have studied the flow of a particle through a hydrocyclone by means of positron emission particle tracking (PEPT) using ^{18}F as radioactive tracer and could track the particle in the cyclone with an accuracy of 0.2 mm/ms.

5. SCALE-UP AND DESIGN

5.1 Scale-up

Preliminary selection of hydrocyclones is usually done from manufacturers' charts. These charts show regions of operation for each cyclone as a function of the inlet pressure and volumetric flow rate. For more in-depth analysis, the operation of hydrocyclones can be captured by three-dimensionless groups, namely the Stokes, Euler, and Reynolds numbers, as indicated in Eqns (8)–(10). In these equations, the velocity v is the mean axial velocity in the body of the hydrocyclone, i.e., the ratio of the volumetric flow to the cross-sectional area of the cylindrical section of the cyclone body.

$$Stk_{50} = \frac{(\rho_s - \rho) v d_{50}^2}{18 \mu D_c} \tag{8}$$

$$Eu = \frac{\Delta P}{\rho v^2 / 2} \tag{9}$$

$$Re = \frac{\rho v D_C}{\mu} \tag{10}$$

Svarovsky (1984) has shown that the product of the Stokes and Euler numbers is approximately constant for geometrically similar hydrocyclones, and as such these equations can serve as a basis for the scale-up and design of hydrocyclones in process circuits.

Broadly speaking, there are essentially three major families of geometrically similar hydrocyclones, viz. the so-called Rietema (1961) and Bradley (1965) hydrocyclones. The geometric proportions of these families of cyclones are indicated in Table 1. The range of sizes of hydrocyclones in industry is comparatively limited, but each cyclone can be operated at different opening sizes for the inlet, underflow, and overflow by means of interchangeable parts (Castilho and Medronho, 2000).

D_c, D_i, and D_o are the diameters of the cyclone, the feed inlet, and the overflow, respectively, $\boldsymbol{L}$ and $\boldsymbol{l}$ are the total length of the cyclone and the length of the vortex finder, respectively, and $\boldsymbol{\theta}$ is the angle of the hydrocyclone cone.

5.2 Hydrocyclone Networks

Satisfactory separation can be achieved in many conventional solid–solid separation systems by means of a single pass through a hydrocyclone. However, under extreme conditions, such as dealing with low density differentials, fine particles, friable particulates, etc., multiple stages of cyclones may be required. The use of multiple stages of hydrocyclones is also necessary where multiple products are required, such as when the dispersed phase is separated into size or density fractions, or when cyclones are used for dewatering in co- or countercurrent washing operations (Williams et al., 1994). Likewise, multiple stages of hydrocyclones are used to improve the efficiency of separation in closed circuit grinding, for example (Heiskanen, 1993). The same equations comprising dimensionless numbers that are used for scale-up can also be used for basic analysis with regard to the design of hydrocyclone networks or clusters.

6. MONITORING AND CONTROL OF HYDROCYCLONES

6.1 Control

To date, the control of hydrocyclones has not enjoyed the same attention as the control of other pieces of processing equipment, in part owing to their simple design and robust operation. Nonetheless, this seems to be changing with the realization that significant gains in plant efficiencies can be realized by better operation of hydrocyclones in process circuits. For example, an important part of control of hydrocyclones is stabilization of the slurry density in the overflow or cut point for feeds with varying solids concentrations and particle size distributions (Neesse et al., 2004b; Hodouin, 2011). Regulation of the volume split has been restricted to single large cyclones fitted with adjustable nozzles. This is not feasible with batteries of smaller cyclones.

6.2 Soft Sensors

Soft sensors are inferential models using other variables correlated to the variable of interest to estimate the value of the latter. In hydrocyclones, nonlinear autoregressive models with exogenous variables (ARX) (Casali et al., 1998), error projection (Sbarbaro et al., 2008), recursive least squares (Du et al., 1997), and artificial neural networks and support vector machines (Sun et al., 2008) have been used to estimate the particle size in the overflow of hydrocyclones in closed grinding circuits.

6.3 Online Monitoring of Hydrocyclones

Over the last 20 years, several approaches have been proposed for the online monitoring of hydrocyclone behaviour or characteristics, as summarized in Table 3. Although some of these approaches have been commercialized, online monitoring of hydrocyclones has not yet been incorporated into process control loops to a large extent and online monitoring technology can therefore not be considered as a mature technology yet. Table 3 gives a summary of online monitoring methods proposed over the last two decades.

6.4 Example of Underflow Monitoring

An example of underflow monitoring of hydrocyclones as discussed by Janse van Vuuren et al. (2011) can be considered briefly here. In this approach, a video camera is used to monitor the spray profile of a hydrocyclone

Table 3 Online monitoring of hydrocyclones

Method	Measured element	References
Mechanical plate	Underflow spray angle	Hulbert (1993)
Load cell gravimetry	Mass of solids in hydrocyclone	Neesse et al. (2004c)
X-ray tomography	Internal distribution of solids	Galvin and Smitham (1994)
Electrical impedance tomography	Internal distribution of solids; air core size and shape	Williams et al. (1997), Gutiérrez et al. (2000)
Electrical resistance tomography	Internal concentration of solids; location, size and shape of air core	Williams et al. (1992), Williams et al. (1995) Williams et al. (1999), West et al. (2000)
Acoustic monitoring	Feed pressure, feed solids concentration; underflow discharge oscillation	Hou et al. (1998, 2002), Neesse et al. (2004c)
Ultrasound	Underflow discharge oscillation	Olson and Waterman (2006)
Laser optical	Underflow discharge profile	Neesse et al. (2004c)
Videography	Underflow discharge spray angle	Petersen et al. (1996), Van Deventer et al. (2003), Janse van Vuuren et al. (2011)
Overflow Pressure	Air core diameter	Krishna et al. (2010)

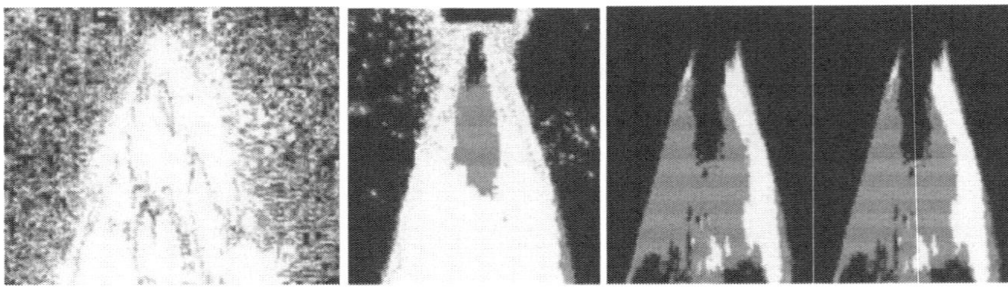

Figure 5 Preprocessing of an image of the underflow of a hydrocyclone. *From Janse van Vuuren et al. (2011); with permission from Elsevier.*

dynamically. Images of the underflow are first enhanced to determine the outline of the flow as best as possible, as shown in Figure 5. This entails removal of the noise in the image on the left in the figure, to finally yield the image on the right.

The width of this spray profile is subsequently measured at a fixed point below the spigot, as a proxy for the angle of the spray profile as indicated in Figure 6). Measurements at different points on the profile can also be obtained for more robust analysis (Petersen et al., 1996).

This information is analysed dynamically from a sequence of images and the underflow widths can then be mapped to a process chart, as shown in Figure 7. On this chart, normal operating conditions are indicated by a region delineated by a 99% confidence limit (the confidence level is adjustable). Image data are continuously mapped to this chart and the operator

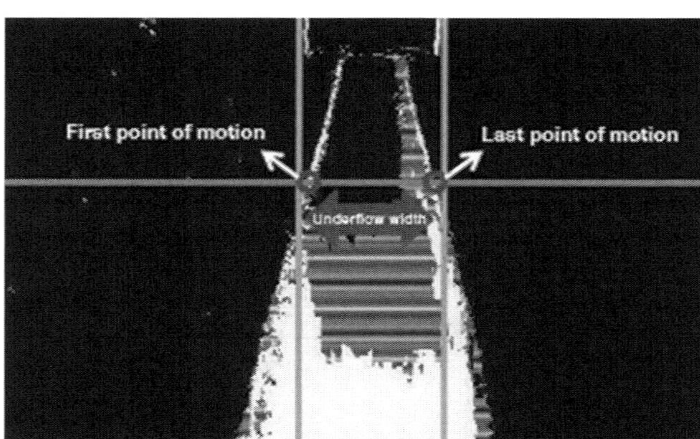

Figure 6 Measurement of the spray profile width from a denoised image of the underflow of a hydrocyclone. *From Janse van Vuuren et al. (2011); with permission from Elsevier.*

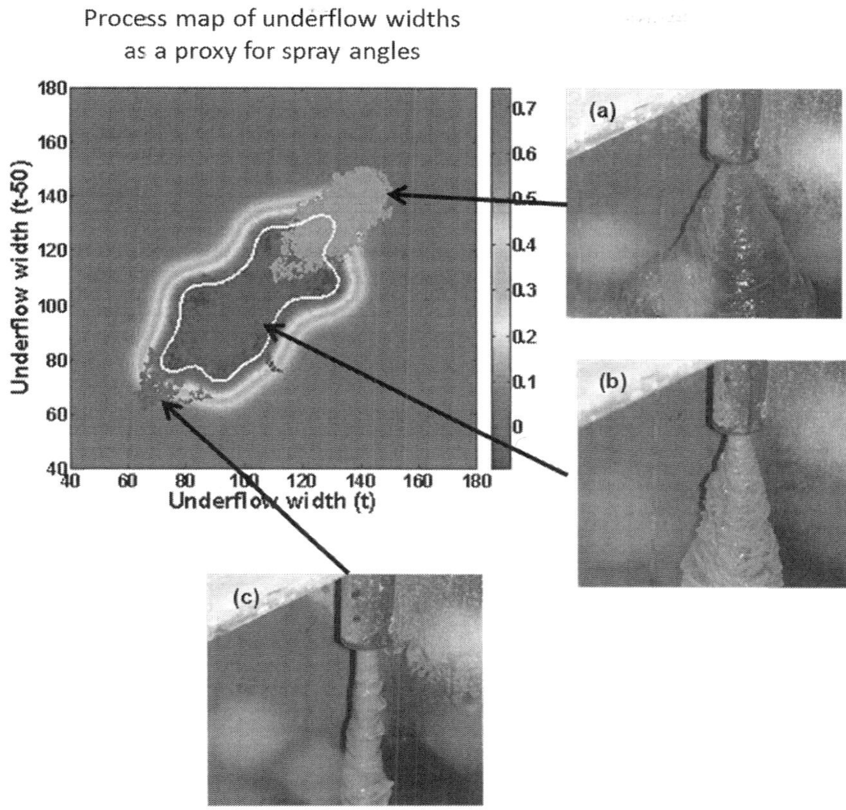

Figure 7 Mapping of image data to a process chart form predictive monitoring of the behaviour of the hydrocyclones.

can then detect problems as before arise (typically overloading of the cyclone leading to roping), since the transition from the normal operating region to abnormal behaviour may be preceded by instability.

7. FUTURE DEVELOPMENTS

The application of hydrocyclones as separation devices is exceedingly diverse, and one of the recent trends is the development of hydrocyclones for the classification of fine particles. However, classification inefficiencies, such as their large bypass, represent a major hurdle that will have to be surmounted before they could be competitive with other devices in this regard. Even so, the potential of very small diameter hydrocyclones in the classification of submicron size particles would be large, if these problems can be solved (Cilliers, 2000).

As indicated in this chapter, the online monitoring of hydrocyclones has attracted considerable attention in recent years. These systems could for the basis for advanced control of hydrocyclone operations, which could translate into significant benefits to plant operations associated with a wide range of products.

Finally, numerical modelling of hydrocyclones continues to advance strongly on the back of concomitant improvements in computer technology, software, and increasing insight into flow processes in hydrocyclones. This will lead to improvement in hydrocyclone design and will extend their range and flexibility even further.

REFERENCES

Asomah, A.K., Napier-Munn, T.J., 1997. An empirical model of hydrocyclones, incorporating angle of cyclone inclination. Miner. Eng. 10 (3), 339–347.

Bamrungsri, P., Puprasert, C., Guigui, C., Marteil, P., Bréant, P., Hébrard, G., 2008. Development of a simple experimental method for the determination of the liquid field velocity in conical and cylindrical hydrocyclones. Chem. Eng. Res. Des. 8 (11), 1263–1270.

Bloor, M.I.G., Ingham, D.B., Laverack, S.D., 1980. An analysis of boundary layer effects in a hydrocyclone. In: Proceedings of the International Conference on Hydrocyclones, Cambridge, Paper 5, 49–62, BHRA Fluid Engineering, Cranfield, 1980.

Bradley, D., Pulling, D.J., 1959. Flow patterns in the hydraulic cyclone and their interpretation in terms of performance. Trans. Inst. Chem. Eng. 37, 34–45.

Bradley, D., 1965. The Hydrocyclone. Pergamon Press Limited, Oxford.

Brennan, M., 2006. CFD simulations of hydrocyclones with an air core comparison between large eddy simulations and a second moment closure. Chem. Eng. Res. Des. 84 (A6), 495–505.

Bretnai, E., 1891. US Patent No. 453, 105.

Casali, A., Gonzalez, G.D., Torres, F., Vallebuona, G., Castelli, L., Gimnez, P., 1998. Particle size distribution soft-sensor for a grinding circuit. Powder Technol. 99, 15–20.

Castilho, L.R., Medronho, R.A., 2000. A simple procedure for design and performance prediction of Bradley and Rietema hydrocyclones. Miner. Eng. 13 (2), 183–191.

Chang, Y.-F., Ilea, C.G., Aasen, Ø.L., Hoffmann, A.C., 2011. Particle flow in a hydrocyclone investigated by positron emission particle tracking. Chem. Eng. Sci. 66 (18), 4203–4211.

Chen, W., Zydek, N., Parma, F., 2000. Evaluation of hydrocyclone models for practical applications. Chem. Eng. J. 80, 295–303.

Cilliers, J.J., Hinde, A.L., 1991. An improved hydrocyclone model for backfill preparation. Miner. Eng. 4 (7–11), 683–693.

Cilliers, J.J., 2000. Hydrocyclones for particle separation. In: II Particle Size Separation. Academic Press.

Coelho, M.A.Z., Medronho, R.A., 2001. A model for performance prediction of hydrocyclones. Chem. Eng. J. 84, 7–14.

Cullivan, J.C., Williams, R.A., Dyakowski, T., Cross, C.R., 2004. New understanding of a hydrocyclone flow field and separation mechanism from computational fluid dynamics. Miner. Eng. 17 (5), 651–660.

Das, A., Miller, J.D., 1996. Swirl flow characteristics and froth phase features in air-sparged hydrocyclone flotation as revealed by X-ray CT analysis. J. Miner. Process. 47 (3–4), 251–274.

Delgadillo, J.A., Rajamani, R.K., 2005. A comparative study of three turbulence closure models for the hydrocyclone problem. Int. J. Miner. Process. 77 (4), 217–230.

Delgadillo, J.A., Rajamani, R.K., 2007. Exploration of hydrocyclone designs using computational fluid dynamics. Int. J. Miner. Process. 84 (1–4), 252–261.

Driessen, M.G., 1951. Theory of Flow in a Cyclone. Revue de L'Industrie Minerale, Numero Special, Saint-Etienne, 449–461.

Du, Y.-G., del Villar, R., Thibault, J., 1997. Neural net-based softsensor for dynamic particle size estimation in grinding circuits. Int. J. Miner. Process 52 (2), 121–135.

Dueck, J., Matvienko, O.V., Neesse, T., 2000. In: Modelling of Hydrodynamics and Separation in a Hydrocyclone. Theoretical Foundations of Chemical Engineering, vol. 34. Kluwer Academic/Plenum Publishers, pp. 428–438.

Dwari, R.K., Biswas, M.N., Meikap, B.C., 2004. Performance characteristics of sand FCC and fly ash in a novel hydrocyclone. Chem. Eng. Sci. 59, 671–684.

Dyakowski, T., Nowakowski, A.F., Kraipech, W., Williams, R.A., 1999. A three dimensional simulation of hydrocyclone behaviour. In: Second International Conference on CFD in the Minerals and Process Industries Melbourne, Australia.

Endres, E., Dueck, J., Neesse, T., 2012. Hydrocyclone classification of particles in the micron range. Miner. Eng. 31, 42–45.

Flament, F., Thibault, J., Hodouin, D., 1993. Neural Networks Based Control of Mineral Grinding Plants. Mineral Engineering 6 (3), 235–249.

Fisher, M.J., Flack, R.D., 2002. Velocity distributions in a hydrocyclone separator. Exp. Fluids 32 (3), 302–312.

Galvin, K.P., Smitham, J.B., 1994. Use of X-rays to determine the distribution of particles in an operating cyclone. Miner. Eng. 7, 1269–1280.

Gutiérrez, J.A., Dyakowski, T., Beck, M.S., Williams, R.A., 2000. Using electrical impedance tomography for controlling hydrocyclone underflow discharge. Powder Technol. 108, 180–184.

Heiskanen, K., 1993. Particle Classification. Chapman and Hall, London, UK.

Hodouin, D., 2011. Methods for automatic control, observation, and optimization in mineral processing plants. J. Process Control 21, 211–225.

Hou, R., Hunt, A., Williams, R.A., 1998. Acoustic monitoring of hydrocyclone performance. Miner. Eng. 11 (11), 1047–1059.

Hou, R., Hunt, A., Williams, R.A., 2002. Acoustic monitoring of hydrocyclones. Powder Technol. 124, 176–187.

Hsieh, K.T., 1988. A Phenomenological Model of the Hydrocyclone (Ph.D. thesis). University of Utah, UT, USA.

Hsieh, K.T., Rajamani, R.K., 1988. Phenomenological model of the hydrocyclone: model development and verification for single-phase flow. Int. J. Miner. Process. 22 (1–4), 223–237.

Hulbert, D.G., 1993. Measurement methods and apparatus for hydrocyclones. In: Mintek (Ed.), EP0522215A3.

Janse van Vuuren, M.J., Aldrich, C., Auret, L., 2011. Videographic monitoring of hydrocyclones. Miner. Eng. 24, 1532–1544.

Karr, C.L., Stanley, D.A., McWhorter, B., 2000. Optimization of hydrocyclone operation using a geno-fuzzy algorithm. Comput. Method. Appl. Mech. Eng. 186, 517–530.

Knowles, S.R., Woods, D.R., Feuerstein, I.A., 1973. The velocity distribution within a hydrocyclone operating without an air-core. Can. J. Chem. Eng. 51, 263–271.

Kraipech, W., Chen, W., Parma, F.J., Dyakowski, T., 2002. Modelling the fish-hook effect of the flow within hydrocyclones. Int. J. Miner. Process. 66 (1–4), 49–65.

Kraipech, W., Chen, W., Dyakowski, T., Nowakowski, A., 2006. The performance of the empirical models on industrial hydrocyclone design. Int. J. Miner. Process. 80, 100–115.

Krishna, V., Sripriya, R., Kumar, V., Chakraborty, S., Meikap, B.C., 2010. Identification and prediction of air core diameter in a hydrocyclone by a novel online sensor based on digital signal processing technique. Chem. Eng. Process. Process Intensif. 49, 165–176.

Lapple, C.E., 1951. Processes use many collector types. Chem. Eng. 58, 144–151.

Leith, D., Licht, W., 1972. Collection efficiency of cyclone type particle collector: a new theoretical approach. AIChE Symp. Series (Air-1971) 68, 196–206.

Lim, E.W., Chen, Y.-R., Wang, C.-H., Wu, R.-M., 2010. Experimental and computational studies of multiphase hydrodynamics in a hydrocyclone separator system. Chem. Eng. Sci. 65, 6415–6424.

Lim, K.S., Kwon, S.B., Lee, K.W., 2003. Characteristics of the collection efficiency for a double inlet cyclone with clean air. Aerosol Sci. 34, 1085–1095.

Lynch, A.J., Rao, T.C., 1975. Modelling and scale-up of hydrocyclone classifiers. In: Proceedings of the 11th International Mineral Processing Congress, Cagliari, pp. 245–269.

Madge, D.N., Romero, J., Strand, W.L., 2004. Hydrocarbon cyclones in hydrophilic oil sand environments. Miner. Eng. 17 (5), 625–636.

Mangadoddy, N., Sripriya, R., Banerjee, P.K., 2005. CFD modelling of hydrocyclone—prediction of cut size. Int. J. Miner. Process. 75, 53–68.

Marins, L.P.M., Duarte, D.G., Loureiro, J.B.R., Moraes, C.A.C., Silva Freire, A.P., 2010. LDA and PIV characterization of the flow in a hydrocyclone without an air-core. J. Pet. Sci. Eng. 70, 168–176.

Monredon, T.C., Hsieh, K.T., Rajamani, R.K., 1992. Fluid flow model of the hydrocyclone: an investigation of device dimensions. Int. J. Miner. Process. 35 (1–2), 65–83.

Murray, H.H., 1980. Major kaolin processing developments. Int. J. Miner. Process. 7 (3), 263–274.

Nageswararao, K., Wiseman, D.M., Napier-Munn, T.J., 2004. Two empirical hydrocyclone models revisited. Miner. Eng. 17, 671–687.

Neesse, T., Dueck, J., Minkov, L., 2004a. Separation of finest particles in hydrocyclones. Miner. Eng. 17, 689–696.

Neesse, T., Golyk, V., Kaniut, P., Reinsch, V., 2004b. Hydrocyclone control in grinding circuits. Miner. Eng. 17, 1237–1240.

Neesse, T., Schneider, M., Golyk, V., Tiefel, H., 2004c. Measuring the operating state of the hydrocyclone. Miner. Eng. 17, 697–703.

Nenu, R.K.T., Yoshida, H., 2009. Comparison of separation performance between single and two inlets hydrocyclones. Adv. Powder Technol. 20, 195–202.

Norimoto, U., Fukui, K., Yoshida, H., 2004. Particle separation performance of hydrocyclone with forced vortex type. J. Soc. Powder Technol. Jpn. 43 (9), 666–675.

Nowakowski, A.F., Cullivan, J.C., Williams, R.A., Dyakowski, T., 2004. Application of CFD to modelling of the flow in hydrocyclones. Is this a realizable option or still a research challenge? Miner. Eng. 17 (5), 661–669.

Olson, T.J., Waterman, R.J., 2006. Hydrocyclone roping Detector and method. In: TUCSON (Ed.), US6983850B2. Krebs Engineers Corporation, United States.

Pericleous, K.A., Rhodes, N., 1986. The hydrocyclone classifier – a numerical approach. Int. J. Miner. Process. 17, 23–43.

Petersen, K.R.P., Aldrich, C., Van Deventer, J.S.J., McInnes, C., Stange, W.W., 1996. Hydrocyclone underflow monitoring using image processing methods. Miner. Eng. 9 (3), 301–316.

Plitt, L.R., 1976. A mathematical model of the hydrocyclone classifier. Can. Inst. Min. Metall. Bull. 60 (776), 114–122.

Puprasert, C., Hebrard, G., Lopez, L., Aurelle, Y., 2004. Potential of using hydrocyclone and hydrocyclone equipped with Grit pot as a pre-treatment in run-off water treatment. Chem. Eng. Process. Process Intensif. 43 (1), 67–83.

Rietema, K., 1961. Performance and design of hydrocyclones – parts I–IV. Chem. Eng. Sci. 15, 298–325.

Sbarbaro, D., Ascencio, P., Espinoza, P., Mujica, F., Cortes, G., 2008. Adaptive soft-sensors for on-line particle size estimation in wet grinding circuits. Control Eng. Pract. 16 (2), 171–178.

Schuetz, S., Mayer, G., Bierdel, M., Piesche, M., 2004. Investigations on the flow and separation behaviour of hydrocyclones using computational fluid dynamics. Int. J. Miner. Process. 73 (2–4), 229–237.

Shepherd, C.B., Lapple, C.E., 1939. Flow pattern and pressure drop in cyclone dust collectors. Ind. Eng. Chem. 31, 972–984.

Silva, A.C., Silva, E.M.S., Matos, J.D.V. Comparison between Plitt's model and its revisions for hydrocyclones simulation. Accessed from: http://www.arber.com.tr/imps2012.org/proceedingsebook/Abstract/absfilAbstractSubmissionFullContent369.pdf.

Slechta, J., Firth, B.A., 1984. Classification of fine coal with a hydrocyclone. Int. J. Miner. Process. 12 (4), 213–237.

Stange, W.W., 1993. Using artificial neural networks for the control of grinding circuits. Miner. Eng. 6 (5), 479–489.

Stegowski, Z., Leclerc, J.-P., 2002. Determination of the solid separation and residence time distributions in an industrial hydrocyclone using radioisotope tracer experiments. Int. J. Miner. Process. 66, 67–77.

Sun, Z., Wang, H., Zhang, Z., 2008. Soft sensing of overflow particle size distributions in hydrocyclones using a combined method. Tsinghua Sci. Technol. 13 (1), 47–53.

Svarovsky, L., 1984. Hydrocyclones. Holt Rinehart and Winston.

Utikar, R., Darmawan, N., Tade, M., Li, Q., Evans, G., Glenny, M., Pareek, V., 2010. Hydrodynamic simulation of cyclone separators. In: Oh, H.W (Ed.), Computational Fluid Dynamics. InTech. Available from. http://www.intechopen.com/books/computational-fluiddynamics/hydrodynamic-simulation-of-cyclone-separators. ISBN: 978-953-7619-59-6.

Van der Walt, T.J., Van Deventer, J.S.J., Barnard, E., 1993. Neural nets for the simulation of mineral processing operations: part II. Applications. Miner. Eng. 6 (11), 1135–1153.

Van Deventer, J.S.J., Feng, D., Petersen, K.R.P., Aldrich, C., 2003. Modelling of hydrocyclone performance based on spray profile analysis. Int. J. Miner Process. 70, 183–203.

Wang, Z.-B., Chu, L.-Y., Chen, W.-M., Wang, S.-G., 2008. Experimental investigation of the motion trajectory of solid particles inside the hydrocyclone by a Lagrange method. Chem. Eng. J. 138 (1–3), 1–9.

West, R.M., Jia, X., Williams, R.A., 2000. Parametric modeling in industrial tomography. Chem. Eng. J. 77 (1–2), 31–36.

White, D.A., 1991. Efficiency curve model for hydrocyclones based on crowding theory. Trans. Inst. Min. Metall. Sect. C: Miner. Process. Extr. Metall. 100, 135–138.

Williams, R.A., Dickin, F.J., Dyakowski, T., Ilyas, O.M., Abdullah, Z., Beck, M.S., 1992. Looking into mineral process plant? Miner. Eng. 5, 867–881.

Williams, R.A., Ilyas, O.M., Dyakowski, T., Dickin, F.J., Gutierrez, J.A., Wang, M., Beck, M.S., Shah, C., Rushton, A., 1995. Air core imaging in cyclonic separators: implications for separator design and modelling. Chem. Eng. J. Biochem. Eng. J. 56, 135–141.

Williams, R.A., Jia, X., West, R.M., Wang, M., Cullivan, J.C., Bond, J., Faulks, I., Dyakowski, T., Wang, S.J., Climpson, N., Kostuch, J.A., Payton, D., 1999. Industrial monitoring of hydrocyclone operation using electrical resistance tomography. Miner. Eng. 12, 1245–1252.

Williams, R.A., Albarran de Garcia Colon, I.L., Lee, M.S., Roldan Villasana, E.J., 1994. Design targeting of hydrocyclone networks. Miner. Eng. 5, 253–256.
Williams, R.A., Dickin, F.J., Gutiérrez, J.A., Dyakowski, T., Beck, M.S., 1997. Using electrical impedance tomography for controlling hydrocyclone underflow discharge. Control Eng. Pract. 5, 253–256.
Woodfield, D., Bickert, G., 2004. Separation of flocs in hydrocyclones - significance of floc breakage and floc hydrodynamics. Int. J. Miner. Process. 73 (2–4), 239–249.
Yalamanchili, M.R., Miller, J.D., 1995. Removal of insoluble slimes from potash ore by air-sparged hydrocyclone flotation. Miner. Eng. 8 (1–2), 169–177.
Yalcin, T., Kaukolin, E., Byers, A., 2003. Axial inlet cyclone for mineral processing applications. Miner. Eng. 16 (12), 1375–1381.
Yang, I.H., Shin, C.B., Kim, T.-H., Kim, S., 2004. A three-dimensional simulation of a hydrocyclone for the sludge separation in water purifying plants and comparison with experimental data. Miner. Eng. 17 (5), 637–641.
Yoshida, H., Norimoto, U., Fukui, K., 2006. Effect of blade rotation on particle classification performance of hydrocyclone. Powder Technol. 164, 103–110.
Yoshioka, N., Hotta, Y., 1955. Liquid cyclones as a hydraulic classifier. Chem. Eng. Jpn. 19, 633–641.
Zhao, B., Shen, H., Kang, Y., 2004. Development of a symmetrical spiral inlet to improve cyclone separator performance. Powder Technol. 145, 47–50.

SECTION ONE

Membrane filters

CHAPTER TWO

Dynamic Filtration with Rotating Disks, and Rotating or Vibrating Membranes

Luhui Ding, Michel Y. Jaffrin and Jianquan Luo
Technological University of Compiegne, Compiegne Cedex, France

Contents

Progress in Filtration and Separation
http://dx.doi.org/10.1016/B978-0-12-384746-1.00002-1

GLOSSARY

CFF Crossflow filtration
CIP Cleaning in place
COD Chemical oxygen demand
DF Dynamic filtration
MF Microfiltration
MSD Multishaft disks
NF Nanofiltration
RDM Rotating disk module
RO Reverse osmosis
SBM Spinning basket module
TMP Transmembrane pressure
UF Ultrafiltration
VMBR Vibrating membrane bioreactor
VRR Volume reduction ratio

Symbols

d Vibration amplitude, m
Es Specific energy consumed per m^3 of permeate, kWh/m^3
F Vibration frequency, Hz
J Permeate flux, $1/L/h/m^2$
p Pressure, bar, Pa
r Local radius, m
R Disk radius, m
γ Shear rate, s^{-1}
μ Dynamic viscosity, Pa s
ν Kinematic viscosity, m^2/s
ρ Density, kg/m^3
ω Angular velocity, rad/s

1. INTRODUCTION

The aim of this chapter is to present a relatively recent technique which permits to increase the performance of membrane filtration as compared to classical techniques known as crossflow filtration in which the fluid circulates along a fixed membrane. This new technique, called dynamic filtration, took its origin after 1990 in the United States with the

VSEP vibrating system of New Logic in California, the SpinTek module with rotating membranes in Huntington Ca, and the DMF rotating disk system of Pall Corp, Mass. Several German companies Bokela, Novoflow, Novoflow, and Canzler built systems with rotating disks and rotating membranes after 2000.

1.1 Differences between Dynamic and Crossflow Filtration

In crossflow filtration (CFF), the fluid to be treated is circulated by a pump along a flat, a spiral wound, a tubular, or a hollow fibre module and the permeate, containing water and small solutes is collected by filtration across the membrane. This process necessitates applying to the feed a pressure higher than that of permeate, together with a tangential fluid velocity high enough to reduce solute accumulation on the membrane which may cause fouling. But a large tangential velocity will induce a large pressure drop along the membrane, reducing the transmembrane pressure (TMP) in the downstream part of the module and decreasing the average permeate flux.

In dynamic or shear-enhanced filtration (DF), the shear rate at the membrane is created by a moving part such as a disk or a rotor (Lee et al., 1995; Mantari et al., 2006) rotating near fixed membranes or by membranes rotating around a shaft in a housing (Kroner et al., 1988) or by azimuthal vibrations of circular membranes stacked around a vertical axis (VSEP, New Logic, Vane et al., 1999) or by longitudinal vibrations of hollow fibres (Beier et al., 2006).

1.2 Advantages and Limitations of Dynamic and Crossflow Filtrations

1.2.1 Dynamic Filtration

The first advantage is that, by using high rotation speeds and large radius membranes, it is possible to generate very high shear rates at the membrane, up to over $3 \cdot 10^5$/s (Ding et al., 2002) which can increase the permeate flux by a large factor as compared to CFF, by reducing concentration polarization and membrane fouling. These high shear rates also increase solute transmission through a microfiltration (MF) membrane by reducing cake formation. Since the transfer of small solutes such as ions and molecules through nanofiltration (NF) or reverse osmosis (RO) membranes is mainly diffusive, this transfer is reduced because solute concentration at membranes has been lowered at high shear rates. This leads to a higher rejection rate by the membrane which is the goal of wastewater treatment by NF or RO (Frappart et al., 2006). So an NF membrane in DF may have almost the same rejection as an RO membrane in CFF, but with a higher permeate flux. Moreover, the permeate flux keeps increasing with TMP until higher pressures at high shear rates since the pressure-limited regime is extended, due to decreased concentration polarization.

Another advantage is that the feed flow needs only to be slightly larger than the permeate flow rate in DF since the membrane shear rate is produced independently from the fluid velocity and the power of feed pumps is lower than in CFF. Rotating membranes or disks consume energy at high speeds, but this is offset by the energy gain on pumps, and one has a choice between using high rotation speeds to produce high permeate fluxes or moderate speeds yielding fluxes similar to those in CFF, but with a large reduction in specific energy consumed per unit volume of permeate.

A third advantage is that DF modules can produce retentates with high dry solutes concentrations of at least 70%, since the fluid velocity near the membrane is small, reducing the pressure drop and energy losses due to viscosity. In multicompartment rotating disk modules such as the SpinTek and the Dyno filter, the retentate keeps increasing since compartments are connected in series avoiding the need for recirculating the fluid though the module.

The drawbacks of DF modules are their high cost and complexity, especially when disks rotate between fixed membranes forming separate compartments. Systems such as the KMPT with ceramic membrane disks rotating around parallel shafts inside a housing are simpler to build and to service than those with separate compartments, but require very precise balancing of the disks to avoid internal vibrations and wear. Another drawback may be the limited membrane area of some modules less than 25 m^2, but at least three of them, the Optifilter CR, the Rotostream, and the VSEP are available with about 150 m^2 of membrane.

1.2.2 Crossflow Filtration

The advantages of CFF are the moderate cost of filtration modules, especially in the case of spiral wound modules and the wide choice of membranes and modules configurations. Their maintenance is also less complex and expensive than that of DF modules and they are better suited for treating very large volumes of fluid. Their drawbacks are that their permeate fluxes are limited when treating viscous and highly charged fluids and for reaching high solid concentrations, of above 50% of dry solids. They also require big and powerful pumps for producing high fluid velocities in large modules.

In fact, there are situations when CFF and DF systems can be complementary, when treating large volumes, while producing a highly concentrated retentate, which is often the case of waste water treatment. The first step of the process, requiring large membrane areas must use CFF modules, like spiral wound ones, while in the final concentration step, the retentate of CFF modules could be further concentrated by DF modules, such as VSEP, capable of handling high solid concentrations.

2. REVIEW OF INDUSTRIAL DYNAMIC FILTRATION MODULES

2.1 VSEP Vibrating Systems

An original concept, known as vibratory shear-enhanced process (VSEP), was invented in 1987. The VSEP uses azimuthal vibrations of membrane to create high shear rate, aiming to control concentration polarization and prevent membrane fouling. The high shear rate at the membrane is produced by the inertia of the retentate which moves out of phase with the membrane and varies sinusoidally with time. The use of resonance permits to minimize the power necessary to produce the vibrations, which is only 9 kW, even for large units of 150 m^2 membrane area (Johnson, 2008). The filter pack may, depending on the model, contain one or more membranes (Figure 1). The Series LP is primarily used for testing

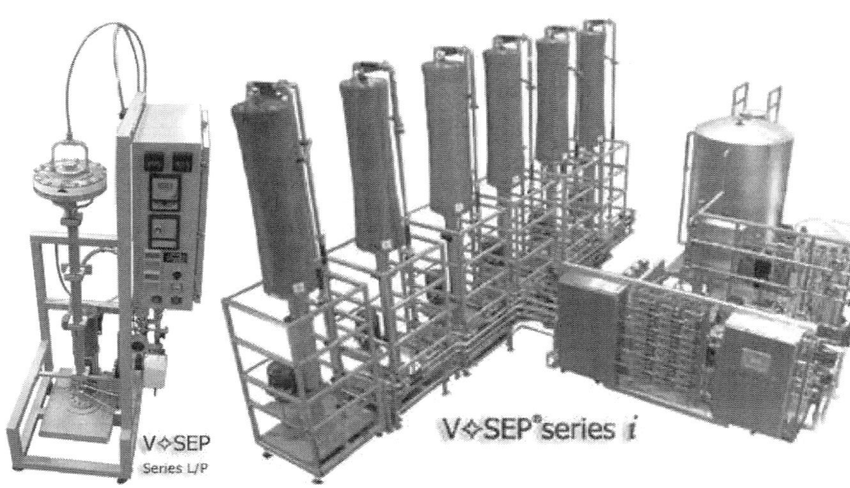

Figure 1 Commercial VSEP modules from New Logic Research. Inc. *Pictures taken from http://www.vsep.com/index.html.*

purposes; however, with flow rates up to 270/Lh, and the Series LP can also be used in small-scale production processes. The most widely used VSEP is the Series i84. With up to 140 m^2 of membrane area in each filter pack, the i84 is the ideal module size to process larger flow rates.

2.2 Multicompartments Systems with Rotating Disks or Rotors between Fixed Membranes

2.2.1 The DYNO Filter, BOKELA

The operation of the DYNO Filter is based on the principle of stirred-cell filtration. The DYNO filter consists of several disk-shaped filter modules that are arranged in series and form a completely closed hermetical chamber system. The slurry flows meander-like from one chamber to the next one and becomes accordingly more concentrated since filtrate is discharged from each chamber. After the last chamber, the concentrate (retentate) is extracted with an automated drain valve. The module structure and industrial equipments of DYNO Filter are shown in Figure 2.

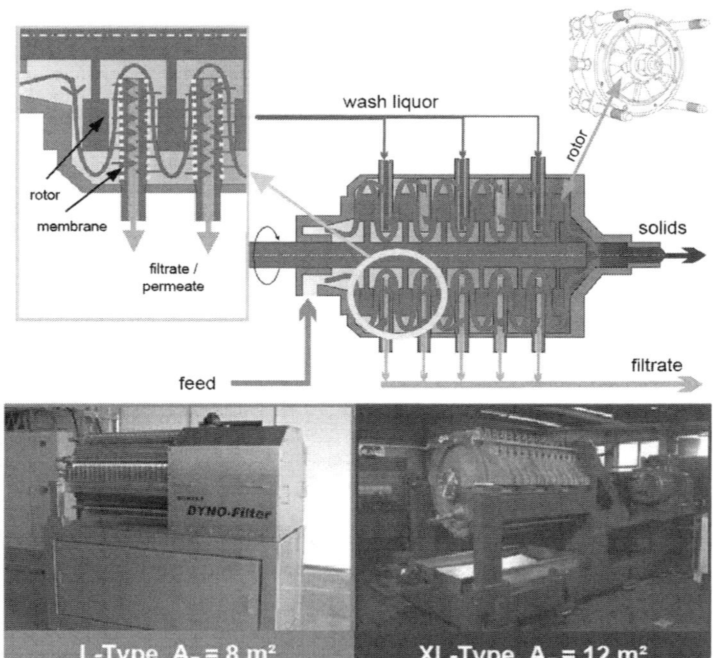

Figure 2 Module structure and industrial equipments of DYNO Filter, BOKELA. *Pictures taken from http://www.bokela.de/de/technologien/cross-flow-filtration.html.*

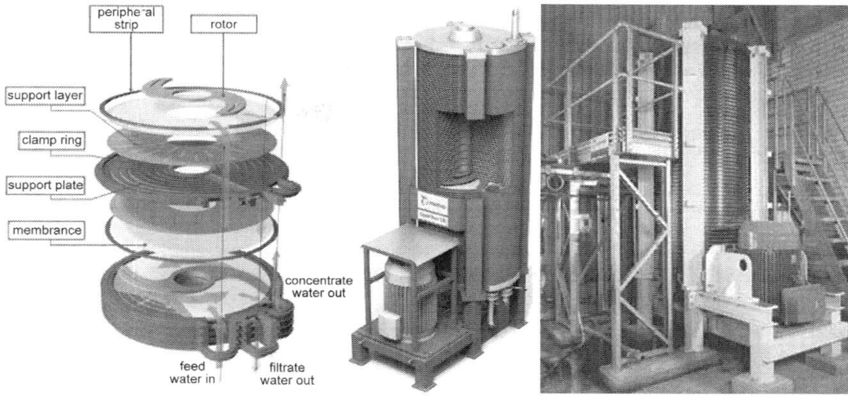

Figure 3 Module structure and industrial equipments of OptiFilter CR, Metso paper. *Pictures taken from http://www.metso.com/MP/marketing/Vault2MP.nsf/BYWID/WID-030410-2256C-2060A*

2.2.2 The Optifilter CR, Metso Paper

The OptiFilter CR is made of flat membranes fastened on both sides of filter cassettes. They are stacked on top of each other. Between each cassette there is a rotor which creates turbulence in order to enhance the filtrate capacity and reduce the fouling effect. Its module structure and industrial equipments are shown in Figure 3. The OptiFilter CR is an advanced device based on fine physical separation and the Cross-rotational technique makes it superior for treating water effluents in the pulp and paper industry. Combining Opti-Filter CR with other membrane technologies makes water-recycle possible in paper industry.

2.2.3 The FMX System, BKT

The FMX is a new antifouling membrane system based on the simple, yet innovative concept of using vortices to prevent fouling of the membrane for high density, high viscosity, and high solid applications. It has a stack structure similar to the OptiFilter CR, but the rotors are replaced by vortex generators (Figure 4). FMX can be used for wastewater treatment, separation and dewatering in manufacturing processes, and recovery (reuse, recycle) applications in various industries.

2.3 Rotating Membrane Systems

2.3.1 Rotary Membrane System, SpinTek

This compact, high shear ultrafilter with rotating membrane disks generates high shear rates that are orders of magnitude greater than conventional UF

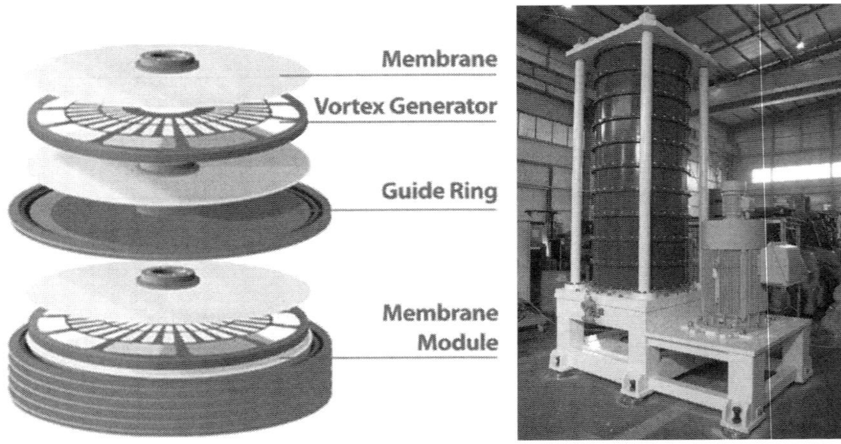

Figure 4 Module structure and industrial equipment of FMX system,BKT. *Pictures taken from http://www.bkt21.com/membrane-and-filtration/anti-fouling-membrane/.*

systems, prevent fouling and provide high, very stable permeate fluxes. This system is now available from SpinTek for filtration and fractionation of high fouling and viscous feed solutions. Successful operation has been performed in >90% recovery of protein from cheese whey, submicron filtration of lube oil, nonsettable solids concentration above 6.4 Pa/s and fine chemical dewatering to above 40% solids. The newest generation product can use polymeric membranes, or ceramic or metallic membranes in various pore sizes and can also be configured to utilize any commercially available flat-sheet polymer membrane. Figure 5 shows a SpinTek industrial module and its associated equipments.

2.3.2 Single-Shaft Disk Filter (SSDF), Novoflow

Dynamic crossflow filtration systems with rotating filter disks are assembled today in multishaft or single-shaft design. In a multishaft module two or more filter stacks with overlapping sectors are rotated or counter rotated, thus creating high shear forces between overlapping disks. In Novoflow's single-shaft disk filter (SSDF) module only one filter stack is rotated on its centre shaft, as shown as Figure 6. Novoflow builds all larger plants in vertical design for reduced foot print and larger filter areas per module. Novoflow's membrane disk technology enables selection of the optimal membrane, drainage, and support material. So filter disks can be optimized for specific applications. Test series with different membrane materials become feasible at reasonable prices for small lots. Novoflow's composite filter disks with NewSeal technology are applied for liquid as well as air filtration and gassing.

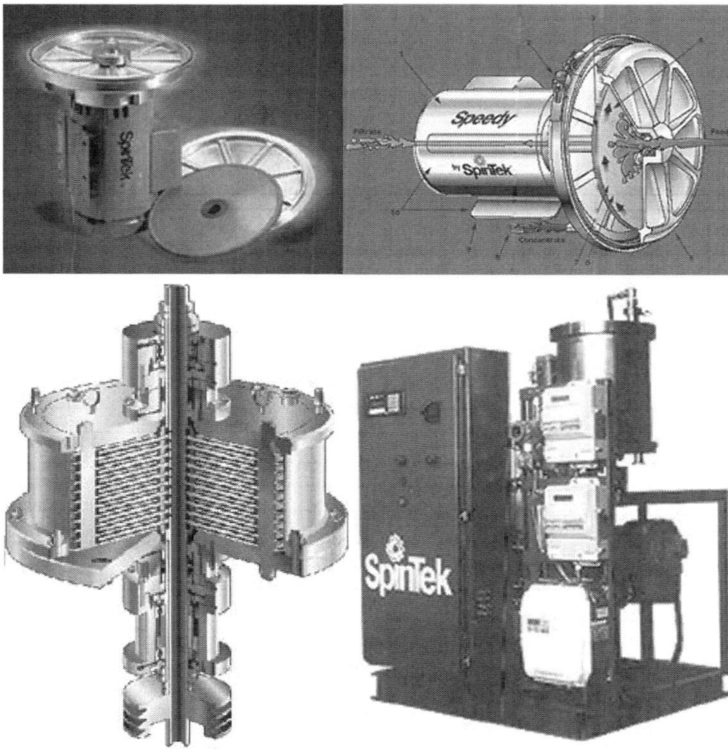

Figure 5 Module structure and industrial equipments of Rotary Membrane System, SpinTek (Speedy T andST-II). *Pictures taken from http://www.spintek.com*

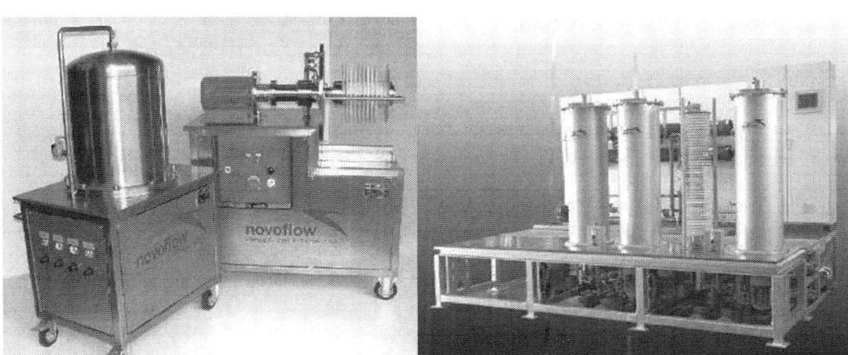

Figure 6 Novoflow filter modules, left: 1×1 m^2; right: 4×10 m^2. *Pictures taken from http://www.novoflow.com/PDF-Filter/SSDF-Filter/ssdf-filter.html*

2.4 Other Developing Dynamic Filtration Modules

Although an industrial multishaft disk (MSD) module, equipped with 31.2 cm diameter ceramic membranes rotating on eight parallel shafts located on a cylinder, has been reported by Jaffrin (2008) (Figure 7, left), its commercial module information still cannot be found in web site of GEA Westfalia Separator group. Ding et al. (2006) investigated an MSD pilot with two parallel hollow shafts rotating at same speed (Figure 7, right), and the ceramic membrane disks from each shaft overlap each over for 15.5% of their surface. He et al. (2007) found that the effect of shear rate increment due to membrane overlapping on permeate flux, could be completely offset by the high concentration increase between two adjacent and overlapping membranes. This pilot was later modified by replacing the ceramic disks on one shaft by metal disks with vanes, and it avoids local over concentration between overlapping membranes, thus increasing permeate flux and saving energy at high rotating speed. Since metal disks are cheaper than ceramic ones, the cost of such systems may be less, for the same output, than for original MSD systems (Jaffrin, 2008; He et al., 2007). Tu and Ding (2010) also employed polymeric membranes to replace ceramic ones, and found that overlapping disks were more efficient to remove cake fouling on polymeric membranes, but their energy consumption were higher.

Figure 7 Industrial MSD module (Westfalia Separator) with eight parallel shafts and 31.2 cm ceramic disks (left); Laboratory MSD pilot in UTC (right).

Compared with SSDF, although the MSD module is more compact, its advantage is marginal in terms of permeate flux, energy, and maintenance cost. That may be the reason why the MSD is not yet commercialized.

A research group in Jadavpur University, India, directed by Bhattacharjee, has done much work about dynamic shear-enhanced membrane filtration. They have designed and developed a rotating disk module (RDM) (Sarkar et al., 2011) and a spinning basket module (SBM) (Sarkar et al., 2012), as illustrated in Figure 8. Their RDM is equipped with two motors with speed-controllers to provide independent rotation of the stirrer and membrane disk. The module can rotate the membrane and the stirrer in opposite direction to provide maximum shear near the membrane (Sarkar et al., 2011). However, their rotating speed was less than 300 rpm for the stirrer and less than 100 rpm for the membrane, which is too low for industrial application. They also installed stationary vanes near membrane surface to increase the shear rate, and found that 45° blade angle vanes having holes on the blade surface produce a maximum permeate flux (Sen et al., 2010). On the other hand, their SBM consists of a hollow basket with four radial arms (which may be increased in a scaled up module) mounted on a central hollow shaft, as shown in Figure 8. The whole system was placed in a stainless steel cylindrical tank and rotated in the direction of the membrane surface (outward normal to the membranes). When flux reaches its steady state, the basket is rotated in the reverse direction (in the direction normal to the impermeable side of the radial arms, Sen et al., 2010). Because of the counter-rotation-induced pressure difference, the accumulated solute is expected to be dispersed from the membrane thereby removing cake fouling, equal to the back flushing operation. However, as expected, due to the high hydraulic resistance of membrane disk, this new device may not be energy efficient.

Liu et al. (2012) have designed a new helical rotating membrane module to enhance turbulence, in order to increase permeate flux and reduce fouling. The membranes at various helical angles have a length of 300 mm and a width of 30 mm. As shown in Figure 9(a), this rotating filtration device included a DC motor, a set of gears, a housing with outlet pipe and a membrane module. The drive gear was mounted on the top of the housing and the membrane module was at the bottom. By adjusting voltage of the DC transformer, the rotating speed of membrane module was controlled at 75–160 rpm. Another rotating flat-sheet membrane system has been produced by the same group, and the membrane module was composed of nine identical flat-sheets in a cylindrical tank with an internal

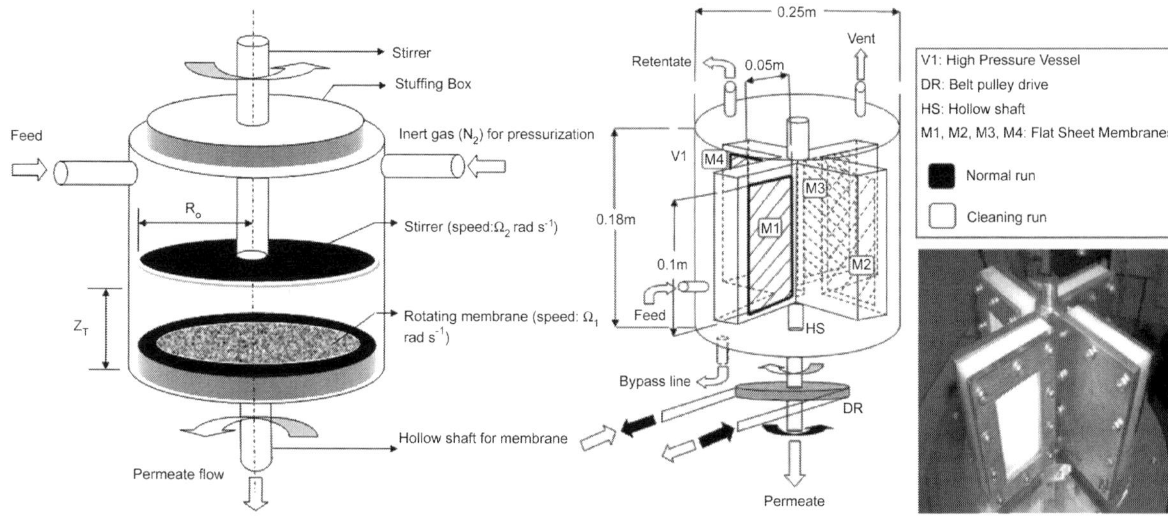

Figure 8 Schematic diagram of RDM (left) and SBM (right) modules in Bhattacharjee'sgroup. *From Sarkar et al. (2011, 2012) with permission*

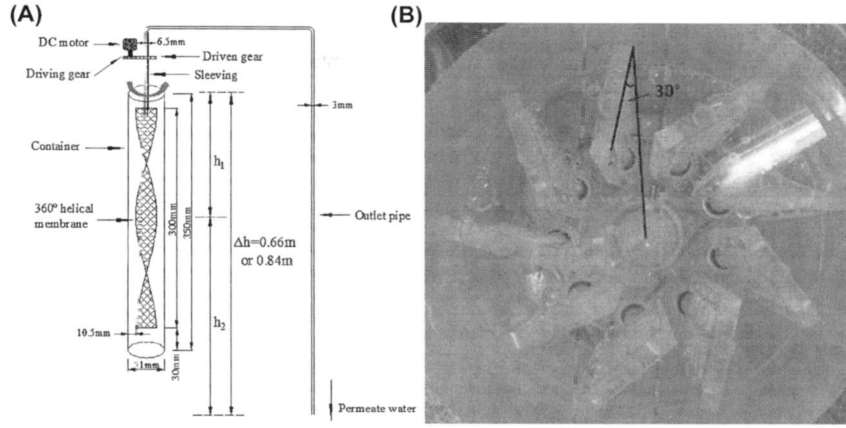

Figure 9 Schematic diagram of rotating helical and flat membranemodules. *From Liu et al. (2012) and Jiang et al. (2013) with permission*

diameter of 240 mm and a total effective volume of 13 L, and the diameter of each plate was 160 mm and the effective height and width of each flat-sheet is 153 and 39 mm, respectively, as shown in Figure 9(b). These two devices may be promising for MBR (Membrane Bio-Reactor) configuration, and needs further development.

A vibrating submersed hollow fibres system is a new configuration for shear-enhanced membrane filtration, and may have a great potential in MBR application. Beier and Jonsson (2009) described a vibrating membrane bioreactor (VMBR), and as shown in Figure 10, consisting of hollow fibres placed vertically in a bundle, and connected to a rod moving up and down due to a "rotation head" fixed to an electric motor. The frequency of the vibrations could be adjusted between 0 and 30 Hz. Another system was reported by Gomaa et al. (2011). Their membrane module included several turbulence promoters and oscillated vertically using an adjustable eccentric driven by a variable speed motor; this arrangement provided a range of oscillation frequencies from 0 to 25 Hz. Bilad et al. (2012) presented a novel magnetically induced membrane vibration (MMV) for fouling control in MBR, and the vibration was created by magnetic attraction/repulsion forces in a "push and pull" mode, and the vibration moves the membrane to the left and the right through a sinusoidal pattern at frequencies between 0 and 60 Hz. As the vibrating engine is integrated in the membrane module and magnetically induced, it is expected to undergo less friction, consume less energy, and have a flexible vibration control.

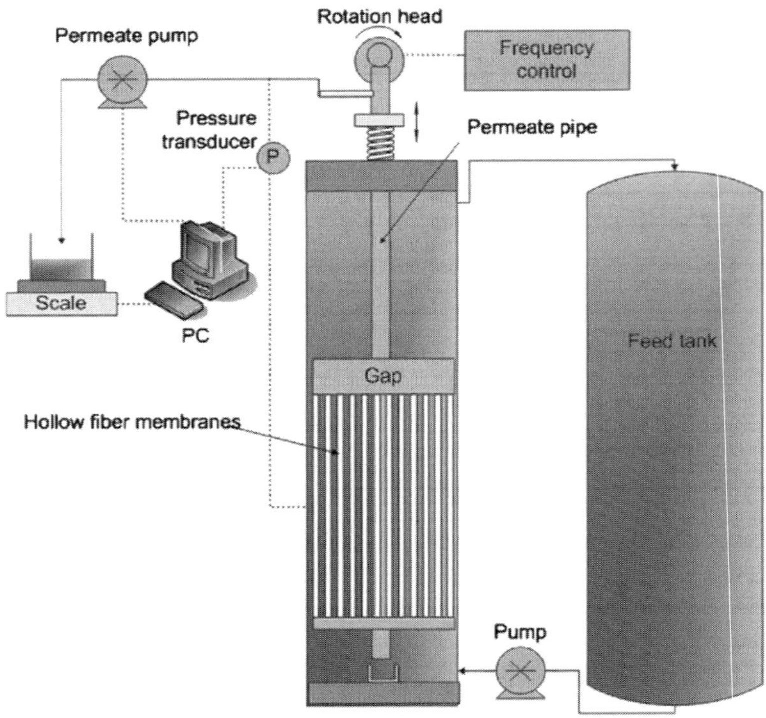

Figure 10 Schematic diagram ofVMBR. *From Beier and Jonsson (2009) with permission*

3. CALCULATIONS OF INTERNAL FLUID DYNAMICS IN VARIOUS DYNAMIC FILTRATION MODULES

3.1 Membrane Shear Rate and Pressure Distribution in Rotating Disk Modules with Fixed Membranes

The internal fluid mechanics of such systems has been investigated by Bouzerar et al. (2000a,b) in a small pilot designed in their laboratory and denoted as UTC RDM. For laminar flow, for small rotation speeds, a solution of Navier Stokes equation gave, for the local shear rate as function of radius r,

$$\gamma_1 = 0.77(k\omega)^{1.5} r\nu^{-0.5} \tag{1}$$

where ω is the disk angular velocity, k a velocity coefficient such that $k\omega$ is the angular velocity of the inviscid core between the disk and the membrane. This coefficient was measured to be 0.42 for a smooth disk, but increased if the disk was equipped with vanes and could exceed 0.82 when the disk was equipped with 8 radial vanes 6 mm high (Brou et al., 2002). In turbulent flow, they proposed for the local shear rate on the stationary membrane, using the Blasius friction coefficient for a flat plate,

$$\gamma_t = 0.0296(k\omega)^{1.8} r^{1.6} \nu^{-0.8} \tag{2a}$$

while the membrane mean shear rate over its radius R is given by

$$\gamma_{tm} = 0.0164(k\omega)^{1.8} R^{1.6} \nu^{-0.8} = 0.55\gamma_{max} \tag{2b}$$

where γ_{max} is the maximum shear rate at disk rim ($r = R_d$)

The pressure distribution in the fluid between the disk and the membrane was given by Bernouilli equation as,

$$p = (1/2)\rho(k\omega r)^2 + p_0 \tag{3}$$

where p_0 is the pressure at centre or when the fluid is at rest. Thus the TMP obtained by integration of Eqn (7) over the disk area is

$$p_{tm} = p_c - 1/4\rho(k\omega R_d)^2 \tag{4}$$

where R_d is the disk radius and p_c the peripheral pressure, measured at a tap in the housing. Calculations of the shear rates were also made by Torras et al. using Fluent software (Torras et al., 2006) who found that shear stresses on the rotating disk and on the fixed membrane were similar and in good agreement with those given by Murkes and Carlson (1998) for a rotating disk.

3.2 Membrane Shear Rate in Vibrating Systems

3.2.1 Unsteady Membrane Shear Rate in VSEP Modules

The flow induced by torsional oscillations of two parallel disks as in the VSEP has been investigated by Rosenblatt (1960). This flow is periodic and the membrane shear rate, which depends upon time as well as upon radius r, has been calculated by Al Akoum et al. (2002). The local shear rate on the membrane is given by

$$\gamma(r,t) = (r/R_2)d(\pi F)^{1.5}\nu^{-0.5}(\cos\omega_1 t - \sin\omega_2 t) \tag{5}$$

where d is the membrane displacement at radius r, itself proportional to r, R_2 is the outer radius, F the oscillations frequency $= \omega_1/2\pi$ and ν is the fluid kinematic viscosity.

The maximum with time of this shear rate at the disk periphery is given by,

$$\gamma_{max1} = 2^{0.5} d_1 (\pi F)^{1.5} \nu^{-0.5} \tag{6}$$

where d_1 is membrane displacement at periphery. The mean shear rate over the membrane area is

$$\gamma_m = \frac{2^{3/2}\left(R_2^3 - R_1^3\right)}{3\pi R_2\left(R_2^2 - R_1^2\right)}\gamma_{\max 1} = 0.330\gamma_{\max 1} \tag{7}$$

The shear rate in vibrating hollow fibre modules has been calculated by Beier et al. (2006) as,

$$\gamma = V_0[\omega_2/(2\nu)]^{0.5}\left[\sin\omega_2 t - \cos\omega_2 t\right] \tag{8}$$

where ω_2 is the angular frequency, d_2 their displacement amplitude, $V_0 = d_2\omega_2$ is the velocity amplitude of the fibres. Its maximum with time, when $\omega_2 t = -\pi/4$ or $3\pi/4$ is

$$\gamma_{\max 2} = 2^{0.5} V_0 d_2[\omega_2/(2\nu)]^{0.5} \tag{9}$$

Since $\pi F = \omega_1/2$, Eqn (11a) may be rewritten as

$$\gamma_{\max 1} = 2^{0.5} d_1(\omega_1/2)^{1.5}\nu^{-0.5} \tag{10}$$

and the ratio of maximum shear rates for these two systems will be

$$\gamma_{\max 1}/\gamma_{\max 2} = (d_1/d_2)(\omega_1/\omega_1)^{1.5} \tag{11}$$

If the membrane displacements and angular frequencies were the same, the hollow fibres module would have a larger mean shear rate on the membrane, as it is uniform inside the fibres, while it decreases toward the centre in the VSEP. But for practical reasons, the vibration amplitude of the fibres is limited to about 3 mm and fibres vibration frequency to 30 Hz (Beier et al., 2006) against 2.5–3 cm and 60 Hz for the VSEP giving a ratio $\gamma_{\max 1}/\gamma_{\max 2}$ equal to 28. A similar system tested by Genkin et al. (2006) had higher vibration amplitude (4 cm) but lower frequency of 2 Hz. Thus, in both systems, the effective shear rate was much lower than in the VSEP.

4. RECENT APPLICATIONS OF DYNAMIC FILTRATION AND INDUSTRIAL CASE STUDIES

4.1 VSEP Applications

4.1.1 MF and UF Applications

Takata et al. (1999) used a small VSEP L series pilot equipped with a single 100 kDa annular membrane of 30 cm diameter and 500 cm^2 area to remove humic substances present in rivers. The TMP was set to 600 kPa, the temperature to 15 °C and the membrane vibration amplitude to 2.5 cm. Permeate fluxes were high, decaying from 420/L/h/m^2 to 300/L/h/m^2 after 25 h of filtration and to 200/L/h/m^2 after 95 h and was restored to 300/L/h/m^2 after chemical cleaning. Feed coagulation by poly aluminium chloride permitted to reject molecules above 30 kDa and increased the flux by 45% after 60 h of filtration.

Low et al. (2004) chose the same VSEP pilot to recycle water from carbon-loaded wastewater generated in a TV picture tubes plant by MF. They used Teflon membranes of 0.07 μm pore size at a TMP of 2.4 bar at a temperature of 28 °C and a vibration amplitude of 2.54 cm. Due to the low carbon concentration of 0.2% wt (about 2/gL) and the high shear rate permeate fluxes were very high, decaying from an initial value of 1800/L/h/m^2 to 520/L/h/m^2 after 5 h of filtration. Other tests performed with a 1.0 μm pore size gave a lower flux, due to internal fouling by carbon particles.

Zouboulis and Petala (2009) filtered landfill leachates wastewaters with the L type VSEP pilot using successively 0.1 μm, 100 and 10 kDa membranes. They confirmed that the VSEP yielded large and stable permeate fluxes of about 150/L/h/m^2 at 100 kDa and 100/L/h/m^2 at 10 kDa while increasing small solutes removal as compared to CFF.

Al Akoum et al. (2002) investigated the internal dynamics of the VSEP L pilot by measuring the membrane displacement amplitude at rim with vibration frequency with an accelerometer placed on its housing. They found that this amplitude rose rapidly from 0.6 cm at 59 Hz to a maximum of 3.0 cm at the resonant frequency of 60.75 Hz. It must be noted that, even with industrial modules of 60 cm, the maximum displacement is the same and New Logic recommends not to exceed 2.5 cm to limit maintenance costs. In MF of yeast suspensions at 20 °C and 30 kPa, the stabilized permeate flux J in 1/L/h/m^2 was found to vary with the mean membrane shear rate γm as

$$J = 0.35\gamma_{\mathrm{m}}^{0.50} \tag{12}$$

While in UF at 50 kDa of bovine albumin solution at 10/gL, 35 °C and 500 kPa, the stabilized permeate flux was higher, but had a similar variation with γ_{m} as

$$J = 3.95\gamma_{\mathrm{m}}^{0.426} \tag{13}$$

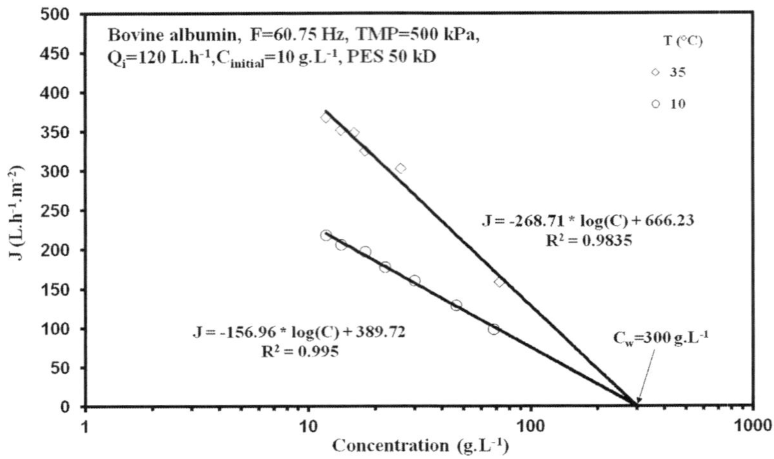

Figure 11 Variation of permeate flux with bovine albumin concentration in UF concentration tests using a VSEP pilot. *From Al Akoum et al. (2002) with permission*

The variation of permeate flux in concentration tests without permeate recycling at $T = 10°$ and 35 °C is shown in Figure 11. These fluxes decayed with increasing albumin concentration according to the logarithmic law of Blatt et al. (1970) for concentration polarization until the albumin gel concentration of 300/gL corresponding to zero flux.

Jaffrin et al. (2004) compared permeate fluxes obtained by a VSEP L pilot and a UTC RDM module equipped with the same 0.2 μm nylon membrane in MF of yeast suspension at 3/gL. These fluxes are plotted as a function of TMP in Figure 12 for the VSEP operated at 60.75 Hz and the RDM with its disk rotating at 2000 rpm with and without radial vanes. Values of maximum shear rates at membrane rims, indicated in the figure, ranged from $0.99 \cdot 10^5$/s for the smooth disk to $3.05 \cdot 10^5$/s for the disk with vanes, while the VSEP had a time maximum shear rate of $1.12 \cdot 10^5$/s. It is interesting to note that, although the RDM and the VSEP have different internal geometry, the ratio of their maximum fluxes (1.13) is equal to the ratio of their shear rates which is 13% higher for the VSEP. This indicates that, in DF, the permeate flux is mainly governed by the shear rate and not by module geometry. When shear rates decreased during concentration tests with the same modules, due to viscosity increase, corresponding permeate fluxes fell on the same line of equation $J = 4.3 \cdot 10^{-6}\ \gamma_{max}^{1.459}$ with a high R^2 correlation coefficient (see Figure 13).

Vane et al. (2002) removed volatile compounds from aqueous solutions using silicone pervaporation membranes in a 10 m^2membrane area VSEP

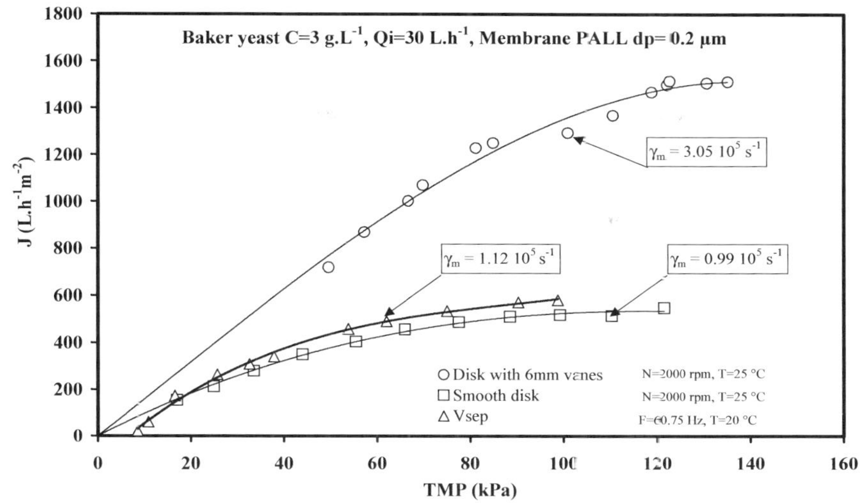

Figure 12 Variation of permeate fluxes with TMP in MF of yeast suspensions with the UTC RDM and two types of disks and with theVSEP. *From Jaffrin et al. (2004) with permission*

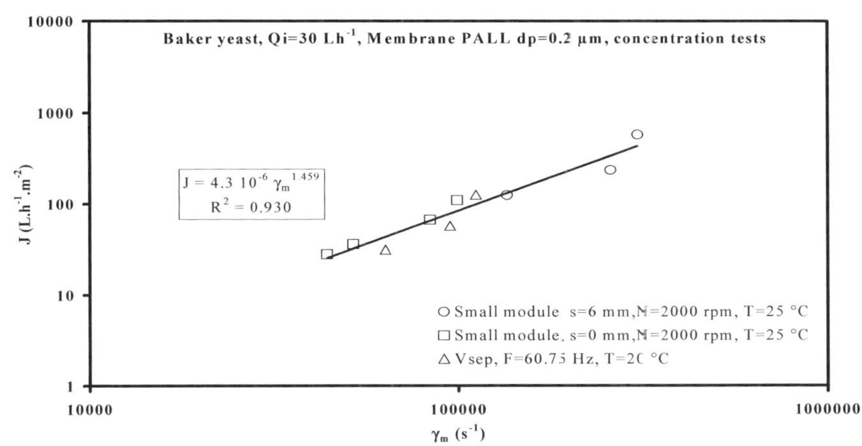

Figure 13 Variation of permeate fluxes with maximum membrane shear rate during concentration tests with RDM andVSEP. *From Jaffrin et al. (2004) with permission*

module CF1010. They noted that the mass transfer was much higher than with spiral wound modules and they further optimized the performance by connecting in series 13 packs of three membranes in parallel in order to avoid a tapering effect of the retentate.

4.1.2 NF and RO Applications

The Kay Bailey Hutchinson desalination plant at El Paso produces 19,000 m^3 per day of concentrated brackish water. This brackish water retentate was further concentrated by a VSEP to reduce the disposal cost, a task which could not be achieved by CFF (Delgado, 2009).

Frappart et al. (2008) treated a model dairy effluent represented by skim milk diluted to 1/3 concentration using a VSEP pilot and a UTC RDM pilot equipped with the same Desal Ag RO membrane and a disk with vanes. The variation of permeate flux with the volume reduction ratio (VRR) during concentration tests is plotted in Figure 14 in semilog coordinates for three frequencies, 60, 60.2 and 60.75 Hz for the VSEP and at 1000 and 2000 rpm for the RDM. Fluxes at initial concentration (VRR = 1) were high (>180/L/h/m^2) and almost the same for the VSEP at 60.75 Hz and the RDM at 2000 rpm were almost the same, although the shear rate was higher for the RDM, but at higher VRR, the RDM flux exceeded the VSEP flux by about 15%. The theoretical maximum VRR, was found by extrapolation to zero flux, to be equal to 12 for both modules. It is interesting to note that the VSEP gave, at 60.75 Hz, a significantly higher flux than the RDM at 1000 rpm for all VRR although its maximum shear rate was, at $1.35 \cdot 10^5$/s only slightly higher than that of RDM ($131 \cdot 10^5$/s). This is probably due to the effect of vibrations resulting in a periodic variation of shear rate

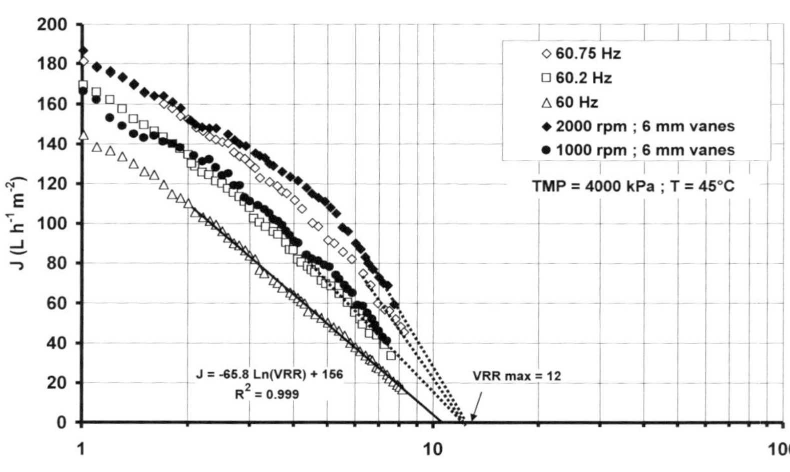

Figure 14 Variation of permeate fluxes in RO of model dairy effluent with a VSEP at various ferequencies and an RDM at 1000 and 2000 rpm. *From Frappart et al. (2008) with permission*

which contribute to raise the flux. This also confirms the efficiency of the VSEP for reaching high solid concentrations.

4.2 Applications of Rotating Disks and Rotating Membranes Modules

4.2.1 MF and UF Applications

One of the first investigations of the rotating disk system (Pall Corp) was carried out by Lee et al. (1995) who utilized a small pilot with a 15 cm diameter disk rotating above a circular flat membrane of 137 cm^2 area and 0.65 μm pores, for MF of yeast suspensions. The shear rate at membrane tip was calculated to be 120,000/s at the maximum speed of 3450 rpm. At this speed, permeate fluxes were higher, by a factor of 10 after 10 min of filtration, than those generated by a flat-sheet module equipped with the same membrane. Frenander and Jönnson (1996) used the same pilot to recover a protein produced by recombinant *Vibrio* cholerae, and when operating at 3300 rpm, a stabilized flux of 180/L/h/m^2 was obtained, three times higher than with normal crossflow filtration (CFF) with a transmission of 97% instead of 70% in CFF. Pessoa and Vitolo (1998) employed a 76 mm diameter disk rotating 4 mm above a flat membrane at the bottom of a tank at speeds ranging from 400 to 1600 rpm for a maximal peripheral speed of 6.4/ms. Using this device, cell harvesting was carried out with several membranes of pore size ranging from 0.04 to 0.45 μm, and a complete transmission was obtained at 1600 rpm and low TMP.

In our laboratory, Bouzerar et al. (2000a,b) designed a 15 cm diameter module with a plastic disk rotating near a fixed circular membrane (Figure 15, denoted as UTC module) and tested different types of disks,

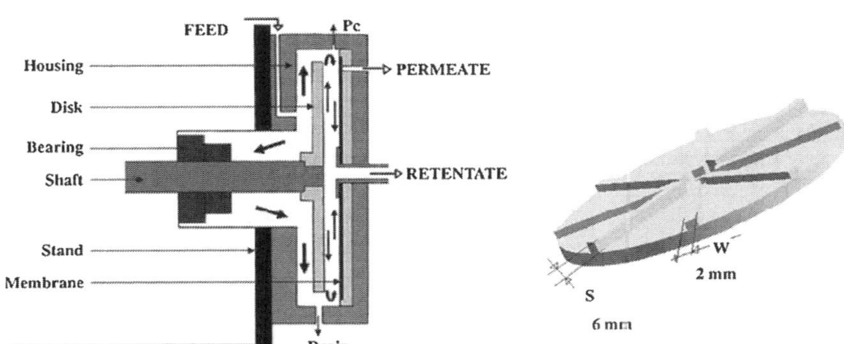

Figure 15 Schematic of UTC rotating disk module with disk equipped withvanes.

smooth and with four or six radial vanes of various heights from 2 to 6 mm. Initial tests were made in MF with $CaCO_3$ mineral suspensions as test fluid. They evaluated the radial distribution of flux using annular membranes and verified that the flux increase with radius was larger at high speed, when flow was turbulent, while at low speeds, the flux reached a maximum at a radius of about 4 cm (Jaffrin et al., 2004). Brou et al. (2002) also measured the permeate fluxes in with various types of disks and confirmed the large increase in flux due to the effect of vanes (Figure 15). The flux increased by 171% at a TMP of 120 kPa when a smooth disk was replaced by a disk with 6 mm vanes.

Harscoat et al. (1999) compared the recovery of glucuronane polysaccharides from fermentation broth using tubular ceramic membranes of 0.5 μm pores and the UTC rotating disk module equipped with a nylon membrane of 0.2 μm pores. Not only stabilized permeate fluxes were higher with the rotating disk module, up to 110/L/h/m^2 for a disk with 2 mm vanes rotating at 1500 rpm, against 55/L/h/m^2 for the tubular membrane, but the glucuronane sieving coefficient remained stable and larger than 90% for the UTC module. Since the high shear stress prevented the formation of a secondary polysaccharides layer, glucuronane mass fluxes were up to 25 times larger with the rotating disk module than with the tubular membrane, after 3 h of filtration. Mellal et al. (2008) also used the UTC module in separation of oligoglucuronans of low degrees of polymerization obtained by enzymatic degradation of a bacterial polysaccharide produced during fermentation by a Sinorhizobium meliloti mutant strain.

Nuortila-Jokinen and Nystrom (1996) have compared a cross rotational CR 500 filter (Raisio Flootek, Finland) pilot in UF of paper mill process waters with PCI tubular polymeric and mineral membranes. They got significantly higher fluxes, by a factor of four–five, with the CR filter rotating at 470 rpm and producing a peripheral rotor speed of 12/ms than tubular membranes at crossflow velocities of 2.1–2.5/ms. After an initial flux decline during the first 8 h, the CR 500 filter could sustain a quasi steady flux around 250–350/L/h/m^2 for close to 100 h. Wu et al. (2008) applied a submerged rotating membrane into a bioreactor with synthetic sewage, and indicated that increasing rotation speed, ratio of aeration intensity to permeate flux, and stoppage time could help to prevent reversible fouling in the operation.

An investigation of MF of oil/water emulsions was made by Viadero et al. (2000) using a Spintek module with rotating ceramic membranes of 0.11 μm pores. They found, during concentration tests, a logarithmic decay of flux with increasing concentration until reaching an oil concentration of 35%. Dal Cin et al. (1998) used a Spintek ST11 L with two rotating UF organic membranes for cutting oil emulsions. They observed that the flux reached a maximum at a certain speed depending upon inlet pressure, due to permeate back pressure. This speed varied from 900 rpm at 310 kPa to 400 rpm at 70 kPa, confirming the flux limitations of rotating membranes at low pressures. The reduction of ionic surfactant (Sodium dodecyl benzene sulfonate) in aqueous solutions was investigated by Moulai-Mostefa et al. (2007) using the UTC module and PES membranes, and they found that at lower pressure (900 kPa), no increase of flux was observed above 500 rpm, indicating that the flux was pressure limited, probably due to strong interactions between surfactant molecules. Li et al. (2009) employed UTC module to recover the linseed oil from oil-in-water (O/W) emulsion by 50 kDa UF membrane, and 76% oil was recovered in supernatant after concentration.

Akoum et al. (2006) also used the UTC rotating disk module for concentrating soy milk proteins while recovering trypsin inhibitors in the permeate by UF. The fluxes at initial concentration were higher with vanes, reaching 98/L/h/m^2 at 2500 rpm and 68/L/h/m^2 at 2000 rpm, than the 37/L/h/m^2 obtained by Berry et al. (1988) in UF of soymilk with hollow fibres. Ding et al. (2003) used the UTC module in UF of skim bovine milk together with a larger 26 cm diameter unit in stainless steel equipped with a 460 cm^2 50 kDa membrane on each side of the disk. The same rotating disk system, together with a Rayflow flat system equipped with the same 40 kDa membrane, were used by Frappart et al. (2011) to separate microalgae from sea water. In concentration tests, the rotating disk module yielded a flux of 80/L/h/m^2 at VRR = 3 versus 35 for the Rayflow. Luo et al. (2013) used the UTC module to clarify raw chicory extract by MF and UF membranes, and found that a moderate rotating speed of 1000 rpm was a good compromise between energy saving and fouling control, while a higher disk rotating speed could be employed at regular intervals for online membrane cleaning, and operating at "threshold flux" in constant flux mode was recommended for chicory juice clarification by high-shear membrane filtration.

Sarkar et al. (2008) described an original system consisting of a membrane disk rotating next to a contra-rotating rotor, and they applied this technique to the recovery of proteins from casein whey, using successively a 30 kDa membrane to concentrate caseins and a 5 kDa membrane to recover lactose in permeate (Sarkar et al., 2009). Fillaudeau et al. (2007) used an RVF module (Profiltra, Boulogne Billancourt, France) with an impeller-shaped rotor, rotating between two membrane disks for clarification of rough beer. The impeller produced TMP variations which vibrated the membranes and possibly contributed to their cleaning. The permeate flux exceeded $250/L/h/m^2$ at 4 °C with a 1.1 μm pore membrane, much higher than with crossflow filtration.

Ding et al., 2006 microfiltered $CaCO_3$ suspensions with a double-shafts MSD pilot (Westfalia separator) equipped with 12 ceramic disks of 9 cm diameter, shown in Figure 16. Their nominal pore size was 0.2 μm and the total membrane area was 0.121 m^2. Disks from each shaft overlap for 15.5% of their surface, and the membrane shear rate is maximum during overlapping. The fluxes keep increasing with TMP until at least 230 kPa and reach $800/L/h/m^2$ at 1930 rpm, but the effect of increasing rotation speed is not as important as with a rotating disk-fixed membrane module, presumably due to the effect of permeate counter pressure at high speed. This pilot was later modified by replacing the ceramic disks on one shaft by metal ones, either smooth or with 2 mm vanes and each shaft was driven by different motors to permit rotating metal disks at different speeds from membranes (He et al., 2007). Figure 3 compares the variations of permeate flux with TMP for these various configurations, at the same speed of 738 rpm for both shafts. The highest flux ($800/L/h/m^2$) was obtained when using metal disks with vanes, and was equivalent to the flux reached at 1930 rpm with 12 ceramic disks (shown in Figure 17). Without metal

Figure 16 Picture of MSD 12 ceramic disks pilot from Westfalia Separator.

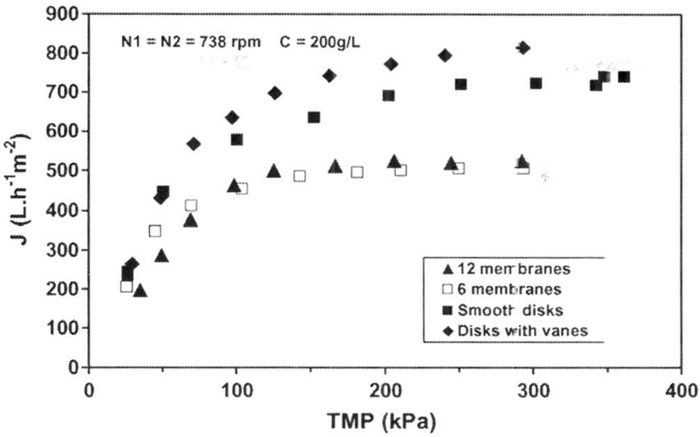

Figure 17 Variation of permeate flux withTMPfor 12 ceramic disks (normal MSDpilot, six ceramic disks only, six ceramic disks and 6 smooth metal disks, six ceramic disks and six metal disks withvanes). *From Ding et al. (2006) with permission*

disks, with either 6 or 12 membranes, the maximum flux was only 500/L/h/m^2. However the total permeate flow rate remains higher with 12 membranes than with 6 membranes and 6 metal disks. These tests showed that the design patented by Feuerpeil et al.(2003) consisting in a large diameter central metal disk surrounded by several ceramic filtrating disks is an interesting alternative to the MSD design with only ceramic disks, as it avoids local over concentration between overlapping membrane Since metal disks are cheaper than ceramic ones, the cost of such systems may be less, for the same output, than for original MSD systems.

Tamneh and Ripperger (2008) compared the performance of an MSD lab pilot in single- and double-shaft configurations to quantify the gain in flux due to overlapping membranes. From electrical power measurements, they concluded that the membrane shear stress in double-shaft configuration was about twice than in single-shaft configuration. This was verified by the absence of cake formation with two shafts and at a speed of 750 rpm, the flux remained steady at 1900/L/h/m^2, while it dropped rapidly to 400/L/h/m^2 with one shaft. Since ceramic membranes for the MSD were only available in limited pore size or cutoffs, Tu and Ding (2010) replaced them by disks equipped with two nylon membranes of same size and pore diameter (0.2 μm) as original ceramic membranes to concentrate $CaCO_3$ suspensions. Maximum permeate fluxes were higher at 300 kPa and 1930 rpm for nylon membranes, reaching 850/L/h/m^2 versus 760/L/h/m^2 for ceramic membranes, due to their higher permeability and hydrophilicity. Espina et al.

(2010) described a two-stage MF-UF process for fractionation of milk proteins using an MSD pilot for extracting casein micelles in MF retentate and 80% of β-Lg proteins in permeate. This permeate was filtrated by a 50 kDa UF membrane in a rotating disk module to recover α-La in permeate, and a 90% transmission and a mean flux of $400/L/h/m^2$ up to VRR = 3 were obtained.

4.2.2 Applications in NF and RO

Since few rotating disk systems can sustain the high pressures more than 1000 kPa, this technique is not widely applied in NF and RO processes. At shear rates of the order of 10^5/s, the reduction in concentration polarization enables the permeate flux to keep increasing with TMP until at least 4000 kPa. Frappart et al. (2006) used the UTC module equipped with a Desal five DK membrane to treat a model of dairy process water, aiming to reduce the chemical oxygen demand (COD) and ionic content. This system was proved to be very efficient, as not only permeate fluxes were much higher than those obtained with spiral wound modules with the same membrane (Balannec et al., 2002), but also the lactose and ion rejections were increased by the reduction in concentration polarization. The permeate COD remained below the allowed limit of 125 mgO_2/L until a VRR of 4.2 even though the initial retentate COD was very high (36,000 mgO_2/ L), while the permeate conductivity surpassed limited value and a two-step NF treatment was required. However, with one single reverse osmosis step, the UTC module permitted to obtain reusable water from the same model effluent (Frappart et al., 2008).

Luo et al. (2010) selected a more permeable NF 270 membrane and compared the results with those of the Desal 5-DK. As seen in Figure 18, the NF270 membrane gives a much higher permeate flux than Desal-5 DK, starting from $450/L/h/m^2$ at VRR = 1 versus only $200/L/h/m^2$ for the 5-DK. But at VRR = 8, the difference was lower, 140 versus 90/ $L/h/m^2$. Luo and Ding (2011) and Luo et al. (2012a) also found that using UTC module with NF270 membrane, the flux of real dairy wastewater was of the same level as that of model effluent in despite of different compositions, and membrane fouling was very sensitive to feed pH in a range of 8–10, while operating at "threshold flux" could minimize the fouling. In two recent papers (Luo et al., 2012b,c), they clarified the mechanisms behind the flux decline when treating dairy wastewater by UTC module. As illustrated in Figure 19, when operating at high shear rate and pressure, after a stable flux period, a slow flux decline caused by surface adsorption

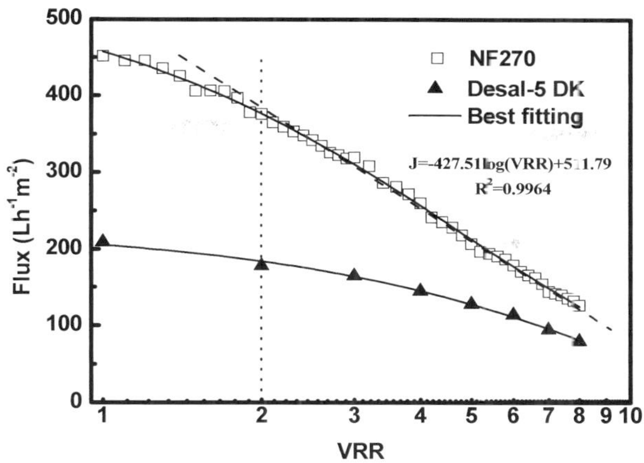

Figure 18 Permeate flux as a function of volume reduction ratio in semilog coordinates for different membranes. Feed: 1:2 diluted skim milk; TMP = 40 bar; rotating speed = 2000 rpm; T = 35 °C. *From Luo et al. (2010) with permission*

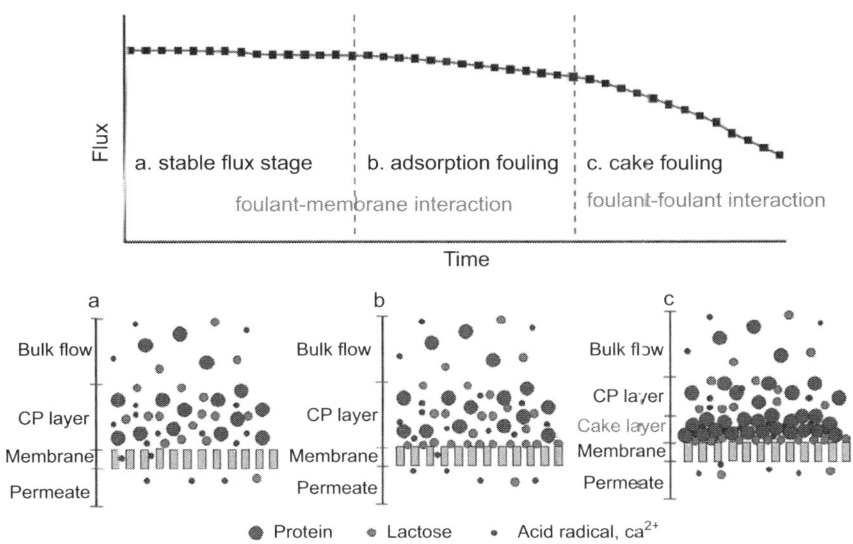

Figure 19 Schematic diagram of fouling mechanism for wastewater treatment by NF under extreme hydraulicconditions. *From Luo et al. (2012b) with permission*

of foulants (lactose, multivalent salt ions, and their aggregates) occurred. In this adsorption fouling stage, pore narrowing and blocking governed by foulant–membrane interaction were the main fouling mechanisms. In absence of chemical cleaning, this adsorption fouling could induce cake fouling formation by proteins-calcium aggregates, resulting in severe flux decline.

Treatment of wastewater containing a cleaning-in-place detergent by using UTC module with NF270 membrane has also been reported by Luo et al. (2012d), where at a rotational speed of 2000 rpm, a pH 7.2, and 25 °C, the permeate flux reached a plateau at $350/Lm^2/h$ above 35 bar while rejection of conductivity and COD were respectively 93% and 97%. Due to the high shear rate, the fouling layer formed in this process was not very compact and could be broken up by water rinse, and thus membrane permeability could be fully recovered without any chemical cleaning. When detergent wastewater was concentrated to VRR = 6 (COD = 237/gL), the flux still remained at $47/Lm^2/h$. The NF permeate might be reused as CIP cleaning water, and the precipitated surfactants in concentrates after sedimentation could be extracted for reuse and the concentration process could be continued to increase recovery of surfactants.

5. DISCUSSION

5.1 Energetic Considerations

DF modules require an additional motor to generate the shear rate on the membrane by a rotor or a rotating disk, or by rotating or vibrating the membrane, but they do not need large and powerful recirculation pumps like CFF modules. Thus, the energy consumed by this additional motor is generally offset by energy savings made on pump motors. In filtration, the relevant energy parameter is the specific energy consumed per m^3 of permeate (*E*s) which is given by the total power consumed Pt divided by the permeate flow rate QF. Even if *P*t is higher for a DF module than for a CFF module of same membrane area, its specific energy will be less if its permeate flux is significantly higher than that of CFF module. This point is illustrated as follows for the VSEP. According to Johnson (2008) the power spent in generating vibrations for a 150 m^2 membrane area VSEP is 9 kW, while the pump power will range from 14 kW in MF to 25 kW in RO because of the higher TMP. From data of Al Akoum (2002) we can estimate the mean permeate flux in concentration tests by MF of baker's yeast to be about $120/Lm^2/h$, giving $QF = 18/m^3/h$. Thus the specific energy will be

$$E_s = (14 + 9)/18 = 1.28\ \text{kWh/m}^3 \tag{14}$$

In the case of reverse osmosis, we can estimate the mean flux during concentration tests of model dairy effluents represented by diluted milk (Frappart et al., 2008) to be about 50/Lm^2/h or QF = 7.5 m^3/h. In that case, the specific energy will be

$$E_s = (25 + 9)/7.5 = 4.53\ \text{kWh/m}^3 \tag{15}$$

Of course, these values do not account for energy spent in heating the fluid and they will slightly increase if the module size is reduced as vibration power decreases less than membrane area in smaller modules.

Unfortunately, we did not find in the literature similar data for industrial rotating disks or rotating membranes modules. Ding et al. (2006) have measured the electrical power consumed by the rotating ceramic disks in a small MSD pilot at different rotation speeds from 738 to 1930 rpm during MF of $CaCO_3$ suspensions. The feed pump power was not included as the motor was oversized for a small pilot. They found that, although the power spent by disk friction increased with rotation speed, the specific energy decreased from 8.0 kWh/m^3 to 4.2 kWh/m^3 as speed increased because the flux increased faster than the power consumed. These values would probably decrease in an industrial module as the relative part of internal friction on shafts would decrease with larger disks. It must also be noted that, even if increasing the permeate flux to high levels may sometimes increase the specific energy, it may still be economically worthwhile as a higher flux permits to decrease the membrane area necessary for a given output and will decrease the investment cost.

5.2 Complementarity of Crossflow and Dynamic Filtrations

It is clear that DF modules cannot replace CFF modules in all circumstances as their cost per m^2 of membrane is generally higher, especially when compared to spiral wound modules and their maintenance probably more expensive. They are presently not suitable for treating very large volume of fluids, as in the case of desalination, wastewater treatment, or dairy applications. But DF modules may be advantageous if the module size is limited because of lack of space, or if the final retentate concentration is too high to be obtained with CFF modules. VSEP modules are often proposed as "end of pipe" treatment, to remove most of BOD, COD, TDS in the permeate and reduce the final volume of discharged retentate in order to lower its cost of disposal. It is cheaper to use CFF in primary MF or UF steps dealing with large volumes, but DF may be more efficient when very high solid concentrations must be obtained by NF or RO, as gains in permeate fluxes at high pressures are higher with these membranes than in MF and UF.

6. CONCLUSIONS

The benefits of high shear rates dynamic filtration for increasing permeate fluxes and membrane selectivity have been well confirmed in the scientific literature by many laboratories as well as in case studies reported by DF modules manufacturers. These benefits are higher in NF and RO than in MF and UF applications as the reduction in concentration polarization extends the pressure limited regime to high TMP exceeding 40 bar, resulting in an additional flux increase and in a diminution of microsolute transmission by the membrane which increase their rejection. The most active company in this field appears to be New Logic Research in California which is the only one selling modules operating at pressure of more than 40 bar, covering NF and RO as well as MF and UF. In addition, their use of resonant frequency permits to maximize the vibration amplitude of their modules with minimal power. It sells its industrial modules worldwide, for a large range of applications, biogas effluent, wastewater, black liquor and landfill leachates treatment, ethanol stillage, polyethylene glycol recovery, processing phosphate fertilizer, and polymer diafiltration.

It is unfortunate that DF modules sold by European manufacturers Bokela, Canzler, KMPT, and Metso Paper are limited to MF and UF. Only Novoflow sells an SSDF unit permitting MF, UF, and NF. The availability of large ceramic NF membrane disks should permit the development of multishaft DF modules with overlapping membranes rotating in a housing like the KMPT or the MSD which should be cheaper to build and to service than multicompartments systems with rotors between fixed membranes.

REFERENCES

Akoum, O., Richfield, D., Jaffrin, M.Y., Ding, L.H., Swart, P., 2006. Recovery of trypsin inhibitor and soy milk concentration by dyanmic filtration. J. Membr. Sci. 279, 291–300.

Al-Akoum, O., Jaffrin, M.Y., Ding, L.H., Paullier, P., Vanhoutte, C., 2002. An hydrodynamic investigation of microfiltration and ultrafiltration in a vibrating membrane module. J. Membr. Sci. 197, 37–52.

Balannec, B., Gesan-Guiziou, G., Chaufer, B., Rabiller-Baudry, M., Daufin, G., 2002. Treatment of dairy process waters by membrane operations for water reuse and milk constituents concentration. Desalination 147, 89–94.

Beier, S.P., Jonsson, G., 2009. A vibrating membrane bioreactor (VMBR): Macromolecular transmission-influence of extracellular polymeric substances. Chem. Eng. Sci. 64, 1436–1444.

Beier, S.P., Guerra, M., Garde, A., Jonsson, G., 2006. Dynamic microfiltration with a vibrating hollow fiber module: filtration of yeast suspensions. J. Membr. Sci. 281, 281–287.

Berry, S.E., Nguyen, M.H., 1988. High rate ultrafiltration of soy milk. Desalination 70, 169–176.

Bilad, M.R., Mezohegyi, G., Declerck, P., Vankelecom, I.F.J., 2012. Novel magnetically induced membrane vibration (MMV) for fouling control in membrane bioreactors. Water Res. 46, 63–72.

Blatt, W.F., David, A., Michaels, A.S., Nelson, L., 1970. Solute polarization and cake formation inmembran eultrafiltraion: cause, consequences and control techniques. In: Flynn, E.F. (Ed.), Membrane Science and Technology. Plenum Press, London, pp. 47–97.

Bouzerar, R., Ding, L.H., Jaffrin, M.Y., 2000a. Local permeate flux-shear-pressure relationship in a rotating disk microfiltration module: implications for global performance. J. Membr. Sci. 170, 127–141.

Bouzerar, R., Jaffrin, M.Y., Ding, L., Paullier, P., 2000b. Influence of geometry and angular velocity on performance of a rotating disk filter. AIChE J. 46, 257–265.

Brou, A., Ding, L.H., Jaffrin, M.Y., 2002. Dynamic microfiltration of yeast suspensions using rotating disks equipped with vanes. J. Membr. Sci. 197, 269–282.

Dal-Cin, M.M., Lick, C.N., Kumar, A., Lealess, S., 1998. Dispersed phase back transport during ultrafiltration of cutting oil emulsions with a spinning disc geometry. J. Membr. Sci. 141, 165–181.

Delgado, G.G., January 1, 2009. Treatment of RO Concentrate Using VSEP Technology. ETD Collection for University of Texas, El Paso. Paper AAI1473859.

Ding, L.H., Akoum, O., Abraham, A., Jaffrin, M.Y., 2002. Milk protein concentration by ultrafiltration with rotating disk modules. Desalination 144, 307–311.

Ding, L.H., Akoum, O., Abraham, A., Jaffrin, M.Y., 2003. High shear skim milk ultrafiltration using rotating disk filtration systems. AIChE J. 49, 2433–2441.

Ding, L.H., Jaffrin, M.Y., Mellal, M., He, G., 2006. Investigation of performances of a multishaft disk (MSD) system with overlapping membranes in microfiltration of mineral suspensions. J. Membr. Sci. 276, 232–240.

Espina, V., Jaffrin, M.Y., Ding, L., Cancino, B., 2010. Fractionation of pasteurized skim milk proteins by dynamic filtration. Food Res. Intern. 43, 1335–1346.

Feuerpeil, H.P., Blase, D., Olapinski, H., 2003. GmbH. German Patent DE 102 39 247 C1.

Fillaudeau, L., Boissier, B., Moreau, A., Blanpain-Avet, P., Ermolaev, S., Jitariouk, N., Gourdon, A., 2007. Investigation of rotating and vibrating filtration for clarification of rough beer. J. Food Eng. 80, 206–217.

Frappart, M., Akoum, O., Ding, L.H., Jaffrin, M.Y., 2006. Treatment of dairy process waters modelled by diluted milk using dynamic nanofiltration with a rotating disk module. J. Membr. Sci. 282, 465–472.

Frappart, M., Jaffrin, M.Y., Ding, L.H., 2008. Reverse osmosis of diluted skim milk: comparison of results obtained from vibratory and rotating disk modules. Sep. Purif. Technol. 60, 321–329.

Frappart, M., Massé, A., Jaffrin, M.Y., Pruvost, J., Jaouen, P., 2011. Influence of hydrodynamics in tangential and dynamic ultrafiltration systems for microalgae separation. Desalination 265, 279–283.

Frenander, U., Jönsson, A.S., 1996. Cell harvesting by cross-flow microfiltration using a shear-enhanced module. Biotech. Bioeng. 52, 397–403.

Genkin, G., Waite, T.D., Fane, A.G., Chang, S., 2006. The effect of vibration and coagulant addition on the filtration performance of submerged hollow fiber membranes. J. Membr. Sci. 281, 726–734.

Gomaa, H.G., Rao, S., Al-Taweel, M., 2001. Flux enhancement using oscillatory motion and turbulence promoters. J. Membr. Sci. 381, 64–73.

Harscoat, C., Jaffrin, M.Y., Bouzerar, R., Courtois, J., 1999. Influence of fermentation conditions and microfiltration process on membrane fouling during recovery of glucuronane polysaccharides from fermentation broths. Biotech. Bioeng. 65, 500–511.

He, G., Ding, L.H., Paullier, P., Jaffrin, M.Y., 2007. Experimental study of a dynamic filtration system with overlapping ceramic membranes and non-permeating disks rotating at independent speeds. J. Membr. Sci. 300, 63–70.

Jaffrin, M.Y., 2008. Dynamic shear-enhanced membrane filtration: a review of rotating disks, rotating membranes and vibrating systems. J. Membr. Sci. 324, 7–25.

Jaffrin, M.Y., Ding, L.H., Akoum, O., Brou, A., 2004. A hydrodynamic comparison between rotating disk and vibratory dynamic filtration systems. J. Membr. Sci. 242, 155–167.

Jiang, T., Zhang, H., Yang, F., Gao, D., Du, H., 2013. Relationships between mechanically induced hydrodynamics and membrane fouling in a novel rotating membrane bioreactor. Desalin. Water Treat. 51, 2850–2861.

Johnson, G., 2008. New Logic Research, USA, Personal Communication (email).

Kroner, K.H., Nissingen, V., 1988. Dynamic filtration of microbial suspensions using an axially rotating filter. J. Membr. Sci. 36, 85–100.

Lee, S.S., Burt, A., Russoti, G., Buckland, B., 1995. Microfiltration of recombinant yeast cells using a rotating disk dynamic filtration system. Biotech. Bioeng. 48, 386–400.

Li, L., Ding, L., Tu, Z., Wan, Y., Clausse, D., Lanoiselle, J.L., 2009. Recovery of linseed oil dispersed within an oil-in-water emulsion using hydrophilic membrane by rotating disk. J. Membr. Sci. 342, 70–79.

Liu, L., Gao, B., Liu, J., Yang, F., 2012. Rotating a helical membrane for turbulence enhancement and fouling reduction. Chem. Eng. J. 181–182, 486–493.

Low, S.C., Han, H.J., Jin, W.X., 2004. Characteristics of vibration membrane in water recovery from fine carbon-loaded wastewater. Desalination 160, 83–99.

Luo, J., Ding, L., 2011. Influence of pH on treatment of dairy wastewater by nanofiltration using shear-enhanced filtration system. Desalination 278, 150–156.

Luo, J., Ding, L., Wan, Y., Paullier, P., Jaffrin, M.J., 2010. Application of NF-RDM (nanofiltration rotating disk membrane) module under extreme hydraulic conditions for the treatment of dairy wastewater. Chem. Eng. J. 163, 307–316.

Luo, J., Ding, L., Wan, Y., Jaffrin, M.Y., 2012a. Threshold flux for shear-enhanced nanofiltration: experimental observation in dairy wastewater treatment. J. Membr. Sci. 409–410, 276–284.

Luo, J., Ding, L., Wan, Y., Paullier, P., Jaffrin, M.Y., 2012b. Fouling behavior of dairy wastewater treatment by nanofiltration under shear-enhanced extreme hydraulic conditions. Sep. Purif. Technol. 88, 79–86.

Luo, J., Cao, W., Ding, L., Zhu, Z., Wan, Y., Jaffrin, M.Y., 2012c. Treatment of dairy effluent by shear-enhanced membrane filtration: the role of foulants. Sep. Purif. Technol. 96, 194–203.

Luo, J., Ding, L., Wan, Y., Jaffrin, M.Y., 2012d. Flux decline control in nanofiltration of detergent wastewater by a shear-enhanced filtration system. Chem. Eng. J. 181–182, 397–402.

Luo, J., Zhu, Z., Ding, L.H., Bals, O., Wan, Y., Jaffrin, M.J., Vorobiev, E., 2013. Flux behaviour in clarification of chicory juice by high-shear membrane filtration: evidence for threshold flux. J. Membr. Sci. 435, 120–129.

Mänttäri, M., Vitikko, K., Nystrom, M., 2006. Nanofiltration of biologically treated effluents from the pulp and paper industry. J. Membr. Sci. 272, 152–160.

Mellal, M., Jaffrin, M.J., Ding, L., Delattre, C., Michaud, P., Courtois, J , 2008. Separation of oligogclucuronans of low degrees of polymerization by using a high shear rotating disk filtration module. Sep. Purif. Technol. 60, 22–29.

Moulai-Mostefa, N., Ding, L.H., Frappart, M., Jaffrin, M.Y., 2007. Treatment of aqueous ionic surfactant solutions by dynamic ultrafiltration. Sep. Sci. Technol. 42, 2583–2594.

Murkes, J., Carlsson, C.G., 1988. In: Crossflow Filtration. John Wiley & Sons.

Nuortila-Jokinen, J., Nyström, M., 1996. Comparison of membrane separation processes in the internal purification of paper mill water. J. Membr. Sci. 119, 99–115.

Pessoa, A., Vitolo, M., 1998. Evaluation of cross-flow microfiltration membrane using a rotary disc-filter. Process. Biochem. 33, 39–45.

Rosenblat, S., 1960. Flow between torsionally oscillating disks. J. Fluid Mech. 8, 388–399.

Sarkar, D., Bhattacharjee, C., 2008. Modeling and analytical simulation of rotating disk ultrafiltration module. J. Membr. Sci. 320, 344–355.

Sarkar, P., Gosh, S., Dutta, S., Sen, D., Bhattacharjee, C., 2009. Effect of different operating parameters on the recovery of proteins from casein whey using a rotating disc membrane ultrafiltration cell. Desalination 249, 5–11.

Sarkar, D., Datta, D., Sen, D., Bhattacharjee, C., 2011. Simulation of continuous stirred rotating disk membrane module: an approach based on surface renewal theory. Chem. Eng. Sci. 66, 2554–2567.

Sarkar, A., Sarkar, D., Bhattacharjee, C., 2012. Design and performance characterization of a new shear enhanced module with inbuilt cleaning arrangement. J. Chem. Technol. Biotechnol. 87, 1121–1130.

Sen, D., Roy, W., Das, L., Sadhu, S., Bhattacharjee, C., 2010. Ultrafiltration of macromolecules using rotating disc membrane module (RDMM) equipped with vanes: effects of turbulence promoter. J. Membr. Sci. 360, 40–47.

Takata, K., Yamamoto, K., Brian, R., Watanabe, Y., 1999. Removal of humic substances with vibratory shear enhanced processing membranes filtration. Water Supply 17, 93–102.

Tamneh, Y., Ripperger, S., 2008. Performance of single and double shaft disk separators. Phys. Sep. Sci. Eng.. ID 508617. http://dx.doi.org/10.1155/2008/508617.

Torras, C., Pallares, J., Garcia-Valls, R., Jaffrin, M.Y., 2006. CFD simulation of a rotating disk flat membrane module. Desalination 200, 453–455.

Tu, Z., Ding, L., 2010. Microfiltration of mineral suspensions using a MSD module with rotating ceramic and polymeric membranes. Sep. Purif. Technol. 73, 363–370.

Vane, L.M., Alvarez, F.R., 2002. Full scale vibrating pervaporation membrane unit:VOC removal from water and surfactant solutions. J. Membr. Sci. 202, 177–193.

Vane, L.M., Alvarez, F.R., Giroux, E.L., 1999. Reduction of concentration polarization in pervaporation using vibrating membrane module. J. Membr. Sci. 153, 233–241.

Viadero, R.C., Masciola, D.A., Reed, B.E., Vaughan Jr., R.L., Jr., 2000. Two-phase limiting flux in high-shear rotary ultrafiltration of oil-in-water emulsions. J. Membr. Sci. 175, 85–96.

Wu, G., Cui, L., Xu, Y., 2008. A novel submerged rotating membrane bioreactor and reversible membrane fouling control. Desalination 228, 255–262.

Zouboulis, A.I., Petala, M.D., 2009. Performance of VSEP vibratory membrane filtration system during the treatment of landfill leachates. Desalination 222, 165–175.

CHAPTER THREE

Membrane Distillation (MD)

Mohamed Essalhi[1] and Mohamed Khayet[1,2]
[1]Department of Applied Physics I, Faculty of Physics, University Complutense of Madrid, Madrid, Spain
[2]Madrid Institute for Advanced Studies of Water (IMDEA Water Institute), Alcalá de Henares, Madrid, Spain

Contents

GLOSSARY

Asymmetric membrane A membrane constituted of two or more nonidentical layers of different structures.

Brine A retentate saline solution with a concentration higher than that of seawater normally more than 50 g/L.

Concentration polarization Accumulation of higher level(s) of nonvolatile solute(s) or lower level(s) of volatile solute(s) close to the membrane surface in comparison to the well-mixed bulk solution in a membrane module.

Fouling Deposition or adsorption of dissolved components on the external membrane surface, at the pore openings, or within the membrane pores reducing the membrane performance.

Membrane casting Preparation of a flat sheet membrane by extending first a polymer solution over a glass plate or a support followed by coagulation in a nonsolvent or solvent evaporation, being the technique phase inversion.

Membrane distillation crystallization Application of MD to process highly concentrated aqueous solution, to recover water and generate the desired supersaturation in the crystallizer where product crystals can be precipitated.

Pore tortuosity Ratio of actual pore length to membrane thickness.

Porosity Void volume fraction or ratio of void space to total membrane volume of a porous membrane.

Retentate A solution containing a higher concentration of rejected solutes by the membrane than the feed solution.

Scaling Precipitation or crystallization of salts at the membrane surface, feed side of the membrane module.

Progress in Filtration and Separation
http://dx.doi.org/10.1016/B978-0-12-384746-1.00003-3

Selectivity Efficiency of solute separation from water by a membrane.

Separation factor A parameter relating the permeate concentration (C_p) to the feed concentration (C_f) of a membrane module ($(1-C_p/C_f) \times 100$).

Temperature polarization The transmembrane temperature is lower than the applied bulk temperature difference between the feed and permeate sides of the membrane module.

1. MEMBRANE DISTILLATION SEPARATION TECHNOLOGY AND ITS VARIANTS

Membrane distillation (MD) is a separation process that involves both nonwetting porous membrane and phase change generally applied for the treatment of solutions in which water is the major component for:

- Separation of water from dissolved solutes and production of distilled/potable water including ultrapure water,
- Concentration of nonvolatile dissolved solutes in aqueous solutions and recycling of valuable materials,
- Removal of volatile solutes from aqueous solution and their concentration in the permeate including separation of azeotropic mixtures, etc.

In this process, separation is carried out based mainly on the two phase changes, evaporation and condensation. Evaporation occurs at the liquid/vapour interfaces formed at one side of nonwetted pores of a hydrophobic membrane, inside the membrane module, whereas condensation step can be taken place inside or outside the membrane module, depending on the MD variant. These involved two phase change phenomena are the origin of the term MD, similar to conventional distillation. Other terms such as thermo-pervaporation, pervaporation, membrane evaporation, capillary distillation, and transmembrane distillation were used before 1986 (Franken and Ripperger, 1988; Smolders and Franken, 1989). However, MD is known since June 1963, when Bodell filed the first patent on MD (Bodell, 1963). However, the first MD publication was done in form of paper four years later in 1967 (Findley, 1967). Since then interest in MD process has been faded quickly because the obtained water production rate was low compared to other processes such as reverse osmosis (RO). In the early of the 1980s MD has recovered much interest when novel membranes with better characteristics and modules became available (Esato and Eiseman, 1975; Cheng, 1981; Gore, 1982; Cheng and Wiersma, 1982, 1983a,b; Carlsson, 1983; Catalogue of Enka AG, 1984; Schneider and van Gassel, 1984; Andersson et al., 1985; Van Gassel and Schneider, 1986).

The membrane required for MD applications must be porous and at least one of its layers must be hydrophobic and must not be wetted by the liquid phase. Only vapour is transported through this hydrophobic layer being the driving force the transmembrane partial pressure gradient. In addition, the membrane must not alter the vapour/liquid equilibrium of the involved compounds and condensation must not occur inside its pores. The hydrophobic nature of the membrane prevents liquid aqueous solutions from entering its pores due to the surface tension forces resulting in the formation of liquid/vapour interfaces at the extremes of each pore. In most cases, to establish the transmembrane driving force and the necessary latent heat for evaporation, the feed aqueous solution is heated to temperatures between 30 and 90 °C (i.e., below the boiling point of the aqueous liquid feed solution). Therefore, simultaneous heat and mass transfers occur in this process.

To establish the necessary driving force, different MD variants can be considered. The differences between them are made only in the permeate side (Figure 1):

1. *Direct contact membrane distillation* (DCMD): An aqueous solution colder than the feed solution is circulated tangentially to the permeate side of the membrane. The transmembrane temperature difference induces the necessary vapour pressure difference. In this case the volatile molecules of the feed aqueous solution (water or volatile organic compounds, VOCs) evaporate at the hot feed liquid/vapour interface, cross the membrane pores in vapour phase, and condense at the cold liquid/vapour interface inside the membrane module. Care must be taken in this case in order to prevent wetting of the pores from the permeate side of the membrane when using VOCs having low surface tension. During the treatment of aqueous solutions containing VOCs, the concentration of the volatile solute in the permeate aqueous solution will increase and will be high compared to the feed aqueous solution. Therefore the risk of membrane pore wetting from the membrane permeate side will be high. In general, DCMD is used for production of distilled/potable water using feed aqueous solutions containing nonvolatile solutes, e.g., desalination.
2. *Sweeping gas membrane distillation* (SGMD): A gas, such as air or nitrogen, sweeps the permeate side of the membrane carrying the evaporated molecules outside the membrane module for condensation. In this MD variant the gas temperature and its hydrostatic pressure are kept below those of the feed aqueous solution. SGMD is used mostly for

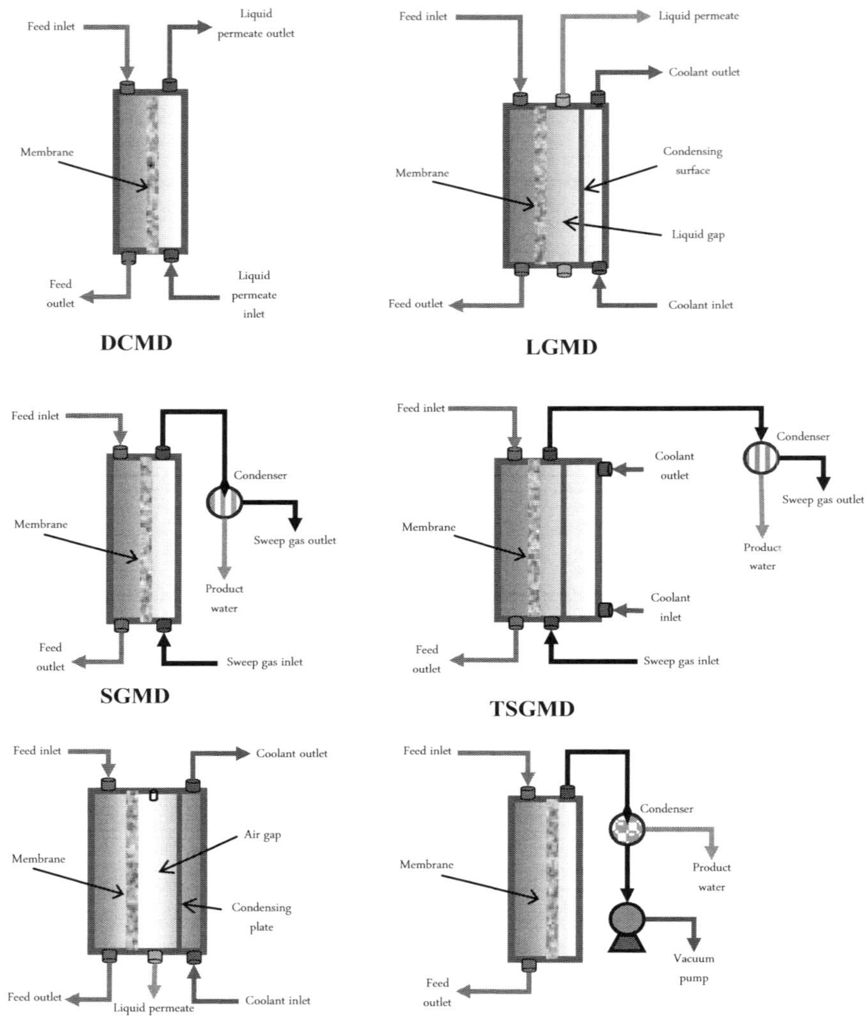

Figure 1 MD process variants: DCMD, SGMD; AGMD; LGMD; TSGMD, and VMD. MD, membrane distillation; DCMD, direct contact membrane distillation; SGMD, sweeping gas membrane distillation; AGMD, air gap membrane distillation; LGMD, liquid gap membrane distillation; TSGMD, thermostatic sweeping gas membrane distillation; VMD, Vacuum Membrane Distillation.

distilled/potable water production, concentration of solutes in the feed membrane side as well as the removal and concentration of VOCs in aqueous solutions. This variant is sometimes called membrane gas stripping or membrane air stripping (MAS) (Mahmud et al., 1998; Juang at al., 2005; Vialadomat at al, 2006). When a dense and selective membrane is used in the membrane module instead of a porous and hydrophobic membrane, the process is termed sweeping gas pervaporation (Calibo et al., 1987; Korngold and Korin, 1993).

3. *Air gap membrane distillation* (AGMD): In this case a cold condensing surface is placed inside the membrane module and a stagnant air gap is interposed between the membrane and the condensation surface to solve the problem of heat loss by conduction through the membrane, which leads to relatively low thermal efficiency of MD. The evaporated volatile molecules cross first the membrane pores and the air gap thickness to finally condense over the cold surface inside the membrane module. The permeate water exits from the bottom part of the membrane module taking advantage of the gravity. Because condensation is carried out over a cold surface rather than directly on the membrane surface, AGMD variant can be applied in fields where the DCMD is limited such as the removal of VOCs from aqueous solutions. In addition, AGMD is also applied for potable/distilled water production and concentration of nonvolatile solutes in the feed aqueous solutions. Because the permeate flux has to overcome the air barrier between the membrane and the condensing surface, it is reduced depending on the effective air gap width.
4. *Liquid gap membrane distillation* (LGMD): This is another variant of MD combining both DCMD and AGMD configurations. In this case the air gap between the membrane and the cold surface is kept to be filled by a stagnant cold liquid solution, frequently the produced distilled water. AGMD module is used but the air gap space between the membrane and the condensing surface is filled with the produced water. The permeate water exits from the top part of the membrane module. This configuration also received the name permeate gap membrane distillation (PGMD). Like DCMD, LGMD is also applied generally for water production and concentration of nonvolatile solutes in the feed side of the membrane module. If the air gap between the membrane and the cold surface is filled with any solid material such as a porous support, sand, or sponge material, the process is called material gap membrane distillation

(MGMD) although this claimed new configuration is either AGMD or LGMD (Francis et al., 2013).

5. *Thermostatic sweeping gas membrane distillation* (TSGMD): This MD variant combines both SGMD and AGMD in order to minimize the temperature of the sweeping gas, which increases considerably along the membrane module length because of the heat transferred from the feed side through the membrane to the permeate side. In SGMD, the gas temperature, the heat transfer rate, and the mass transport through the membrane change during the gas progression along the membrane module. The presence of the cold wall in the permeate side reduces the increase of the sweeping gas temperature resulting in an enhancement of the driving force and the water production rate as a consequence. TSGMD can also be applied for distilled/potable water production, concentration of the nonvolatile solutes in the feed aqueous solution, and concentration of VOCs in aqueous solutions.
6. *Vacuum Membrane Distillation* (VMD): In this case vacuum or a low pressure is applied in the permeate side of the membrane module by means of a vacuum pump. The downstream pressure is maintained below the saturation pressure of volatile molecules to be separated from the feed aqueous solution. External condensers are needed to collect the permeate. At laboratory scale, nitrogen liquid cold traps are often used when a very low downstream pressure is applied. VMD is generally used for separation of VOCs from water. Membranes having smaller pore size (i.e., less than 0.45 μm) than in the other MD variants are used providing that in VMD the risk of pore wetting is high. When a dense and selective membrane is used in the membrane module instead of a porous and hydrophobic membrane this process is termed pervaporation (Khayet and Matsuura, 2004).

For a given application, an adequate MD configuration is selected among the above cited MD variants taking into consideration the type of the feed solution, their benefits, and drawbacks. Currently, MD technology is gaining an increasing importance in membrane processes (Figure 2) and becomes more attractive than other popular separation processes because MD:

- exhibits high rejection factors (near 100%) of nonvolatile solutes present in water such as ions, macromolecules, colloids, cells, etc.
- can be operated at lower temperatures than conventional distillation so that waste heat and/or alternative energy sources, such as solar and geothermal can be used.

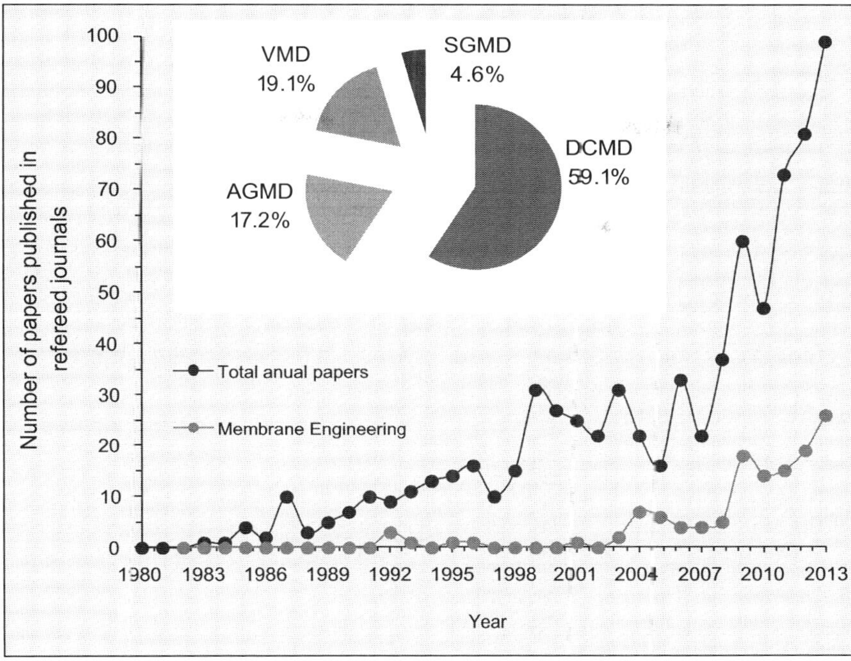

Figure 2 Growth of MD activity up to December 31, 2013 represented as a plot of number of papers published in refereed journals per year, annual published studies on membrane design and fabrication for MD applications and percentages of the studied MD variants. MD, membrane distillation.

- can be operated at lower operating hydrostatic pressures than conventional pressure-driven membrane processes used in filtration and liquid separation (i.e., reverse osmosis, RO; nanofiltration, NF; ultrafiltration, UF; microfiltration, MF).
- can be combined with other processes in integrated systems (i.e., pressure-driven membrane processes; forward osmosis, FO, etc.).
- uses membranes with less demanding membrane mechanical properties and reduced chemical interactions with process solution.
- it needs smaller spaces compared to conventional distillation processes.
- can be used in applications where other processes cannot be applied or their applications are very expensive (e.g., treatment of aqueous solutions with high osmotic pressures or with a solute concentration near saturated solution).
- requires less pretreatment steps compared to other membrane processes.
- exhibits less fouling propensity compared to other pressure-driven membrane processes.

In spite of all these cited advantages, the main limitation of MD is the drawback of membrane pore wetting. This can be avoided by using membranes with high liquid entry pressure (*LEP*) of feed solution inside the membrane pores (i.e., high hydrophobicity membranes, high water contact angles) and small maximum pore size, and the feed aqueous solution when containing solutes with low surface tension must be sufficiently dilute. This limits MD for certain applications such as the separation of organic/organic solutions and the treatment of highly concentrated aqueous solutions with surfactants, alcohols, and VOCs in general. The present MD technology is still need to be improved for its adequate industrial implementation in different separation applications. Multistaged MD systems, development of adequate membranes and membrane modules for MD, improvement of economical and energy efficiency of MD systems are actual proposed research areas.

As can be seen in Figure 2, interest on MD has increased significantly. Among the published papers in International Journals up to December 31, 2013, the most used MD variant is DCMD with 59.1% of the MD published studies because in this configuration condensation phenomenon is carried out inside the membrane module leading in this way to a simple operation mode. However, compared to the other MD variants the air entrapped within the pores of a membrane used in DCMD results in a high mass transfer inefficiency, while the heat loss by conduction through the membrane, which is considered heat loss in MD is high. In contrast, SGMD is the least studied MD variant with only 4.6% of the MD published studies because this MD variant requires external condensers to collect the permeate and a source for gas circulation. However, SGMD combines a relatively low conductive heat loss through the membrane with a reduced mass transfer resistance. As occurred in AGMD variant, in SGMD there is a gas barrier that reduces the heat loss by conduction through the membrane. Nevertheless, compared to AGMD variant the gas in SGMD sweeps the membrane resulting in higher mass transfer coefficients and therefore higher permeate fluxes. It is worth noting that the calculated percentages of the two MD variants TSGMD and LGMD is negligible. In practically all the published MD studies, commercial microporous hydrophobic membranes available in capillary or in flat sheet forms, have been used although these membranes were prepared initially for other purposes, for example microfiltration (MF). Some commercial membranes commonly used in MD are presented in Table 1.

A membrane for MD application has to meet the following requirements simultaneously (Khayet and Matsuura, 2011):

Table 1 Some commercial membranes used in MD (membrane thickness, δ; mean pore size, d_p; porosity, ε; liquid entry pressure of water in the membrane pores, LEP_w)

Membrane type	Membrane name	Manufacturer	Material	δ (μm)	d_p (μm)	*E* (%)	LEP_w (kPa)
Flat sheet membranes	TF200	Gelman	PTFE/PP[1]	178	0.20	80	282
	TF450				0.45		138
	TF1000				1.00		48
	GVHP	Millipore	PVDF[2]	110	0.22	75	204
	HVHP			140	0.45		105
	Gore		PTFE	64	0.2	90	368
				77	0.45	89	288
			PTFE/PP[1]	184	0.2	44	463
	Enka		PP	100	0.1	75	–
	3MA	3M Corporation		91	0.29[3]	66	–
	3MB			81	0.40[3]	76	
Capillary membranes	Accurel® S6/2 MD020CP2N[4]	AkzoNobel Microdyn		450	0.2	70	140
	MD020TP2N	Enka Microdyn		1550	0.2	75	–
	Accurel® BFMF 06-30-33[5]	Enka A.G. Euro-Sep		200	0.2	70	

[1]Flat sheet polytetrafluoroethylene, PTFE, membranes supported by polypropylene, PP, or polyethylene, PE.
[2]Flat sheet polyvinylidene fluoride, PVDF, membranes.
[3]Maximum pore size.
[4]Shell-and-tube capillary membrane module: Filtration area: 0.1 m^2, inner capillary diameter: 1.8 mm; length of capillaries: 470 mm.
[5]Shell-and-tube capillary membrane module: Filtration area: 0.3 m^2, inner capillary diameter: 0.33 mm, length of capillaries: 200 mm.
Reprinted from Khayet (2011), copyright (2011), with permission from Elsevier.

- good thermal stability for temperatures ranging from ambient temperature up to the boiling temperature of water,
- high chemical resistance to a wide range of aqueous solutions,
- high permeability membranes taking into account that there is an increase of the permeate flux with the increase of the membrane pore size and porosity, and with a decrease of the membrane thickness and pore tortuosity;
- high *LEP*, which is the minimum transmembrane hydrostatic pressure that is applied on the membrane before liquid solution penetrates into the pores. The *LEP* is characteristic of each membrane. It is high for membranes prepared with a high hydrophobicity material (i.e., large water contact angle) and a small maximum pore size. However, when using a membrane with a small maximum pore size, the *LEP* is high but the permeability of the membrane is low.
- narrow pore size distribution,
- low thermal conductivity because the heat transferred by conduction through the membrane from the feed to the permeate side is heat loss in MD. This conductive heat loss is greater for thinner membranes. However, using thicker membranes contradicts the achievement of high permeability. Therefore, a compromise exists between a membrane having a high permeability and a low heat transfer by conduction.

It is worth noting that only 16.8% of the MD published studies are focused on membrane engineering for MD, i.e., design and fabrication of membranes specifically for MD (Figure 2). Very few authors have considered the possibility of manufacturing novel membranes and membrane module designs specifically for MD applications (Khayet and Matsuura, 2011; Khayet et al., 2006a; Khayet et al., 2006b; Khayet, 2011). Hydrophobic porous membranes can be prepared by different techniques depending on the properties of the used materials that should be selected according to criteria including compatibility with the feed aqueous solutions, cost, ease of fabrication and assembly, useful operating temperatures, and thermal conductivity, which must be as low as possible. Microporous membranes can be made by sintering, stretching, phase inversion, or thermally induced phase separation (TIPS) (Khayet and Matsuura, 2011; Mulder, 1996; Lloyd, 1990). The MD membrane can be a single hydrophobic layer (i.e., conventional and most used membrane type), a composite porous bilayered hydrophobic/hydrophilic membrane or composite trilayered hydrophilic/hydrophobic/hydrophilic or hydrophobic/hydrophilic/hydrophobic porous membranes. Both supported and unsupported membranes with different structures are used in MD and their pore sizes ranges between 10 nm and 1 μm. Recently nanostructured, hybrid, and exotic membranes are designed for MD applications (Khayet and Matsuura, 2011; Khayet, 2011; Essalhi and Khayet, 2013). As can be seen in Figure 2, there is a growing interest on membrane engineering for MD and therefore the future of MD is promising as a consequence.

2. MD MODULES AND FLUID FLOW

Various types of membrane modules were designed for each MD variant and tested in different systems and applications. There are three major MD module configurations, which are plate-and-frame module, shell-and-tube or tubular module, and spiral wound membrane module. Figure 3 shows schematic diagrams of these modules. Different types of membranes were packed in a large variety of module configurations and tested in MD systems.

The permeate flux obtained in the MD process is also affected significantly by the module design, the MD configuration, and its operating conditions not only the membrane itself. In addition to the previously mentioned membrane requirement, a good module to be used in MD must:

- exhibit a high packing density (i.e., high membrane surface area) with an optimized size and a high membrane module performance (i.e., high permeability and high separation factor),
- use housing with high resistance to chemicals, pressures, and temperatures (i.e., high thermal stability),
- pack properly the membrane in potting resins, free of cracks, and with a good adhesion,
- permit its drying in case of membrane wetting problem as well as easy inspection and defects reparation,
- allow high feed and permeate flow rates tangentially to the membrane or in cross-flow mode including baffles, spacers, and/or turbulent promoters in order to increase the heat and mass transfer coefficients, reduce the effects of the both the temperature polarization (i.e., thickness of the thermal boundary layer) and concentration polarization (i.e., thickness of the concentration boundary layer), and increase the thermal efficiency,
- provide high mass and heat transfer rates between the bulk solutions and the membrane surface,

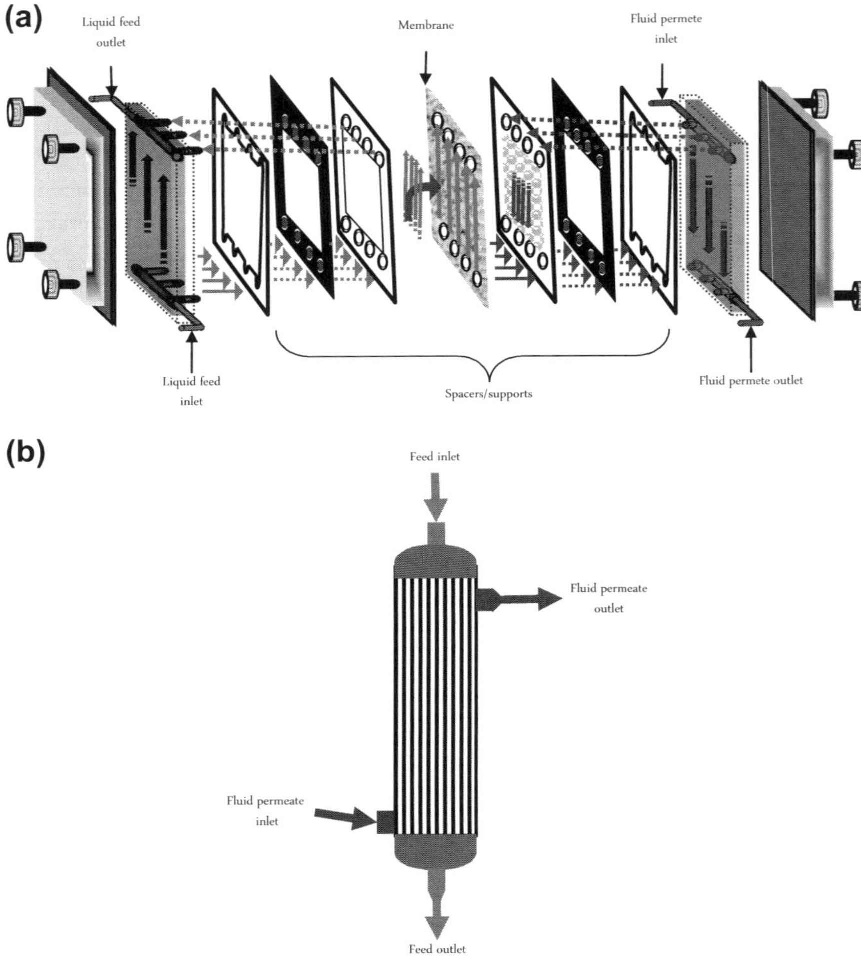

Figure 3 MD modules: (a) plate-and-frame; (b) shell-and-tube or tubular, and (c) spiral wound membrane module for AGMD or LGMD variants. MD, membrane distillation; AGMD, air gap membrane distillation; LGMD, liquid gap membrane distillation. *Reprinted from Winter et al. (2011), copyright (2011), with permission from Elsevier; Reprinted from Winter et al. (2012), copyright (2012), with permission from Elsevier.*

- satisfy low pressure drop along the membrane module length to prevent excessively high transmembrane hydrostatic pressures that may cause wetting of membrane pores,
- guarantee uniform flows throughout the whole membrane module avoiding dead corners and channel formation,

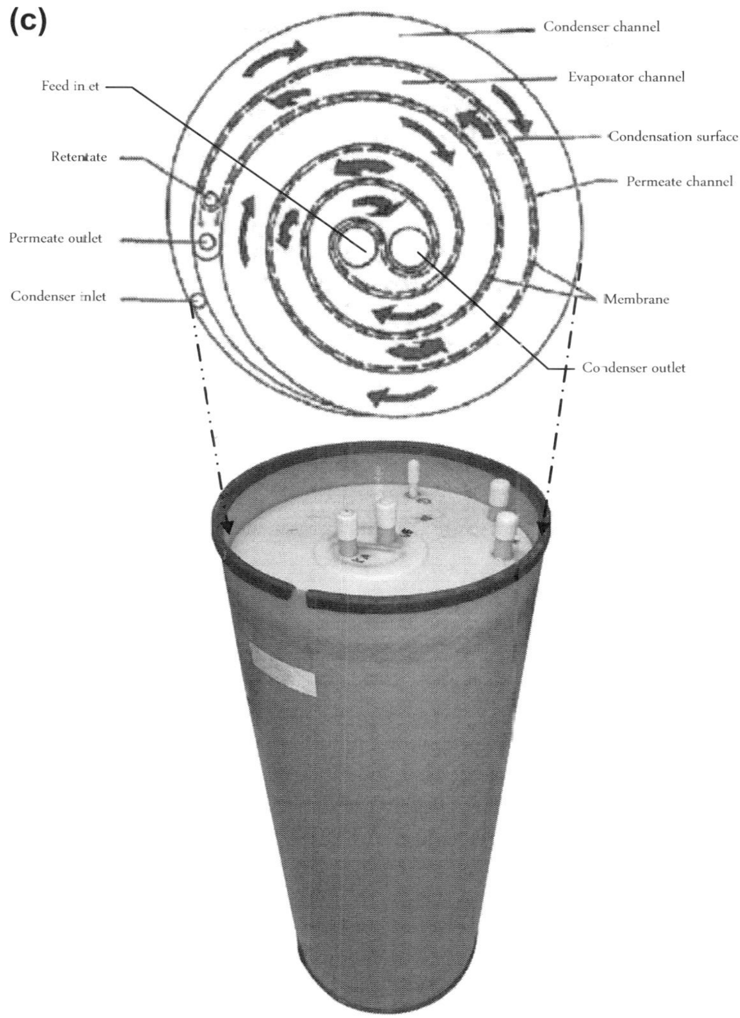

Figure 3 cont'd

- guarantee low heat loss to the environment and if possible a good heat recovery system,
- avoid erosion problems by using for example plastic materials,
- contain a membrane support if necessary, that must be chosen to be strong enough to prevent deflection or rupture of the membrane,
- be properly designed allowing an easy cleaning and membrane replacement, with low scaling, low fouling, etc.

It is to be noted that the choice of a membrane module for each MD configuration is usually determined by both economic and operative conditions. Most of laboratory scale membrane modules are plate-and-frame modules designed for use with flat sheet membranes due to their versatility and simplicity in fabrication, as compared to the spiral wound or shell-and-tube membrane modules. In fact, flat sheet membranes can easily be removed from a plate-and-frame module for their examination, cleaning or replacement and the same module can be used to test different membranes. However, tubular or shell-and-tube membrane modules fabricated using capillary or hollow fiber membranes are more attractive than plate-and-frame modules fabricated with flat sheet membranes because much higher membrane surface area to module volume ratio can be packed. The packing density of plate-and-frame membrane modules can vary between 100 and 400 m^2/m^3 depending on the number of membrane sheets (Khayet and Matsuura, 2011). On the other hand, a large number of membrane capillaries or hollow fibers can be packed in the modules with packing densities of about 600–1200 m^2/m^3 (Khayet and Matsuura, 2011). In the case of hollow fiber membranes the inner diameters are even smaller, 50–100 μm, and thousands of hollow fibers can be packed in shell and tube membrane modules with very high packing densities, which may reach 3000 m^2/m^3. Capillary or hollow fiber membranes do not require any support and are an integrated part of the module. The main inconvenient is the membranes in these last modules cannot be replaced easily in case the membrane pores are wetted by liquid solutions. Capillary membranes were also assembled in plate-and-frame membrane modules in cross-flow mode to reduce the temperature polarization effect by increasing the heat transfer coefficients (Li and Sirkar, 2004). Considerably enhanced water production rate in both DCMD and VMD configurations were achieved. Flat sheet membranes were also arranged in spiral wound modules as shown in Figure 3 and the membrane packing density normally ranges between 300 and 1000 m^2/m^3, depending on the channel height. Spiral wound modules for MD has been first proposed in 1982 by Gore & Associated Co. (Gore, 1982) and then by Hanbury and Hodgkiess three years later (Hanbury and Hodgkiess, 1985). Later, commercial spiral wound membrane modules were used in DCMD (Zakrewaska-Trznadel et al., 1999), AGMD, or LGMD variants with an integrated heat recovery for the design of solar-powered desalination system (Tarnacki et al., 2012).

The majority of the designed membrane modules for MD were more academically orientated than industrially. Several attempts of commercialization have failed due to difficulties in engineering aspects. Reliability of the membrane module is still a serious issue in MD and each configuration imposes certain fluid dynamic conditions on both feed and permeate sides. Actually, the availability of industrial MD modules is one of the limitations for MD process industrial implementation. A historical survey together with different designs of membrane modules for MD was described in details in (Khayet and Matsuura, 2011). Nowadays, the most relevant companies and research institutions in the world developing MD membrane modules and applying MD on a commercial scale are:

- **Scarab Development AB and XZero AB (Sweden):** Scarab Development AB was founded in 1973 in order to exploit low temperature distillation technologies. Plate-and-frame AGMD module design was patented by Scarab in 1981 (Figure 4(a)). The Swedish company XZero acquired the license to use Scarab's technology in semiconductor industry for producing ultrapure water systems with zero liquid discharge (www.Xzero.se). The AGMD modules have been tested by different institutions in Sandia National Laboratory in the United States (Liu, 2004), by the University of Texas at El Paso sponsored by the US Bureau of Reclamation (Walton et al., 2004; Lu et al., 2001), using solar thermal collectors in Spain and Mexico under the frame of MEDESOL project (Gálvez et al., 2009) and in the Royal Institute of Technology (Department of Energy Technology) in Stockholm (Sweden) for water purification in cogeneration plants (Kullab and Martin, 2011).
- **Fraunhofer Institute for Solar Energy System(ISE) and SolarSpring GmbH (Germany):** In 2003 Fraunhofer ISE began developing spiral wound AGMD modules with different sizes and with an integrated heat recovery for different solar-powered desalination systems installed in the Island of Gran Canaria in Spain, Jordan, Egypt, Mexico, Pantelleria in Italy, and Amarika in Namibia (Figures 3(c) and 4(b)) (Tarnacki et al., 2012; Koschikowski et al., 2005, 2009; Banat et al., 2007a; Banat et al., 2007b; Fath et al., 2008; Raluy et al., 2012; Schwantes et al., 2013). These modules were also considered for LGMD. SolarSpring GmbH, based in Freiburg, Germany, was founded in 2009 as a spin-off of Fraunhofer ISE. Its

(a)

(b)

Figure 4 Commercial modules: (a) Scarab AB and XZero AB (www.scarab.se; www.Xzero.se; Liu, 2004; Kullab and Martin, 2011), (b) Fraunhofer ISE *(Reprinted from Winter et al. (2011), copyright (2011), with permission from Elsevier)*, (c) Memstill *(Reprinted from Jansen et al. (2013), copyright (2013), with permission from Elsevier; Reprinted from Hanemaaijer et al. (2006), copyright (2006), with permission from Elsevier)*, and (d) Memsys *(Reprinted from Zhao et al. (2013), copyright (2013), with permission from Elsevier; Reprinted from Ong et al. (2012), copyright (2012), with permission from Elsevier)*.

overall objective is the design and installations of decentralized autonomous systems for remote areas.

- **Memstill, TNO (Netherlands Organisation for Applied Scientific Research), Keppel Seghers (Belgium), and Aquastill (Netherlands):** Memstill technology initiated its technology

(c)

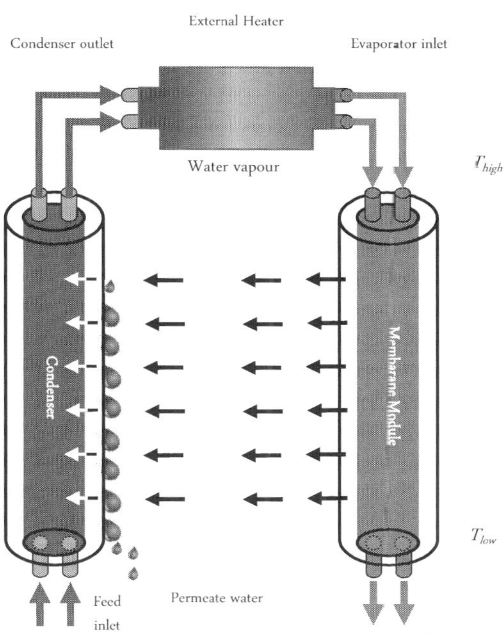

Figure 4 cont'd

(d)

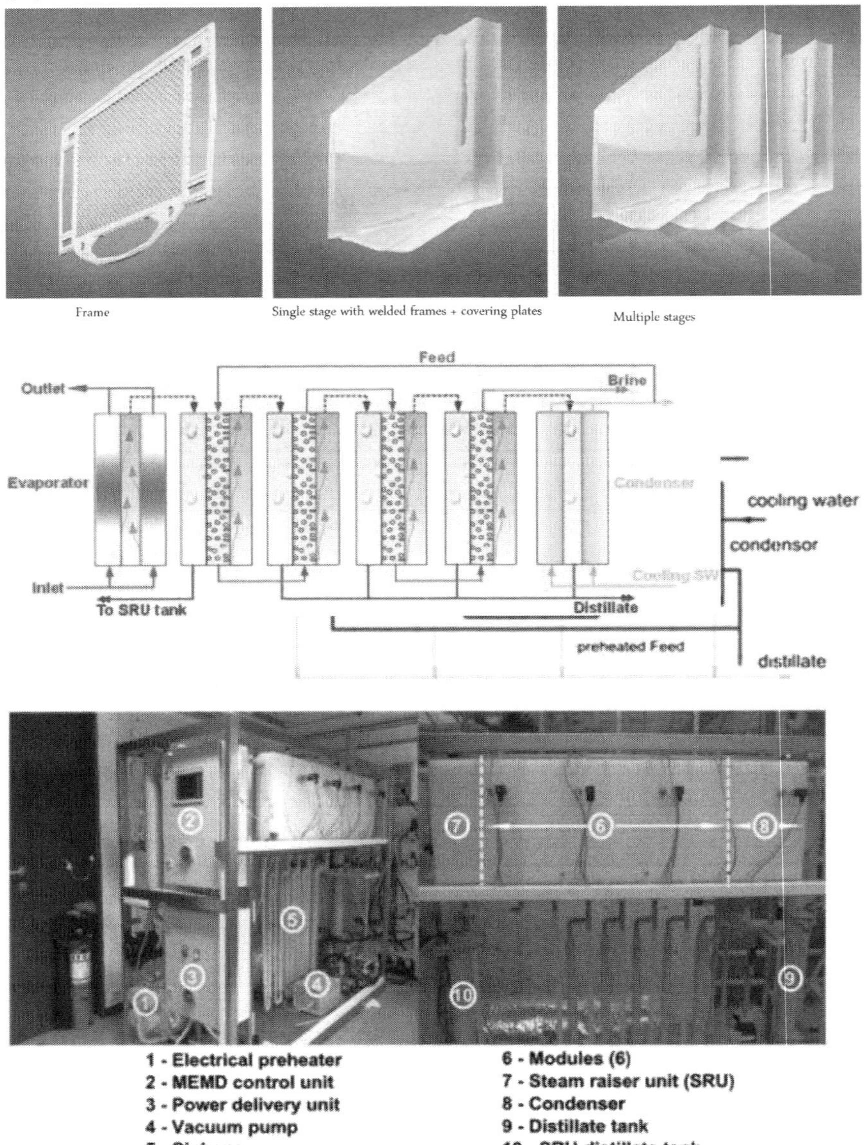

Figure 4 cont'd

development in 1999 and emerged during 2006. It was developed by a consortium including TNO (Netherlands Organisation for Applied Scientific Research) and Keppel Seghers Belgium N.V. (formerly known as Seghers Keppel Technology N.V.). The design consists of an AGMD module, in which the cold saline water flows through a condenser with nonpermeable walls, increasing its temperature due to the condensing permeate, and then passes through a heat exchanger where additional heat is added before entering in direct contact with the membrane (Dotremont et al., 2008) Figure 4(c): first pilot modules with 300 m^2 of membrane). The technology was patented by TNO in 1999 and 2005 and licensed to Aquastill (founded in 2008) and Kepple Seghers. Memstill pilot plants have been operating since 2006 in Singapore, in Belgium (BASF Antwerp), and in the Netherlands (Jansen, et al., 2010). In Singapore (Jurong Island), the plans of Memstill were to operate at 100 m^3 per day on a petroleum refinery. It was claimed a thermal energy consumption as low as 56–100 kWh/m^3 with a Gained Output Ratio (GOR) as high as 11.2 calculated for feed temperatures of 80–90 °C with an electrical energy of 0.75 kWh/m^3 (Tarnacki et al., 2012). In 2008, 2009, there was a large investment to reduce the cost of this type of MD modules (Jansen et al., 2010). It was observed that since 2012 Aquastill offers AGMD, DCMD, and LGMD installations.

- **Memsys (Germany, Singapore) and Aquaver (The Netherlands):** The Memsys system is based on vacuum enhanced multieffect AGMD variant incorporating heat recovery and recycling in a plat-and-frame membrane module (Figure 4(d)). The stages are composed of alternative hydrophobic membrane and foil (c, PP) frames. Each stage recovers the heat of condensation providing a multiple-effect design while the distillate is produced in each evaporation/condensation stage and in the condenser. The module components are made of PP, which eliminates corrosion and scaling and allows large-scale cost efficient production. The company's module production started in 2010 and their pilot plants have been installed in Singapore, Australia, and India among others (Jansen et al., 2010, 2013). In Singapore, Memsys Clearwater Pte Ltd and Nanyang Technological University (NTU) are collaborating on the treatment of water contaminated with oil. In November 2011, Memsys and Aquaver (part of Ecover Group) agree on exclusive license agreement to cooperate worldwide on small-scale units for potable water supply and process water applications. In 2012, Memsys awarded a grant from the Environment and Water Industry Programme Office (EWI) to build a 50 m^3/day MD test system at Senoko Power Plant, the largest power supplier in Singapore. General Electric Co. (GE) and Memsys Clearwater Pte Ltd (Germany and Singapore) have entered into an agreement to develop Memsys's MD technology for the unconventional resource applications including shale gas, coal seam gas, and other fuels recovered by hydraulic fracturing (high saline produced water). With Concord Enviro Systems Pvt Ltd of Mumbai (India), Germany's Memsys have signed a global license agreement for cooperation on treating molasses wastewater from sugar industries (Concord Enviro Systems, 2012).

Other compact modules including solar MD modules were presented by different research institutions (Khayet and Matsuura, 2011; Cipollina et al., 2012; Hausmann et al., 2012; Hogan et al., 1990; Chen and Ho, 2010; Kim et al., 2013). In addition to the development of novel MD membranes, researchers have investigated strategies to improve the MD process such as optimizing MD operating parameters (i.e., flow rates and temperatures) and designing novel modules to reduce temperature polarization, concentration polarization, fouling, scaling, and pressure drop and therefore enhance permeate flux (Li and Sirkar, 2004; Cath et al., 2004; Martinez and Rodriquez-Maroto, 2007; Teoh et al., 2008; Yang et al., 2011; He et al., 2011; Martinez-Diez and Vázquez-González, 2000; Qtaishat et al., 2008).

The permeate flux in MD increases with the increase of the feed flow rate or permeate flow rate in the MD module channels when these are operated under laminar or transitional flow hydrodynamic regimes, whereas it tends to an asymptotic value when the flow regime is turbulent. As the flow rate increases the thickness of the thermal and/or the concentration boundary layers in the membrane module channels become thin and results in low effects of temperature polarization and concentration polarization. For example, in SGMD variant the sweeping gas flow rate is together with the feed temperature are the important parameters controlling the permeate flux of MD (Khayet et al., 2002). The main temperature polarization in SGMD is located in the air phase and permeate flux in the SGMD process is mostly controlled by the heat transfer through the air boundary layer.

Computational fluid dynamics (CFD) techniques were applied for optimum design of different MD membrane modules (Cipollina et al., 2009; Yu et al., 2012; Shakaib et al., 2012; Xu et al., 2009; Yang et al., 2012). Thermo-fluid dynamics of an MD module showed that spacers

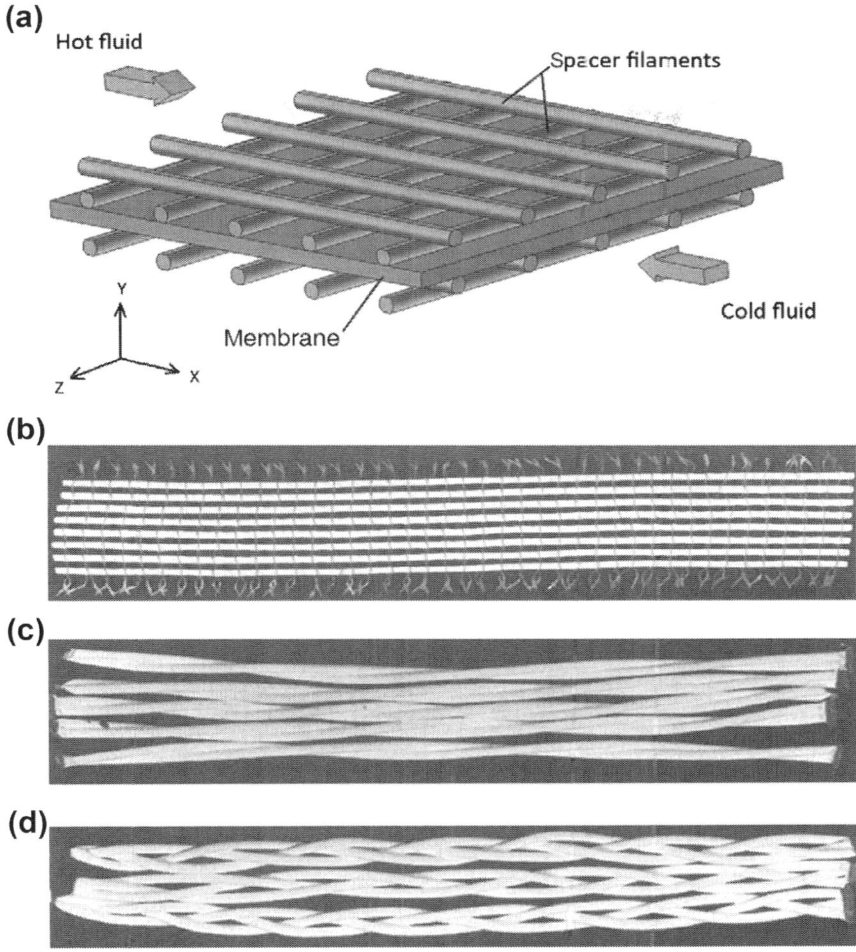

Figure 5 Spacer-filled MD channel (a) *(Reprinted from Shakaib et al. (2012), copyright (2012), with permission from Elsevier)*; membrane fibers arranged in fabric (b), twisted (c) and braided (d) configurations *(Reprinted from Schneider et al. (1988), copyright (1988), with permission from Elsevier)*; different module design and hollow fiber configurations (e) *(Reprinted from Teoh et al. (2008), copyright (2008), with permission from Elsevier)*; and Images of possible baffles for capillaries or hollow fiber membrane modules (f) *(Reprinted from Ahmad and Mariadas (2004), copyright (2004), with permission from Elsevier)*.

can significantly affect temperature gradients within its channel, permitting to design an optimal spacer (Figure 5) (Teoh et al., 2008; Shakaib et al., 2012; Yang et al., 2012; Ahmad and Mariadas, 2004). It was also suggested that by adding baffles to the modules the fluid dynamics

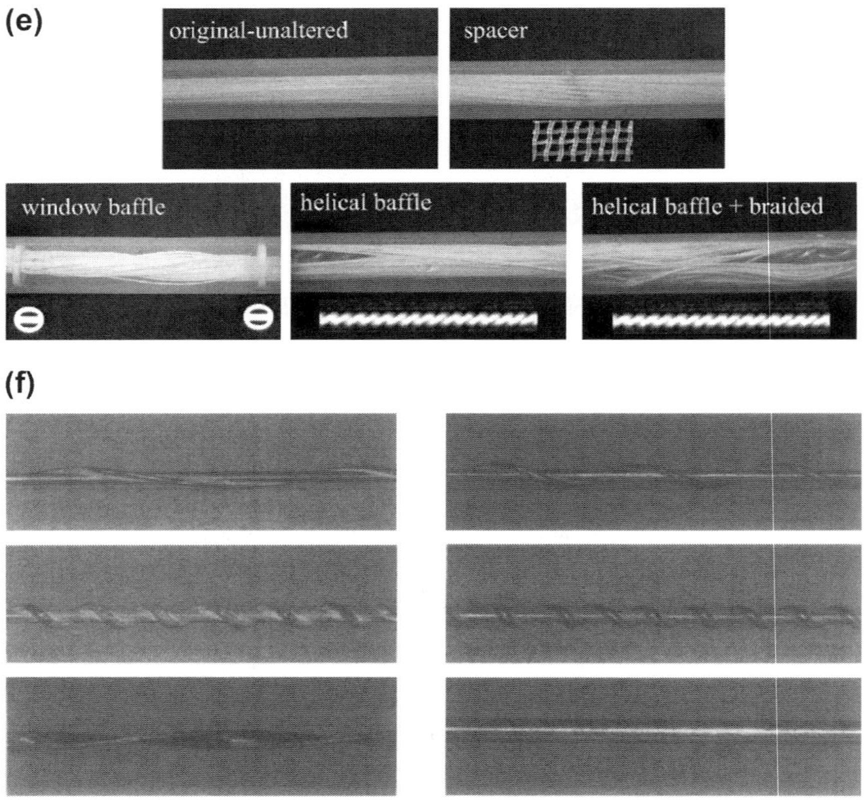

Figure 5 cont'd

may reduce the temperature polarization effect responsible for low permeate fluxes in MD modules (Figure 5). Inserting baffles in membrane modules create fluid instabilities in the liquid flow and the formed vortices improve the mixing between the boundary layers (i.e., temperature and concentration layers) of the membrane. Furthermore, if the fibers are twisted or braided (Figure 5) instead of been arranged straight or in a fabric, more turbulent and uniform flow outside the fibers can be produced leading to an enhancement of both heat and mass transfer coefficients in the shell side of the membrane (Teoh et al., 2008; Schneider et al., 1988). Bundles with twisted or braided fibers act as static mixers around the fibers. However, special care must be taken because the used spacers or baffles filled channels have an electrical energy penalty because of the increase of the backpressure. (Yang et al., 2012; Al-Sharif et al., 2013; Wang and Cussler, 1993).

3. MD APPLICATIONS: FILTRATION AND SEPARATION

MD technology is gaining an increasing importance in separation processes and it is currently applied for environmental, chemical, petrochemical, metallurgical, food, pharmaceutical, and biotechnology industries. In general, the MD typical applications are summarized in Figure 6.

- **Brackish water and seawater desalination**: This is the most considered MD application for water production because the obtained salt rejection factor is very close to 100% (i.e., practically a total rejection of salts). In the case of a feed aqueous solution containing nonvolatile components, either electrolytes (sodium chloride, NaCl; potassium chloride, KCl; lithium bromide, LiBr; etc.) or nonelectrolytes (glucose, sucrose, fructose, etc.) only water molecules flow through the membrane pores in vapor phase. It must be pointed out that 20% of the world's population has inadequate access to drinking water although over two-thirds of the planet is covered with water (99.3% of the total water is either too salty as seawater or inaccessible as ice caps). Moreover, water is potable only when it contains less than 500 ppm of salt. Table 2 presents

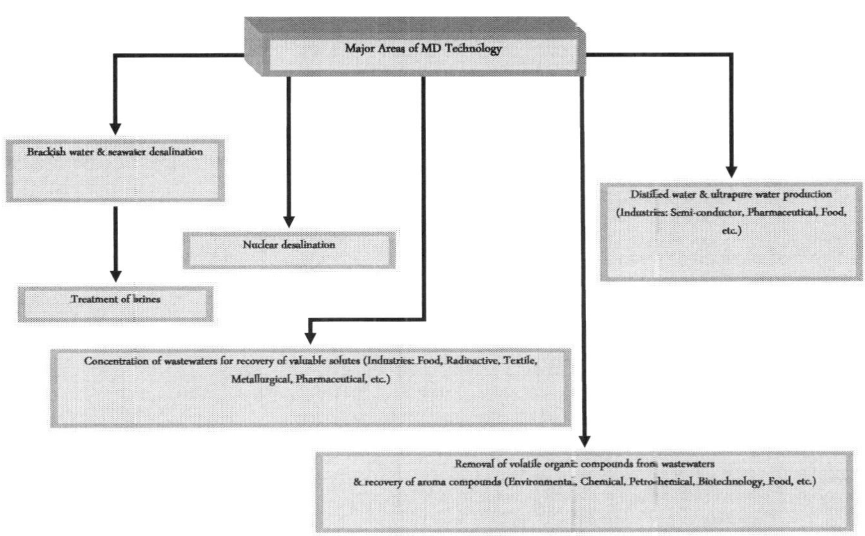

Figure 6 Typical field applications of MD. MD, membrane distillation.

Table 2 Reported permeate fluxes (J_w) and salt rejection factors (α) of different membranes used for desalination by different MD variants. Feed temperature (T_f), permeate temperature in DCMD (T_p), temperature difference between feed and permeate in DCMD (ΔT), cooling temperature in AGMD (T_c), liquid flow rate (Q_f), liquid circulation velocity (v_f), air gap width in AGMD (a), electrical conductivity of the permeate (ψ_p), and electrical conductivity of the feed (ψ_f)

Membrane	J_w (10^{-3} kg/m^2 s)	Observation	References
GVHP	13.52	DCMD: Distilled water as feed; $T_f = 90.7$ °C; $T_p = 19.7$ °C.	Khayet et al. (2010)
	9.00	DCMD: Distilled water as feed; $T_f = 70$ °C; $T_p = 20$ °C.	Phattaranawik et al. (2003)
	0.89	DCMD: 3 g/L NaCl; $\Delta T = 10$ °C; $T_f = 51.9$ °C.	Khayet et al. (2007)
	0.83	DCMD: 64.5 g/L NaCl; $\Delta T = 10$ °C; $T = 52.7$ °C.	Khayet et al. (2007)
	2.28	AGMD: $T_f = 50$ °C, $T_c = 20$ °C, $Q_f = 70$ L/h, a=1.8 mm, distilled water.	Izquierdo-Gil et al. (1999)
HVHP	18.61	Deareation DCMD; $T_f = 80$ °C; $T_p = 21$ °C distilled water.	Schofield et al. (1990)
	16.39	NaCl (14 g/L)	
	11.11	NaCl (25 g/L)	
	10.80	DCMD: Distilled water as feed; $T_f = 70$ °C; $T_p = 20$ °C.	Phattaranawik et al. (2003)
	7.31	AGMD: $a = 0.8$ m, $T_c = 7$ °C $T_f = 82$ °C, tapwater ($\psi_f = 297$ μS/cm).	Izquierdo-Gil et al. (1999)
	1.94	$T_f = 52$ °C, tap water ($\psi_f = 297$ μS/cm, 99% salt rejection).	
	1.67	$T_f = 52$ °C, seawater model solution ($\psi_f = 37.6$ mS/cm, $\psi_p = 1100$ μS/cm).	

TF200	18.69	DCMD: Distilled water as feed; $T_f = 80.1\ ^\circ C$; $T_p = 20.1\ ^\circ C$.	Khayet et al. (2010)
	2.90	DCMD: 1.9 g/L NaCl; $\Delta T = 10\ ^\circ C$; $T_f = 52.2\ ^\circ C$.	Khayet et al. (2007)
	2.23	DCMD: 64.5 g/L NaCl; $\Delta T = 10\ ^\circ C$; $T_f = 52.7\ ^\circ C$.	Khayet et al. (2007)
	1.31	AGMD: $T_f = 70\ ^\circ C$, $T_c = 30\ ^\circ C$, distilled water as feed.	Gostoli and Sarti (1989)
TF450	14.19	AGMD: $T_f = 71\ ^\circ C$, $T_c = 13.9\ ^\circ C$, 30 g/L NaCl, $a = 3$ mm, $Q_f = 205$ L/h, $\alpha = 99.92\%$.	Khayet and Cojocaru (2012b)
	13.11	AGMD: $T_f = 71\ ^\circ C$, $T_c = 13.9\ ^\circ C$, 30 g/L NaCl, $a = 5.6$ mm, $Q_f = 183$ L/h, $\alpha = 99.98\%$.	Khayet and Cojocaru (2012a)
	8.67	AGMD: $T_f = 59\ ^\circ C$, $T_c = 13.9\ ^\circ C$, 30 g/L NaCl, $a = 5.6$ mm, $Q_f = 205$ L/h, $\alpha = 99.98\%$.	Khayet and Cojocaru (2012a)
3MA	25.2 22.5 19.8	DCMD: $T_f = 74\ ^\circ C$; $T_p = 20\ ^\circ C$; distilled water as feed 19.5 g/L; 42.2 g/L.	Lawson and Lloyd (1996)
3MB	21.6	DCMD: Distilled water as feed; $T_f = 70\ ^\circ C$; $T_p = 20\ ^\circ C$.	Lawson and Lloyd (1996)

MD, membrane distillation; DCMD, direct contact membrane distillation; AGMD, air gap membrane distillation.

as an example the permeate fluxes of some membranes commonly used for desalination by different MD variants.

It was observed that the temperature of the feed solution is the most significant MD operating parameter controlling the MD permeate flux. Generally, it is admitted that there is an exponential increase of the MD permeate flux with the feed temperature because the partial vapor pressure increases exponentially with the temperature following an Arrhenius type of dependence ($J \propto e^{-A/T}$ where J is the permeate flux, T is the absolute temperature, and A is a constant) (García-Payo et al., 2000). Remember that the transmembrane vapor pressure is the driving force of mass transfer in MD. A linear increase of the MD permeate flux with the vapor pressure difference between the feed and permeate was obtained (Banat and Simandl, 1994, 1998; Kurokawa et al., 1990). The MD flux is lower for higher permeate temperature and higher nonvolatile solute concentration of the feed solution (Khayet and Matsuura, 2011; Khayet, 2011; Lawson, and Lloyd, 1997; El-Bourawi et al., 2006). Moreover, the permeate flux was found to be greater for membranes having larger pore sizes. In AGMD variant it was observed a decrease of the permeate flux with the increase of the air gap thickness between the membrane and the condensing surface because of the increase of the resistance to mass transfer in the air gap of the membrane module (Banat and Simandl, 1994, 1998; Jonsson et al., 1985; Kimura et al., 1987).

- **Treatment of concentrated brines:** MD is proposed to solve the problem of brine disposal although the MD permeate flux declines as the concentration of salt in water is increased due to the decrease of the partial vapour pressure of the salt feed solution (Khayet and Matsuura, 2011; Al-khudhiri et al., 2012; Wu and Drioli, 1989; Mariah et al., 2006). It was demonstrated that MD can be applied for processing high salinity aqueous solutions and concentrated brines derived from other separation processes such as RO and NF instead of their discharge to the environment. It must be pointed out that pressure-driven membrane processes are not able to treat concentrated brines (i.e., >75 g/L of salt) because of their high osmotic pressures. Therefore, various hybrid processes integrating MD as a final stage with other separation pressures such as pressure-driven membrane processes and crystallization were proposed looking at zero salty water discharge (Bouguecha and Dhahbi, 2003; Ji et al., 2010; Mericq et al., 2010; Edwie and Chung, 2012; Chen et al., 2014). Pure water is recovered by MD while the resulting saturated salt aqueous solutions are used in crystallization process from which precipitated solids can be produced and enabling the formation of high quality crystals. This technology is also termed membrane distillation crystallization (MDC). Recently some researches start on the development of membranes for MDC (Edwie and Chung, 2012).

- **Concentration of wastewaters containing valuable compounds:** MD technology has been applied successfully for the treatment of wastewaters derived from different origins for recovery of valuable compounds and production of water less hazardous to the environment. The types of treated wastewaters are pharmaceutical wastewater containing taurine, concentration of biological solutions (bovine plasma, bovine blood, protein), metallurgical wastewater, textile wastewater contaminated with dyes, wastewater reclamation in space, olive oil mill wastewater for polyphenols recovery, waters contaminated with boron, arsenic, heavy metals, ammonia (NH_3), coolant liquid (i.e., glycols), humic acid, acid solutions rich in specific compounds for example the concentration of hydrogen iodide (HI) and sulfuric acid aqueous solutions of interests in hydrogen energy production from water, radioactive wastewater solutions (i.e., nuclear desalination), brine and other impotable water sources, etc. (Khayet and Matsuura, 2011; Zakrewska-Trznadel et al., 1999; Wu et al., 1991; Calabró et al., 1991; Cath et al., 2005; Caputo et al., 2007; El-Abbassi et al., 2012; El-Abbassi et al., 2013; Gryta and Karakulski, 1999; Khayet et al., 2004; Qu et al., 2009; Rincón et al., 1999; Tomaszewska et al., 1995; Zolotarev et al., 1994; Hou et al., 2010). One of the advantages of using MD technology is the possibility to operate at low temperatures of feed aqueous solutions. Therefore, MD was proved to be successfully applicable in fields where high temperatures result in degradation of the valuable compounds present in food wastewaters (i.e., concentration of milk, concentration of must, which is the juice obtained from grape pressing containing sugars and various aroma compounds and for the concentration of many other types of juices including orange juice, mandarin juice, apple juice, sugarcane juice, etc.) (Kimura et al., 1987; Drioli et al., 1992; Calabró et al., 1994; Bandini and Sarti, 2002; Gunko et al., 2006; Lukanin et al., 2003; Nene et al., 2002).
- **Removal of trace volatile organic compounds (VOCs) and dissolved gases from wastewaters:** Various dilute mixtures containing VOCs at different concentrations were tested by different MD variants for VOCs extraction from water. These are alcohols (i.e., methanol,

ethanol, isopropanol, and *n-butanol*), halogenated VOCs (i.e., chloroform, trichloroethylene, and tetrachloroethylene), benzene, acetone, acetonitrile, ethyl acetate, methyl acetate, methyl tert-butyl ether, etc. (Calibo et al., 1987; Khayet and Matsuura, 2011, 2004, 2001; Gostoli and Sarti, 1989; Hoffman et al., 1987; Fujii et al., 1992a,b; Bandini et al., 1992, 1997; Sarti et al., 1993., Banat and Simandl, 1996; Couffin et al., 1998; Bandini and Sarti, 1999; García-Payo et al., 2000; Izquierdo-Gil and Jonsson, 2003; Urtiaga et al., 2000). Such applications are appropriate for environmental, chemical, petrochemical, and biotechnology industries. Successful applications were also observed in food processing for recovery of volatile aroma compounds from juices (Bagger-Jorgensen et al., 2004), for ethanol recovery from fermentation broth (Gostoli and Sarti, 1989; Udriot et al., 1994; Banat and Simandl, 1999), and for breaking azeotropic mixtures (e.g., hydrochloric acid/water, propionic acid/water, and formic acid/water azeotrope mixtures (Udriot et al., 1994; García-Payo et al., 2000)). In MD, both water and VOCs are transported through the pores of the membrane and therefore the vapor pressure in the permeate side must be adequately chosen to reach good selectivities. MD was also proposed as an alternative separation technology for extraction of dissolved gases in water such as oxygen and ammonia, but only very few studies were reported in this field (Bandini et al., 1992; Ding et al., 2006; El-Bourawi et al., 2007; García-Payo et al., 2002).

- **Distilled and ultrapure water production:** MD is based on membrane evaporation and condensation phenomena. Therefore, when using adequate membranes for MD with narrow pore size distribution (see section 1), the produced water exhibits a high quality providing that a high rejection factor of nonvolatile solutes, very close to 100%, are achieved. A high quality water with an electrical conductivity as low as 0.8 μS/cm with 0.6 ppm TDS (total dissolved solids) was produced by MD (Karakulski et al., 2002). Because the produced water is very pure it is suitable for use in medical, pharmaceutical, and semiconductor industrial sectors (Khayet and Matsuura, 2011; Liu, 2004; Weyl, 1967).

Most of the above cited MD applications are reviewed in the recently published book by Khayet and Matsuura (Khayet and Matsuura, 2011). Furthermore, different propositions to improve product quantity and quality and reduce energy consumption were reported including integrated MD technology to other conventional processes such as distillation systems (i.e., multieffect distillers), to pressure-driven membrane processes (RO, UF, NF, forward osmosis, FO), to alternative energy sources such as solar and geothermal energy, and also to nuclear installations where waste heat can be recovered (Tarnacki et al., 2012; Walton et al., 2004; Lu et al., 2001; Gálvez et al., 2009; Koschikowski et al., 2005, 2009; Banat et al., 2007a,b; Fath et al., 2008; Raluy et al., 2012; Schwantes et al., 2013; Dotremont et al., 2008; Jansen et al., 2010; Jansen et al., 2010, 2013; Heinzl et al., 2012; Memsys–NTU partnership, 2011; Concord Enviro Systems, 2012; Zhao et al., 2013; Ong et al., 2012; Cipollina et al., 2012; Hausmann et al., 2012; Hogan et al., 1990; Chen and Ho, 2010; Kim et al., 2013; Li and Sikar, 2004; Bouguecha and Dhahbi, 2003; Zakrewska-Trznadel et al., 2001; De Andrés et al., 1998; Calabró et al., 1990, 1994; Drioli et al., 1999; Criscuoli and Drioli, 1999; Gryta et al., 2001a; Criscuoli et al., 2002; Rommel et al., 2008; Banat and Jwaied, 2010; Wange et al., 2009, 2011; Suareza et al., 2010; Mericq et al., 2011; Ding et al., 2005; Mertinetti et al., 2009). Solar ponds and solar collectors can be used to provide heat (solar thermal) or electrical energy (solar photovoltaic panels) requirements to operate an MD plant. Figure 7 shows as an example a solar powered MD plant.

It is worth quoting that since the first publication in the field of solar assisted MD by Hogan et al. (Hausmann et al., 2012) in 1990 using flat plate solar collectors, various research studies were reported in the MD literature incorporating different solar devices to MD modules and tested in different countries around the world (Tarnacki et al., 2012; Walton et al., 2004; Lu et al., 2001; Gálvez et al., 2009; Kullab and Martin, 2011; Banat et al., 2007a,b; Fath et al., 2008; Koschikowski et al., 2009; Raluy et al., 2012; Schwantes et al., 2013; Cipollina et al., 2012; Hogan et al., 1990; Chen and Ho, 2010; Kim et al., 2013; Rommel et al., 2008; Banat and Jwaied, 2010; Wang et al., 2009; Suareza et al., 2010; Mericq et al., 2011; Ding et al., 2005). However, up to now few studies are published on MD economics, energy analysis, and costs evaluations (Khayet, 2013). Wide dispersed and confusing water production costs (WP_{Cost}) and specific energy consumption (E_C) analysis are reported. The E_C varies from about 1 to 9000 kWh/m^3 while the WP_{Cost} varies from \$0.3/m^3 to \$130/m^3. These scattered values are due to the different type and size of the MD systems, type of feed processed water, energy source, energy recovery systems, cost of energy, economic analysis procedure, etc. Some commercial MD applications are still under evaluation due to the high energy consumption, high costs of membrane modules, difficulties with long time operation, and membrane wettability among others.

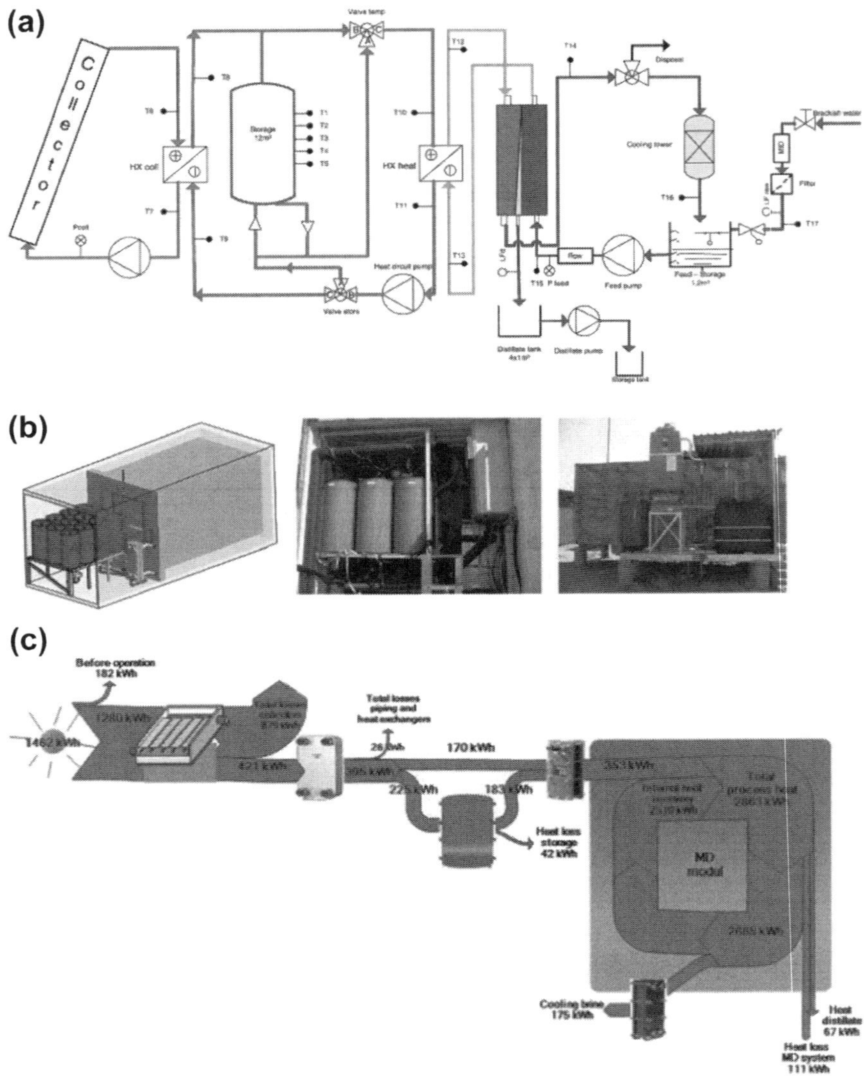

Figure 7 Schema (a), images (b), and heat flows and losses in a typical sunny day (c) of a solar AGMD plant installed in Amarika (Namibia) in 2010 (by Fraunhofer ISE) (4 m^3/day at alternating temperatures 65–80 °C, 12 modules, 168 m^2 membrane, raw water, drilling well 28,000 ppm, solar thermal flat plate collectors single glazed 232 m^2, 12 m^3 integrated heat storage, and brine cooling tower). *Reprinted from Schwantes et al. (2013) copyright (2013), with permission from Elsevier.*

Recently, a value of 27.6 kJ/kg (7.67 kWh/m^3) was reported as the theoretical minimum energy consumption of single-pass MD associated with a heat recovery system (i.e., heat exchanger) proposed for seawater desalination at 60 °C. For RO with a typical recovery of 50% this value is much lower, around 3.18 kJ/kg (0.88 kWh/m^3), although RO uses high cost electric energy (Lin et al., 2014; Elimelech and Philip, 2011). Nevertheless, MD is claimed to be an attractive low cost process for clean water production when low grade thermal energy is available.

4. TIPS, REMARKS, AND FUTURE DIRECTIONS IN MD

After more than 45 years of hard and continuous researches, recently MD technology begins to acquire industrial interests boosted by some companies such as Memsys, Memstill, Scarab Development AB, Keppel Seghers, and Fraunhofer ISE among others. Still MD researchers are looking for identification of new applications of MD process including integrated MD systems to other separation processes. In addition, few research studies are reported on long term MD performance and membrane fouling (i.e., deposition of particles, colloids, emulsions, suspensions, macromolecules, etc.) and microorganism growth on membrane surface (i.e., membrane biofouling) although fouling phenomena in MD are significantly lower than those faced in other pressure-driven membrane separation processes (Van Gassel and Schneider, 1986; Li and Sirkar, 2004; Schneider et al., 1988; Banat and Simandl, 1994, 1998; Kimura et al., 1987; Sakai et al., 1986, 1988; Ortiz de Zárate et al., 1998; Calabró et al., 1994; Karakulski et al., 2002; Gryta et al., 2001a; Gryta et al., 2001b; Drioli and Wu, 1985; Gryta, 2000, 2002, 2005; Khayet and Mengual, 2004; Srisurichan et al., 2005). Membrane fouling and scaling in MD can lead to wetting of the membrane pores and reduce the effective membrane area. As consequence the MD water production rate together with the water quality and the separation or rejection factors are reduced. MD researchers are now discussing various issues such as the energy consumption, especially that of the recent proposed commercial MD plants, the water production cost, the difficulties faced with long-term operation, the simultaneous risk of membrane wetting, scaling, and fouling. Among the areas that are roughly studied are membrane engineering for preparation of improved and novel membranes, membrane modules designed specifically for MD applications, and optimized coupling of renewable energy systems to MD plants.

ACKNOWLEDGMENTS

M. Essalhi is thankful to Middle East Desalination Research Centre (MEDRC) for the grant (Project 06-AS-02). The authors gratefully acknowledge the financial support of the I + D + I Project MAT2010-19249 (Spanish Ministry of Science and Innovation).

REFERENCES

Ahmad, A.L., Mariadas, A., 2004. Baffled microfiltration membrane and its fouling control for feed water of desalination. Desalination 168, 223–230.

Al-khudhiri, A., Darwish, N., Hilal, N., 2012. Treatment of high salinity solutions: application of air gap membrane distillation. Desalination 287, 55–60.

Al-Sharif, S., Albeirutty, M., Cipollina, A., Micale, G., 2013. Modelling flow and heat transfer in spacer-filled membrane distillation channels using open source CFD code. Desalination 311, 103–112.

Andersson, S.I., Kjellander, N., Rodesjo, B., 1985. Design and field tests of a new membrane distillation desalination process. Desalination 56, 345–354.

Bagger-Jorgensen, R., Meyer, A.S., Varming, C., Jonsson, G., 2004. Recovery of volatile aroma compounds from black currant juice by vacuum membrane distillation. J. Food Eng. 64, 23–31.

Banat, F., Jwaied, N., 2010. Autonomous membrane distillation pilot plant unit driven by solar energy: experiences and lessons learned. Int. J. Sustain. Water Environ. Syst. 1, 21–24.

Banat, F.A., Simandl, J., 1994. Theoretical and experimental study in membrane distillation. Desalination 95, 39–52.

Banat, F.A., Simandl, J., 1996. Removal of benzene traces from contaminated water by vacuum membrane distillation. Chem. Eng. Sci. 51, 1257–1265.

Banat, F.A., Simandl, J., 1998. Desalination by membrane distillation: a parametric study. Sep. Sci. Technol. 33, 201–226.

Banat, F.A., Simandl, J., 1999. Membrane distillation for dilute methanol: separation from aqueous streams. J. Membr. Sci. 163, 333–348.

Banat, F., Jwaied, N., Rommel, M., Koschikowski, J., Wieghaus, M., 2007a. Desalination by a "compact SMADES" autonomous solar-powered membrane distillation unit. Desalination 217, 29–37.

Banat, F., Jwaied, N., Rommel, M., Koschikowski, J., Wieghaus, M., 2007b. Performance evaluation of the "large SMADES" autonomous desalination solar-driven membrane distillation plant in Aqaba, Jordan. Desalination 217, 17–28.

Bandini, S., Sarti, G.C., 1999. Heat and mass transport resistances in vacuum membrane distillation per drop. AIChE J. Vol. 45, 1422–1433.

Bandini, S., Sarti, G.C., 2002. Concentration of must through vacuum membrane distillation. Desalination 149, 253–259.

Bandini, S., Gostoli, C., Sarti, G.C., 1992. Separation efficiency in vacuum membrane distillation. J. Membr. Sci. 73, 217–229.

Bandini, S., Saavedra, A., Sarti, G.C., 1997. Vacuum membrane distillation: experiments and modeling. AIChE J. 43, 398–408.

Bodell, B.R., 1963. Silicone Rubber Vapor Diffusion in Saline Water Distillation. United States Patent Serial No. 285,032.

Bouguecha, S., Dhahbi, M., 2003. Fluidized bed crystalliser and air gap membrane distillation as a solution to geothermal water desalination. Desalination 152, 237–244.

Calabró, V., Pantano, G., Kang, M., Molinari, R., Drioli, E., 1990. Experimental study on integrated membrane processes in the treatment of solutions simulating textile effluents. Energy and exergy analysis. Desalination 78, 257–277.

Calabró, V., Drioli, E., Matera, F., 1991. Membrane distillation in the textile wastewater treatment. Desalination 83, 209–224.

Calabró, V., Jiao, B.L., Drioli, E., 1994. Theoretical and experimental study on membrane distillation in the concentration of orange juice. Ind. Eng. Chem. Res. 33, 1803–1808.

Calibo, R.L., Matsumura, M., Takahashi, J., Kataoka, H., 1987. Ethanol stripping by pervaporation using porous PTFE membrane. J. Ferment. Technol. 65, 665–674.

Caputo, G., Felici, C., Tarquini, P., Giaconia, A., Sau, S., 2007. Membrane distillation of HI/H_2O and H_2SO_4/H_2O mixtures for the sulphur-iodine thermochemical process. Int. J. Hydrogen Energy 32, 4736–4743.

Carlsson, L., 1983. The new generation in sea water desalination: SU membrane distillation system. Desalination 45, 221–222.

Catalogue of Enka AG., June 1984. Presented at Europe-Japan Joint Congress on Membranes and Membrane Processes. In: Drioli, E., Nakagaki, M. (Eds.). Plenum Press, NY, USA, Stresa, Italy.

Cath, T.Y., Adams, V.D., Childress, A.E., 2004. Experimental study of desalination using direct contact membrane distillation: a new approach to flux enhancement. J. Membr. Sci. 228, 5–16.

Cath, T.Y., Adams, D., Childress, A.E., 2005. Membrane contactor processes for wastewater reclamation in space, II. Combined direct osmosis, osmotic distillation, and membrane distillation for treatment of metabolic wastewater. J. Membr. Sci. 257, 111–119.

Chen, T.C., Ho, C.D., 2010. Immediate assisted solar direct contact membrane distillation in saline water desalination. J. Membr. Sci. 358, 122–130.

Chen, G., Lu, Y., Krantz, W.B., Wang, R., Fane, A.G., 2014. Optimization of operating conditions for a continuous membrane distillation crystallization process with zero salty water discharge. J. Membr. Sci. 450, 1–11.

Cheng, D.Y., 1981. Method and Apparatus for Distillation. United States Patent Serial No. 4,265,713.

Cheng, D.Y., Wiersma, S.J., 1982. Composite Membrane for a Membrane Distillation System. United States Patent Serial No. 4,316,772.

Cheng, D.Y., Wiersma, S.J., 1983a. Apparatus and Method for Thermal Membrane Distillation. United States Patent Serial No. 4,419,187.

Cheng, D.Y., Wiersma, S.J., 1983b. Composite Membrane for a Membrane Distillation System, United States Patent Serial No. 4,419,242.

Cipollina, A., Miceli, A.D., Koschikowski, J., Micale, G., Rizzuti, L., 2009. CFD simulation of a membrane distillation module channel. Desalin. Water Treat. 6, 177–183.

Cipollina, A., Di Sparti, M.G., Tamburini, A., Micale, G., 2012. Development of a membrane distillation module for solar energy seawater desalination. Chem. Eng. Res. Des. 90, 2101–2121.

Concord Enviro Systems uses Memsys'technolog to treat molasses wastewater. Membr. Technol 11, 2012, 1–16.

Couffin, N., Cabassud, C., Lahoussine-Turcaud, V., 1998. A new process to remove halogenated VOCs for drinking water production: vacuum membrane distillation. Desalination 117, 233–245.

Criscuoli, A., Drioli, E., 1999. Energetic and exegetic analysis of an integrated membrane desalination system. Desalination 124, 243–249.

Criscuoli, A., Drioli, E., Capuano, A., Memoli, B., Andreucci, V.E., 2002. Human plasma ultrafiltrate purification by membrane distillation: process optimisation and evaluation of its possible application on-line. Desalination 147, 147–148.

De Andrés, M.C., Doria, J., Khayet, M., Peña, L., Mengual, J.I., 1998. Coupling of a membrane distillation module to a multieffect distiller for pure water production. Desalination 115, 71–81.

Ding, Z., Liu, L., El-Bourawi, M.S., Ma, R., 2005. Analysis of a solar-powered membrane distillation system. Desalination 172, 27–40.

Ding, Z., Liu, L., Li, Z., Ma, R., Yang, Z., 2006. Experimental study of ammonia removal from water by membrane distillation (MD): the comparison of three configurations. J. Membr. Sci. 286, 93–103.

Dotremont, C., Kregersman, B., Puttemans, S., Ho, P., Hanemaaijer, J., 2008. Memstill: A Near-future Technology for Sea Water Desalination. ICOM 2008, Honolulu, Hawaii, 12–18 July, http://www.icom2008.org/viewpaper.cfm?ID=2050.

Drioli, E., Wu, Y., 1985. Membrane distillation: an experimental study. Desalination 53, 339–346.

Drioli, E., Jiao, B.L., Calabró, V., 1992. The preliminary study on the concentration of orange juice by membrane distillation. Proc. Int. Soc. Citric. 3, 1140–1144.

Drioli, E., Laganá, F., Criscuoli, A., Barbieri, G., 1999. Integrated membrane operations in desalination processes. Desalination 122, 141–145.

Edwie, F., Chung, T.S., 2012. Development of hollow fiber membranes for water and salt recovery from highly concentrated brine via direct contact membrane distillation and crystallization. J. Membr. Sci. 412-422, 111–123.

El-Abbassi, A., Kiai, H., Hafidi, A., García-Payo, M.C., Khayet, M., 2012. Treatment of olive mill wastewater by membrane distillation using polytetrafluoroethylene membranes. Sep. Purif. Technol. 98, 55–61.

El-Abbassi, A., Hafidi, A., Khayet, M., García-Payo, M.C., 2013. Integrated direct contact membrane distillation for olive mil wastewater treatment. Desalination 323, 31–38.

El-Bourawi, M.S., Ding, Z., Ma, R., Khayet, M., 2006. A framework for better understanding membrane distillation separation process: review. J. Membr. Sci. 285, 4–29.

El-Bourawi, M.S., et al., 2007. Application of vacuum membrane distillation for ammonia removal. J. Membr. Sci. 301, 200–209.

Elimelech, M., Phillip, W.A., 2011. The future of seawater desalination: energy, technology, and the environment. Science 333, 712–717.

Esato, K., Eiseman, B., 1975. Experimental evaluation of Gore-Tex membrane oxygenator. J. Thorac. Cardiovasc. Surg. 69, 690–697.

Essalhi, M., Khayet, M., 2013. Self-sustained webs of polyvinylidene fluoride electrospun nanofibers at different electrospinning times: 1. Desalination by direct contact membrane distillation. J. Membr. Sci. 433, 167–179.

Fath, H.E.S., et al., 2008. PV and thermally driven small-scale, stand-alone solar desalination systems with very low maintenance needs. Desalination. 225, 58–69.

Findley, M.E., 1967. Vaporization through porous membranes. Ind. Eng. Chem. Process. Des. Dev. 6, 226–237.

Francis, L., Ghaffour, N., Alsaadi, A.A., Amy, G.L., 2013. Material gap membrane distillation: a new design for water vapor flux enhancement. J. Membr. Sci. 448, 240–247.

Franken, A.C.M., Ripperger, S., 1988. Terminology for Membrane Distillation. European Society of Membrane Science and Technology. Issued January.

Fujii, Y., Kigoshi, S., Iwatani, H., Aoyama, M., 1992a. Selectivity and characteristics of direct contact membrane distillation type experiment: I. Permeability and selectivity through dried hydrophobic fine porous membranes. J. Membr. Sci. 72, 53–72.

Fujii, Y., Kigoshi, S., Iwatani, H., Aoyama, M., Fusaoka, Y., 1992b. Selectivity and characteristics of direct contact membrane distillation type experiment: II. Membrane treatment and selectivity increase. J. Membr. Sci. 72, 73–89.

Gálvez, J.B., García-Rodríguez, L., Martín-Mateos, I., 2009. Seawater desalination by an innovative solar-powered membrane distillation system: the medesol project. Desalination 246, 567–576.

García-Payo, M.C., Izquierdo-Gil, M.A., Fernández-Pineda, C., 2000. Air gap membrane distillation of aqueous alcohol solutions. J. Membr. Sci. 169, 61–80.
García-Payo, M.C., Rivier, C.A., Marison, I.W., von Stockar, U., 2002 Separation of binary mixtures by thermostatic sweeping gas membrane distillation: II. Experimental results with aqueous formic acid solutions. J. Membr. Sci. 198, 197–210.
Gore, D.W., 1982. Gore-Tex membrane distillation. In: Proc. Of the 10th Ann. Convention of the Water Supply Improvement Assoc., Honolulu, USA, July 25–29.
Gostoli, C., Sarti, G.C., 1989. Separation of liquid mixtures by membrane distillation. J. Membr. Sci. 41, 211–224.
Gryta, M., 2000. Concentration of saline wastewater from the production of heparin. Desalination 129, 35–45.
Gryta, M., 2002. The assessment of microorganism growth in the membrane distillation system. Desalination 142, 79–88.
Gryta, M., 2005. Long-term performance of membrane distillation process. J. Membr. Sci. 265, 153–159.
Gryta, M., Karakulski, K., 1999. The application of membrane distillation for the concentration of oil-water emulsions. Desalination 121, 23–29.
Gryta, M., Karakulski, K., Morawski, A.W., 2001a. Purification of oily wastewater by hybrid UF/MD. Water Res. 35, 3665–3669.
Gryta, M., Tomaszewska, M., Grzechulska, J., Morawski, A.W , 2001b. Membrane distillation of NaCl solution containing natural organic matter. J. Membr. Sci. 181, 279–287.
Gunko, S., Verbych, S., Bryk, M., Hilal, N., 2006. Concentration of apple juice using direct contact membrane distillation. Desalination 190, 117–124.
Hanbury, W.T., Hodgkiess, T., 1985. Membrane distillation: an assessment. Desalination 56, 287–297.
Hanemaaijer, J.H., et al., 2006. Memstill membrane distillation – a future desalination technology. Desalination 199, 175–176.
Hausmann, A., Sanciolo, P., Vasiljevic, T., Weeks, M., Duke, M , 2012. Integration of membrane distillation into heat paths of industrial processes. Chem. Eng. J. 211–212, 378–387.
He, K., Hwang, H.J., Woo, M.W., Moon, I.S., 2011. Production of drinking water from saline water by direct contact membrane distillation (DCMD). J. Ind. Eng. Chem. 17, 41–48.
Heinzl, W., Büttner, S., Lange, G., 2012. Industrialized modules for MED desalination with polymer surfaces. Desalin. Water Treat. 42, 177–180.
Hoffman, E., et al., 1987. Evaporation of alcohol/water mixtures through hydrophobic porous membranes. J. Membr. Sci. 34, 199–206.
Hogan, P.A., Sudjito, Fane, A.G., Morrison, G.L., 1990. Desalination by solar heated membrane distillation. Desalination 81, 81–90.
Hou, D., et al., 2010. Boron removal from aqueous solution by direct contact membrane distillation. J. Hazard. Mater. 177, 613–619.
Izquierdo-Gil, M.A., Jonsson, G., 2003. Factors affecting flux and ethanol separation performance in vacuum membrane distillation (VMD). J. Membr. Sci. 214, 113–130.
Izquierdo-Gil, M.A., García-Payo, Fernández-Pineda, C., 1999. Air gap membrane distillation of sucrose aqueous solutions. J. Membr. Sci. 155, 291–307.
Jansen, A., et al., 2010. Pilot plants prove feasibility of a new desalination technique. Asian Water 26, 22–26.
Jansen, A.E., Assink, J.W., Hanemaaijer, J.H., van Medevoort, J., van Sonsbeek, E., 2013. Development and pilot testing of full-scale membrane distillation modules for deployment of waste heat. Desalination 323, 55–65.

Ji, X., et al., 2010. Membrane distillation-crystallization of seawater reverse osmosis brines. Sep. Purif. Technol. 71, 76–82.

Jonsson, A.S., Wimmerstedt, R., Harrysson, A.C., 1985. Membrane distillation: a theoretical study of evaporation through microporous membranes. Desalination 56, 237–249.

Juang, R.S., Lin, S.H., Yang, M.C., 2005. Mass transfer analysis on air stripping of VOCs from water in microporous hollow fibers. J. Membr. Sci. 255, 79–87.

Karakulski, K., Gryta, M., Morawski, A., 2002. Membrane processes used for potable water quality improvement. Desalination 145, 315–319.

Khayet, M., 2011. Membranes and theoretical modeling of membrane distillation: a review. Adv. Colloid Interface. Sci. 164 (1–2), 56–88.

Khayet, M., 2013. Solar desalination by membrane distillation: Dispersion in energy consumption analysis and water production costs (a review). Desalination 308, 89–101.

Khayet, M., Cojocaru, C., 2012a. Air gap membrane distillation: desalination, modelling and optimization. Desalination 287, 138–145.

Khayet, M., Cojocaru, C., 2012b. Artificial neural network modelling and optimization of desalination by air gap membrane distillation. Sep. Purif. Technol. 86, 171–182.

Khayet, M., Matsuura, T., 2001. Preparation and characterization of polyvinylidene fluoride membranes for membrane distillation. Ind. Eng. Chem. Res. 40, 5710–5718.

Khayet, M., Matsuura, T., 2004. Pervaporation and vacuum membrane distillation processes: modeling and experiments. AIChE J. 50, 1697–1712.

Khayet, M., Matsuura, T., 2011. Membrane Distillation: Principles and Applications. Elsevier, The Netherlands.

Khayet, M., Mengual, J.I., 2004. Effect of salt concentration during the treatment of humic acid solutions by membrane distillation. Desalination 168, 373–381.

Khayet, M., Godino, M.P., Mengual, J.I., 2002. Thermal boundary layers in sweeping gas membrane distillation processes. AIChE J. 48, 1488–1497.

Khayet, M., Velázquez, A., Mengual, J.I., 2004. Direct contact membrane distillation of humic acid solutions. J. Membr. Sci. 240, 123–128.

Khayet, M., Matsuura, T., Mengual, J.I., Qtaishat, M., 2006a. Design of novel direct contact membrane distillation membranes. Desalination 192, 105–111.

Khayet, M., Mengual, J.I., Zakrzewska-Trznadel, G., 2006b. Direct contact membrane distillation for nuclear desalination. Part. II. Experiments with radioactive solutions. Int. J. Nucl. Desalin. 2, 56–73.

Khayet, M., Cojocaru, C., García-Payo, M.C., 2007. Application of response surface methodology and experimental design in direct contact membrane distillation. Ind. Eng. Chem. Res. 46, 5673–5685.

Khayet, M., Imdakm, A.O., Matsuura, T., 2010. Monte Carlo simulation and experimental heat and mass transfer in direct contact membrane distillation. Int. J. Heat Mass Transfer 53, 1249–1259.

Kim, Y.D., Thu, K., Ghaffour, N., Ng, K.C., 2013. Performance investigation of a solar-assisted contact membrane distillation system. J. Membr. Sci. 427, 345–364.

Kimura, S., Nakao, S., Shimantani, S., 1987. Transport phenomena in membrane distillation. J. Membr. Sci. 33, 285–298.

Korngold, E., Korin, E., 1993. Air sweep water pervaporation with hollow fiber membranes. Desalination 91, 187–197.

Koschikowski, J., Rommel, M., Wieghaus, M., 6–12 August 2005. Solar Thermal Membrane Distillation for Small Scale Desalination Plants. ISES World Congress, Orlando.

Koschikowski, J., et al., 2009. Experimental investigations on solar driven stand-alone membrane distillation systems for remote areas. Desalination 248, 125–131.

Kullab, A., Martin, A., 2011. Membrane distillation and applications for water purification in thermal cogeneration plants. Sep. Purif. Technol. 76, 231–237.

Kurokawa, H., Ebara, K., Kuroda, O., Takahashi, S., 1990. vapor permeate characteristics of membrane distillation. Sep. Sci. Technol. 25, 1349–1359.
Lawson, K.W., Lloyd, D.R., 1996. Membrane distillation: II. Direct contact MD. J. Membr. Sci. 120, 123–133.
Lawson, K.W., Lloyd, D.R., 1997. Review: membrane distillation. J. Membr. Sci. 124, 1–25.
Li, B., Sirkar, K.K., 2004. Novel membrane and device for direct contact membrane distillation-based desalination process. Ind. Eng. Chem. Res. 43, 5300–5309.
Lin, S., Yip, N.Y., Elimelech, M., 2014. Direct contact membrane distillation with heat recovery: thermodynamic insights from module scale modelling. J. Membr. Sci. 453, 498–515.
Liu, C., 2004. Polygeneration of Electricity, Heat and Ultrapure Water for the Semiconductor Industry, Heat & Power Technology, Department of Energy Technology, Royal Institute of Technology, Stockholm, Sweden, Master of Science Thesis.
Lloyd, D.R., 1990. Microporous membrane formation via thermally-induced phase separation, part 1, solid-liquid phase separation. J. Membr. Sci. 52, 239–261.
Lu, H., Walton, J.C., Swift, A.H.P., 2001. Desalination coupled with salinity-gradient solar ponds. Desalination 136, 13–23.
Lukanin, O.S., Gunko, S.M., Bryk, M.T., Nigmatullin, R.R., 2003. The effect of content of apple juice biopolymers on the concentration by membrane distillation. J. Food Eng. 60, 275–280.
Mahmud, H., Kumar, A., Narbaitz, R.M., Matsuura, T., 1998. Membrane air stripping: a process for removal of organics from aqueous solutions. Sep. Sci. Technol. 33, 2241–2255.
Mariah, L., et al., 2006. Membrane distillation of concentrated brines-Role of water activities in the evaluation of driving force. J. Membr. Sci. 280, 937–947.
Martinetti, C.R., Childress, A.E., Cath, T.Y., 2009. High recovery of concentrated RO brines using forward osmosis and membrane distillation. J. Membr. Sci. 331, 31–39.
Martinez, L., Rodriquez-Maroto, J.M., 2007. Effects of membrane and module design improvements on flux in direct contact membrane distillation. Desalination 205, 97–103.
Martinez-Diez, L., Vázquez-González, M.I., 2000. A method to evaluate coefficients affecting flux in membrane distillation. J. Membr. Sci. 173, 225–234.
Memsys-NTU partnership aims to enhance water desalination technology. Membr. Technol. 2, 2011, 1–16.
Mericq, J.P., Laborie, S., Cabassud, C., 2010. Vacuum membrane distillation of seawater reverse osmosis brines. Water Res. 44, 5260–5273.
Mericq, J., Laborie, S., Cabassud, C., 2011. Evaluation of systems coupling vacuum membrane distillation and solar energy for seawater desalination. Chem. Eng. J. 166, 596–606.
Mulder, M., 1996. Basic Principles of Membrane Technology. Kluwer Academic Publishers, Dordrecht, The Netherlands.
Nene, S., Kaur, S., Sumod, K.K., Joshi, B., Raghavarao, K.S.M.S., 2002. Membrane distillation for the concentration of raw-cane sugar syrup and membrane clarified sugarcane juice. Desalination 147, 157–160.
Ong, C.L., Escher, W., Paredes, S., Khalil, A.S.G., Michel, B., 2012. A novel concept of energy reuse from high concentration photovoltaic thermal (HCPVT) system for desalination. Desalination 295, 70–81.
Ortiz de Zárate, J.M., Rincón, C., Mengual, J.I., 1998. Concentration of bovine serum albumin aqueous solutions by membrane distillation. Sep. Sci. Technol. 33, 283–296.
Phattaranawik, J., Jiraratananon, R., Fane, A.G., 2003. Effect of pore size distribution and air flux on mass transport in direct contact membrane distillation. J. Membr. Sci. 215, 75–85.
Qtaishat, M., Matsuura, T., Kruczek, B., Khayet, M., 2008. Heat and mass transfer analysis in direct contact membrane distillation. Desalination 219, 272–292.

Qu, D., Wang, J., Hou, D., Luan, Z., Fan, B., Zhao, C., 2009. Experimental study of arsenic removal by direct contact membrane distillation. J. Hazard. Mater. 163, 874–879.

Raluy, R.G., et al., 2012. Operational experience of a solar membrane distillation demonstration plant in Pozo Izquierdo-Gran Canaria Island (Spain). Desalination 290, 1–13.

Rincón, C., Ortiz de Zárate, J.M., Mengual, J.I., 1999. Separation of water and glycols by direct contact membrane distillation. J. Membr. Sci. 158, 155–165.

Rommel, M., Wieghaus, M., Koschikowski, J., 2008. Solar powered desalination: an autonomous water supply. Desalination 3, 22–24.

Sakai, K., et al., 1986. Extraction of solute-free water from blood by MD. Am. Soc. Artif. Intern. Organs 32, 397–400.

Sakai, K., Koyano, T., Muroi, T., Tamura, M., 1988. Effects of temperature and concentration polarization on water vapour permeability for blood in membrane distillation. Chem. Eng. J. 38, B33–B39.

Sarti, G.C., Gostoli, C., Bandini, S., 1993. Extraction of organic components from aqueous streams by vacuum membrane distillation. J. Membr. Sci. 80, 21–33.

www.scarab.se, Scarab Development AB, Stockholm, Sweden.

Schneider, K., van Gassel, T.J., 1984. Membrandestillation. Chem. Ing. Tech. 56, 514–521.

Schneider, K., Hölz, W., Wollbeck, R., 1988. Membranes and modules for transmembrane distillation. J. Membr. Sci. 39, 25–42.

Schofield, R.W., Fane, A.G., Fell, C.J.D., Macoun, R., 1990. Factors affecting flux in membrane distillation. Desalination 77, 279–294.

Schwantes, R., et al., 2013. Membrane distillation: solar and waste heat driven demonstration plants for desalination. Desalination 323, 93–106.

Shakaib, M., Hasani, S.M.F., Ahmed, I., Yunus, R.M., 2012. A CFD study on the effect of spacer orientation on temperature polarization in membrane distillation modules. Desalination 284, 332–340.

Smolders, K., Franken, A.C.M., 1989. Terminology for membrane distillation. Desalination 72, 249–262.

Srisurichan, S., Jiraratananon, R., Fane, A.G., 2005. Humic acid fouling in the membrane distillation process. Desalination 174, 63–72.

Suareza, F., Tyler, S., Childress, A., 2010. A theoretical study of a direct contact membrane distillation system coupled to a salt-gradient solar pond for terminal lakes reclamation. Water Res. 44, 4601–4615.

Tarnacki, K., Meneses, M., Melin, T., van Medevoort, J., Jansen, A., 2012. Environmental assessment of desalination processes: reverse osmosis and memstill®. Desalination 296, 69–80.

Teoh, M.M., Bonyadi, S., Chung, T.S., 2008. Investigation of different hollow fiber module designs for flux enhancement in the membrane distillation process. J. Membr. Sci. 311, 371–379.

Tomaszewska, M., Gryta, M., Morawski, A.W., 1995. Study on the concentration of acids by membrane distillation. J. Membr. Sci. 102, 113–122.

Udriot, H., Araque, A., von Stockar, U., 1994. Azeotropic mixtures may be broken by membrane distillation. Cheng. Eng. J. 54, 87–93.

Urtiaga, A.M., Ruiz, G., Ortiz, I., 2000. Kinetic analysis of the vacuum membrane distillation of chloroform form aqueous solutions. J. Membr. Sci. 165, 99–110.

Van Gassel, T.J., Schneider, K., 1986. An energy-efficient membrane distillation process. In: Drioli, E., et al. (Eds.), Membranes and Membrane Processes. Plenum Press, New York, pp. 343–348.

Viladomat, F.G., Souchon, I., Athès, V., Marin, M., 2006. Membrane air-stripping for aroma compounds. J. Membr. Sci. 277, 129–136.

Walton, J., Lu, H., Turner, C., Solis, S., Hein, H., 2004. Solar and Waste Heat Desalination by Membrane Distillation. Desalination and Water Purification Research and Development Program Report No. 81. Bureau of Reclamation, Denver, CO, USA.

Wang, K.L., Cussler, E.L., 1993. Baffled membrane modules made with hollow fiber fabric. J. Membr. Sci. 85, 265–278.

Wang, X., Zhang, L., Yang, H., Chen, H., 2009. Feasibility research of potable water production via solar-heated hollow fiber membrane distillation system. Desalination 247, 403–411.

Wang, K.Y., Teoh, M.M., Nugroho, A., Chung, T.S., 2011. Integrated forward osmosis-membrane distillation (FO-MD) hybrid system for the concentration of protein. Chem. Eng. Sci. 66, 2421–2430.

Weyl, P.K., 1967. Recovery of demineralized water from saline waters. United States Patent US3340186 A, 5 Sep.

Winter, D., Koschikowski, J., Wieghaus, M., 2011. Desalination using membrane distillation: experimental studies on full scale spiral wound modules. J. Membr. Sci. 375, 104–112.

Winter, D., Koschikowski, J., Ripperger, S., 2012. Desalination using membrane distillation: flux enhancement by feed water deaeration on spiral-wound modules. J. Membr. Sci. 423–424, 215–224.

Wu, Y., Drioli, E., 1989. The behaviour of membrane distillation of concentrated aqueous solution. Water Treat. 4, 399–415.

Wu, Y., Kong, Y., Liu, J., Zhang, J., Xu, J., 1991. An experimental study on membrane distillation-crystallization for treating waste water in taurine production. Desalination 80, 235–242.

Xu, Z., Pan, Y., Yu, Y., 2009. CFD simulation on membrane distillation of NaCl solution. Front. Chem. Eng. China 3, 293–297.

www.Xzero.se.

Yang, X., Wang, R., Shi, L., Fane, A.G., Debowski, M., 2011. Performance improvement of PVDF hollow fiber-based membrane distillation process. J. Membr. Sci. 369, 437–447.

Yang, X., Yu, H., Wang, R., Fane, A.G., 2012. Analysis of the effect of turbulence promoters in hollow fiber membrane distillation modules by computational fluid dynamic (CFD) simulations. J. Membr. Sci. 415-416, 758–769.

Yu, H., Yang, X., Wang, R., Fane, A.G., 2012. Analysis of heat and mass transfer by CFD for performance enhancement in direct contact membrane distillation. J. Membr. Sci. 405–406, 38–47.

Zakrewska-Trznadel, G., Harasimowicz, M., Chmielewski, A.G., 1999. Concentration of radioactive components in liquid low-level radioactive waste by membrane distillation. J. Membr. Sci. 163, 257–264.

Zakrewska-Trznadel, G., Harasimowicz, M., Chmielewski, A.G., 2001. Membrane processes in nuclear technology-application for liquid radioactive waste treatment. Sep. Purif. Technol. 22–23, 617–625.

Zhao, K., et al., 2013. Experimental study of the memsys vacuum-multi-effect-membrane-distillation (V-MEMD) module. Desalination 323, 150–160.

Zolotarev, P.P., Urgozov, V.V., Volkina, I.B., Nikulin, V.N., 1994. Treatment of waste water for removing heavy metals by membrane distillation. J. Hazard. Mater. 37, 77–82.

CHAPTER FOUR

Pervaporation

Bart Van der Bruggen and Patricia Luis
Department of Chemical Engineering, Laboratory of Applied Physical Chemistry and Environmental Technology, KU Leuven, Leuven, Belgium

Contents

1. INTRODUCTION

One of the most remarkable membrane separation processes, without any doubt, is pervaporation. It is the only membrane process in which a phase transition occurs during transport through a dense or microporous matrix, leading to the peculiar feature of an undefined phase condition within the membrane itself. The phase transition occurs due to the vacuum (or, more generally, the low partial pressure) at the permeate side. Selectivity is obtained by selective dissolution of components from a liquid mixture in the membrane material and differences in diffusion through the membrane. This allows separation of solvents on the basis of their affinity for the membrane material, in contrast to classical distillation where separation occurs as a result of differences in volatility. Thus, azeotropic mixtures and mixtures with a relative volatility close to one can be separated by pervaporation, on condition that they have a clear difference in properties determining their transport through pervaporation membranes. As further described in this chapter, the main characteristics that are decisive for rejection by or transport through the membrane are the polarity of the solvent, and its molecular size.

Progress in Filtration and Separation
http://dx.doi.org/10.1016/B978-0-12-384746-1.00004-5

Because mixtures of organic solvents with water have a large difference in these parameters when water is to be removed, this is an obvious application for pervaporation. Similarly, the removal of an organic compound from an aqueous solution is also feasible. On the other hand, organic–organic separations are much more challenging since they have a smaller difference in polarity and molecular size.

The oldest reference to pervaporation in ISI's Web of Science dates back to 1956, and reports on the use of pervaporation for separation of aqueous alcoholic mixtures (Heisler et al., 1956). However, the origins of pervaporation are older. Kober first introduced the term pervaporation in 1917 in a publication reporting on the selective permeation of water from aqueous solutions of albumin and toluene through cellulose nitrate films (Kober, 1917). He observed that a liquid in a collodion bag, which was suspended in the air, evaporated, although the bag was tightly closed. This brought him to introduce the term pervaporation as a contraction of "permselective evaporation." In general, the definition of pervaporation can be derived from this observation as a process in which one or more components of a liquid mixture selectively permeate through a dense or microporous membrane, driven by a difference in partial vapour pressure, leaving the membrane as a vapour, and being recovered in a condensed form as a liquid. Even before Kober, a similar observation was made by Kahlenberg, who reported in 1906 on the separation of a mixture of a hydrocarbon and an alcohol through a rubber membrane (Kahlenberg, 1906). Later, in 1935 Farber recognized the potential of pervaporation for separating and concentrating protein and enzyme solutions, using cellophane (Farber, 1935). However, Binning and co-workers at American Oil Company (Amoco) were the first to carry out systematic studies to determine the principles and understand the potential of pervaporation. Between 1958 and 1962, several patents and applications by Binning and his group appeared, mostly related to their research on the separation of hydrocarbon mixtures through a nonporous polyethylene film (Binning and James, 1958a; Binning et al., 1961, 1962). Experiments were carried out at high temperatures up to 150 °C, at which the permeation rates were high enough to permit practical applications. They demonstrated that there was a difference in permeation rate between paraffins and aromatics and olefins, which permeated faster. In general, linear hydrocarbons proved to permeate faster than branched isomers. This brought them to potential applications for improvement of gasoline, since it would allow to increase the octane number by applying pervaporation.

In addition, they also reported the separation of other mixtures such as methanol/benzene, alcohols/water and pyridine/water by pervaporation (Binning and James, 1958b,c; Binning, 1961).

The breakthrough of pervaporation came in the 1970s, when the potential for many applications became apparent. In 1976, in one of the first issues of Journal of Membrane Science, azeotropic mixtures were considered for the first time (Aptel et al., 1976). Due to its simplicity in operation (i.e., automation), the possibility of working at elevated temperatures and the reduced amounts of chemicals needed, pervaporation was found to be a good alternative for conventional energy consuming processes such as distillation, or complex methods involving secondary materials cycles, such as extraction. In addition, challenging separations such as benzene-cyclohexane (Rautenbach and Albrecht, 1980) and 1,3 butadiene-isobutene (Vasse et al., 1975). Pervaporation had potential as an energy-saving separation method, which was crucial for its development in the years of the oil crisis in the 1970s. Slow permeation rates, the traditional impediments for membrane separations, were improved by the development of asymmetric membranes. This has catalyzed studies on pervaporation in the 1970s, leading to significant breakthroughs after roughly a decade of intensive studies, in the 1980s. Worth mentioning is the development of thin film polyvinyl alcohol-polyacrylonitrile (PVA/PAN) composite membranes by GFT (Gesellschaft fur Trenntechnik GmbH, since 1997 Sulzer Chemtech).

Pilot-scale industrial applications appeared already in the early 1980s. The first operating pilot plant was started in 1982 as a demonstration unit for alcohol dehydration, to produce 1200 L per day with a purity of 99.2 weight percent ethanol. One of the successes of this plant, in addition to the product purity, was the low energy consumption compared with that of conventional dehydration of ternary azeotropic distillation after addition of benzene (Feng and Huang, 1997; Kujawski, 2000; Jonquières et al., 2002). Plants with larger production capacities, ranging from 2000 to 15,000 L per day, appeared not much later during the 1980s. Over 50 plants were installed by GFT in Europe and the USA for the dehydration of ethanol and other organic solvents. The first large-scale pervaporation unit, used for the dehydration of ethanol to a concentration of 99.8 wt%, with a capacity of 150,000 L per day, was installed in 1988 in the sugar refinery of Bethéniville, France (Feng and Huang, 1997; Kujawski, 2000; Jonquières et al., 2002).

Dehydration was successful in more pilot-scale and full-scale plants. In 2000, inorganic NaA zeolite membranes were for the first time installed in a large-scale plant for the dewatering of ethanol to 99.8 wt% (Morigami et al., 2001). Dehydration was further used in pervaporation membrane reactors for equilibrium reactions in which water was obtained as a by-product. By removing water from the reaction broth, the yield of the reaction could be increased, according to the principle of Le Châtelier-Braun. This allows even a full conversion. In 1994 a plant was installed in which water is continuously removed from a reaction mixture in order to shift the reaction equilibrium towards the wanted product (Jonquières et al., 2002).

Parallel to dehydration, hydrophobic pervaporation was also developed, mainly using polydimethylsiloxane (PDMS) membranes. These were used for extraction of organic compounds from aqueous streams (Lipnizki et al., 1999a,b; Peng et al., 2003), for example, in the case of removal of trichloroethylene from groundwater using spiral-wound PDMS membranes in a plant in California, USA, in 1993 (Feng and Huang, 1997; Kujawski, 2000; Jonquières et al., 2002).

Apart from hydrophilic membranes for dehydration and hydrophobic membranes for extraction of organic compounds, a third membrane type was developed specifically for organic–organic separations. These membranes are denoted as "organoselective" or "organophilic." The initially targeted industrial application was the purification of ethyl tert-butyl ether (ETBE), a fuel octane enhancer used in Europe, to replace formerly applied lead derivatives. The feasibility of pervaporation for the purification of ETBE were demonstrated at pilot-scale by the Institut Français du Pétrole, and enabled the commercialization of the first organoselective membranes for the removal of methanol or ethanol from purely organic mixtures (Smitha et al., 2004).

Evidently, hybrid configurations are logical key applications for pervaporation nowadays (together with solvent dehydration, for which pervaporation can be considered a standard process with proven performance). Because the separation mechanism in pervaporation is based on completely different principles than for example, distillation, a combination of pervaporation and distillation yields an efficient hybrid method in which both processes are used for that part of the overall separation where they perform best (Lipnizki et al., 1999a,b). Another successful tandem, as already indicated, is the combination of pervaporation with a reactor for (bio)chemical conversions; although it requires some more expertise than a plug-and-play dehydration, this is one of the areas where a substantial further growth is to be expected (Van der Bruggen et al., 2010).

Pervaporation is closely related to vapour permeation, a membrane separation process in which a vapour mixture is separated by dense or microporous membranes due to a difference in vapour pressure. The only difference with pervaporation is in the feed phase (vapor vs liquid); separation mechanisms and even membranes are usually quite the same. This is related to the "absence" of a well-defined phase during transport through the membranes, making the feed condition irrelevant (although it would of course still determine the driving force, i.e., the difference in partial pressure). Applications of both processes were reviewed in 2002, including an overview of patented applications, and suppliers for PV and VP membranes and membrane separation systems (Jonquières et al., 2002). Since then, new applications were developed particularly for organic–organic separations and pervaporation membrane reactors, but the reference applications are still the same today. Research efforts indicate, however, a far greater potential for pervaporation as the standard technology for difficult organic–organic separations in the pharmaceutical and (petro)chemical industry (Smitha et al., 2004). One of the key issues to be resolved is the development of a reliable mass transport model to predict membrane performance and to aid in the production of sufficiently stable, cheap and high performance membranes. This will be outlined in the following sections.

In terms of membrane structures, pervaporation is closely related to gas separation and organic solvent nanofiltration, two processes in which similar membrane types are used (Van der Bruggen et al., 2006). The difference with pervaporation is in the driving force: gas separation and organic solvent nanofiltration require a pressure gradient, and do not involve a phase transition. Membranes, however, are essentially the same, but optimization of membrane structures and morphologies is carried out in a different way. Crosslinking, for example, is more important for polymeric pervaporation membranes, in view of the gradient between the liquid feed (which causes swelling at the feed side) and the permeate side held at vacuum (in unswollen, dry conditions). This lowers the permeability, so that a careful balance between membrane stability and permeability is needed for pervaporation membranes. This has triggered research on membrane synthesis in particular, yielding a vast amount of reports on preparation of pervaporation membranes using a wide range of polymers, and ceramic materials (Bolto et al., 2011).

2. FUNDAMENTALS OF PERVAPORATION

The selectivity in pervaporation originates from different affinities for the membrane material among components of the feed mixture, and differences in permeation rate by diffusion. During transport, evaporation of the permeating compounds occurs; the heat of vaporization has to be supplied by the feed, which causes a temperature drop. As shown in Figure 1, heat may be supplied by intermediate heat exchangers installed between a number of pervaporation modules in series, for an industrial set-up of pervaporation.

In the absence of coupling effects, transport through a pervaporation membrane is typically described by the solution-diffusion model (Wijmans and Baker, 1995). This model describes mass transport of a fluid through a dense, nonporous membrane as:

$$\text{Permeability}\ (P) = \text{Solubility}\ (S) \times \text{Diffusivity}\ (D)$$

The solubility of a compound is the amount sorbed by the membrane under equilibrium conditions, and hence, a thermodynamic parameter, in contrast to the diffusivity, which is a kinetic parameter quantifying the rate of permeation through the membrane. This can be described as:

$$J_i = -L_i \frac{d\mu_i}{dx}$$

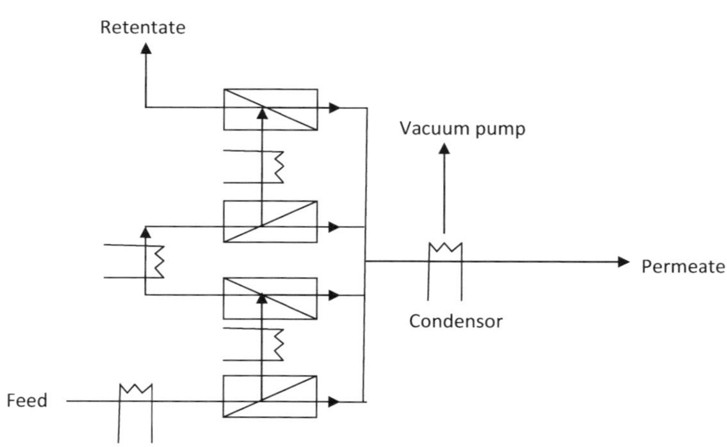

Figure 1 Pervaporation modules in series with intermediate heat exchangers.

where J_i is the flux of component i, L_i is the permeability of component i, and μ_i is the chemical potential of component i.

The chemical potential can be expressed as follows:

$$d\mu_i = RT\text{dln}(\gamma_i n_i) + v_i dp$$

where γ_i is the activity coefficient of component i, n_i the molar concentration, v_i the molar volume, and p the pressure.

The flux in pervaporation is also often expressed as

$$J_i = -D_{im}\nabla c_i$$

where ∇c_i is the molar concentration gradient of component i over the membrane and D_{im} is the Fickian diffusivity of i in the membrane.

This leads to the following simple expression:

$$J_i = D_{im}\left(c_{im}^f - c_{im}^p\right)$$

where c_{im}^f and c_{im}^p are the concentrations of component i inside the membrane on the feed side and on the permeate side, respectively.

The concentration inside the membrane at the membrane interface may be obtained by Henry's law:

$$c_{im} = S_{im}p_i$$

Where S_{im} is the adsorption coefficient or the solubility, a measure of the amount of penetrant sorbed by the membrane under equilibrium conditions and p_i is the partial pressure of species i in the phase adjoining the interface.

The resulting equation is the classical representation of the solution-diffusion transport mechanism:

$$J_i = \frac{S_{im}D_{im}}{L}\left(p_i^f - p_i^p\right)$$

$p_i^f - p_i^p = \Delta p_i$ is the partial pressure difference over the membrane, which is, calculated as (Wijmans and Baker, 1995):

$$p_i^f = x_i\gamma_i p_i^{\text{vap}}$$

and

$$p_i^p = y_i P_i^p$$

where x_i is the molar fraction of component i in a multicomponent feed, γ_i the activity coefficient at temperature T(K), P_i^{vap} the vapour pressure of pure i at temperature T(K), y_i the molar fraction of component i in the permeate, and P^p the total permeate pressure. The activity coefficients can be calculated with the UNIFAC equation. For pure species the partial pressure difference Δp_i reduces to the vapour pressure P_i^{vap} (since $P^p << P_i^f$), which can be calculated with the Antoine equation:

$$P^{\text{vap}} = A \exp\left(\frac{-\Delta H^{\text{vap}}}{RT}\right)$$

where A is the preexponential factor and ΔH^{vap} the heat of vaporization.

The temperature dependency of the flux is composed of a sorption related factor, a diffusion related factor, and the heat of vaporization, which can be written as:

$$J_i = S_{i0} \exp\left(\frac{-\Delta H_S}{RT}\right) D_{i0} \exp\left(\frac{-E_D}{RT}\right) A_i \exp\left(\frac{-\Delta H^{\text{vap}}}{RT}\right)$$

or

$$J_i = F_{i0} \exp\left(\frac{-E_F}{RT}\right) A_i \exp\left(\frac{-\Delta H^{\text{vap}}}{RT}\right)$$

where S_{i0}, D_{i0}, and F_{i0} are the preexponential factors, ΔH_s is the enthalpy of adsorption, E_D the activation energy for diffusion, and $E_F = E_D + \Delta H_s$ the activation energy for permeability.

Thus, the total flux follows an Arrhenius (exponential) type of relation:

$$J_i = J_0 \exp\left(\frac{-E_J}{RT}\right)$$

where $E_J = E_F + \Delta H^{\text{vap}} = E_D + \Delta H_s + \Delta H^{\text{vap}}$ is the activation energy for flux and J_0 is the preexponential factor.

The difference in activation energy for flux and permeability is the heat of vaporization ΔH^{vap}. Therefore, the most general and objective way to compare membrane systems relates to the energy difference E_F, which is a combination of the activation energy for diffusion (E_D), and the enthalpy of adsorption (ΔH_s). Since adsorption processes are exothermic, the enthalpy of adsorption is negative.

Diffusive transport through a pervaporation membrane may occur by surface diffusion (pore sizes below 2 nm) or by molecular sieving (micropore or configurational diffusion, pore sizes below 1 nm) (de Lange et al., 1995; Burggraaf and Cot, 1996; Beuscher and Gooding, 1998; Gilron and Soffer, 2002; Wood et al., 2002). For surface diffusion, E_F can be positive or negative, depending on whether the transport activation energy exceeds the enthalpy of adsorption or not. When $E_F < 0$, transport will decrease with increasing temperature and when $E_F > 0$, it will increase with increasing temperature. Transport through molecular sieving always increases with temperature ($E_F > 0$).

The solution-diffusion equations are a simplification, because it can be argued whether the boundary conditions that allow the application of Henry's law are met. Using a Langmuir adsorption isotherm and a concentration dependent diffusivity would solve this.

In order to take concentration effects and dragging effects into account, more complex adsorption-diffusion models based on the Maxwell–Stefan theory are required (Krishna, 1990; Krishna and Wesselingh, 1997; Taylor and Krishna, 1993). This is particularly important for inorganic membranes, where sorption is in fact a surface adsorption mechanism, and transport can be described as hopping from one adsorption position to the next one. In the literature, several derivations of Maxwell–Stefan adsorption-diffusion models have been published to describe transport in gas separation and pervaporation applications using both polymeric (Heintz and Stephan, 1994; Ghoreyshi et al., 2002; Izák et al., 2003) and inorganic membranes (Van den Broeke and Krishna, 1995; Van den Broeke et al., 1999a,b; Van den Broeke et al., 1999a,b; Krishna & Van den Broeke, 1995; Krishna, 2000; Kapteijn et al., 2000; Pera-Titus et al., 2006; de Bruijn et al., 2003; Yu et al., 2007; Lee, 2007; Krishna & van Baten, 2008). These models assume that transport of a component i is not only the result of a gradient in chemical potential of that component, but that transport is also influenced by the presence of the membrane, and of the gradients of all the other components j in the mixture.

The generalized Maxwell–Stefan theory assumes that movement of a species in a multicomponent mixture is caused by a driving force, which is balanced by the friction of this species with other species and its surroundings (Bettens, 2008). The Maxwell-Stefan equation can be generally written as:

$$-\frac{x_i}{RT}\nabla_{T,P}\mu_i = \sum_{j=1}^{n}\frac{x_i x_j\left(u_i - u_j\right)}{Đ_{ij}}$$

where μ_i represents the chemical potential of species i, $\nabla\mu_i$ represents the gradient of the chemical potential of species i, $(u_i - u_j)$ the relative velocity between species i and j, $Đ_{ij}$ the Maxwell–Stefan diffusivity between species i and j, and x_i (x_j) the molar fraction of i (j).

For a binary mixture, the equation becomes:

$$-\frac{\theta_1}{RT}\nabla_{T,P}\mu_1 = \frac{\theta_1\theta_2}{Đ_{12}^S}(u_1 - u_2) + \frac{\theta_1\theta_M}{Đ_{1M}^S}(u_1 - u_M)$$

The first term on the right-hand side expresses the friction exerted between both adsorbed components 1 and 2. The second term expresses the friction between component 1 and the membrane's pore wall. $Đ_{12}^S$ and $Đ_{1M}^S$ are the corresponding Maxwell–Stefan surface diffusivities. Interchanging 1 and 2 provides the analogous equation for component 2.

θ is the fractional coverage of the surface:

$$\theta_i = \frac{q_i}{q_{i,\text{sat}}}$$

For the membrane, θ_M represents the fraction of available adsorption sites. Because it is not well defined, it is often integrated in $Đ_{1M}^S$ by replacing $\frac{\theta_M}{Đ_{1M}^S}$ by $\frac{1}{Đ_{1M}^{S'}}$ (Krishna and Wesselingh, 1997; Taylor and Krishna, 1993).

Since fluxes should be considered here with respect to the membrane, $u_M = 0$. Further eliminating the velocities using the definition of the molar fluxes ($i = 1,2$):

$$J_i = \rho q_i u_i$$

with ρ the density of the membrane material, yields:

$$-\frac{\rho\theta_1}{RT}\nabla_{T,P}\mu_1 = \frac{\theta_1\theta_2}{Đ_{12}^S}\left(\frac{J_1}{q_{1,\text{sat}}\theta_1} - \frac{J_2}{q_{2,\text{sat}}\theta_2}\right) + \frac{\theta_1}{Đ_{1M}^{S'}}\left(\frac{J_1}{q_{1,\text{sat}}\theta_1}\right)$$

The gradient in chemical potential can be related to the gradients in fractional coverage and to the thermodynamic correction factors Γ (Krishna and Wesselingh, 1997; Taylor and Krishna, 1993):

$$-\frac{\theta_1}{RT}\nabla_{T,P}\mu_1 = \Gamma_{11}\nabla_{T,P}\theta_1 + \Gamma_{12}\nabla_{T,P}\theta_2$$

with

$$\Gamma_{1j} = \theta_1\frac{\partial \ln p_1}{\partial \theta_j}$$

where p_1 is the partial pressure of component 1.

The thermodynamic factors are determined by the adsorption isotherm chosen to relate the surface coverage θ_1 to the partial pressure p_1. Adsorption of a binary mixture can be described by the extended Langmuir model (Bettens, 2008):

$$\theta_i = \frac{q_i}{q_{i,\text{sat}}} = \frac{b_i p_i}{1 + \sum_{j=1}^{2} b_j p_j}$$

If the saturation loadings $q_{i,\text{sat}}$ for both components are the same, the thermodynamic factors can be obtained analytically:

$$\Gamma_{11} = \frac{1 - \theta_2}{1 - \theta_1 - \theta_2}$$

and

$$\Gamma_{12} = \frac{\theta_1}{1 - \theta_1 - \theta_2}$$

The final form of the Maxwell–Stefan equations in matrix form for a binary mixture is then:

$$-\rho[q_{\text{sat}}][\Gamma](\nabla\theta) = [B](J)$$

or

$$(J) = -\rho[q_{\text{sat}}][B]^{-1}[\Gamma](\nabla\theta)$$

The matrices are as follows:

$$(J) = \begin{pmatrix} J_1 \\ J_2 \end{pmatrix}$$

$$[q_{\text{sat}}] = \begin{bmatrix} q_{1,\text{sat}} & 0 \\ 0 & q_{2,\text{sat}} \end{bmatrix}$$

$$[\Gamma] = \frac{1}{1 - \theta_1 - \theta_2} \begin{bmatrix} 1 - \theta_2 & \theta_1 \\ \theta_2 & 1 - \theta_1 \end{bmatrix}$$

$$[B] = \begin{bmatrix} \dfrac{1}{Đ_{1M}^{S'}} + \dfrac{\theta_2}{Đ_{12}^{S'}} & -\dfrac{\theta_1}{Đ_{12}^{S}} \\ -\dfrac{\theta_2}{Đ_{12}^{S}} & \dfrac{1}{Đ_{2M}^{S'}} + \dfrac{\theta_1}{Đ_{12}^{S'}} \end{bmatrix}$$

$$[B]^{-1} = \frac{\begin{bmatrix} \frac{1}{Đ_{2M}^{S'}} + \frac{\theta_1}{Đ_{12}^{S}} & \frac{\theta_1}{Đ_{12}^{S}} \\ \frac{\theta_2}{Đ_{12}^{S}} & \frac{1}{Đ_{1M}^{S'}} + \frac{\theta_2}{Đ_{12}^{S}} \end{bmatrix}}{\frac{\theta_1}{Đ_{1M}^{S'} Đ_{12}^{S}} + \frac{\theta_2}{Đ_{2M}^{S'} Đ_{12}^{S}} + \frac{1}{Đ_{1M}^{S'} Đ_{2M}^{S'}}}$$

$$(\nabla\theta) = \begin{pmatrix} \nabla\theta_1 \\ \nabla\theta_2 \end{pmatrix}$$

The solution for J_1 and J_2 can be expressed as follows:

$$J_1 = -\frac{\rho q_{1,\text{sat}} Đ_{1M}^{S'} \left[\left(\Gamma_{11} + \theta_1 \frac{Đ_{2M}^{S'}}{Đ_{12}^{S}} (\Gamma_{11} + \Gamma_{21}) \nabla\theta_1 \right) + \left(\Gamma_{12} + \theta_1 \frac{Đ_{2M}^{S'}}{Đ_{12}^{S}} (\Gamma_{12} + \Gamma_{22}) \right) \nabla\theta_2 \right]}{1 + \theta_1 \frac{Đ_{2M}^{S'}}{Đ_{12}^{S}} + \theta_2 \frac{Đ_{1M}^{S'}}{Đ_{12}^{S}}}$$

$$J_2 = -\frac{\rho q_{2,\text{sat}} Đ_{2M}^{S'} \left[\left(\Gamma_{22} + \theta_2 \frac{Đ_{1M}^{S'}}{Đ_{12}^{S}} (\Gamma_{22} + \Gamma_{12}) \nabla\theta_2 \right) + \left(\Gamma_{21} + \theta_2 \frac{Đ_{1M}^{S'}}{Đ_{12}^{S}} (\Gamma_{21} + \Gamma_{11}) \right) \nabla\theta_1 \right]}{1 + \theta_1 \frac{Đ_{2M}^{S'}}{Đ_{12}^{S}} + \theta_2 \frac{Đ_{1M}^{S'}}{Đ_{12}^{S}}}$$

For pure species, these equations reduce to ($i = 1,2$):

$$J_i = -\frac{\rho q_{i,\text{sat}} Đ_{iM}^{S'} \nabla\theta_i}{1 - \theta_i}$$

The main remaining difficulty in these equations is the gradient of the fractional coverage, which is difficult to estimate. A possible simplification is to assume that the gradient of fractional coverage is constant and that the fractional coverage itself can be averaged (De Bruijn et al., 2007):

$$\overline{\theta_1} = \frac{\theta_1^f + \theta_1^p}{2}$$

Or using the logarithmic average:

$$\overline{\theta_1} = 1 - \frac{(1 - \theta_1^p) - \left(1 - \theta_1^f\right)}{\ln\left(\frac{1 - \theta_1^p}{1 - \theta_1^f}\right)}$$

In the Maxwell–Stefan expressions for the fluxes, three Maxwell–Stefan diffusivities have to be determined: $Đ_{12}^{S}$, $Đ_{1M}^{S'}$, and $Đ_{2M}^{S'}$. For the counter-exchange Maxwell–Stefan diffusivity, the Vignes equation is often used:

$$Đ_{12}^{S} = Đ_{1M}^{S'\frac{\theta_1}{\theta_1+\theta_2}} Đ_{2M}^{S'\frac{\theta_2}{\theta_1+\theta_2}}$$

Alternatively, when no interactions between species occur, the corresponding counter-exchange coefficient can be assumed infinite ($D_{12}^{S} = \infty$). This is denoted as single-file diffusion (Van den Broeke et al., 1999a,b; Van den Broeke et al., 1999a,b); components are not able to pass each other in the narrow pores.

The single-species Maxwell–Stefan diffusivities can either be considered to be independent of the amount adsorbed:

$$Đ_{1M}^{S'} = Đ_{1M}^{S'}(0)$$

and

$$Đ_{2M}^{S'} = Đ_{2M}^{S'}(0)$$

Alternatively, the single-species Maxwell–Stefan diffusivities can be described as a function of the amount adsorbed:

$$Đ_{1M}^{S'} = Đ_{1M}^{S'}(0)(1 - \theta_1 - \theta_2)$$

and

$$Đ_{2M}^{S'} = Đ_{2M}^{S'}(0)(1 - \theta_1 - \theta_2)$$

The selectivity of a pervaporation membrane is calculated as a dimensionless separation factor α. For a binary mixture consisting of components A and B, the separation factor is:

$$\alpha_{\mathrm{A,B}} = \frac{y_{\mathrm{A}}/x_{\mathrm{A}}}{y_{\mathrm{B}}/x_{\mathrm{B}}}$$

where y_{A} and y_{B} are the permeate compositions and x_{A} ad x_{B} are the feed or retentate compositions. These compositions can be described by means of mole fractions, mass fractions or volume fractions. The separation factor is chosen in such a way that its value is greater than unity and so that component A permeates preferentially. If $\alpha = 1$, no separation can be achieved.

The separation factor used in pervaporation is similar to the relative volatility used for distillation. Interestingly, their values can be quite different for the same binary system. As an example, a diagram showing the vapour concentrations in distillation as a function of their corresponding liquid concentrations in equilibrium (McCabe–Thiele diagram) is shown for the binary mixture ethanol-water (Kujawski, 2000). For pervaporation, the diagram gives the composition of the feed mixture on the *x*-axis; the *y*-axis shows the composition of the permeate. The membrane in the Figure 2 is a hydrophilic membrane: water is passing preferentially through the membrane for any water fraction (the curve is always above the diagonal) and is enriched in the permeate. The highest separation factor is observed when the water fraction is low. Thus, pervaporation is best suited for the removal of a minor component from a mixture with high selectivity. In distillation, the more volatile ethanol is enriched in the vapour phase, at least at high water concentrations (the curve is below the diagonal). With decreasing water in the feed, the vapour liquid equilibrium shows azeotropic behaviour (the azeotropic point crosses the diagonal).

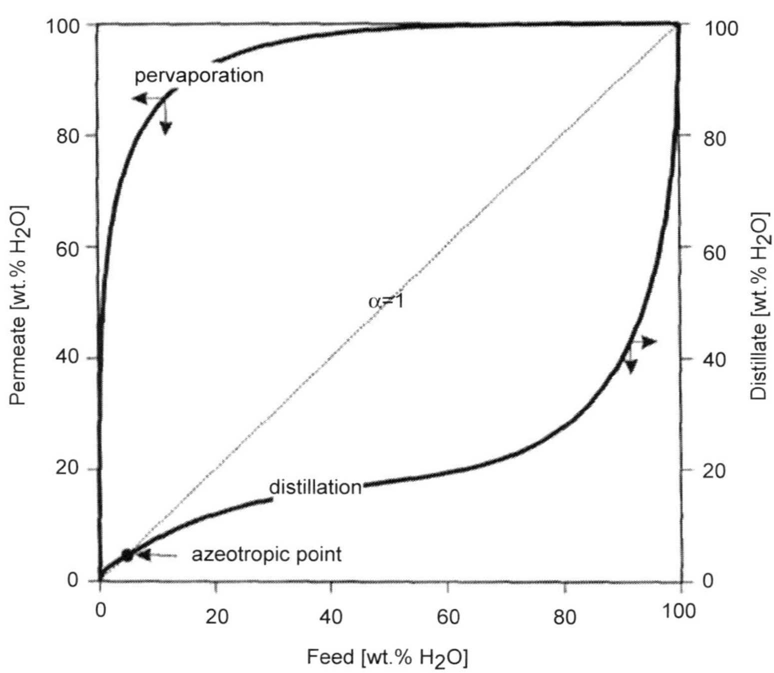

Figure 2 Comparison between selectivity in pervaporation (upper curve) and distillation (lower curve) (Kujawski, 2000).

Although the separation factor is commonly used to describe pervaporation (and will be further used in this chapter as well, because all data can be found in this logic), Baker et al. (2010) point out that this may lead to misinterpretations; a better definition would be the ratio of permeabilities or permeances, as also applied for gas separation. This is referred to as selectivity:

$$\alpha_{ij} = \frac{P_i^G}{P_j^G} = \frac{P_i^G/l}{P_j^G/l}$$

where P is the permeability of component I or j, and l the membrane thickness.

An alternative way to quantify selectivity is by the dimensionless enrichment factor β, which is defined as the ratio of the fraction (molar or mass) of component A in the permeate to the fraction of component A in the feed:

$$\beta_A = \frac{y_A}{x_A}$$

The enrichment factor is, however, concentration dependent.

Usually, there is a trade-off between the permeation flux and selectivity: when the selectivity increases, the permeability decreases and vice versa. A pragmatic combined parameter is the pervaporation separation index (PSI), originally defined to measure of the separation ability of a membrane as (Huang and Yeom, 1990):

$$\mathrm{PSI} = J_{tot}\alpha_{A,B}$$

The PSI is still frequently used in this way. However, a broken membrane with a very high flux and no separation ($\alpha_{A,B} = 1$) would have a very high PSI, which makes no sense. Therefore, the definition was adjusted as follows (Teli et al., 2007):

$$\mathrm{PSI} = J_{tot}\left(\alpha_{A,B} - 1\right)$$

Unfortunately, one can find both versions in publications, although the "new" definition should definitely be used uniformly.

3. PERVAPORATION MEMBRANES

Membrane preparation techniques and methods used to deposit thin layers upon a support have been discussed thoroughly by Mulder (1996). Membranes used in pervaporation are usually dense or microporous asymmetric polymeric membranes, which allow a high flux in combination with a high selectivity (and thus, a high PSI). They have an asymmetric or composite structure: a thin microporous top layer and a porous sublayer made from the same or a different material. The first layer enables an effective separation while maintaining a high flux. The second sublayer gives the membrane the desired mechanical strength. It can be a single plate, tube, hollow-fibre or honeycomb structure. The major requirement for the sublayer is that it needs to have an open structure to minimize resistance to vapour transport and to avoid capillary condensation. On the other hand, if the pores in the sublayer are too large, it is difficult to apply a thin selective layer directly upon the support. Hence, a multilayer membrane is often considered to meet specific applications. Often the support layer is similar to an ultrafiltration membrane (made of e.g., polyacrylonitrile, polyetherimide, polysulfone, polyethersulfone, or polyvinylidene fluoride, with pores in the range of c. 20–50 nm), cast on a porous support layer consisting of a woven or nonwoven fabric (micrometre pore range). The top layer, coated on this structure, is a thin dense separating layer with thickness ranging between 0.5 and 5 μm. Coating can be carried out in many ways; a common method is by evaporation of a solution of the polymer in an appropriate solvent, which is spread onto the support layer.

The selection of the material strictly depends on the application. Figure 3 gives an overview of different pervaporation application areas (Lipnizki et al., 1999a,b).

The target compound determines the type of membrane material for the desired separation. For dehydration applications, hydrophilic membranes are used, while for removal of organics from an aqueous mixture, hydrophobic membranes are applicable. In target organophilic pervaporation, where two or more organics must be separated, the choice of the membrane material is strongly dependent on the hydrophobicity of the target compound. A general overview of membrane materials, categorized according to the application, is given in Table 1 (Satyanarayana et al., 2004; Fleming, 1989; Yoshioka et al., 2000; Polyakov et al., 2004).

Commonly used polymer membranes for dewatering applications are polyvinyl alcohol (PVA) (Kujawski, 2000; Van Baelen et al., 2004), polyimide (PI) (Kim et al., 2000), chitosan (Kanti et al., 2004), and sodium alginate (Kanti et al., 2004). To optimize their performance and to reduce excessive swelling, the polymer membranes are grafted, blended, or crosslinked by special agents and very often, inorganic fillers (zeolite, metal oxide,

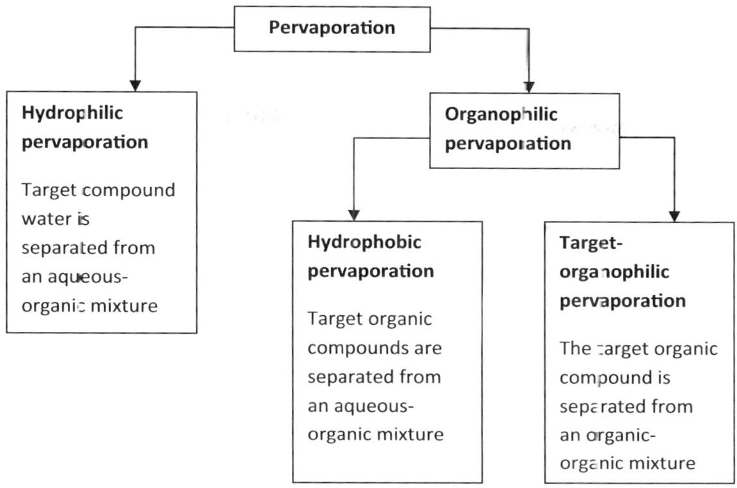

Figure 3 Schematic overview of application areas for pervaporation (Lipnizki et al., 1999a,b).

silica) are added (Jonquières et al., 2002). A wide range of other polymers have been reported as hydrophilic membrane materials, among which polyurethane/polyvinyl alcohol blends (Yao et al., 2012), the polysaccharide Konjac glucomannan (Li et al., 2012), phosphorylated chitosan (Sunitha et al., 2012), and many other.

The standard material for hydrophobic pervaporation membranes is polydimethylsiloxane (PDMS). PDMS can be either used to fabricate hollow fibre, tubular, unsupported sheet, or thin layer supported sheet membranes. Selectivity factors with pure PDMS membranes for ethanol/water mixtures range from 4 to 12. In the continuous search for better materials, poly-1-trymetylsilyl-1-propylene (PTMSP) has shown to give higher selectivity factors, under similar conditions, ranging from 9 to 26 (Volkov et al., 2004; Nagase et al., 1991; Claes et al., 2011).

Both PDMS and its modification polyoctylmethylsiloxane (POMS) were found to show very good results for VOC recovery and the removal of different organic compounds from aqueous solutions (Baker et al., 1997; Liang et al., 2004; Sampranpiboon et al., 2000; Boeddeker and Bengtson, 1989; Lipsi and Cote, 1990; Raghunath and Hwang, 1992). Several other elastomers such as ethylene-propylene-diene terpolymer (EPDM), nitrile butadiene rubber (NBR), polyether-block-polyamide (PEBA) were reported to show promising results (Panek and Konieczny,

Table 1 Membrane materials and their applications in pervaporation (Satyanarayana et al., 2004; Fleming, 1989; Yoshioka et al., 2000; Polyakov et al., 2004)

Pervaporation Type	Hydrophilic	Hydrophobic	Organophilic
Membrane materials	Polyvinyl alcohol (PVA) Polyvinylalcohol/ polyacrylonitrile (PVA/PAN) Cellulose acetate Polysulfone Carboxylmethyl-cellulose Polyetherimide (PEI) Chitosan	Polydimethylsiloxane (PDMS) Polyoctylsiloxane (POMS) Polyether-block-polymide (PEBA) Ethylene-propylene diene terpolymers (EPDM) Poly-1-trymetylsilyl-1-propylene (PTMSP) Polypropylene (PP) Polybutadiene (PB) Hydrophobic zeolites	Polydimethylsiloxane (PDMS) Polyether-block-polymide (PEBA) Polypropylene (PP) Polyethylene (PE) Polyvinyl alcohol (PVA) Polyetherimide (PEI)
Applications	Breaking of azeotropes of binary mixtures Dehydration of alcohols	Removal of organics from water Removal of alcohol from alcoholic beverages Recovery of aromatic compounds in food technology	Separation of (m)ethanol from (m)ethyl tert-butyl ether Separation of benzene and cyclohexane

2007, 2008; Meuleman et al., 2001). Since the 1990s research was carried out in order to find better or more cost-effective materials. For example, in 1999 Dutta and Sikdar discovered a good separation performance for the removal of trichloroethylene (TCE) and 1,1,1-trichloroethane (TCA) from water for a membrane made of styrene-butadiene-styrene block copolymer (SBS) (Ganapathi-Desai and Sikdar, 2000). An additional advantage of SBS is that the membrane is composed of only one material, which reduces manufacturing time and cost in comparison to composite PDMS membranes. Crosslinked and carbon black filled natural rubber membranes were also used for the removal of organics from water (Singha and Ray, 2012).

Organophilic or organoselective polymers may be made of a wide variety of materials, depending on the specific separation. For example, membranes that have been studied for the separation of benzene and cyclohexane include polypropylene (PP), polyvinylidene fluoride (PVDF) and polymethyl methacrylate (PMMA) (Smitha et al., 2004; Villaluenga and Mohammadi, 2000). Aromatic polyimide and polybenzoxazole membranes were used for the fractionation of toluene/n-heptane and benzene/n-heptane mixtures (Ribeiro et al., 2012), and were found to be selective towards the aromatic hydrocarbon. Permeabilities could be manipulated through the chemical structure of the diamine used in the polycondensation method applied to prepare the films; this allowed a change by approximately four orders of magnitude in permeability of an individual compound. Other materials for often very challenging separations by pervaporation have been reported as well, such as the use of divinyl benzene crosslinked hydroxy terminated polybutadiene-based polyurethaneurea membranes for the separation of *o*-xylene from *p*-xylene (Gu et al., 2012). A separation factor of two was measured for this application.

A new approach for organophilic membranes synthesis is the use of molecular surface engineering (MSE), applied by the German manufacturer PolyAn. The strategy of MSEis to fill the pores of a porous membrane with a functional polymer, which eventually yields a dense structure with specific selective transport properties. MSE membranes are tailor-made for well-defined organic–organic separations.

Inorganic membranes offer superior chemical and thermal stability compared to polymers. Water has been removed from organics with microporous silica (Van Veen et al., 2001), titania (Sekulic et al., 2002) and zirconia membranes (Sekulic et al., 2002). Furthermore, zeolites were studied for this purpose. These include ZSM-5 (Bowen et al., 2004), mordenite (Bowen et al., 2004), A-, X-, Y-, and T-type zeolite membranes (Morigami et al., 2001; Okamoto et al., 2001; Sommer and Melin, 2005; Ying et al., 2004). Despite their superior stability compared to polymers, the stability of microporous silica membranes in acids and bases is limited to pH values from 3 to 11. Under strong acidic conditions the silica layer is dissolved, whereas in a basic environment the sublayer is unstable. Moreover, in dehydration experiments, water fluxes tend to decline gradually in time and the membrane loses selectivity. In contrast, titania, zirconia and methylated silica membranes are more stable (Campaniello et al., 2004). The stability of zeolites depends on the silicium to aluminium framework ratio.

For removal of organics from aqueous solutions, inorganic membranes based on hydrophobic zeolites were found to have higher separation factors and fluxes on laboratory scale than polymeric membranes (Bowen et al., 2004). Silicalite-1 was the most studied zeolite for this application. For silicalite-based membranes, ethanol/water selectivities under same working conditions as above, range from 12 up to values over 100 (Lin et al., 2001, 2003; Matsuda et al., 2002). In addition, all fluxes met or exceeded those obtained for the thinnest PDMS films. Silicalite-1 (aluminium free) is quite stable, whereas A-type zeolites with a silicium to aluminium ratio of one are very sensitive against the influence of mineral and organic acids at a pH even just slightly below neutral.

Unfortunately, the high cost of fabrication inorganic membranes is limiting its applications. For these reasons, the interest went into the direction of manufacturing PDMS membranes with incorporated zeolite fillers. In addition to the nature of the fillers, other important parameters affecting the membrane performance are the loading of the particles, the particle size, the zeolite type and the membrane casting conditions. These mixed matrix membranes showed also better performance characteristics (ethanol/water selectivities up to 60) than unfilled PDMS membranes under the same working conditions, with only a little cost increase (Jia et al., 1992; Chen et al., 1998; Vankelecom et al., 1995; Li and Wang, 1996). Mixed matrix membranes based on PDMS with other fillers such as silica have also been considered (Shirazi et al., 2012). It was found that incorporating silica nanoparticles significantly improves the selectivity of the membrane, because the polymer chains are rigidified and the polymer free volume decreases. However, this is at the expense of the permeability, which decreases.

Hydrophilic mixed matrix membranes were also studied. For example, zeolite 4A and ZSM-5 were incorporated in Matrimid 5218 polyimide (Mosleh et al., 2012); sodium carboxymethyl cellulose/poly (vinyl alcohol) mixed matrix membranes were filled with zeolite 4A (Prasad et al., 2011); and carbon nanotubes were studied as fillers in polyvinyl alcohol nanocomposite membranes (Shirazi et al., 2011). The high number of publications on mixed matrix membranes for pervaporation (among other uses) shows that there is much scientific interest in such membrane types; nevertheless, commercial applications remain to be seen.

4. HYDROPHILIC PERVAPORATION: APPLICATIONS IN DEHYDRATION

The most classical application of pervaporation as a stand-alone process is the removal of small fractions of water from organic solvents by using a hydrophilic membrane. The success of this application is related to two factors:

1. The ease of separation: because differential transport in pervaporation membranes is related to differences in molecular size and polarity, the separation between water and an organic solvent is beneficial, and high separation factors (and high PSI values) can be obtained rather easily.
2. Process economics: in sharp contrast with wastewater, organic solvents represent an economic value, depending on the specific compound but always positive and sometimes high. This justifies investments in technologies for purification.

The most interesting range of applications is in the removal of moderate water fractions from an organic solvent. Very low water fractions in the permeate are technically possible, but at high cost since the driving force for permeation for the removal of small concentrations of water is low. In practice, it can be assumed that a purity of 99.8% is economically and technically feasible. The typical reference solvent for dehydration is isopropanol, for which separation factors are high (sometimes ethanol is taken as the reference, but with lower separation factors). For easier separations, the purity of the permeate can be higher.

When large water fractions are to be removed, the process cost increases because larger fluxes are needed, which translates into larger investment costs (membrane surface area related) and higher operational costs (energy related). Therefore, water fractions below 10% are to be preferred.

Ethanol dehydration is assumably the most important application for solvent dehydration. However, because the general process comprises more than only the removal of a relatively low water fraction, (bio)ethanol purification will be considered separately further in this chapter.

Dehydration of isopropanol is also among the most studied applications of hydrophilic pervaporation. Isopropanol forms an azeotrope with water at a concentration of 87.9% and dehydration is therefore a very interesting application, since it is difficult to achieve with a classical distillation (Figure 4) (Bruschke, 1995). Shao and Huang (2007) state that for a mixture of water and isopropanol containing 10 wt% water, the maximum single plate separation factor (isopropanol to water) in distillation is about 2, whereas a pervaporation membrane can offer a one-through separation factor for water over isopropanol of 2000–10,000. The performance of distillation is of course related to the proximity of the azeotrope, but the separation factors are indicative.

Typical applications make use of PVA membranes. However, crosslinking of the membrane is an important parameter to consider. Burshe et al. (Burshe et al., 1997) studied the effect of crosslinking with citric acid, adipic acid, maleic acid, glutaraldehyde, and glyoxal, and concluded that the type of crosslinker has an important impact on the permeation characteristics of

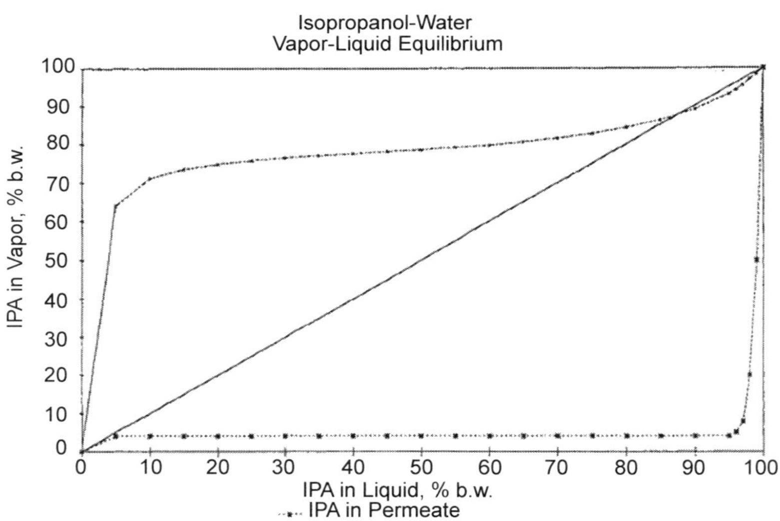

Figure 4 Comparison of distillation (upper curve) and pervaporation (lower curve) for separation of isopropanol and water. The horizontal axis displays the isopropanol concentration in the feed for pervaporation and in the liquid phase for distillation. The vertical axis displays the isopropanol concentration in the permeate for pervaporation and in the vapour phase for distillation (Bruschke, 1995).

the membranes. A better crosslinking, however, not only increases the separation factor but also decreases the permeability. For carboxymethylated PVA membranes, a separation factor of 362 was observed for a flux of 831 g/m^2 h (feed temperature 80 °C, 85% isopropanol) (Nam et al., 1999). More recent efforts yielded separation factors that surpass this even by an order of magnitude, but with lower fluxes: for sulfonated PVA membranes, a separation factor of 3452 was reported with a flux of 35.1 g/m^2 h (40 °C, 10 wt%) (Rachipudi et al., 2011). A similar effect is obtained by using fillers; using 2 wt% of carbon nanotubes in PVA, a selectivity of c. 1800 was obtained for a flux of 80 g/m^2 h (10 wt% water in IPA at 30 °C) (Shirazi et al., 2011). An even more impressive separation factor of 30,000 was obtained for mixed matrix membranes of PVA and poly(vinyl pyrrilidone) blends loaded with 4% phosphomolybdic acid, again with a modest flux of 36 g/m^2 h (10 wt% water in IPA at 30 °C) (Mali et al., 2011). Higher fluxes with reasonably good separation factors can be found as well; for example, by blending PVA with a soluble polyelectrolyte complex made from poly(diallyldimethylammonium chloride) and sodium carboxymethyl cellulose, a separation factor of c. 1000 was obtained with a flux of 1350 g/m^2 h (10 wt% water in IPA at 70 °C) (Zhao et al., 2009). Note, however, that the temperature in this case is much higher, so that the actual improvement is lower than it seems. High fluxes are also obtained when PVA is blended with chitosan, typically at high fractions of chitosan (~75%) (Svang-Ariyaskul et al., 2006). Many more publications of different PVA-based membranes for isopropanol-water separations can be found in the literature, usually with fluxes and separation factors in the same range as described above.

Materials other than PVA have also been considered for dehydration of isopropanol, and a wide selection of polymers and hybrid structures is available. A logical candidate, often studied, is the polymer chitosan (Nawawi and Huang, 1997; Choudhari et al., 2007; Zielinska et al., 2011; Lee et al., 1998; Liu et al., 2005; Devi et al., 2005; Moon et al., 1999; Liu et al., 2007, 2011; Choudhari and Kariduraganavar, 2009). Separation factors for chitosan are typically lower than for (modified) PVA, but fluxes tend to be higher. Lee et al. (1998) reported a separation factor of 150 for a flux of 1800 g/m^2 h. Blending chitosan with gamma-(glycidyloxypropyl) trimethoxysilane (5%) improves the performance to a separation factor of nearly 700 and a flux of 1730 g/m^2 h (although this was measured for a water fraction of 30%) (Liu et al., 2005). In another procedure (Liu et al., 2005), chitosan membranes were crosslinked with toluene-2,4-diisocyanate, yielding a separation factor of 472 for a flux of 390 g/m^2 h (temperature 30 °C, 8.4 wt% water). Interestingly, Devi et al. (2005) measured the concentration dependency of flux and separation factor. Higher water fractions had a substantially lower separation factor, decreasing sharply between 30 and 40 wt% (from 190 to 12), whereas fluxes steadily increase. This compromises a fair comparison between different procedures for chitosan modification. Nevertheless, a trend can be seen, which is consistent with the analysis of Devi et al. (2005), who compared various modifications reported in the literature and conclude that fluxes are in the range between three and 200 g/m^2 h, with selectivities below 1000. One exception is the work of Moon et al. (1999), who obtained a separation factor of c. 2500 and a flux of 554 g/m^2 h for a composite membrane of NaAlg and chitosan. Nevertheless, this mainly shows the blending effect with polymers other than chitosan. Similar results can be found for many other blends; for example, for composite chitosan/poly(tetrafluoroethylene) (PTFE) membranes, a permeation flux of 1730 g/m^2 h and a separation factor of 775 were obtained at 70 °C for a 70 wt% isopropanol aqueous solution (Liu et al., 2007). The addition titanate nanotubes, as studied by Liu et al. (2011), may further increase the performance of chitosan membranes: chitosan membranes with 6 wt% poly(aspartic acid) modified titanate nanotubes were found to have a permeation flux and separation factor of 1498 g/m^2 h and 6237, respectively (90% isopropanol, 80 °C). More good results for the use of nanoparticles in chitosan membranes can be found in the literature; the most remarkable example is a chitosan membrane with 10% sodium montmorillonite as filler material, which had a flux of 142 g/m^2 h and a separation factor of 15,000 at 30 °C and 10 wt% water (Choudhari and Kariduraganavar, 2009).

Among other reported materials for isopropanol dehydration, modified sodium alginate (NaAlg) membranes are prominent; for example, filled with zeolites (Sudhakar et al., 2011). Sodium alginate membranes and their use in pervaporation have been described extensively by Bhat and Aminabhavi (2007), who underline the versatility to modify and tune their structures to achieve a desired separation. They review literature data on grafted NaAlg, blends of NaAlg, hybrid composites of NaAlg, mixed matrix membranes of NaAlg and modified NaAlg including blends, grafts, filled matrices, and composites. For more details on this membrane type, the reader is referred to Bhat and Aminabhavi (2007).

Polyimide membranes are further candidates for isopropanol dehydration. Zeolite 4A and ZSM-5 filled polyimide (Matrimid 5218) membranes were found to have separation factors of 8991, 3904, and 1276, respectively, for zeolite 4A fillers, ZSM-5, and for neat Matrimid 5218 membranes (Mosleh et al., 2012).

Among other membrane materials for dehydration of isopropanol, reports can be found on sulfonated poly(ether ether ketone) (SPEEK) (Huang et al., 2001), poly(tetrafluoroethylene)/polyamide (Liu et al., 2008), Torlon (R) 4000 TF polyamide-imide (PAI) and Ultem (R) 1010 polyetherimide (PEI) (Wang et al., 2008), and ceramic materials including NaA type zeolite membranes, which have a higher temperature and mechanical stability than polymeric membranes (Van Hoof et al., 2006), and amorphous silica (Verkerk et al., 2001). Ceramic membranes are less studied than polymeric membranes for isopropanol-water separations, presumably because their main advantages, a high chemical and thermal resistance combined with excellent separation factors, are not essential here: polymeric membranes have sufficient resistance and have been shown to have good separation factors.

Applications for dehydration of solvents other than isopropanol have been reviewed by Chapman et al. (Chapman et al., 2008a,b) (the review includes isopropanol as well, but is of interest here mainly because of its broad coverage of various solvents). The authors include both ceramic membranes (inorganic, zeolitic, and hybrid membranes) and polymeric membranes; interestingly, they are classified according to the membrane material and not according to the application, which yields a good overview of the potential of various materials for separation of (mainly) ethanol and isopropanol; the potential of membranes for the dehydration of other commonly used organics such as acetic acid, tetrahydrofuran (THF) and acetone is then further described. Most other organic compounds can be dehydrated in a similar way than isopropanol (and ethanol), although separation factors can differ significantly. In general, they tend to be higher compared to isopropanol, because isopropanol is relatively polar and small and therefore, not the easiest compound from which to separate water. Dehydration of THF, for example, which forms an azeotrope with water at 95 wt% THF, was studied using a blend membrane of PVA/PVP crosslinked with maleic acid (Sunitha et al., 2009); an infinite selectivity for dehydration of 97% THF at 30 °C was reported. Fluxes tend to be in the same order of magnitude as for isopropanol, since the membranes are water selective and the fluxes mainly depend on the water fraction. In this case, a low flux of 7 g/m^2 h was reported, which is presumably related to the low water fraction. A measurable separation factor (2140) with substantial flux (170 g/m^2 h at 30 °C and 5 wt% water) was obtained for a chitosan membrane with incorporated NaY zeolite (Varghese et al., 2009). Again, it shows that chitosan-based membranes tend to have higher fluxes and lower separation factors. Three different copolymers of acrylamide with increasing amount of 2-hydroxyethyl methacrylate were studied by Ray and Ray (2008), and showed quite high fluxes and a moderate separation factor. Further down the selectivity scale is polyaniline (Chapman et al., 2008a,b), which proved to have a reasonable flux (600–800 g/m^2 h at 55 °C and a water concentration of 1–5 wt%) but the average separation factor was only 43.8. Ceramic membranes, however, which may be of interest in this case in view of their better chemical stability, have been shown to have a remarkably high separation factor going up to 20,000 at a flux of 980 g/m^2 h (feed containing 7 wt% water at 55 °C) (Urtiaga et al., 2003). In spite of their higher cost, ceramic membranes may be a robust solution for this separation, with a very good performance.

Similarly, dehydration of *N,N*-dimethylformamide (DMF) and 1,4-dioxane may be considered. Again, ceramic membranes are of interest (Ten Elshof et al., 2003). Surprisingly, however, not many studies consider ceramic membranes for these separations. Polymeric membranes were also studied for DMF dehydration; for example, crosslinked composite membranes of NaAlg filled with clinoptilolite had a separation factor of 23 (Cahya et al., 2011). The stability of these membranes for longer durations might not be secured. Dehydration of 1,4-dioxane (which forms an azeotrope with water at 82.1 wt% dioxane) has a somewhat better perspective. PVA and zeolite T (15 wt%) mixed matrix composite membranes were studied for mixtures containing 10 wt% water; a separation factor of 1689 was obtained with a flux of 213 g/m^2 h, a clearly better performance than plain PVA membranes (Veerapur et al., 2008).

Caprolactam dehydration was considered in a similar way, using for example, crosslinked PVA membranes (Zhang et al., 2007). Furthermore, several other solvents can be studied in a similar way; of particular interest are those solvents that have an azeotrope with water at relatively low water fractions. These include tert-butanol, methyl ethyl ketone, acetonitrile, ethylene chloride, and n-propyl acetate, among others. In addition, close boiling mixtures are of interest; well-known examples are acetic acid-water and acetone-water, two very challenging separations. Acetone dehydration was studied by using polybenzimidazole membranes (Shi et al., 2012), but remains difficult to achieve. Acetic acid dehydration, in a similar way, is still a standing problem, although attempts have been made by developing polyphenylsulphone membranes, among others (Jullok et al., 2011).

5. HYDROPHOBIC PERVAPORATION

Hydrophobic pervaporation, i.e., the selective removal of organic compounds from an aqueous solution by pervaporation, is less applied than dehydration even though it is technically on a similar level. This is related to the applications: removal of organics such as the VOC fraction from wastewater, for example, is feasible but not economically viable. Some studies on this subject were made, however, such as the use of hydrophobic pervaporation for industrial wastewater streams of water-chloroform, and of water-methyl-isobutyl ketone (Lipnizki and Field, 2002). When the extracted solvent can be recovered, as studied here, the process proves to be also economically interesting. For the wastewater streams considered, it was possible to recover over 99% of the organic components at concentrations of more than 98 wt%. For organics with limited solubility in water, a hybrid process combining pervaporation with a decanter system may be of interest (Field and Lobo, 2003). In this system, the pervaporation unit produces permeate that after condensation gives two liquid phases. The organic phase is relative pure and concentrated (and available for reuse), whereas the aqueous phase can be recycled into the feed stream (Field and Lobo, 2003). Hybrid systems specifically for the system aniline-water were suggested by Mecki and Lichtenthaler (Mecki and Lichtenthaler, 1996). Other studies are limited to toluene-water systems and phenol removal from wastewater (Satyanarayana et al., 2004; Panek and Konieczny, 2009; Moraes et al., 2007).

Applications in other fields include recovery of aroma compounds in the food industry. This was considered in an early publication by Beaumelle et al. (1992), who considered permeation of propanol, ethyl acetate and ethyl butyrate as model aroma compounds through PDMS membranes. Later studies focused either on development of applications, or on membrane improvements. Standard membranes are made of PDMS as the separating layer, but by using fillers, their performance can be improved. This has been shown by Vankelecom et al. (1997), who used carbon black, in situ methylated silica and silylated silica as fillers to maximize fluxes and selectivities in the pervaporation of aqueous ethanol, tertiary butyl alcohol, or aroma solutions. However, apart from some rare studies describing e.g., the use of hydroxy terminated polybutadiene based polyurethaneurea membranes for aroma recovery (Bai et al., 2007), not many efforts have been done (or: were needed) to improve standard PDMS membranes. These have been successfully used in several similar applications of aroma recovery with different products, such as strawberry aroma's (Isci et al., 2006), dairy aroma components (Baudot and Marin, 1996), etheric oil aroma components (Mauz et al., 1996), apple juice aroma's (Borjesson et al., 1996), pomegranate aroma compounds (Raisi and Aroujalian, 2011), and bergamot essential oil (Figoli et al., 2006).

An evident and prominent application of hydrophobic pervaporation is the use in product recovery in bioconversions, in ethanol production, in the acetone-butanol-ethanol (ABE) process, and butanol recovery. The specific case of ethanol purification will be discussed in a separate chapter because it relates to hydrophobic as well as hydrophilic pervaporation. However, in view of the benefits of butanol compared to ethanol, studies increasingly focus on butanol instead of ethanol. For example, Ikegami et al. (2011) separated butanol obtained in ABE conversions by using silicone rubber-coated silicalite membranes. Evidently, this requires a new approach to membrane materials and selectivities, which may stimulate further research on the development of novel hydrophobic pervaporation membranes that go beyond the classical PDMS membranes. Silica filled poly[1-(trimethylsilyl)-1-propyne] (PTMSP) membranes are also potential candidates for such separations (Claes et al., 2012). In another study, a tri-layer composite membrane consisting of an active layer made of PDMS and dual support layers of high porosity polyethylene and high mechanical stiffness perforated metal was studied (Li et al., 2010). It can be expected that more research on novel membrane materials and membrane structures will emerge; new approaches such as the use of liquid membranes have been already proposed as well (Thongsukmak and Sirkar, 2007).

Other applications of hydrophobic pervaporation are rather scattered and not well developed, in spite of their potential. Apart from applications in biotechnology, which seem to come closer to a breakthrough, azeotropic separations can be considered as well. However, azeotropic separations that concern the removal of a smaller fraction of an organic compound from water are of less importance, and occur less than the reverse. Some azeotropes with water at low concentrations are with benzyl alcohol, cyclohexanol and furfuryl alcohol; in such cases, hydrophobic pervaporation would be straightforward. Lipnizki et al. argue that the slow development of hydrophobic pervaporation is in part related to fact that the majority of studies have been with binary feed systems, while real feed mixtures are often multicomponent and might contain impermeable components, such as salts and sugars. Thus, the real performance of hydrophobic pervaporation is not yet entirely understood in process conditions. This may be a part of the explanation; what is clear is that it is not related to insufficient performance of the membranes (at least not in general).

6. ORGANOPHILIC PERVAPORATION

A newly emerging application area is organophilic pervaporation for the separation of aromatic/aliphatic, aromatic/alicyclic and isomeric mixtures. Applications until now are still scarce, which is mainly due to the limited availability of organophilic pervaporation membranes. The separation layer has to be tuned for almost each distinct application (Claes, 2012). Of specific interest are applications in separating olefins from paraffins such as propene/propane, aliphatic hydrocarbons containing benzene or toluene, and xylene isomers. The easiest separations, which are already operational, use hydrophilic membranes for separating ethanol or methanol from mixtures containing hydrocarbons, ethers and esters. An overview of applications was given by Smitha et al. (2004) and is summarized in Figure 5. For a detailed overview of membranes and applications in the four areas shown in Figure 5, the reader is referred to Smitha et al. (2004). They conclude that pervaporation can compete with well-established separation technologies for organic–organic separations, due to distinctive advantages like economy, safety and its ecofriendly nature. They refer to the increasing number of patents and publications in scientific journals on this subject. Nevertheless, they also mention the need for process economics to be better evaluated, and suggest the use of hybrid process combining pervaporation with conventional separation methods like distillation for organic–organic separations.

The use of molecular surface engineering, applied by PolyAn (Germany), is a potential breakthrough (Matuschewski and Schedler, 2008), which allows more targeted separations. One of these is desulfurization of jet fuel (Pasel et al., 2012), which was studied for a duration of over 500 h. Apart from PolyAn membranes, Teflon AF2400 is another interesting membrane material that has been applied for organic–organic separations in binary mixtures containing acetone, chloroform and methanol (Polyakov et al., 2006) and mixtures of chloromethanes (Polyakov et al., 2004). Furthermore,

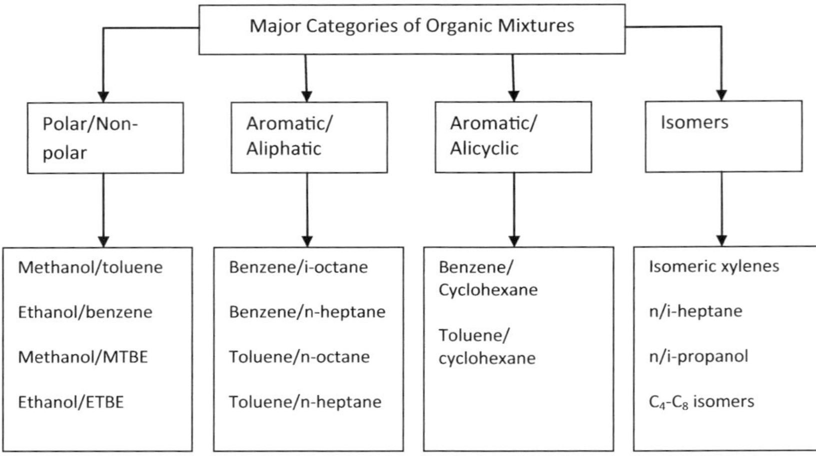

Figure 5 Main application areas of organophilic pervaporation. *Reprinted with permission from Smitha et al. (2004).*

fluorinated siloxane-imide block copolymer membranes have been reported (Jiang et al., 2011).

An application intermediate between hydrophobic and organophilic pervaporation is the recovery of aroma compounds from fermentation media, which has been mentioned earlier for hydrophobic applications. Nevertheless, the use of organophilic membranes may allow fractionation of the aromas according to their chemical nature (Brazinha and Crespo, 2009).

Organophilic pervaporation can of course also be used in the separation of organic–organic azeotropes, for example, the dimethyl carbonate/methanol system (Wang et al., 2011), and isopropanol-benzene and isopropanol-toluene (Mandal and Pangarkar, 2003).

7. HYBRID SYSTEMS

Hybridization of processes in which pervaporation is applied in conjunction with another separation method is a very promising approach that combines the advantages of classical processes (e.g., distillation) and membrane separations. Pervaporation has a clear advantage for concentrations around the azeotropic composition. The application of pervaporation avoids the use of entrainers (in extractive or azeotropic distillation) or pressure variations (in pressure-swing distillation), and reduces the number of trays. On the other hand, the separation factors for mixtures at concentrations not near the azeotrope might be insufficient to compete cost-wise with classical separation processes. As already mentioned, pervaporation is the most cost-effective when operated in a moderate concentration range. Permeation of larger quantities requires larger installations (with larger surface areas for larger fluxes), while for a higher purity of permeates, driving forces become too small so that again large installations are needed, which are not economical. A hybrid configuration uses the best of both worlds and is often the most realistic scenario.

Lipnizki et al. (1999a,b) reviewed pervaporation-based hybrid processes with the belief that the optimal process design is unlikely to consist solely of pervaporation. Several examples are given, with much attention for ethanol-water separation (see further in this chapter) and isopropanol-water. A suggested configuration for the latter separation is shown in Figure 6 (Bruschke, 1995). The system shown here is typical: distillation is used for the concentration range where water fractions are large, whereas pervaporation is used to overcome the azeotrope. Remark that the recycle stream of 10 wt% isopropanol is an essential part of such hybrid configuration. More complex schemes can be developed, based on a multiple pervaporation or using a second (small) distillation column for polishing of the permeate.

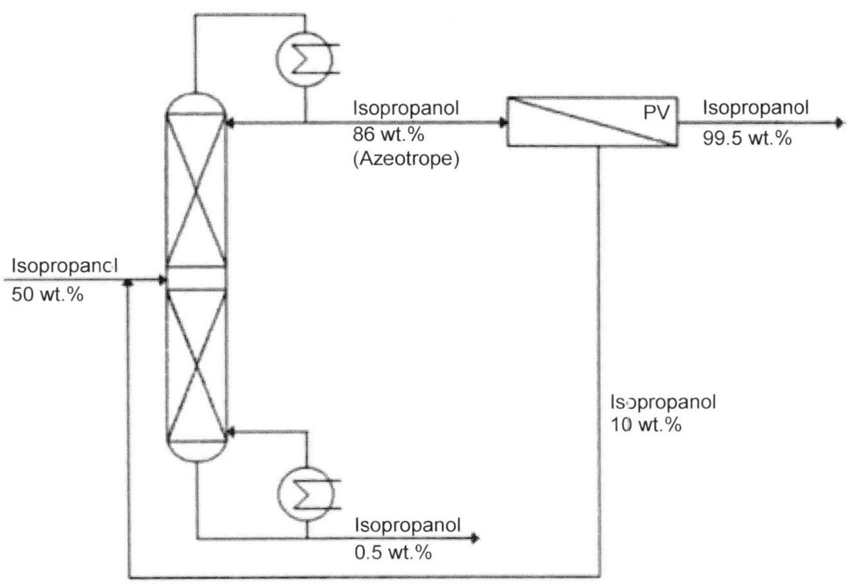

Figure 6 Flow sheet for hybrid separation of isopropanol and water using distillation and pervaporation. *Adapted from Bruschke (1995).*

Other examples given by Lipnizki et al. (1999a,b) include dimethyl acetal production (hydrophilic pervaporation, azeotrope with water), dehydration of acetic acid (with suggestion of a hybrid with reverse osmosis; no azeotrope but low relative volatility), benzene-cyclohexane (using extractive distillation in a hybrid with organophilic pervaporation), carboxyl acid ester and dimethylcarbonate production (azeotrope with methanol, organophilic pervaporation in hybrid with distillation), and ethyl tert-butyl ether production (azeotrope with C_4 fraction; organophilic pervaporation in hybrid with distillation). As can be seen, many hybrid solutions require the use of organophilic pervaporation; advances in membranes for such separations (see above) would indeed open up many applications.

Van Hoof et al. (2004) made an economic comparison between (azeotropic) distillation and different combined systems combining distillation with pervaporation for the dehydration of isopropanol. The systems they proposed are shown in Figure 7 and Figure 8 (Van Hoof et al., 2004).

In Figure 7, the feed mixture is distilled to obtain a top product with composition close to the azeotrope at 83.0 wt% isopropanol. The top product is further dehydrated by a pervaporation unit, using polymeric membranes. Remark that the recycle stream necessary for a hybrid configuration is not considered here (the permeate of the pervaporation unit is obtained as a second aqueous product along with the bottom product of the distillation column).

Figure 8 proposes a more complex configuration in which a second distillation column is integrated in the hybrid configuration, so that the pervaporation is used only close to the azeotropic point. In this configuration, hybridization is still incomplete (the permeate obtained from the pervaporation should be recycled as well).

Interestingly, the authors make calculations for polymeric as well as for ceramic membranes. The performance of the ceramic membranes was better, which was mainly due to a loss of selectivity at low water concentrations. This effect is assumably due to competition in transport and is not usually considered, although the effect of coupled diffusion is well known (Van Baelen et al., 2005). Ceramic membranes suffer less from this competition effect at low water fractions and have therefore a strategic advantage. A second advantage of ceramic membranes, the better temperature resistance, is here less important; it is also mentioned that acid resistance might be reduced, which could be an impediment for implementation.

For the configurations in Figures 7 and 8, it was found that the hybrid system distillation-pervaporation was the most interesting process from

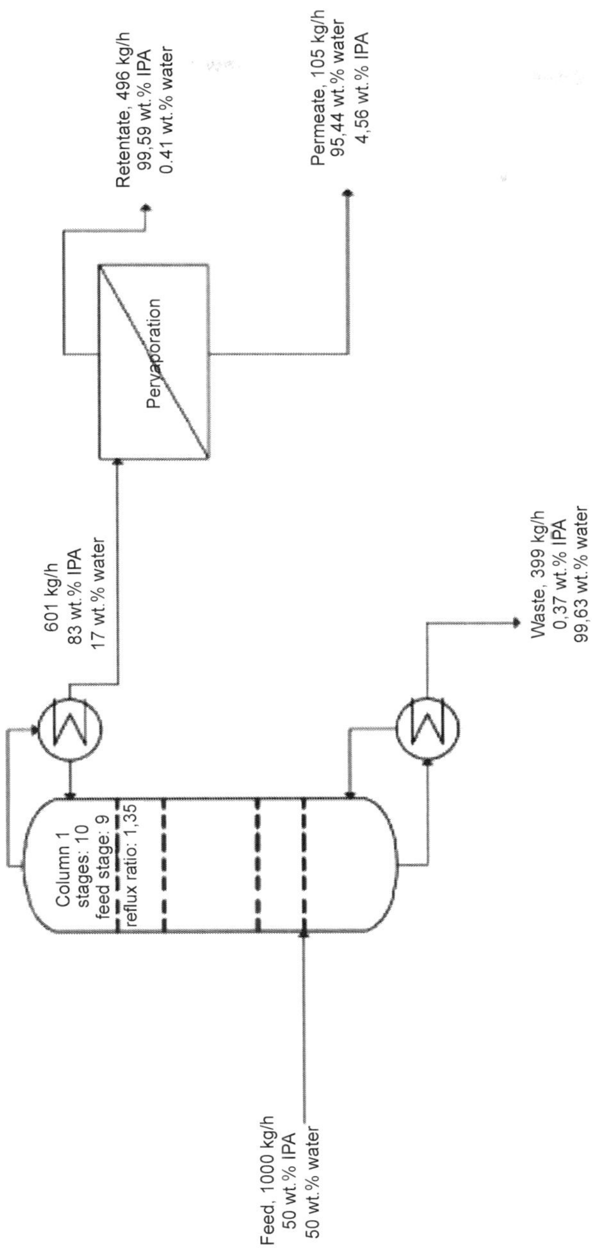

Figure 7 Combination of distillation and pervaporation for separation of isopropanol and water (Van Hoof et al., 2004).

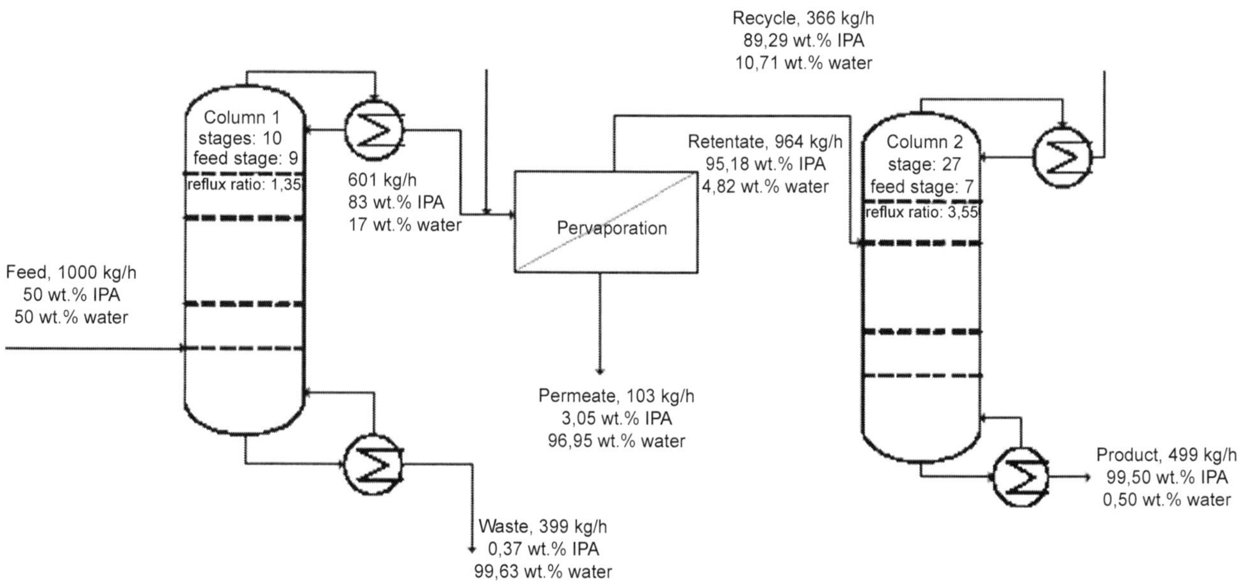

Figure 8 Hybrid process for isopropanol-water separation proposed by Van Hoof et al. (2004). The pervaporation permeate is not recycled in this configuration.

economic point of view. When ceramic membranes are used, the integration could lead to a saving in total costs of 49% compared to azeotropic distillation.

8. ETHANOL PURIFICATION AND PRODUCTION OF BIO-ETHANOL

An interesting case for application of pervaporation is the purification of ethanol. Production of ethanol based on bioconversions typically yields aqueous products with ethanol fraction of 5 wt%, or even below. Lignocellulosic sources are considered the most promising, when all aspects are taken into account, in spite of the low(er) concentration of ethanol produced in the broth. This increases the challenge for further processing and sharpens the demands for cost-effective purification processes. Distillation, vapour permeation, and pervaporation are potential candidates to purify bio-ethanol (Vane, 2005). A classical process scheme is shown in Figure 9 (Haelssig et al., 2012). A centrifuge is used for solids separation. Then, the aqueous ethanol is sent to a steam stripping column, denoted as the beer column, which can operate with a reboiler or direct steam injection. The

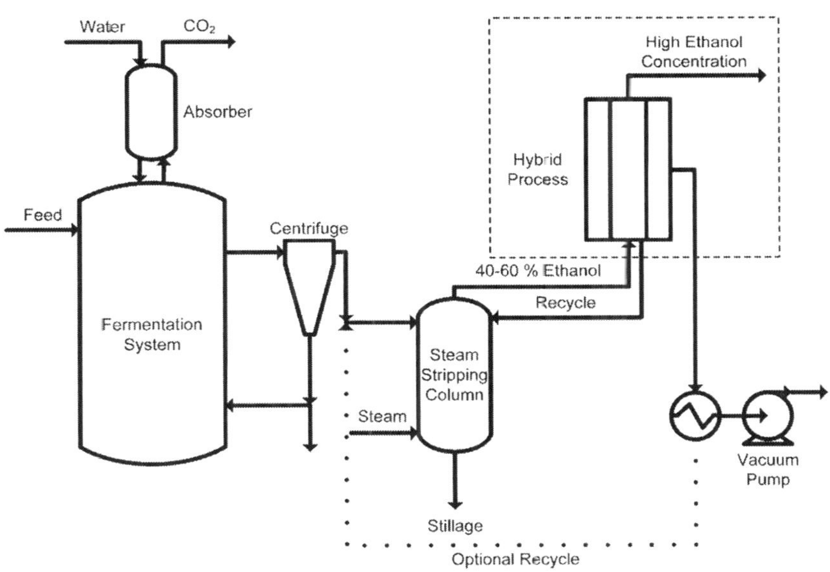

Figure 9 Classical scheme for production of ethanol from biomass. *Reprinted from Haelssig et al. (2012) with permission of Elsevier.*

vapour phase distillate stream coming from this column usually contains between 40 and 60 wt% ethanol (depending on the design of the beer column as well as the feed composition). The distillate is then sent to a rectifying column, which increases the ethanol concentration to near the ethanol-water azeotrope. The concentration of the distillate stream leaving the rectifying column varies depending on the design of the column (i.e., reflux ratio and number of stages), but cannot exceed the composition of the ethanol–water azeotrope. Thus, the distillate concentration is between 90 and 94 wt% ethanol. This distillate stream is dehydrated in pressure swing adsorption, pervaporation or vapour permeation, which produces anhydrous ethanol. The bottom product leaving the rectifying column is water, but has a substantial concentration of ethanol, and is either recycled to the beer column or sent to a separate side stripping column to maintain a high ethanol recovery.

Distillation is a nonpreferred option, although multicolumn distillation followed by molecular sieve adsorption is currently still the standard method for producing fuel-grade ethanol from dilute fermentation broths in modern corn-to-ethanol facilities (Vane, 2008). Combining distillation with for example, vapour permeation allows to significantly reduce the energy cost (Huang et al., 2010). Vapour permeation and pervaporation have potential, but have to provide sufficiently high permeabilities and separation factors to make the purification feasible. Stand-alone single-step or multistep pervaporation can be considered, or a hybrid approach combining membrane separation with distillation (or other processes such as pressure swing adsorption, which is applicable for dehydration of 87 wt% ethanol (Jeong et al., 2012), or extractive and azeotropic distillation in dividing-wall columns (Kiss and Suszwalak, 2012)). An interesting hybridization is the system shown in Figure 10, in which distillation is carried out using falling film evaporation, while at the same time permeating water through a hydrophilic membrane (Haelssig et al., 2012). This advanced hybrid goes beyond the classical integration of two separate units with interconnection, since the two processes are now carried out in one device.

In theory, it should be possible to obtain pure ethanol in a single-step pervaporation, using a hydrophobic membrane with preferential ethanol transport. The separation required for this is calculated as:

$$\alpha = \frac{c_{\text{eth},p}/c_{\text{eth},f}}{c_{w,p}/c_{w,f}}$$

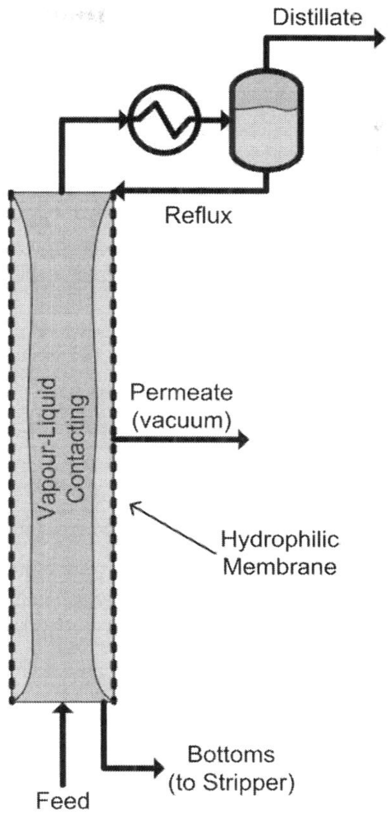

Figure 10 Hybrid distillation—pervaporation applied in a single device. *Reprinted from Haelssig et al. (2012) with permission of Elsevier.*

The ethanol concentration in the feed can vary considerably, depending on process conditions and the strain used. For a rough calculation, a concentration of 5 wt% is assumed here, although concentrations can be substantially lower or higher (Pejin et al., 2012). The water concentration can be approximated as $1 - c_{eth,f}$, ignoring all other components in the broth. It should be taken into account that this would yield only a rough approximation; nevertheless, it can be calculated that a change of some percents would not have a large influence.

The relation between $c_{eth,p}$ and α can be calculated with these assumptions as:

$$c_{eth,p} = \frac{500\ \alpha}{95 + 5\ \alpha}$$

An average, nonoptimized pervaporation membrane has a typical selectivity of five; this yields an ethanol concentration of only 21%. Optimized membranes, such as a thin PDMS layer on a polyetherimide support, would yield a separation factor of 7–9 (Lee et al., 2012). Ceramic membranes hydrophobic mesoporous silica have similar separation factors, reported to be close to 8 (Jin et al., 2011). The best membranes currently available have a separation factor of c. 20 (e.g., reported value of 18.3 for a PTMSP membrane with silica filler (Claes et al., 2012)). The ethanol concentration in the product is then considerably higher (51%), although this is still insufficient. It can be shown that a selectivity of at least 500 is needed to overcome the azeotropic point, which is not realistic. Figure 11 shows the permeate concentration as a function of the selectivity of a hypothetical membrane. For an overview of ethanol selectivities of pervaporation membrane materials, the reader is further referred to Peng et al. (2011).

More realistically, a two-step pervaporation for ethanol purification can be considered. In general, the first distillation column for purification of bioethanol has a distillate concentration of c. 37%. A single-step hydrophobic pervaporation unit to replace this distillation column should yield the same concentration. Assuming that a selectivity of 20 is realistic, an initial ethanol concentration of 3% is sufficient to obtain 37% purity. This, however, does not improve the upstream section of the purification. Using the same membrane with a selectivity of 20, the permeate concentration is only 92%.

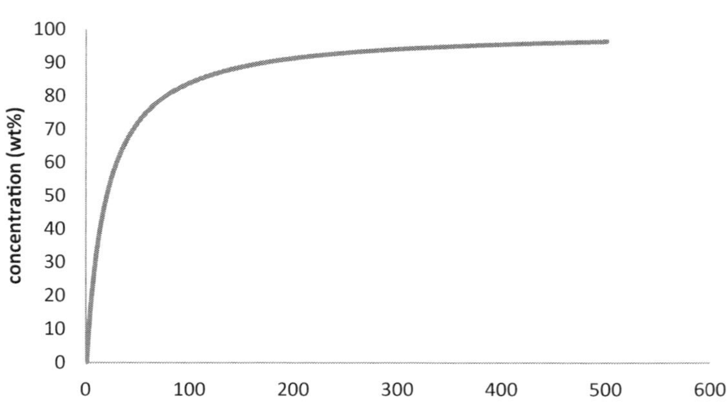

Figure 11 Concentration of ethanol in the permeate as a function of the selectivity of the membrane, assuming an initial ethanol concentration of 5 wt%.

A further extension of the multistep approach is a three-step pervaporation. It is evident that a good purity of the permeate is difficult to achieve with hydrophobic pervaporation, either in one step or in two steps. Thus, it is clear that a third unit, making use of hydrophilic membranes, is needed. This depends on the desired purity. Based on the 92% purity that was shown to be realistic for hydrophobic pervaporation, it can be calculated that a purity of 99.5% in a three-step approach is feasible. Hydrophilic pervaporation for ethanol dehydration is a much easier process than hydrophobic pervaporation; very high separation factors reaching easily the hundreds can be obtained, with extreme values over 10,000. For a detailed review of separation factors for ethanol dehydration as a function of the membrane material, the reader is referred to Bolto et al. (2011).

Apart from lignocellulosic ethanol production, another issue to consider is fractionation of all fermentation products in ABE fermentation. ABE fermentation is an anaerobic process in which a strain of bacteria from the *Clostridium* family is used, such as *Clostridium* acetobutylicum or *Clostridium* beijerinckii. Apart from ethanol, also acetone and butanol can be considered. There is increasing interest in particularly butanol in view of its higher value and energy content. Ong et al. (2011) studied separation of acetone and ethanol, which is a challenging separation because of the small difference in properties. They found separation factors up to 4 with classical PDMS membranes. Thus, to make ABE fermentation profitable on large scale, more work should be done on advanced separations for product fractionation; this has not been considered extensively. Zhou et al. (2011) considered a combined process of adsorption on silicalite-1 and hydrophobic pervaporation; acetone-ethanol separation proved to be the most difficult aspect.

A final remark is to be made on the method for determination of fluxes and selectivities, particularly in lignocellulosic and ABE conversions. It has been shown that fermentation by-products have an influence on the purification of ethanol from water during hydrophobic pervaporation; in particular, sugars and salts were found to increase the membrane performance while 2,3-butanediol decreased the ethanol flux and selectivity factor (Chovau et al., 2011). Thus, care should be taken when translating lab scale results to industrial applications.

9. PERVAPORATION MEMBRANE REACTORS

For a detailed overview of the use of pervaporation in membrane reactors, the reader is referred to Van der Bruggen et al. (2010). The combination of a chemical (equilibrium) reaction with simultaneous removal of water as a side product by pervaporation is an interesting application that has been demonstrated in many cases, of which the majority is related to esterification reactions (Kita et al., 1988; David et al., 1991a,b; Parulekar, 2007; Benedict et al., 2003, 2006; Datta and Tsai, 1995; Waldburger and Widmer, 1996; Krupiczka and Koszorz, 1999; Tanna and Mayadevi, 2007; Lim et al., 2002; Zhu et al., 1996; Dams and Krug, 1991; Brüschke et al., 1995; Shah and Ritchie, 2005; Bernal et al., 2002; Coronas and Santamaria, 1999; Tanaka et al., 2001).

Esterifications are organic transformations that can be considered among the most important in organic synthesis. Esterification reactions involve a carboxylic acid and an alcohol, with the following reaction:

$$\mathrm{RCOOH} + \mathrm{R'OH} \leftrightarrow \mathrm{RCOOR'} + \mathrm{H_2O}$$

The reaction is slow in the absence of a catalyst; usually an acid catalyst such as sulphuric acid or polymeric sulfonic acids, or a resin such as Amberlyst type resins. The yield of the ester RCOOR′ is limited by the reaction equilibrium. However, as expressed by the principle of Le Châtelier-Braun, when one of the reaction products (right-hand side) is removed, the equilibrium will shift to the right. This results in a higher yield of the ester. As explained in the previous sections, water removal is the showcase reaction for pervaporation and thus, pervaporation membrane reactors for esterification are successful in many conversions. Nevertheless, the application is more complex than dehydration of a pure solvent, since at least three products other than water are involved. The alcohol, and to a lesser extent the carboxylic acid, might be a challenge. The most common esterification reaction is the reaction of ethanol with acetic acid to produce ethyl acetate. Thus, water has to be removed from ethanol, acetic acid, and ethyl acetate. As demonstrated above, dehydration of ethanol is quite feasible, with high separation factors. Separation between acetic acid and water is somewhat more difficult, although good results have been obtained lately (Jullok et al., 2011). In any case, the separation is sufficient for attaining a yield of ethyl acetate of nearly 100%. A large number of pervaporation membranes and reactor configurations have been proposed in various publications (Van der Bruggen et al., 2010).

In addition to the reaction of ethanol and acetic acid, a wide range of esterification reactions coupled to pervaporation can be found in the literature. The most challenging among these is methyl acetate synthesis from methanol and acetic acid. This conversion was described by e.g.,

Assabumrungrat et al. (2003a,b), who compare three modes of operation, namely a semi-batch reactor, a plug flow reactor, and a continuous stirred tank. The hybrid design itself was found feasible, and the limiting factor is not the dehydration by pervaporation. This catalyzes new research on reactor design (in addition to the continuous search for improved membrane materials). One approach is the development of microreactors, in which the reaction and separation are as closely connected as possible; examples are packed-bed membrane reactors and catalytic membrane reactors. Miniaturization is a way to improve heat and mass transfer by two combined effects: (1) the diffusion distance within the reactor is much lower and (2) the interfacial area per unit reactor volume is larger (Jensen, 2001). A second pathway is the development of new membrane structures with catalysts integrated in the membranes, applied in e.g., the active zeolite membrane reactor proposed by Bernal et al. (2002), on the basis of an acid zeolite, H-ZSM-5.

The use of pervaporation in other membrane-assisted reactions is less explored. Nevertheless, the same principles can be applied to any equilibrium reaction in which the side product can be selectively removed by pervaporation. When the by-product is water, it can be easily separated from the reaction products by using a hydrophilic pervaporation membrane. Examples are the production of dimethyl urea (DMU) and methyl isobutyl ketone (MIBK) (Heroin et al., 1991; Staudt-Bickel and Lichtenthaler, 1996a,b). MIBK, a solvent for paints and protective coatings, is produced in two ways in a three step procedure with reactions as shown in Figure 12 (Staudt-Bickel and Lichtenthaler, 1996). The first reactions are a base-catalyzed aldol condensation of acetone and an acid catalyzed dehydration, followed by a hydrogenation of the resulting unsaturated ketone with a metal catalyst. Unfortunately, the first step is limited by the equilibrium of the condensation reaction, and yields in the last two steps are rather low. An alternative one step procedure uses a palladium/acid organic ion exchanger as a bifunctional catalyst (Wöllner and Engelhardt, 1968). Due to the high polarity of the sulfonated organic ion exchanger catalyst, water formed in the reaction accumulates in the pores of the ion exchanger and hinders the accessibility of the active sites at the palladium surface to the less polar organic reactant mesityl oxide. This compromises the reactor yield drastically. Removal of water would solve the problems with the catalyst and enhance the conversion of acetone. This can be done with hydrophilic crosslinked PVA membranes in a hybrid configuration (Staudt-Bickel and Lichtenthaler, 1996).

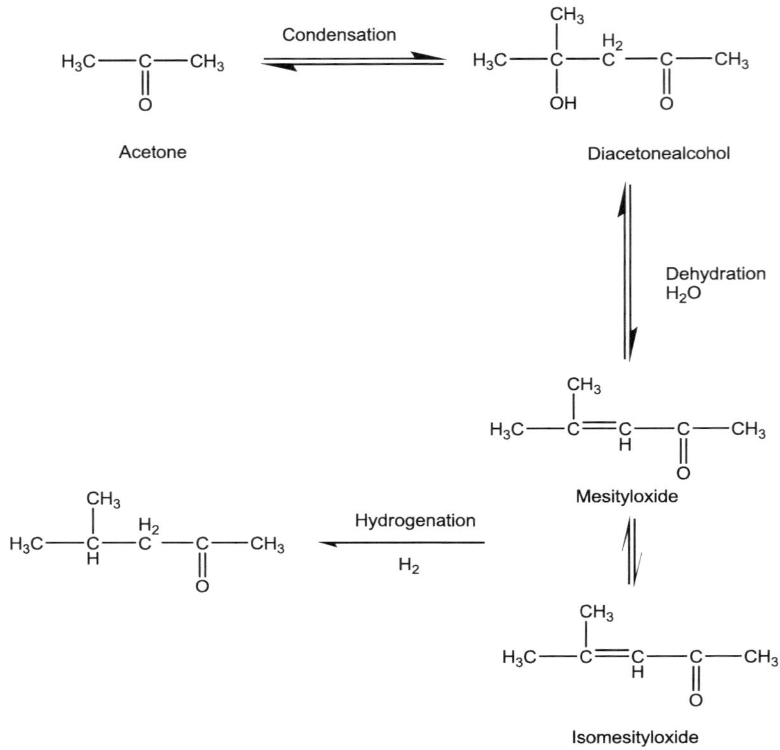

Figure 12 Synthesis of methyl isobutyl ketone (MIBK) using a three-step procedure. *Reproduced with permission from Staudt-Bickel and Lichtenthaler (1996).*

The synthesis of tertiary ether is another typical example of equilibrium-limited reactions, with catalytic activity strongly inhibited by the presence of water and generally low conversions due to limits imposed by thermodynamics (Van der Bruggen et al., 2010). The synthesis of ethyl tert-butyl ether (ETBE) from tert-butyl alcohol and ethanol (Yang and Goto, 1997) and the synthesis of methyl tert-butyl ether (MTBE) from tert-butyl alcohol and methanol (Matouq et al., 1994) are often studied, and can be enhanced by using pervaporation combined with a reactor. In general, all equilibrium reactions involving water as by-product can be carried out in a similar way. Progress in organophilic membranes, as described above, also allows to apply the same principles to synthesis procedures involving methanol or ethanol as the by-product, as is the case in transesterification reactions.

Pervaporation membrane reactors have another large application field in bioconversions. In the fermentation broth, the conversion is limited by the concentration of produced compounds (for example, of ethanol), so that bioreactors are normally operated at a very low efficiency. By removing the product continuously with a pervaporation membrane, the yield can be drastically increased. The most prominent example of such reactors is the production of bio-ethanol, which has been described in the previous section. Other possible extracts from fermentation reactors are aroma components, which can be removed by using hydrophobic membranes (Schafer and Crespo, 1999).

In general, the concept of pervaporation membrane reactors can be applied to any reaction with conversion limitations, provided that a pervaporation membrane is sufficiently selective; therefore, many more applications can be carried out with currently available membranes, and further growth can be expected following new developments in selective membranes, particularly in the field of organophilic separations.

10. CONCLUSIONS

Pervaporation is a mature technology today, with several applications with proven performance. For solvent dehydration, pervaporation has become a standard method because (1) the selectivity for water is very high with state-of-the-art membranes and (2) the concentration range for dehydration, in which a relatively small amount of water has to be removed, is ideal for pervaporation. Thus, while dehydration is quite feasible for a wide range of solvents (with isopropanol among the most studied compounds, along with ethanol), research challenges mainly remain in the development of robust processes for dehydration of a wide range of solvents that form azeotropes with water at relatively low water fractions, not only by developing new materials but also considering their application on a larger (time)scale.

Hydrophobic pervaporation and particularly organophilic pervaporation have a very good prognosis, but fewer applications have made it to industrial scale so far. Organophilic pervaporation is a new development, and it is evident that the separation challenge for the membranes is much larger than for any other application. Thus, research in this area focuses on improving the membrane selectivity for challenging separations, and the estimation of the potential of organophilic pervaporation for a wide range of separations. For hydrophobic pervaporation, the slower development is only partly related to the separation performance. A better ethanol selectivity would certainly catalyze large-scale applications; nevertheless, economic feasibility is often a dominating issue, which depends on the amount and value of the solvent to be recycled. For lignocellulosic bioethanol, the most realistic configuration for upgrading the classical distillation-based purification is in a three-step pervaporation with two hydrophobic and one hydrophilic units. A purity of 99.5% can be achieved with a selectivity of 20 in hydrophobic pervaporation.

Hybrid configurations with pervaporation are interesting in two ways. The combination with classical separation technologies, often distillation, is successful because of the complementarity between pervaporation and methods based on phase equilibria. Pervaporation can operate in the optimal concentration range, where often distillation is impeded by an azeotrope. The second hybrid configuration is the pervaporation membrane reactor, in which pervaporation is used for removal of the product or the by-product of a conversion. Equilibrium reactions involving water as the by-product are indicators of the success of pervaporation membrane reactors; esterifications are a typical example. The potential for developing new applications is very high, with novel membrane types but even with currently available state-of-the-art membranes.

REFERENCES

Aptel, P., Challard, N., Cuny, J., Neel, J., 1976. Application of pervaporation process to separate azeotropic mixtures. J. Membr. Sci. 1 (3), 271–287.

Assabumrungrat, S., Phongpatthanapanich, J., Praserthdam, P., Tagawa, T., Goto, S., 2003a. Theoretical study on the synthesis of methyl acetate from methanol and acetic acid in pervaporation membrane reactors: effect of continuous flow modes. Chem. Eng. J. 95, 57–65.

Assabumrungrat, S., Kiatkittipong, W., Praserthdam, P., Goto, S., 2003b. Simulation of pervaporation membrane reactors for liquid phase synthesis of ethyl tert-butyl ether from tert-butyl alcohol and ethanol. Catal. Today 79–80, 249–257.

Bai, Y., Qian, J.W., Yin, J., Zhai, Z.B., Yang, Y., 2007. HTPB-based polyurethaneurea membranes for recovery of aroma compounds from aqueous solution by pervaporation. J. Appl. Polym. Sci. 104 (1), 552–559.

Baker, R.W., Athayde, A.L., Daniels, R., Le, M., Pinnau, I., Ly, J.H., Wijmans, J.G., Kaschemekat, J.H., Helm, V.D., 1997. Development of pervaporation to recover and reuse volatile organic compounds from industrial waste streams. DOA/AL/98769-1, Report to US Department of Energy.

Baker, R.W., Wijmans, J.G., Huang, Y., 2010. Permeability, permeance and selectivity: a preferred way of reporting pervaporation performance data. J. Membr. Sci. 348, 346–352.

Baudot, A., Marin, M., 1996. Dairy aroma compounds recovery by pervaporation. J. Membr. Sci. 120 (2), 207–220.

Beaumelle, D., Marin, M., Gibert, H., 1992. Pervaporation of aroma compounds in water ethanol mixtures – experimental-analysis of mass-transfer. J. Food Eng. 16 (4), 293–307.

Benedict, D.J., Parulekar, S.J., Tsai, S.P., 2003. Esterification of lactic acid and ethanol with/without pervaporation. Ind. Eng. Chem. Res. 42 (11), 2282–2291.

Benedict, D.J., Parulekar, S.J., Tsai, S.P., 2006. Pervaporation-assisted esterification of lactic and succinic acids with downstream ester recovery. J. Membr. Sci. 281, 435–445.

Bernal, M.P., Coronas, J., Menendez, M., Santamaria, J., 2002. Coupling of reaction and separation at the microscopic level: esterification processes in a H-ZSM-5 membrane reactor. Chem. Eng. Sci. 57, 1557–1562.

Bettens, B., August 2008. Pervaporation through microporous silica membranes (Ph.D. thesis). KU Leuven, Belgium.

Beuscher, U., Gooding, C.H., 1998. The permeation of binary gas mixtures through support structures of composite membranes. J. Membr. Sci. 150, 57–73.

Bhat, S.D., Aminabhavi, T.M., 2007. Pervaporation separation using sodium alginate and its modified membranes – a review. Sep. Purif. Rev. 36 (3–4), 203–229.

Binning, R.C., James, F.E., 1958a. Permeation. A new commercial separation tool. Refin. Eng. 30 (6), C14–C15.

Binning, R.C., James, F.E., 1958b. How to separate by membrane permeation. Pet. Refin. 37 (5), 214.

Binning, R.C., James, F.E., 1958c. Permeation: a new way to separate mixtures. Oil Gas J. 56, 104.

Binning, R.C., Lee, R.J., Jennings, J.F., Martin, E.C., 1961. Separation of liquid mixtures by permeation. Ind. Eng. Chem. 53, 45–50.

Binning, R.C., Lee, R.J., Jennings J.F., Martin, E.C., 1962. U.S. Patent 3,035,060.

Binning, R.C., April 25, 1961. U.S. Patent 2,981,680.

Boeddeker, K.W., Bengtson, G., 1989. In: Huang, R.Y.M. (Ed.), Pervaporation: Membrane Separation Processes. Elsevier Science Publisher, Amsterdam, pp. 437–460.

Bolto, B., Hoang, M., Xie, Z.L., 2011. A review of membrane selection for the dehydration of aqueous ethanol by pervaporation. Chem. Eng. Proc. Intens 50 (3), 227–235.

Borjesson, J., Karlsson, H.O.E., Tragardh, G., 1996. Pervaporation of a model apple juice aroma solution: comparison of membrane performance. J. Membr. Sci. 119 (2), 229–239.

Bowen, T.C., Noble, R.D., Falconer, J.L., 2004. Fundamentals and applications of pervaporation through zeolite membranes. J. Membr. Sci. 245, 1–33.

Brazinha, C., Crespo, J.G., 2009. Aroma recovery from hydro alcoholic solutions by organophilic pervaporation: modelling of fractionation by condensation. J. Membr. Sci. 341 (1–2), 109–121.

Brüschke, H.E.A., Ellinghorst, G., Schneider, W.H., 1995. Optimization of a coupled reaction – pervaporation process. In: Bakish, R. (Ed.), Proceedings of the Seventh International Conference on Pervaporation Processes in the Chemical Industry. Bakish Material Corporation, Eaglewood, NJ, USA, pp. 310–320.

Bruschke, H., 1995. Industrial application of membrane separation processes. Pure Appl. Chem. 67 (6), 993–1002.

Burggraaf, A.J., Cot, L., 1996. Fundamentals of Inorganic Membrane Science and Technology. In: Membrane Science and Technology Series 4. Elsevier, Amsterdam.

Burshe, M.C., Netke, S.A., Sawant, S.B., Joshi, J.B., Pangarkar, V.G., 1997. Pervaporative dehydration of organic solvents. Sep. Sci. Technol. 32 (8), 1335–1349.

Cahya, S., Sanli, O., Camurlu, E., 2011. Crosslinked sodium alginate and sodium alginate-clinoptilolite (natural zeolite) composite membranes for pervaporation separation of dimethylformamide-water mixtures: a comparative study. Desal. Water Treat. 25 (1–3), 297–309.

Campaniello, J., Engelen, C.W.R., Haije, W.G., Pex, P.P.A.C., Vente, J.F., 2004. Long term pervaporation performance of microporous methylated silica membranes. Chem. Commun., 834.

Chapman, P.D., Oliveira, T., Livingston, A.G., Li, K., 2008a. Membranes for the dehydration of solvents by pervaporation. J. Membr. Sci. 318 (1–2), 5–37.

Chapman, P.D., Loh, X.X., Livingston, A.G., Li, K., 2008b. Polyaniline membranes for the dehydration of tetrahydrofuran by pervaporation. J. Membr. Sci. 309 (1–2), 102–111.

Chen, X., Ping, Z.H., Long, Y.C., 1998. Separation properties of alcohol-water mixture through silicalite-1-filled silicone rubber membranes by pervaporation. J. Appl. Polym. Sci. 67, 629–636.

Choudhari, S.K., Kariduraganavar, M.Y., 2009. Development of novel composite membranes using quaternized chitosan and Na^+-MMT clay for the pervaporation dehydration of isopropanol. J. Coll. Interf. Sci. 338 (1), 111–120.

Choudhari, S.K., Kittur, A.A., Kulkarni, S.S., Kariduraganavar, M.Y., 2007. Development of novel blocked diisocyanate crosslinked chitosan membranes for pervaporation separation of water-isopropanol mixtures. J. Membr. Sci. 302 (1–2), 197–206.

Chovau, S., Gaykawad, S., Straathof, A.J.J., Van der Bruggen, B., 2011. Influence of fermentation by-products on the purification of ethanol from water using pervaporation. Bioresour. Technol. 102, 1669–1674.

Claes, S., Vandezande, P., Mullens, S., Van Bael, M.K., Maurer, F.H.J., 2011. Enhanced performance in pervaporation of supercritical carbon dioxide treated poly1-(trimethylsilyl)-1-propyne pervaporation membranes. J. Membr. Sci. 382, 177–185.

Claes, S., Vandezande, P., Mullens, S., De Sitter, K., Peeters, R., Van Bael, M.K., 2012. Preparation and benchmarking of thin film supported PTMSP-silica pervaporation membranes. J. Membr. Sci. 389, 265–271.

Claes, S., 2012. Development, modification and optimization of poly[1-(trimethylsilyl)-1-propyne] pervaporation membranes (Ph.D. thesis). University of Hasselt, Belgium.

Coronas, J., Santamaria, J., 1999. Separations using zeolite membranes. Sep. Purif. Meth. 28 (2), 127–177.

Dams, A., Krug, J., 1991. Pervaporation aided esterification – alternatives in plant extension for an existing chemical process. In: Bakish, R. (Ed.), Proceedings of the Fifth International Conference on Pervaporation Processes in the Chemical Industry. Bakish Material Corporation, Eaglewood, NJ, USA, pp. 338–348.

Datta, R., Tsai, S., 1995. Esterification of fermentation-derived acids via pervaporation. U.S. Patent 5,723,639.

David, M.O., Gref, R., Nguyen, T.Q., Neel, J., 1991a. Pervaporation-esterification coupling. 1. Basic kinetic model. Chem. Eng. Res. Des. 69 (4), 335–340.

David, M.O., Gref, R., Nguyen, T.Q., Neel, J., 1991b. Pervaporation-esterification coupling. 2. Modelling of the influence of different operating parameters. Trans. Inst. Chem. Eng 69, 341–346.

de Bruijn, F.T., Sun, L., Olujic, Ž., Jansens, P.J., Kapteijn, F., 2003. Influence of the support layer on the flux limitation in pervaporation. J. Membr. Sci. 223, 141–156.

De Bruijn, F.T., Groß, J., Olujic, Ž., Jansens, P.J., Kapteijn, F., 2007. On the driving force of methanol pervaporation through a microporous methylated silica membrane. Ind. Eng. Chem. Res. 46 (12), 4091–4099.

de Lange, R.S.A., Hekkink, J.H.A., Keizer, K., Burggraaf, A.J., 1995. Permeation and separation studies on microporous sol-gel modified ceramic membranes. Micropor. Mater. 4, 169–186.

Devi, D.A., Smitha, B., Sridhar, S., Aminabhavi, T.M., 2005. Pervaporation separation of isopropanol/water mixtures through crosslinked chitosan membranes. J. Membr. Sci. 262 (1–2), 91–99.

Farber, L., 1935. Application of pervaporation. Science 82, 158.

Feng, X., Huang, R.Y.M., 1997. Liquid separation by membrane pervaporation: a review. Ind. Eng. Chem. Res. 36, 1048–1066.

Field, R.W., Lobo, V., 2003. Hydrophobic pervaporation – toward a shortcut method for the pervaporation-decanter system. In: Li, N.N., Drioli, E., Ho, W.S.W.,

Lipscomb, G.G. (Eds.), Advanced Membrane Technology, 984. Annals of the New York Academy of Sciences, pp. 401–410.
Figoli, A., Donato, L., Carnevale, R., Tundis, R., Statti, G.A., Menichini, F., Drioli, E., 2006. Bergamot essential oil extraction by pervaporation. Desalination 193 (1–3), 160–165.
Fleming, H.L., 1989. Dehydration of organic/aqueous mixtures by membrane pervaporation. In: Kampen, W. (Ed.), Proceedings of International Conference on Fuel Alcohols and Chemicals. K-Engineering, Charlotte, NC.
Ganapathi-Desai, S., Sikdar, S.K., 2000. A polymer-ceramic composite membrane for recovering volatile organic compounds from wastewaters by pervaporation. Clean Prod. Proc. 2, 140–148.
Ghoreyshi, S.A.A., Farhadpour, F.A., Soltanieh, M., 2002. Multicomponent transport across nonporous polymeric membranes. Desalination 144, 93–101.
Gilron, J., Soffer, A., 2002. Knudsen diffusion in microporous carbon membranes with molecular sieving character. J. Membr. Sci. 209, 339–352.
Gu, J., Zhang, C.F., Bai, Y.X., Zhang, L., Sun, Y.P., Chen, H.L., 2012. Divinyl benzene cross-linked HTPB-based polyurethaneurea membranes for separation of *p*-/*o*-xylene mixtures by pervaporation. J. Appl. Polym. Sci. 123 (4) 1968–1976.
Haelssig, J.B., Tremblay, A.Y., Thibault, J., 2012. A new hybrid membrane separation process for enhanced ethanol recovery: process description and numerical studies. Chem. Eng. Sci. 68 (1), 492–505.
Heintz, A., Stephan, W., 1994. A generalized solution-diffusion model of the pervaporation process through composite membranes, Part II. Concentration polarization, coupled diffusion and the influence of the porous support layer. J. Membr. Sci. 89, 153–169.
Heisler, E.G., Hunter, A.S., Siciliano, J., Treadway, R.H., 1956. Solute and temperature effects in the pervaporation of aqueous alcoholic solutions. Science 124 (3211), 77–79.
Heroin, C., Spiske, L., Hefner, W., 1991. Dehydration in the synthesis of dimethylurea by pervaporation. In: Bakish, R. (Ed.), Proceedings of the Fifth International Conference on Pervaporation Processes in the Chemical Industry. Bakish Material Corporation, Englewood, NJ, USA, pp. 349–361.
Huang, R.Y.M., Yeom, C.K., 1990. Pervaporationseparation of aqueous mixtures using crosslinked poly(vinyl alcohol)(PVA). II. Permeation of ethanol-water mixtures. J. Membr. Sci. 51 (3), 273–292.
Huang, R.Y.M., Shao, P., Feng, X., Burns, C.M., 2001. Pervaporation separation of water/isopropanol mixture using sulfonated poly(ether ether ketone) (SPEEK) membranes: transport mechanism and separation performance. J. Membr. Sci. 192 (1–2), 115–127.
Huang, Y., Baker, R.W., Vane, L.M., 2010. Low-Energy distillation-membrane separation process. Ind. Eng. Chem. Res. 49 (8), 3760–3768.
Ikegami, T., Negishi, H., Sakaki, K., 2011. Selective separation of *n*-butanol from aqueous solutions by pervaporation using silicone rubber-coated silicalite membranes. J. Chem. Technol. Biotechnol. 86 (6), 845–851.
Isci, A., Sahin, S., Sumnu, G., 2006. Recovery of strawberry aroma compounds by pervaporation. J. Food Eng. 75 (1), 36–42.
Izák, P., Bartovská, L., Friess, K., Sípek, M., Uchytil, P., 2003. Description of binary liquid mixtures transport through non-porous membrane by modified Maxwell–Stefan equations. J. Membr. Sci. 214 (2), 293–309.
Jensen, K.F., 2001. Microreaction engineering – is small better? Chem. Eng. Sci. 56 (2), 293–303.
Jeong, J.S., Jeon, H., Ko, K.M., Chung, B., Choi, G.W., 2012. Production of anhydrous ethanol using various PSA (Pressure Swing Adsorption) processes in pilot plant. Renew. Energ. 42, 41–45.

Jia, M.-D., Peinemann, K.-V., Behling, R.-D., 1992. Preparation and characterization of thin-film zeolite-PDMS composite membranes. J. Membr. Sci. 73, 119–128.

Jiang, X., Gu, J.A., Shen, Y., Wang, S.G., Tian, X.Z., 2011. New fluorinated siloxane-imide block copolymer membranes for application in organophilic pervaporation. Desalination 265 (1–3), 74–80.

Jin, T., Ma, Y., Matsuda, W., Masuda, Y., Nakajima, M., Ninomiya, K., Hiraoka, T., Daiko, Y., Yazawa, T., 2011. Ethanol separation from ethanol aqueous solution by pervaporation using hydrophobic mesoporous silica membranes. J. Ceram. Soc. Jap. 119 (1391), 549–556.

Jonquières, A., Clément, R., Lochon, P., Néel, J., Dresch, M., Chrétien, B., 2002. Industrial state-of-the-art of pervaporation and vapour permeation in the western countries. J. Membr. Sci. 206, 87–117.

Jullok, N., Darvishmanesh, S., Luis, P., Van der Bruggen, B., 2011. The potential of pervaporation for separation of acetic acid and water mixtures using polyphenylsulfone membranes. Chem. Eng. J. 175, 306–315.

Kahlenberg, L., 1906. On the nature of the process of osmosis and osmotic pressure with observations concerning dialysis. J. Phys. Chem. 10, 141–209.

Kanti, P., Srigowri, K., Madhuri, J., Smitha, B., Sridhar, S., 2004. Dehydration of ethanol through blend membranes of chitosan and sodium alginate by pervaporation. Sep. Purif. Technol. 40 (3), 259–266.

Kapteijn, F., Moulijn, J.A., Krishna, R., 2000. The generalized Maxwell-Stefan model for diffusion in zeolites: sorbate molecules with different saturation loadings. Chem. Eng. Sci. 55, 2923–2930.

Kim, J.-H., Lee, K.-H., Kim, S.Y., 2000. Pervaporation separation of water from ethanol through polyimide composite membranes. J. Membr. Sci. 169 (1), 81–93.

Kiss, A.A., Suszwalak, D.J.P.C., 2012. Enhanced bioethanol dehydration by extractive and azeotropic distillation in dividing-wall columns. Sep. Purif. Technol. 86, 70–78.

Kita, H., Sasaki, S., Tanaka, K., Okamoto, K., Yamamoto, M., 1988. Esterification of carboxylic acid with ethanol accompanied by pervaporation. Chem. Lett. 12, 2025–2028.

Kober, P.A., 1917. Pervaporation, perstillation, and percrystallisation. J. Am. Chem. Soc. 39, 9444.

Krishna, R., van Baten, J.M., 2008. Insights into diffusion of gases in zeolites gained from molecular dynamics simulations. Micropor. Mesopor. Mat. 109 (1–3), 91–108.

Krishna, R., Van den Broeke, L.J.P., 1995. The Maxwell-Stefan description of mass transport across zeolite membranes. Chem. Eng. J. Biochem. Eng. J. 57 (2), 155–162.

Krishna, R., Wesselingh, J.A., 1997. The Maxwell-Stefan approach to mass transfer. Chem. Eng. Sci. 52 (6), 861–911.

Krishna, R., 1990. Multicomponent surface diffusion of adsorbed species: a description based on the generalized Maxwell–Stefan equations. Chem. Eng. Sci. 45 (7), 1779–1791.

Krishna, R., 2000. Diffusion of binary mixtures in zeolites: molecular dynamics simulations versus Maxwell–Stefan theory. Chem. Phys. Lett. 326 (5–6), 477–484.

Krupiczka, R., Koszorz, Z., 1999. Activity-based model of the hybrid process of an esterification reaction coupled with pervaporation. Sep. Purif. Technol. 16, 55–59.

Kujawski, W., 2000. Application of pervaporation and vapour permeation in environmental protection. J. Environ. Stud. 9, 13–26.

Lee, Y.M., Nam, S.Y., Woo, D.J., 1998. Pervaporation performance of beta-chitosan membrane for water/alcohol mixtures. J. Polym. Eng 18 (1–2), 131–146.

Lee, H.J., Cho, E.J., Kim, Y.G., Choi, I.S., Bae, H.J., 2012. Production of anhydrous ethanol using various PSA (Pressure Swing Adsorption) processes in pilot plant. Bioresour. Technol. 109, 110–115.

Lee, S.C., 2007. Prediction of permeation behavior of CO_2 and CH_4 through silicalite-1 membranes in single-component or binary mixture systems using occupancy dependent Maxwell–Stefan diffusivities. J. Membr. Sci. 306 (1–2), 267–276.

Li, X., Wang, S., 1996. Some characteristics of pervaporation for dilute ethanol-water mixtures by alcohol-permselective membranes. Sep. Sci. Technol. 31, 2867–2873.

Li, S.Y., Srivastava, R., Parnas, R.S., 2010. Separation of 1-butanol by pervaporation using a novel tri-layer PDMS composite membrane. J. Membr. Sci. 363 (1–2), 287–294.

Li, Q., Yu, P., Lin, Y.W., Zhu, T.R., Luo, Y.B., 2012. A novel membrane on pervaporation performance for dehydration of caprolactam solution. J. Ind. Eng. Chem. 18 (2), 604–610.

Liang, L., Dickson, J.M., Jiang, J., Brook, M.A., 2004. Effect of low flow rate on pervaporation of 1,2-dichloroethane with novel polydimethylsiloxane composite membranes. J. Membr. Sci. 231, 71–79.

Lim, S.Y., Park, B., Hung, F., Sahimi, M., Tsotsis, T.T., 2002. Design issues of pervaporation membrane reactors for esterification. Chem. Eng. Sci. 57, 4933–4946.

Lin, X., Kita, H., Okamoto, K., 2001. Silicalite membrane preparation, characterization, and separation performance. Ind. Eng. Chem. Res. 40, 4069–4078.

Lin, X., Chen, X., Kita, H., Okamoto, K., 2003. Synthesis of silicalite tubular membranes by in situ crystallization. AlChE J. 49, 237–247.

Lipnizki, F., Field, R.W., 2002. Hydrophobic pervaporation for environmental applications: process optimization and integration. Environ. Prog. 21 (4), 265–272.

Lipnizki, F., Hausmanns, S., Field, R.W., 2004. Influence of impermeable components on the permeation of aqueous 1-propanol mixtures in hydrophobic pervaporation. J. Membr. Sci. 228 (2), 129–138.

Lipnizki, F., Hausmanns, S., Ten, P.-K., Field, R.W., Laufenberg, G., 1999a. Organophilic pervaporation: prospects and performance. Chem. Eng. J. 73 (2), 113–129.

Lipnizki, F., Field, R.W., Ten, P.-K., 1999b. Pervaporation-based hybrid process: a review of process design, applications and economics. J. Membr. Sci. 153 (2), 183–210.

Lipsi, C., Cote, P., 1990. The use of pervaporation for the removal organic contaminants from water. Environ. Prog. 9, 254–261.

Liu, Y.L., Lee, K.R., Lai, J.Y., 2005. Crosslinked organic-inorganic hybrid chitosan membranes for pervaporation dehydration of isopropanol-water mixtures with a long-term stability. J. Membr. Sci. 251 (1–2), 233–238.

Liu, Y.L., Yu, C.H., Lee, K.R., Lai, J.Y., 2007. Chitosan/poly(tetrafluoroethylene) composite membranes using in pervaporation dehydration processes. J. Membr. Sci. 287 (2), 230–236.

Liu, Y.L., Yu, C.H., Lai, J.Y., 2008. Poly(tetrafluoroethylene)/polyamide thin-film composite membranes via interfacial polymerization for pervaporation dehydration on an isopropanol aqueous solution. J. Membr. Sci. 315 (1–2), 106–115.

Liu, G., Yang, D., Zhu, Y.Y., Ma, J., Nie, M.C., Jiang, Z.Y., 2011. Titanate nanotubes-embedded chitosan nanocomposite membranes with high isopropanol dehydration performance. Chem. Eng. Sci. 66 (18), 4221–4228.

Mali, M.G., Magalad, V.T., Gokavi, G.S., Aminabhavi, T.M., Raju, K.V.S.N., 2011. Pervaporation separation of isopropanol–water mixtures using mixed matrix blend membranes of poly(vinyl alcohol)/poly(vinyl pyrrolidone) loaded with phosphomolybdic acid. J. Appl. Polym. Sci. 121 (2), 711–719.

Mandal, S., Pangarkar, V.G., 2003. Effect of membrane morphology in pervaporative separation of isopropyl alcohol-aromatic mixtures – a thermodynamic approach to membrane selection. J. Appl. Polym. Sci. 90 (14), 3912–3921.

Matouq, M., Tagawa, T., Goto, S., 1994. Combined process for production of methyl tert-butyl ether from tert-butyl alcohol and methanol. J. Chem. Eng. Jpn. 27, 302–306.

Matsuda, H., Yanagishita, H., Negishi, H., Kitamoto, D., Ikegami, T., Haraya, K., Nakane, K., Idemoto, Y., Koura, N., Sano, T., 2002. Improvement of ethanol selectivity of silicalite membrane in pervaporation by silicone rubber coating. J. Membr. Sci. 210, 433–437.

Matuschewski, H., Schedler, U., 2008. MSE – modified membranes in organophilic pervaporation for aromatics/aliphatics separation. Desalination 224 (1–3), 124–131.
Mauz, M., Kimmerle, K., Hulser, D.F., 1996. Concentration of native etheric oil aroma components by pervaporation. J. Membr. Sci. 118 (2), 145–150.
Mecki, K., Lichtenthaler, R.N., 1996. Hybrid process using pervaporation for the removal of organics from process and waste water. J. Membr. Sci. 113 (1), 81–86.
Meuleman, E.E.B., Willemsen, J., Mulder, M.H.V., Strathmann, H., 2001. EPDM as a selective membrane material in pervaporation. J. Membr. Sci. 188, 235–249.
Moon, G.Y., Pal, R., Huang, R.Y.M., 1999. Novel two-ply composite membranes of chitosan and sodium alginate for the pervaporation dehydration of isopropanol and ethanol. J. Membr. Sci. 156, 17–27.
Moraes, E.B., Alvarez, M.E.T., Perioto, F.R., Wolf-Maciel, M.R., 2007. Modeling and simulation for pervaporation process: an alternative for removing phenol from wastewater. In: Pierucci, S. (Ed.), ICHEAP-9: 9th Int. Conf. on Chem. Proc. Eng., Pts 1-3, vol. 17. Chemical Engineering Transactions, pp. 1621–1626.
Morigami, Y., Kondo, M., Abe, J., Kita, H., Okamoto, K., 2001. The first large-scale pervaporation plant using tubular-type module with zeolite NaA membrane. Sep. Purif. Technol. 25 (1–3), 251–260.
Mosleh, S., Khosravi, T., Bakhtiari, O., Mohammadi, T., 2012. Zeolite filled polyimide membranes for dehydration of isopropanol through pervaporation process. Chem. Eng. Res. Des. 90 (3A), 433–441.
Mulder, M.H.V., 1996. Basic Principles of Membrane Technology. Kluwer Academic, Dordrecht, The Netherlands.
Nagase, Y., Takamura, Y., Matsui, K., 1991. Chemical modification of poly(substitutedacetylene): V. Alkylsilylation of poly(1-trimethylsilyl-1-propyne) and improved liquid separating property at pervaporation. J. Appl. Polym. Sci. 42, 185–190.
Nam, S.Y., Chun, H.J., Lee, Y.M., 1999. Pervaporation separation of water-isopropanol mixture using carboxymethylated poly(vinyl alcohol) composite membranes. J. Appl. Polym. Sci. 72 (2), 241–249.
Nawawi, M.G.M., Huang, R.Y.M., 1997. Pervaporation dehydration of isopropanol with chitosan membranes. J. Membr. Sci. 124 (1), 53–62.
Okamoto, K., Kita, H., Horii, K., Tanaka, K., Kondo, M., 2001. Zeolite NaA membrane: preparation, single-gas ppermeation, and pervaporation and vapor permeation of water/organic liquid mixtures. Ind. Eng. Chem. Res. 40 (1), 163–175.
Ong, Y.K., Widjojo, N., Chung, T.S., 2011. Fundamentals of semi-crystalline poly(vinylidene fluoride) membrane formation and its prospects for biofuel (ethanol and acetone) separation via pervaporation. J. Membr. Sci. 378 (1–2), 149–162.
Panek, D., Konieczny, K., 2007. Preparation and applying the membranes with carbon black to pervaporation of toluene from the diluted aqueous solutions. Sep. Purif. Technol. 57, 507–512.
Panek, D., Konieczny, K., 2008. Applying filled and unfilled polyether-block-amide membranes to separation of toluene from wastewaters by pervaporation. Desalination 222, 280–285.
Panek, D., Konieczny, K., 2009. Pervaporative separation of toluene from wastewaters by use of filled and unfilled poly(dimethylosiloxane) (PDMS) membranes. Desalination 241 (1–3), 197–200.
Parulekar, S.J., 2007. Analysis of pervaporation-aided esterification of organic acids. Ind. Eng. Chem. Res. 46 (25), 8490–8504.
Pasel, J., Wang, Y., Hurter, S., Dahl, R., Peters, R., Schedler, U., Matuschewski, H., 2012. Desulfurization of jet fuel by pervaporation. J. Membr. Sci. 390, 12–22.
Pejin, D.J., Mojovic, L.V., Pejevic, J.D., Grujic, O.S., Markov, S.L., Nikolic, S.B., Markovic, M.N., 2012. Increase in bioethanol production yield from triticale by

simultaneous saccharification and fermentation with application of ultrasound. J. Chem. Technol. Biotechnol. 87 (2), 170–176.

Peng, M., Vane, L.M., Liu, S.X., 2003. Recent advances in VOCs removal from water by pervaporation. J. Hazard. Mater. 98 (1–3), 69–90.

Peng, P., Shi, B.L., Lan, Y.Q., 2011. A review of membrane materials for ethanol recovery by pervaporation. Sep. Sci. Technol. 46 (2), 234–246.

Pera-Titus, M., Llorens, J., Tejero, J., Cunill, F., 2006. Description of the pervaporation dehydration performance of A-type zeolite membranes: a modeling approach based on the Maxwell–Stefan theory. Catal. Today 118 (1–2), 73–84.

Polyakov, A.M., Starannikova, L.E., Yampolskii, Y.P., 2004. Amorphous teflons AF as organophilic pervaporation materials: separation of mixtures of chloromethanes. J. Membr. Sci. 238, 21–23.

Polyakov, A., Bondarenko, G., Tokarev, A., Yampolskii, Y., 2006. Intermolecular interactions in target organophilic pervaporation through the films of amorphous teflon AF2400. J. Membr. Sci. 277 (1–2), 108–119.

Prasad, C.V., Swamy, B.Y., Sudhakar, H., Sobharani, T., Sudhakar, K., Subha, M.C.S., Rao, K.C., 2011. Preparation and characterization of 4A zeolite-filled mixed matrix membranes for pervaporation dehydration of isopropyl alcohol. J. Appl. Polym. Sci. 121 (3), 1521–1529.

Rachipudi, P.S., Kariduraganavar, M.Y., Kittur, A.A., Sajjan, A.M., 2011. Synthesis and characterization of sulfonated-poly(vinyl alcohol) membranes for the pervaporation dehydration of isopropanol. J. Membr. Sci. 383 (1–2), 224–234.

Raghunath, B., Hwang, S.-T., 1992. Effect of boundary layer mass transfer resistance in the pervaporation of dilute organics. J. Membr. Sci. 65, 147–161.

Raisi, A., Aroujalian, A., 2011. Aroma compound recovery by hydrophobic pervaporation: the effect of membrane thickness and coupling phenomena. Sep. Purif. Technol. 82, 53–62.

Rautenbach, R., Albrecht, R., 1980. Separation of organic binary mixtures by pervaporation. J. Membr. Sci. 7 (2), 203–223.

Ray, S., Ray, S.K., 2008. Dehydration of tetrahydrofuran (THF) by pervaporation using crosslinked copolymer membranes. Chem. Eng. Proc. 47 (9–10), 1620–1630.

Ribeiro, C.P., Freeman, B.D., Kalika, D.S., Kalakkunnath, S., 2012. Aromatic polyimide and polybenzoxazole membranes for the fractionation of aromatic/aliphatic hydrocarbons by pervaporation. J. Membr. Sci. 390, 182–193.

Sampranpiboon, P., Jiraratananon, R., Uttapap, D., Feng, X., Huang, R.Y.M., 2000. Separation of aroma compounds from aqueous solutions by pervaporation using polyoctylmethylsiloxane (POMS) and polydimethyl siloxane (PDMS) membranes. J. Membr. Sci. 174, 55–65.

Satyanarayana, S.V., Sharma, A., Bhattacharya, P.K., 2004. Composite membranes for hydrophobic pervaporation: study with the toluene water system. Chem. Eng. J. 102, 171–184.

Schafer, T., Crespo, J.G., August 22–27, 1999. Extraction of aromas from active fermentation reactors by pervaporation. In: BelafiBako, K., Gubicza, L., Mulder, M. (Eds.), Integration of Membrane Processes into Bioconversions, XVIth European Membrane Society Annual Summer School on Integration of Membrane Processes into Bioconversions. Veszprem, Hungary.

Sekulic, J., Luiten, M.W.J., ten Elshof, J.E., Benes, N.E., Keizer, K., 2002. Microporous silica and doped silica membrane for alcohol dehydration by pervaporation. Desalination 148 (1–3), 19–23.

Shah, T.N., Ritchie, S.M.C., 2005. Esterification catalysis using functionalized membranes. Appl. Catal. A Gen. 296, 12–20.

Shao, P., Huang, R.Y.M., 2007. Polymeric membrane pervaporation. J. Membr. Sci. 287 (2), 162–179.

Shi, G.M., Wang, Y., Chung, T.S., 2012. Dual-layer PBI/P84 hollow fibers for pervaporation dehydration of acetone. AIChE J. 58 (4), 1133–1145.

Shirazi, Y., Tofighy, M.A., Mohammadi, T., 2011. Synthesis and characterization of carbon nanotubes/poly vinyl alcohol nanocomposite membranes for dehydration of isopropanol. J. Membr. Sci. 378, 551–561.

Shirazi, Y., Ghadimi, A., Mohammadi, T., 2012. Recovery of alcohols from water using polydimethylsiloxane-silica nanocomposite membranes: characterization and pervaporation performance. J. Appl. Polym. Sci. 124 (4), 2871–2882.

Singha, N.R., Ray, S.K., 2012. Removal of pyridine from water by pervaporation using crosslinked and filled natural rubber membranes. J. Appl. Polym. Sci. 124 (1), E99–E107.

Smitha, B., Suhanya, D., Sridhar, S., Ramakrishna, M., 2004. Separation of organic–organic mixtures by pervaporation—a review. J. Membr. Sci. 241 (1), 1–21.

Sommer, S., Melin, T., 2005. Performance evaluation of microporous inorganic membranes in the dehydration of industrial solvents. Chem. Eng. Proc. 44 (10), 1138–1156.

Staudt-Bickel, C., Lichtenthaler, R.N., 1996a. Integration of pervaporation for the removal of water in the production process of methylisobutylketone. J. Membr. Sci. 111, 135–141.

Staudt-Bickel, C., Lichtenthaler, R.N., 1996b. Integration of pervaporation for the removal of water in the production process of methylisobutylketone (MIBK). In: Proceedings of the International Congress on Membranes and Membrane Processes (ICOM), Yokohama, Japan, August 18–23, pp. 394–395.

Sudhakar, H., Prasad, C.V., Sunitha, K., Rao, K.C., Subha, M.C.S., Sridhar, S., 2011. Pervaporation separation of IPA-water mixtures through 4A zeolite-filled sodium alginate membranes. J. Appl. Polym. Sci. 121 (5), 2717–2725.

Sunitha, K., Kumar, Y.V.L.R., Sridhar, S., 2009. Effect of PVP loading on pervaporation performance of poly(vinyl alcohol) membranes for THF/water mixtures. J. Mater. Sci. 44 (23), 6280–6285.

Sunitha, K., Satyanarayana, S.V., Sridhar, S., 2012. Phosphorylated chitosan membranes for the separation of ethanol-water mixtures by pervaporation. Carbohydr. Polym. 87 (2), 1569–1574.

Svang-Ariyaskul, A., Huang, R.Y.M., Douglas, P.L., Pal, R., Feng, X., Chen, P., Liu, L., 2006. Blended chitosan and polyvinyl alcohol membranes for the pervaporation dehydration of isopropanol. J. Membr. Sci. 280 (1–2), 815–823.

Tanaka, K., Yoshikawa, R., Ying, C., Kita, H., Okamoto, K., 2001. Application of zeolite membranes to esterification reactions. Catal. Today 67 (1–3), 121–125.

Tanna, N.P., Mayadevi, S., 2007. Analysis of a membrane reactor: influence of membrane characteristics and operating conditions. Int. J. Chem. React. Eng 5, A5.

Taylor, R., Krishna, R., 1993. Multicomponent Mass Transfer. Wiley.

Teli, S.B., Gokavi, G.S., Sairam, M., Aminabhavi, T.M., 2007. Mixed matrix membranes of poly(vinyl alcohol) loaded with phosphomolybdic heteropolyacid for the pervaporation-separation of water–isopropanol mixtures. Coll. Surf. A Physicochem. Eng. Asp. 301, 55–62.

Ten Elshof, J.E., Abadal, C.R., Sekulic, J., Chowdhury, S.R., Blank, D.H.A., 2003. Transport mechanisms of water and organic solvents through microporous silica in the pervaporation of binary liquids. Micropor. Mesopor. Mater. 65 (2–3), 197–208.

Thongsukmak, A., Sirkar, K.K., 2007. Pervaporation membranes highly selective for solvents present in fermentation broths. J. Membr. Sci. 302 (1–2), 45–58.

Urtiaga, A., Gorri, E.D., Casado, C., Ortiz, I., 2003. Pervaporative dehydration of industrial solvents using a zeolite NaA commercial membrane. Sep. Purif. Technol. 32 (1–3), 207–213.

Van Baelen, D., Reyniers, A., Van der Bruggen, B., Vandecasteele, C., Degreve, J., 2004. Pervaporation of binary and ternary mixtures of water with methanol and/or ethanol. Sep. Sci. Technol. 39 (3), 559–576.

Van Baelen, D., Van der Bruggen, B., Van den Dungen, K., Degreve, J., Vandecasteele, C., 2005. Pervaporation of water-alcohol mixtures and acetic acid-water mixtures. Chem. Eng. Sci. 60 (6), 1583–1590.

Van den Broeke, L.J.P., Krishna, R., 1995. Experimental verification of the Maxwell-Stefan theory for micropore diffusion. Chem. Eng. Sci. 50 (16), 2507–2522.

Van den Broeke, L.J.P., Bakker, W.J.W., Kapteijn, F., Moulijn, J.A., 1999a. Transport and separation properties of a silicalite-1 membrane—I. Operating conditions. Chem. Eng. Sci. 54 (2), 245–258.

Van den Broeke, L.J.P., Kapteijn, F., Moulijn, J.A., 1999b. Transport and separation properties of a silicalite-1 membrane—II. Variable separation factor. Chem. Eng. Sci. 54 (2), 259–269.

Van der Bruggen, B., Jansen, J.C., Figoli, A., Geens, J., Boussu, K., Drioli, E., 2006. Characteristics and performance of a 'universal' membrane suitable for gas separation, pervaporation and nanofiltration applications. J. Phys. Chem. B 110 (28), 13799–13808.

Van der Bruggen, B., Drioli, E., Giorno, L., 2010. Pervaporation membrane reactors. In: Comprehensive Membrane Science and Engineering, vol. 3. Academic Press, Oxford, pp. 135–163.

Van Hoof, V., Van den Abeele, L., Buekenhoudt, A., Dotremont, C., Leysen, R., 2004. Economic comparison between azeotropic distillation and different hybrid systems combining distillation with pervaporation for the dehydration of isopropanol. Sep. Purif. Technol. 37 (1), 33–49.

Van Hoof, V., Dotremont, C., Buekenhoudt, A., 2006. Performance of Mitsui NaA type zeolite membranes for the dehydration of organic solvents in comparison with commercial polymeric pervaporation membranes. Sep. Purif. Technol. 48 (3), 304–309.

Van Veen, H.M., Van Delft, Y.C., Engelen, C.W.R., Pex, P.P.A.C., 2001. Dewatering of organics by pervaporation with silica membranes. Sep. Purif. Technol. 22-23, 361–366.

Vane, L.M., 2005. A review of pervaporation for product recovery from biomass fermentation processes. J. Chem. Technol. Biotechnol. 80 (6), 603–629.

Vane, L.M., 2008. Separation technologies for the recovery and dehydration of alcohols from fermentation broths. Biofuels Bioprod. Biorefin. 2 (6), 553–588.

Vankelecom, I.F.J., Depre, D., De Beukelaer, S., Uytterhoeven, J.B., 1995. Influence of zeolites in PDMS membranes of water/alcohol mixtures. J. Phys. Chem. 99, 13193–13197.

Vankelecom, I.F.J., DeKinderen, J., Dewitte, B.M., Uytterhoeven, J.B., 1997. Incorporation of hydrophobic porous fillers in PDMS membranes for use in pervaporation. J. Phys. Chem. B 101 (26), 5182–5185.

Varghese, J.G., Kittur, A.A., Kariduraganavar, M.Y., 2009. Dehydration of THF-water mixtures using zeolite-incorporated polymeric membranes. J. Appl. Polym. Sci. 111 (5), 2408–2418.

Vasse, F., Brun, J.P., Guillou, M., 1975. Pervaporation of binary-mixtures of 1,3 butadiene and isobutene through nitrile rubber films – effect of upstream conditions. Comptes Rendus 281 (24), 1073–1075.

Veerapur, R.S., Patil, M.B., Gudasi, K.B., Aminabhavi, T.M., 2008. Poly(vinyl alcohol)-zeolite T mixed matrix composite membranes for pervaporation separation of water+1,4-dioxane mixtures. Sep. Purif. Technol. 58 (3), 377–385.

Verkerk, A.W., Van Male, P., Vorstman, M.A.G., Keurentjes, J.T.F., 2001. Properties of high flux ceramic pervaporation membranes for dehydration of alcohol/water mixtures. Sep. Purif. Technol. 22–23, 689–695.

Villaluenga, J.P.G., Mohammadi, A.T., 2000. A review on the separation of benzene/cyclohexane mixtures by pervaporation processes. J. Membr. Sci. 169 (2), 159–174.

Volkov, V.V., Fadeev, A.G., Khotimsky, V.S., Litvinova, E.G., Selinskaya, Y.A., McMillan, J.D., Kelley, S.S., 2004. Effects of synthesis conditions on the pervaporation properties of poly[1-(trimethylsilyl)-1-propyne] useful for membrane bioreactors. J. Appl. Polym. Sci. 91, 2271–2277.

Waldburger, R.M., Widmer, F., 1996. Membrane reactors in chemical production processes and the application to the pervaporation-assisted esterification. Chem. Eng. Technol. 19 (2), 117–126.

Wang, Y., Jiang, L.Y., Matsuura, T., Chung, T.S., Goh, S.H., 2008. Investigation of the fundamental differences between polyamide-imide (PAI) and polyetherimide (PEI) membranes for isopropanol dehydration via pervaporation. J. Membr. Sci. 318 (1–2), 217–226.

Wang, L., Han, X.L., Li, J.D., Zhan, X., Chen, J., 2011. Separation of azeotropic dimethylcarbonate/Methanol mixtures by pervaporation: sorption and diffusion behaviors in the pure and nano silica filled PDMS membranes. Sep. Sci. Technol. 46 (9), 1396–1405.

Wijmans, J.G., Baker, R.W., 1995. The solution-diffusion model: a review. J. Membr. Sci. 107, 1–21.

Wöllner, J., Engelhardt, F., 1968. Process for the single stage manufacture of methyl isobutyl ketone. U.S. Patent 3,405,178.

Wood, J., Gladden, L.F., Keil, F.J., 2002. Modelling diffusion and reaction accompagnied by capillary condensation using three-dimensional pore networks. Part 2. Dusty gas model and general reaction kinetics. Chem. Eng. Sci. 57, 3047–3059.

Yang, B.L., Goto, S., 1997. Pervaporation with reactive distillation for the production of ethyl tert-butyl ether. Sep. Sci. Technol. 32 (5), 971–981.

Yao, L.L., Wu, C.M., Yang, Z.J., Qiu, W.Q., Cui, P., Xu, T.W., 2012. Waterborne polyurethane/poly(vinyl alcohol) membranes: preparation, characterization, and potential application for pervaporation. J. Appl. Polym. Sci. 124, E216–E224.

Ying, C., Kita, H., Okamoto, K.-I., 2004. Zeolite T membrane: preparation, characterization, pervaporation of water/organic liquid mixtures and acid stability. J. Membr. Sci. 236 (1–2), 17–27.

Yoshioka, T., Fujiume, J., Murakami, A., 2000. Pervaporation separation of MeOH/MTBE through agarose membranes. J. Membr. Sci. 178, 75–78.

Yu, M., Falconer, J.L., Noble, R.D., Krishna, R., 2007. Modeling transient permeation of polar organic mixtures through a MFI zeolite membrane using the Maxwell–Stefan equations. J. Membr. Sci. 293 (1–2), 167–173.

Zhang, L., Yu, P., Luo, Y.B., 2007. Dehydration of caprolactam-water mixtures through cross-linked PVA composite pervaporation membranes. J. Membr. Sci. 306 (1–2), 93–102.

Zhao, Q., Qian, J., An, Q., Zhu, M.H., Yin, M.J., Sun, Z.W., 2009. Poly(vinyl alcohol)/polyelectrolyte complex blend membrane for pervaporation dehydration of isopropanol. J. Membr. Sci. 343 (1–2), 53–61.

Zhou, H.L., Su, Y., Chen, X.R., Wan, Y.H., 2011. Separation of acetone, butanol and ethanol (ABE) from dilute aqueous solutions by silicalite-1/PDMS hybrid pervaporation membranes. Sep. Purif. Technol. 79 (3), 375–384.

Zhu, Y., Minet, G., Tsotsis, T.T., 1996. A continuous pervaporation membrane reactor for the study of esterification reactions using composite polymeric/ceramic membrane. Chem. Eng. Sci. 51, 4103–4113.

Zielinska, K., Kujawski, W., Chostenko, A.G., 2011. Chitosan hydrogel membranes for pervaporative dehydration of alcohols. Sep. Purif. Technol. 83, 114–120.

CHAPTER FIVE

Liquid – Membrane Filters

Aloke Kumar Ghoshal and Prabirkumar Saha
Department of Chemical Engineering, Indian Institute of Technology Guwahati, Guwahati, India

Contents

Progress in Filtration and Separation
http://dx.doi.org/10.1016/B978-0-12-384746-1.00005-7

GLOSSARY

Bioactive Having direct effects on a living organism
Carrier One that transports or conveys
Environmentally benign Not harmful to the environment or atmosphere or surroundings
Extraction Method of separation of a solute from one phase into the other
Stripping Method of removal of a solute from one phase into the other
Membrane A barrier that selectively transport solutes
Transport Movement from one place to the other

1. INTRODUCTION

In recent times, membrane-based technology has emerged as an efficient process for pre-concentration and separation of chemicals in aqueous solutions (Bunt and Waanders, 2008; Chakrabarty, 2010). A membrane can be defined essentially as a barrier, which separates two fluid phases and restricts transport of various components between these phases in a selective manner. A solute typically diffuses from one phase to the other across the membrane by means of a driving force (concentration gradient, pressure gradient, electrostatic, or chemical potential difference). A membrane can be homogenous or heterogeneous, symmetric, or asymmetric in structure, solid, or liquid. The solid membrane-based separation faces the problem of low flux rate, low selectivity, and large size equipment is needed for their operation (Kamiński and Kwapiński, 2000). The liquid membrane (LM) based separation, on the other hand, is an effective method for selective separation and pre-concentration of trace quantity of solutes and it has been applied in various fields, such as metallurgy, biotechnology, environmental sciences, and material sciences, etc. (Araki and Tsukube, 1990).

The concept of LM was proposed way back in the beginning of last century; when Nernst and Riesenfeld (1901) studied the properties of systems consisting of an oil layer separating electrolyte solutions. LM is a homogeneous, thin film of liquid (membrane phase) interposed between two other liquid phases, viz. feed (or source) phase and receiving (or strip) phase. Feed phase typically contains solute(s) that need to be transported across the thin film LM to the strip phase. The transport of solute across the LM occurs by means of diffusion due to concentration difference. The main advantage of LM over solid membrane is the higher solubility and diffusivity coefficients of compounds in a liquid medium than in a solid one. The membrane liquid must be chosen in such a way that it is immiscible with the feed/receiving phases. Depending on the type of application, the source/membrane/strip phase combination can have either aqueous/organic/aqueous configuration or organic/aqueous/organic configuration.

2. THEORETICAL BACKGROUND OF SOLUTE TRANSPORT THROUGH LM

Solute containing aqueous phase applied for LM-based separation is mostly an ionic solution, where pressure and temperature are usually constant. According to thermodynamics, Gibb's free energy must decrease in any spontaneously occurring irreversible state change. Gibb's free energy is an extensive parameter of a system, i.e., it increases linearly with the size of the process. As it is inconvenient to keep track of size of the process in most calculations, the electrochemical potential is used in lieu of Gibb's free energy. Electrochemical potential is the contribution of 1 mol of the ith constituent of the process to its molar free energy. If n_i is the number of moles of the ith constituent in the process, then

$$G = \sum_i \mu_i^e n_i \tag{1}$$

where μ_i^e is the electrochemical potential of species i (in kJ/kmol). Consider the case of two aqueous phases separated by a membrane. The electrochemical potentials of a solute in the two phases (A and B) are μ_A^e and μ_B^e, respectively. Transport of solutes occurs from regions of higher electrochemical potential to regions of lower potential, i.e., $\mu_B^e < \mu_A^e$.

In generic terms, electrochemical potential can be expressed as a combination of chemical potential and electrostatic potential,

$$\mu_i^e = \mu_i^c + z_i F\phi \tag{2}$$

where μ_i^c is the chemical potential of the species i (in kJ/kmol), z_i is the valency (charge) of the ion i (dimensionless), F is Faraday's Constant (9.65×10^7 C/kmol), ϕ is the local electrostatic potential (kV). In the special case of an uncharged atom, $z_i = 0$ and so $\mu_i^e = \mu_i^c$. The chemical potential of a species can further be denoted as

$$\mu_i^c = \mu_i^{c,\mathrm{std}} + RT \ln C_i \tag{3}$$

where, $\mu_i^{c,\mathrm{std}}$ is the chemical potential in a given standard state. When an ion is acted on by an electric field, it tends to move with a drift velocity (v) that is assumed to be proportional to the force (E) provided by the field. Then

$$u_i = \frac{v}{E} \tag{4}$$

where, u_i is the mobility of the ion. The force (E) that drives ions in solution is assumed to be the spatial gradient of the electrochemical potential and is given by the following equation.

$$\begin{aligned} E &= -\frac{\mathrm{d}\mu_i^e}{\mathrm{d}x} = -\frac{\mathrm{d}}{\mathrm{d}x}\{\mu_i^c + z_i F\phi\} = -\frac{\mathrm{d}}{\mathrm{d}x}\left\{\mu_i^{c,\mathrm{std}} + RT \ln C_{i,x} + z_i \mathrm{F}\phi\right\} \\ &= -RT\frac{\mathrm{d}C_{i,x}}{C_{i,x}\mathrm{d}x} - z_i\mathrm{F}\frac{\mathrm{d}\phi}{\mathrm{d}x} \end{aligned} \tag{5}$$

Again the flux J_i of the ion is the number of moles of ion passing through a unit area per second and is given by the product of the concentration of the ion and its drift velocity. Thus the ion flux driven by an electric field is

$$\begin{aligned} J_i &= C_{i,x}v = C_{i,x}u_iE = C_{i,x}u_i\left\{-RT\frac{\mathrm{d}C_{i,x}}{C_{i,x}\mathrm{d}x} - z_i\mathrm{F}\frac{\mathrm{d}\phi}{\mathrm{d}x}\right\} \\ &= -u_iRT\frac{\mathrm{d}C_{i,x}}{\mathrm{d}x} - z_i\mathrm{F}C_{i,x}u_i\frac{\mathrm{d}\phi}{\mathrm{d}x} \end{aligned} \tag{6}$$

The above equation is named as the *Nernst–Planck* equation. If there is no electrostatic potential, the above equation is reduced to Fick's law of diffusion,

$$J_i = -u_iRT\frac{\mathrm{d}C_{i,x}}{\mathrm{d}x} = -D\frac{\mathrm{d}C_{i,x}}{\mathrm{d}x} \tag{7}$$

where $D = u_iRT$ is the simple diffusivity of uncharged atom through the LM. For a constant concentration gradient, the Nerst–Planck Equation is re-written as

$$J_i = -\frac{D(C_{i,2} - C_{i,1})}{L} - z_i F\left\{C_{i,1} + \frac{(C_{i,2} - C_{i,1})}{L}x\right\}u_i\frac{d\phi}{dx} \tag{8}$$

Correspondingly the Fick's law of diffusion (in case of no electrostatic potential) takes the form of

$$J_i = -\frac{D(C_{i,2} - C_{i,1})}{L} \tag{9}$$

The transport through LM is highly dependent on the viscosity of the membrane phase as diffusivity according to Stokes Einstein equation, is inversely proportional to viscosity (Mulder, 1991), i.e.,

$$D = \frac{kT}{6\pi\eta r} \tag{10}$$

where, k is the Boltzmann constant, η is the viscosity of the organic phase, and r is the molecule radius. Hence, the viscosity of the membrane phase should be low for higher transport.

3. MECHANISM OF TRANSPORT OF SOLUTE IN LM-BASED SEPARATION

The transport mechanisms in LM-based separation can be categorized into two major types, viz. passive transport and active transport.

3.1 Passive Transport

Passive transport is a movement of ions/atoms/molecules across membranes that is driven by the growth of entropy of the system and does not require any chemical energy. Passive transport through LM involves solution-diffusion or facilitated diffusion mechanisms.

Solution-diffusion is the phenomenon by which components first get dissolved and then moves towards lower concentration zone due to concentration gradient (Figure 1). Diffusion continues until this gradient is

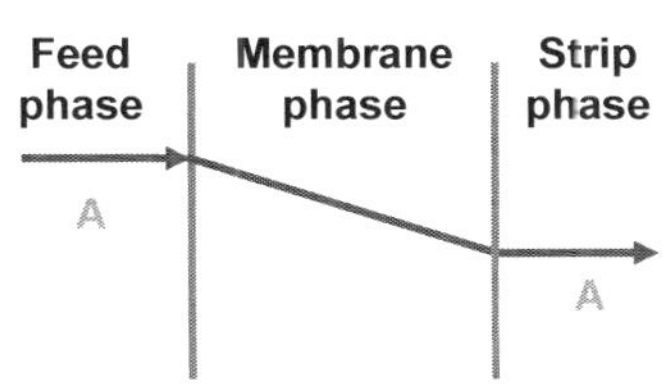

Figure 1 Ordinary diffusive transport of component A through liquid membrane.

eliminated (Ellinghorst et al., 1989). The rate of mass transfer in this case is usually low and depends on the solubility of solute in the organic as well as strip phase.

Facilitated diffusion (or carrier-mediated diffusion) is the movement of molecules across the cell membrane via special carrier agents embedded within the LM. Solute molecules bind with the carrier agents, and the complex move through the membrane. Facilitated diffusion is also a passive process as the solutes move down the concentration gradient without using energy. To increase the rate of mass transfer or efficiency of the LM separation, a carrier agent is added to the membrane phase. The carrier should be soluble only in membrane phase and should have the ability to form complex reversibly with the solute. The mechanism is represented schematically through Figure 2(a). This is called the uniport mechanism because a single component is transferred through the LM. Here the transport of component A is enhanced by the presence of the carrier molecule C. The carrier C forms a complex AC at the feed/membrane interface. Complex AC then diffuses through the membrane due to concentration gradient across the membrane

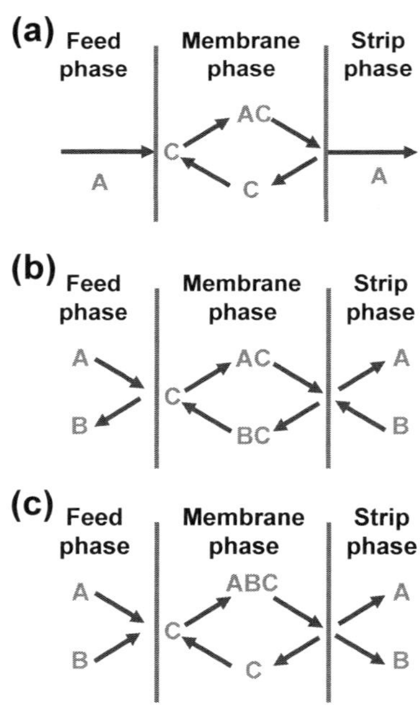

Figure 2 (a) Mechanism of carrier mediated (or facilitated) transport in liquid membrane (LM) with mobile carrier. (b) Mechanism of coupled counter transport in LM. (c) Mechanism of coupled co-transport in LM.

and releases the solute A at the membrane/strip interface. The free carrier C then diffuses back to the feed/membrane interface due to concentration gradient and the cycle continues. In this case two processes occur simultaneously. Part of component A is transported by free diffusion (i.e., solution-diffusion mechanism) whilst other part is transported due to the formation of solute–carrier complex that enhances the solubility of the solute A in the membrane phase. Hence, the transport rate is increased. One basic feature of carrier-mediated transport is that the complexation reaction must be reversible. Otherwise solute transport would stop when all the carrier molecules would have formed complex with the solute. Secondly, the affinity between the carrier and solute should not be very strong or very weak. A strong complex, i.e., one exhibiting high affinity between the carrier and solute may result in slow release at the membrane/strip interface while a weak complex, i.e., one exhibiting low affinity between the carrier and solute would yield limited facilitation. Therefore, there should be optimum bond energies of this reversible complex. This bond energy is recommended to be in the range of $1–5 \times 10^4$ kJ/kmol (Mulder, 1991). Another feature of facilitated transport is the occurrence of two different processes, i.e., chemical reaction and mass transfer at the same time.

3.2 Active Transport

In active transport, the movement of a substance across a membrane occurs against its concentration gradient (from low to high concentration). This is known as uphill transport. Kedem (Kedem and Katchalsky, 1961; Kedem, and Essig, 1965) has proposed a more general definition. According to him, active transport is accomplished only by the cross-coupling of the flux of species, *i* with that of other species or the chemical reaction, and the driving force is supplied by the free energy change of the coupled processes (Araki and Tsukube, 1990). There are two types of active transport: primary and secondary.

Primary active transport, also called direct active transport, directly uses chemical energy (such as from adenosine triphosphate or ATP in case of cell membrane) to transport all species of solutes across a membrane against their concentration gradient. Uptake of glucose in the human intestines is an example of primary active transport. Other sources of energy for primary active transport are redox energy (chemical reaction such as oxidation and reduction) and photon energy (light). An example of primary active transport using redox energy is the mitochondrial electron transport chain that uses the reduction energy of NADH (nicotinamide adenine dinucleotide, reduced form) to move protons across the inner mitochondrial membrane against their concentration gradient. An example of primary active transport using light energy is the proteins involved in photosynthesis.

Secondary active transport, on the other hand, allows one solute to move downhill (along its electrochemical potential gradient) in order to yield enough entropic energy to drive the transport of the other solute uphill (from a low concentration region to a high one). This is also known as coupled transport, as opposed to noncoupled or uniport transport where transport of a single component is facilitated. There are two main forms of coupled transport: antiport and symport. In antiport two species of ion or other solutes are pumped in opposite directions across a membrane (Figure 2(b)) and in symport transport two species move in the same direction (Figure 2(c)).

Cytosis is an active transport mechanism for the movement of large quantities of molecules into and out of biological cells. There are two types of cytosis: exocytosis and endocytosis. A cell directs the contents of secretory vesicles out of the cell membrane in exocytosis. The vesicles fuse with the cell membrane and their content, usually protein, is released out of the cell. In endocytosis, a cell absorbs molecules, such as proteins, from outside the cell by engulfing it with the cell membrane.

4. TYPES OF TRANSPORT OF SOLUTE IN LM-BASED SEPARATION

There are four basic types of transport systems, viz. cation transport, anion transport, neutral transport and switchable transport, each of which has its own mechanisms and carrier types. In each of these types, it has to be noticed that regardless of mechanism of their complex formation, charge-neutrality must be maintained.

4.1 Cation Transport

Cation transport can occur in two ways, symport or antiport, as shown in Figures 3 and 4. In the symport configuration, a neutral carrier moves the target cation and co-transported anion together across the membrane. This occurs in several stages as described below:

- Target cation and co-transported anion diffuse from bulk of feed phase to the feed/membrane interface.
- At the feed/membrane interface, the ions are complexed with the carrier.

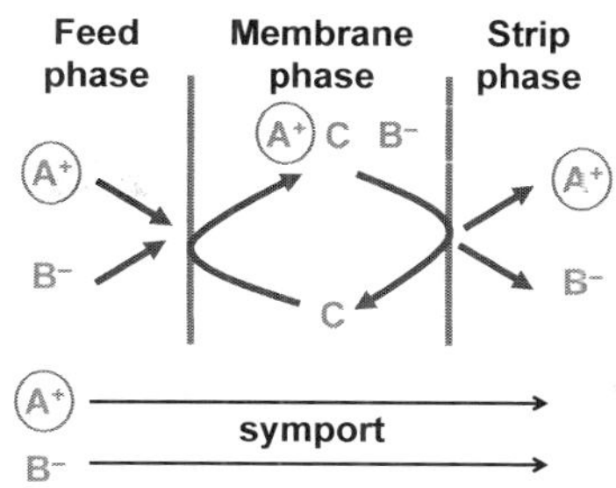

Figure 3 Cationic symport.

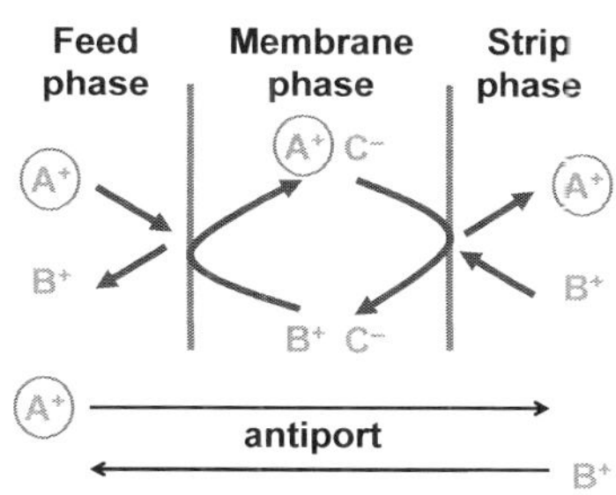

Figure 4 Cationic antiport.

- That complex diffuses across the membrane.
- Decomplexation occurs at the strip–membrane interface.
- The ions are released to the strip phase which eventually diffuse to the bulk of strip phase
- The carrier diffuses back across the membrane, to the feed–membrane interface.

Removal of Cu^{++} through LM using 2-hydroxy-5-dodecylbenzaldehyde (2H5DBA) as a carrier is a typical example of such type of transport. The carrier contains hydroxyl and aldehyde groups as ion-binding groups that facilitate the transport. The complexation (extraction) and decomplexation (stripping) reactions are given below (Monlinari et al., 2006):

$$\text{Extraction}: \quad Cu^{++}_{(aq)} + 2HA_{(org)} \rightleftharpoons CuA_{2(org)} + 2H^{+}_{(aq)} \tag{11}$$

$$\text{Stripping}: \quad CuA_{2(org)} + 2H^{+} \rightleftharpoons 2HA + Cu^{++} \tag{12}$$

where, HA is the acidic extractant.

For the antiport transport, an anionic carrier is used and the stages are bit different:

- Target cation diffuses from bulk of feed phase to the feed–membrane interface.
- "Counter transportable" cation diffuses from bulk of strip phase to the strip–membrane interface.
- At the feed–membrane interface, the anionic carrier forms a neutral complex with the target cation.
- The complex diffuses across the membrane.
- Cation-exchange reaction occurs at the strip–membrane interface, whereby target cation is released to the strip phase and "counter transportable" cation forms a new neutral complex with the anionic carrier.
- The new neutral complex is counter-transported (diffuses back) across the membrane.
- Decomplexation reaction occurs whereby counter-transported cation is released to the feed phase and the anionic carrier gets ready to repeat the cycle.

Transportation of Co^{++} through LM using di-(2-ethylhaxyl)-phosphoric acid (D2EHPA) as a carrier is a typical example of such type of transport where H^+ acts as a counter-transported ion. The relevant reaction is given below (Leon and Guzman, 2005):

$$Co^{++}_{(aq)} + 2(HR)_{2(org)} \rightleftharpoons CoR_2(HR)_{2(org)} + 2H^+ \qquad (13)$$

The transport mechanism is coupled counter-ion transport with cobalt (II) and proton moving in the opposite direction.

4.2 Anion Transport

As it can be evident in Figures 5 and 6, anionic transport is similar in mechanism to the cationic transport, except use of a cationic carrier instead of an anionic one in the antiport configuration.

Typical example of symport anion transport is the extraction of Hg (II) ions through LM in presence of tri-octylamine (ToA) as a carrier (Li et al., 1996). The transport steps are given as below:

$$\begin{aligned} \text{Extraction:}\quad & Hg^{++} + 4Cl^- \rightleftharpoons HgCl_4^{-2} \\ & R_3N_{(org)} + H^+Cl^-_{(aq)} \rightleftharpoons R_3NHCl_{(org)} \\ & 2R_3NH^+Cl^-_{(org)} + HgCl_4^{--} \rightleftharpoons (R_3NH)_2HgCl_{4(org)} + 2Cl^-_{(aq)} \end{aligned} \qquad (14)$$

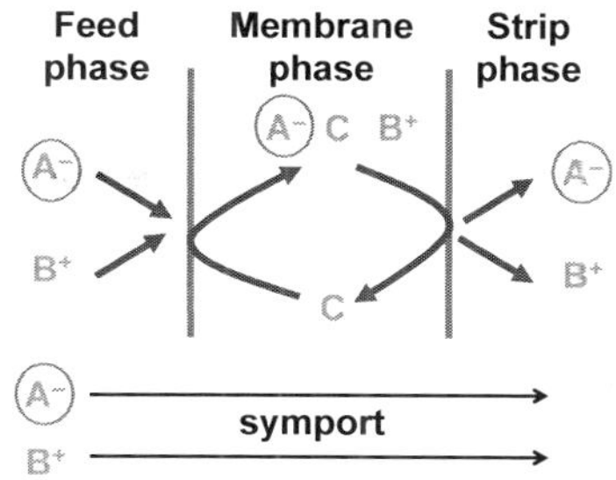

Figure 5 Anionic symport.

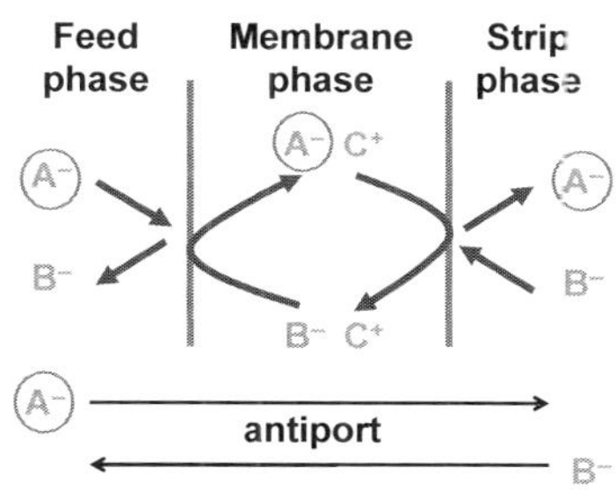

Figure 6 Anionic antiport.

$$\text{Stripping: } (R_3NH)_2HgCl_{4(org)} + OH^- \rightleftharpoons R_3N_{(org)} + HgCl^{--}_{4(aq)} + 2H_2O \quad (15)$$

Typical example of antiport anion transport is the separation of ligno-sulphonate (LS) through LM (Kontturi et al., 1990a). The transport steps are given as below:

$$\begin{aligned}\text{Extraction: } & R_3N + HCl \rightleftharpoons R_3NH^+Cl^- \\ & R_3NH^+Cl^- + LSNa_n \rightleftharpoons R_3NHLSNa_{(n-1)} + NaCl\end{aligned} \quad (16)$$

$$\text{Stripping: } R_3NHLSNa_{(n-1)} + NaCl \rightleftharpoons R_3NH^+Cl^- + LSNa_n \quad (17)$$

A schematic representation of the above transport mechanism is available elsewhere (Chakraborty, 2010).

4.3 Neutral Transport

In this type of transport, a neutral molecule is transported through an LM using suitable carrier. It is channelized via uniport mechanism. Figure 7 presents a schematic of this type of transport. Transport of oxygen from a

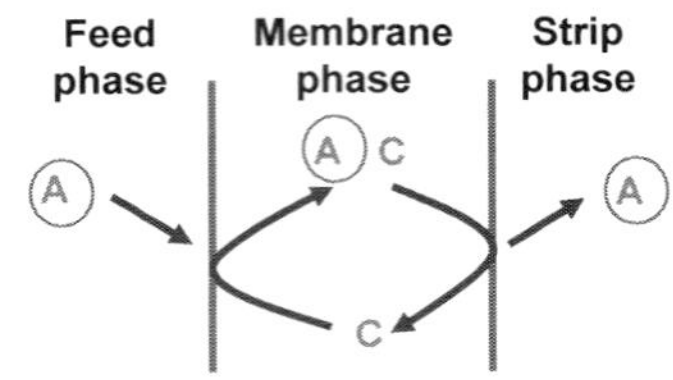

Figure 7 Neutral transport.

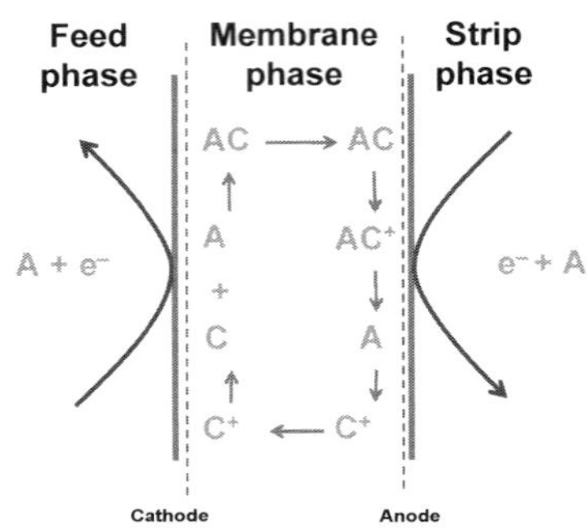

Figure 8 Switchable transport.

mixture of oxygen and nitrogen through a water film in presence of cobalt as a carrier is a good example of neutral transport.

4.4 Switchable Transport

The use of photochemistry and electrochemistry has been proved to be increasing the rates at which the carrier complexes dissociate, which would, therefore, increase the transport rate. A sample of this is shown in Figure 8. Essentially, the switchable transport system works in addition to the regular transport system:

$$AC - e \rightleftharpoons AC^{+} \rightarrow A + C^{+} \tag{18}$$

Only the second step of the reaction is accelerated.

5. CARRIER

Carriers are reagents that play an important catalytic role in the LM system. Small amount of carrier is sufficient for the above purpose because of the small solvent inventory associated with the membrane and also because of their nonvolatile nature.

5.1 Advantages of Carriers

Carriers attribute to selective separation and concentration of specific solutes with high flux of transport. Carriers are characterized by

- ability to quick bind and release particular solute(s)
- ability of selective and reversible binding of component(s) in the solution
- lack of ability to bind with solvent
- lack of ability to coalesce
- nontoxicity.

Careful integration of a suitable carrier into the LM phase is very important in order to enhance the effective solubility of the species to be transported in the membrane phase.

5.2 Types of Carrier

Carriers are categorized into three major types, viz. acidic, basic and neutral, primarily on the basis of their functional groups.

5.2.1 Acidic Carrier

The acidic carriers are most effective for extracting the cations as the carriers do form complexed salt with cations with the exchange of protons. The acidic carriers do have COOH, P(OH), SO_3H or chelating groups. A typical example of acidic carrier-mediated transport is presented by the following equation (Gherrou et al., 2002):

$$Cu^{++} + 2C_{16}H_{34}O_3P(OH) \rightleftharpoons C_{16}H_{34}O_3P(OCu) + 2H^+ \tag{19}$$

where, copper is extracted from its aqueous solution with di(2-ethylhexyl) phosphoric acid.

5.2.2 Basic Carrier

The alkaline/basic carriers are used for the extraction of the anionic metal complexes. Amines are ideal examples of such carriers. Typical example of basic carrier-mediated transport is presented by the following equation (Warey, 2007):

$$CdCl^{--}_{4(aq)} + 2H^{+}_{(aq)} + 2R_3N_{(org)} \rightleftharpoons ((R_3NH)_2)CdCl_{4(org)} \tag{20}$$

where Cd (II) is transported in the form of $CdCl_4^{--}$ from the aqueous acidic solution by the carrier trioctylamine (TOA). The TOA (R_3N) accepts a proton to form a positively charged species and cadmium ion is extracted as $((R_3NH)_2)CdCl_4$.

5.2.3 Neutral Carrier

Neutral carriers are generally used as cation carriers in LMs for the selective transport of different metal ions. They carry the metal ions by encapsulating them within their cavity. The extraction efficiency of such neutral carriers depends on the size of their cavity and the size of the inserted ions. When such carriers are used as metal ion carriers in the membrane phase, the concurrent transport of an anion with the cation occurs across the LM. Organic phosphoryl compounds and macrocyclic molecules are the most widely used neutral carriers in LM processes. Typical example of such type of carrier-mediated transport is presented by the following equation (Bachiri et al., 1996):

$$Ag^+_{(aq)} + NO^-_{3(aq)} + C_{20}H_{36}O_6 \rightleftharpoons Ag\left(C_{20}H_{36}O_6\right)NO_{3(org)} \qquad (21)$$

where, silver ion is co-transported along with the nitrate anion from the aqueous solution by the neutral carrier di-cyclohexanone-18-crown-6 ($C_{20}H_{36}O_6$).

5.3 Choice of Carrier

The selection of a right carrier is the key factor in all types of carrier-mediated LM. The carrier should be very specific and should have high selectivity towards the solute to be removed. The carrier may be mobile or fixed. If it is dissolved in the liquid it is called mobile carrier. On the other hand, the carrier can be bound chemically or physically (fixed carrier) to a solid polymer, which is termed as polymer inclusion membrane. In mobile carrier system, the carrier solute complex diffuses across the membrane whereas in fixed carrier system the solute jumps from one site to the other. The diffusivity in mobile carrier is much higher than the fixed carrier system. The most frequently used carriers include phospho-organic compounds, crown ethers, hydro-oximes and amines (Kamiński and Kwapiński, 2000). Typical carriers used in various LM processes are listed in Table 1. Various combination of LM systems employed in the extraction of different materials in anionic form are presented in Table 2.

Table 1 Various carriers used in various LM processes

Carrier Type	Carriers	Application in the extraction of	Literature cited
Acidic	Acetylacetone	Cr (III)	Sekine et al., 1988
	Benzoylacetone	Ur (VI)	Rajan and Shinde, 1996
	β-Hydroxime	Cu (II)	Sengupta et al., 1990
	D2EHPA[di(2-ethylhaxyl) phosphoric acid]	Ag (I), Cu (II)	Gherrou et al., 2002
	Cyanox 272 (R_2POH)	Co (II)	Swain et al., 2007
	Trioctylphosphinoxide	As (V)	Pérez et al., 2007
Basic	Primene JMT [$(C_8H_{17})NH_2$]	Mo (VI)	Neková and Schrotterova, 1999
	Trioctylamine [$(C_8H_{17})_3N$]	Cr (VI)	Kumbasar 2008
	Trinonylamine [$(C_9H_{19})_3N$]	Lignosulfonate	Kontturi et al., 1990a
	Trilaurylamine	Lignosulfonate	Kontturi et al., 1990b
	Alamine 336 [$R_3N(R:C_8-C_{10})$]	Citric acid	Yordanov and Boyadzhiev, 2004
	Aliquat 336 [$CH_3(C_8C_{17})_3N$]	Cd (II)	Lv et al., 2007
Neutral	Di-benzo-18-crown-6	Na^+, Ur (VI)	Goyette et al., 2003
	Di-cyclohexyle-18-crown-6	Hg (II)	Jabbari et al., 2001
	Cryptana (2,2,1) [N_2O_5]	Ag (I), Cu (II) and Zn (II)	Arous et al., 2004
	Cryptand(2,2,2) [N_2O_6]	Ag (I), Cu (II) and Zn (II)	Arous et al., 2004

Table 2 LM system employed in the extraction of anions

Materials extracted	Carrier used	Diluent	Stripping agent	References
$HgCl_4^{2-}$	Trioctylamine (TOA)	Toluene	NaOH	Li et al., 1996
Hg(picrate)$^-$	Tetrathia-12-crown-4	Chlorofom	EDTA	Shamsipur et al., 2002
Hg(picrate)$^-$	Bis-calixarene-nitrile derivative	Chlorofom	H_2O	Alpoguz et al., 2002
$Cr_2O_7^{2-}$	Cyanex 923	Xylene	NaCl	Alguacil et al., 2005
$Cr_2O_7^{2-}$	Tri *n*-butylphosphate	Hexane	NaOH	Muthuraman et al., 2009
Lignosulfonate	TOA	Decanol	NaOH	Kontturi et al., 1990
Lignosulfonate	Trilaurylamine	Decanol	NaOH	Kontturi et al., 1990
NO_3^-	TOA	Kerosene	NaOH	Lin and Long, 1997

6. SOLVENTS

Ideally the solvent phase of the LM setup should be the thinnest possible strip that provides least resistance for solute transport, yet is stable enough to resist accidental channellingchannelling between feed and strip phases. Mass transfer across membrane phase is inversely proportional to its viscosity.

6.1 Choice of Solvents

Selection of suitable solvent (for membrane phase) is one of the keys issues of LM-based separation. Volatile solvents are not preferred due to instability of the LM for its quicker evaporation alongside the environmental pollution. Low viscous, nonvolatile solvents are ideal for LM applications. The efficiency of solvents is characterized by its distribution coefficient. The distribution coefficient is defined as the ratio of the amounts of target solute present in the organic phase and the aqueous phase after equilibrium is reached. A very high distribution coefficient indicates excellent extraction and vice versa. Strong extractant will not release the solute easily at the strip–membrane interface. Therefore, thorough investigation is necessary to identify a suitable solvent–carrier combination that would give the best possible separation for a target solute.

6.2 Types of Solvents

Solvents can be classified into three major types:

- *Conventional organic chemicals*: dichloroethane, *n*-heptane, kerosene, dodecane, xylene, chloroform, toluene, carbon tetrachloride, etc.
- *Environmentally benign solvents*: coconut oil, sunflower oil, mustered oil, soybean oil, linseed oil, olive oil, sesame oil, etc.
- *Ionic liquids (IL)*: butyl-3-methylimidazolium hexafluorophosphate ([BMIM] [PF6] or [C4MIM] [PF6]), 1-butyl-3-methyl-imidazolium bis(trifluoromethanesulfonyl)imide [BMIM] [Tf2N], etc.

Most of the conventional organic chemicals are toxic, flammable and volatile in nature. In recent times, the research activities on extraction processes have shifted towards echo-friendly solvents, due to the stricter environmental regulations (Chakrabarty et al., 2010a,b). The environmentally benign solvents, such as vegetable oils, are also termed as green solvents. ILs may be another suitable alternative for the conventional volatile organic solvents, because of their novel properties (Chakrabarty, 2010). An IL is a salt which remains in molten state at room temperature. While ordinary liquids such as water are predominantly made of electrically neutral molecules, an IL is largely made of ions and short-lived ion pairs. They are colourless, designable, nonvolatile, nonflammable, highly solvating, water-immiscible and noncoordinating. The thermal stability and miscibility of ILs depend on the anion and the length of the alkyl chain in cation and some other properties, such as viscosity, surface tension and density. IL does not require any carrier agent for facilitation of solute transport.

7. TYPES OF LM

Based on the configurations, LMs are broadly classified into three types: bulk liquid membrane (BLM), emulsion liquid membrane (ELM) and supported liquid membrane (SLM). According to the geometry SLM is again of three types, flat sheet, hollow fibre and spiral wound. Besides these, one more LM is included in the LM family (Figure 9), i.e., electrostatic pseudo LM.

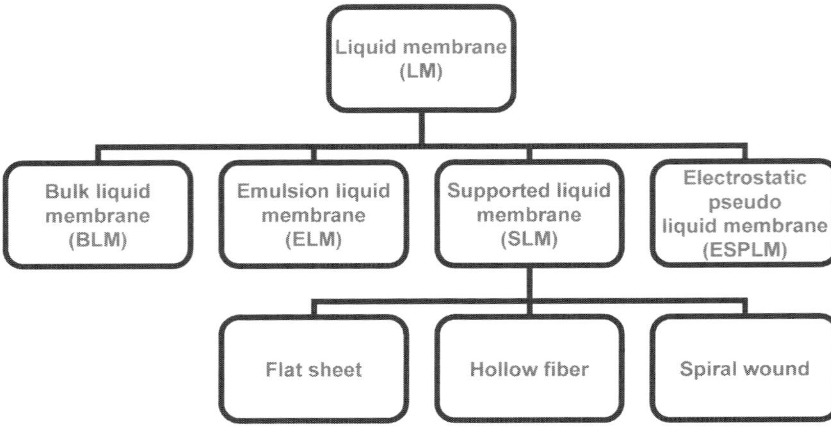

Figure 9 Family of liquid membranes.

7.1 Bulk Liquid Membrane

BLM is the simplest of all LMs where all three phases of the LM unit are placed in a vessel in bulk quantities. Two aqueous phases are separated with a solid barrier (e.g. a glass wall) and the LM shares an interface with both these aqueous phases as shown in the schematic representations (Figure 10). Figure 10(a) shows the case when membrane phase is heavier

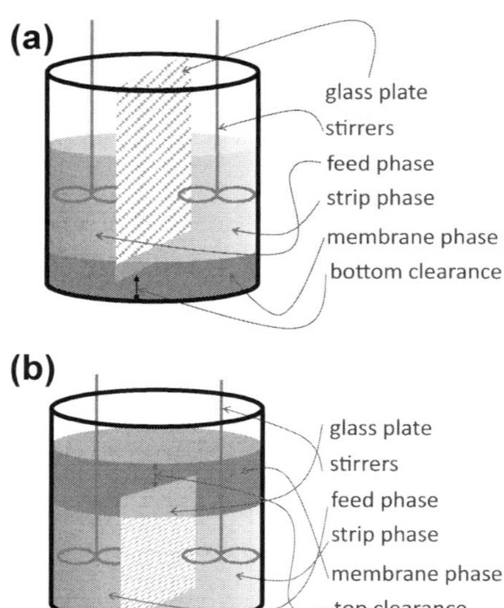

Figure 10 (a) Schematic of bulk liquid membrane (BLM) when membrane liquid is heavier than both feed and strip phases. (b) Schematic of BLM when membrane liquid is lighter than both feed and strip phases.

than both feed and strip phases whereas Figure 10(b) indicates otherwise. BLM is basically used as an important and relevant lab scale study because experimentation using BLM helps understanding the feasibility of separation using LM technique for any system of concern.

7.1.1 Merits and Demerits

Various important equilibrium and kinetic parameters, useful for mass transfer studies, can also be estimated through simple experimental set-up like BLM. However, applicability of BLM in higher scale application suffers due to lower value of surface area to volume ratio. For the same reason, scaling up of the lab scale BLM to a pilot/commercial scale happens to be practically infeasible.

7.1.2 Stability

BLMs are quite stable if the fabrication of the setup is done carefully. Care should be taken so that there is no leakage between the compartments. A simple "dye test" can be useful in detecting possible channelling between the compartments (Chakrabarty et al., 2009b). Moreover stirring of the aqueous phases should be optimally maintained so that the mixing is proper yet the chance of accidental mixing of two aqueous phases is less.

7.2 Emulsion Liquid Membrane

ELM (also known as liquid surfactant membrane) is prepared by forming an emulsion between two immiscible phases (membrane phase and strip phase) and dispersing the emulsion into a third continuous phase (feed phase) by agitation. Therefore, ELM is also referred to as double emulsion. A simplified diagram of ELM is shown in Figure 11. The membrane in ELM is the liquid phase that separates the encapsulated internal droplets

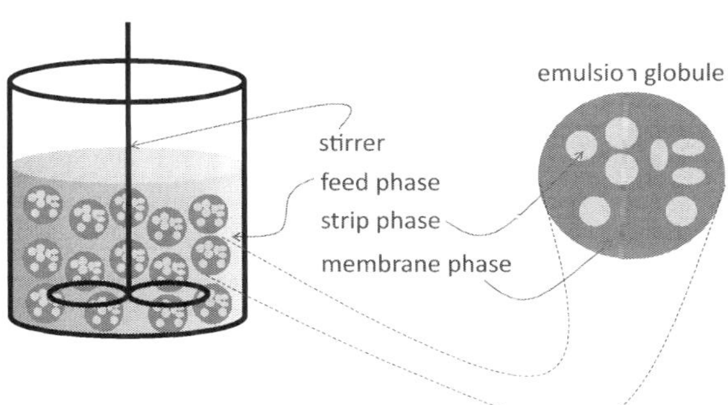

Figure 11 Schematic of emulsion liquid membrane.

in emulsion from the continuous phase. The encapsulated internal phase and the continuous phase are miscible with each other, but a stable membrane phase keeps them separated. The globules of emulsion are typically 10 μm to 1 mm diameter in size. The encapsulated internal droplets of receiving phases have the typical diameter of 1–10 μm. ELM is of two types, water in oil in water (W/O/W) type or oil in water in oil (O/W/O) type. In W/O/W type, oil acts as the membrane phase and in O/W/O type water acts as the membrane phase. ELM was invented by Norman N. Li in 1968 and this process has already been used in industrial scale (Li, 1968; Ho and Sirkar, 2001).

7.2.1 Merits and Demerits

ELM process provides large mass transfer area because large numbers of emulsion globules can be dispersed in a small volume of continuous phase. Therefore, rapid mass transfer occurs in ELM process from external continuous phase to dispersed internal phase. The other advantages of ELMs are (Othman et al., 2004)

- High diffusive flux and low energy consumption,
- Simple configuration/equipment,
- Zero disposal,
- Recycling of material,
- Operational simplicity, and
- High efficiency.

Although an ELM process offers high mass transfer efficiency, it has some limitations too. Demulsification step is usually energy intensive. In addition, in ELM process, surfactant is employed to maintain membrane integrity. Once the surfactant is dissolved in the solution it is difficult to remove it from the solution. This makes the process complicated. In order to recover the receiving phase and in order to replenish the carrier, the emulsion has to be demulsified. This is a difficult task, and consumes a good amount of energy. Emulsion swelling is another troublesome aspect in ELM system, which dilutes the internal phase and makes it difficult to recycle (Ho and Sirkar, 2001).

7.2.2 Stability

In order to maintain the stability of the emulsion during extraction process, the membrane phase is enriched with some surfactants and/or additives as stabilizer. An ionized surfactant has relatively high water solubility and thus generally makes W/O/W type emulsion. A nonionic surfactant, however, can be used to make either type of emulsions. An appropriate selection of surfactant can be carried out by using its hydrophylic-lipophylic balance (HLB) value. HLB is the percentage of hydrophilic functional groups in the surfactant molecule divided by five Surfactants with a low HLB are more lipids affine and thus, tend to make an C/W/O emulsion while those with a high HLB are more hydrophilic in nature and thus, tend to make a W/O/W emulsion. A blend of two or more nonionic surfactants yields better stability in emulsion than a single surfactant does. Internal droplets in emulsion with 1–3 μm in diameter yield good emulsion stability (Zhou, 1996). Kopp suggested (Marr and Kopp, 1982) the following set of guidelines for the formation of stable water in oil emulsions (Noble and Way, 1987).

- Organic phase soluble surfactant concentration should be between 0.1 and 5 wt.%
- Organic phase viscosity should be between 30 and 1000 mPa s
- Volume ratio of the internal phase to membrane phase should be between 0.2 and 2.
- Volume ratio of internal phase to continuous external phase should be between 0.2 and 0.05.
- Volume ratio of continuous phase to emulsion phase should be between 1 and 40.
- Surfactant HLB value should be between 6 and 8.

7.3 Supported Liquid Membrane

SLM or commonly known as immobilized LM is an LM setup in which the membrane liquid (organic or aqueous) is immobilized within the pores of porous solid membrane that separates feed and strip phases. An SLM consists of three main parts, viz. support material, solvent (membrane liquid) and carrier agent. The membrane phase is prepared by dissolving the carrier agent into the solvent. The porous support material serves as a supporting framework for the membrane phase. The porous support can be inorganic or organic (polymer) with suitable chemical properties and mechanical stability. The surface porosity and overall porosity of such support materials should be high in order to obtain a high permeation rate. Permeation rate also depends upon the thickness of membrane because the flux is inversely proportional to the membrane thickness (Mulder, 1991). Therefore, support should be as thin as possible to obtain high flux. Based on the geometry of the supports, SLM is in general of three types, viz. flat sheet, hollow fibre, and spiral wound.

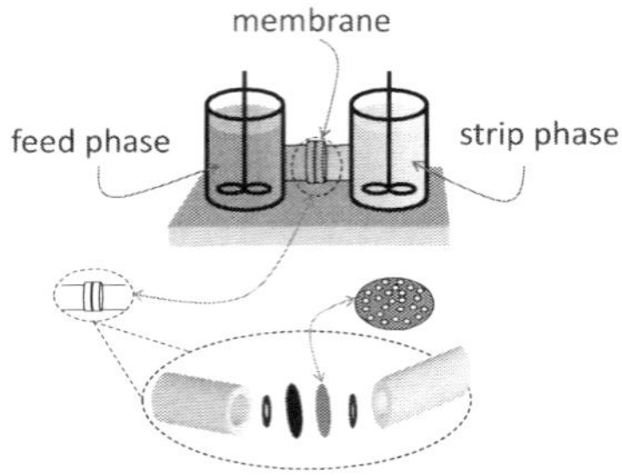

Figure 12 Schematic of flat sheet supported liquid membrane.

7.3.1 Flat Sheet SLM

Flat sheet SLM, as the name implies, has support material in flat sheet form. It is simple in structure, having low cost, but requires large space. A schematic of flat sheet SLM is shown in Figure 12.

7.3.2 Hollow Fibre SLM

Hollow fibre SLM (HFSLM) acts similar to a shell and tube heat exchanger, thus compact with increased mass transfer area per unit volume (Ho and Sirkar, 2001). A schematic of HFSLM is shown in Figure 13. It has an outer shell and there are large numbers of thin porous fibres in the shape of tubes with ID 0.5–1.0 mm (Warey, 2007) running inside the shell along its length, all in neatly ordered rows. The pores of the fibres are filled with the membrane phase. The feed phase and strip phase are separately passed through the channels (fibres and shell).

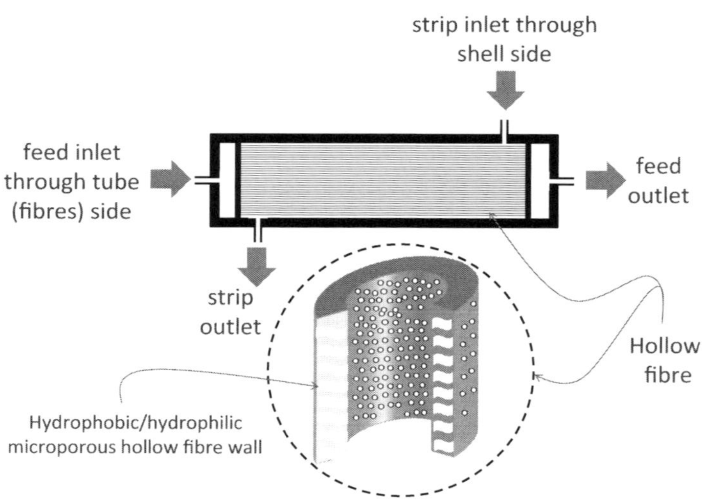

Figure 13 Schematic of hollow fibre supported liquid membrane.

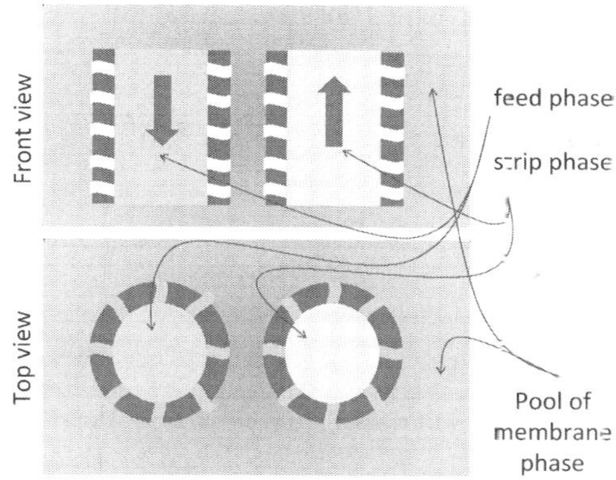

Figure 14 Schematic of Hollow Fibre Contained Liquid Membrane.

7.3.3 Hollow fibre Contained SLM

Hollow-fibre contained SLM (HFCSLM) consists of a cylindrical shell as shown in Figure 14. The shell side is filled with the membrane liquid connected to an external membrane liquid reservoir. Two channels of hollow porous fibres, with aqueous phases (i.e., feed and strip phases) flowing through them, are immersed in the stationary membrane phase of the shell side. Pressure in the shell side is maintained higher than the pressure of both feed and strip solution so that the liquid of aqueous phases does not leach out of the fibres. In this type of configuration, loss of membrane liquid is automatically replenished from the external membrane liquid reservoir.

7.3.4 Spiral Wound Membrane

The spiral wound membrane as shown in Figure 15 is in fact a plate and frame system wrapped around a central perforated collection pipe, in a

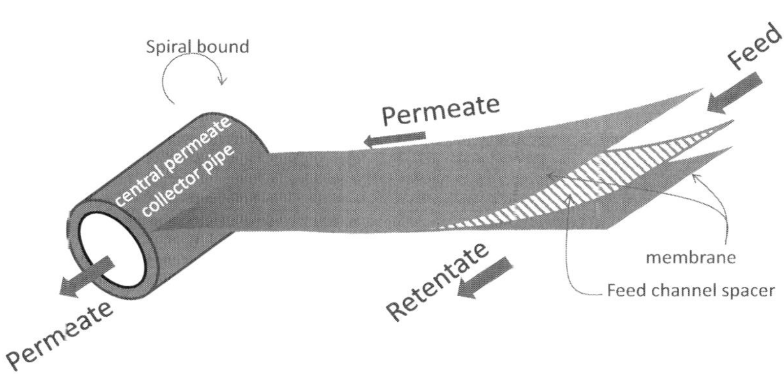

Figure 15 Schematic of spiral wound supported liquid membrane.

similar fashion to a sandwich roll (Ho and Sirkar, 2001). The sandwich is made of four layers; a membrane, a feed channel, another membrane, and a strip channel which forces all the separated materials towards the perforated tube in the centre. The feed flows axially through the cylindrical module parallel along the central pipe, whereas, the permeate flows radially toward the central pipe. This type of SLM is an intermediate between flat sheet and hollow fibre membrane in terms of stability and surface area per unit volume.

7.4 Electrostatic Pseudo Liquid Membrane

Electrostatic pseudo liquid membrane (ESPLM) is a combination of an electrostatic technique and the principle of LM, developed by Gu in 1988 (Gu, 1988; Ho and Sirkar, 2001). An ESPLM tank, as shown in Figure 16, is filled with continuous organic phase. The tank is divided into two compartments, viz. extraction cell and stripping cell, which are separated with two narrowly spaced porous baffles that contains an oil layer between them that acts as an LM. The cells have inlet and outlet pipes that allow flow of aqueous phases (feed solution and stripping solution) through them. Oil layer allows the transport of carrier complex from the extraction cell to the stripping cell and regenerated carrier from the stripping cell to the extraction cell, while mixing of the aqueous phases is prevented. A high voltage (AC) electrostatic field is applied simultaneously across the extraction and stripping cells. Under the electric field, the aqueous phases are dispersed into large numbers of droplets in the continuous organic phase. In the extraction cell, the solute in the aqueous droplets is

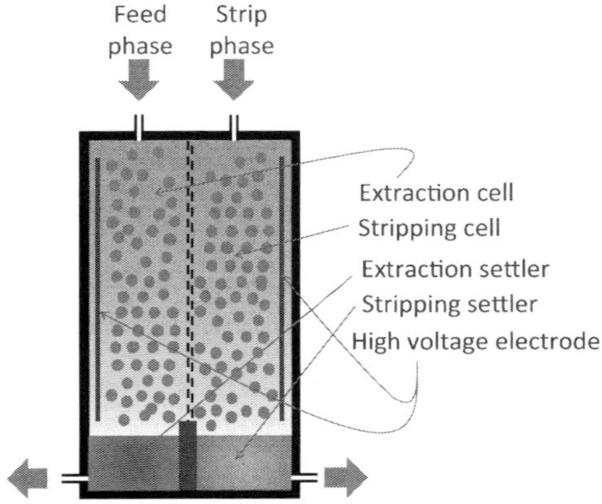

Figure 16 Schematic of electrostatic pseudo liquid membrane.

extracted into the organic phase (Gu, 1990). The complex formed in the extraction cell, driven by its own concentration gradient, diffuses through the perforated baffle plate into the stripping cell. The extractant is regenerated after the solute is stripped off in the stripping cell and diffuse back to the extraction cell through the perforated baffle plate.

7.5 Hollow Fibre Renewal Liquid Membrane

Hollow fibre renewal liquid membrane (HFRLM) setup (Zhongqi et al., 2007) is based on the surface renewal theory. HFRLM setup, as shown in Figure 17, consists of hydrophobic and porous hollow fibres, the pores of which are filled with the organic phase. The lumen side of hollow fibres is filled with aqueous phase in which organic phase droplets are dispersed uniformly and the shell side of the fibres is filled with another aqueous phase. As the fibres are hydrophobic and organo-affine, a thin organic film is developed in the lumen side of the fibres that acts as an LM. The surface of the LM layer is peeled off due to the sheer force of the flowing fluid through the lumen, however, the surface gets renewed with the organic droplets present in the aqueous phase of the lumen side. The renewal of the surface reduces the mass transfer resistance in the aqueous boundary layer in the lumen side and thereby accelerates the mass transfer rate. Hence, the additional organic phase in the lumen side helps maintaining the continuity of the LM layer and replenishment of the loss of the membrane liquid.

7.5.1 *Merits and Demerits*

SLM-based separation process has several advantages such as low capital, operating, maintenance and energy costs, high separation factor, low extractant consumption and easy to scale-up. However, the major drawback of SLM is the instability of the liquid film due to the gradual loss of membrane solution by dissolution and spontaneous emulsification during the operation.

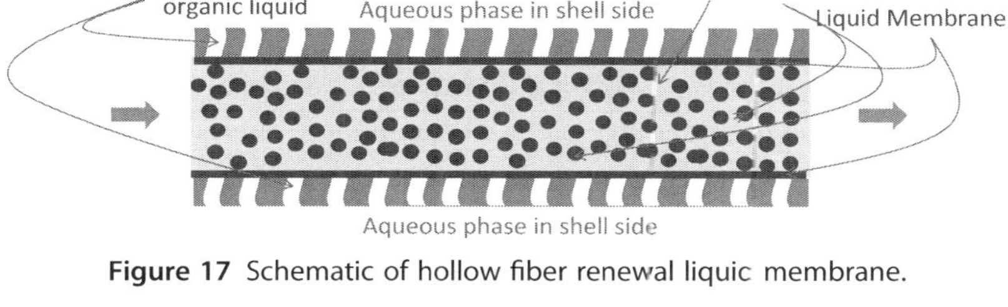

Figure 17 Schematic of hollow fiber renewal liquic membrane.

HFCSLM suffers from the problems like pore fouling, difficulty in cleaning of fibres between uses, high pressure drop and high cost.

Though ESPLM is a simple process, with easy operation, high flux, negligible leakage and swelling, and low energy consumption, it is only effective when the continuous phase is an organic solution (Ho and Sirkar, 2001).

Merits of the HFRLM are (Zhongqi et al., 2007):

- Mass transfer rate is increased because of the renewal effect of LM and high membrane surface area.
- Longer stability.
- Expensive solvents can be used because consumption of extractant is low.
- Energy consumption, capital investment, maintenance cost and operation cost are all less.
- Scale-up is easy as hollow fiber device is modular.
- Easy to operate.
- This technique can be applied for simultaneous extraction and stripping processes using either LM structure (W/O/W systems or O/W/O systems) with fibre being either hydrophobic or hydrophilic, depending on the LM structure.

7.5.2 Stability

Stable SLMs can be achieved by using a membrane liquid that has high interfacial tension yet lower surface tension than the critical surface tension of the polymeric solid support. It may be advisable to use polytetrafluoroethylene membranes with a pore size of the order of 0.1 μm (Takeuchi et al., 1987). Hence, aliphatic hydrocarbons of higher boiling point are suitable membrane solvents. Such solvents have the additional advantage of being chemically more inert to polymeric solids. Again a surface-active carrier agent reduces the stability of the SLM as a result of the lower solvent–water interfacial tension.

Care should be taken during fabrication of ESPLM setup so that there is no channelling of dispersion droplet between the compartments. The perforation in the baffles should be done properly.

HFRLM is quite stable by virtue of its concept of design. As opposed to the normal SLMs where loss of membrane liquid is a major concern, the operation of HFRLM integrates the replenishment of LM layer as the separation process progresses. Hence, the stability of HFRLM is established as long as supply of dispersed organic phase is uninterrupted.

8. OPERATIONAL ISSUES RELATED TO LM-BASED SEPARATION UNIT

Operational issues of LM-based separation unit are discussed in this section. Typically a two-phase extraction study yields an initial idea about choice of solvent, carrier and favourable operational conditions. The conditions are fine tuned in the subsequent three phase studies.

8.1 Two Phase Study

Two-phase study is basically an extraction equilibrium study which helps one to select a suitable membrane–carrier combination. For a W/O/W type LM operation, an aqueous solution of target solute is prepared by dissolving the required quantity of such solute in water. Carrier is added in required amount of solvent in an Erlenmeyer flask. The mixing of carrier in solvent is done in shaking incubator to make the membrane (organic) phase homogeneous. The pH of aqueous solution is maintained by adding acid/base, as and when needed. Small quantities of both membrane and aqueous phases are the mixed in an Erlenmeyer flask and the mixture is then well-stirred in a shaking incubator at certain temperature and stirring speed for a certain time of extraction to reach the equilibrium. The mixture is allowed to remain undisturbed for some time (about 12 h) in order to separate the phases. The clear sample of aqueous solution is collected for further tests. Concentration of solute remaining in aqueous phase is determined with the help of a pregenerated calibration curve. The quantity of solute transferred to the membrane phase is calculated by the mass balance, with fair assumption that no solute is lost during experimentation. Experiments should be repeated to ensure the repeatability of the results. Distribution coefficient (*a.k.a.* separation factor) of solute is then calculated as the ratio of solute in membrane phase to solute in the aqueous phase at equilibrium. A high distribution coefficient of the solute between the membrane phase and the aqueous feed phase necessarily means better transport of solute from feed phase to membrane phase. Effects of various physical parameters such as temperature, pH, and carrier concentration in the membrane, etc., on the equilibrium distribution of solute (distribution coefficient) are examined with the selected solvents. Stripping (reverse extraction) experiments are carried out with the membrane phase carrying the solute and fresh stripping phase to understand the solute releasing characteristics of the membrane phase as well as to optimize the strip phase conditions that involves pH, type and concentration of stripping solute in strip phase, etc.

8.2 Three Phase Study

In the three phase study, the effects of various process variables such as stirring speed, carrier concentration and feed and strip phase concentration on the transport of solute are studied.

In a BLM set up, two compartments of the reactor houses the feed and strip phases in such a way that the level of these liquids remain well below the top of the wall. The membrane phase liquid is then poured from the top in such a way that the height of the membrane phase clears the top of the glass wall and thus creates a bridge between the feed and strip phases for possible mass transfer of solutes. On the other hand, if the membrane phase is heavier than the feed/strip phases, the reactor is divided into two parts from the top with a glass wall at the middle with sufficient clearance at the bottom. The membrane phase is placed at the bottom of the reactor in such a way that the liquid level of the membrane phase rises beyond the bottom clearance of the glass wall. Two compartments are thereby formed due to such arrangement which can house the feed and strip phases. Care is taken so that there is no leakage/accidental mixing of these phases in both the cases. Continuous stirring ensures the aqueous solutions to be well mixed and the bulk concentration to be uniform throughout, and nevertheless, care is taken so that interface between aqueous phases and membrane phase was not disturbed.

In SLM setup, the micro-porous polymeric support is impregnated with the organic phase by immersing the support in the liquid for sufficiently longer time (about 24 h). The pores in the polymeric support get filled with the liquid due to capillary action. The support is taken out of the liquid and the liquid on the surface of the support is allowed to drip for few seconds before it is placed in the permeation cell. It is important to decide upon an ideal stirring speed in order to ensure good mixing and uniform bulk concentration inside the aqueous phases. Stirring minimizes concentration polarization at the membrane interface too. The pH value of the feed phase is continuously monitored and adjusted by adding appropriate acid/base as and when required. Small quantities of both the aqueous phases are collected periodically for further analysis. The solute content of the aqueous phase is determined using appropriate analytical instruments.

In ELM setup, the organic phase is first prepared by dissolving required quantity of carrier and surfactants in the solvent. To prepare a stable W/O (water in oil) emulsion, the internal or strip phase is added drop wise to this organic phase under stirring condition until the required volume ratio of organic phase to internal phase is obtained. The blended solution is then kept in a sonicator until a stable milky white W/O emulsion is obtained. Temperature is maintained during sonication to avoid demulsification. Measured quantity of freshly prepared W/O emulsion is then poured slowly into aqueous feed phase (continuous) in the ELM set-up. The solution is then stirred by a motor driven mechanical stirrer to disperse the emulsion phase. The speed of the stirrer is regulated by a voltage regulator. Samples are collected periodically from the continuous phase and allowed to settle by gravity for few minutes to separate the emulsion phase from the continuous feed phase. The emulsion phase, being heavier settles at the bottom. The upper layer of feed phase is filtered and analyzed for measuring concentration of solute. The effects of process variables on the transport of solute are evaluated by measuring the concentration of solute in both feed and strip phase after sufficient time for calculation of extraction and recovery, respectively.

9. CASE STUDY

In recent times, LM-based separation technology has been studied in lab-scale experiments for transport and pre-concentration of various types of solutes, which can broadly be classified in three major types: (1) organic pollutants (2) inorganic heavy metals (3) precious bioactive compounds. The issue which is common to all these types is that they are available in trace quantities in their respective sources. The common aim of all these case studies is to detect the best combination of solvent, carrier, stripping agent, emulsifier, supporting material, etc., and also the optimum operating conditions, viz. pH, temperature, stirring rate, concentrations of feed phase, strip phase, carriers, etc., that would yield the maximum transport and recovery of target solute from feed phase to strip phase. Most of the LM-based processes use organic solvents as the diluents. One common problem, seldom encountered with LMs (particularly with SLM), is the instability of the membrane due to loss of liquid that oozes out from the membrane pores. Organic solvents are generally volatile in nature and they do have some degree of toxicity. The volatility of the solvent thereby increases its loss and that eventually makes the LM more unstable. Moreover, the toxicity due to the strayed solvent is never desirable. Thus, it is essential also to look for environment-friendly solvent for successful operation of LM-based unit. Vegetable oils are ideal replacement for the conventional volatile organic diluents. These oils are less volatile and nontoxic in nature. Hence, with an obligation towards Green Technology, identification of a vegetable oil would provide environmentally and physiologically benign alternative for otherwise conventional organic solvents for LMs. Following sections present notes on the above issues for all three major classifications as above.

9.1 Organic Pollutant

There are various types of organic pollutants available in domestic as well as industrial environment which can be separated using LM-based technology. Present discussion focuses on separation aspects of one such pollutant namely, LS.

9.1.1 Lignosulfonate

Lignin (Figure 18(a)) is a complex chemical compound most commonly derived from wood. Highly lignified wood is durable and therefore, a good raw material for many applications including pulp and paper industry. Lignin is responsible for yellowing of paper with age. It must be removed from the pulp before high-quality bleached paper can be manufactured. In sulfite pulping, lignin is removed from wood pulp as sulfonates, viz. lignosulfonate (LS). LS, a polydispersed polyelectrolyte and a spherical macromolecule, have two kinds of functional groups (SO_3^{-2} and OH^-). The model structure of LS is shown in Figure 18(b) as a sodium salt of LS (sodium lignosulphonate). Extensive usage of commercial grade LS includes production of vanillin, animal feed pellet binders, dispersant for dyes, pesticides, carbon black, etc. LS find its application in boiler and cooling tower water treatment where it prevents scale deposition and also in industrial cleaning applications where it acts as dirt dispersant. Separation of LS from paper mill effluent thereby serves dual purpose, using paper mill waste as a source of LS in one hand and an effective wastewater treatment on the other. A detailed procedure of separation of LS using LM-based techniques has been provided in Chakrabarty (2010), Shaik et al. (2010), Chakrabarty et al. (2009a,b, 2010a,c).

9.1.1.1 Mechanism of Transport

Separation of LS is possible through either of symport and antiport transport mechanism. In order for the complex to be formed at the feed-membrane interface, protonation should occur at the feed phase. Hence, the feed phase should be maintained in acidic condition by adding HCl. Consequently the stripping phase should be maintained in alkaline condition. A strip phase

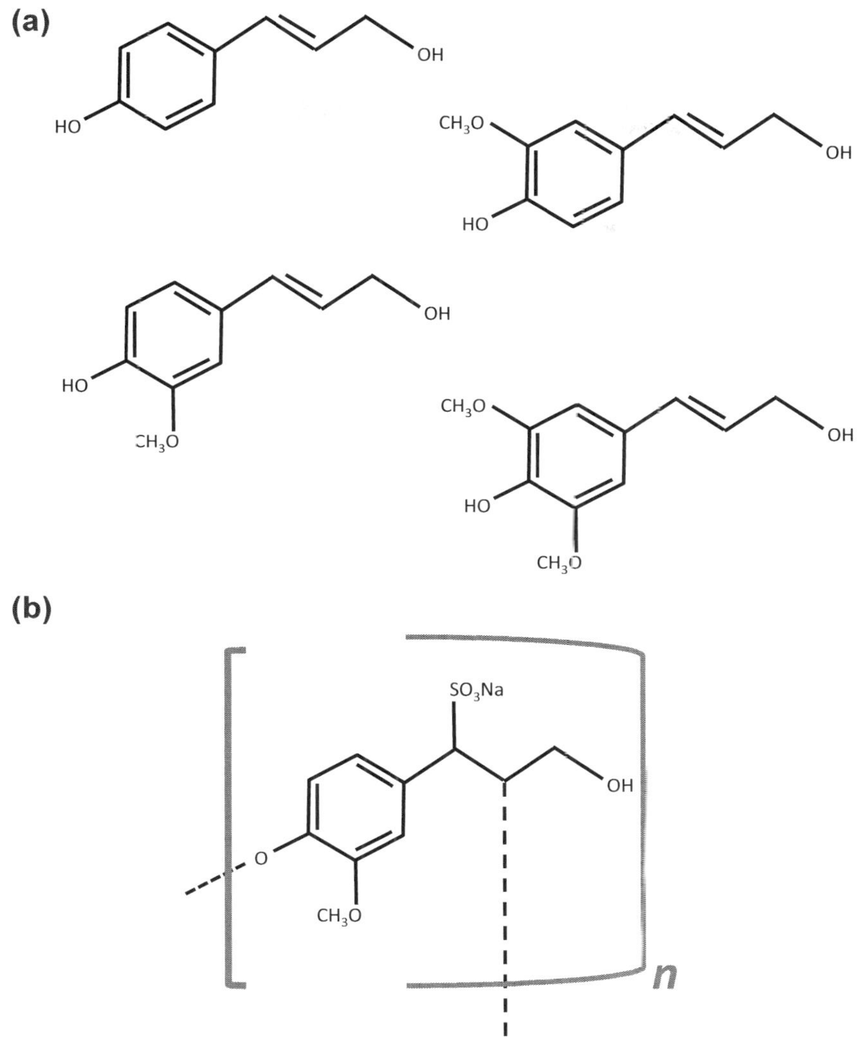

Figure 18 (a) Various structures of monolignols. (b) Unit structure of a typical lignosulphonate macromolecule.

containing NaOH demonstrates symport mechanism (Figure 19(a)) whereas that containing NaCl demonstrates antiport mechanism (Figure 19(b)). LS ion moves uphill in both the cases whereas H^+ and Cl^- move downhill in symport and antiport mechanisms respectively. TOA is used as a carrier agent for both these cases (Chakrabarty et al., 2009b). The reaction at the feed-membrane interface follows a two-step mechanism which is same for both types of transport. The reactions are shown as follows:

$$\text{Extraction}: R_3N + HCl \rightleftharpoons R_3NH^+Cl^- \tag{22}$$

$$\text{Extraction}: R_3NH^+Cl^- + LSNa_n \rightleftharpoons R_3NHLSNa_{(n-1)} + NaCl \tag{23}$$

$$\text{Stripping}: R_3NHLSNa_{(n-1)} + NaOH \rightleftharpoons R_3N + H_2O + LSNa_n \tag{24}$$

or

$$\text{Stripping}: R_3NHLSNa_{(n-1)} + NaCl \rightleftharpoons R_3NH^+Cl^- + LSNa_n \tag{25}$$

where, R_3N is TOA in organic phase, $LSNa_n$ is sodium lignosulphonate and $R_3NHLSNa_{(n-1)}$ is the complex that is formed as a result of reaction at feed/membrane interface. The overall extraction reaction can be written as

$$R_3N + HCl + LSNa_n \rightleftharpoons R_3NHLSNa_{(n-1)} + NaCl \tag{26}$$

The equilibrium constant for the above reaction can be written as

$$K_{eq} = \frac{[R_3NHLSNa_{(n-1)}][NaCl]}{[R_3N][HCl][LSNa_n]} \tag{27}$$

After rearrangement the above equation takes the form,

$$[LSNa_n] = \frac{1}{[R_3N]} \times \frac{[R_3NHLSNa_{(n-1)}][NaCl]}{K_{eq}[HCl]} \tag{28}$$

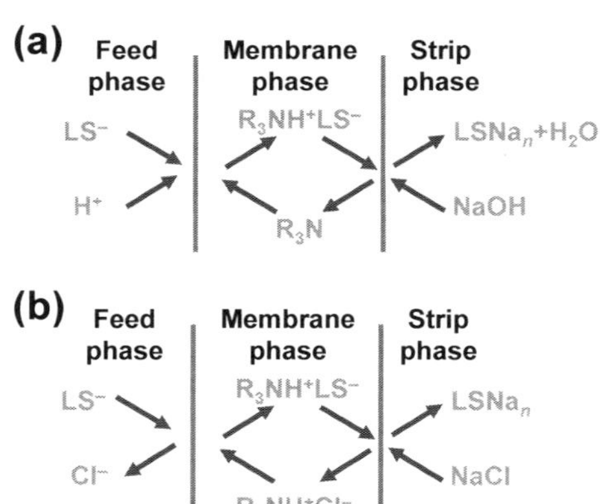

Figure 19 (a) Schematic of symport mode of transport of lignosulphonate (LS). (b) Schematic of antiport mode of transport of LS.

or

$$\log\{\mathrm{LSNa}_n\} = (-1)\log\{\mathrm{R_3N}\} + \log\left\{\frac{[\mathrm{R_3NHLSNa}_{(n-1)}][\mathrm{NaCl}]}{K_{\mathrm{eq}}[\mathrm{HCl}]}\right\} \tag{29}$$

At a constant pH, [HCl] remains constant. It also can be assumed that the variation in [NaCl] would be negligible within a small concentration range of TOA. Therefore, [NaCl] is considered as constant. Hence, a plot of log $\{\mathrm{LSNa}_n\}$ versus $\log\{\mathrm{R_3N}\}$ yields a straight line with the slope of (−1). Chakrabarty et al. (2009a) reported the experimentally evaluated slope to be (−0.866) which is very close to the theoretical value, taking the experimental error into consideration.

9.1.1.2 Choice of Solvent

Dichloroethane ($C_2H_4Cl_2$) is found to be a good solvent for the extraction of LS with a distribution coefficient of 21.5 followed by carbon tetrachloride (8.5) and chloroform (1.25). Other organic solvents show negligible extraction of LS. It has also been observed that the solubility of LS is very low in absence of carrier agent even for dichloroethane. Hence, it is evident that mere solution diffusion is not enough for LS transport; some facilitation is required for it. On the other hand, it was observed that coconut oil could be used as an important alternative for solvent in lieu of toxic organic solvent (Chakrabarty et al., 2010b). It was also observed that LS is highly soluble in coconut oil even without any carrier agent. This fact augments the philosophy of green technology in LM-based operation. Nevertheless, addition of carrier agent (TOA) reduces the time for reaching equilibrium extraction.

9.1.1.3 LM Configurations

Chakrabarty (2010) performed LM-based separation studies in all three configurations, i.e., BLM, SLM and ELM. The results on BLM established the utility of such separation technique, whereas the SLM and ELM provided some more insight to the applicability as well as scalability of LM-based techniques in industrial level operation.

9.1.1.3.1 SLM Configuration Polyvinyldenefluoride proves to be an ideal support material for the SLM with 0.2 μm pore size. It was also noticed that a larger pore size (say 0.45 μm) leads to stability issue for the LM as the capillary action inside the pores is insufficient to hold the liquid (Chakrabarty et al., 2009a). Porosity (ε) and thickness (l) of support membrane can be measured using scanning electron microscope. The tortuosity (τ) can be calculated as

$$\tau = \frac{1 + V_p}{1 - V_p} \tag{30}$$

where, $V_p = 1 - \varepsilon$, is the volume fraction of the polymeric framework (Alguacil et al., 2005).

A permeation model has been developed (Chakrabarty et al., 2009a) based on the following assumptions:

- The mass transfer of LS complex across the SLM occurs through diffusion mechanism only.
- The chemical reaction takes place only at the feed–membrane and membrane–strip interfaces, and not inside the pores. The rate of reaction is very high compared to the rate of diffusion process. Hence, the interfacial flux due to chemical reaction is neglected.

Thus, diffusion flux of LS (J_{aq}) from the bulk of the feed phase to the feed/membrane interface and diffusion flux of LS-TOA complex (J_{org}) from the feed/membrane interface to the membrane–strip interface can be written as follows:

$$J_{aq} = \Delta aq^{-1}\left\{[LSNa_n]_f - [LSNa_n]_{fm}\right\} \tag{31}$$

$$J_{org} = \Delta org^{-1}\left\{\left[R_3NHLSNa_{(n-1)}\right]_{fm} - \left[R_3NHLSNa_{(n-1)}\right]_{ms}\right\} \tag{32}$$

where, $[LSNa_n]_f$ and $[LSNa_n]_{fm}$ are concentrations of LS at bulk of the feed phase and the feed/membrane interface respectively, whereas $[R_3NHLSNa_{(n-1)}]_{fm}$ and $[R_3NHLSNa_{(n-1)}]_{ms}$ are the concentrations of LS-TOA complex at feed/membrane and membrane/strip interfaces respectively. Alonoso et al. (2006) argued that $\left[R_3NHLSNa_{(n-1)}\right]_{fm} \gg \left[R_3NHLSNa_{(n-1)}\right]_{ms}$ and hence, Eqn (32) was be approximated as

$$J_{org} = \Delta org^{-1}\left\{\left[R_3NHLSNa_{(n-1)}\right]_{fm}\right\} \tag{33}$$

Moreover at steady state condition the fluxes of diffusion at aqueous phase and organic phase are equal. Hence,

$$J_{org} = J_{aq} = J \tag{34}$$

Resistances in aqueous and organic phases are denoted by Δaq and Δorg respectively, which can further be expressed as

$$\Delta \mathrm{aq}^{-1} = \frac{D_{\mathrm{LS}}}{\delta} \tag{35}$$

$$\Delta \mathrm{org}^{-1} = \frac{\varepsilon D_{\mathrm{LS-TOA}}}{l\tau^2} \tag{36}$$

where, D_{LS} and $D_{\mathrm{LS\text{-}TOA}}$ are diffusivities of LS and LS-TOA complex respectively. From Eqn (28) it can be concluded that

$$[\mathrm{LSNa}_n]_{\mathrm{fm}} = \frac{\left[\mathrm{R_3NHLSNa}_{(n-1)}\right]_{\mathrm{fm}}\left[\mathrm{NaCl}\right]}{K_{\mathrm{eq}}\left[\mathrm{HCl}\right]\left[\mathrm{R_3N}\right]} \tag{37}$$

Using Eqns (33), (34) and (42), one obtains

$$J = \Delta \mathrm{aq}^{-1}\left\{[\mathrm{LSNa}_n]_{\mathrm{f}} - \frac{\left(\frac{J}{\Delta \mathrm{org}^{-1}}\right)[\mathrm{NaCl}]}{K_{\mathrm{eq}}[\mathrm{HCl}][\mathrm{R_3N}]}\right\} \tag{38}$$

Rearranging the above,

$$P = \frac{J}{[\mathrm{LSNa}_n]_{\mathrm{f}}} = \frac{K_{\mathrm{eq}}\left[\mathrm{HCl}\right]\left[\mathrm{R_3N}\right]}{(\Delta \mathrm{org})\left[\mathrm{NaCl}\right] + (\Delta \mathrm{aq})K_{\mathrm{eq}}\left[\mathrm{HCl}\right]\left[\mathrm{R_3N}\right]} \tag{39}$$

where, P is the permeability coefficient. Again rearranging the above and using Eqn (37),

$$\begin{aligned}\frac{1}{P} &= \frac{(\Delta \mathrm{org})\left[\mathrm{NaCl}\right] + (\Delta \mathrm{aq})K_{\mathrm{eq}}\left[\mathrm{HCl}\right]\left[\mathrm{R_3N}\right]}{K_{\mathrm{eq}}\left[\mathrm{HCl}\right]\left[\mathrm{R_3N}\right]}\\ &= (\Delta \mathrm{org})\frac{1}{\left\{\frac{K_{\mathrm{eq}}[\mathrm{HCl}][\mathrm{R_3N}]}{[\mathrm{NaCl}]}\right\}} + (\Delta \mathrm{aq})\\ &= (\Delta \mathrm{org})\frac{1}{\left\{\frac{\left[\mathrm{R_3NHLSNa}_{(n-1)}\right]_{\mathrm{fm}}}{[\mathrm{LSNa}_n]_{\mathrm{fm}}}\right\}} + (\Delta \mathrm{aq}) = (\Delta \mathrm{org})\frac{1}{m} + (\Delta \mathrm{aq})\end{aligned} \tag{40}$$

where, m is the distribution of LS at various carrier concentrations and at a constant pH maintained during the experiments. Again, J can be expressed as

$$J = \frac{V_f}{A}\left(-\frac{\mathrm{d}[\mathrm{LSNa}_n]}{\mathrm{d}t}\right) \tag{41}$$

combining Eqns (39) and (41), one obtains

$$P = \frac{J}{[\mathrm{LSNa}_n]} = \frac{V_f}{A[\mathrm{LSNa}_n]}\left(-\frac{\mathrm{d}[\mathrm{LSNa}_n]}{\mathrm{d}t}\right) \tag{42}$$

or

$$\frac{\mathrm{d}[\mathrm{LSNa}_n]}{[\mathrm{LSNa}_n]} = -\frac{PA}{V_\mathrm{f}}\mathrm{d}t \tag{43}$$

Simple integrations of Eqn (43) for the specified duration of experiment would yield

$$P = \frac{V_\mathrm{f}}{At}\ln\left\{\frac{[\mathrm{LSNa}_n]_\mathrm{f}\big|_0}{[\mathrm{LSNa}_n]_\mathrm{f}\big|_t}\right\} \tag{44}$$

where, t is the elapsed time during the experiment, $[\mathrm{LSNa}_n]_\mathrm{f}\big|_0$ and $[\mathrm{LSNa}_n]_\mathrm{f}\big|_t$ are the concentrations of LS at initial time and at time t, respectively. Thus, by plotting P^{-1} versus m^{-1} of the Eqn (40) for various $[R_3N]$ at constant pH a straight line with slope (Δorg) and intercept (Δaq) would be obtained. The mass transfer coefficients for aqueous and organic phases can thereby be found. Chakrabarty et al. (2009a) reported the values of D_{LS} and $D_{\mathrm{LS\text{-}TOA}}$ to be 1.67×10^{-9} and 66.8×10^{-9} m^2/s respectively.

9.1.1.3.2 ELM Configuration The ELM module for LS separation was developed (Chakrabarty, 2010) with a suitable nonionic surfactant such as sorbitan monooleate (SPAN 80) or poly ethylene glycol of molecular weight 20,000 (PEG 20000). The external (continuous) phase of the emulsion is the source phase of LS in these experiments. Hence, the LS transport occurs from external phase to internal phase. The organic phase was first prepared by dissolving required quantity of carrier agent (TOA) and surfactants (PEG 20000 or SPAN 80) in the solvent (dichloroethane). To prepare a stable W/O (water in oil) emulsion, the internal or strip phase (aqueous NaOH solution) was added drop wise to this organic phase under stirring (1000 rpm) condition until the required volume ratio (*a.k.a.* phase ratio) of organic phase to internal phase was obtained. The blended solution was then kept in a sonicator for 1–3 h until a stable milky white W/O emulsion was obtained. Sonication reduces intermolecular interaction by using sound energy to buzz the molecules uniformly. During sonication, temperature was maintained in such a way so that it does not exceed 40 °C to avoid demulsification. Thereafter, the W/O emulsion was dispersed in continuous (source) phase at a proper feed to emulsion ratio (*a.k.a.* treat ratio) with proper stirring. The resultant emulsion took the form of the desired ELM and it was a W/O/W emulsion. It is observed that with increase of phase ratio beyond 1:1 the transport of LS decreases. With the increase of phase ratio the absolute amount of each component in the membrane phase is raised and hence, extraction should increase (Shen et al., 1996). Though at higher phase ratio, capacity of the membrane phase increases, at the same time membrane thickness and viscosity of the emulsion phase also increases. This increases the membrane phase resistance to transport and also hampers the intermixing in the organic membrane phase (Boyadzhiev and Bezenshek, 1983). This phenomenon becomes more predominant particularly for macromolecules like LS. Hence, the merit of incorporation of relatively more membrane volume could not be extracted at higher phase ratio. It may be referred from Chakrabarty (2010) that with increase of phase ratio, extraction of LS remains more or less same during 20–30 min of experiment. However, a phase ratio of 1.5:1 is more justified. On the other hand, the influence of treat ratio is almost same for various treat ratios 50, 25 and 20 and the extraction decreases when the treat ratios are lowered further (below 40:3). This behaviour can be explained from the following two opposite phenomena occurring simultaneously.

- Decrease of treat ratio in turn increases the membrane and internal phase amount, amount of total carrier in the membrane phase and membrane phase surface area for transport. This would enhance the permeation and stripping of LS (Jee et al., 1985; Boyadzhiev and Bezenshek, 1983).
- Increased coalescence of emulsion globules with increase in emulsion volume reduces the mass transfer area and hence, the rate of extraction. In that case, higher extraction time is needed. With increase in extraction time emulsion breakage increases (Jee et al., 1985). This, in turn, reduces the permeability and stripping of LS.

9.1.1.4 Optimum Operating Conditions

It is understood that the feed phase should be maintained at acidic condition for enabling sufficient protonation. It is experimentally detected that the feed phase pH for LS transport should be 2. However, too much protonation ($pH < 2$) is not favourable for LS transport because of the following two reasons

- In highly acidic condition LS, molecules with higher molar mass on average, is not ionized (Kontturi and Kontturi, 1987) and therefore, reaction shown by Eqn (23) is not favoured.
- Increase in chloride ion at lower pH inhibits the formation of amine complex, as reaction shown by Eqn (22) is at equilibrium.

Increase of temperature enhances the distribution coefficient (m) of LS initially till it reaches an optimum at 313 K. However, the distribution coefficient drops sharply above 313 K. The initial increase in the distribution coefficient is possibly due to the fact that increase in temperature favours reaction at feed/membrane interface. However, effective charge of LS drops to zero above 313 K (Kontturi and Kontturi, 1987; Kontturi et al., 1990). Hence, the transport of LS due to carrier (reaction shown by Eqn (23)) drops sharply above 313 K as it loses its charge and is found to be almost zero at relatively higher temperature of 323 K.

Increase in carrier concentration enhances the equilibrium distribution coefficient and an optimum is attained at 2 vol% TOA. Value of distribution coefficient is very low (<1) in absence of a carrier. This indicates poor solubility of the LS in the membrane liquid. Once the carrier is added to the membrane phase, LS–amine complex is formed which is soluble in the membrane liquid. The maximum distribution coefficient at 2 vol% TOA indicates saturation capacity of the membrane phase for the complex. With further increase in the carrier concentration, the saturation capacity of the membrane phase for the complex is marginally reduced due to presence of more amount of carrier. However, it is worth mentioning here that this optimum carrier concentration will vary depending on the initial feed phase concentration of LS, the ratio of feed phase to membrane phase volumes, and other physicochemical properties like temperature, pH, etc.

It is observed that stirring of feed phase greatly increases the extraction operation, whereas stirring of strip phase has practically no effect on stripping operation. Hence, effects of hydrodynamic conditions on the LS transport are visible in the extraction operation, but not in stripping operation.

The influence of initial feed concentration on the transport of LS was studied by Chakrabarty et al. (2009b). With increase in initial feed concentration the initial rate of the transport of LS is higher but the transport continues for a longer time. At the same time, with increase in initial feed concentration recovery of LS also increases. In antiport mode of transfer of LS using NaCl as stripping agent it is observed that though the extraction of LS is 98%, its recovery is only 5–10%. This might be due to the ineffectiveness of the stripping agent (NaCl) to decomplex the LS–amine complex. As the rate of stripping is slow compared to the rate of extraction, a level of saturation is attained in the membrane phase at high initial feed concentration and hence, more time is needed for achieving high percentage of extraction (>90 wt%).

In SLM operation, increase in NaOH concentration in the stripping phase up to 0.5 M enhances the recovery of LS (Chakrabarty et al.,

2009a). However, further increase in NaOH concentration shows adverse effect on LS transport. This phenomenon is different from the one observed by Chakrabarty et al. (2009b) with the separation of LS through BLM, where recovery of LS increased and never showed any adverse effect with increase of NaOH concentration. The decrease of recovery of LS at high concentration in SLM may be due to following two reasons:

- Process performance is considerably affected by the presence of a difference in ionic strengths of feed and strip phases. The increase of recovery of LS from 0.25 to 0.5 M NaOH concentration reveals that the ionic concentrations in the two phases (feed and strip) are balanced within this range. However, beyond 0.5 M NaOH the difference in ionic strength increases and creates an osmotic pressure differential, which eventually results in decrease of LS flux.
- According to the progressive wetting mechanism, the aqueous phase from low interfacial tension side progressively penetrates the membrane pores (Zha et al., 1995). With increase in NaOH concentration in strip side (0.1–1 M) the interfacial tension decreases. Therefore, strip solution may penetrate the support pores causing the loss of membrane liquid and thereby reducing the LS flux.

9.2 Heavy Metals

The major sources of effluent, containing heavy metals (viz. Hg, Pb, Cd, etc.), are industries such as chloro-alkali, pulp and paper, battery, fertilizer, mining and combustion, and rubber processing. Major problem in the separation of heavy metals from industrial effluent is their availability in trace quantities. The separation and purification of trace amount of solute from their dilute solutions has been an interesting problem for the scientists and engineers for decades. There have been several literatures working on the LM-based separation of such heavy metals like Shamsipur et al. (2002) for mercury; Castro et al. (2004) for copper; Mortaheb et al. (2010), Bhatluri et al. (in press) and He et al. (2000) for cadmium. Detailed investigations on applicability of LM technology has been reported by Shaik et al. (2010) and Chakrabarty et al. (2010b,c) for mercury. A brief account on the removal of mercury from wastewater by LM is presented below.

9.2.1 Mercury

Natural erosion of soil and ore deposits is the source of mercury in surface and underground water. Industries like chloro-alkali, oil refining, electrical, rubber processing, battery, paint, fluorescent lamp, fertilizer, pharmaceuticals, plastics and pulp and paper are typical sources of mercury pollution (Lopes et al., 2008;

Senkal and Yavuz, 2007; Rajesh and Hari, 2008). Maximum concentration of mercury in wastewater from various sources is about 10 mg/l (Zambrano et al., 2004). Mercury (II) is usually more toxic than other forms because of higher solubility in water and high tendency for binding to the proteins (Uludag et al. 1997). According to European environmental regulations (Ghodbane and Hamdaoui, 2008), maximum level of mercury in drinking water and wastewater is restricted to 1 and 5 μg/l, respectively. Thus, thorough investigation on development of technology for removal of mercury from water as well as wastewater is very important as well as appropriate.

9.2.1.1 Mechanism of Transport

TOA has been found to be the most promising carrier agent and has been employed to transport Hg (II) where TOA reacts with the Hg (II) only when it is in anionic complex form (Shaik et al., 2010; Chakrabarty et al., 2010b,c). Transport of mercury and H^+ ions occurs in the same direction and is called as coupled co-transport. The reaction mechanism is as follows.

9.2.1.1.1 Extraction Reaction Complexation of Hg (II) by chloride ions in the feed solution

$$Hg^{+2} + 4Cl^- \leftrightarrow HgCl_4^{-2} \tag{45}$$

Reaction of TOA (denoted as R_3N) at the feed/membrane interface with HCl in the feed solution:

$$R_3N_{(org)} + H^+Cl^-_{(aq)} \Leftrightarrow R_3NHCl_{(org)} \tag{46}$$

In feed, $HgCl_4^{2-}$ exchange with Cl^- of $R_3NH^+Cl^-$ in the membrane phase:

$$2R_3NH^+Cl^-_{(org)} + HgCl_4^{2-} \Leftrightarrow (R_3NH)_2HgCl_{4(org)} + 2Cl^-_{(aq)} \tag{47}$$

9.2.1.1.2 Stripping Reaction

$$(R_3NH)_2HgCl_{4(org)} + Na^+OH^- \Leftrightarrow R_3N_{(org)} + HgCl^{2-}_{4(aq)} + 2H_2O \tag{48}$$

9.2.1.2 Choice of Solvent

Various solvents, viz. *n*-heptane, hexane, toluene, cyclohexane and dichloroethane, were employed and tested for the extraction of Hg (II) by (Shaik et al., 2010; Chakrabarty et al., 2010b,c). It is observed that transport of mercury is not significant when *n*-heptane and toluene were used as membrane. Significant separation was observed with hexane, followed by dichloroethane and cyclohexane. Since vapour pressures of hexane and cyclohexane at 25 °C are 201 and 130 mbar respectively, they prove to be more volatile than the dichloroethane having vapour pressure 105 mbar at 25 °C. Therefore, hexane and cyclohexane have not been considered and dichloroethane-TOA duo has been selected as the solvent–carrier combination in the BLM for further experiments. The estimated distribution coefficient in dichloroethane with and without carrier (TOA) is found to be 2.94 and 26.97 respectively.

9.2.1.3 LM Configurations

I have already discussed earlier that the actual transport phenomena in LMs are quite complex and are strongly influenced by feed pH, carrier concentration and physiochemical properties of carrier in the membrane phase. When TOA (carrier) is imparted to the membrane phase, formation of Hg (II)-amine complex (Eqn (47)) takes place in the membrane which thereby increases the mass transfer through the interface of feed-membrane and eventually yields higher distribution coefficient. Concentration of carrier, TOA is found to be optimum at its concentration 1.5 vol%, which again depends upon the ratio of membrane phase volume to feed phase volume and the initial feed phase concentration.

The reaction mechanism reported above (Eqns (45)–(48)) indicates that acidic condition is required for the protonation of TOA as it binds with Hg (II) in anionic form only. Distribution coefficient of Hg (II) is found to be the maximum at pH 2.5. Extraction is not effective at very low pH as concentration in the feed solution forms H_2HgCl_4 which does not dissociate enough to form adequate $HgCl_4^{-2}$ that forms the complex with the cation. At much higher pH, the protonation is inadequate due to less availability of proton. Details of various LM configurations, their operations, merits and demerits, and stability have already been discussed in detail in earlier sections. In this section, the results of study of mercury separation by BLM and SLM methods (Shaik et al., 2010; Chakrabarty et al., 2010b,c) have been highlighted and reasons are discussed.

9.2.1.3.1 BLM Configuration The BLM separation studies was carried out with carrier concentration of 1.5 vol% and feed pH 2.5. Aqueous NaOH solution was used as strip phase and it is found that up to 94% extraction was possible. Experiments suggested that both percentage extraction and recovery were more for 0.5 M NaOH strip solution.

The effect of initial feed concentration was studied using concentration range of 200 ppb to 4 ppm. Lower feed phase concentration of Hg (II) leads to a higher recovery but with increased concentration of Hg (II) recovery decreases due to the slower diffusion of complex from feed to strip phases. At lower initial feed concentration, diffusion rate of the complex through the membrane is higher and amount of Hg (II) is inadequate for precipitation reaction at the strip-side. But at higher concentration, adequate supply of Hg (II) enhances the precipitation reaction (forming HgO) and therefore, the recovery decreases.

9.2.1.3.2 SLM Configuration The BLM method shows that the LM comprising of TOA and dichloroethane is a suitable combination for the separation of mercury. Results of a systematic investigation on the transport of mercury from aqueous solution through an SLM containing TOA as carrier and dichloroethane as solvent are discussed in this section. Nylon 6, 6 has been used as the support and aqueous NaOH as the strip phase. Influence of various parameters such as feed pH, strip concentration, carrier concentration and feed concentration on the transport of mercury were studied.

Effect of pH on the extraction and recovery of mercury has already been discussed earlier and pH of 2.5 yielded the best results for SLM also.

Strip phase concentration played an important role in the transport of mercury through SLM. Detailed study with various concentration of NaOH suggested that in case of SLM the recovery as well as extraction of mercury reaches maximum at NaOH concentration of 0.1 M. The value is lower than BLM value. This is because of yellow coloured precipitate of HgO as discussed earlier for which membrane pores get clogged and transport of mercury gets affected.

Experiments with variation of initial feed concentration suggested that optimum feed phase pH changes with the feed concentration and the membrane under optimum feed phase pH is equally effective both at high and very low feed concentrations.

9.2.1.4 Optimum Operating Conditions

TOA-dichloroethane is proved to be a suitable solvent–carrier combination for the separation of mercury. Extraction of mercury is a highly pH dependent with optimum at pH of 2.5. The SLM configuration Nylon 6,6-TOA-dichloroethane demonstrated efficient separation (about 81%) capability of mercury from its aqueous solution.

9.3 Bioactive Compounds

Plant extracts containing the bioactive compounds are getting increased attention for use as herbal medicines due to their low or no side effects. Since such extracts do contain some unwanted components, it becomes necessary to separate these target components from their respective sources and use them at required dosage. Separation, purification or pre-concentration of bioactive compounds such as antibiotics and amino acids are usually carried out by solvent extraction method. In such situation solvent requirement is very high and the same hazardous volatile organic solvents create pollution to the environment. LM technology with environmentally benign solvent can be a suitable alternative for such application. Important antibiotics such as penicillin, cephalosporin, etc., and their derivatives are separated from the hydrolysates and reaction broths by LM with Aliquat 336 as carrier and *n*-heptane-kerosene (1:1) as diluents (Sahoo and Dutta, 1998). Wang and Juang (2002) used acidic extractant, di (2-ethylhexyl) phosphoric acid (D2EHPA) for extraction of amino acids using ELM. The present discussion focuses on separation of one such very commonly used bioactive compound, catechin. Manna et al. (2012) have reported such study in details.

9.3.1 Catechin

Tea plant (Camellia sinensis) is a rich source of catechin (±C), a pharmacological substance that is effective for a number of health protection problems. Five different catechin derivatives namely catechin (C), epicatechin (EC), epicatechingallate (ECG), epigallocatechin (EGC) and epigallocatechingallate (EGCG) (Figure 20) are found in various proportions in the extract from tea leaves and flowers. The amount of various catechins varies depending on the plant species, climate and cultivation conditions. Typically, in 1 g of solid green tea leaves, Indian black tea contains 3.4 mg catechin (+C), 4.86 mg epicatechin (-EC) and 14.99 mg total catechin compounds (5 nos) (Jin et al., 2006). Normally, the amounts of polyphenols, catechins (±C) and epigallocatechingallate (EGCG) are 37–56, 30–42 and 10–13% of green tea, respectively (Jin et al., 2006). Other sources of catechins include berries, olive oil, chocolate/cocoa, coffee, pomegranates, popcorn, fruits and other vegetables.

9.3.1.1 Mechanism of Transport

The chiral mixture of catechin and epicatechin (called as (±)-catechin) has been used as solute to be extracted by the neutral extractant, TBP (tributyl phosphate) (Manna et al., 2012). The possible reaction mechanism is the

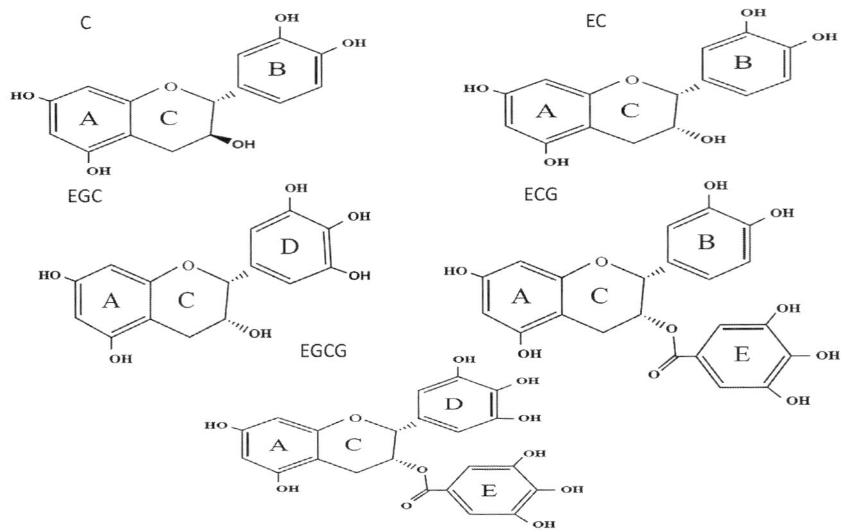

Figure 20 Various catechin derivatives.

hydrogen bonding that makes a complex. Figure 21 is indicative of the possible sites of formation of linkages via *H*-bonding between TBP and catechin during extraction. Thus, the reaction can be described as follows.

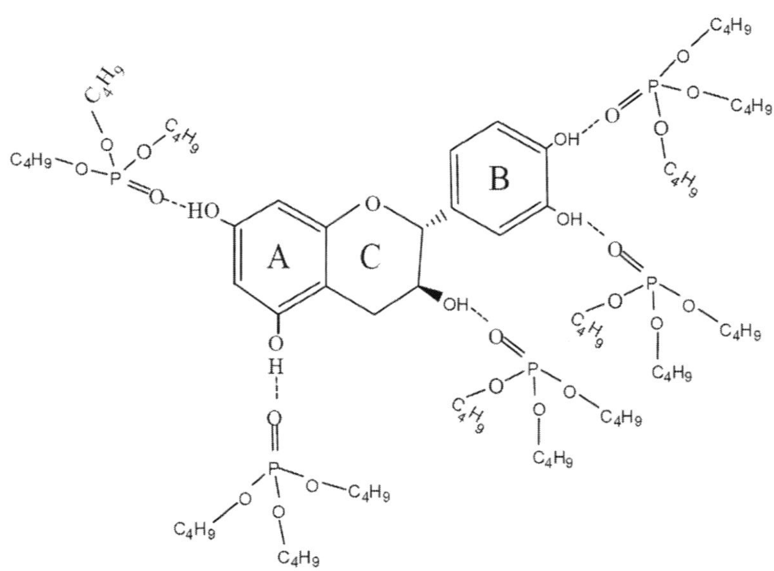

Figure 21 Possible sites of formation of linkages via *H*-bonding between tributyl phosphate and catechin.

$$[\text{CatOH}]_{\text{aq}} + x\ \text{TBP}_{\text{org}} \rightleftharpoons (\text{CatOH}.x\ \text{TBP})_{\text{org}} \tag{49}$$

with the equilibrium constant, K_{ex} by:

$$K_{ex} = \frac{[\text{CatOH}.x\text{TBP}]_{\text{org}}}{[\text{CatOH}]_{\text{aq}}[\text{TBP}]_{\text{org}}^{x}} \tag{50}$$

Equation (50) is re-written to find out the extraction reaction stoichiometry as follows:

$$\log[\text{CatOH}]_{\text{aq}} = \log\left(\frac{[\text{CatOH}.x\text{TBP}]}{K_{ex}}\right) - x\log[\text{TBP}]_{\text{org}} \tag{51}$$

where, $[\text{CatOH}]_{\text{aq}}$, $[\text{TBP}]_{\text{org}}$ and $[\text{CatOH.TBP}]_{\text{org}}$ represent catechin in aqueous feed phase, TBP in organic phase and the complex between catechin and TBP in organic phase, respectively. Catechin distribution coefficient between membrane phase containing various concentrations of carriers and aqueous phase was determined in two-phase equilibrium study. According to Eqn (51) and experimental data on variation of $[\text{CatOH}]_{\text{aq}}$ with $[\text{TBP}]_{\text{org}}$, x can be approximated as nearly equal to 2. Mass transfer of catechin across the membrane is described considering only diffusion parameters with the assumption that the interfacial reaction is very fast and hence, the concentration at the interface should be equal to the equilibrium concentration. Transport mechanism of catechin is schematically represented through Figure 22.

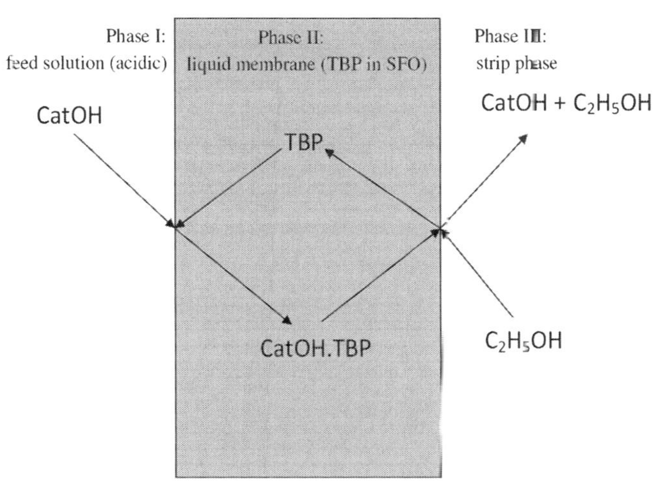

Figure 22 Catechin transport mechanism. TBP, tributyl phosphate; SFO, sunflower oil.

9.3.1.2 Choice of Solvent

Experiments with various combinations of solvents and carrier suggested sunflower oil (SFO) containing TBP to be the best solvent–carrier combination. Catechin extraction was 78% in SFO containing 0.55 M TBP followed by 75% in soya bean oil,74% in coconut oil and 34% in mustard oil. Carrier in membrane phase is necessary for extraction of catechin. No extraction of catechin was observed in organic solvents like *iso*-octane, *n*-heptane and *n*-decane and vegetable oils in absence of any carrier agent. The carrier makes the solute–carrier complex which is soluble in the membrane phase and facilitates transport of solute. The LM with 0.55 M TBP in SFO has the highest distribution coefficient (3.55) and the same has been used by Manna et al., (2012) for further experimental work.

9.3.1.3 LM Configurations

Catechin, a polyphenol and weak acid with a dissociation constant (pKa value) of 8.6431, remains almost undissociated upto 7.0. TBP, the neutral carrier, extracted catechin in molecular state. Therefore, extraction is more or less constant in acidic pH but drastically reduced at pH beyond 7.0. In alkaline pH, catechin dissociates to its anionic state leading to drastic reduction in extraction by the neutral extractor, TBP. Hence, further studies were carried out at neutral pH.

At low carrier concentration the number of carrier molecules at feed-membrane interface is stoichiometrically less than catechin needed for complexation of solute-carrier. Therefore, distribution coefficient initially increases efficiently with the increase of carriers and then with a declining slope. Thus, carrier concentration in the range of 0.44–0.73 M is the most suitable and 0.55 M TBP has been used in further studies. Two-phase study revealed that most of the catechin is extracted quite early (within ~5 h) and maximum extraction, 78.5%, is reached within 24 h. The section further focuses on the results of BLM-based separation of catechin as reported by Manna et al. (2012).

9.3.1.3.1 BLM Configuration BLM experiments were carried out varying the process variables like feed and strip phase concentration on the transport of catechin. Feed phase pH was maintained at 4.0 to ensure catechin in its molecular state.

Recovery of a target solute using LM technology is strongly dependent on the choice of a suitable stripping agent. Experiments with various stripping agents such as ethanol, *iso*-propanol, sodium hydroxide (NaOH), sodium chloride (NaCl) in their aqueous solutions suggested that ethanol and *iso*-propanol have nearly the same recovery efficiency but no or little catechin was recovered by sodium hydroxide and sodium chloride. Ethanol, due to its low cost and easy availability, is been selected as stripping agent.

Extraction of catechin attained a maximum of 70% and recovery in stripping phase attains a maximum of 44% for 0.36 M TBP in SFO. Thus, the optimum carrier concentration is lower than that was obtained in two-phase study. This is because of the fact that carrier diffuses back to the feed-membrane interface after being freed from its complex releasing the solute at the membrane–strip interface. Hence, any concentration higher than the optimum one does not yield any better extraction and recovery is decreased due its resistance on transport of the complex through the membrane.

BLM Experiments with pH 4.0 of feed phase and 0.36 M carrier in membrane phase suggested that 0.2 M ethanol in stripping phase is optimum for both extraction and recovery of catechin. Both extraction and recovery start declining beyond ethanol concentration of 0.2 M due to crowding of excess ethanol in the stripping phase which inhibits the catechin transportation from membrane–strip interface to the bulk strip phase.

Extraction decreases very slowly from 25 to 100 mg/l initial catechin concentration and then rapidly due to concentration polarization of the solute in the feed-membrane interface with respect to the carrier concentration. The optimum initial catechin concentration was found to be in the range of 100–150 mg/l for carrier concentration of 0.36 M.

A maximum flux of 5.7×10^{-9} g/cm^2 s across feed side interface at 2 h and 4.9×10^{-9} g/cm^2 s at 5 h across strip side interface was obtained for conditions of initial feed concentration = 100 mg/l, feed phase pH = 4.0, stirring speed = 400 rpm, strip phase concentration = 0.2 M. As assumed, complexation takes place instantaneously in the feed-side interface leading to high flux of solute (complex) through the feed side interface up to 2 h due to high concentration gradient of complex between two interfaces. However, because of low diffusivity of catechin through the membrane, maximum flux through strip–side interface occurs at 5 h.

9.3.1.4 Optimum Operation Conditions

LM-based separation is found to be an efficient way to extract bioactive compound like catechin. TBP–SFO–ethanol is found to be good carrier–solvent–stripping agent combination. The optimum conditions for the BLM studies are pH = 4.0, carrier concentration = 0.36 M, ethanol concentration = 0.2 M, stirring speed = 400 rpm and initial feed concentration = 100 mg/l. The feed phase must be acidic to ensure the catechin in its molecular state for extraction by a neutral carrier like TBP. The study demonstrates the promise of LM technology through use of physiologically benign vegetable oil for the recovery of a medicinal substance indicating an enormous scope for pharmaceutical industries.

REFERENCES

Alguacil, F.J., Coedo, A.G., Dorado, M.T., 2005. Transport of chromium(VI) through a Cyanex 923-xylene flat sheet supported liquid membrane. Hydrometallurgy 57, 51–56.

Alonoso, M., Delgado, A.L., Sastre, A.M., Alguacil, F.J., 2006. Kinetic modeling of the facilitated transport of cadmium using Cyanex 923 as ionophore. Chem. Eng. J. 118, 213–219.

Alpoguz, H.K., Memon, S., Ersoz, M., Yilmaz, M., 2002. Transport of Hg^{2+} through bulk liquid membrane using bis-calix[4]arene nitril derivative as carrier: kinetic analysis. New. J. Chem. 26, 477–480.

Araki, T., Tsukube, H., 1990. Liquid Membranes: Chemical Applications. CRC Press, Boca Raton, FL.

Arous, O., Gherrou, A., Kerdjoudj, H., 2004. Removal of Ag (l), Cu (II) and Zn (ll) ions with a supported liquid membrane containing cryptands as carrier. Desalination 161, 295–303.

Bachiri, A.E., Hag, A., Burgard, M., 1996. Recovery of silver nitrate by transport across a liquid membrane containing dicyclohexano-18-crown-6 as a carrier. J. Membr. Sci. 121, 159–168.

Bhatluri, K.K., Manna, M.S., Saha, P., Ghoshal, A.K., 2013. Separation of Cd (II) from its aqueous solution using environmentally benign vegetable oil as liquid membrane. Asia-Pacific J. Chem. Eng. 8, 775-785.

Boyadzhiev, L., Bezenshek, E., 1983. Carrier mediated extraction: application of double emulsion technique for mercury removal from wastewater. J. Membr. Sci. 14, 13–18.

Bunt, J.R., Waanders, F.B., 2008. Trace element behaviour in the Sasol–Lurgi MK IV FBDB gasifier. Part 1 – the volatile elements: Hg, As, Se, Cd and Pb. Fuel 87 (12), 2374–2387.

Castro, M.D.G., Riano, M.D.G., Vargas, M.G., 2004. Model experiment to test the use of a liquid membrane for separation and pre-concentration of copper from natural water. Anal. Chim. Acta 506, 81–86.

Chakrabarty, K., 2010. Liquid Membrane Based Technology for Removal of Pollutants from Wastewater (Ph.D. dissertation). Indian Institute of Technology Guwahati, Assam, India.

Chakrabarty, K., Saha, P., Ghoshal, A.K., 2010a. Separation of lignosulfonate from its aqueous solution using emulsion liquid membrane. J. Membr. Sci. 360, 34–39.

Chakrabarty, K., Saha, P., Ghoshal, A.K., 2010b. Separation of mercury from its aqueous solution through supported liquid membrane using environmentally benign diluent. J. Membr. Sci. 350, 395–401.

Chakrabarty, K., Saha, P., Ghoshal, A.K., 2010c. Simultaneous separation of mercury and lignosulfonate from aqueous solution using supported liquid membrane. J. Membr. Sci. 346, 37–44.

Chakrabarty, K., Saha, P., Ghoshal, A.K., 2009a. Separation of lignosulfonate from its aqueous solution using supported liquid membrane. J. Membr. Sci. 340, 84–91.

Chakrabarty, K., Krishna, K.V., Saha, P., Ghoshal, A.K., 2009b. Extraction and recovery of lignosulfonate from its aqueous solution using bulk liquid membrane. J. Membr. Sci. 330, 135–144.

Ellinghorst, G., Goetz, B., Niemoeller, A., Scholz, H., Brueschke, H.E.A., Tusel, G., 1989. Method for Production of Solution Diffusion Membranes and Their Application for Pervaporation, United States Patent no. 4865743, dated September 12, 1989.

Gherrou, A., Kerdjoudj, H., Molinari, R., Drioli, E., 2002. Removal of silver and copper ions from acidic thiourea solutions with a supported liquid membrane containing D2EHPA as carrier. Sep. Purif. Technol. 28, 235–244.

Ghodbane, I., Hamdaoui, O., 2008. Removal of mercury (II) from aqueous media using eucalyptus bark kinetic and equilibrium studies. J. Hazard. Mater. 160, 301–309.

Goyette, M.L., Longin, T.L., Noble, R.D., Koval, C.A., 2003. Selective photofacilitated transport of sodium ions through liquid membranes: key factors in experimental design, transport results and comparison with a mathematical model. J. Membr. Sci. 212, 225–235.

Gu, Z.M., 1988. Electrostatic pseudo-liquid membrane separation technology. Huagong Xuebao 6, 667.

Gu, Z.M., 1990. A new liquid membrane technology–electrostatic pseudo liquid membrane. J. Membr. Sci. 52 (1), 77–88.

He, D., Ma, M., 2000. Kinetics of cadmium (II) transport through a liquid membrane containing tricapryl amine in xylene. Sep. Sci. Technol. 35, 1573–1585.

Ho, W.S.W., Sirkar, K.K., 2001. Membrane Handbook. Kluwer academic publishers, Dordrecht.

Jabbari, A., Esmaeili, M., Shamsipur, M., 2001. Selective transport of mercury as $HgCl_4^{2-}$ through a bulk liquid membrane using K^+-dicyclohexyl-18-crown-6 as carrier. Sep. Purif. Tech. 24, 139–145.

Jee, C.F., Long, T.B., Xia, X.M., Jin, Q.Q., Ying, Z.L., 1985. A study on a two components liquid membrane system. J. Membr. Sci. 23, 137–154.

Jin, Y., Jin, C.H., Row, K.H., 2006. Separation of catechin compounds from different teas. Biotechnol. J. 1, 209–213.

Kamiński, W., Kwapiński, W., 2000. Applicability of liquid membranes in environmental protection. Pol. J. Environ. Stud. 9, 37–43.

Kedem, O., Essig, A., 1965. Isotope flows and flux ratios in biological membranes. J. Gen. Physiol. 48 (6), 1047–1070.

Kedem, O., Katchalsky, A., 1961. A physical interpretation of the phenomenological coefficients of membrane permeability. J. Gen. Physiol. 45 (1), 143–179.

Kontturi, A.K., Kontturi, K., Niinikoski, P., Sundholm, G. 1990a. Extraction of a polyelectrolyte using SLM –II extraction and fraction of lignosulfonate. Acta Chem. Scand. 44, 883–891.

Kontturi, A.K., Kontturi, K., 1987. A method for determination of ionic diffusion coefficients of polydisperse polyelectrolyte. J. Colloid Interface Sci. 120, 256–262.

Kontturi, A.K., Kontturi, K., Niinikoski, P., Sundholm, G., 1990b. Extraction of a polyelectrolyte using SLM –I. Acta Chem. Scand. 44, 879–882.

Kumbasar, R.A., 2008. Selective separation of chromium(VI) from acidic solutions containing various metal ions through emulsion liquid membrane using trioctylamine as extractant. Sep. Purif. Tech. 64, 56–62.

Leon, G., Guzman, M.A., 2005. Kinetic study of the effect of carrier and stripping agent concentrations on the facilitated transport of cobalt through bulk liquid membranes. Desalination 184, 79–87.

Li, N., 1968. Separating Hydrocarbons with Liquid Membranes, US Patent 3410794.

Li, Q., Liu, Q., Wei, X., 1996. Separation study of mercury through and emulsion membrane. Talanta 43, 1837–1842.

Lin, C.C., Long, R.L., 1997. Removal of nitric acid by emulsion liquid membrane: experimental results and model prediction. J. Membr. Sci. 134, 33–45.

Lopes, C.B., Lito, P.F., Otero, M., Lin, Z., Rocha, J., Silva, C.M., Pereira, E., Duarte, A.C., 2008. Mercury removal with titanosilicate ETS-4: batch experiments and modeling. Micropor. Mesopor. Mater. 115, 98–105.

Lv, J., Yang, Q., Jiang, J., Chung, T.S., 2007. Exploration of heavy metal ions transmembrane flux enhancement across a supported liquid membrane by appropriate carrier selection. Chem. Eng. Sci. 62, 6032–6039.

Manna, M.S., Bhatluri, K.K., Saha, P., Ghoshal, A.K., 2012. Transportation of catechin (±C) using physiologically benign vegetable oil as liquid membrane. Ind. Eng. Chem. Res. 51, 15207–15216.

Marr, R., Kopp, A., 1982. Liquid membrane technology – a survey of phenomena, mechanisms and models. Int. Chem. Eng. 22 (1), 44–60.

Monlinari, R., Poerio, T., Argurio, P., 2006. Selective removal of Cu^{++} versus Ni^{++}, Zn^{++} and Mn^{++} by using a new carrier in a supported liquid membrane. J. Membr. Sci. 280, 470–477.

Mortaheb, H.R., Zolfaghari, A., Mokhtarani, B., Amini, M.H., Mandanipour, V., 2010. Study on removal of cadmium by hybrid liquid membrane process. J. Hazard. Mater. 177, 660–667.

Mulder, M., 1991. Basic Principles of Membrane Technology. Kluwer Academic Publishers, Dordrecht.

Muthuraman, G., Teng, T.T., Leh, C.P., Norli, I., 2009. Use of bulk liquid membrane for the removal of chromium (VI) from aqueous acidic solution with tri-n-butyl phosphate as a carrier. Desalination 249, 884–890.

Nernst, W., Riesenfeld, E.H., 1901. Ueber elektrolytische erscheinungen an der grenzfläche zweier lösungsmittel. Nachr. Königliche Ges. Wiss. Gött. 1, 54–61.

Neková, P., Schrotterova, D., 1999. Extraction of Mo (VI) by primene JMT. Solvent Extr. Ion. Exch. 17, 163–175.

Noble, R.D., Way, J.D., 1987. Liquid Membrane Technology. National Bureau of Standards, Centre for Chemical Engineering, Boulder.

Othman, N., Goto, M., Mat, H., 2004. Liquid Membrane Technology for Precious Metals Recovery from Industrial Waste. University Teknologi Malaysia (Johor) and Kyushu University (Japan).

Pérez Carrera, A, y Fernández Cirelli, A., 2007. Problemática del arsénico en la llanura sudeste de la provincia de Córdoba. Biotransferencia a leche bovina. Revista Investigación Veterinaria (INVET) 9 (1), 123–125.

Rajan, M., Shinde, V.M., 1996. Synergistic extraction and separation study of urenium (VI). J. Radioanal. Nucl. Chem. 203, 169–176.

Rajesh, N., Hari, M.S., 2008. Spectrophotometric determination of inorganic mercury (II) after preconcentration of its diphenylthiocarbazone complex on a cellulose column. Spectrochim. Acta Part A 70, 1104–1108.

Sahoo, G.C., Dutta, N.N., 1998. Studies on Emulsion Liquid Membrane Extraction of Cephalexin. J. Membr. Sci. 145, 15–26.

Sekine, T., Inaba, K., Aikawa, H., 1988. Rate of solvent extraction of Cr (III) with acetylacetone into carbon tetrachloride. Bull. Chem. Soc. Jpn. 61, 1131–1134.

Sengupta, A., Raghuraman, B., Sirkar, K.K., 1990. Liquid membranes for flue gas desulfurization. J. Membr. Sci. 51 (1–2), 105–126.

Senkal, B.F., Yavuz, E., 2007. Sulfonamide based polymeric sorbents for selective mercury extraction. React. Funct. Polym. 67, 1465–1470.

Shaik, A., Chakrabarty, K., Saha, P., Ghoshal, A.K., 2010. Separation of Hg (II) from its aqueous solution using bulk liquid membrane. Ind. Eng. Chem. Res. 49, 2889–2894.

Shamsipur, M., Mashhadizadeh, M.H., Azimi, G., 2002. Highly selective and efficient transport of mercury (II) ions across a bulk liquid membrane containing tetrathia-12-crown-4 as a specific ion carrier. Sep. Purif. Tech. 27, 155–161.

Shen, J.Q., Yin, W.P., Zhao, Y.X., Yu, L.J., 1996. Extraction of alanine using emulsion liquid membranes featuring a cationic carrier. J. Membr. Sci. 120, 45–53.

Swain, B., Jeong, J., Chun, J., Gae, L., Lee, H., 2007. Extraction of Co (II) by supported liquid membrane and solvent extraction using Cyanex-272 as an extractant: a comparison study. J. Membr. Sci. 288, 139–148.

Takeuchi, H., Takahashi, K., Goto, W., 1987. Some observations on the stability of supported liquid membranes. J. Membr. Sci. 34, 19–31.

Uludag, Y., Ozbelge, H., Yilmaz, L., 1997. Removal of mercury from aqueous solutions via polymer enhanced ultrafiltration. J. Membr. Sci. 129, 93–99.

Wang, Y.Y., Juang, R.S., 2002. Amino acid separation with D2EHPA by solvent extraction and liquid surfactant membranes. J. Membr. Sci. 207, 241–252.

Warey, P.B., 2007. New Research on Hazardous Materials. Nova Publishers, New York.

Yordanov, B., Boyadzhiev, L., 2004. Pertraction of citric acid by means of emulsion liquid membranes. J. Membr. Sci. 238, 191–197.

Zambrano, J.B., Laborie, S., Viers, P., Rakib, M., Durand, G., 2004. Mercury removal and recovery from aqueous solutions by coupled complexation–ultrafiltration and electrolysis. J. Membr. Sci. 229, 179–186.

Zha, F.F., Fane, A.G., Fell, C.J.D., 1995. Instability mechanism of supported liquid membranes in phenol transport process. J. Membr. Sci. 107, 59–74.

Zhongqi, R., Zhang, W., Liu, Y.M., Dai, Y., Cui, C., 2007. New liquid membrane technology for simultaneous extraction and stripping of copper (II) from wastewater. Chem. Eng. Sci. 62, 6090–6101.

Zhou, D.W., 1996. Emulsion Liquid Membrane Removal of Arsenic and Strontium from Wastewater an Experimental and Theoretical Study (Ph.D. dissertation). New Jersey Institute of Technology.

CHAPTER SIX

Electrodialysis

Yoshinobu Tanaka
IEM Research, Ushiku-shi, Ibraki, Japan

Contents

Progress in Filtration and Separation
http://dx.doi.org/10.1016/B978-0-12-384746-1.00006-9

1. INTRODUCTION

Ion-exchange membrane electrodialysis is a process to separate or extract ionic species in an electrolyte solution. The largest scale application of the electrodialysis is the demineralization of saline water. Other applications are desalination and reuse of sewage or industrial waste, recovery of useful components from a metal surface treatment process, refining of amino acid solution, desalination of milk, whey, sugar liquor, concentration of seawater, organic acid, etc. Electrodialysis is the fundamental technology based on the ion-exchange membranes, and it is applied to the succeeding technology such as bipolar membrane electrodialysis, electrodeionization, electrolysis, diffusion dialysis, fuel cell, etc.

Looking back over the development in the history of on ion-exchange membrane electrodialysis, notable achievement is the theory of mass transport; the Nernst–Planck equation developed by Planck (1890). Donnan (1924) presented the membrane equilibrium theory. Michaelis and Fujita (1925) studied permselectivity of biological membranes using colloidal membranes. Teorell (1935) and Meyer and Sievers (1936) discussed membrane phenomena and revealed the mechanism of membrane potential, electric conductivity, transport number, etc. Meyer and Straus (1940) electrodialyzed a KCl solution and realized doubled KCl concentration. Wyllie and Patnode (1950) and Juda and McRae (1950) invented artificial ion-exchange membranes and the membranes were applied to electrodialysis for desalting saline water. Ion-exchange membrane electrodialysis is now one of the basic technologies in saline water desalination industry.

This chapter introduces computer simulation of an ion-exchange membrane electrodialysis process for saline water desalination. Electrodialysis operation is classified into (1) continuous (single-pass), (2) batch, and (3) feed-and-bleed (Schaffer and Minz, 1966; Yawataya, 1986). In the continuous operation, a feeding solution flows from a feed source to an electrodialyzer and a desalted solution is produced continuously. In order to increase the desalting ratio, the continuous process is formed in multistages. In the batch operation, the desalted solution is recycled through a circulation tank and an electrodialyzer. The solution in the circulation tank is gradually desalted and finally the desalted solution is taken out from the tank. In the feed-and-bleed operation, a feeding solution is recycled between a circulation tank and an electrodialyzer continuously. The recycle rate is determined as desired from the targeted concentration changes of the desalted solution.

An electrodialyzer and a circulation tank are the main units working in the electrodialysis operation. The function of the electrodialyzer is computed with the continuous program which is developed for analyzing the performance of the continuous electrodialysis operation with no circulation tank. The batch and feed-and-bleed operations are carried out with the electrodialyzer and the circulation tank. The circulation tank is an effective working unit in the batch and feed-and-bleed operations. The computer program of the batch and feed-and-bleed operations consists of (1) the continuous program which describes the function of the electrodialyzer and (2) the batch and feed-and-bleed programs which describe the function of the circulation tank (Tanaka, 2015).

The program functions as a pilot plant operation and computes the performance of the electrodialyzer such as; ion and solution flux across a membrane pair, cell voltage, current density, energy consumption, water recovery, limiting current density, pressure drop, etc. The electrodialysis operation is classified into constant current mode, constant voltage mode, and constant salt concentration mode. The program for each mode is integrated into spreadsheets and the computation is carried out with a trial-and-error calculation by inputting electrodialyzer specifications and operating conditions and adjusting control keys. The algorithm is incorporated in common software (Excel). It can be computed with use of several pieces of papers, an ink cartridge and an ordinary hardware (Windows) within about 10 min. There are no specific hardware requirements. The spreadsheets are included in separate web sites (Tanaka, 2015). Readers can operate the program using the web sites (Companion sites) and discuss the performance of the electrodialysis process. The continuous program is the most fundamental one. The batch and feed-and-bleed program are established based on the continuous program. Thus the continuous program is explained definitely by arranging equations systematically in Section 3. The batch and feed-and-bleed program are explained respectively in Sections 4 and 5. All equations in the programs are integrated in the companion sites (Tanaka, 2015). The equations in the program are described using the cgs unit system because the equations are developed based on the fundamental principles of electrochemistry. However, in order to apply these equations to practical operations, it is necessary to replace the cgs units with practical units in the final stage. For examples, salt concentration is computed using the fundamental cgs unit eq/cm^3, but it is converted to practical unit $mg/dm^3 = ppm$. Current density is converted from A/cm^2 to A/dm^2. Energy consumption is converted from Ws/cm^3 to kWh/m^3. Because of the above equation development for improving the program availability, the units in this chapter are not unified and are mixed.

2. ELECTRODIALYZER

Many ion-exchange membranes are integrated in an electrodialyzer. Electrodialyzers are arranged to form an electrodialysis process. The electrodialyzer is classified to a sheet flow type operated with an one directional direct current and a tortuous flow type operated with a two directional direct current. The electrodialysis (ED) process is operated with an one directional current. The electrodialysis reversal (EDR) process is operated with periodic reversal of two directional currents. Both processes have extremely unique features and are applied according to the requirements for the process.

2.1 Structure of an Electrodialyzer and Process Operations

The basic structure of an electrodialyzer consists of stacks in which cation-exchange membranes, anion-exchange membranes, gaskets (desalting cells and concentrating cells) are arranged alternately (Figure 1). Fastening frames are put on both outsides of the stack which is fastened up together through crossbars setting in the frames. The deformation of the membranes is prevented by regulating hydrostatic pressure in the fastening frames. Inlet manifold slots and outlet manifold slots are prepared at the bottoms and heads of the gaskets, respectively. Spacers are incorporated with the gaskets to prevent the contact of cation-exchange membranes with anion-exchange membranes and to mix the solution. Many stacks are arranged through the fastening

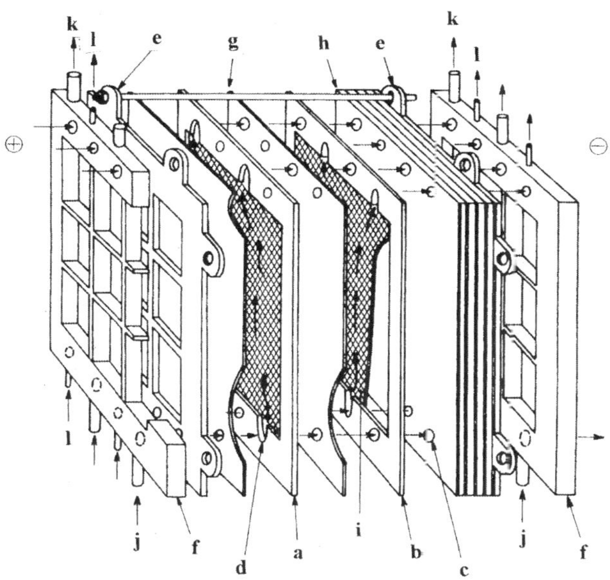

Figure 1 Structure of a stack (Sheet flow type). (a) desalting cell, (b) concentrating cell, (c) manifold, (d) slot, (e) fastening frame, (f) feeding frame, (g) cation exchange membrane, (h) anion exchange membrane, (i) spacer, (j) feeding solution, (k) desalted solution, and (l) concentrated solution. *(Azechi, 1980).*

frames, and electrode cells are put on both ends of the electrodialyzer, which are fastened by a press putting on the outsides of electrode cells (Figure 2).

An electrolyte solution to be desalted is supplied from solution feeding frames to entrance manifolds, flows through entrance slots, current passing portions and exit slots, and discharged from exit manifolds to the outside of the stack (Figures 1 and 2). A concentrated solution is supplied to concentrating cells with a part of a feeding solution in a circulation flow system, and discharged to the outside of the stack through an overflow extracting system.

Effective membrane area is in the range from less than 0.5 m^2 to about maximum 2 m^2. In order to reduce energy consumption, it is desirable to decrease the electric resistance of the membrane and gasket thickness. Gasket material is selected from synthesized rubber, polyethylene, polypropylene, polyvinyl chloride and ethylene–vinyl acetate copolymer, etc. The spacer is usually incorporated with the gasket and a solution flows dispersing along the spacer net.

A continuous, a batch, or a feed-and-bleed electrodialysis system is formed at first. A feeding solution is filtered and supplied to the electrodialyzer via a filtered solution tank. A part of the filtered solution is supplied to electrode chambers and a concentrated solution tank. Titanium is adopted as the anode material. In order to avoid membrane destruction due to Cl_2 and HClO generated by an anode reaction, a perfluorinated ion-exchange membrane is inserted between the cathode chamber and the adjacent stack. A washed solution from the anode chamber is mixed with the filtered feeding solution to suppress the growth of microorganisms in the feeding solution. An HCl solution is supplied to the cathode chamber to neutralize OH^- ions generated by the cathode reaction. A washing system is provided

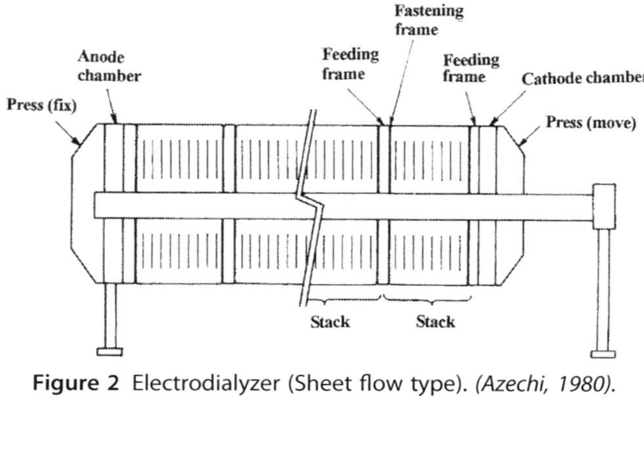

Figure 2 Electrodialyzer (Sheet flow type). *(Azechi, 1980).*

for washing the inside of desalting cells by acid or chemical reagents and dissolving adhered substances. Some times, fine organisms pass through the filter and breed in the electrodialyzer. These troubles are avoided by disassembling and washing the electrodialyzer.

2.2 Parts of an Electrodialyzer

The electrodialyzer is composed of the parts as follows.

2.2.1 Fastening Frame

Maximum 2000 pairs of membranes are arranged between electrodes in an electrodialyzer, so as to let disassembling and assembling works be easy. The membrane array is divided further into several stacks consisting of 50–400 pairs. Fastening frames are fixed by bolts on both ends of the stack. The fastening frame is usually served as a solution feeding frame, so that a desalting and a concentrating solution are supplied to each gasket cell from the feeding frame incorporated in every stack. Material of the fastening frame is selected from polyvinyl chloride, polypropylene, and rubber-lined iron, etc.

2.2.2 Solution Feeding Frame

A solution feeding frame is integrated for feeding solutions to each desalting and concentrating cell. Manifold holes are prepared at corresponding positions of the holes fitted in the gasket. Solutions are usually supplied through the manifolds to each stack, but as the case may be supplied to each plural stack.

2.2.3 Gasket

The shape of the gasket is presented in Figure 3. A solution is supplied from the inlet manifold put at the bottom, flows through the slot, and is fed into

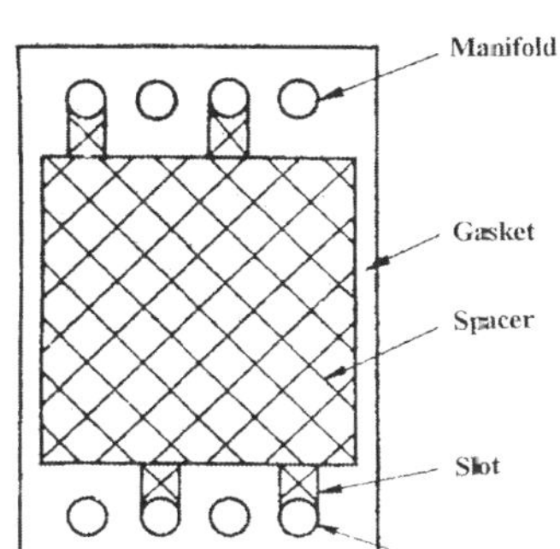

Figure 3 Shape of the gasket. *(Urabe and Doi, 1978).*

the current passing portion. Then the solution is discharged through the outlet slot to the manifold fitted to the head. The gasket has the following functions: (1) prevents solution leakage from the inside to the outside of the electrodialyzer, (2) adjusts the distance between a cation-exchange membrane and an anion-exchange membrane, and (3) prevents solution leakage between a desalting cell and a concentrating cell occurring at a slot section. In order to prevent the solution leakage, it is desirable to adopt a soft material for the gasket. On the other hand, it is desirable to adopt a hard and stable material to avoid dimension changes during a long-term operation. The material of the gasket is selected from rubber, ethyl-vinyl acetate copolymer, polyvinyl chloride, polyethylene, etc. The thickness of the gasket is in the range of 0.5–2.0 mm.

2.2.4 Slot

It is important to reduce the inside solution leakage, which arises through pinholes and cracks in the membranes or through gaps due to the membrane deformation at the slot as shown in Figure 4. In order to prevent these troubles, a lot of devices are proposed as exemplified in Figure 5 in which (a) decrease the width of the slot, (b) bend the slot, and (c) insert the support in the slot.

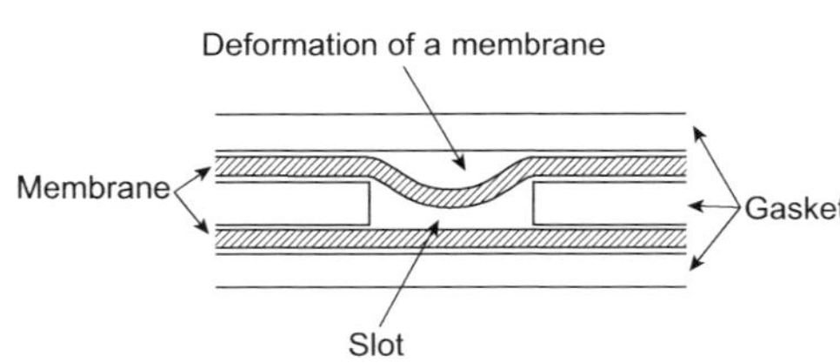

Figure 4 Deformation of an ion-exchange membrane. *(Urabe and Doi, 1978).*

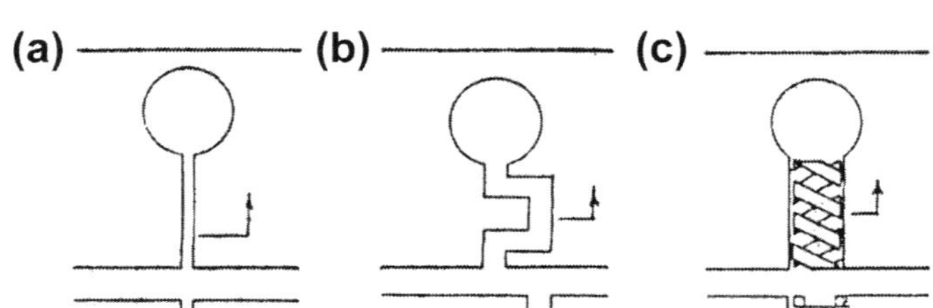

Figure 5 Structure of slots. *(Urabe and Doi, 1978).*

2.2.5 Spacer

The function of a spacer is to keep the distance between the membranes. In addition, the spacer increases the limiting current density due to solution disturbance. The spacer is selected taking account of the requirements such as (1) low friction head loss, (2) low electric current screening effect, (3) easy air discharge, and (4) less blocking of a flow-pass caused by the precipitation of fine particles suspended in a feeding solution. The structures of a spacer are classified in Figure 6, as (a) expanded polyvinyl chloride, (b) wave porous plate, (c) diagonal net, (d) mikoshiro texture, and (e) honeycomb net.

2.2.6 Electrode and Electrode Chamber

Platinum plated titanium, graphite, or magnetite is used for anode material and stainless or iron is used for cathode material. The shape of electrodes is classified into net, bar, and flat. A partition is inserted between an electrode chamber and a stack for preventing the mixing of solutions. In an anode chamber, oxidizing substances such as chlorine gas evolve. An ion-exchange membrane (hydrocarbon membrane) is easily deteriorated by contact with the oxidizing substances, so it is necessary to use

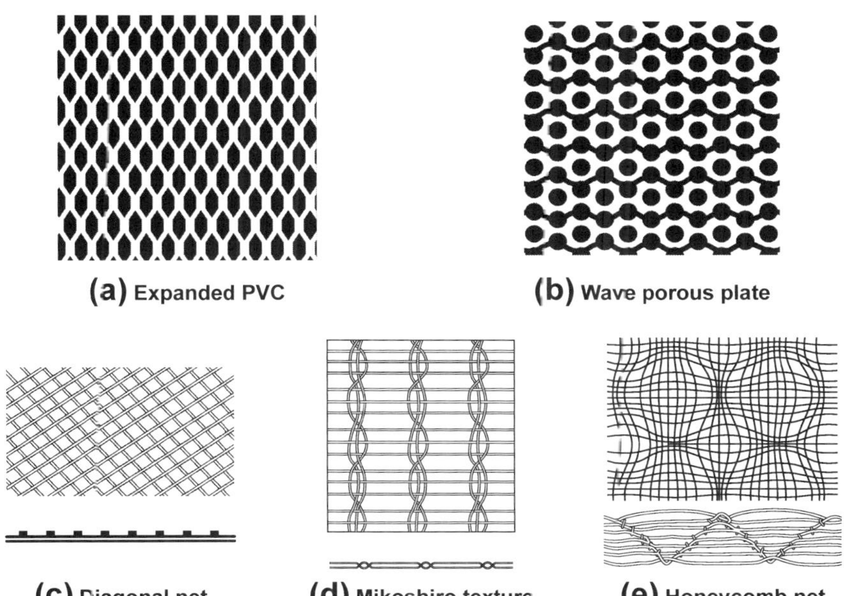

Figure 6 Structure of spacers. *(Urabe and Doi, 1978).*

two sheets of partitions and put a buffer chamber between the two partitions. Material of the partition is an ion-exchange membrane (fluorocarbon membrane), an asbestos sheet, or a battery partition.

An acid solution is added into a cathode solution and the electrodialyzer is operated under controlling pH of the cathode solution for preventing the precipitation of magnesium hydroxides in the cathode chamber. A feeding solution or a concentrated solution is supplied into the electrode chamber. The concentration of oxidizing substances in the anode solution is reduced by adding sodium sulfite or sodium thiosulfate into the solution being discharged. Sometimes, a sodium sulfate solution is supplied to an anode and a cathode chamber, achieving the neutralization by mixing the effluent of both chambers.

2.2.7 Press

An oil pressure press is usually used to adjust the pressure to 5–10 kg/cm^2.

2.3 Requirements for Improving the Performance of an Electrodialyzer

In order to improve the performance of an electrodialyzer, membrane characteristics should be naturally improved. At the same time, the circumstances in an electrodialyzer in which the membranes work should be better. The following sections describe the definite problems lowering the circumstances and performance of an electrodialyzer.

2.3.1 Solution Velocity Distribution between Desalting Cells

In an electrodialyzer, ion-exchange membranes, desalting cells, and concentrating cells are arranged alternately and a solution is supplied into desalting cells. In this flow system, the solution velocity distribution in desalting cells does not become uniform. This phenomenon causes the concentration distribution, electric resistance distribution and current density distribution in the electrodialyzer, and gives rise to the decrease of the limiting current density of the electrodialyzer. In order to operate the electrodialyzer stably, it becomes necessary to make the solution velocities between the desalting cells uniform.

2.3.2 Solution Leakage in an Electrodialyzer

The dimensions of all parts of an electrodialyzer are not always consistent with the figures in the specifications. Small pinholes can open in an electrodialyzer because the mechanical strength of ion-exchange membranes is relatively low. Gaps may occur between the materials composing the electrodialyzer in the assembling works of an electrodialyzer. If a pressure difference between the desalting cells and concentrating cells exists in these circumstances, solutions leak through the membranes and the leak lowers the performance of the electrodialyzer. In order to avoid these troubles, we have to remove the pinholes and gaps in the electrodialyzer and control the pressure difference between desalting cells and concentrating cells.

2.3.3 Distance between Membranes

Decrease of the distance between the membranes brings about the decrease of electrical resistance and energy consumption. On the other hand, it brings about the increase of friction loss of solution flow, blocking of the materials suspended in a feeding solution and increase of pumping motive power. Accordingly, it becomes necessary to realize the optimum distance between the membranes. The optimum distance is decided further taking account of electric resistance of ion-exchange membranes and that of electrolyte solutions in desalting and concentrating cells.

2.3.4 Spacer

Main functions of a spacer is to create space between a cation-exchange membrane and an anion-exchange membrane. When solution velocity and the Reynolds number are decreased, hydrodynamic pattern exhibits laminar flow, which means that disturbing effect of the spacer is low. In order to increase the limiting current density, turbulent flow should be induced by increasing the Reynolds number.

2.3.5 Electric Current Leakage

A part of electric current flows through slots and manifolds causing ineffective current leakage. Current leakage is increased by the increase of the number of cell pairs integrated in a stack and the increase of sectional area of slots and manifolds. These events, however, related to the solution velocity distribution between the cells.

2.3.6 Simplicity of Structure of an Electrodialyzer

Disassembling and assembling works are peculiar features in operating an electrodialyzer. Excellent durability of ion-exchange membranes is owing to careful treatment in this work. So, the simplicity of the structure is a requirement for handling this work.

3. CONTINUOUS (SINGLE-PASS) ELECTRODIALYSIS PROGRAM

3.1 Overview

The continuous process is applicable to large-scale electrodialysis and its performance was so far investigated as follows. Belfort and Daly (1970) constructed optimization routine for a continuous electrodialysis plant. The algorithm was applied to the Office of Saline Water test bed plant at Webster, South Dakota and compared the actual cost and operating conditions. Avriel and Zeligher (1972) developed a mathematical model for preliminary engineering design and economical evaluation of a continuous electrodialysis plant. Detailed cost computations were performed resulting in capital investment and annual operating costs. Lee et al. (2002) developed a computer simulation program for describing a continuous flow process and estimated investment and operation costs. Further, electrodialysis plant was designed and optimized in terms of overall costs and the different parameters. Moon et al. (2004) investigated ionic transport across membranes based on the one- and two-dimensional continuous electrodialysis modeling using the principles of electrochemistry, transport phenomena, and thermodynamics. Fidaleo and Moresi (2005) simulated mass transfer, mass balance, potential drop, and limiting current density in a continuous operation based on the Nernst–Planck equation. Sadrzadeh et al. (2007) modeled continuous flow desalination starting from a differential equation of steady-state mass balance and giving salt concentration in dilute compartments or separation percent for various voltages, flow rates, and feed concentrations. Nikonenko et al. (2008) described electrodialysis or electrodeionization characteristics (mass transfer coefficient, Sherwood number, degree of desalination, and others). The reasonability of the functions was discussed with experimental measurements. Brauns et al. (2009) developed a simulation model through solver software. Experimental verification of the software was performed using industrial type pilot plant. Limiting current density was theoretically evaluated in the model calculations for design purpose and corrected with the experimental results. Tanaka developed a program for computing the performance of a continuous constant current mode seawater concentrating process. In this research, the reliability of the program was discussed by comparing the calculated data with experimental data (Tanaka, 2011b) and industrial data (Tanaka, 2003, 2010). The principles of seawater concentration are fundamentally equivalent to saline water desalination. The program developed in the seawater concentration process is applicable to saline water desalination by taking account of the difference of operating conditions in both processes. The above program was applied to saline water desalination (Tanaka, 2012a, 2015), which is described in this section.

3.2 Continuous Electrodialysis Process

The continuous electrodialysis process (single stage) is illustrated in Figure 7. The electrodialyzer is incorporated with the desalting cells, concentrating cells in the stack marked with gray. Figure 8 illustrates the structure of the desalting and concentrating cell. Number of cells is N for desalting cells and $N + 1$ for concentrating cells. The anode and cathode cells are placed at both outsides of the stack and an electric current is passed through the electrodes. A feeding solution of salt concentration C'_{in} is supplied to desalting cells at linear velocity of u'_{in} at the inlets. C'_{in} and u'_{in} are decreased to respectively C'_{out} and u'_{out} at the outlets of desalting cells. A part of the feeding solution is also supplied to concentrating cells for preventing scale formation in the concentrating cells. The salt concentration and linear velocity in concentrating cells are respectively $C''_{\text{in}} = C'_{\text{in}}$ and u''_{in} at the inlets and C''_{out} and u''_{out} at the outlets. Partition cells are incorporated between the stack and electrode cells for preventing the influence of electrode reactions to the performance of the electrodialyzer. A part of a concentrated solution flowing from the outlets of concentrating cells is supplied to electrode cells and partition cells.

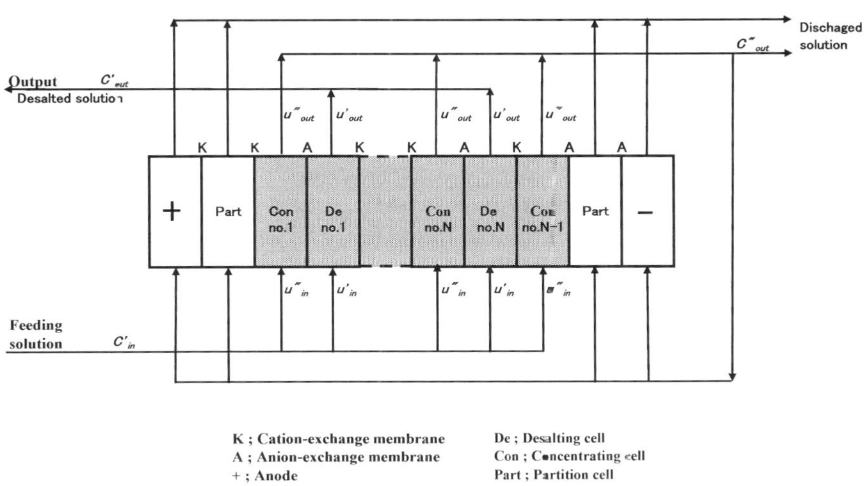

Figure 7 Continuous electrodialysis process. *(Tanaka, 2012a).*

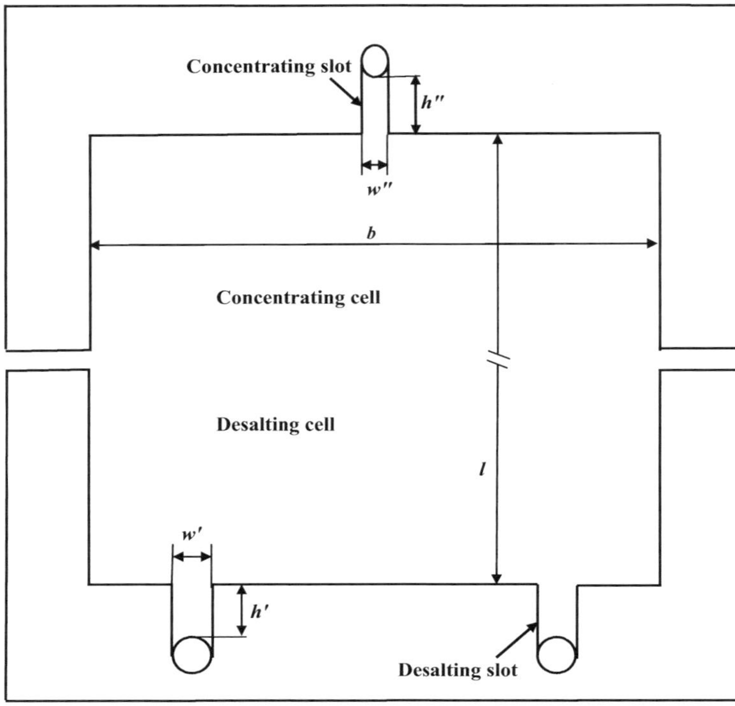

Figure 8 Structure of a desalting cell and a concentrating cell. *(Tanaka, 2012a).*

Specifications of an electrodialyzer and its operating conditions are described below.

1. Flow-pass thickness in a desalting and a concentrating cell; a (cm)
2. Flow-pass width in a desalting and a concentrating cell; b (cm)
3. Flow-pass length in a desalting and a concentrating cell; l (cm)
4. Membrane area; $S = bl$ (cm^2)
5. Number of stacks in an electrodialyzer: 1
6. Number of desalting cells, concentrating cells, and membrane pairs integrated in a stack; N, $N+1$, N
7. Probe electrodes are inserted into concentrating cells integrated at the ends of a stack for measuring cell voltage
8. Average current density; I/S (A/cm^2)
9. Salt concentration at the inlets of desalting cells; C'_{in} (eq/cm^3)

10. Salt concentration at the inlets of concentrating cells; $C''_{in} = C'_{in}$ (eq/cm^3)
11. Linear velocity at the inlets of desalting and concentrating cells; u'_{in}, u''_{in} (cm/s)
12. Temperature; T (°C)
13. Standard deviation of the normal distribution of solution velocity ratio; σ
14. Flow system in desalting and concentrating cells; single-pass flow
15. Diagonal net spacers are integrated in desalting and concentrating cells and in desalting and concentrating slots. Dimensions of a spacer are:
    Diameter of a spacer rod d_S = half thickness of the cell $a/2$
    Distance between the rods χ
    Crossing angle of the rods θ
16. Number and dimension of the cells and slots in a cell pair are:

Number	Number	Thickness	Width	Length
Desalting cell	1	a	b	l
Concentrating cell	1	a	b	l
Desalting slot	n'	a	w'	h'
Concentrating slot	n''	a	w''	h''

3.3 Mass Transport in an Electrodialyzer

Mass transport in an electrodialyzer is illustrated in Figure 9. A salt solution (raw salt solution, concentration: C'_{in}) is supplied to the inlets of desalting cells (De) at average linear velocity of u'_{in}. For preventing scale formation in concentrating cells, a part of a raw salt solution is supplied also to the inlets of concentrating cells (Con) at the average linear velocity of u''_{in}. By supplying an electric current I, ions and solutions are transferred from desalting cells to concentrating cells across an ion-exchange membrane pair and their flux is defined by J_S and J_V respectively. In desalting (concentrating) cells, salt concentration is decreased (increased) from C'_{in} ($C''_{in} = C'_{in}$) under applied average current density I/S and reaches average salt concentration C'_{out} (C''_{out}) at the outlets of desalting (concentrating) cells. Salt concentration change in desalting cells causes current density change along the flow-pass from i_{in} at the inlets to i_{out} at the outlets. The current density becomes i at x distance from the inlets of desalting cells. $I/S, J_S, J_V, C'_p, C''_p, u'_p$ and u''_p are altogether the values at $x = pl$ distance from the inlets of desalting and concentrating cells. V_{in}, V_{out} and V_p are voltage difference between electrodes respectively at the inlets ($x = 0$), the outlets ($x = l$), and $x = pl$ of desalting cells ($V_{in} = V_p = V_{out}$).

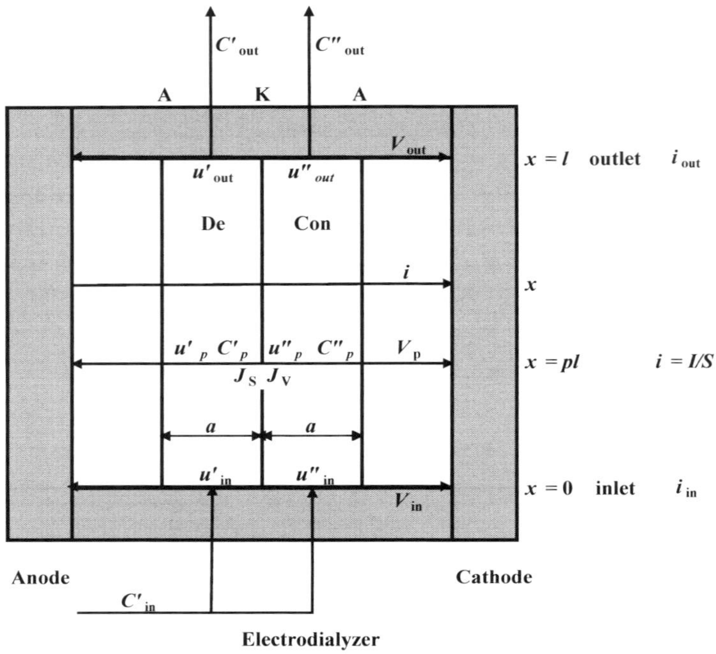

De: Desalting cell, Con: Concentrating cell
K: Cation exchage membrane, A: Anion exchange membrane
J_S, J_V: **Fluxes of ions and solutions across membrane pairs at** $x = pl$
C'_p, C''_p: **Electrolyte concentration in desalting and concentrating cells at** $x = pl$
u'_p, u''_p: **Linear velocity in desalting cells and concentrating cells at** $x = pl$
$V_{in} = V_p = V_{out}$

Figure 9 Mass transport in desalting cells and concentrating cells. *(Tanaka, 2012a).*

3.4 Overall Mass Transport Equation and Membrane Characteristics

Fluxes of ions J_S and a solution J_V across an ion-exchange membrane pair at $x = pl$ distance from the inlets of desalting and concentrating cells are expressed by the following overall mass transport equation (Tanaka, 2006).

$$\begin{aligned} J_s &= \lambda(I/S) - \mu\left(C''_p - C'_p\right) \\ &= (t_K + t_A - 1)(I/S)/F - \mu\left(C''_p - C'_p\right) = \eta(I/S)/F \end{aligned} \tag{1}$$

$$J_V = \phi(I/S) + \rho\left(C_p'' - C_p'\right) \tag{2}$$

where λ (eq/C) is the overall transport number, μ (cm/s) is the overall solute permeability, ϕ (cm^3/C) is the overall electro-osmotic permeability and ρ (cm^4/eq s) is the overall hydraulic permeability. t is the transport number of counter-ions in the membrane.η is current efficiency and F is the Faraday constant.

ρ versus λ, μ, and ϕ plots are given by the following empirical equations.

$$\lambda = 9.208 \times 10^{-6} + 1.914 \times 10^{-5}\rho \tag{3}$$

$$\mu = 2.005 \times 10^{-4}\rho \tag{4}$$

$$\phi = 3.768 \times 10^{-3}\rho^{0.2} - 1.019 \times 10^{-2}\rho \tag{5}$$

ρ versus alternating current electric resistance of an ion-exchange membrane pair $r_{alter,K} + r_{alter,A}$ is expressed by the following equation.

$$r_{alter} = r_{alter,K} + r_{alter,A} = 1.2323\rho^{-(1/3)} \tag{6}$$

The relationship between ρ and solution temperature T (°C) is approximated by the following equation (Tanaka, 2011a).

$$\rho = 3.421 \times 10^{-3} + 3.333 \times 10^{-4}T \tag{7}$$

The above equations mean that the leading parameter ρ represents all of the overall membrane characteristics and that λ, μ, ϕ, $r_{alter,K} + r_{aler,A}$, and ρ are determined by setting T.

3.5 Salt Concentration and Linear Velocity in Desalting Cells

Salt solutions are supplied to desalting cells and pass through the cells by single-pass flow.

This phenomenon is expressed by the following equations.

$$q'_{in}\left(\mathrm{cm}^3/\mathrm{s}\right) = ab \times u'_{in}(\mathrm{cm/s}) \tag{8}$$

$$q'_{out}\left(cm^3/s\right) = q'_{in} - q \tag{9}$$

$$q\left(\mathrm{cm}^3/\mathrm{s}\right) = blJ_V \tag{10}$$

$$u'_{out}\left(\mathrm{cm/s}\right) = q'_{out}/ab \tag{11}$$

Assuming the linear velocities to be changed linearly in desalting cells, average linear velocity at $x = pl$ distance from the inlets of desalting cells u'_p is

$$u'_{\mathrm{p}} = u'_{\mathrm{in}} - p\left(u'_{\mathrm{in}} - u'_{\mathrm{out}}\right) = u'_{\mathrm{in}} - p(q/ab) \tag{12}$$

p in Eqn (12) is calculated using Eqns (67), (80) and (81) in "Section 3.10 Current density distribution" in the course of the trial-and-error calculation. C'_{out} is introduced as follows.

$$C'_{\mathrm{out}}\left(\mathrm{eq/cm^3}\right) = C'_{\mathrm{in}} - \left(\frac{\eta l}{aF}\right)\left(\frac{1}{u'_{\mathrm{p}}}\right)\left(\frac{I}{S}\right) \tag{13}$$

Linear velocities in desalting cells are not uniform between the cells. The limiting current density $(I/S)_{\mathrm{lim}}$ is strongly influenced by the solution velocity distribution between desalting cells. Solution velocity ratio in desalting cells is defined as:

$$\xi = \frac{u - \overline{u}}{\overline{u}} \tag{14}$$

u is the linear velocity in each desalting cell and $\overline{u}$ is the average linear velocity in desalting cells integrated in a stack. We assume here that the frequencies Y_j of ξ in group j; ξ_j is expressed by the normal distribution (standard deviation; σ) (Tanaka, 2005). We divide all ξ_j among $0 - n$.

$$\xi_j = -3\sigma\left(1 - \frac{j}{n/2}\right), \quad j = 0 - n \tag{15}$$

Linear velocity at $x = pl$ distance from the inlet of the jth desalting cell is

$$u'_{\mathrm{p},j} = u'_{\mathrm{p}}\left(\xi_j + 1\right) \tag{16}$$

Accordingly

$$C'_{\mathrm{out},j} = C'_{\mathrm{in},j} - \left(\frac{\eta l}{aF}\right) \times \frac{1}{u'_{\mathrm{p}}\left(\xi_j + 1\right)}\left(\frac{I}{S}\right) \tag{17}$$

in which,

$$C'_{\mathrm{in},j} = C'_{\mathrm{in}}$$

Salt concentration at $x = pl$ distance from the inlets of desalting cells C'_{p} is computed by summing up $C'_{\mathrm{p},j}$ between $j = 0$ and $j = n$ and calculating the average of the summation for N cells.

$$C'_{\mathrm{p}} = \frac{1}{N} \sum_{j=0}^{n} Y_j C'_{\mathrm{p},j} \tag{18}$$

in which

$$C'_{\mathrm{p},j} = C'_{\mathrm{in},j} - \left(\frac{\eta l}{aF}\right) \times \frac{1}{u'_{\mathrm{p}/2}(\xi_j + 1)} \left\{a_1 p + (a_2/2)p^2 + (a_3/3)p^3\right\} \tag{19}$$

3.6 Salt Concentration and Linear Velocity in Concentrating Cells

For preventing scale formation in concentrating cells, salt solutions supplied to the desalting cells are also fed to the inlets of concentrating cells at the average linear velocity of u''_{in}. and they pass through the concentrating cells by single-pass flow. Linear velocities are also not uniform between concentrating cells as in the desalting cells. However, the solution velocity distribution in concentrating cells never influences to the limiting current density, so we assume the uniform velocity distribution between concentrating cells. The equations are introduced as follows.

The material balance of solutions and ions transferring from desalting cells to concentrating cells at $x = pl$ is given by the following equations (cf. Eqns (1) and (2)).

Solutions

$$abN\left(u''_{\mathrm{out}} - u''_{\mathrm{in}}\right) = SNJ_{\mathrm{V}} = lbN\left\{\phi\left(\frac{I}{S}\right) + \rho\left(C''_{\mathrm{p}} - C'_{\mathrm{p}}\right)\right\} \tag{20}$$

Ions

$$abN\left(C''_{\mathrm{out}}u''_{\mathrm{out}} - C''_{\mathrm{in}}u''_{\mathrm{in}}\right) = SNJ_{\mathrm{S}} = lbN\left\{\lambda\left(\frac{I}{S}\right) - \mu\left(C''_{\mathrm{p}} - C'_{\mathrm{p}}\right)\right\} \tag{21}$$

Assuming linear salt concentration and solution velocity changes in concentrating cells, C''_{out}, C''_{p}, u''_{p} and u''_{p} are introduced from Eqns (20) and (21) as follows.

$$C''_{\mathrm{out}} = \frac{\sqrt{M^2 + 4LN} - M}{2L} \tag{22}$$

$$C''_{\mathrm{p}} = C''_{\mathrm{in}} + p\left(C''_{\mathrm{out}} - C''_{\mathrm{in}}\right) \quad C''_{\mathrm{in}} = C'_{\mathrm{in}} \tag{23}$$

$$u''_{\text{out}} = \frac{1}{2}\left(\sqrt{M^2 + 4LN} + M\right) \tag{24}$$

$$u''_{\text{p}} = u''_{\text{in}} + p\left(u''_{\text{in}} + u''_{\text{in}}\right) \tag{25}$$

in which

$$L = \frac{l}{a}\rho\, p \tag{26}$$

$$M = \frac{l}{a}\left[\phi\left(\frac{I}{S}\right) - \rho\left\{C'_{\text{p}} - (1-p)C''_{\text{in}}\right\}\right] + u''_{\text{in}} \tag{27}$$

$$N = C''_{\text{in}}u''_{\text{in}} + \frac{l}{a}\left\{\lambda\left(\frac{I}{S}\right) - \mu\left(C''_{\text{p}} - C'_{\text{p}}\right)\right\} \quad C''_{\text{in}} = C'_{\text{in}} \tag{28}$$

Salt concentration of a solution passing through the membrane pair at $x = pl$; $C''^{\text{b}}_{\text{p}}$ is introduced from Eqns (1) and (2) as follows.

$$C''^{\text{b}}_{\text{p}} = \frac{1}{2\rho}\left(\sqrt{A^2 + 4\rho B} - A\right) \tag{29}$$

$$A = \phi(I/S) + \mu - \rho C'_{\text{p}} \tag{30}$$

$$B = \lambda(I/S) + \mu C'_{\text{p}} \tag{31}$$

3.7 Physical Properties of Solutions in Desalting and Concentrating Cells

The unit of salt concentration C (eq/cm^3) described in Sections 3.5 and 3.6 is changed to C_1 (eq/dm^3), C_2 (salt g/solution dm^3), and C_3 (salt g/solution kg) using the following equations.

$$C_1\left(\text{eq/cm}^3\right) = C\left(\text{eq/cm}^3\right) \times 10^3 \tag{32}$$

$$C_2\left(\text{g salt/dm}^3\text{ solution}\right) = 57.87C_1 \tag{33}$$

$$C_3(\text{g salt/kg solution}) = \frac{\sqrt{Q^2 + 4PC_2} - Q}{2P} \tag{34}$$

$$P = (7.881 - 1.368 \times 10^{-2}T + 8.978 \times 10^{-5}T^2) \times 10^{-4} \tag{35}$$

$$Q = 1.001 - 1.101 \times 10^{-4}T - 3.356 \times 10^{-6}T^2 \tag{36}$$

Density d (kg/dm^3) (Sato and Matsuo, 1974), specific electric conductance κ (S/cm) (Akiyama, 1992) and electrolyte (NaCl) activation coefficient γ of an electrolyte solution (Harned and Owen, 1998) are given by the following functions of temperature T (°C) and salt concentration C_3 (g salt/kg solution).

$$\begin{aligned} d = & \, 1.001 - 1.101 \times 10^{-4} T - 3.356 \times 10^{-6} T^2 \\ & + \left(7.881 - 1.368 \times 10^{-2} T + 8.978 \times 10^{-5} T^2\right) \times 10^{-4} C_3 \end{aligned} \tag{37}$$

$$\begin{aligned} k = & \left(0.9383 + 3.463 \times 10^{-2} \mathrm{T}\right) \times 10^{-3} C_3 - \left(1.655 + 3.863 \times 10^{-2} \mathrm{T}\right) \\ & \times 10^{-6} C_3^2 - \left(1.344 + 3.160 \times 10^{-2} T\right) \times 10^{-9} C_3^3 \end{aligned} \tag{38}$$

$$\gamma = 0.5927 + 0.4355 C_3^{-0.5} - 7.201 \times 10^{-5} C_3 + 3.503 \times 10^{-6} C_3^2 \tag{39}$$

3.8 Electric Resistance of an Ion-Exchange Membrane Pair and Solutions in Desalting and Concentrating Cells

The alternating current electric resistance of an ion-exchange membrane r_{alter} is measured conventionally at 25 °C by passing a 1000 Hz alternating current.

For measuring the direct current electric resistance of an ion-exchange membrane, the membrane is set in a two-cell apparatus and a low-concentration NaCl solution (specific conductivity, κ' (S/cm)) is supplied into both cells. Direct current electric resistance of the membrane r^*_{dire} (Ωcm^2) is measured at 25 °C by passing a direct current. The relationship between κ' and $r^*_{\text{dire}}/r_{\text{alter}}$ is expressed by the following empirical equation (Tanaka et al., 2003).

$$\log\left(\frac{r^*_{\text{dire}}}{r_{\text{alter}}}\right) = 0.3380 + 0.6386 \log\kappa' + 0.2961 (\log\kappa')^2 \tag{40}$$

A low-concentration NaCl solution (specific conductivity, κ') is supplied to the desalting side and a high-concentration NaCl solution (specific conductivity, κ'') is supplied to the concentrating side of the above apparatus. The direct current electric resistance of the membrane r_{dire} is measured at

25 °C by passing a direct current and subtracting the effect of membrane potential. The empirical relationship between κ''/κ' and $r_{\text{dire}}/r^*_{\text{dire}}$:

$$\frac{r_{\text{dire}}}{r^*_{\text{dire}}} = 1.000 - 0.1359\log\left(\frac{\kappa''}{\kappa'}\right) \tag{41}$$

Direct electric current of a membrane pair r_{memb} is estimated by multiplying Eqn (40) by Eqn (41):

$$r_{\text{memb}} = r_K + r_A = \left(\frac{r^*_{\text{dire}}}{r_{\text{alter}}}\right) \times \left(\frac{r_{\text{dire}}}{r^*_{\text{dire}}}\right) \times r_{\text{alter}} = r_{\text{dire}} \tag{42}$$

Electric resistance of a desalting cell r' and of a concentrating cell r'' in an electrodialyzer are given:

$$r' = \frac{a}{(1-\varepsilon)\kappa'} \tag{43}$$

$$r' = \frac{a}{(1-\varepsilon)\kappa''} \tag{44}$$

ε defines an electric current screening effect of a diagonal spacer and it is determined by the volume ratio of spacer rods in a desalting and concentrating cell as follows.

$$\varepsilon = \frac{\pi\ a}{8\chi\ \sin\theta} \tag{45}$$

in which χ is the distance between spacer rods and θ is the crossing angle of the rods. Structure of the diagonal spacer is given in Figure 10.

3.9 Pressure Drop in Desalting and Concentrating Cells

Hydrodynamic diameter of a desalting or a concentrating cell; $d_{\text{H,cell}}$ and that of a desalting or a concentrating slot; $d_{\text{H,slot}}$ incorporated with a diagonal

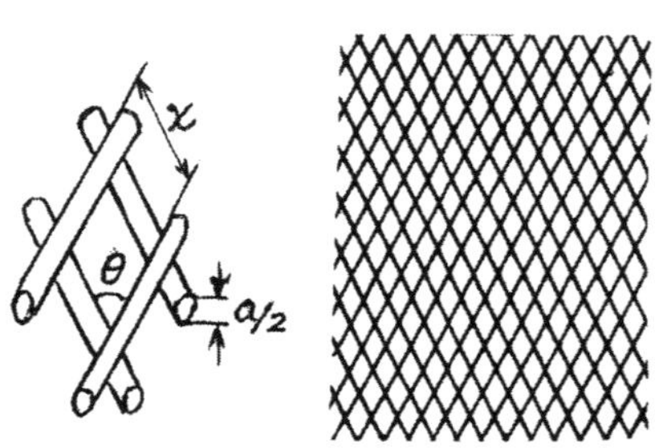

Figure 10 Structure of a diagonal net spacer. *(Tanaka, 2011b).*

net spacer are expressed by the following equation (Zimmerer and Kotte, 1996) (Tsiakis and Papageorgiou, 2005).

$$d_{\mathrm{H,cell}} = \frac{8 - \pi \frac{a}{\chi}}{4\left(\frac{1}{b} + \frac{1}{a}\right) + 2\pi\left(1 - \frac{a}{4b}\right)\frac{1}{\chi}} \tag{46}$$

$$d_{\mathrm{H,slot}} = \frac{8 - \pi \frac{a}{\chi}}{4\left(\frac{1}{w} + \frac{1}{a}\right) + 2\pi\left(1 - \frac{a}{4w}\right)\frac{1}{\chi}} \tag{47}$$

in which w is the flow-pass width in the slot (Figure 1).

Pressure difference between the inlet and the outlet of the cell in a desalting or a concentrating cell; pressure drop in the cell ΔP_{cell} and that in a desalting or a concentrating slot; pressure drop in the slot ΔP_{slot} are

$$\Delta P_{\mathrm{cell}}(\mathrm{Pa}) = \frac{3.2\mu \; l u_{\mathrm{cell}}}{\left(d_{\mathrm{H,cell}}\right)^2} \tag{48}$$

$$\Delta P_{\mathrm{slot}}(\mathrm{Pa}) = \frac{3.2\mu \; h u_{\mathrm{slot}}}{\left(d_{\mathrm{H,slot}}\right)^2} \tag{49}$$

where h is the flow-pass length in the slot (Figure 8), u_{cell} is linear velocity in the desalting or concentrating cell, and u_{slot} is linear velocity in the slot. μ (g/cm/s) is the viscosity coefficient of a solution and it is expressed by the following function of temperature T (°C) and electrolyte concentration C (g salt/kg solution) of a solution (Akiyama, 1992).

$$\begin{aligned}\mu = {} & 1.200 \times 10^{-2} - 1.224 \times 10^{-4} T + \left(2.107 \times 10^{-5} - 1.529\right. \\ & \left. \times 10^{-7} T\right) C + \left(-1.392 \times 10^{-8} + 1.123 \times 10^{-10} T\right) C^2 + \left(5.819\right. \\ & \left. \times 10^{-10} - 6.769 \times 10^{-12} T\right) C^3\end{aligned} \tag{50}$$

u_{cell} and u_{slot} are linear velocity in the cell and slot. ΔP (m) is calculated using $\Delta P\ (\mathrm{m}) = 1.01972 \times 10^{-4}\ \Delta P(\mathrm{Pa})$.

3.10 Current Density Distribution

3.10.1 Quadratic Current Density Distribution Equation

In an electrodialysis system, current density is decreased due to the salt concentration decrease in desalting cells along a flow-pass. This phenomenon influences to the limiting current density of an electrodialyzer $(I/S)_{\mathrm{lim}}$ (cf. Section 3.12). The current density distribution is assumed to be approximated by the following quadratic equation expressed at x/l distance from the inlet of a desalting cell.

$$i = a_1 + a_2\left(\frac{x}{l}\right) + a_3\left(\frac{x}{l}\right)^2 \tag{51}$$

To determine a_1, a_2 and a_3 in Eqn (51), three-dimensional simultaneous equations are set up (Tanaka, 2000, 2002) as described in the succeeding sections.

3.10.2 The First Three-Dimensional Simultaneous Equation ($V_{in} = V_p$) for Computing ζ_{inout}

When a large number of membrane pairs are integrated in an electrodialyzer, ohmic loss of electrodes incorporated in the electrodialyzer is negligible compared to the values between the electrodes. This event is expressed by Eqn (52) which means that the voltage difference between the electrodes at the entrance of desalting cells V_{in} is equal to the value at the exits V_{out}.

$$V_{\mathrm{in}} = V_{\mathrm{out}} \quad \text{The first equation} \tag{52}$$

$$V_{\mathrm{in}} = \mathrm{A}_1 i_{\mathrm{in}} + \mathrm{A}_2 \tag{53}$$

$$V_{\mathrm{out}} = \mathrm{B}_1 i_{\mathrm{out}} + \mathrm{B}_2 \tag{54}$$

A_1, A_2, B_1, and B_2 are calculated for N membrane pairs and for solution velocity ratio ξ_j between $j = 0$ and $j = n$ in a stack.

Electric resistance at $x = 0$ (inlets)

$$\mathrm{A}_1 = \left(r'_{\mathrm{in}} + r''_{\mathrm{in}} + r_{\mathrm{memb,in}}\right)N \tag{55}$$

Membrane potential at $x = 0$ (inlets)

$$A_2 = 2(t_K + t_A - 1)\left(\frac{RT}{F}\right)N \ln\left(\frac{\gamma''_{\mathrm{in}} C''_{\mathrm{in}}}{\gamma'_{\mathrm{in}} C'_{\mathrm{in}}}\right) \tag{56}$$

Electric resistance at $x = l$ (outlets)

$$B_1 = \sum_{\mathrm{j}=0}^{\mathrm{n}} Y_{\mathrm{j}} r'_{\mathrm{out,j}} + r'' N + \sum_{\mathrm{j}=0}^{\mathrm{n}} Y_{\mathrm{j}} r_{\mathrm{memb,out,j}} \tag{57}$$

Membrane potential at $x = l$ (outlets)

$$B_2 = 2(t_K + t_A - 1)\left(\frac{RT}{F}\right)\sum_{j=0}^{n} Y_j\left(\frac{\gamma''_{out}C''_{out}}{\gamma'_{out}C'_{out}}\right) \tag{58}$$

$$Z_1 = \frac{B_1}{A_1} \tag{59}$$

$$Z_2 = -\frac{A_2 - B_2}{A_1} \tag{60}$$

$$\alpha_1 = \left(\frac{I}{S}\right) - Z_2 \tag{61}$$

$$\alpha_2 = -2\left\{3\left(\frac{I}{S}\right) - 2Z_2\right\} \tag{62}$$

$$\alpha_3 = 3\left\{2\left(\frac{I}{S}\right) - Z_2\right\} \tag{63}$$

$$\beta_1 = Z_1\left(\frac{I}{S}\right) \tag{64}$$

$$\beta_2 = -2(2Z_1 + 1)\left(\frac{I}{S}\right) \tag{65}$$

$$\beta_3 = 3(Z_1 + 1)\left(\frac{I}{S}\right) \tag{66}$$

$$\zeta_{inout} = \frac{\alpha_1 + \alpha_2 p + \alpha_3 p^2}{\beta_1 + \beta_2 p + \beta_3 p^2} \tag{67}$$

p predetermined is substituted to Eqn (67) to calculate temporary ζ_{inout}.

3.10.3 *The Second Three-Dimensional Simultaneous Equation ($V_{in} = V_p$) for Computing ζ_{inp}*

Equations (68) is naturally introduced from Eqn (52).

$$V_{in} = V_p \quad \text{The second equation} \tag{68}$$

$$V_{\text{in}} = \text{A}_1 i_{\text{in}} + \text{A}_2 = \text{ Eqn } (53) \tag{69}$$

$$V_{\text{p}} = C_1\left(\frac{I}{S}\right) + C_2 \tag{70}$$

$$\text{A}_1 = \text{Eqn } (55) \tag{71}$$

$$\text{A}_2 = \text{Eqn } (56) \tag{72}$$

Electric resistance at $x = pl$

$$C_1 = \sum_{j=0}^{\text{n}} Y_j r'_{\text{p},j} + r'' N + \sum_{j=0}^{\text{n}} Y_j r_{\text{memb,p},j} \tag{73}$$

Membrane potential at $x = pl$

$$C_2 = 2(t_{\text{K}} + t_{\text{A}} - 1)\left(\frac{RT}{F}\right)\sum_{j=0}^{n} Y_j\left(\frac{\gamma''_{\text{p}} C''_{\text{p}}}{\gamma'_{\text{p}} C'_{\text{p}}}\right) \tag{74}$$

$$Z_3 = \frac{C_1}{A_1} \tag{75}$$

$$Z_4 = -\frac{A_2 - C_2}{A_1} \tag{76}$$

$$\gamma_1 = (Z_3 - 1)\left(\frac{I}{S}\right) + Z_4 \tag{77}$$

$$\gamma_2 = 2\left\{(3 - 2Z_3)\left(\frac{I}{S}\right) - 2Z_4\right\} \tag{78}$$

$$\gamma_3 = 3\left\{(Z_3 - 2)\left(\frac{I}{S}\right) + Z_4\right\} \tag{79}$$

$$\zeta_{\text{inp}} = \frac{\gamma_1 + \gamma_2 p + \gamma_3 p^2}{\left\{(2p - 3p^2)\left(\frac{I}{S}\right)\right\}} \tag{80}$$

ζ_{inp} is calculated by substituting predetermined p which was substituted to Eqn (67)

3.10.4 The Third Three-Dimensional Simultaneous Equation ($\zeta_{inout} = \zeta_{inp}$) for Computing p

p is computed by adjusting p with trial-and-error calculation to realize Eqn (81).

$$\zeta_{\rm inout} = \zeta_{\rm inp} \quad \text{The third equation} \tag{81}$$

3.10.5 Coefficients in the Current Density Distribution Equation and Current Density Nonuniformity Coefficient

Using Eqn (52), the coefficients in the quadratic current density distribution equation (Eqn (51)) are introduced as follows.

$$a_1 = Z_1 \zeta_{\rm out}\left(\frac{I}{S}\right) + Z_2 \tag{82}$$

$$a_2 = 2\left[\{3 - (2Z_1 + 1)\zeta_{\rm out}\}\left(\frac{I}{S}\right) - 2Z_2\right] \tag{83}$$

$$a_3 = -3\left[\{2 - (Z_1 + 1)\zeta_{\rm out}\}\left(\frac{I}{S}\right) - Z_2\right] \tag{84}$$

$\zeta_{\rm in}$ and $\zeta_{\rm cut}$ included in Eqns (82)–(84) are respectively inlet and outlet current density non-uniformity coefficient defined by:

$$\zeta_{\rm in} = \frac{i_{\rm in}}{I/S} = \frac{a_1}{I/S} \tag{85}$$

$$\zeta_{\rm out} = \frac{i_{\rm out}}{I/S} = \frac{a_1 + a_2 + a_3}{I/S} \tag{86}$$

3.11 Cell Voltage, Energy Consumption, Water Recovery, Desalting Ratio

From Eqns (85) and (86), current density at the inlet; $i_{\rm in}$ and outlet; $i_{\rm out}$ of the desalting cell are:

$$i_{\rm in} = (I/S)\zeta_{\rm in} \tag{87}$$

$$i_{\rm out} = (I/S)\zeta_{\rm out} \tag{88}$$

Ohmic voltage and membrane voltage at the inlet of the desalting cell ($V_{\Omega,\rm in}$ and $V_{\rm memb,in}$) and those at the outlet of the desalting cell ($V_{\Omega,\rm out}$ and $V_{\rm memb,out}$) are:

$$V_{\Omega,\rm in} = \left(r'_{\rm in} + r_{\rm memb,in} + r''\right) i_{\rm in} \tag{89}$$

$$V_{\text{memb,in}} = 2(t_{\text{K}} + t_{\text{A}} - 1) \times \left(\frac{RT}{F}\right) \ln \frac{\gamma''_{\text{in}} C''_{\text{in}}}{\gamma'_{\text{in}} C'_{\text{in}}} \tag{90}$$

$$V_{\Omega,\text{out}} = \left(\sum_{j=0}^{n} Y_j r'_{\text{out},j} + \sum_{j=0}^{n} Y_j r_{\text{memb,out},j} + r''_{\text{out}} N\right) \times i_{\text{out}} \times \left(\frac{1}{N}\right) \tag{91}$$

$$V_{\text{memb,out}} = 2(t_{\text{K}} + t_{\text{A}} - 1)\left(\frac{RT}{F}\right) \sum_{j=0}^{n} \ln \frac{\gamma''_{\text{out}} C''_{\text{out}}}{\gamma'_{\text{out},j} C'_{\text{out},j}} \times \left(\frac{1}{N}\right) \tag{92}$$

Cell voltage V_{cell} is introduced from Eqn (52) as shown in Eqn (93).

$$V_{\text{cell}}(V/\text{pair}) = V_{\Omega,\text{in}} + V_{\text{memb,in}} = V_{\Omega,\text{out}} + V_{\text{memb,out}} \tag{93}$$

Energy consumption E is expressed by the following equation.

$$E(\text{kWh/m}^3) = \frac{V_{\text{cell}} I}{q'_{\text{out}}} \times 3600 \times 10^3 \tag{94}$$

q'_{out} is the solution volume; output of the desalted solution (cm^3/s cell) (Eqn (9)).

Water recovery Re is:

$$\text{Re} = \frac{q'_{\text{out}}}{q'_{\text{in}} + q''_{\text{in}}} \tag{95}$$

Desalting ratio α is:

$$\alpha = \left(1 - \frac{C'_{\text{out}}}{C'_{\text{in}}}\right) \tag{96}$$

3.12 Limiting Current Density

Limiting current density of a cation-exchange membrane is less than that of an anion-exchange membrane, because the mobility of counter-ions in a solution for a cation-exchange membrane is less than that for an anion-exchange membrane. So the limiting current density of an ion-exchange membrane integrated in an electrodialyzer i_{lim} is given by the following empirical equation established for a cation-exchange membrane (Tanaka, 2005, 2013).

$$i_{\text{lim}}(\text{A/cm}^2) = \left\{l_1 + l_2\left(\frac{T}{25}\right) + l_3\left(\frac{T}{25}\right)^2\right\}(m_1 + m_2 u'_{\text{out}})\, C'^{\,n_1 + n_2 u'_{\text{out}}}_{\text{out}} \tag{97}$$

The coefficients in Eqn (97) were measured experimentally as follows:

$$\begin{aligned} &l_1 = 0.5950, l_2 = 0.2731, l_3 = 0.1310, m_1 = 83.50, m_2 = 24.00, \\ &n_1 = 0.7846, n_2 = 8.612 \times 10^{-3} \end{aligned} \tag{98}$$

When current density reaches the limit of a cation-exchange membrane i_{lim} at the outlet of a desalting cell in which linear velocity becomes the least among $u'_{\text{out}} : u'^{\#}_{\text{out}}$, the average current density applied to an electrodialyzer is defined as its limiting current density $(I/S)_{\text{lim}}$ which is expressed by Eqn (99) introduced from Eqns (88) and (97).

$$\begin{aligned} \left(\frac{I}{S}\right)_{\text{lim}} &= \frac{i_{\text{lim}}}{\zeta_{\text{out}}} \\ &= \left\{ l_1 + l_2\left(\frac{T}{25}\right) + l_3\left(\frac{T}{25}\right)^2 \right\} \frac{\left(m_1 + m_2 u'^{\#}_{\text{out}}\right)}{\zeta_{\text{out}}} \left(C'^{\#}_{\text{out}}\right)^{n_1 + n_2 u'^{\#}_{\text{out}}} \end{aligned} \tag{99}$$

in which $C'^{\#}_{\text{out}}$ is C'_{out} at $u' = u'^{\#}_{\text{out}}$.

In Eqn (99), $u'^{\#}_{\text{out}}$ is nearly equal to $u'^{\#}_{\text{in}}$. Substituting $u'^{\#}_{\text{in}} = u'^{\#}_{\text{out}}$ in Eqn (99) leads to:

$$\left(\frac{I}{S}\right)_{\text{lim}} = \left\{ l_1 + l_2\left(\frac{T}{25}\right) + l_3\left(\frac{T}{25}\right)^2 \right\} \frac{m_1 + m_2 u'^{\#}_{\text{in}}}{\zeta_{\text{out}}} \left(C'^{\#}_{\text{out}}\right)^{n_1 + n_2 u'^{\#}_{\text{in}}} \tag{100}$$

in which, $C'^{\#}_{\text{in}}$ is C'_{in} at $u' = u'^{\#}_{\text{in}}$ which is given by;

$$u'^{\#}_{\text{in}} = u'_{\text{in}}(1 - 3\sigma) \tag{101}$$

σ is the standard deviation of the normal distribution of solution velocity ratio defined by Eqn (14). The influence of σ to general performance of the electrodialyzer is relatively few (Tanaka, 2009), but it influences to the limiting current density of the electrodialyzer.

The relationship between $(I/S)_{\text{lim}}$ and $C'^{\#}_{\text{out}}$ is also introduced as follows

$$\left(\frac{I}{S}\right)_{\text{lim}} = \left(\frac{a}{\lambda l}\right) u'^{\#}_{\text{in}} \left(C'_{\text{in}} - C'^{\#}_{\text{out}}\right) \tag{102}$$

Putting Eqn (100) = Eqn (102):

$$Z_1 = \left\{ l_1 + l_2\left(\frac{T}{25}\right) + l_3\left(\frac{T}{25}\right)^2 \right\} \frac{\left(C'^{\#}_{\text{out}}\right)^{n_1+n_2 u'^{\#}_{\text{in}}}}{C'_{\text{in}} - C'^{\#}_{\text{out}}} \tag{103}$$

$$Z_2 = \left(\frac{a\zeta_{\text{out}}}{\lambda l}\right)\left(\frac{u'^{\#}_{\text{in}}}{m_1 + m_2 u'^{\#}_{\text{in}}}\right) \tag{104}$$

$$Z_1 = Z_2 \tag{105}$$

Limiting current density of an electrodialyzer $(I/S)_{\text{lim}}$ is computed with trial and error calculation by substituting control key $C'^{\#*}_{\text{out}}$ for $C'^{\#}_{\text{out}}$ in Eqn (103) to realize $Z_1 = Z_2$ (Eqn (105)) for determining $C'^{\#}_{\text{out}}$. $(I/S)_{\text{lim}}$ is calculated by substituting $C'^{\#}_{\text{out}}$into Eqn (100).

3.13 Electrodialysis Program

3.13.1 Constant Electric Current Continuous Program

The program is shown in Figure 11 and operated in the Companion site (Tanaka, 2015).

3.13.2 Constant Voltage Continuous Program

The program is shown in Figure 12 and operated in the Companion site (Tanaka, 2015).

3.13.3 Constant Salt Concentration Continuous Program

The program is shown in the chart Figure 13 and operated in the Companion site (Tanaka, 2015).

3.13.4 Limiting Current Density Program

The program is shown in Figure 14. It is included in the Companion site (Tanaka, 2015).

The above programs are integrated into the companion sites (web sites), which include all equations described in Sections 3.4–3.12 (Tanaka, 2015).

3.14 Saline Water Desalination

Specification and operating conditions of the electrodialyzer inputting into the program are presented in Table 1.

Ion flux J_S and solution flux J_V are plotted against current density I/S (Figure 15, constant current operation), cell voltage V_{cell} (Figure 16, constant voltage operation), and salt concentration at the outlets of desalting cells C'_{out} (Figure 17, constant salt concentration operation).

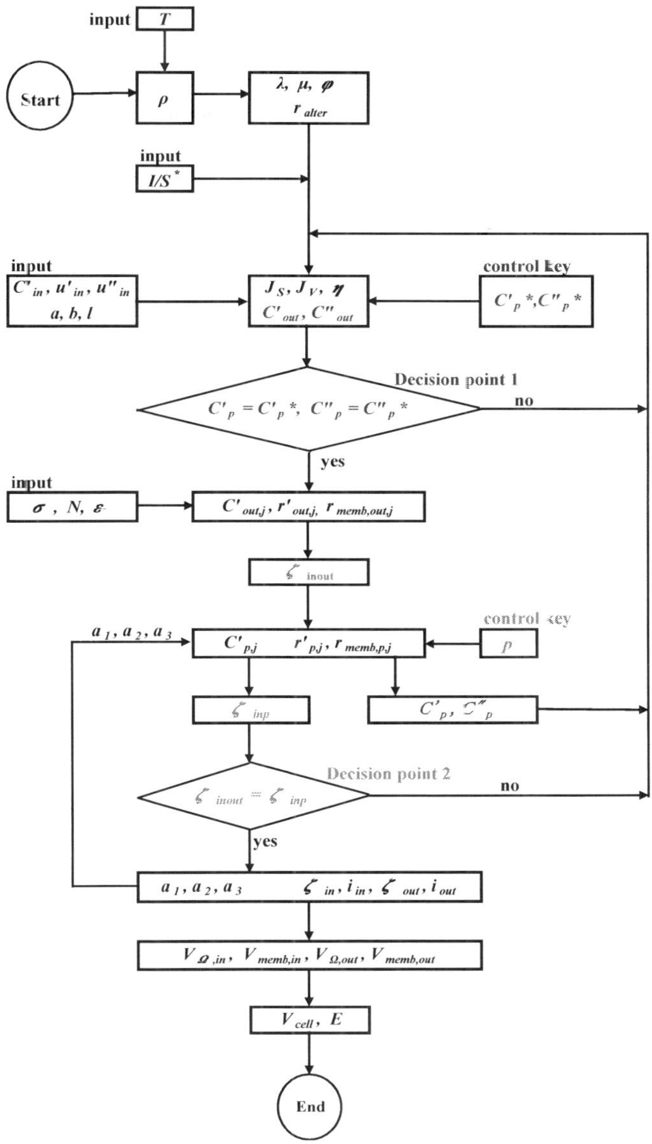

Figure 11 Constant electric current continuous electrodialysis program chart.

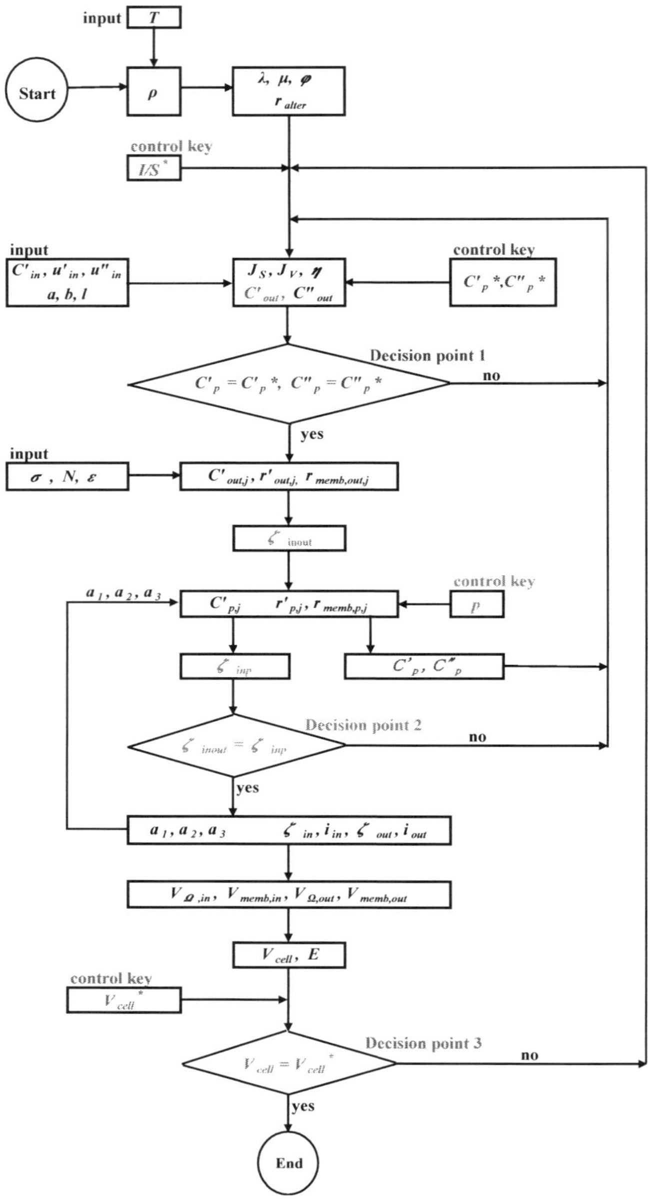

Figure 12 Constant voltage continuous electrodialysis program chart.

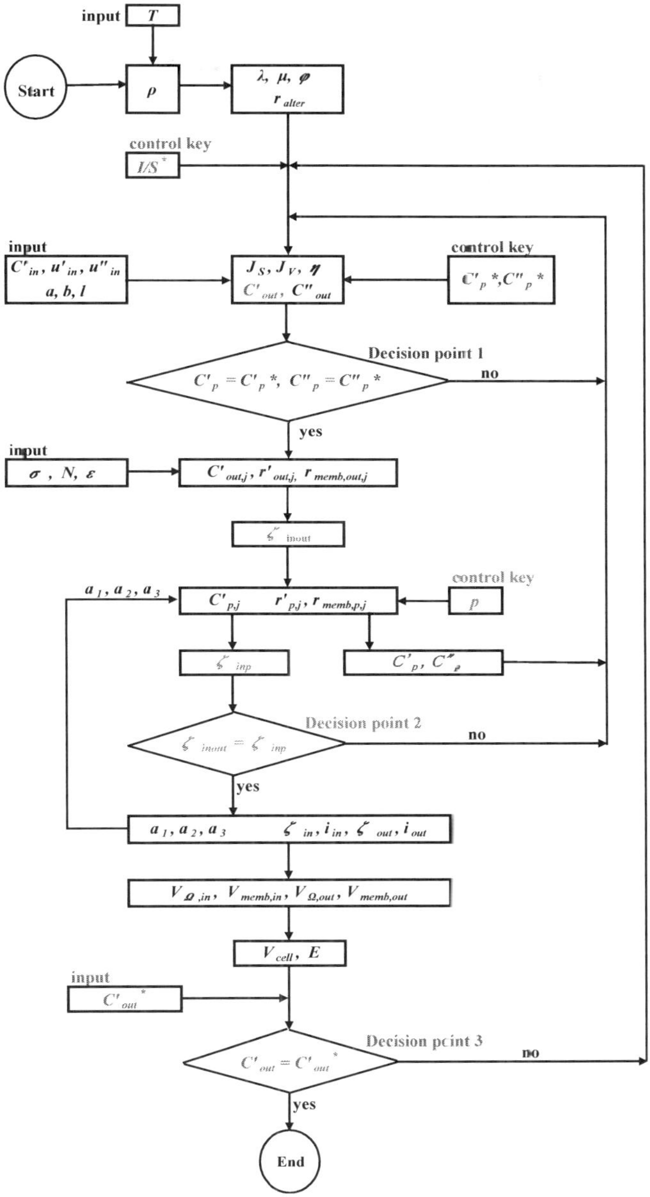

Figure 13 Constant salt concentration continuous electrodialysis program chart.

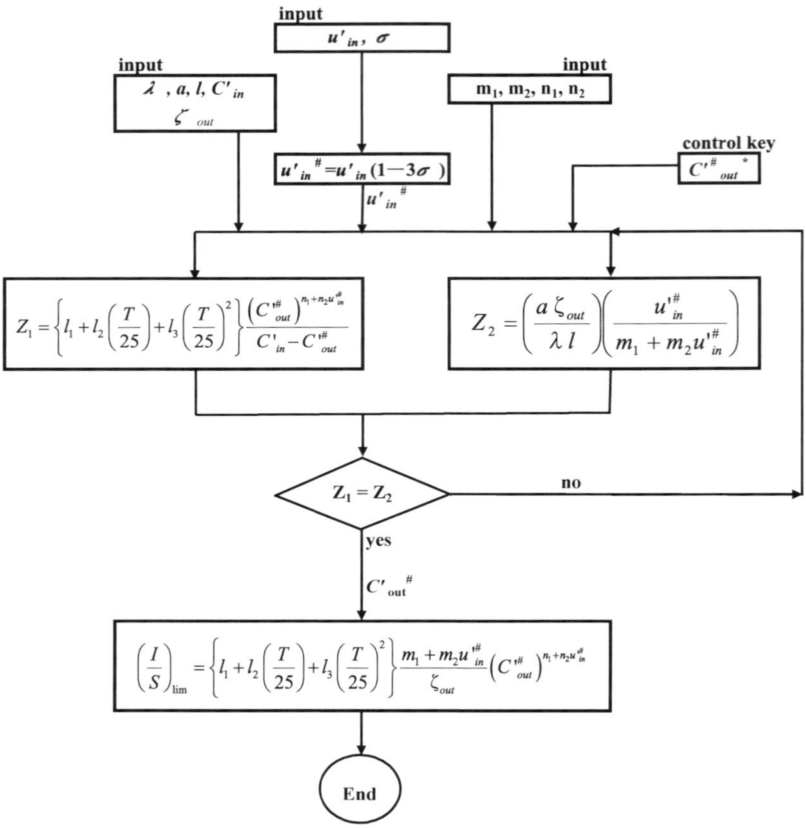

Figure 14 Limiting current density program chart. Superscript # means the least value in an electrodialyzer.

Figure 18 shows I/S versus energy consumption E and desalting ratio α in the constant current operation. Figure 19 shows V_{cell} versus E and α in the constant voltage operation. Figure 20 shows salt concentration at the outlets of desalting cells C'_{out} versus E and α in the constant salt concentration operation.

Figure 21 shows relationship (for constant voltage operation) between V_{cell} and pressure drop in the total desalting cell $\Delta P'_{\text{total}}$, desalting cell $\Delta P'_{\text{cell}}$, total concentrating cell $\Delta P''_{\text{total}}$ and concentrating cell $\Delta P''_{\text{cell}}$. The pressure drop in the desalting slot $\Delta P'_{\text{slot}}$ and in the concentrating slot $\Delta P''_{\text{slot}}$ are calculated as; $\Delta P'_{\text{slot}} = \Delta P'_{\text{total}} - \Delta P'_{\text{cell}}$, and $\Delta P''_{\text{slot}} = \Delta P''_{\text{total}} - \Delta P''_{\text{cell}}$.

Table 1 Electrodialysis specifications and operating conditions

Folw-pass thickness of desalting and concentrating cells a	0.05 cm
Flow-pass width desalting and concentrating cells b	100 cm
Flow-pass length of desalting and concentrating cells l	100 cm
Number of cell pairs N	300
Salt concentration at the inlets of desalting cells C'_{in}	2000 mg/dm^3
Salt concentration at the inlets of concentrating cells cells C''_{in}	2000 mg/dm3
Linear velocity at the inlets of desalting cells u'_{in}	10 cm/s
Linear velocity at the inlets of concentrating cells u'_{in}	1 cm/s
Temperature T	25 °C
Standard deviation of solution velocity ratio in desalting cells σ	0.1
Distance between spacer rod χ	0.3 cm
Crossing angle of the rod θ	π/3 rad
Number of slots in desalting cells n'	5
Number of slots in concentrating cells n''	2
Width of slots in desalting cells w'	4 cm
Width of slots in concentrating cells w''	2 cm
Length of slots in desalting cells h'	4 cm
Length of slots in concentrating cells h''	4 cm

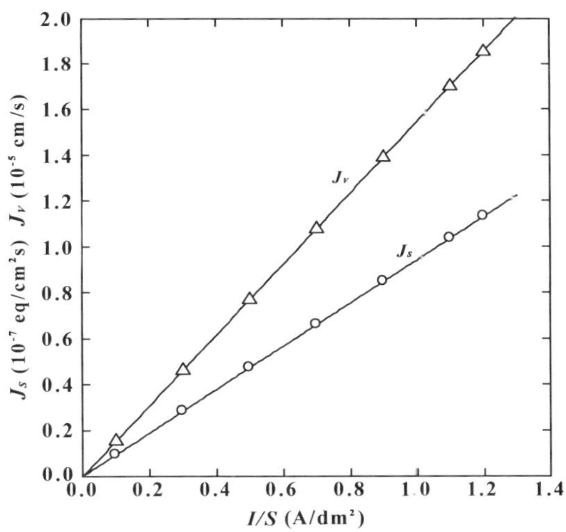

Figure 15 Current density vs. ion flux and solution flux. Constant electric current operation.

Figure 22 gives limiting current density $(I/S)_{\text{lim}}$ versus V_{cell} (for constant voltage operation) and C'_{out} (for constant salt concentration operation). $(I/S)_{\text{lim}}$ is computed assuming that the standard deviation of the normal distribution of the solution velocity ratio $\sigma = 0.1$ (cf. Eqns (14) and (15) and Table 1) which functions as a safety factor for operating the

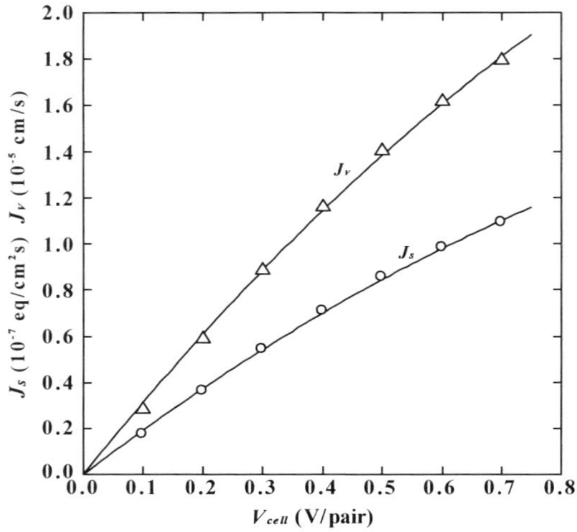

Figure 16 Cell voltage vs. Ion flux and solution flux. Constant voltage operation.

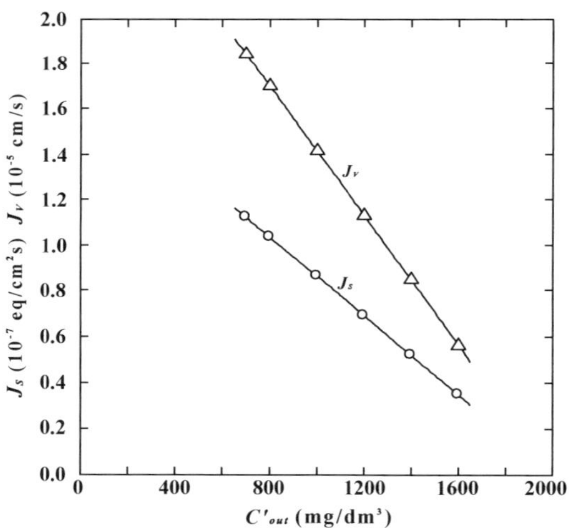

Figure 17 Salt concentration vs. ion flux and solution flux. Constant salt concentration operation.

electrodialyzer stably (Tanaka, 2012b). $(I/S)_{\mathrm{lim}}$ is plotted against I/S for constant electric current operation in Figure 23 (mark ○). Figure 23 includes also I/S versus $(I/S)_{\mathrm{lim}}$ (for constant voltage operation; mark △) and I/S versus $(I/S)_{\mathrm{lim}}$ (for constant salt concentration; mark □). Data for three type operations are plotted on the same line. The real limiting current density $(I/S)_{\mathrm{lim,real}} = 1.144\ \mathrm{A/dm^2}$ is

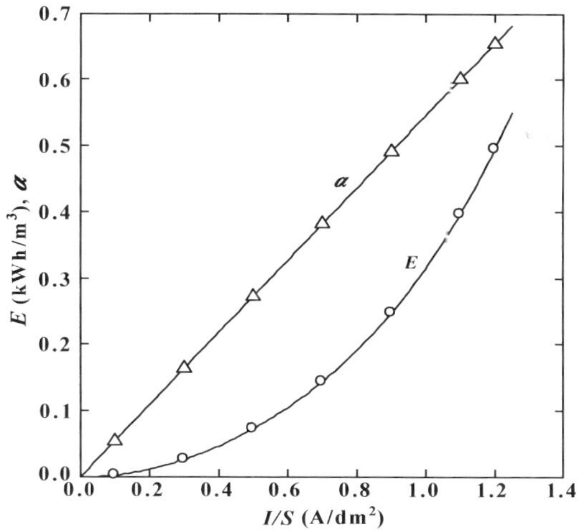

Figure 18 Current density vs. energy consumption and desalting ratio. Constant electric current operation.

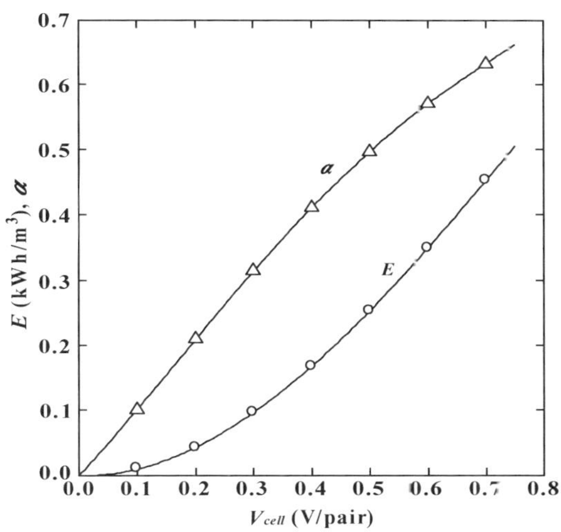

Figure 19 Cell voltage vs. energy consumption and desalting ratio. Constant voltage operation.

determined from the intersection between the I/S versus $(I/S)_{\mathrm{lim}}$ plots and the $I/S = (I/S)_{\mathrm{lim}}$ line in Figure 23. Incorporating the $(I/S)_{\mathrm{lim,real}}$ into Figure 22, limiting cell voltage $V_{\mathrm{cell,lim}}$ and limiting salt concentration $C'_{\mathrm{out,lim}}$ at $(I/S)_{\mathrm{lim,real}}$ are computed as; $V_{\mathrm{cell,lim}} = 0.688$ V/pair and $C'_{\mathrm{out,lim}} = 750$ mg/dm^3.

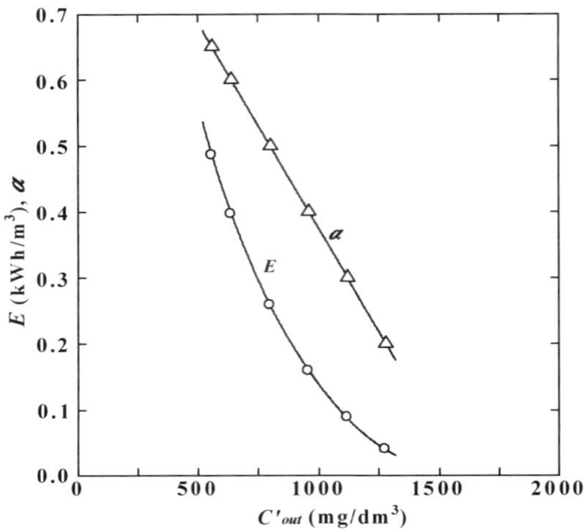

Figure 20 Salt concentration vs energy consumption and desalting ratio. Constant salt concentration operation.

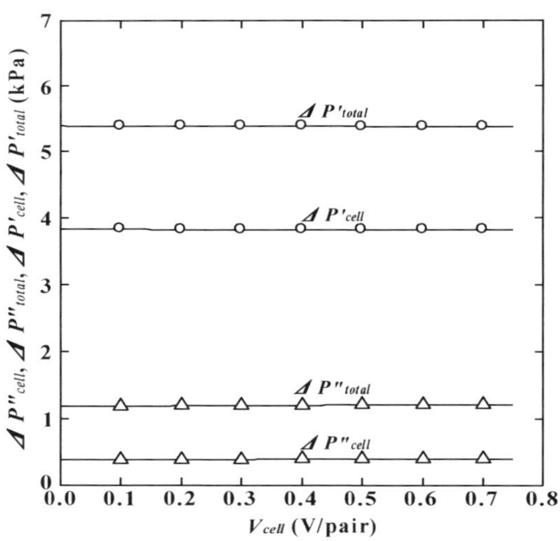

Figure 21 Cell voltage vs pressure drop in a desalting cell and in a concentrating cell. Constant voltage operation.

3.15 Drinking Water Production in the Two-Stage Operation

The threshold of acceptable aesthetic criteria of human drinking water is 500 mg/dm^3. So, the salt concentration of the drinking water is assumed to be 400 mg/dm^3 in this section. In order to produce drinking water

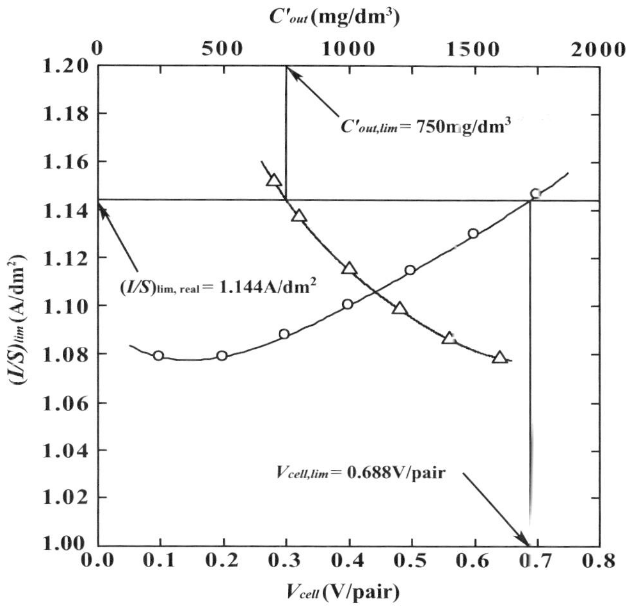

Figure 22 Limiting current density. ○ Constant voltage operation. △ Constant salt concentration operation.

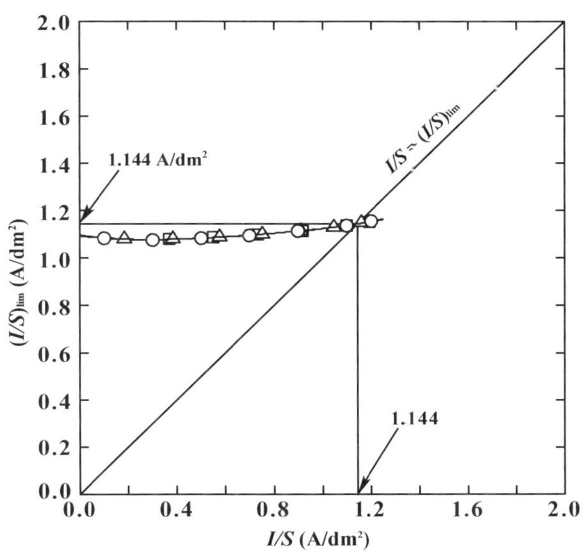

Figure 23 Real limiting current density. ○ Constant electric current operation △ Constant voltage operation □ Constant salt concentration operation.

with the continuous process, a single stage process (Figure 7) is generally arranged to form a multistage process. Figure 24 shows salt concentration and linear velocity changes in desalting cells in the two-stage continuous electrodialysis process. In the first stage, salt concentration and linear velocity are $C'_{in,I}$ and $u'_{in,I}$ at the inlets and $C'_{out,I}$ and $u'_{out,I}$ at the outlets. In the second stage, these parameters become $C'_{in,II} = C'_{out,I}$ and $u'_{in,II} = u'_{out,I}$ at the inlets and $C'_{out,II}$ and $u'_{out,II}$ at the outlets. The computation program for the multistage operation is established by arranging the programs (Figures 11–13) and companion sites (Tanaka, 2015) to form the multiple-process.

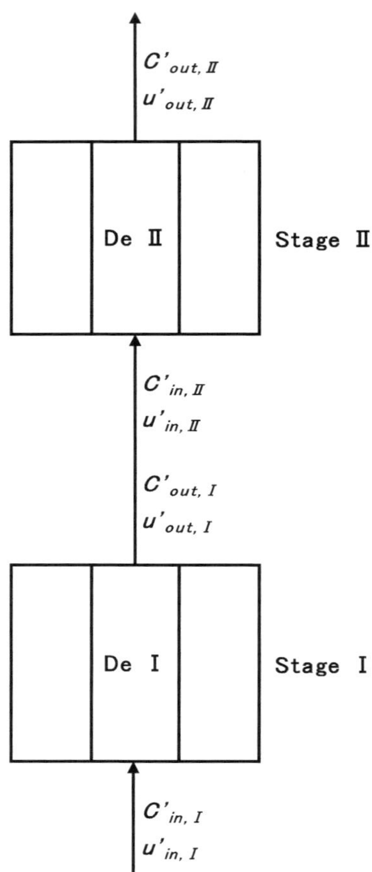

Figure 24 Two-stage electrodialysis process De: Desalting cells. *(Tanaka, 2012a).*

In this section, a salt solution is assumed to be supplied to a two-stage electrodialysis process. The specifications and operating conditions of each stage are equivalent to the list in Table 1. Salt concentration at the inlets of the first stage $C'_{\text{in,I}}$ and the outlets of the second stage $C'_{\text{out,II}}$ are; $C'_{\text{in,I}} = 2000\ \text{mg/dm}^3$ and $C'_{\text{out,II}} = 400\ \text{mg/dm}^3$. Salt concentration of a solution at the junction between the stage I and sage II is defined as $C'_{\text{junc}} = C'_{\text{out,I}} = C'_{\text{in,II}}$ and the computer simulation is performed with constant salt concentration mode (Figure 13) by changing C'_{junc}.

Figure 25 gives C'_{junc} versus current density I/S, cell voltage V_{cell} and energy consumption E in both stages. I/S, V_{cell} and E are decreased in the first stage and increased in the second stage with the increase of C'_{junc}. Total energy consumption in the two stage operation $E_{\text{total}} = E_{\text{I}} + E_{\text{II}}$ takes minimum value 0.452 kWh/m^3 at $C'_{\text{junc}} = 1044\ \text{mg/dm}^3$.

Figure 26 shows the current density I/S versus limiting current density $(I/S)_{\text{lim}}$. Real limiting current density is determined from the intersection between the I/S versus $(I/S)_{\text{lim}}$ plots and the $I/S = (I/S)_{\text{lim}}$ line drawn in the figure. In stage I, I/S becomes larger than $(I/S)_{\text{lim}}$ when $C'_{\text{junc}} = 700\ \text{mg/dm}^3$ is inputted. In this situation, current density i surpasses the limiting value of a cation-exchange membrane i_{lim} at the outlet of a

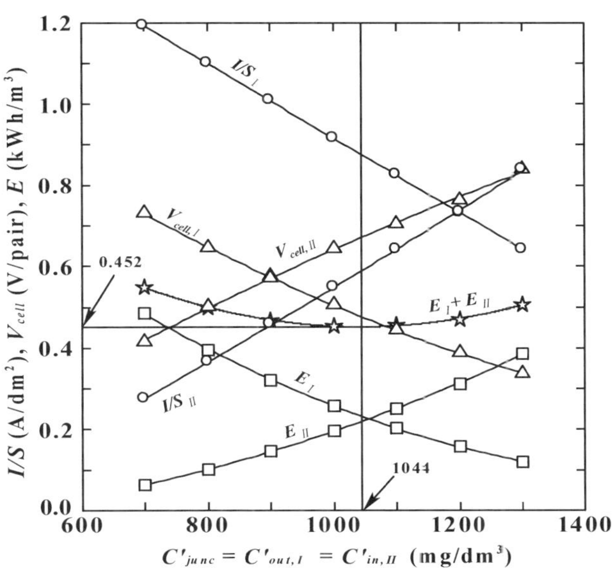

Figure 25 Current density, cell voltage and energy consumption in the two-stage operation for drinking water production. $C'_{\text{in,I}} = 2000$ mg/dm^3, $C'_{\text{out,II}} = 400$ mg/dm^3.

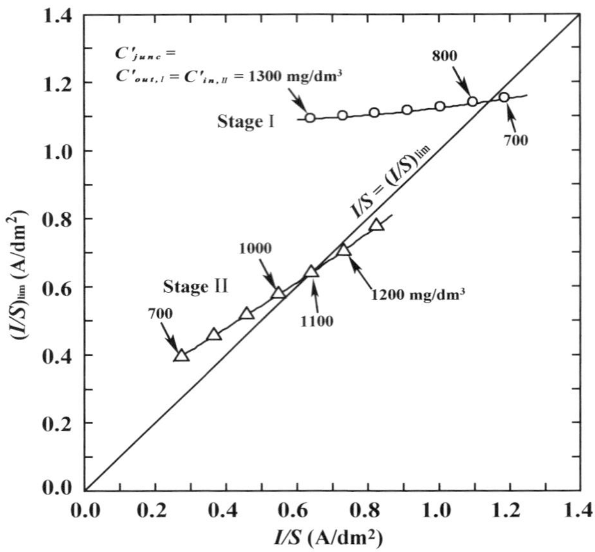

Figure 26 Limiting current density in the two-stage process. $C'_{\text{in,I}} = 2000$ mg/dm^3, $C'_{\text{out,II}} = 400$ mg/dm^3.

desalting cell in which linear velocity becomes the least among u'_{out} (cf. Section 3.12). An electric current can be passed in the computation, because very small amount of ions is remaining still in bulk at the outlet of the desalting cell in spite of at over limiting current density. The computation can not be carried out when $C'_{\text{junc}} = 600$ mg/dm^3 is inputted in the program, because the salt concentration in the bulk $C'^{\#}_{\text{out}}$ in Eqn (99) becomes zero. In stage II, I/S surpasses $(I/S)_{\text{lim}}$ at $C'_{\text{junc}} = 1200$ and 1300 mg/dm^3. The computation can not be performed at $C'_{\text{junc}} = 1400$ mg/dm^3 because $C'^{\#}_{\text{out}}$ becomes zero.

From Figure 26, C'_{junc} should be kept to $800 < C'_{\text{junc}} < 1000$ mg/dm^3 for operating the above two-stage electrodialysis process stably. $C'_{\text{junc}} = 1044$ mg/dm^3 computed in Figure 25 is economically feasible but it is not recommended in order to operate the process stably.

4. BATCH ELECTRODIALYSIS PROGRAM

4.1 Overview

The batch process is applicable to the operation of every kind of small- or middle-scale electrodialysis operations; Kusakari et al. (1977)

installed a commercial multistage batch system electrodialysis plant for desalination of brackish groundwater at Hatsushima, Atami, Shizuoka Pref. Japan. Tani et al. (1978) developed a seawater desalination batch mode unit operated in a vessel. Rapp and Pfromm (1998a) removed chloride from the chemical recovery cycle of a kraft pulp mill. Elidaoui et al. (2006) demineralized sugar liquor in a beet sugar manufacturing pilot plant. Banasiak et al. (2007) investigated the removal of fluoride and nitrate from brackish groundwater. Walha et el. (2007) demineralized brackish ground water by nanofiltration, reverse osmosis, and electrodialysis in Tunisia and compared their energy consumption. Kabay et al. removed nitrate (2007), fluoride (2008a), and boron (2008b) in a salt solution with a batch electrodialyzer.

The performance of a batch process has been discussed from various view points; Parulekar (1998) investigated energy consumption of batch operation such as (1) constant current, (2) constant voltage, (3) constant current followed by constant voltage, (4) constant voltage followed by constant current, and (5) operation with time-variant current and voltage. Demircioglu et al. (2001) introduced equations expressing the ionic mass balance around a dilute circulation tank and discussed energy consumption in a batch process. Ahmed et al. (2002) developed a mathematical model to predict changes in contaminants with time, and to estimate contaminant fluxes of migration, diffusion, and convection in a laboratory-scale batch electrodialysis cell for the regeneration of contaminated hard-chrome plating baths. Moon et al. (2004) predicted the performance of a one- and two-dimensional batch electrodialysis process based on the fundamental principles of electrochemistry, transport phenomena, and thermodynamics. Ortiz et al. (2005) developed a mathematical model for a batch process and discussed mass balance, ohmic drop, and membrane potential.

4.2 Batch Electrodialysis Process

Figure 27, illustrates the batch electrodialysis process, in which the valve V_1 is opened at first and a definite volume (Q_0) of a raw salt solution (salt concentration C^0) is supplied to the circulation tank. Next, V_1 is closed, V_2 is opened and the solution is supplied to the inlets of desalting cells (De) integrated into the stack (marked gray) in the electrodialyzer. The solution flowing out from the outlets of desalting cells is returned to the circulation tank. The raw feeding solution is also supplied to the inlets of concentrating cells (Con) for preventing scale formation in the concentrating cells.

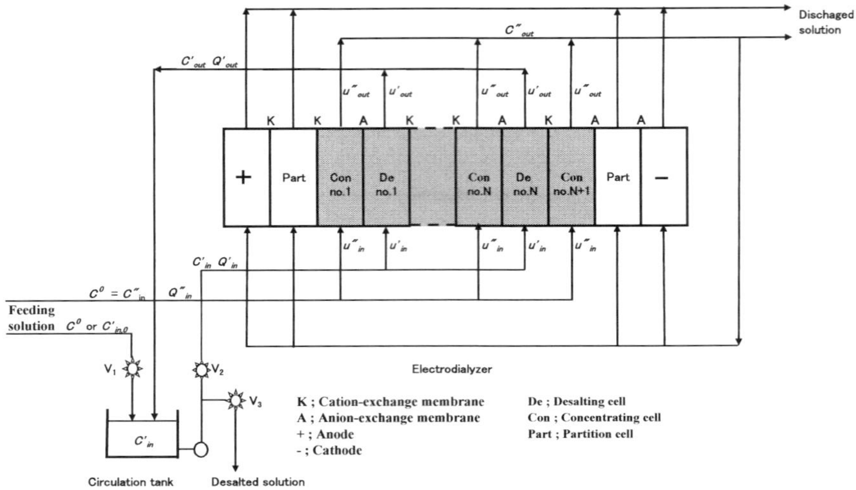

Figure 27 Batch electrodialysis process. *(Tanaka, 2013).*

A part of the solution flowing out from the outlets of concentrating cells is further supplied to electrode cells and partition cells (Part), which are integrated for preventing the influence of electrode reactions to the performance of the electrodialyzer. Then, the solution is discharged to the outside of the process with the residual solution flowing out from the outlets of concentrating cells. Next, the electrodialysis is proceeded applying constant voltage between electrodes and circulating the salt solution between the circulation tank and desalting cells. After the salt concentration in desalting cells decreased to a targeted value, V_2 is closed, V_3 is opened and the desalted solution (the output of the process) is extracted to the outside of the process.

4.3 Performance of an Electrodialyzer

In order to discuss the performance of the batch process, we have to define the performance of an electrodialyzer by determining the relationship between salt concentration at the inlets of desalting cells C'_{in}; cell voltage V_{cell} and the five parameters (salt concentration at the outlets of desalting cells C'_{out}; salt concentration at the outlets of concentrating cells C''_{out}; linear velocity at the outlets of desalting cells u''_{out}; current density I/S; ion flux across a membrane pair J_S). The above relationship is calculated from the constant voltage continuous electrodialysis program (Tanaka, 2013) according to Figure 12 (Section 3.13.2) and Companion site (Tanaka, 2015).

Five parameters (C'_{out}, C''_{out}, u''_{out}, I/S, JS,) are calculated for an electrodialyzer specifications and operating conditions presented in Table 1 (Excepting C'_{in} and C''_{in}). They are plotted against C'_{in} taking V_{cell} as a parameter and shown respectively in Figures 28, 29, 30, 31 and 32. The five parameters are expressed by the following five functions of C'_{in} and V_{cell}.

$$C'_{out} = \left(1 - \alpha V_{cell}^{0.9}\right) C'_{in} \tag{106}$$

$$C''_{out} = C^0 - \beta_1 V_{cell}^{0.9} + \beta_2 V_{cell}^{0.9} C_{in}'^{0.8} \tag{107}$$

$$u'_{out} = \gamma_1 + \gamma_2 V_{cell}^{1.1} - \gamma_3 V_{cell}^{0.9} C_{in}'^{0.8} \tag{108}$$

$$I/S = -\delta_1 V_{cell} + \delta_2 V_{cell}^{0.9} + C_{in}'^{0.8} \tag{109}$$

$$J_s = -\varepsilon_1 V_{cell} + \varepsilon_2 V_{cell}^{0.9} C_{in}'^{0.8} \tag{110}$$

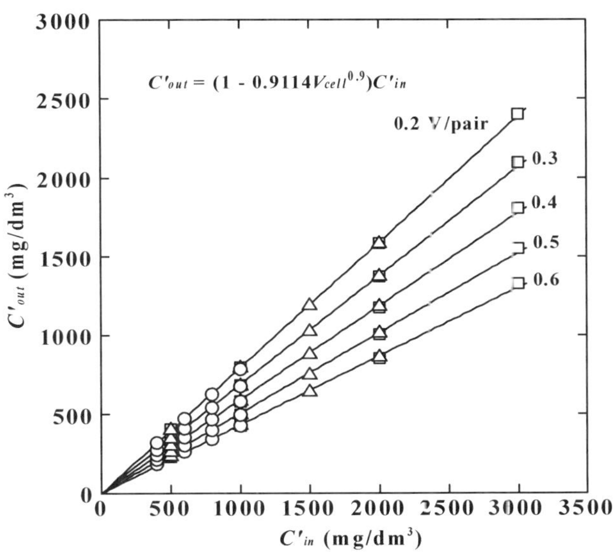

Figure 28 C'_{in} vs salt concentration at the outlets of desalting cells. $C^0 = 1000$ (○), 2000 (△), 3000 (□) mg/dm^3.

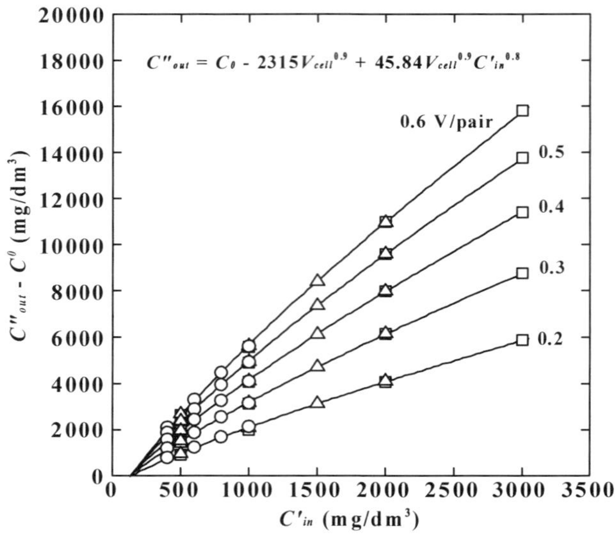

Figure 29 C'_{in} vs salt concentration at the outlets of concentrating cells. $C^0 = 1000$ (○), 2000 (△), 3000 (□) mg/dm^3.

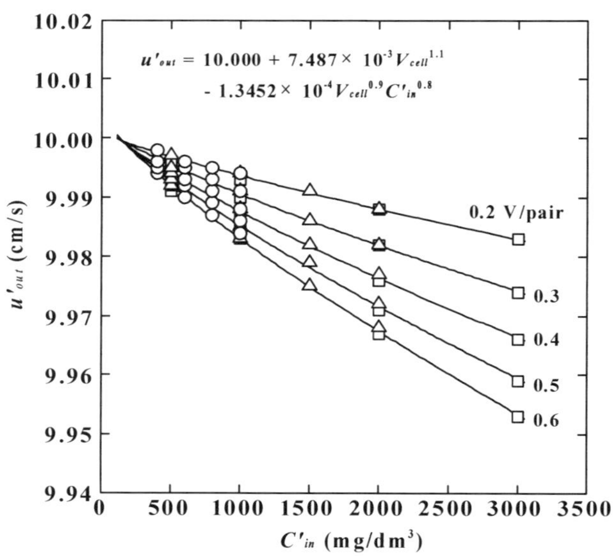

Figure 30 C'_{in} vs linear velocity at the outlets of desalting cells. $C^0 = 1000$ (○), 2000 (△), 3000 (□) mg/dm^3.

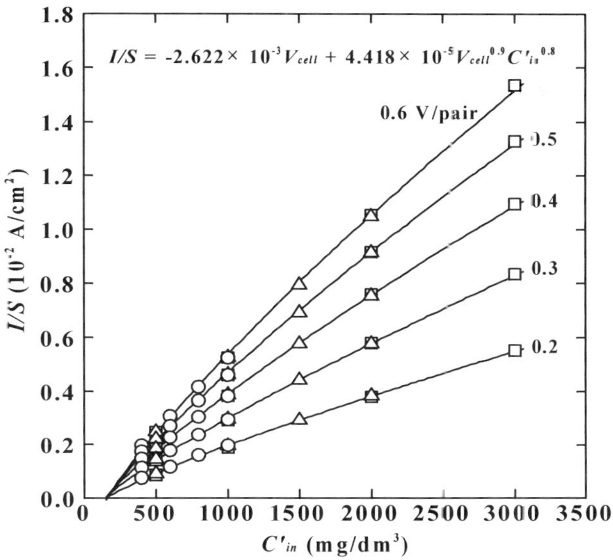

Figure 31 C'_{in} vs current density. $C^0 = 1000$ (○), 2000 (△), 3000 (□) mg/dm^3.

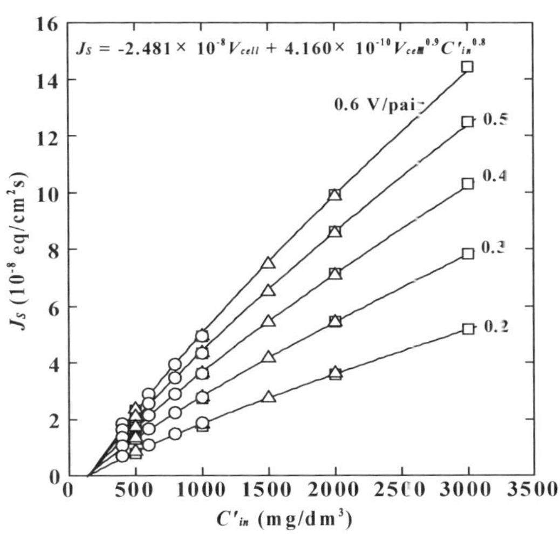

Figure 32 C'_{in} vs ion flux. $C^0 = 1000$ (○), 2000 (△), 3000 (□) mg/dm^3.

Ten coefficients (α, β_1, β_2, γ_1, γ_2, γ_3, δ_1, δ_2, ε_1, ε_2) appearing in Eqns (106)–(110) are shown in Figures 28–32. It should be noted that these coefficients are calculated on the basis of Table 1. Number of cell pairs $N = 300$ pairs is inputted but it does not influence the computed results.

4.4 Performance of the Batch Process

The equations are developed by incorporating the imaginary feeding cell; Fe and shutter; Sh at the outlet of the circulation tank as illustrated in Figure 33 in which definite volume (Q_0 dm^3) of a solution (salt concentration: $C^0 = C'_{\text{in},0}$ mg/dm^3) is prepared at first (Tanaka, 2013, 2015).

We assume here that the operation time t is divided at an interval of Δt minutes. The shutter repeats an instantaneous imaginary open/shut operation and supplies a definite flow rate (Q'_{in}) of a solution from the circulation tank to the desalting cells through the feeding cell at an interval of Δt minutes.

$$Q'_{\text{in}}\left(\text{dm}^3/\Delta t\text{min}\right) = abNu'_{\text{in}} \times \left(60\Delta ts\right) \times 10^{-3} \tag{111}$$

$$Q'_{\text{out}}\left(\text{dm}^3/\Delta t\text{min}\right) = abNu'_{\text{out}} \times \left(60\Delta ts\right) \times 10^{-3} \tag{112}$$

Before the step calculation;

$C'_{\text{out},0}$ is computed from Eqn (106).

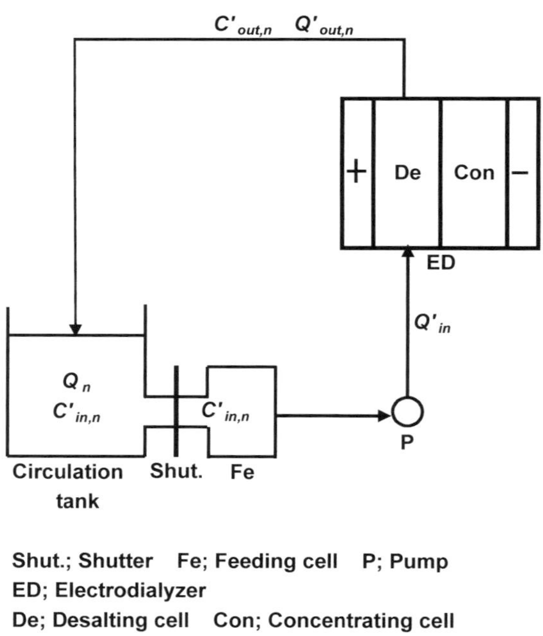

Figure 33 Instanteneous open/shut solution feed batch operation. *(Tanaka, 2013).*

$u'_{out,0}$ is calculated from Eqn (108).

$Q'_{out,0}$ is computed from Eqns (108) and (112).

$C''_{out,0}$ is calculated from Eqn (107).

Steps during operation time are numbered as $n = 1$ ($t = \Delta t$ min), $n = 2$ ($t = 2\Delta t$ min), $n = 3$ ($t = 3\Delta t$ min), $n = n$ ($t = n\Delta t$ min). The solution in the circulation tank is mixed vigorously and its concentration is the same to the concentration in the feeding cell and at the inlets of the electrodialyzer (desalting cells) C'_{in}. Table 2 gives salt concentration changes in the circulation tank (in the feeding cell) ($C'_{in,n}$) and at the outlets of the electrodialyzer ($C'_{out,n}$) and solution volume in the circulation tank (Q_n) in each step n. The mass balance and volume balance in the circulation tank in each step is introduced as follows.

$n = 1$

Mass balance:

$$C'_{in,1} = \frac{C'_{in,0} Q_0 - C'_{in,0} Q'_{in} + C'_{out,0} Q'_{out,0}}{Q_1} \tag{113}$$

Volume balance:

$$Q_1 = Q_0 - Q'_{in} + Q'_{out,0} \tag{114}$$

$n = 2$

Mass balance:

$$C'_{in,2} = \frac{C'_{in,1} Q_1 - C'_{in,1} Q'_{in} + C'_{out,1} Q'_{out,1}}{Q_2} \tag{115}$$

Table 2 Changes of salt concentration and solution volume with time in the batch operation

	Time	Circulation tank		Feeding cell (ED inlet)	ED outlet	
n	min	mg/dm^3	dm^3	mg/dm^3	mg/dm^3	dm^3/Δt min
0	0	$C'_{in,0}$	Q_0	$C'_{in,0}$	$C'_{out,0}$	$C'_{out,0}$
1	Δt	$C'_{in,1}$	Q_1	$C'_{in,1}$	$C'_{out,1}$	$C'_{out,1}$
2	$2\Delta t$	$C'_{in,2}$	Q_2	$C'_{in,2}$	$C'_{out,2}$	$C'_{out,2}$
n	$n\Delta t$	$C'_{in,n}$	Q_n	$C'_{in,n}$	$C'_{out,n}$	$C'_{out,n}$
$n+1$	$(n+1)\Delta t$	$C'_{in,n+1}$	Q_{n+1}	$C'_{in,n+1}$	$C'_{out,n+1}$	$C'_{out,n+1}$

Volume balance:

$$Q_2 = Q_1 - Q'_{\text{in}} + Q'_{\text{out},1} \tag{116}$$

$C'_{\text{in},1}$ and Q_1 are calculated using Eqns (113) and (114). $C'_{\text{out},1}$ and $Q'_{\text{out},1}$ are calculated using Eqns (106), (108), and (112).

$n = n$.

Mass balance

$$C'_{\text{in},n} = \frac{C'_{\text{in},n-1} Q_{n-1} - C'_{\text{in},n-1} Q'_{\text{in}} + C'_{\text{out},n-1} Q'_{\text{out},n-1}}{Q_n} \tag{117}$$

Volume balance

$$Q_n = Q_{n-1} - Q'_{\text{in}} + Q'_{\text{out},n-1} \tag{118}$$

$C''_{\text{out},n}$ and I/S_n are computed substituting $C'_{\text{in},n}$ and V_{cell} into Eqns (107) and (109). I/S_n and $J_{\text{S,n}}$ are computed using Eqns (109) and (110). $(I/S)_{\text{average}}$ is computed using the following equation.

$$\left(\frac{I}{S}\right)_{\text{average}} = \frac{\sum_{n=1}^{n} I/S_n}{n} \tag{119}$$

Current efficiency η_n, desalting ratio α_n, water recovery Re_n, and energy consumption E_n are computed as follows.

$$\eta_n = \frac{F\sum_{n=1}^{n} J_{\text{S},n}}{\sum_{n=1}^{n} I/S_n} \tag{120}$$

$$\alpha_n = 1 - \frac{C'_{\text{in},n}}{C'_{\text{in},0}} \tag{121}$$

$$\text{Re}_n = \frac{Q_n}{Q_0 + Q''_{\text{in}}(\text{dm}^3/\text{min}) \times t(\text{min})} \tag{122}$$

$$Q''_{\text{in}}\left(\text{dm}^3/\text{min}\right) = ab(N+1)u''_{\text{in}} \times (60\ \text{s/min}) \times 10^{-3} \tag{123}$$

$$E_n\left(\text{kWh/m}^3\right) = \frac{S\left(\text{cm}^2\right)NV_{\text{cell}}(I/S)_{\text{average}}\left(\text{A/cm}^2\right) \times 10^{-3}}{\frac{Q_n(\text{dm}^3)}{t(\text{min})} \times 60(\text{min}/h) \times 10^{-3}} \tag{124}$$

The computation is carried out by inputting the data in Table 1 into the spreadsheet software (Microsoft Excel) and the performance changes with time in the batch process are calculated. In order to improve the precision of computation, Δt should be as small as possible taking into account the values of N and Q_0. The following computation is carried out by inputting $\Delta t = 4$ min.

4.5 Saline Water Desalination

The fundamental parameters described in the site are fixed based on Table 1 (Excepting C'_{in} and C''_{in}) in all computation in this program. Computation is carried out by inputting and changing C^0 $(= C_{in,0})$, Q_0, V_{cell}, N and Δt optionally. The performance of the batch process is plotted against operation time t. The followings are the performance of the operation plotted against t calculated by changing $C^0 (= C'_{in,0})$ and V_{cell} as parameters and putting $N = 300$ pairs and $Q_0 = 30{,}000$ dm^3. The computation can be carried out in the spread sheet integrated into the companion site (Tanaka, 2015).

Figure 34 shows changes of C'_{in} with time. C'_{in} is equivalent to the salt concentration in the circulation tank. So it corresponds to the salt concentration of product water. It equals to $C_{in,0}$ at $t = 0$ and decreases toward lower concentration. The decreasing rate is increased with V_{cell}. I/S

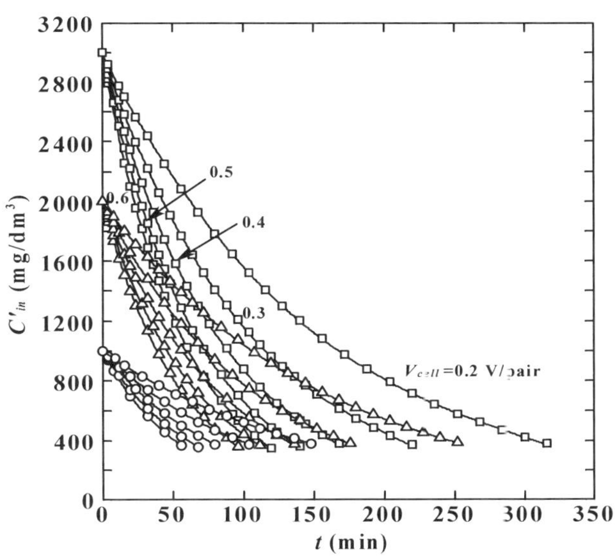

Figure 34 Changes of salt concentration with time at the inlets of desalting cells. $C^0 = 1000$ (○), 2000 (△), 3000 (□) mg/dm^3. $Q_0 = 30$ m^3, $N = 300$ pairs.

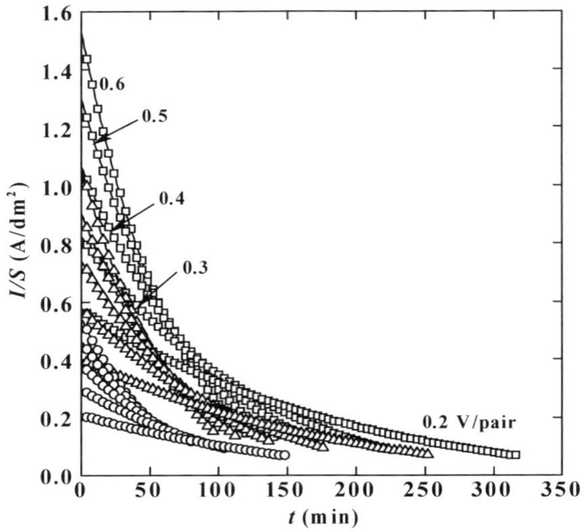

Figure 35 Changes of current density with time. $C^0 = 1000$ (○), 2000 (△), 3000 (□) mg/dm^3. $Q_0 = 30$ m^3, $N = 300$ pairs.

increases with C^0 at $t = 0$ and decreases with t, and decreasing rate is increased with V_{cell} (Figure 35). Figures 34 and 35 show that the desalination is progressed in the first half time and the purification is proceeded in the latter time to produce the product water. C''_{out} is increased with C^0 and decreased with t (Figure 36) due to current density decrease (Figure 35). Solution volume in the circulation tank Q corresponds to the output of the batch process. Q is slightly decreased with the increase of t and V_{cell} (Figure 37). Energy consumption E is increased with t, V_{cell} and C^0 (Figure 38). Current efficiency η is decreased slightly with the increase of t and V_{cell} and decrease of C^0 (Figure 39). Desalting ratio α increases from 0 to 1 with the increase of t, increases with V_{cell} and independent of C^0 (Figure 40). Water recovery Re is decreased with the increase of t and independent of C^0 and V_{cell} (Figure 41).

4.6 Drinking Water Production

4.6.1 One-Stage Operation

Drinking water is assumed to be produced according the process described in Section 4.4. Companion site presents the spreadsheet which can be operated by changing ($C^0(= C''_{\mathrm{in},0})$, Q_0, V_{cell}), and Δt optionally (Tanaka, 2015).

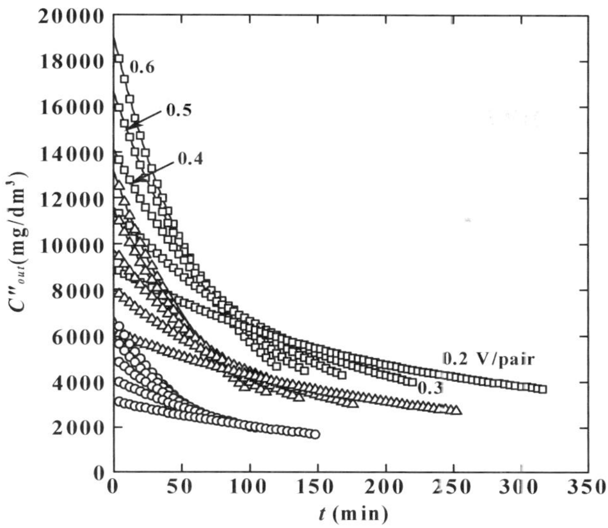

Figure 36 Changes of salt concentration with time at the outlets of concentrating cells. $C^0 = 1000$ (○), 2000 (△), 3000 (□) mg/dm^3. $Q_0 = 30$ m^3, $N = 300$ pairs.

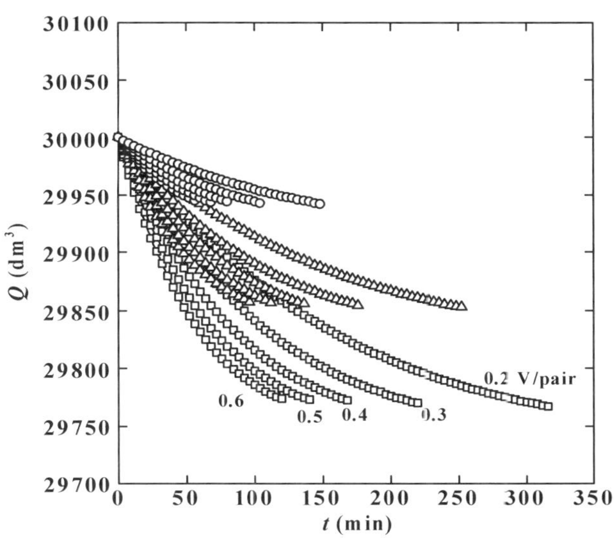

Figure 37 Changes of solution volume with time in the circulation tank. $C^0 = 1000$ (○), 2000 (△), 3000 (□) mg/dm^3. $Q_0 = 30$ m^3, $N = 300$ pairs.

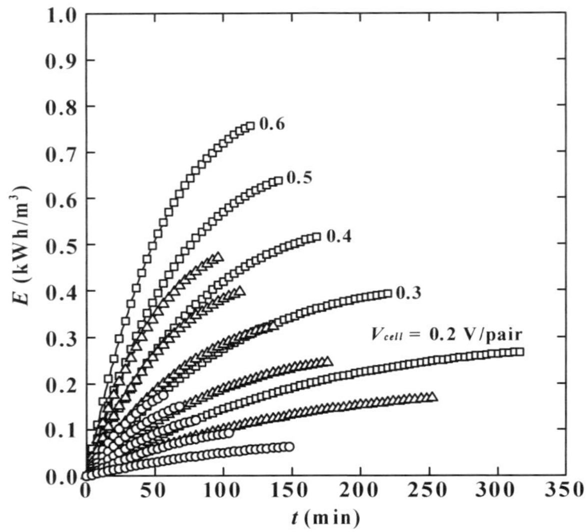

Figure 38 Changes of energy consumption with time. $C^0 = 1000$ (○), 2000 (△), 3000 (□) mg/dm^3. $Q_0 = 30$ m^3, $N = 300$ pairs.

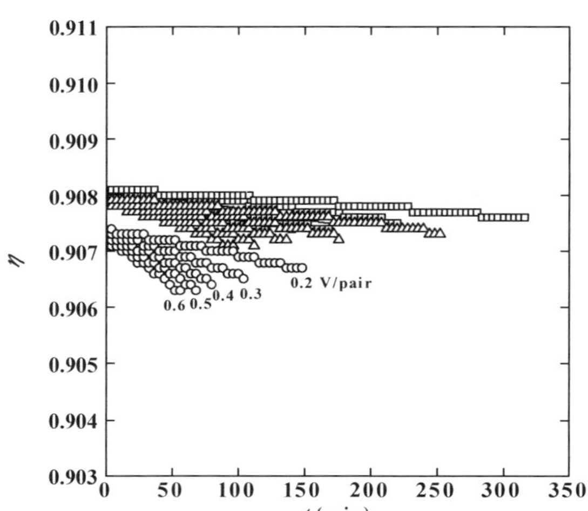

Figure 39 Changes of current efficiency with time. $C^0 = 1000$ (○), 2000 (△), 3000 (□) mg/dm^3. $Q_0 = 30$ m^3, $N = 300$ pairs.

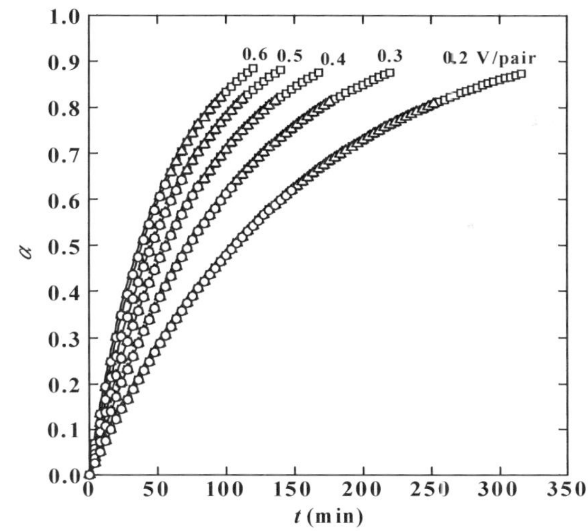

Figure 40 Changes of desalting ratio with time. $C^0 = 1000$ (○), 2000 (△), 3000 (□) mg/dm^3. $Q_0 = 30$ m^3, $N = 300$ pairs.

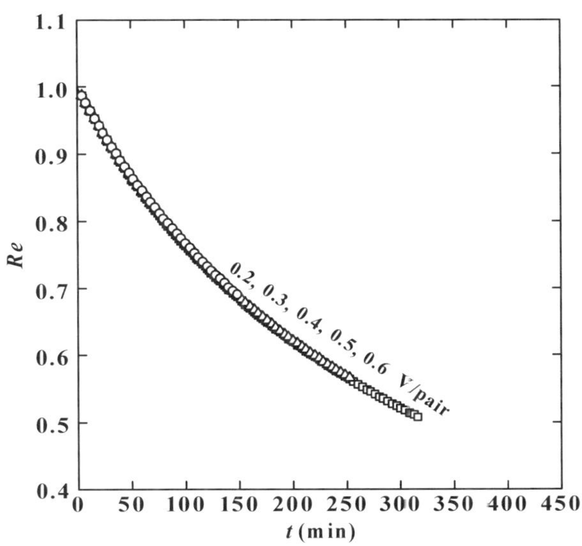

Figure 41 Changes of water recovery with time. $C^0 = 1000$ (○), 2000 (△), 3000 (□) mg/dm^3. $Q_0 = 30$ m^3, $N = 300$ pairs.

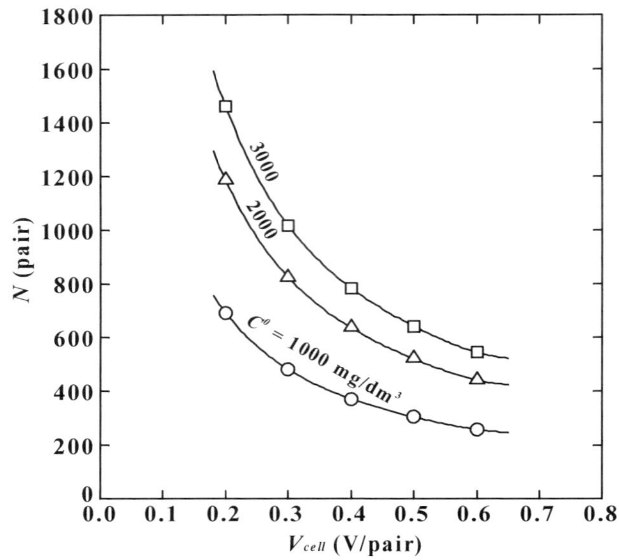

Figure 42 Cell voltage vs cell pair number integrated in an electrodialyzer. $C'_{in}(t = 60\text{min}) = 400$ mg/dm^3, $Q_0 = 30$ m^3.

Computation is carried out by changing as C^0 ($= C_{in,0}$) $= 1000$, 2000, and 3000 mg/dm^3, $V_{cell} = 0.3$, 0.4, 0.5, and 0.6 V/pair and setting $Q_0 = 30{,}000$ dm^3 and $\Delta t = 4$ min. Number of cell pairs N is calculated by adjusting it to realize $C'_{in} = 400$ mg/dm^3 at $t = 60$ min (1 h, $n = 15$). Figure 42 shows the relationship between N and V_{cell}. Figure 43 shows the relationship between energy consumption and cell voltage.

4.6.2 Two-Stage Operation

A salt solution is supplied to the first-stage in the two-stage batch electrodialysis process in Figure 44. Number of desalting cells is N_I in the first-stage and N_{II} in the second stage. The solution is electrodialyzed applying constant cell voltage V_{cell} (V/pair) to produce drinking water in the second-stage. In Fig. 44, the salt concentration of the solution supplied from stage I to stage II is $C'_{in,I}$ which is referred as junction salt concentration C'_{junc}. Figure 45 exemplifies the relationship between C'_{junc} and N, I/S, and E in each stage calculated putting $C^0 = 2000$ mg/dm^3, $V_{cell} = 0.4$ V/pair, $Q_0 = 30$ m^3 and $\Delta t = 4$ min. The plots at $C'_{junc} = 400$ and 2000 mg/dm^3 correspond to N, $I/S_{average}$ and E in the one-stage operation integrated with N_I and N_{II} alone. $N_I + N_{II}$ and $E_I + E_{II}$ in the two-stage operation are larger than the values in the

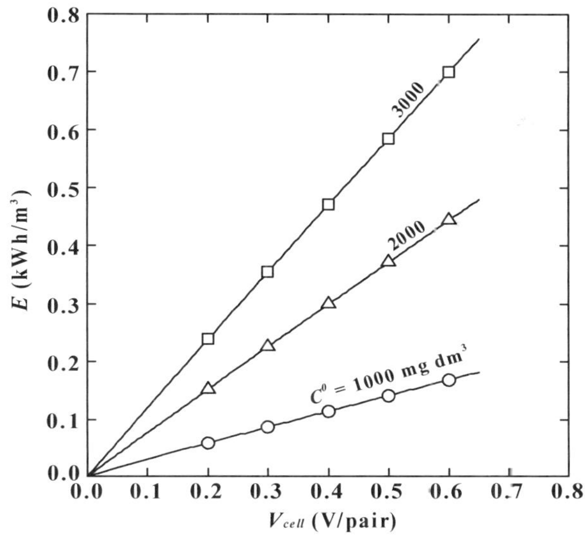

Figure 43 Cell voltage vs energy consumption. $C'_{in}(t = 60\text{ min}) = 400$ mg/dm^3, $Q_0 = 30$ m^3.

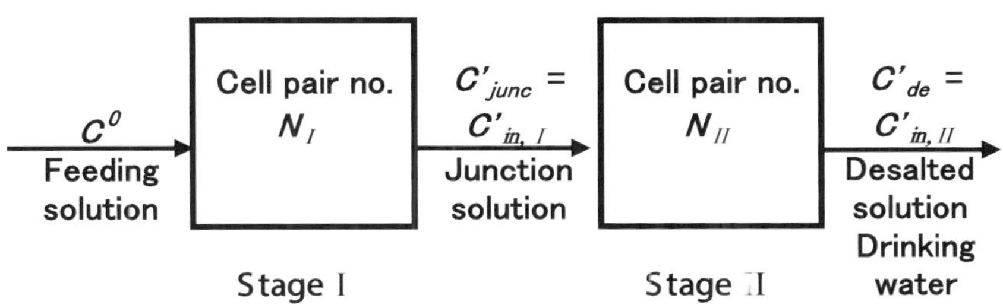

Figure 44 Two-stage process.

one-stage operation. Thus the multistage operation is assumed to be disadvantageous comparing to the single-stage operation. Current density in the first stage is larger than the value in the second stage; $I/S_{\text{average I}} > I/S_{\text{average II}}$ which means that the desalination is progressed in the first half stage and the purification is proceeded in the latter stage to produce acceptable drinking water. Total average current density $(I/S)_{\text{total, average}}$ plotted in Figure 45 is calculated by the following equation.

$$\left(\frac{I}{S}\right)_{\text{total,average}} = \frac{N_{\text{I}}(I/S)_{\text{average,I}} + N_{\text{II}}(I/S)_{\text{average,II}}}{N_{\text{I}} + N_{\text{II}}} \tag{125}$$

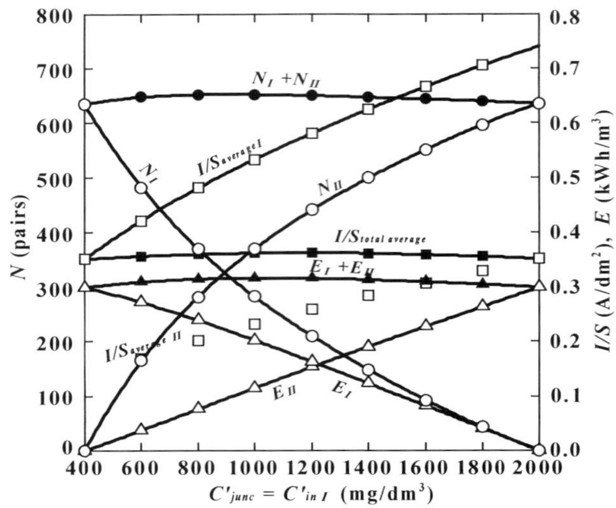

Figure 45 C'_{junc} vs cell pair number, current density and energy consumption. $C^0 = 2000$ mg/dm^3, $C'_{\text{de}} = 400$ mg/dm^3, $V_{\text{cell}} = 0.4$ V/pair, $t = 1$ h (each stage), $Q_0 = 30$ m^3.

4.7 Limiting Current Density

Limiting current density $(I/S)_{\text{lim}}$ is expressed by Eqn (100). The program for calculating $(I/S)_{\text{lim}}$ is presented in Figure 14 and included in Companion sites (Tanaka, 2015). Current density I/S and $(I/S)_{\text{lim}}$ are plotted against C'_{in} at $C^0 = 3000$ mg/dm^3 taking V_{cell} as a parameter in Figure 46. Almost the same plots are obtained at $C^0 = 2000$ and 1000 mg/dm^3. I/S is recognized to be less than $(I/S)_{\text{lim}}$; $I/S < (I/S)_{\text{lim}}$, so the electrodialysis is assumed to be operated stably. Linear velocities in desalting cells are not uniform between the cells and are expressed by the normal distribution. The above $(I/S)_{\text{lim}}$ is computed assuming that the standard deviation of the normal distribution of the solution velocity ratio $\sigma = 0.1$ (safety factor) for operating the electrodialyzer stably. However, if V_{cell} exceeds 0.6 V/pair, I/S is estimated to exceed $(I/S)_{\text{lim}}$. In order to evaluate the real limiting current density of the electrodialyzer $(I/S)_{\text{lim,real}}$, $(I/S)_{\text{lim}}$ is plotted against I/S taking V_{cell}, C^0 and salt concentration at the inlets of desalting cells C'_{in} as parameters (Figure 47). $(I/S)_{\text{lim,real}}$ is determined from the intersections between the $(I/S)_{\text{lim}}$ plots and the $I/S = (I/S)_{\text{lim}}$ line in Figure 47. From the intersection, the relationship between C'_{in} and $(I/S)_{\text{lim,real}}$ is introduced as shown in Figure 48. $(I/S)_{\text{lim,real}}$ is computed on the basis of the electrodialyzer specifications defined in Table 1 and expressed by the function of C'_{in}. Figure 48 suggests that the limiting current density is the performance of the electrodialyzer.

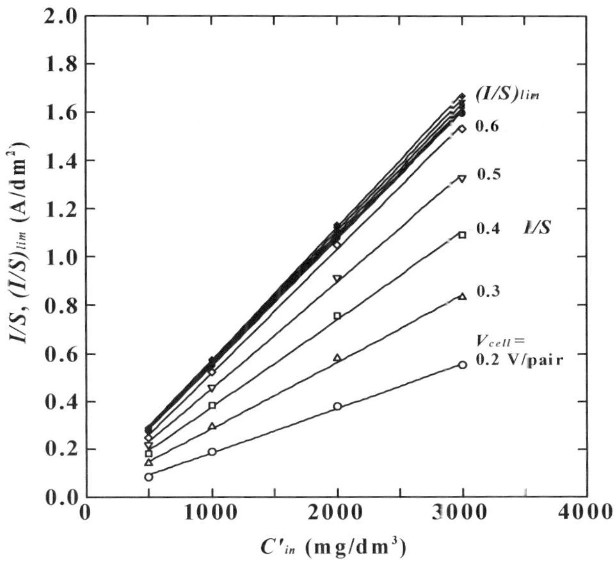

Figure 46 Limiting current density. Filled: $(I/S)_{\text{lim}}$ Open: I/S. $C^0 = 3000$ mg/dm^3.

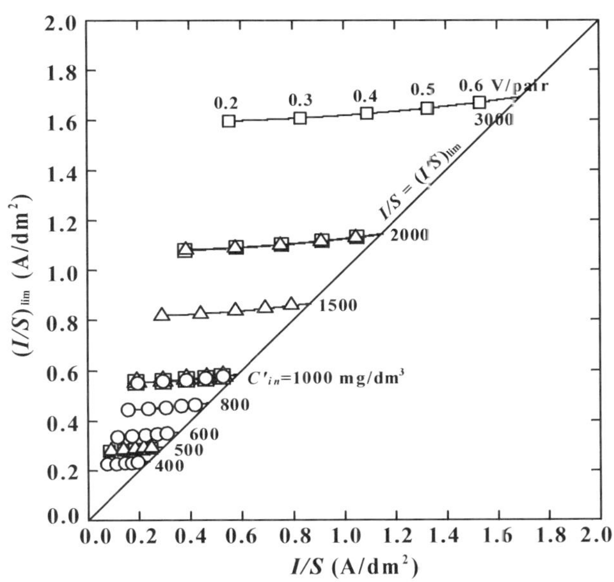

Figure 47 Real limiting current density. $C^0 = 1000$ (○), 2000 (△), 3000 (□) mg/dm^3.

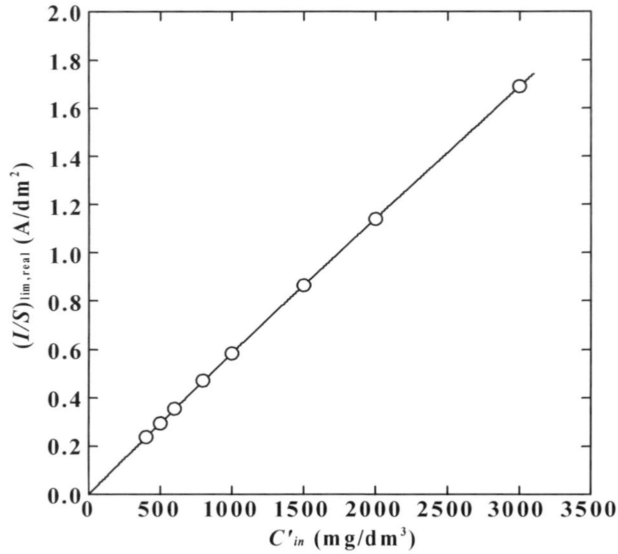

Figure 48 Relationship between C'_{in} and $(I/S)_{lim,real}$.

5. FEED-AND-BLEED ELECTRODIALYSIS PROGRAM

5.1 Overview

The feed-and-bleed operation is widely performed so far. It is applicable to middle-scale electrodialysis. For instance, Asahi Chemical Co. developed a feed-and-bleed electrodialysis plant (2500 m^3/day) for producing potable water in Ohoshima island, Tokyo (Koga and Mitsugami, 1978). Tokuyama Inc. developed the technology to establish a closed system by means of electrodialytic reuse of waste water in a plating process (Matsunaga, 1986). Asahi Glass Co. constructed large-scale blackish water desalination plants in Kashima Power Station (2000 m^3/day) and Kashima South Joint Power Station (12,000 m^3/day), Ibaraki Pref. Japan (Kawahara, 1994). Thompson et al. (1997) separated sulfide from hydroxide in kraft white liquor with an electrodialysis system operated in both the batch and feed-and-bleed modes, producing sulfide-rich white liquor. Rapp and Pfromm (1998b) removed chloride in a kraft pulping process with a feed-and-bleed operation incorporated with monovalent-selective anion exchange membranes. Ryabtsev et al. (2001) developed a two-stage feed-and-bleed set-up for desalination of underground saline water. The first stage involves galvanostatic regime at increased current density, and the second stage profound in a potentiostatic regime resulting in desalted water. Schoeman et al. (2005) electrodialyzed a hazardous leachate from an industrial landfill site to decrease TDS concentration and effluent volume. Electrodialysis pilot tests were conducted in the batch and feed-and-bleed modes to develop process design criteria for full-scale application. Zhang et al. (2011) electrodialyzed the reverse osmosis concentrate to reduce the volume of salty water discharge and improve the overall water recovery. Ion transport mechanisms in the batch and feed-and-bleed mode were studied to monitor the effluent water compositions. Tran et al. (2012) treated reverse osmosis concentrate with the batch and feed-and-bleed mode electrodialysis to realize high recovery of the reverse osmosis system. A pellet reactor was used before electrodialysis treatment to prevent scale precipitation. Tanaka developed the feed-and-bleed program for saline water desalination (2014, 2015). This section explains the program taking account of the continuous program described in Section 3.

5.2 Feed-and-Bleed Process

A raw salt solution (salt concentration; C_0') is assumed to be desalinated by the feed-and-bleed electrodialysis process illustrated in Figure 49. The electrodialyzer is incorporated with desalting cells and concentrating cells in the stack marked with gray. Number of cells is N for desalting cells, $N + 1$ for

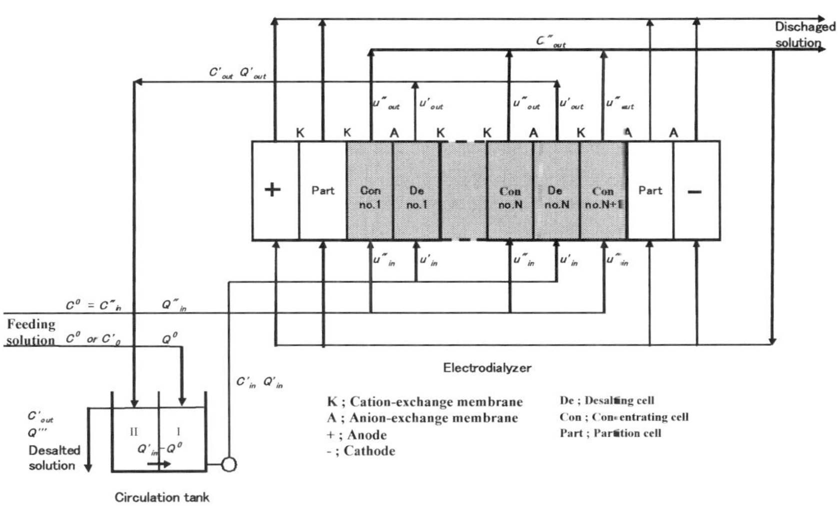

Figure 49 Feed-and-bleed electrodialysis process. *(Tanaka, 2014).*

concentrating cells and N for cell pairs. The anode and cathode cells are placed at both outsides of the stack and an electric current is supplied between the electrodes. The circulation tank is separated to the compartment I and compartment II and the salt solution is supplied to the electrodialyzer through the circulation tank.

The raw salt solution (salt concentration; C^0, volume flow rate; Q^0) is supplied to the compartment I in the circulation tank. It is further supplied to desalting cells in the electrodialyzer $(C'_{\rm in}, Q'_{\rm in})$. Linear velocity in the desalting cells is $u'_{\rm in}$ at the inlets and $u'_{\rm out}$ at the outlets. The solution flowing out from the desalting cells is returned to the compartment II $(C'_{\rm out}, Q'_{\rm out})$. A part of the desalted solution $(C'_{\rm out}, Q''')$ is extracted from the compartment II to the outside of the process. Another part of the desalted solution $(C'_{\rm out}, Q_{\rm move} = Q'_{\rm in} - Q^0)$ moves from the compartment II to the compartment I. The process is operated keeping $Q_{\rm move} > 0$.

The raw salt solution is also supplied to concentrating cells in the electrodialyzer $(C^0 = C''_{\rm in}, Q''_{in})$ for preventing scale formation in the cells. Linear velocity in the concentrating cells is $u''_{\rm in}$ at the inlets and $u''_{\rm out}$ at the outlets. A part of the solution extracted from the outlets of concentrating cells is further supplied to the electrode cells and the partition cells (Part), which are equipped for preventing the influence of electrode reactions to the performance of the electrodialyzer. Then, the solution is discharged to the outside of the process with the residual solution flowing out from the outlets of concentrating cells.

5.3 Mass Balance and Energy Consumption in the Feed-and-Bleed Operation

The performance of the feed-and-bleed operation is understandable by discussing the following phenomena generating in the circulation tank in Figure 49.

In the circulation tank, the volume rate of a solution flowing into desalting cells $Q'_{\rm in}$ and concentrating cells $Q''_{\rm in}$ are defined as;

$$Q'_{\rm in} = abNu'_{\rm in} \tag{126}$$

$$Q''_{\rm in} = ab(N+1)u''_{\rm in} \tag{127}$$

The volume rate of a solution flowing out from desalting cells $Q'_{\rm out}$ and that of a solution flowing out from the compartment II in the circulation tank Q''' are

$$Q'_{\rm out} = Q'_{\rm in} - {\rm SNJ_V} \tag{128}$$

$$Q''' = Q^0 - SNJ_V \tag{129}$$

The material balance in the compartment I is

$$C'_{in}Q'_{in} = C'_0Q^0 + C'_{out}\left(Q'_{in} - Q^0\right) \tag{130}$$

Feeding solution volume flow rate Q^0 is introduced from Eqn (130) as follows

$$Q^0 = \frac{C'_{in} - C'_{out}}{C'_0 - C'_{out}}Q'_{in} \tag{131}$$

In the circulation tank, the volume rate $Q_{move} = Q'_{in} - Q^0$ of a solution moves from the compartment II to the compartment I. In order to operate the batch electrodialysis process, Q^0 must be less than Q'_{in};

$$Q_{move} = Q'_{in} - Q^0 > 0 \tag{132}$$

Energy consumption E, water recovery Re, and desalting ratio α are given as follows.

$$E = \frac{IV_{cell}N}{Q'''} = \left(\frac{1}{Q'''}\right)\left(\frac{I}{S}\right)SV_{cell}N \tag{133}$$

$$\mathrm{Re} = \frac{Q'''}{Q^0 + Q''_{in}} \tag{134}$$

$$\alpha = 1 - \frac{C'_{out}}{C^0} \tag{135}$$

5.4 Electrodialysis Program of the Feed-and-Bleed Operation

Figure 50 gives the electrodialysis program chart of the single-stage feed-and-bleed process operated under a constant voltage. In this program, the steps from "Start" to "decision point 4" are defined as the constant salt concentration and constant voltage continuous steps, which are computed with the continuous program (Section 3, Tanaka, 2012a). The decision point 1 functions as the constant salt concentration continuous operation (Section 3.13.3, Figure 13). However, the decision point 4 functions as the constant voltage continuous operation (Section 3.13.2, Figure 12), so that the overall process comes to be operated under constant voltage. Limiting current density $(I/S)_{lim}$ is the performance of the electrodialyzer and it is computed after

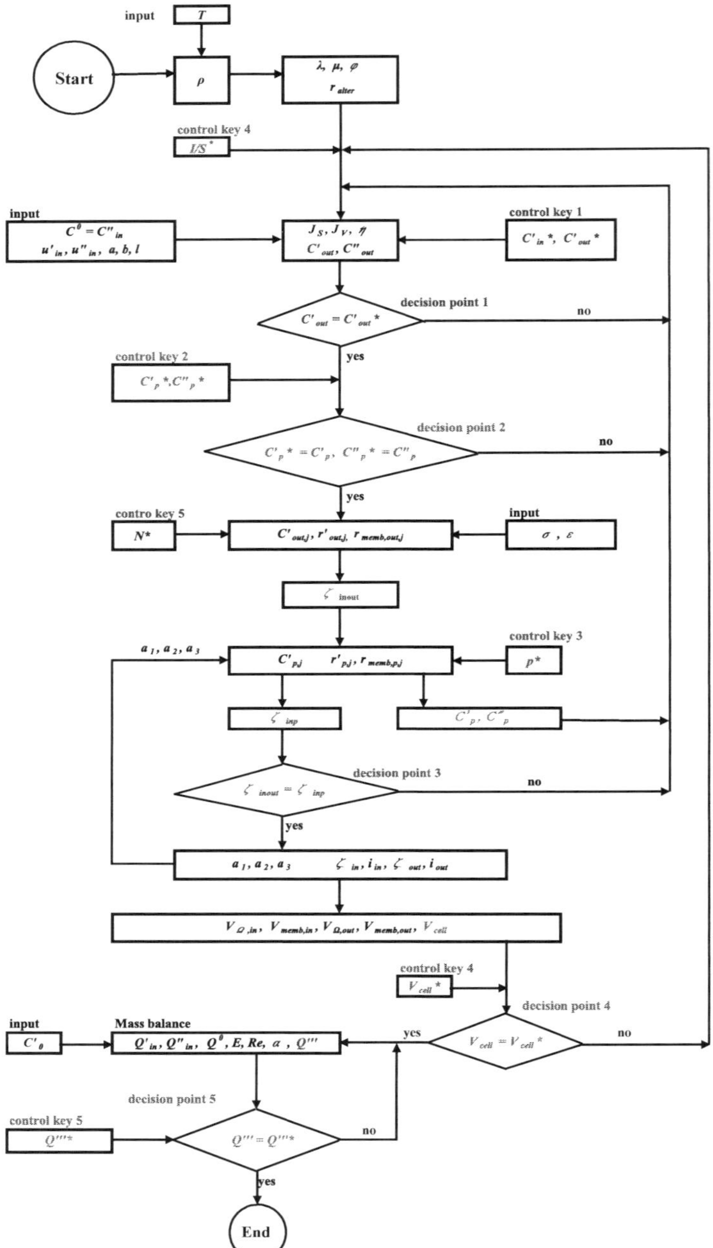

Figure 50 Feed-and-bleed electrodialysis program chart. *(Tanaka, 2014).*

the continuous steps by inputting the information in mass transport in the electrodialyzer into the limiting current program (Section 3.13.4, Figure 14). In Figure 50, the feed-and-bleed steps including the "decision point 5" are added after the above continuous steps. Computation in the feed-and-bleed steps is carried out based on the mass balance and energy consumption (Section 5.3) after the above computation in the continuous steps is finished. Cell pair number N, energy consumption E, water recovery Re, desalting ratio α, and Q_{move} are computed in the feed-and-bleed steps.

The computation can be carried out in the spreadsheets (Microsoft Excel) presented in the Companion site (Tanaka, 2015).

In the feed-and-bleed process in Figure 49, the concentrating cells are supplied with a raw salt concentration (concentration C^0). Thus at the inlets of concentrating cells, $C^0 = C''_{in}$ holds. In the single-stage feed-and-bleed process, the circulation tank is supplied with a raw salt solution (concentration C^0). However, if C^0 is high, it is advantageous to establish the multi-stage feed-and-bleed process and the desalted solution extracted from the previous stage (concentration C'_{out}) should be supplied to the circulation tank in the next stage.

5.5 Specifications and Operating Conditions of the Electrodialyzer

The fundamental specifications and operating conditions of the electrodialyzer is defined in Table 1 (Excepting C'_{in} and C''_{in}). We produce the drinking water (salt concentration 400 mg/dm^3) in the single-stage constant voltage feed-and-bleed operation under the following conditions.

Cell voltage $V_{cell} = 0.2, 0.3, 0.4, 0.5, 0.6$ V/pair.

Salt concentration of a feeding solution $C^0 = 1000, 2000, 3000$ mg/dm^3 (ionic constituent ratio is the same to that of seawater)

Salt concentration of a desalted solution $C'_{out} = 400$ mg/dm^3 (drinking water)

Output of a desalted solution $Q''' = 10, 20, 30, 40, 50$ m^3/h.

5.6 Drinking Water Production

5.6.1 One-Stage Operation

Figure 51 shows the relationship between cell voltage V_{cell} and current density I/S taking salt concentration of a feeding solution C^0 as a parameter. I/S increases with V_{cell} and this phenomenon is understandable. However, it is noticed that I/S is slightly decreased with the increase of C^0.

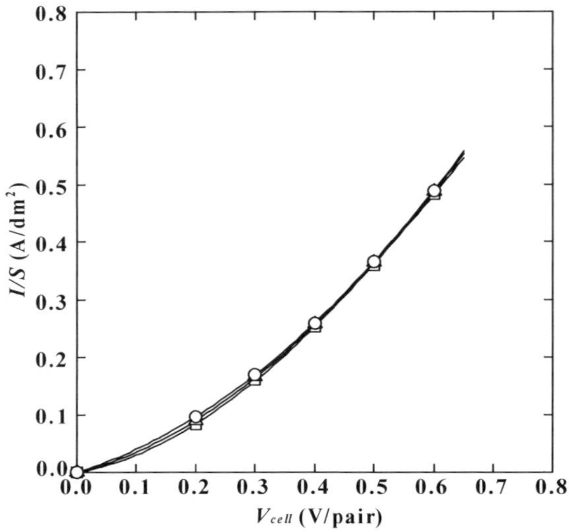

Figure 51 Cell voltage vs current density. $C^0 = 1000$ (○), 2000 (△), 3000 (□) mg/dm^3.

This phenomenon seems to be unconventional because electric resistance is decreased with the increase of C^0, so I/S is estimated to be increased with C^0. This phenomenon is commented in the following:

In this electrodialysis system, (a) C'_{out} is predetermined as 400 mg/dm^3 and (b) A raw solution (salt concentration C^0) is supplied also to concentrating cells ($C''_{\text{in}} = C^0$). At under an applied cell voltage V_{cell}, if I/S is increased with C^0 due to the decrease of electric resistance of desalting cells, concentrating cells and membranes, C'_{out} is possibly decreased to less than 400 mg/dm^3. In order to maintain $C'_{\text{out}} = 400$ mg/dm^3 in this circumstance, C'_{in} must be decreased with the increase of C^0 in order to reduce I/S (Figure 51). The decrease of C'_{in} is the only one phenomenological measure to solve the problem. For understanding this phenomenon, C'_{in} is plotted against V_{cell} and shown in Figure 52. In this figure, C'_{in} is slightly decreased with the increase of C^0.

Membrane pair number N is plotted against water production rate Q''' keeping $C^0 = 2000$ mg/dm^3 and exemplified in Figure 53. N is increased with the increase of Q''' and decrease of V_{cell}. Keeping $Q''' = 30$ m^3/h, N versus V_{cell} is exemplified in Figure 54. N is increased with the increase of C^0 and decrease of V_{cell}. Figure 55 exemplifies energy consumption E versus V_{cell} setting $Q''' = 30$ m^3/h . E is increased with V_{cell} and C^0. Figure 56 exemplifies water recovery Re at $Q''' = 30$ m^3/h. Re increases with the

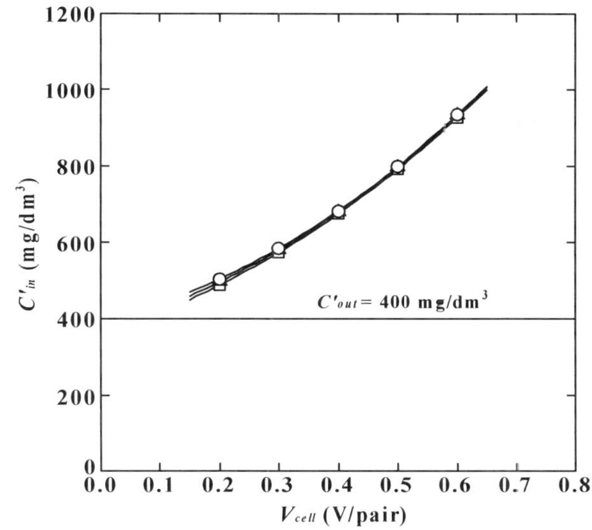

Figure 52 Salt concentration at the inlets of desalting cells. $C^0 = 1000$ (○), 2000 (△), 3000 (□) mg/dm^3.

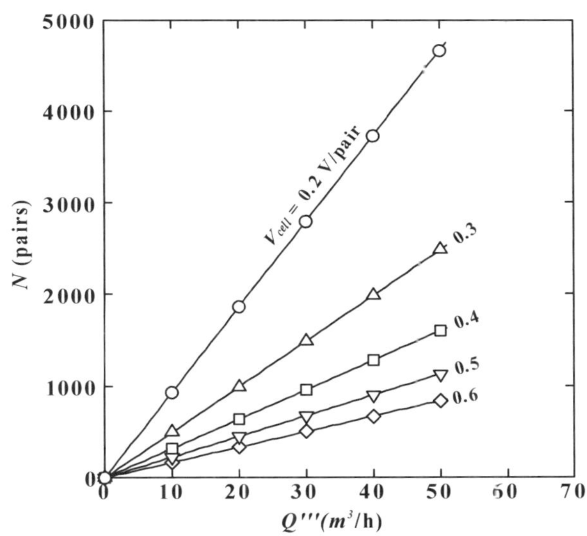

Figure 53 Water production rate vs cell pair number. $C^C = 2000$ mg/dm^3.

decrease of C^0 and the increase of V_{cell} (Figure 57). Re is independent of water production rate Q'''. Desalination ratio α is 0.6 ($C^0 = 1000$ mg/dm^3), 0.8 (2000 mg/dm^3) and 0.8667 (3000 mg/dm^3) because the salt concentration of the desalted solution C'_{out} is adjusted to 400 mg/dm^3.

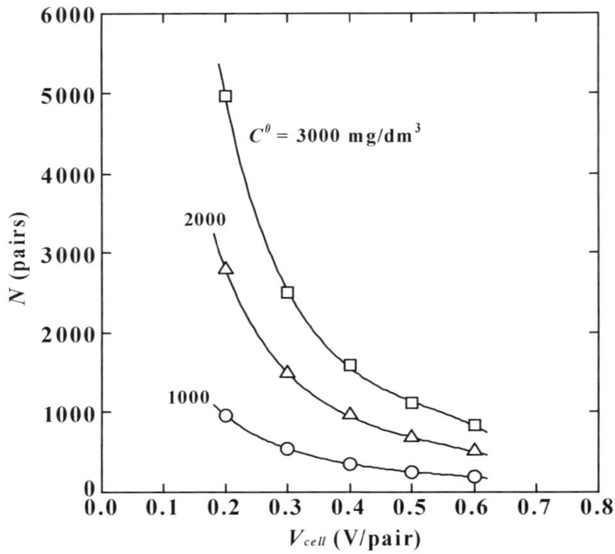

Figure 54 Cell voltage vs cell pair number. $Q''' = 30$ m^3/h.

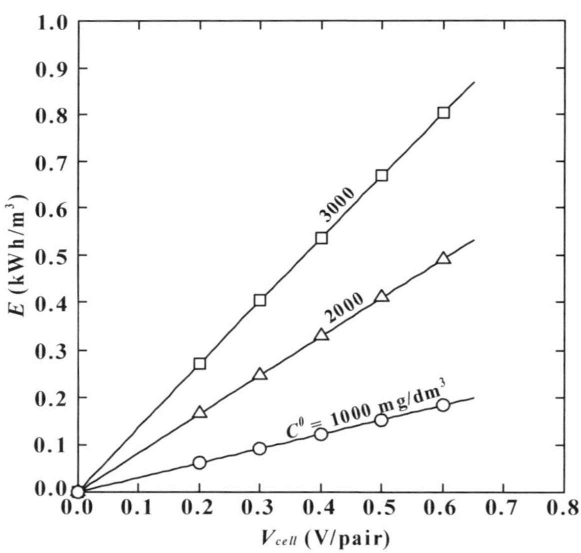

Figure 55 Cell voltage vs energy consumption. $Q''' = 30$ m^3/h.

In the circulation tank, the solution must move from the compartment II to the compartment I. Thus the process must be operated in the condition defined by Eqn (132). Figure 58 gives the relationship between V_{cell} and Q_{move} at $Q''' = 30$ m^3/h. When $C^0 = 1000$ mg/dm^3, Q_{move} becomes

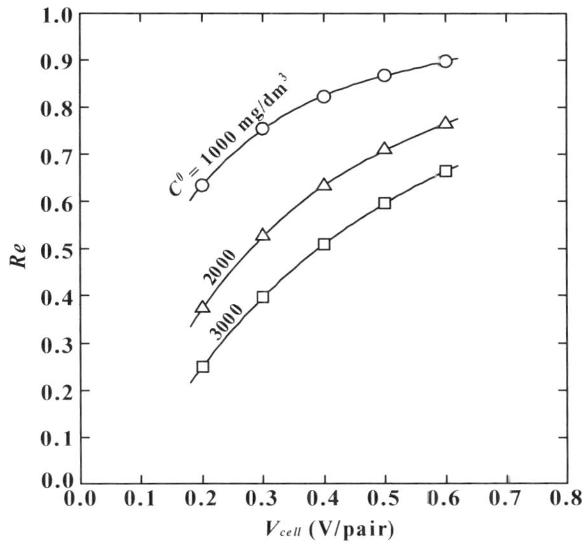

Figure 56 Cell voltage vs water recovery. $Q''' = 30$ m^3/h.

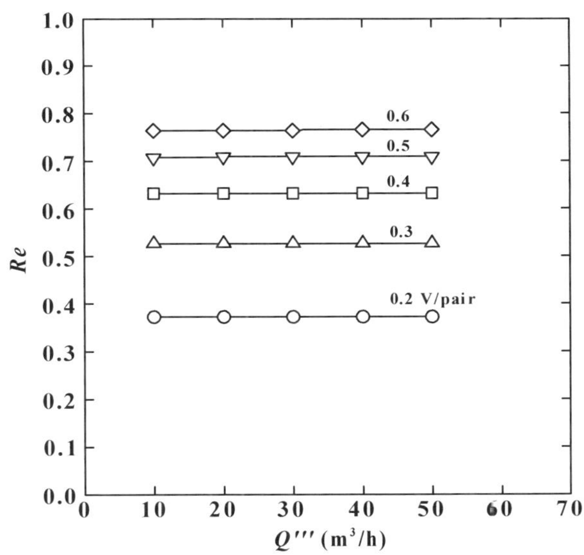

Figure 57 Water production rate vs water recovery. $C^0 = 2000$ mg/dm^3.

zero at $V_{\text{cell}} = 0.674$ V/pair. So, in order to operate the process according to Eqn (132), cell voltage must be kept as $V_{\text{cell}} < 0.674$ V/pair. When the salt concentration of the feeding solution $C^0 = 2000$ and 3000 mg/dm^3, Q_{move} does not reach zero in the range of V_{cell} applied in Figure 58; so the system is

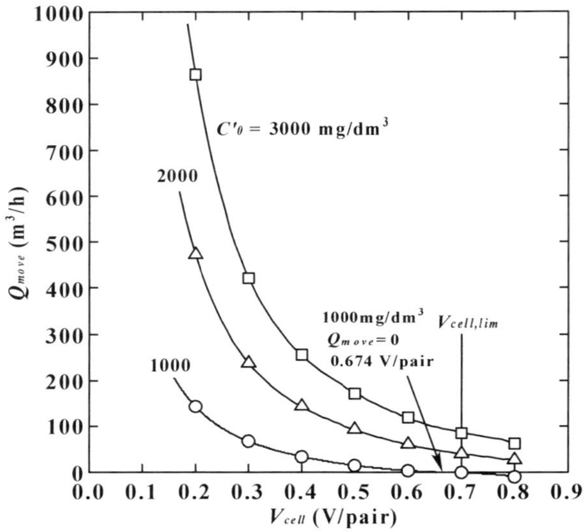

Figure 58 Cell voltage vs Q_{move}. $Q''' = 30$ m^3/h.

estimated to be operated stably. However, limiting cell voltage $V_{cell,lim}$ appears at 0.7 V/pair (Figure 59). Thus the electrodialysis should be operated at $V_{cell} < 0.7$ V/pair.

Limiting current density $(I/S)_{lim}$ is plotted against I/S and shown in Fig. 59. The real limiting current density $(I/S)_{lim,real}$ is determined from the intersection between the plots and the $I/S = (I/S)_{lim}$ line indicated in the figure and it is estimated as $(I/S)_{lim,real} = 0.637$ A/dm^2 ($C^0 = 1000$ mg/dm^3), 0.635 A/dm^2 (2000 mg/dm^3), and 0.634 A/dm^2 (3000 mg/dm^3). Limiting cell voltage $V_{cell,lim}$ applied under the $(I/S)_{lim,real}$ is estimated to be 0.7 V/pair for $C^0 = 1000$, 2000, and 3000 mg/dm^3.

5.6.2 Two-Stage Operation

The performance of the two-stage process is computed by arranging the program for the single-stage process (Figure 49) in series. A salt solution ($C'_{\mathrm{I}} = 2000$ and 3000 mg/dm^3) is assumed to be supplied to the two-stage feed-and-bleed process. Integrated cell pair number is N_{I} in the stage I and N_{II} in the stage II. Applying constant cell voltage (0.3, 0.4, 0.5 V/pair) to each stage, the salt solution is electrodialyzed to produce drinking water at the exit of the stage II ($C'_{out,\mathrm{II}} = 400$ mg/dm^3). Salt concentration of a desalted solution at the exit of the stage I; $C'_{out,\mathrm{I}}$ is equivalent to the salt concentration at the inlet of the stage II; $C'_{0,\mathrm{II}}$, i.e., $C'_{out,\mathrm{I}} = C'_{0,\mathrm{II}}$ which is

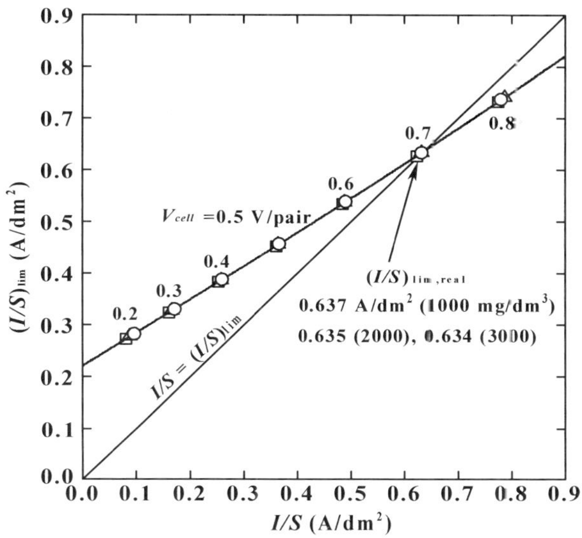

Figure 59 Limiting current density. $C^0 = 1000$ (○), 2000 (△), 3000 (□) mg/dm^3.

defined as the junction salt concentration C'_{junc} at the junction of the stage I and the stage II. The performance of the feed-and-bleed process is evaluated by plotting the performance against C'_{junc} keeping $Q''' = 30\ \text{m}^3/\text{h}$. The following figures show the performance at $C_{\text{I}}^0 = 2000\ \text{mg/dm}^3$

Figure 60 shows C'_{junc} versus Q_{move} in both stages. When the salt solution ($C_{\text{I}}^0 = 2000\ \text{mg/dm}^3$) is supplied to the first stage of the process, the following ranges of C'_{junc} are determined from Fig. 60 to satisfy Eqn (132); $C'_{\text{junc}} = 581 - 1371\ \text{mg/dm}^3$ ($V_{\text{cell}} = 0.3$ V/pair), 680–1179 mg/dm^3 (0.4 V/pair), and 796–1008 mg/dm^3 (0.5 V/pair).

For $C_{\text{I}}^0 = 3000\ \text{mg/dm}^3$, the following C'_{junc} range satisfies Eqn (132); $C'_{\text{junc}} = 575 - 2068\ \text{mg/dm}^3$ ($V_{\text{cell}} = 0.3$ V/pair), 675–1807 mg/dm^3 (0.4 V/pair), 792–1551 mg/dm^3 (0.5 V/pair).

Figure 61 shows number of cell pairs integrated into stage I; N_{I}, and stage II; N_{II}. N_{I} decreases and N_{II} increases with the increase of C'_{junc}. Total cell number $N_{\text{I}} + N_{\text{II}}$ becomes minimum at $N_{\text{I}} = N_{\text{II}}$. So, it is advantageous to integrate equal number of cell pairs into stage I and stage II. Suitable C'_{junc} realizing $N_{\text{I}} = N_{\text{I}}$ is; For $C_{\text{I}}^0 = 2000\ \text{mg/dm}^3$ (Figure 61), $C'_{\text{junc}} = 883.6\ \text{mg/dm}^3$ ($V_{\text{cell}} = 0.3$ V/pair), 889.4 mg/dm^3 (0.4 V/pair), 893.3 mg/dm^3 (0.5 V/pair). For $C_{\text{I}}^0 = 3000\ \text{mg/dm}^3$, $C'_{\text{junc}} = 1073.9\ \text{mg/dm}^3$ ($V_{\text{cell}} = 0.3$ V/pair), 1086.6 mg/dm^3 (0.4 V/pair), 1094.6 mg/dm^3 (0.5 V/pair).

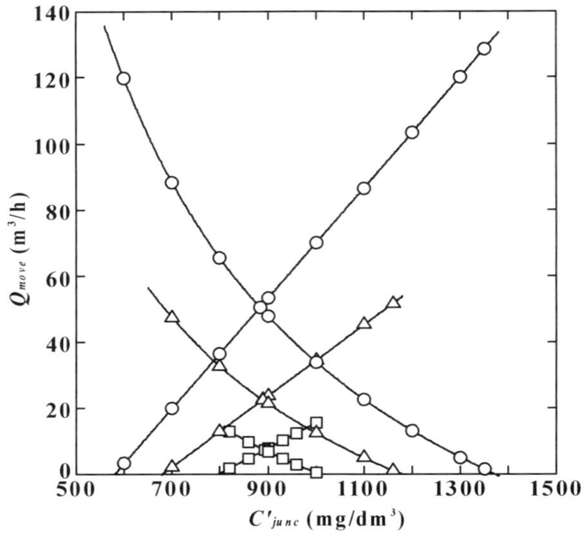

Figure 60 C'_{junc} vs Q_{move}. ○ 0.3, △ 0.4, □ 0.5 V/pair. $C_{I}^{0} = 2000$ mg/dm^3, $C_{out,II}^{0} = 400$ mg/dm^3, $Q''' = 30$ m^3/h.

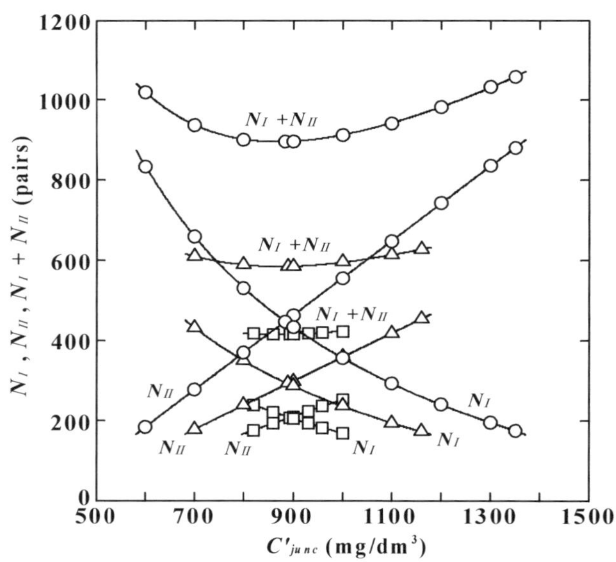

Figure 61 C'_{junc} vs cell pair number. ○ 0.3, △ 0.4, □ 0.5 V/pair. $C_{I}^{0} = 2000$ mg/dm^3, $C_{out,II}^{0} = 400$ mg/dm^3, $Q''' = 30$ m^3/h.

Figure 62 shows current density I/S. Average current density $(I/S)_{\text{average}}$ is calculated from the current density in stage I $(I/S)_{\text{I}}$ and that in stage II $(I/S)_{\text{II}}$ using the following equation.

$$\left(\frac{I}{S}\right)_{\text{average}} = \frac{\left(\frac{I}{S}\right)_{\text{I}} N_{\text{I}} + \left(\frac{I}{S}\right)_{\text{II}} N_{\text{II}}}{N_{\text{I}} + N_{\text{II}}} \quad (136)$$

$(I/S)_{\text{I}}$ is larger than $(I/S)_{\text{II}}$. $(I/S)_{\text{I}}$ increases with C'_{junc}. $(I/S)_{\text{II}}$ is scarcely influenced by C'_{junc}. $(I/S)_{\text{average}}$ becomes maximum when total cell number $N_{\text{I}} + N_{\text{II}}$ becomes minimum at $N_{\text{I}} = N_{\text{II}}$ (cf. Figure 61)

Figure 63 shows energy consumption in stage I; E_{I} and in stage II; E_{II}. E_{I} decreases and E_{II} increases with the increase of C'_{junc}. Total energy consumption $E_{\text{I}} + E_{\text{II}}$ is scarcely influence by C'_{junc} and almost constant.

Figure 64 shows water recovery Re_{I} and Re_{II} in each stage and total water recovery Re_{total} for both stages. They are computed using the following equations.

$$\text{Re}_{\text{I}} = \frac{Q'''_{\text{I}}}{Q^0_{\text{I}} + Q''_{\text{in,I}}} \qquad \text{Re}_{\text{II}} = \frac{Q'''_{\text{I}}}{Q^0_{\text{II}} + Q''_{\text{in,II}}} \quad (137)$$

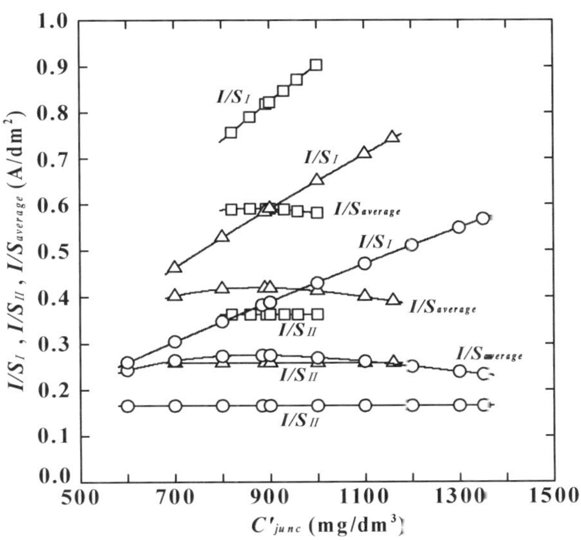

Figure 62 C'_{junc} vs current density. ○ 0.3, △ 0.4, □ 0.5 V/pair. $C^0_{\text{I}} = 2000$ mg/dm^3, $C^0_{\text{out,II}} = 400$ mg/dm^3, $Q''' = 30$ m^3/h.

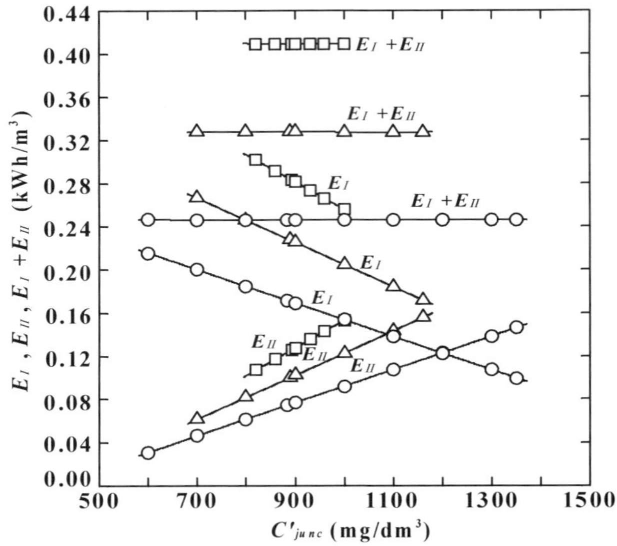

Figure 63 C'_{junc} vs energy consumption. ○ 0.3, △ 0.4, □ 0.5 V/pair. $C_I^0 = 2000$ mg/dm^3, $C_{out,II}^0 = 400$ mg/dm^3, $Q''' = 30$ m^3/h.

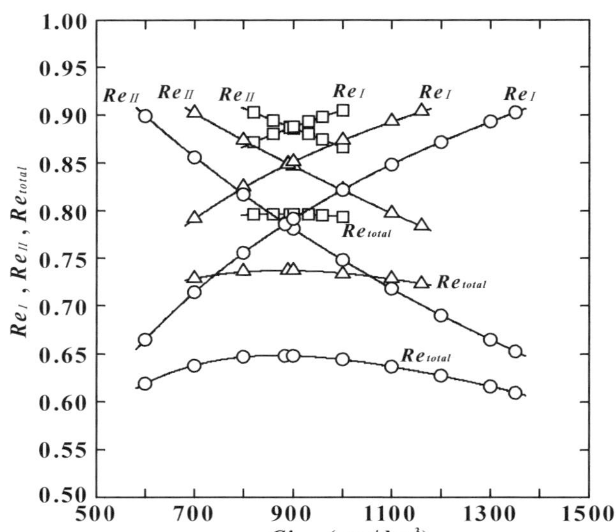

Figure 64 C'_{junc} vs water recovery. ○ 0.3, △ 0.4, □ 0.5 V/pair. $C_I^0 = 2000$ mg/dm^3, $C_{out,II}^0 = 400$ mg/dm^3, $Q''' = 30$ m^3/h.

$$\mathrm{Re_{total}} = \frac{Q'''_{\mathrm{II}}}{Q^0_{\mathrm{I}} + Q''_{\mathrm{in,I}} + Q''_{\mathrm{ir,II}}} \tag{138}$$

$\mathrm{Re_I}$ increases and $\mathrm{Re_{II}}$ decreases with the increase of C'_{junc}. $\mathrm{Re_{total}}$ becomes maximum at $N_{\mathrm{I}} = N_{\mathrm{II}}$.

Limiting current density $(I/S)_{\mathrm{lim}}$ is plotted against current density I/S and shown in Figure 65. The real limiting current density $(I/S)_{\mathrm{lim,real}}$ is obtained from the intersection between the plots and the $I/S = (I/S)_{\mathrm{lim}}$ line in the figure and it is estimated for the first and the second stages as follows.

$C^0 = 2000$ mg/dm^3; 1.325 A/dm^2 (1st stage), 0.622 A/dm^2 (2nd stage).

$C^0 = 3000$ mg/dm^3; 1.615 A/dm^2 (1st stage), 0.622 A/dm^2 (2nd stage)

Limiting cell voltage $V_{\mathrm{cell,lim}}$ applied at $(I/S)_{\mathrm{lim,real}}$ is estimated to be about 0.7 V/pair for both stages, which is also suggested in the one-stage operation (cf. Figure 59).

In conclusion, total cell pair number becomes minimum when the cell pair number in each stage is the same. Saline water is desalinated considerably in the first half stage under an increased current density. The desalinated solution is purified in the latter stage under a decreased electric current to produce acceptable drinking water. The above functions in the multistage operation reduce total cell pair number integrated in the process. The energy consumption and limiting cell voltage are equivalent in the single-stage

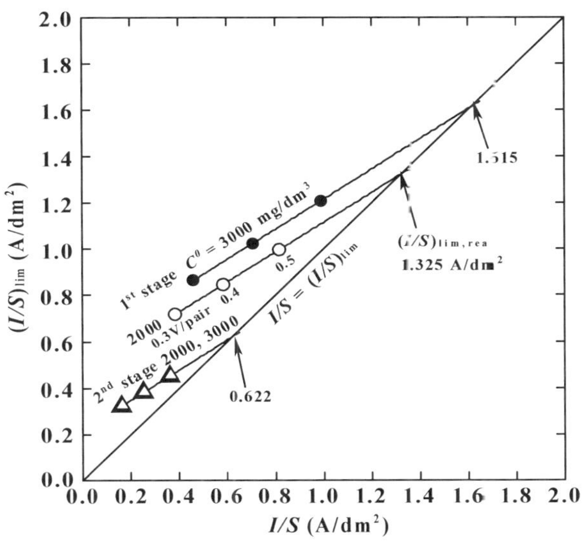

Figure 65 Limiting current density.

operation and multistage operation. However in a two-stage system, energy for pumping and recirculation will be approximately doubled compared to the value in a one-stage system. The multistage operation is advantageous to desalinate a higher salt concentration solution.

REFERENCES

Ahmed, M.I., Chang, H.T., Selman, J.R., Holsen, T.M., 2002. Electrochemical chromic acid regeneration process: fitting of membrane transport properties. J. Membr. Sci. 197, 63–74.

Akiyama, M., 1992. Physical Properties of Salt Brine, Sea Water Science Research Laboratory Technical Report. Japan Tobacco & Salt Public Corp.

Avriel, M., Zeligher, N., 1972. A computer method for engineering and economic evaluation of electrodialysis plant. Desalination 10, 113–146.

Azechi, S., 1980. Bull. Soc. Sea Water Sci., Jpn. 34.

Banasiak, L.J., Kruttschnitt, T.W., Schafer, A.I., 2007. Desalination using electrodialysis as a function of voltage and salt concentration. Desalination 205, 38–46.

Belfort, G., Daly, J.A., 1970. Optimization of an electrodialysis plant. Desalination 8, 153–166.

Brauns, E., De Wilde, W., Van den Bosch, B., Lens, P., Pinoy, L., Empsten, M., 2009. On the experimental verification of an electrodialysis simulation model for optimal stack configuration design through solver software. Desalination 249, 1030–1038.

Demircioglu, M., Kabay, N., Ersoz, E., Kurucaovali, I., Safak, C., Gizli, N., 2001. Cost comparison and efficiency modeling in the electrodialysis of brine. Desalination 136, 317–323.

Donnan, F.G., 1924. The theory of membrane equilibria. Chem. Rev. 1, 73–90.

Eimidaoui, A., Chay, L., Tahaikt, M., Menkouchi Sahli, M.A., Taky, M., Tiyal, F., Khalidi, A., Alaoui Hafidi, My R., 2006. Demineralisation for beet sugar solutions using an electrodialysis pilot plant to reduce malassigenic ions. Desalination 189, 209–214.

Fidaleo, M., Moresi, M., 2005. Optimal strategy to model the electrodialytic recovery of a strong electrolyte. J. Membr. Sci. 260, 90–111.

Harned, H.S., Qwen, B.B., 1998. Physical Chemistry of Electrolyte Solutions. Reinhold p. 726.

Juda, W., McRae, W.A., 1950. Coherent ion-exchange gels and membranes. J. Am. Chem. Soc. 72, 1044. Ion-exchange materials and method of making and using the same, USP at 2,636,851.

Kabay, N., Yuksel, M., Samata, S., Arar, O., Yuksel, U., 2007. Removal of nitrate from ground water by a hybrid process combining electrodialysis and ion exchange process. Sep. Sci. Technol. 42, 2615–2627.

Kabay, N., Arar, O., Samatya, S., Yuksel, U., Yuksel, M., 2008a. Separation of fluoride from aqueous solution by electrodialysis: effect of process parameters and other ionic species. J. Hazard. Mater. 153, 107–113.

Kabay, N., Arar, O., Acar, F., Ghazal, A., Yuksel, U., Yuksel, M., 2008b. Removal of boron from water by electrodialysis: effect of feed characteristics and interfering ions. Desalination 223, 63–72.

Kawahara, T., 1994. Construction and operation experience of a large-scale electrodialysis water desalination plant. Desalination 96, 341–348.

Koga, S., Mitsugami, Y., 1978. Supply of potable water from brackish water by electro-dialysis desalination process at Ohoshima island in Tokyo prefecture. Ind. Water 239, 41–47.

Kusakari, K., Kawamata, F., Matsumoto, N., Saeki, H., Terada, Y., 1977. Electrodialysis plant at Hatsushima. Desalination 21, 45–50.

Lee, H.J., Sarfert, F., Strathmann, H., Moon, S.H., 2002. Designing of an electrodialysis desalination plant. Desalination 142, 267–286.

Matsunaga, Y., 1986. Reuse of Waste Water by Electrodialytic Treatment, Industrial Application of Ion Exchange Membranes, vol. 1, Research Group of Electrodialysis and Membrane Separation Technology. Soc. Sea Water Sci. Jpn., pp. 188–196.

Meyers, K.H., Sievers, J.F., 1936. La permeabilite des membranes I, Theorie de la permeabilite ionique. Helv. Chim. Acta 19, 649–664. La permeabilite des membranes II, Essis avec des membranes selectives artifielles, Hel. Chim. Acta 19, 665–677.

Meyers, K.H., Straus, W., 1940. La permeabilite des membranes VI, Sur le passage du courant electrique a travers des membranes selectives. Helv. Chim. Acta 23, 795–800.

Michaelis, L., Fujita, A., 1925. Permselectivity of biological membranes. Biochem. Z. 158, 28–37, 161, 47–60, 164, 23–30.

Moon, P., Sandi, G., Stevens, D., Kizilel, R., 2004. Computational modeling of ionic transport in continuous and batch electrodialysis. Sep. Sci. Technol. 29, 2531–2555.

Nikonenko, V.V., Pismenskaya, N.D., Itoshin, A.G., Zabolotsky, V.I., Shudrenko, A.A., 2008. Description of mass transfer characteristics of ED and EDI apparatus by using the similarity theory and compartmentation method. Chem. Eng. Proc. 47, 1118–1127.

Ortiz, J.M., Sotoca, J.A., Exposito, E., Gallud, F., Garcia-Garcia, V., Montiel, V., Aldaz, A., 2005. Brackish water desalination by electrodialysis: batch recirculation operating modeling. J. Membr. Sci. 252, 65–75.

Parulekar, S.J., 1998. Optimal current and voltage trajectories for minimum energy consumption in batch electrodialysis. J. Membr. Sci. 148, 91–103.

Plank, M., 1890. Ueber die potentialdifferenz zwichen zwei verdunnten losungen binarer electrolyte. Ann. Der Phys. u Chem. 40, 561–576.

Rapp, H.J., Pfromm, P.H., 1998a. Electrodialysis for chloride removal from the chemical recovery cycle of a kraft pulp mill. J. Membr. Sci. 146, 249–261.

Rapp, H.J., Pfromm, P.H., 1998b. Electrodialysis field test for selective chloride removal from the chemical recovery cycle of a kraft pulp mill. Ind. Eng. Chem. Res. 37, 4761–4767.

Ryabstev, A.D., Kotsupalo, N.P., Titarenko, V.I., Igumenov, I.K., Gelfond, N.V., Fe dotova, N.E., Moroziva, N.B., Shipachev, V.A., Tibilov, A.S., 2001. Development of two-stage electrodiaslysis set-up for economical desalination of artesian and surface waters of sea type. Desalination 137, 207–214.

Sadrzadeh, M., Kaviani, A., Mohammadi, T., 2007. Mathematical modeling of desalination by elctrodialysis. Desalination 206, 534–549.

Sato, K., Matsuo, T., 1974. Seawater Handbook. Soc. Sea Water Sci. Jpn., p. 110.

Schoeman, J.J., Steyn, A., Makgae, M., 2005. Evaluation of electrodialysis for the treatment of an industrial solid waste leachate. Desalination 186, 273–289.

Shaffer, L.H., Minz, M.S., 1966. Electrodialysis. In: Spiegler, K.S. (Ed.), Principles of Desalination. Academic Press, New York, London, pp. 200–239.

Tanaka, Y., 2000. Current density distribution and limiting current density in ion-exchange membrane electrodialusis. J. Membr. Sci. 173, 179–190.

Tanaka, Y., 2002. Current density distribution, limiting current density and saturation current density in an ion-exchange membrane electrodialyzer. J. Membr. Sci. 210, 65–75.

Tanaka, Y., 2003. Mass transport and energy consumption in ion-exchange membrane electrodialysis of seawater. J. Membr. Sci. 215, 265–279.

Tanaka, Y., 2005. Limiting current density of an ion-exchange membrane and of an electrodialyzer. J. Membr. Sci. 266, 6–17.

Tanaka, Y., 2006. Irreversible thermodynamics and overall mass transport. J. Membr. Sci. 281, 517–531.

Tanaka, Y., 2009. A computer simulation of batch ion exchange membrane electrodialysis for saline water desalination. Desalination 249, 1039–1047.

Tanaka, Y., 2010. A computer simulation of ion exchange membrane electrodialysis for concentration of seawater. Membr. Water Treat. 1, 13–37.

Tanaka, Y., 2011a. Ion-exchange membrane electrodialysis for saline water desalination and its application to seawater concentration. Ind. Eng. Chem. Res. 50, 7494–7503.

Tanaka, Y., 2011b. Ion-exchange membrane electrodialysis of saline water and its numerical analysis. Ind. Eng. Chem. Res. 50, 10765–10777.

Tanaka, Y., 2012a. Ion-exchange membrane electrodialysis program and its application to multi-stage continuous saline water desalination. Desalination 301, 10–25.

Tanaka, Y., 2012b. Continuous ion-exchange membrane electrodialysis for drinking water production: numerical analysis. In: Proc. The 2012 World Congress on Advances in Civil, Environmental, and Material Research, P. 141, August 26-30, Seoul, Korea.

Tanaka, Y., 2013. Development of a computer simulation program of batch ion-exchange membrane electrodialysis for saline water desalination. Desalination 320, 118–133.

Tanaka, Y., 2014. Development of a computer simulation program of feed-and-bleed ion-exchange membrane electrodialysis for saline water desalination. Desalination 342, 126–138.

Tanaka, Y., 2015. Ion Exchange Membranes, Fundamentals and Applications, 2nd edition. Elsevier, to be published.

Tanaka, Y., Ehara, R., Itoi, S., Goto, T., 2003. Ion-exchange membrane electrodialytic salt production using brine discharged from a reverse osmosis seawater desalination plant. J. Membr. Sci. 222, 71–86.

Tani, Y., Doi, K., Terada, Y., Yokota, M., Wakayama, M., 1978. Electrodialysis seawater desalination unit in a vessel. Ind. Water 239, 86–89.

Teorell, T., 1935. An attempt to formulate a quantitative theory of membrane permeability. Proc. Soc. Exp. Biol. Med. 33, 282–285.

Thompson, R., Paleologou, M., Wong, R.Y., Berry, R.M., 1997. Separation of sulphide from hydroxide in white liquor by electrodiaslysis. J. Pulp Pap. Sci. 23, J182–J187.

Tran, A.T.K., Zhang, Y., Jullok, N., Meesschaert, B., Pinoy, L., Van der Bruggen, B., 2012. RO concentrate treatment by a hybrid system consisting a pellet reactor and electrodialysis. Chem. Eng. Sci. 79, 228–238.

Tsiakis, P., Papageorgiou, 2005. Optimal design of an electrodialysis brackish water desalination. Desalination 173, 173–186.

Urabe, S., Doi, K., 1978. Ind. Water, 239.

Walha, K., Amar, R.B., Firdaous, L., Quemeneur, F., Jaouen, P., 2007. Brackish groundwater treatment by nanofiltration, reverse osmosis and electrodialysis in Tunisia. Performance and cost comparison. Desalination 207, 95–106.

Wyllie, R., Patnode, H.W., 1950. The development of membranes from artificial cation-exchange materials with particular reference to the determination of sodium-ion activity. J. Phys. Chem. 54, 204.

Yawataya, T., 1986. Ion Exchange Membranes for Engineers. Kyoritsu Shuppan Co. Ltd., Tokyo, pp. 94–98.

Zhang, Y., Ghyselbrecht, K., Meesschaert, B., Pinoy, L., Van der Brugen, B., 2011. Electrodialysis on RO concentyrate to improve water recovery in wastewater reclamation. J. Membr. Sci. 378, 101–110.

Zimmerer, C.C., Kotte, V., 1996. Effects of spacer geometry on pressure drop, mass transfer, mixing behavior, and residence time distribution. Desalination 104, 129–134.

SECTION TWO

Force field assisted separators

CHAPTER SEVEN

Magnetic Techniques for Mineral Processing

Luzheng Chen[1] and Dahe Xiong[2]
[1]State Key Laboratory of Complex Nonferrous Metal Resources Clean Utilization, Kunming University of Science and Technology, Kunming, Yunnan, China
[2]SLon Magnetic Separator Co., Ltd., Ganzhou, Jiangxi, China

Contents

1. INTRODUCTION TO MAGNETIC SEPARATION

Magnetic separations in the field of mineral processing are mainly used for concentration of magnetic components and for removal of magnetic impurities, under dry or wet conditions. This process is achieved by passing

Progress in Filtration and Separation
http://dx.doi.org/10.1016/B978-0-12-384746-1.00007-0

a slurry or mixture containing magnetic and nonmagnetic particles through a nonhomogenous magnetic field, which leads to the selective capture of magnetic particles from the slurry or mixture.

All materials on earth display certain magnetic properties in the presence of an external magnetic field; however, only a small number of minerals are sensitive to magnetic field and separable with magnetic techniques. Magnetic minerals, according to the magnitude of their magnetic susceptibilities, are usually classified into two basic groups, i.e., strongly and weakly magnetic minerals. Strongly magnetic minerals such as magnetite and vanadic titano-magnetite are recoverable in a low-intensity magnetic separator at a magnetic induction lower than 0.2 T, with a relatively low magnetic field gradient smaller than 5 T/m. Weakly magnetic minerals such as hematite, limonite, ilmenite, wolframite, and manganese are recoverable in a high-intensity or high-gradient magnetic separator at a strong magnetic induction higher than 0.6 T, with a high magnetic field gradient larger than 50 T/m. The mass magnetic susceptibilities of these two groups of magnetic minerals are generally defined in the ranges of greater than 10^{-4} m^3/kg and 10^{-7} m^3/kg, respectively; however, such definitions will be extended, with the innovations and developments in magnetic techniques.

The mass magnetic susceptibilities of typical magnetic minerals recoverable with magnetic techniques are illustrated in Table 1 (Wang, 2008).

Although various classification schemes exist for magnetic separators, based upon the magnitude of magnetic induction and its magnetic field gradient, magnetic separators are classified into three types and they are:

- Low-intensity magnetic separators. They are primarily used for concentration of strongly magnetic minerals with high magnetic susceptibility, under wet conditions for fine particles, or under dry conditions for coarse particles. Low-intensity magnetic separators include magnetic pulley, drum magnetic separator, column magnetic separator, screen magnetic separator, etc.
- High-intensity magnetic separators. They are used for concentration of weakly magnetic minerals with low magnetic susceptibility, under wet conditions for relatively fine and coarse particles, or under dry conditions for sufficiently large particles. High-intensity magnetic separators include magnetic pulley, disc magnetic separator, roll magnetic separator and drum magnetic separator, etc.
- High-gradient magnetic separators. They are practically required for the recovery of fine weakly magnetic minerals, under dry or wet conditions. There are various high-gradient magnetic separators reported to be

Table 1 Mass magnetic susceptibilities of typically magnetic minerals

Minerals	Chemical composition	Mass magnetic susceptibility ($\times 10^{-6}$, m^3/kg)	Content of magnetic component (%)
Magnetite	Fe_3O_4	1156	Fe, 72.40
Vanadic titano-magnetite	$(TiO_2, V_2O_5)Fe_3O_4$	917	Fe, 63.70; TiO_2, 6.91; V_2O_5, 0.90
Hematite	Fe_2O_3	0.60–2.16	Fe, 69.94
Specularite	Fe_2O_3	3.70	Fe, 69.94
Limonite	$mFe_2O_3 \cdot nH_2O$	0.31–1.00	Fe, 48.00–62.90
Siderite	$FeCO_3$	0.70–1.50	Fe, 48.28
Chromite	$FeO \cdot Cr_2O_3$	0.63–0.88	Fe, 25.00; Cr_2O_3, 67.91
Pyrrhotite	Fe_7S_8	57.00	Fe, 63.53
Ilmenite	$FeTiO_3$	0.34–5.00	Fe, 33.33; TiO_2, 52.66
Wolframite	(Fe, Mn)WO_4	0.49	Fe, 15.60; WO_3, 76.58
Manganite	$Mn_2O_3 \cdot H_2O$	0.35–1.02	Mn, 62.50
Pyrolusite	MnO_2	0.34	Mn, 63.22
Psilomelane	$Mn \cdot MnO \cdot MnO_2 \cdot nH_2O$	0.30–0.62	Various

applicable; however, until today, pulsating high-gradient magnetic separator has achieved a wide range of applications in the processing of weakly magnetic minerals. Moreover, vibrating high-gradient magnetic separation has shown a prospective potential for the purification of nonmetallic ores such as quartz, feldspar, etc.

It is well known that, in the magnetic separation of magnetic particles, the dominant force that differentiates magnetic particles from nonmagnetic particles is the magnetic force $\vec{F}_m$, which is given by Svoboda and Fujita (2003):

$$\vec{F}_m = \frac{\kappa}{\mu_0} VB\vec{\nabla}B \tag{1}$$

where κ is the volumetric magnetic susceptibility of magnetic particle, μ_0 is the magnetic permeability of vacuum, V is the volume of the particle, B is the external magnetic induction, and $\vec{\nabla}B$ is the magnetic field gradient of the magnetic induction.

It is clear that the magnetic force acting upon the particle is determined by the magnetic induction and the field gradient, and is in the direction of the gradient. In a homogeneous magnetic field, the magnetic force upon particle is zero, as ∇B is zero. Therefore, an optimally designed magnet in a magnetic separator is crucially important for the generation of field gradient appropriate for magnetic separations.

In a magnetic separator, the competing forces ($\vec{F}_c$) as schematically shown in Figure 1 are also acting on particles. They are, among others, mainly hydrodynamic drag and gravity. As the capture of magnetic particles from slurry or mixture occurs, in most cases, in a laminar flow field, the hydrodynamic force $\vec{F}_d$ upon the particles may be written as (Xiong et al., 1998):

$$\vec{F}_d = 6\pi\eta d\vec{v} \tag{2}$$

where η is the dynamic viscosity of slurry or mixture flow, d is the radius of particle, and $\vec{v}$ is the relative velocity of particle with respect to slurry or mixture flow.

Magnetic particles will be separated from nonmagnetic or less magnetic particles if only the magnetic force upon the particles is greater than the sum of competing forces.

Some weakly magnetic minerals such as hematite and limonite can be transformed into strongly magnetic magnetite, through roasting treatment, for example (Liu, 1994):

Transformation of iron oxide ores such as hematite, limonite, and specularite into magnetite through reductive roasting, at a temperature of 550–570 °C:

$$3\,Fe_2O_3 + C \rightarrow 2\,Fe_3O_4 + CO\uparrow \tag{3}$$

Transformation of siderite into magnetite through nonaerated roasting, at a temperature of 400–570 °C:

$$3\,Fe_2CO_3 \rightarrow Fe_3O_4 + 2\,CO_2\uparrow + CO\uparrow \tag{4}$$

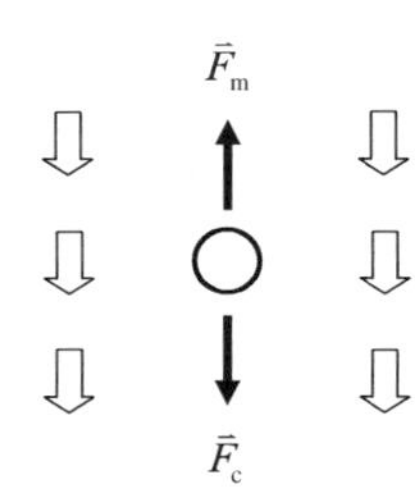

Figure 1 Schematic diagram of a magnetic separation process.

And transformation of pyrite into magnetite through oxidative roasting, at a temperature not higher than 200 °C:

$$7\ FeS_2 + 6\ O_2 \rightarrow Fe_7S_8 + 6\ SO_2\uparrow \tag{5}$$

$$3\ Fe_7S_8 + 3\ 8O_2 \rightarrow 7\ Fe_3O_4 + 24\ SO_2 \tag{6}$$

By roasting treatment, the above weakly magnetic iron ores become strongly magnetic and they are separable in a low-intensity magnetic separator, but, at the expense of environment and cost.

In this chapter of magnetic techniques for mineral processing, theoretical principle of magnetic separations and three types of magnetic separators, i.e., low-intensity, high-intensity and high-gradient magnetic separators, and their technological innovations and operational practices are reviewed. In particular, pulsating high-gradient magnetic separation, a landmark progress in the treatment of fine weakly magnetic minerals such as iron oxide ores and ilmenite, and its continual developments and operational variables are expatiated. Practical cases in the processing of typically magnetic ores, i.e., magnetite, hematite, limonite, and ilmenite and nonmetallic ores where magnetic techniques are well embodied, are introduced. At the end of this chapter, the merits and limitations of the current magnetic techniques and the future trends in the research and development of potential magnetic techniques are discussed.

2. MAGNETIC SEPARATION TECHNIQUES

2.1 Low-Intensity

There are several types of low-intensity magnetic separators in operation in the field of mineral processing; but, with the advent and applications of neodymium iron boron (NbFeB) permanent magnets in magnetic equipments, the permanent magnet-based magnetic separator has mostly replaced the electromagnet-based one and has become one of the most widely used type in the concentration of magnetic minerals.

Low-intensity magnetic separators are basically limited to the concentration of strongly magnetic minerals such as magnetite, titano-magnetite and vanadic titano-magnetite, and to the removal of iron tramps and ferromagnetic impurities.

2.1.1 Magnetic Pulley

Magnetic pulley presents an easy and simple way to remove iron tramps and ferromagnetic impurities from sufficiently coarse dry materials moving on a

Figure 2 Operation of a magnetic pulley (separator).

conveyor belt. In practice, it is used to protect processing machinery in a mineral processing plant. Magnetic pulley is usually suspended over the conveyor belt or installed at the end of the belt.

In recent years, there is a tendency to convert magnetic pulley into a magnetic separator, as show in Figure 2. Material is carried over the belt and the magnetic pulley holds magnetic particles until they leave the magnetic region, while nonmagnetic particles are thrown over the pulley. Magnetic pulleys may be manufactured in permanent or electromagnetic magnets, and are available for continuous and discontinuous operations.

There are two basic designs for magnets in magnetic pulleys, that is, axial poles with polarities alternating along the circumference of the pulley, and radial poles with polarities alternating across the width of the pulley (Svoboda and Fujita, 2003). The axial polarity design is suited for large particles and is preferable to achieve a high-quality magnetic product as the tumbling motion of particles over magnet polarities facilitates the release of entrained nonmagnetic particles, while the radial one is suited for fine particles as it has a more uniform magnetic field distribution, in comparison with that of the axial polarity, and thus is preferable to achieve a high recovery of magnetic particles.

Magnetic pulleys are manufactured in a wide range of diameters from 500 to 2400 mm, and its selection should be matched with the width of the belt. They are installed at a certain distance from the materials, at which magnetic induction varies from 100 to 300 mT, based on the magnetic property and particle size of materials to be separated.

2.1.2 Drum Magnetic Separator

Drum magnetic separators are one of the most widely applied types of magnetic separators and they are operated under wet conditions, for

Figure 3 Wet drum magnetic separator (left) and operation for magnetite (right).

concentration of fine or relatively coarse, strongly magnetic particles from slurry. The dominant areas, where the drum magnetic separators are used, are the processing of magnetite, titano-magnetite, and vanadic titano-magnetite ores. Figure 3 shows the wet drum magnetic separator and its operation for vanadic titano-magnetite. The technical specifications of applied wet drum magnetic separators are illustrated in Table 2.

The drum is partially submerged in a round tank and slurry is carried across the lower surface of the drum. Magnetic particles are attracted onto the drum surface from slurry and brought out of tank with the rotation of drum, to where they are flushed down as magnetic concentrate with water. Nonmagnetic particles are discharged at the bottom of the tank as tailings.

The design of drum magnetic separators is basically the same, with permanent magnet blocks arranged in a stationary position inside a nonmagnetic rotating drum. However, there are three basic designs for the tank of drum, according to the directions of drum rotation and slurry flow, that is, concurrent, counterrotation, and countercurrent (Svoboda, 2004).

In the concurrent design, slurry is fed through feed box at one side of drum and flows in the direction of drum rotation, with nonmagnetic particles discharged at the bottom of tank through the tailings discharge opening. In practice, this design is mainly used as rougher and cleaner, as a result of its high selectivity to magnetic particles.

In the counterrotation design, slurry is fed through a special feed box to the lower side of drum and flows in the opposite direction of drum rotation. Since slurry flows along the whole magnetic arc of drum, this design is characteristic of high recovery for magnetic particles and is mainly applied as rougher and scavenger.

The countercurrent design combines the advantages of concurrent and counterrotation patterns in the characteristics of separation, and thus is

Table 2 Technical specifications of wet low-intensity drum magnetic separators

Drum (mm)		Drum rotation speed (r/min)	Magnetic induction (mT)	Feed solids concentration (%)	Processing capacity		Drum driver (Kw)	Mass weight (t)
Diameter	Length				Solids (t/h)	Slurry (m^3/h)		
600	1200	35		20–30	10–20	35–55	1.5	1.25
	1800	35		20–30	15–30	40–65	1.5	1.3
750	1200	35		20–30	15–30	40–65	2.2	2.0
	1800	35		20–30	20–40	55–110	3	2.45
900	1800	25		25–35	20–45	90–160	3	2.8
	2400	25		25–35	30–55	120–190	4	3.2
1050	1800	20		25–35	45–60	130–200	4	3.3
	2100	20	80–200	25–35	55–80	150–240	4	3.8
	2400	20		25–35	60–100	160–280	4	4.3
	3000	20		25–35	80–120	240–400	4	4.8
1200	1800	19		25–35	60–75	160–250	4	4.7
	2400	19		25–35	70–110	220–380	4	5.2
	3000	19		25–35	80–140	260–460	5.5	5.7
	3600	19		25–35	100–160	280–540	7.5	8.0
	4500	19		25–35	130–190	360–640	15	11.0

capable of achieving a high-grade concentrate with high recovery. In this design, slurry is fed at the near bottom of drum and nonmagnetic particles flow out of the opposite end of tank through a tailings discharge overflow, which also acts as a slurry level controller. The countercurrent based drum magnetic separator is well suited for the treatment of fine materials.

Wet drum magnetic separators of special design may be used as a magnetic thickener for thickening purpose of magnetic concentrates, usually at the end of a magnetic separation process, or as a high-intensity magnetic separator, for the removal of very fine ferromagnetic particles and iron corrosions produced in the grinding process of a nonmetallic ore processing flowsheet, at a magnetic induction reaching as high as 0.6 T.

With the increase in diameter and length of the drum, the solids throughput of the separators increases. But, it is believed that a drum diameter around 1200 mm is optimum, beyond which the throughput does not necessarily increase in a corresponding manner to the drum diameter, and even at the expense of separation performance.

2.1.3 Column Magnetic Separator

Column magnetic separator is developed on the principle of magnetic agglomeration of strongly magnetic minerals such as magnetite in a low-intensity magnetic field, and is primarily used as cleaner to produce a sufficiently high-grade magnetic concentrate. As shown in Figure 4, slurry is fed at the top of column and flows downward inside the column, where magnetic particles agglomerate with the size of agglomerations increasingly growing. The agglomerations settle down to the bottom of column and are

Figure 4 Operation of column magnetic separators for cleaning of magnetite.

discharged through an adjustable valve, under the combined actions of magnetic force and gravity in the downward direction, and hydrodynamic drag in the upward direction from the fluidizing water near the bottom of column. Nonmagnetic particles flow out of the column through the overflow launder installed near the top of column. Figure 4 shows the operation of column magnetic separators in the cleaning of magnetite. The technical specifications of applied column magnetic separators are illustrated in Table 3.

The solids throughput of column magnetic separators varies from 3 to 150 t/h by mass weight and is primarily used for cleaning purposes of strongly magnetic minerals such as magnetite, as a result of its high selectivity to magnetic particles. Table 4 illustrates the cleaning performance of column magnetic separators for different magnetite ores.

Column magnetic separators of special design may be used as magnetic thickener, with the solids concentration of magnetic concentrate reaching 60–70% by mass weight.

2.1.4 Screen Magnetic Separator

Screen magnetic separator presents another magnetic technique for the cleaning of strongly magnetic minerals such as magnetite, based on the principle of magnetic agglomeration (Li, 2005).

Table 3 Technical specifications of column magnetic separators

Column (mm)		Magnetic induction (mT)	Feed size (mm)	Solids throughput (t/h)	Water consumption (m^3/h)
Diameter	Height				
260	1600			2–3	5–10
400	2600			5–8	10–20
500	2900	0–500	0–0.2	10–15	20–30
600	3145–3480			15–20	25–35
700	3200			20–30	30–40

Table 4 Performance of column magnetic separators for cleaning of different magnetite ores

Diameter of column (mm)	Feed grade (% Fe)	Concentrate			Tailings grade (% Fe)
		Mass weight (%)	Grade (% Fe)	Recovery (% Fe)	
200	61.99	91.79	65.76	97.37	19.86
400	60.18	81.07	66.18	89.15	34.49
600	51.60	74.33	67.00	96.51	7.02

Figure 5 Operation of screen magnetic separators for cleaning of magnetite.

As shown in Figure 5, slurry is fed from the top of an inclined screen in a low-intensity magnetic field, with the mesh size of screen sufficiently larger than those of particles in slurry. As the slurry flows down the above surface of screen, magnetic particles agglomerate with the size of agglomerations increasingly growing and roll down as magnetic concentrate at the lower end of screen. The less- or nonmagnetic particles pass through the screen as tailings. Figure 5 shows the operation of screen magnetic separators for cleaning of magnetite.

This magnetic technique may achieve a similar separation selectivity in the cleaning of magnetite, to the column magnetic separators.

2.2 High-Intensity

High-intensity magnetic separators, including the conventional cross-belt and rotating disc magnetic separators, etc., are used for beneficiation of weakly magnetic ores, at a much higher magnetic induction than that in a low-intensity magnetic separator. High-intensity magnetic separators are available under dry and wet conditions, and dry high-intensity magnetic separators have been commercially applied more than a century ago for sufficiently coarse magnetic particles. For the treatment of fine weakly magnetic particles, wet high-intensity magnetic separators are practically required, and are more and more increasingly replaced by high-gradient magnetic separators, due to the developments and applications of high-gradient magnetic separators in the past three decades.

Although high-gradient magnetic separators fall in the scope of high-intensity magnetic separation, it is introduced more detail in a single section in this chapter.

2.2.1 Magnetic Pulley

Magnetic pulleys are conventionally manufactured for a specific installation to remove iron tramps and ferromagnetic impurities of sufficiently large size under dry conditions, as mentioned earlier. With the increasing exploitation of low-grade magnetic ores, such as magnetite, vanadic titano-magnetite and hematite ores in the recent decade, magnetic pulleys are more and more converted into a high-intensity magnetic separator, for the preconcentration treatment of low-grade magnetic ores. The preconcentration of such ores significantly improves the feed grade and the economy of processing plants.

Figure 6 shows a dry high-intensity magnetic pulley as separator and its operation for preconcentration of a low-grade magnetic ore. High-intensity magnetic pulleys are manufactured in a wide range of magnetic inductions from 0.6 to 1.5 T even higher, in correspondence to the magnetic susceptibility and particle size of materials to be treated.

Table 5 illustrates the technical specifications of applied dry high-intensity magnetic pulleys.

2.2.2 Disc and Roll Magnetic Separators

In the past, cross-belt and rotating disc high-intensity magnetic separators were used for concentration of relatively coarse weakly magnetic particles such as wolframite and ilmenite, etc., under dry conditions. In the operation of these two magnetic separators, material is distributed onto the moving conveyor belt in a very thin layer, through a vibrating feeder. Such magnetic separators are not effective even inapplicable for the treatment of fine materials.

With the increasing reduction in liberation size of valuable components in magnetic ores, the conventional cross-belt and rotating disc high-intensity magnetic separators are almost replaced by gravity and flotation, particularly

Figure 6 High-intensity magnetic pulley (left) and operation for preconcentration of low-grade magnetic ore (right).

Table 5 Technical specifications of dry high-intensity magnetic pulleys

Drum (mm)		Drum rotation	Magnetic induction	Feed size	Solids throughput	Mass weight
Diameter	Length	(r/min)	(T)	(%)	(t/h)	(t)
500	600			10–40	20–40	0.7
	750			10–40	30–50	0.8
	950			10–40	35–55	1.0
600	750			10–150	35–55	1.0
	950			10–150	50–60	1.4
	1200			10–150	80–100	1.8
800	950	0–120	0.6–1.5	10–200	50–80	2.2
	1200	(varies with		10–200	65–100	2.5
	1400	diameter)		10–200	80–120	2.7
	1600			10–200	80–120	4.35
1000	1200			10–300	70–120	4.2
	1400			10–300	80–150	4.8
	1600			10–300	150–200	5.4
1200	1400			≤350	100–200	5.45
	1600			≤350	150–250	6.8
1400	1600			≤350	200–350	10.0

by high-gradient magnetic separators, as a result of its effectiveness to fine materials and high solids throughput. In the recent years, however, a wet permanent disc high-intensity magnetic separator as shown in left Figure 7 seems applicable in recovering fine magnetic particles from tailings. In this disc separator, slurry is fed across a round tank, in which vertically rotating discs with permanent magnet blocks pick up fine magnetic particles, and they are brought up and scraped down by rotating scrapers, near the top of discs. Nonmagnetic particles are discharged at the bottom of tank.

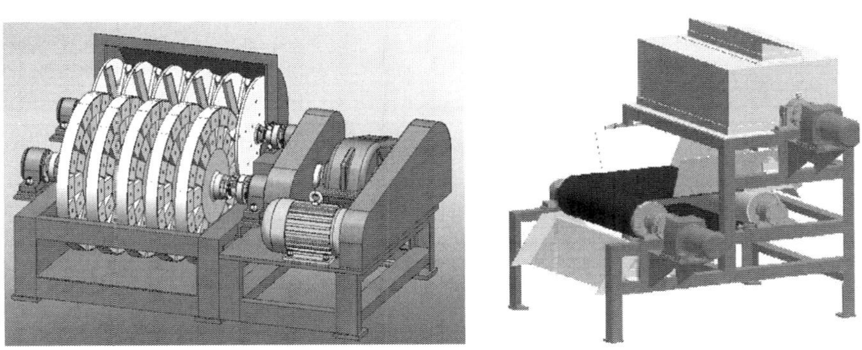

Figure 7 High-intensity disc (left) and roll (right) magnetic separators.

And, a dry high-intensity roll magnetic separator as shown in right Figure 7 is replacing the conventional roll magnetic separators and is used for concentration of relatively coarse magnetic particles. The design of such a roll magnetic separator is similar to that of the conventional roll magnetic separator, but it achieves a higher magnetic induction and its installation requires a much smaller occupation for space.

2.2.3 Drum Magnetic Separator

The availability and affordability of permanent magnets of high-magnetic power result in the design and manufacture of high-intensity drum magnetic separators, and it is used to remove fine ferromagnetic impurities from nonmetallic ores.

The design of high-intensity drum magnetic separator is somewhat different from that of low-intensity one. In the high-intensity drum magnetic separators, a plurality of small magnet blocks with polarities alternating in small intervals are used to achieve a high-magnetic induction of 0.6–0.8 T on the surface of drum. Such a magnet design is believed to require less magnet material, in comparison to the conventional ones (Svoboda, 2004). The configuration of wet high-intensity drum magnetic separator is almost the same as the wet low-intensity one, as shown in Figure 3.

High-intensity drum magnetic separator may be used for concentration of magnetic particles from coarse materials, under dry conditions.

2.3 High-Gradient

2.3.1 Brief Review on High-Gradient

High-gradient magnetic separation has achieved remarkable progress and wide applications since Jones in 1955 achieved a high-gradient magnetic field in a magnetized matrix (Svoboda, 2004). By using magnetic matrix, magnetic force upon magnetic particles is remarkably increased, resulting in the significantly improved recovery for fine weakly magnetic particles.

However, it has undergone a long period in the development of high-gradient magnetic separators, from cyclic to continuous operation, and from the horizontal design philosophy of separating ring to the vertical one. At the same time, magnetic matrix, as carrier for the capture of magnetic particles in a high-gradient magnetic separator, has evolved from the early used fibers and balls to meshes, and to the present rod poles. Now, the rod matrix made of rod poles has proven the most applicable medium in the magnetic capture of fine weakly magnetic particles from slurry in a high-gradient magnetic separator, due to its high operational reliability, simplified combinatorial optimization, and resistance to clogging (Chen et al., 2013a).

2.3.2 Pulsating High-Gradient

In the conventional high-gradient magnetic separations of fine weakly magnetic ores, the mechanical entraining of non-magnetic particles and the clogging of magnetic matrix are the biggest problems. The entraining of nonmagnetic particles deteriorates the selectivity of separation and the clogging makes the maintenance of separators a hard task. Since 1986, pulsating high-gradient magnetic separation was proposed and developed to solve the above problems, and they are now widely applied for concentration of fine weakly magnetic minerals such as hematite, limonite, siderite, etc. (Xiong et al., 1998), and for purification of nonmetallic ores such as quartz, feldspar, and kaolin.

2.3.2.1 Principle of Pulsating High-Gradient Magnetic Separation

The principle of pulsating high-gradient magnetic separation is explicable through a cyclic pilot-scale pulsating high-gradient magnetic separator as schematically shown in Figure 8 (Chen et al., 2009). This pilot-scale separator is fed periodically and rod poles made of magnetic stainless steel are used as separating matrix.

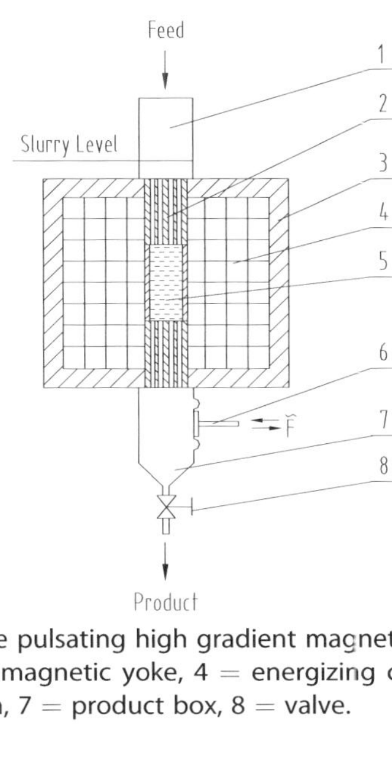

Figure 8 Cyclic pilot-scale pulsating high gradient magnetic separator: 1 = feed box, 2 = magnetic pole, 3 = magnetic yoke, 4 = energizing coils, 5 = magnetic matrix, 6 = pulsating mechanism, 7 = product box, 8 = valve.

While the pilot-scale separator is operated, a direct current flows through the energizing coils and a magnetic field is built up in the separating zone. Firstly, the separating zone is filled with flowing water so that the pulsating energy can be transmitted to the separating zone, in which the level of water and its flow rate are adjustable through the valve below the pulsating mechanism. Then, slurry is fed into the matrix in the separating zone through feed box. Magnetic particles are attracted from slurry onto the surface of matrix, while nonmagnetic particles pass through the matrix and go out through the product box as tailings, under the combined actions of slurry pulsation, gravity, and hydrodynamic drag. The pulsating mechanism drives the slurry in the separating zone up and down, keeping particles in the matrix in a loose state so that magnetic particles can be more easily captured by matrix and nonmagnetic particles can be more easily dragged out of the matrix. This magnetic separator achieves a superior performance to traditional separators, due to the innovative introduction of slurry pulsation in the separating process and the use of rod matrix (Zeng and Xiong, 2003).

The most essential features in pulsating high-gradient magnetic separator, differing from other high-gradient magnetic separators, is a slurry in the matrix exposed to pulsation to loose particles, which is achieved by the pulsating mechanism below the lower magnetic yoke, as shown in Figure 8.

The pulsation of slurry determines the behavior of slurry and particles in the separating zone. While the separating zone is filled with slurry and the pulsating mechanism is working, the pulsating energy is transmitted to the separating zone and drives the slurry up and down. The pulsating velocity $\tilde{v}$, maximum pulsating velocity $\tilde{v}_{\max}$, and average pulsating velocity $\bar{v}$ of slurry are respectively written as (Xiong et al., 1998):

$$\tilde{v} = \frac{1}{2} S\omega \sin(\omega t) \tag{7}$$

$$\tilde{v}_{\max} = \frac{1}{2} S\omega \tag{8}$$

$$\bar{v} = \frac{1}{2\pi} \int_{0}^{2\pi} \left| \frac{1}{2} S\omega \sin(\omega t) \right| d(\omega t) = \frac{S\omega}{\pi} \tag{9}$$

where S is the pulsating stroke, ω is the angle speed of pulsation, and t is the time variable, respectively.

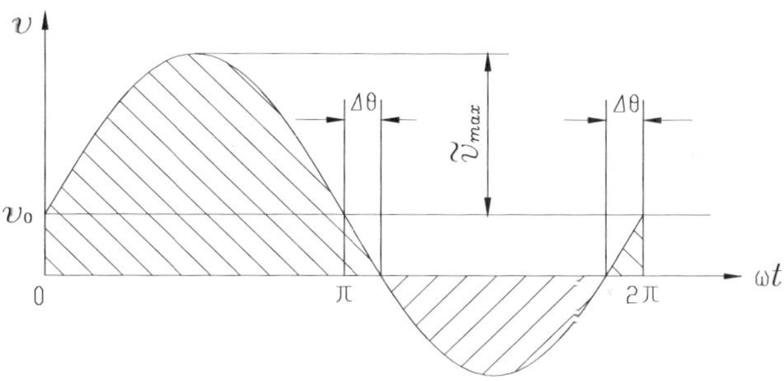

Figure 9 Velocity curve of slurry in a cycle of pulsating high gradient magnetic separation.

As shown in Figure 9, the actual velocity of slurry in the separating zone is the sum of feed velocity and pulsating velocity:

$$v = v_0 + \tilde{v} = v_0 + \frac{1}{2}S\omega\sin(\omega t) = v_0 + \tilde{v}_{\max}\sin(\omega t) \tag{10}$$

From Eqn (10), when $\tilde{v}_{\max} > v_0$, the velocity of slurry is periodically reversed and the slurry moves up and down. As shown in Figure 9, the shadowed area below the horizontal axis indicates that slurry moves upwards, while that above the axis indicates that slurry moves downwards; in such a case, the pulsating slurry produces a relaxing effect on the particles in matrix, and the mechanical entrapments due to bridge effect are reduced or destroyed. But, if $\tilde{v}_{\max} \leq v_0$, slurry flows downwards and nonmagnetic particles may be trapped in the matrix and the matrix may be mechanically clogged, resulting in the deterioration of separation performance. It is clear that it is crucially important that $\tilde{v}_{\max} > v_0$ is met in pulsating high-gradient magnetic separation.

The pulsation of slurry increases the layers or distance of matrix, in which particles pass through, thus improves the collision efficiency between matrix and particles and consequently the recovery of magnetic particles. From Figure 10, when $\tilde{v}_{\max} \leq v_0$ or the pulsating velocity is very small, slurry flows downwards with no reversion upward, and the particles pass n layers of matrix, at a distance equal to the matrix depth L_0, that is:

$$n = n_0 = \frac{L_0}{\Delta L} \tag{11}$$

where n is the layers of particles passing through matrix, n_0 is the layers of matrix, and ΔL is the spacing between two neighbouring layers in matrix.

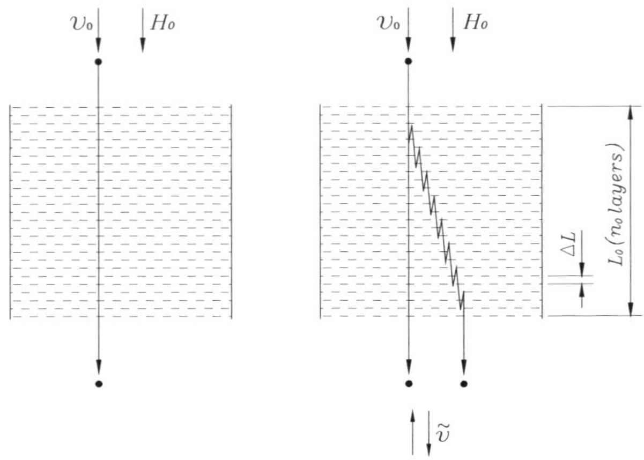

Figure 10 Trajectories of a particle passing through a matrix depth under nonpulsating (left) and pulsating (right) conditions.

As the pulsating velocity increases and $\tilde{v}_{\max} > v_0$ is met, the velocity direction of slurry reverses periodically, and the slurry moves up and down. In such a case, the actual layers and distance of particles passing through the matrix increases as shown in Figure 10, where, for visibility, the particle is artificially deflected from the passing route through the matrix.

The actual distance of particles through matrix in a pulsating cycle in the separating zone as shown in Figure 10 is calculated as:

$$L_T = \int_0^T |v_0 + \tilde{v}_{\max}(\omega t)| dt \tag{12}$$

where L_T is the actual distance of particles passing through matrix in a pulsating cycle of T.

When $\tilde{v}_{\max} \leq v_0$, the particles pass through the depth of matrix with no reversion upward, and the actual distance is L_0. However, under $\tilde{v}_{\max} > v_0$ condition, the slurry moves up and down and the actual distance is integrated from Eqn (12):

$$L_T = \frac{2}{\pi} v_0 T \cdot \left[\arcsin \frac{v_0}{\tilde{v}_{\max}} + \sqrt{\left(\frac{\tilde{v}_{\max}}{v_0}\right)^2 - 1} \right] \tag{13}$$

From Eqn (13), the actual distance of particles passing through matrix may be considerably increased to a high level if only $\tilde{v}_{\max} > v_0$ is met.

And when $\tilde{v}_{max} \leq v_0$, particles pass n_0 layers of matrix. However, when $\tilde{v}_{max} > v_0$ is met, the actual layers of particles through matrix is calculated by dividing L_T with ΔL:

$$n = \frac{2}{\pi} n_0 \cdot \left[\arcsin \frac{v_0}{\tilde{v}_{max}} + \sqrt{\left(\frac{\tilde{v}_{max}}{v_0} \right)^2 - 1} \right] \tag{14}$$

It can be seen that the actual layers increase as the maximum pulsating velocity increases. For instance, when the maximum pulsating velocity $\tilde{v}_{max}$ is 10 times of v_0, the actual layers is increased to 7 times of n_0, significantly increasing the chance of matrix capturing magnetic particles from slurry.

The pulsation of slurry produces a considerable improvement on the separation performance of a pulsating high-gradient magnetic separator, and the effect of slurry pulsation on the grade of magnetic concentrate is given by Oberteuffer (1974):

$$G_m = \frac{G_{max}}{1 + A_{nm} K \frac{F_i}{F_c}} \tag{15}$$

The effect of slurry pulsation on the recovery of magnetic particles is expressed as (Xiong et al., 1998):

$$R = R_{max} \cdot [1 - (1 - p)^n] \tag{16}$$

where G_m is the grade of magnetic concentrate, G_{max} is the maximum grade of magnetic concentrate, A_{nm} is the mass ratio of nonmagnetic particles to magnetic particles in feed, K is constant, F_i is the interactive force between magnetic and nonmagnetic particles, F_c is the competing forces acting upon particles, R is the recovery of magnetic particles, R_{max} is the maximum recovery of magnetic particles, and p is the capture possibility of a particle passing through a single layer of matrix.

From Eqn (15), the concentrate grade increases as the competing forces increase. As the pulsation of slurry produces an additional competing force on particles in pulsating high-gradient magnetic separation, so that it improves the concentrate grade; moreover, the improved concentrate grade is also benefited from the loose state of particles in the separating matrix, as a result of slurry pulsation wherein. It can be seen from Eqn (16), the recovery increases with increase in the actual layers of particles passing through matrix.

The pulsation of slurry produces a different capturing model of matrix for magnetic particles from that with no pulsation (Xiong et al., 1998), as shown in Figure 11.

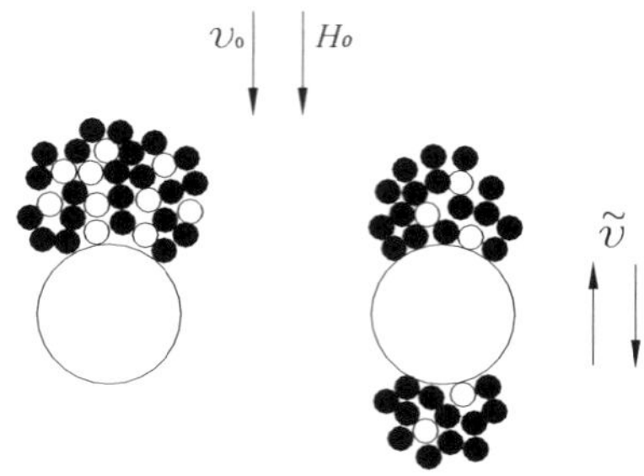

Figure 11 Capture of magnetic particles onto matrix under nonpulsating (left) and pulsating (right) conditions.

Under nonpulsating condition, slurry flows downward and the matrix mainly captures magnetic particles on the upper surface of magnetic elements of matrix. But, under pulsating condition, slurry moves up and down in the matrix and the matrix captures magnetic particles on both the upper and lower surfaces of magnetic elements in the matrix. From this point of view, the pulsation of slurry increases the capture area of matrix for magnetic particles, and thus improves the solids throughput of a pulsating high-gradient magnetic separator.

2.3.2.2 SLon Pulsating High-Gradient Magnetic Separator

SLon pulsating high-gradient magnetic separator as shown in Figure 12 has been innovatively developing since 1986, based on the principle of pulsating

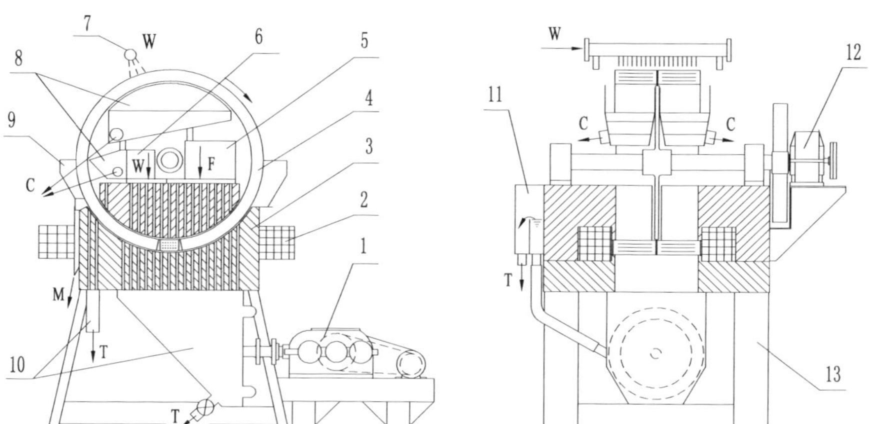

Figure 12 SLon pulsating high gradient magnetic separator: 1 = pulsating mechanism, 2 = energizing coils, 3 = magnetic yoke, 4 = separating ring, 5 = feed box, 6 = wash water box, 7 = concentrate flush, 8 = concentrate boxes, 9 = middling chute, 10 = tailings boxes, 11 = slurry level box, 12 = ring driver, 13 = support frame, F = feed, W = water, C = concentrate, M = middling, T = tailings.

high-gradient magnetic separation, and it is now the mostly applied high-gradient magnetic separator over the world, widely used for concentration of fine weakly magnetic ores such as hematite, ilmenite, etc., and for purification of nonmetallic ores such as quartz, feldspar, etc.

It can be seen from Figure 12, SLon pulsating high-gradient magnetic separator mainly consists of pulsating mechanism, energizing coils, magnetic yoke, separating ring, and feed and product boxes. In most cases, rods matrix made of magnetic stainless steel is used as magnetic matrix.

While working, a direct current flows through the energizing coils and a magnetic field is built up in the separating zone. The separating ring with magnetic matrix rotates around its horizontal axis. Slurry is fed from feed box and enters into the matrix moving through the separating zone. Magnetic particles are attracted from slurry onto the surface of matrix and brought to the top of separating ring, where magnetic field is negligible, and is flushed into concentrate boxes. Nonmagnetic particles flow through the matrix and enter into the tailings boxes under the combined actions of pulsating, gravity, and hydrodynamic drag. As the flushing direction of water is opposite to that of feed for each matrix pile, coarse particles will be flushed out without necessity to pass through the matrix depth, so that the clogging of matrix is avoidable. Moreover, the pulsating mechanism below the magnetic yoke drives the slurry in the separating zone up and down, keeping particles in the zone in a loose state, thus magnetic particles will be more easily captured by matrix and nonmangetic particles will be more easily dragged out as tailings.

Obviously, the vertical rotation of ring prevents matrix from being clogged and the pulsation of slurry improves the selectivity of separation. Due to these two innovative designs, SLon magnetic separators possess such merits as high efficiency, high reliability and flexibility, etc., in comparison to conventional high-gradient magnetic separators.

There are more than 3000 of SLon magnetic separators in operation in the field of mineral processing. It treats materials in a wide range of solids throughput from 100g per batch in SLon-100 cyclic pilot-scale magnetic separator to several hundred tons per hour in SLon full-scale magnetic separators. The technical specifications of SLon magnetic separators are listed in Table 6.

Figure 13 shows the largest SLon-4000 full-scale pulsating high-gradient magnetic separator and its operation for concentration of fine ilmenite from tailings at Panshihua Iron and Steel Company in southwest Sichuan province of China. It achieved a much superior performance to conventional separators, at a solids throughput reaching 570 t/h.

Table 6 Technical specifications of SLon pulsating high gradient magnetic separators.

SLon separators	SLon-100	SLon-500	SLon-750	SLon-1000	SLon-1500	SLon-1750	SLon-2000	SLon-2250	SLon-2500	SLon-3000	SLon-3500	SLon-4000
Ring diameter (mm)	100 (matrix height)	500	750	1000	1500	1750	2000	2250	2500	3000	3500	4000
Magnetic induction (T)	1.2 (1.75)	1.0	1.0	1.0	1.0	1.0	1.0	1.0	1.0	1.0	1.0	1.0
Pulsating stroke (mm)	0–30	0–50	0–50	0–20	0–30	0–30	0–30	0–30	0–30	0–30	0–30	0–30
Pulsating frequency (r/min)	0–600	0–400	0–400	0–300	0–300	0–300	0–300	0–300	0–300	0–300	0–300	0–300
Energizing current (A)	1200	1200	1100	650	950	1200	1200	1400	1400	1600		1700
Energizing voltage (V)	16.4	8.3	12	26.5	28	31	35	36	45	57		60
Energizing power (Kw)	19.7	10	13	17	27	37	42	51	63	91		102
Ring power (Kw)		0.18	0.55	1.1	3	4	5.5	7.5	11	18.5		37
Ring rotation speed (r/min)		0.3–3	0.3–3	2–4	2–4	3–4	3–4	3–4	3–4	3–4	3–4	3-4

power (Kw)												
Feed size (mm)	−1.3	−1.0	−1.0	−1.2	−1.2	−1.2	−1.2	−1.2	−1.2	−1.2	−1.2	−1.2
Feed solids concentration (%)	5−40	10−40	10−40	10−40	10−40	10−40	10−40	10−40	10−40	10−40	10−40	10−40
Feed solids (t/h)	100-600g	0.03−0.125	0.06−0.25	4−7	20−30	30−50	50−80	80−120	100−150	150−250	250−400	350 550
Feed volume flow rate (m^3/h)		0.25−0.5	0.5−1.0	12.5−20	50−100	75−150	100−200	160−300	200−400	350−650		750−1400
Water pressure (MPa)	0.1−0.2	0.1−0.2	0.1−0.2	0.2−0.3	0.2−0.3	0.2−0.3	0.2−0.3	0.2−0.4	0.2−0.4	0.2−0.4	0.2−0.4	0.2−0.4
Flush water consumption (m^3/h)		0.75−1.5	1.5−2.5	8−14	40−60	60−100	80−120	120−170	160−220	240−400		560−800
Mass weight of mathine (t)	1.15	1.5	3	6	20	35	50	70	105	175	200	380
Dimension (L × W × H, mm)	1600 × 800 × 1600	1800 × 1400 × 1320	2000 × 1360 × 1680	2700 × 2000 × 2400	3700 × 2900 × 3200	3900 × 3300 × 3800	4200 × 3550 × 4200	4580 × 4300 × 4590	5800 × 5000 × 5400	6600 × 5300 × 6400		8700 × 7400 × 8500
Remarks	cyclic pilot-scale	Continuous semi-full-scale						Continuous full-scale				

A higher magnetic induction of 1.75 T is achievable in SLon-100 pilot-scale magnetic separator of special design and 1.3 T is achievable in SLon full-scale magnetic separators.

Figure 13 The largest SLon-4000 full-scale pulsating high gradient magnetic separator (left) and operation for concentration of fine ilmenite from tailings (right).

2.3.2.3 Operation of SLon Magnetic Separator

The main operative variables of SLon magnetic separators include magnetic induction, pulsating stroke, pulsating frequency, and selectivity of rod matrix. Moreover, the diameter of ring determines the electric efficiency of the separators. Under any conditions, the performance of the separatorrs is closely related to the characteristics of feed.

Magnetic induction: Magnetic induction is the direct and dominant force which controls the number of magnetic particles going into concentrate, thereby generating a significant effect on the performance. In conventional high-gradient magnetic separators, an increase in magnetic induction decreases the grade of magnetic concentrate but improves the recovery of magnetic particles. However, in SLon pulsating high-gradient magnetic separator, the increase in magnetic induction improves both the concentrate grade and the recovery of magnetic particles, due to the fact that the pulsation of slurry improves the separation selectivity and the capture probability of magnetic particles.

The finer is the material, a higher magnetic induction is required to overcome the competing forces upon particles. But, the induction is not the only variable to recover fine magnetic particles, as the magnetic force acting on a magnetic particle, to a great extent, depends on the field gradient of matrix. Therefore, the optimum magnetic induction is connected to the selectivity of matrix.

Pulsating stroke and frequency: The most essential feature in SLon magnetic separators is the pulsation of slurry in the separating zone, which produces a significant effect on the separation performance. Figure 14 shows the separation results of SLon-100 cyclic pilot-scale magnetic separator in concentrating fine hematite tailings assaying 28.76% Fe, as a function of pulsating

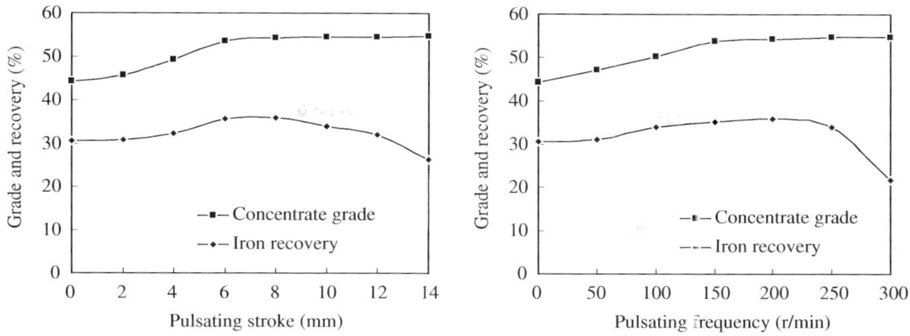

Figure 14 Effect of pulsating stroke (left) and frequency (right) on performance of SLon-100 magnetic separator in concentrating fine hematite tailings.

stroke and frequency, respectively. It is noted that more than 70% by mass weight of the material is smaller than 37 μm.

It can be seen that the pulsating stroke and frequency have significant effects on the separation performance of SLon-100 magnetic separator. As the stroke and frequency increase, concentrate grade increases and approaches the maximums after they reach 6 mm and 150 r/min, respectively; and with increase in the stroke and frequency, the recovery increases and respectively reaches the peak, and then begins to decrease when the stroke and frequency reach excessively high levels.

It is crucial that the pulsating energy is adequate enough so that slurry in the separating zone moves up and down and produces a relaxing effect on the particles in matrix, thereby improving the selectivity of separation and eliminating the mechanical entraining of nonmagnetic particles (Chen et al., 2009). The bigger is the stroke and the higher is the frequency, the more adequate is the pulsating energy in the separating zone.

From Figure 14, when the stroke and frequency are excessively small, the separation performance is improved slightly, due to the inadequate pulsating energy to relax particles in the matrix. The pulsating energy becomes more adequate as the stroke and frequency increase, and slurry in the matrix reverses periodically in the separating zone; in such a case, the actual distance and layers of particles passing through the matrix increase and there are more chances for the magnetic particles to be captured as concentrate, thereby improving the performance greatly. But, the performance decreases after the stroke and frequency reach excessively high levels, and this is due to the fact that fine magnetic particles are subjected to a too strong hydrodynamic competing force and are dragged out as tailings.

The basic principle in choosing the stroke and frequency is based on the nature of materials, and as well as is related to the magnetic induction applied. For a given material, a larger stroke or a higher frequency is usually required if a high enough magnetic induction is applied; and for a set magnetic induction, a fine material tolerates a gentle pulsating intensity. In practice, the stroke is usually fixed while the frequency is adjustable, in response to the variation in feed properties.

Selectivity of rod matrix: Except for the introduction of slurry pulsation in the separating zone of SLon magnetic separators, the rod matrix plays a vital role in the wide applications of the separators. Compared with fine matrix such as meshes and coarse matrix such as balls, rod matrix as shown in Figure 15 has such advantages as high operational reliability, simple combinatorial optimization, and resistance to clogging.

The selectivity of rod matrix determines the magnetic field gradient on the surface of matrix and subsequently the magnetic force imposed on a magnetic particle, which produces a significant effect on the performance of SLon magnetic separators. The finer is the matrix, the higher is the field gradient and the stronger is the magnetic force upon a magnetic particle. The selectivity of rod matrix on the separation performance of SLon-100 cyclic pilot-scale magnetic separator in concentrating fine hematite tailings is illustrative in Table 7.

Selectivity of ring diameter: SLon magnetic separators are termed with the diameter of separating ring and it has a decisive control on the efficiency

Figure 15 Rod matrix widely applied in SLon pulsating high gradient magnetic separators.

Table 7 Comparison of separation performance between 1 mm and 2 mm rod matrix

Rod diameter (mm)	Concentrate mass (%)	Concentrate grade (% Fe)	Iron recovery (%)
1	20.62	52.36	37.54
2	13.16	52.46	24.00

of energy consumptions of the separators. With scale-up of SLon magnetic separators, the energy consumptions decrease. Figure 16 shows the unit electric consumption of the separators for per ton of material, i.e., unit electric consumption, as a function of the diameter of separating ring.

It is clear that the scale-up of SLon magnetic separators reduces energy consumptions. The unit electric consumption reduces from 0.85 to 0.64 and to 0.39 KWh, as the diameter of separating ring is scaled-up from 2000 to 3000 and to 4000 mm. Thus, the selectivity of large-scale separators is extremely economic for a high throughput of processing plant.

2.3.3 Vibrating High-Gradient

The interfusion of iron impurities is a major problem in the utilization of nonmetallic ores, such as kaolin, feldspar, quartz, etc. Such impurities affect the brightness of clay products and reduce its commercial values (Chen et al., 2012). In practice, wet high-gradient magnetic separation is used for the purification of nonmetallic ores, as it is more applicable to have the motion of fine magnetic particles under control, thus an effective separation is achievable in a wet high-gradient magnetic separator. However, wet high-gradient magnetic separation has such demerits as necessity of dry treatment after

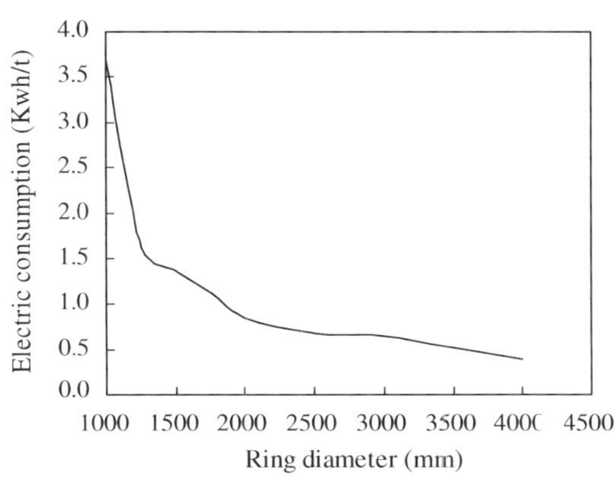

Figure 16 Effect of ring diameter on unit electric energy consumption of SLon magnetic separators.

separation, difficulty of running the process in cold or water-deficient regions, etc.

The most difficulty in applying a dry high-gradient magnetic separation lies in the fact that the mechanical entrapment of nonmagnetic particles in the powder coagulations can not be effectively overcome; this, however, in most cases is the most important aspect of the selectivity problem preventing dry high-gradient magnetic separation from applications. In a dry high-gradient magnetic separation, powder coagulations caused by particle interaction is considerable and tends to cause a blockage in magnetic matrix, which results in the deterioration in the separation performance of dry high-gradient magnetic separation.

2.3.3.1 SLon Vibrating High-Gradient Magnetic Separator

A SLon-1000 dry vibrating high-gradient magnetic separator is reported for purification of nonmetallic ores (Chen et al., 2012). As shown in Figure 17, the separator mainly consists of magnetic yoke, separating ring, energizing coils, ring vibrating mechanism, feed vibrating mechanism, and a shaft joint of rubber. Rod matrix is used as magnetic matrix.

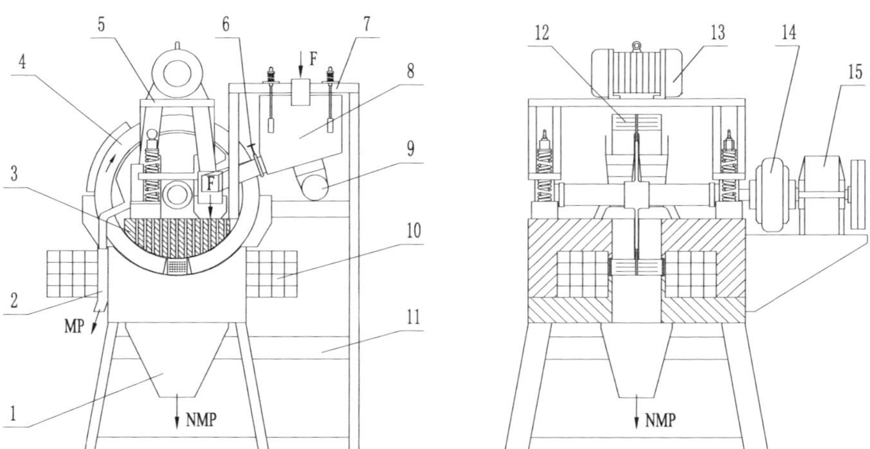

Figure 17 SLon-1000 dry vibrating high gradient magnetic separator. 1 = collecting box of nonmagnetic product, 2 = launder of magnetic product, 3 = magnetic yoke, 4 = separating ring, 5 = ring vibrating mechanism, 6 = feed distributor, 7 = feed vibrating mechanism, 8 = feed box, 9 = motor of feed vibrating mechanism, 10 = energizing coils, 11 = support frame, 12 = magnetic matrix, 13 = vibrating motor of ring vibrating mechanism, 14 = shaft joint of rubber, 15 = ring driver. F = Feed, MP = Magnetic product, NMP = Nonmagnetic product.

While the separator is operated, a direct current flows through the energizing coils and a magnetic field is built up in the separating zone. The separating ring with magnetic matrix positioned on the ring vibrating mechanism rotates around its horizontal axis in a vibrating manner. The material is evenly fed into the matrix in the separating zone, in a very thin layer through feed distributor, which is installed under the vibrating feed mechanism. Magnetic particles are captured from the material onto matrix, brought to the top of ring where the magnetic field is negligible, and fall into the launder of magnetic product below as magnetic product. Nonmagnetic particles, however, pass through the matrix and fall into the collecting box of nonmagnetic product, under the gravity and ring vibration.

Table 8 lists the main technical parameters of SLon-1000 vibrating magnetic separator.

The separator uniquely combines a ring vibrating mechanism that makes the ring vibrate during its rotation, and a feed vibrating mechanism that makes the material evenly pass through matrix, in a very thin layer and without blockage. The rotation torque from the ring driver is stably transmitted to the vibrating ring, through a shaft joint made of rubber in the middle of the ring axis (Liu, 2007).

2.3.3.2 Application of SLon Dry Vibrating Magnetic Separator

The SLon-1000 dry vibrating high-gradient magnetic separator is used for purification of kaolin, with results compared to SLon-1500 wet pulsating high-gradient magnetic separator as illustrated in Table 9. The iron impurities in the kaolin are finely disseminated in the form of hematite and

Table 8 Technical parameters of SLon-1000 dry vibrating high gradient magnetic separator

Technical parameters	Range	Technical parameters	Range
Diameter of ring (mm)	1000	Ring driver (Kw)	1.1
Rotation speed of ring (r/min)	0–3	Power of vibrating ring mechanism (Kw)	1.5
Magnetic induction (T)	1.0	Power of vibrating feed mechanism (Kw)	0.18
Vibrating amplitude of ring (mm)	2–4	Material throughput (t/h)	2–3
Vibrating frequency of ring (1/min)	0–1450	Mass weight (t)	7.6
Energizing power (Kw)	17	Dimension of machine (L × W × H) (mm)	1910 × 2320 × 2708

Table 9 Performance comparison of SLon wet and dry magnetic separators in purificating kaolin

Magnetic separators	Feed grade (% Fe_2O_3)	Nonmagnetic product: Mass weight (%)	Nonmagnetic product: Grade (% Fe_2O_3)	Magnetic product: Mass weight (%)	Magnetic product: Grade (% Fe_2O_3)	Removal rate of Fe_2O_3 (%)
SLon-1500 wet pulsating separator	0.65	92.42	0.44	7.58	3.21	37.44
SLon-1000 dry vibrating separator	0.73	84.56	0.50	15.44	1.99	42.08
Differences	−0.08	+7.86	−0.06	−7.86	+1.22	−4.64

Table 10 Performance of SLon-1000 vibrating magnetic separator in purifying fine quartz

Feed grade (% Fe_2O_3)	Nonmagnetic product: Mass weight (%)	Nonmagnetic product: Grade (% Fe_2O_3)	Magnetic product: Mass weight (%)	Magnetic product: Grade (% Fe_2O_3)	Removal rate of Fe_2O_3 (%)
0.08	97.40	0.04	2.60	1.58	51.30

limonite. From Table 9, compared to SLon wet pulsating high-gradient magnetic separator, the dry magnetic separator achieves a higher iron removal rate but a lower mass weight of nonmagnetic product. As the dry vibrating separation has inferiority in magnetic selectivity and an inherent sensitivity to form powder coagulations to the wet one, the dry vibrating separator is effective in purifying the kaolin material.

Table 10 shows the separation performance of SLon-1000 dry vibrating high-gradient magnetic separator in purifying fine quartz with a particle size below 74 μm by around 90%. It can be seen that it achieves a high mass weight of nonmagnetic product, with the iron removal rate reaching 51.30%.

3. CASE STUDIES ON MAGNETIC SEPARATIONS

In this section, several cases of magnetic techniques for the treatment of typically magnetic ores are introduced, and they are magnetite, hematite, ilmenite, limonite, and nonmetallic ores. In the processing of these magnetic ores, magnetic techniques are applied and their advantages over gravity and flotation methods are well reflected.

3.1 Magnetite and Hematite Ores

Anshan type lean hematite ore, located in Anshan area of northeast China, is one of the mostly typical iron ores of super-large reserve. In the past, this low-grade iron ore was crushed and classified into coarse and fine fractions, with the coarse fraction reductively roasted and the fine fraction processed in a gravity-high-intensity magnetic separation-flotation flowsheet. This flowsheet produces a relatively low-grade concentrate assaying around 63% Fe at a low recovery. In the past decade, with development of SLon pulsating high-gradient magnetic separator and flotation reagents, a high-gradient magnetic separation–reverse flotation flowsheet is proposed and applied to treat the Anshan type lean hematite ore, as shown in Figure 18.

In the new flowsheet, the ore is crushed, ground, and classified into coarse and fine fractions by hydrocyclones. The coarse fraction is fed into spirals to produce a qualified high-grade concentrate, and its tailings are successively processed by low-intensity magnetic and high-gradient magnetic separations, with a relatively high-grade concentrate achieved and returned to the grinding-classification system to achieve full liberation. The fine fraction, as shown in Figure 18, is processed with low-intensity magnetic separation to achieve a magnetite concentrate and its tailings is processed with SLon pulsating high-gradient magnetic separators as shown in Figure 19

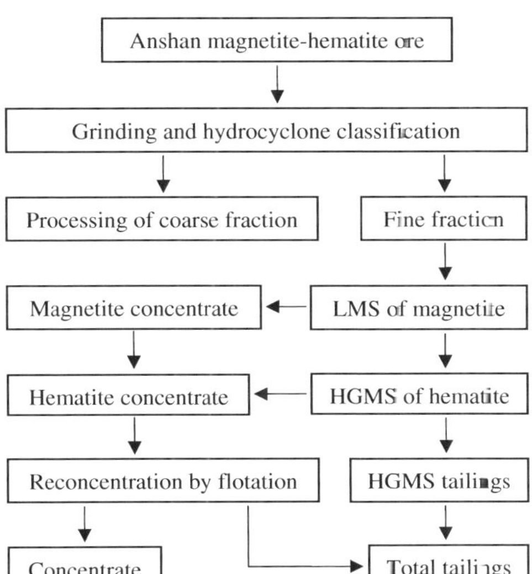

Figure 18 Processing flowsheet of Anshan type lean hematite ore (fine fraction).

Figure 19 Operation of SLon pulsating magnetic separators in Qidashan processing plant for concentration of Anshan type lean hematite ore.

to achieve a hematite concentrate. These two concentrates are combined and upgraded in a reverse flotation flowsheet to produce a high-grade concentrate assaying 67.98%. This concentrate is combined with that from the coarse fraction and produces a final concentrate assaying 67.50% at a high recovery approaching 80%.

Table 11 illustrates the processing results of this typically lean hematite from Qidashan processing plant of Anshan Iron and Steel Company.

Apparently, in the new flowsheet, gangues are effectively discarded as tailings at an extremely low grade through SLon pulsating high gradient magnetic separators, which creates a vital role in the whole flowsheet, for the production of high-grade concentrate at a high recovery.

Due to the successful application of high-gradient magnetic separation–reverse flotation flowsheet in processing Anshan type lean hematite ore in Anshan Iron and Steel Company, high-gradient magnetic separation is

Table 11 Processing results from Qidashan plant for Anshan type lean hematite ore (fine fraction)

Products	Mass weight (%)	Grade (% Fe)	Iron recovery (%)	Remarks
Feed	100.00	25.61	100.00	Fine fraction
Magnetite concentrate	28.52	54.26	60.43	Wet drum LMS
Hematite concentrate	19.53	28.21	21.51	SLON PHGMS
HGMS tailings	51.95	8.90	18.06	
Total tailings	74.50	11.11	32.31	
Concentrate	25.50	67.98	67.69	Magnetite + hematite

now widely applied over the world, for the primary concentration of hematite ore.

3.2 Limonite Ore

There are abundant limonite resources, for instance, in Yunnan province of southwest China, and the main characteristics of such magnetic ores are low density, fine dissemination or inclusion with gangues, small magnetic susceptibility, and high sensitivity to argillation during its grinding process. Thus, its effective utilization, in most cases, becomes difficult even impossible with gravity and flotation. With the development of magnetic techniques, more and more limonite resources are deploited, and used as raw material in the iron and steel industry. Figure 20 shows the processing flowsheet of magnetic separations for a fine limonite from a leaching residue of gold ore in Yunnan province, with results of the flowsheet illustrated in Table 12.

It can be seen that, after removal of strongly magnetic magnetites with wet low-intensity drum magnetic separation, the limonite is effectively recovered with SLon pulsating high-gradient magnetic separator from the residue.

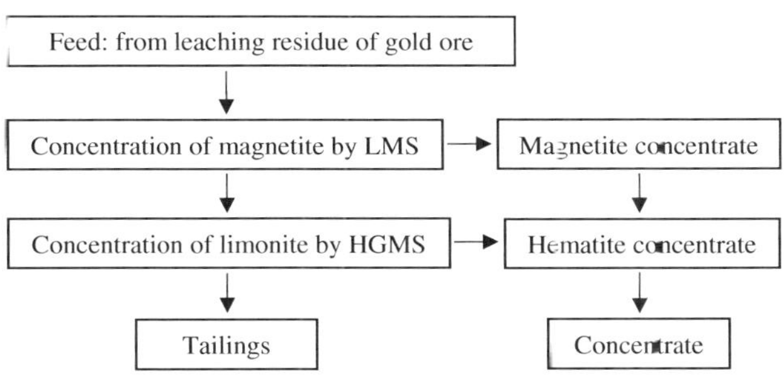

Figure 20 Processing flowsheet of fine limonite from leaching residue of gold ore.

Table 12 Processing results of limonite from leaching residue of gold ore

Products	Mass weight (%)	Grade (% Fe)	Iron recovery (%)	Remarks
Feed	100.00	48.52	100.00	From gold leaching
Magnetite concentrate	30.07	63.57	39.39	Wet drum LMS
Limonite concentrate	33.81	50.36	35.09	SLon PHGMS
Tailings	36.12	34.28	25.52	
Concentrate	63.88	56.58	74.48	Magnetite + limonite

3.3 Ilmenite Ore

Panzhihua ilmenite, located in Sichuan province of southwest China, is the largest ilmenite reserve in the world, with an estimated reserve of 870 million tons. This deposit accounts for more than 90% of the total titanium resource of China and over 35% of the world. In the present processing flowsheet, the ore is first processed with wet low-intensity drum magnetic separator to concentrate strongly magnetic vanadic titanomagnetite, with its tailings classified into coarse and fine fractions; and then the ilmenite is respectively recovered from the two fractions, by high-gradient magnetic separation-reverse flotation flowsheet. Figure 21 shows the processing flowsheet for the fine fraction in the ilmenite processing plant of Panshihua Iron and Steel Company, with separation results listed in Table 13.

It can be seen that high-gradient magnetic separation plays a crucially important role in the recovery of fine ilmenite from the ore. In the flowsheet as shown in Figure 21, the feed is firstly processed by wet low-intensity drum magnetic separation to concentrate vanadic titano-magnetite, and then is effectively processed through an SLon pulsating high-gradient magnetic separation flowsheet, in which 67.87% by mass weight of the feed is discarded as tailings, at a very low grade of 3.02% TiO_2.

It is apparent that high-gradient magnetic separation plays the most significant role in the flowsheet for the production of qualified high-grade ilmenite concentrate, by providing a primary ilmenite concentrate to the reverse flotation flowsheet, at a low mass flow rate.

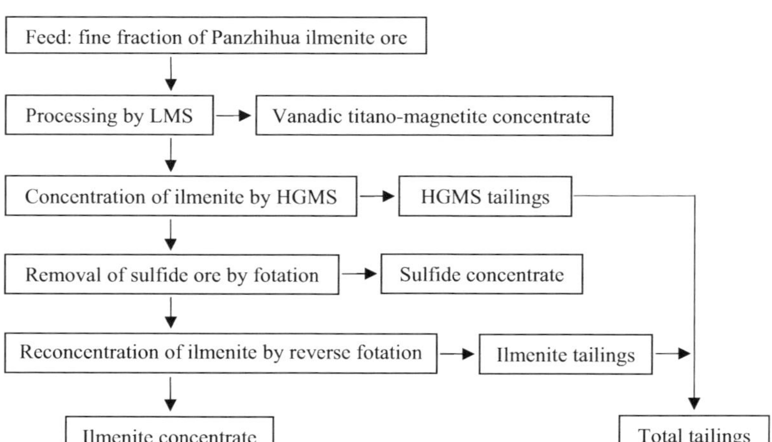

Figure 21 Processing flowsheet of Panzhihua ilmenite ore (fine fraction).

Table 13 Processing results of fine ilmenite of Panshihua ilmenite ore (fine fraction)

Products	Mass weight (%)	Grade (% TiO_2)	TiO_2 recovery (%)	Remarks
Feed	100.00	8.53	100.00	Fine fraction
Vanadic titano-magnetite concentrate	2.62	12.62	3.88	Wet drum LMS
HGMS tailings	67.87	3.02	24.03	SLon PHGMS
Sulfide concentrate	0.89	9.58	1.00	
Ilmentite concentrate	10.11	47.46	56.25	
Ilemenite tailings	18.51	6.84	14.84	
Total tailings	86.38	3.94	39.87	

Table 14 Performances of high gradient magnetic separation in purifying nonmetallic ores

Non-metallic ores	Feed grade (% Fe_2O_3)	Concentrate grade (% Fe_2O_3)	Removal rate (% Fe_2O_3)	Processing method
Quartz	0.22	0.06	91.27	SLon PHGMS
Feldspar	1.48	0.26	84.77	SLon PHGMS
Kaolin	0.72	0.43	87.68	SLon PHGMS
Nepheline	1.69	0.179	91.52	SLon PHGMS
Andalusite	4.32	0.49	93.10	SLon PHGMS

3.4 Nonmetallic Ores

Magnetic techniques have become the mostly used method for purification of nonmetallic ores, mainly due to its effectiveness, simple operation, renewability, and low operating cost. Table 14 shows the performances of SLon pulsating high-gradient magnetic separators in purifying the typically nonmetallic ores in China. From Table 14, high-gradient magnetic separation is well suited for the purification of nonmetallic ores, at a high removal rate for magnetic impurities.

4. FUTURE TRENDS IN MAGNETIC SEPARATION

Magnetic techniques are now extensively applied in the field of mineral processing, for concentration of magnetic components or for removal of magnetic impurities with tremendous successes. But, the mineral industry is encountering such difficulties as declines in grade and liberation size of valuable components and increase of deleterious components in magnetic ores, as well as the increasing compulsion to cut costs and avoid environmental pollution during the mining process. Such challenges have to be considered in the design of new magnetic techniques for environmental and sustainable requirements.

In the past decade, the increasing exploitation of low-grade and finely liberated magnetic ores, which are previously discarded as tailings, has become a prospective proposition (Watson and Beharrell, 2006), with the increasing demand for raw materials. Magnetite and hematite ores, for instance, with grades lower than 15% Fe and 25% Fe, respectively, are being profitably exploited, as a result of the sharp increase for raw materials by iron and steel industries. In the exploitation of such magnetic ores, in the most cases, preconcentration treatment is inevitably required to increase the feed grade and the economic efficiency of processing flowsheets; this, however, results in considerable loss of magnetic values into tailings, due to insufficient magnetic techniques available to preconcentrate such ores. From this point of view, the design and development of new magnetic techniques suited for the preconcentration purpose of low-grade and fine-sized magnetic ores is an urgent problem calling for solution.

In contrast to flotation methods, magnetic separation is a powerful technique for the manipulation of magnetic particles, and does not require chemicals that are responsible for high operating costs and environmental concerns. However, in the processing of low-grade magnetic ores, flotation is practically required in the upgrading of primary concentrates to produce qualified high-grade concentrates. This outcome is due to the strong capturing effect of magnetic force upon low-grade intergrowths, which deteriorates the quality of magnetic product. In the recent years, attentions have arisen in the development of new magnetic techniques such as centrifugal high-gradient magnetic separation (Chen et al., 2013b) or other physical methods, for the production of qualified high-grade concentrates, and this requires an improved understanding on the theoretical and operational principles of magnetic separations, and needs the innovative design of magnetic equipments with higher selectivity to magnetic particles.

Theoretically, the technological significance of superconducting high-gradient magnetic separation is considerable, due to its advantages of low energy consumption and small mass weight of equipments. But, until today, the technical need for such a magnetic technique is negligible in the treatment of magnetically metallic ores, as there is no obvious need for magnetic inductions higher than 1.3 T in the field of mineral processing. Superconducting high-gradient magnetic separation is still dominantly restricted in the removal of very fine magnetic iron and titanium oxides from kaolin and in the treatment of wastewater.

Nevertheless, this current status in superconducting high-gradient magnetic separation does not fade its prospective superiority over conventional magnetic techniques. With the continual depletion of the world's mineral resources, the recovery of very fine magnetic values from discarded tailings and finely disseminated ores necessitate the development of magnetic techniques available with higher magnetic inductions.

5. CONCLUSIONS

Magnetic techniques in various types, as a powerful and inherent technique for the manipulation of magnetic particles, is now widely applied in the field of mineral processing with tremendous successes, due to its effectiveness, simple operation, low operating costs, and environmental sustainability. Numerous efforts have been expanding, in order to convert this superiority into practice, with a wealth of innovative ideas and designs.

Although its tremendous successes in the field of mineral processing, the mineral industry is encountering new difficulties, which have to be considered in the design of new magnetic techniques, for environmental and sustainable requirements, and this requires an improved understanding on the theoretical and operational principles of magnetic techniques.

REFERENCES

Chen, L.Z., Liao, G.P., Qian, Z.H., Chen, J., 2012. Vibrating high gradient magnetic separation for purificaiton of iron impurities under dry condition. Int. J. Mineral Process. 102–103, 136–140.

Chen, L.Z., Qian, Z.H., Wen, S.M., Huang, S.W., 2013a. High gradient magnetic separation of ultrafine particles with rod matrix. Miner. Process. Extr. Metall. Rev. 34 (5), 340–347.

Chen, L.Z., Xiong, D.H., Huang, H.C., 2009. Pulsating high gradient magnetic separation of fine hematite from tailings. Miner. Metall. Process. 26 (3), 163–168.

Chen, L.Z., Xu, G.D., Huang, J.X., 2013b. Principle of cyclic centrifugal high gradient magnetic separation pilot-scale test for fine ilmenite. J. Kunming Univ. Sci. Technol. 38 (1), 28–31.

Li, Y.G., 2005. Application results of magnetic field screen in concentrators. China Min. Mag. 14 (7), 63–65, 68. (in Chinese).

Liu, S.Y., 1994. Science in magnetic and electric separation. Changsha. Central South University Press (in Chinese).

Liu, X.M., 2007. Analysis and practice of SLon-1000 dry vibration high gradient magnetic separator. Min. Process. Equip. 35 (12), 103–1056 (in Chinese).

Obereuffer, J.A., 1974. Magnetic separation: a review of principles, devices and applications. IEEE Trans. 2, 223–238.

Svoboda, J., 2004. Magnetic Techniques for the Treatment of Materials. Kluwer Academic Publishers, Dordrecht, Netherlands.

Svoboda, J., Fujita, T., 2003. Recent developments in magnetic methods of material separation. Miner. Eng. 16 (9), 785–792.

Wang, C.R., 2008. Magnetic and Electric Separation. Metallurgical Industry Press, Beijing (in Chinese).

Watson, J.H.P., Beharrell, P.A., 2006. Extracting values from mine dumps and tailings. Miner. Eng. 19 (15), 1580–1587.

Xiong, D.H., Liu, S.Y., Chen, J., 1998. New technology of pulsating high gradient magnetic separation. Int. J. Mineral Process. 54 (2), 111–127.

Zeng, W.Q., Xiong, D.H., 2003. The latest application of SLon vertical ring and pulsating high gradient magnetic separator. Mine. Eng. 16 (6), 563–565.

LIST OF RELEVANT WEB SITES

Information on column magnetic separator: www.asjyf.com.cn.

Information of SLon pulsating high gradient magnetic separator: www.slon.com.cn.

CHAPTER EIGHT

Electric (Electro/Dielectro-Phoretic)—Force Field Assisted Separators

Morgane Citeau[1], Olivier Larue[2] and Eugene Vorobiev[3]
[1]CREOL, Pessac, France
[2]Solvay, Center for Research & Technology of Shanghai, Xin Zhuang Industrial Zone, Shanghai, China
[3]Département de Génie des Procédés Industriels, Laboratoire Transformations Intégrées de la Matière Renouvelable, Université de Technologie de Compiègne, Centre de Recherche de Royallieu, Compiègne, France

Contents

Progress in Filtration and Separation
http://dx.doi.org/10.1016/B978-0-12-384746-1.00008-2

NOMENCLATURE

A	Surface area (m^2)
C	Particle concentration (mol/m^3)
D	Liquid dielectric constant (F/m)
∇E	Electric field strength gradient (V/m)
E	Electric field intensity (V/m)
E_{cr}	Critical electric field strength (V/m)
$\vec{F}_{DEP}$	Dielectrophoresis force (−)
$\vec{F}_{TW-DEP}$	Travelling-wave dielectrophoretic force (−)
F	Faraday number (96,500 C/mol)
f	Frequency of the applied A.C. field (Hz)
f_{CM}	Clausius-Mossotti factor (−)
i	The imaginary vector (−)
I	Amperage (A)
J	Permeate flux of the liquid (m/s)
m_{water}	Collected water weight (kg)
m_{sludge}	Sludge weight (kg)
n_{e^-}	Number of electron mole (mol)
Δp_E	Electroosmotic pressure (Pa)
Δp_H	Hydraulic pressure (Pa)
q	Superficial liquid velocity (m/s)
q_P	Pressure induced superficial liquid velocity (m/s)
q_E	Electroosmotic superficial liquid velocity (m/s)
r	Radius of spherical particle (m^2)
R_{cell}	Electrical resistance of the material between the electrodes (Ω)
R_m	Filter cloth resistance (1/m)
t	Time (s)
u_e	Electrophoretic mobility of the particle (m^2/V s)
u_{DEP}	Dielectrophoretic mobility of the particle (m^2/V s)
$U_{A/eq}$	Electrochemical potentials at anode (V)
$U_{C/eq}$	Electrochemical potentials at cathode (V)

V_L	Filtrate volume (m^3)
V_P	Particle volume (m^3)
w	Ratio between solid content and filtrate volume (kg DS/m^3)

Greek Letters

α_{av}	average specific filter cake resistance (m/kg)
ε	porosity (−)
$\varepsilon^\star$	complex permittivity (C/V m)
ε_0	permittivity of the free space (8.854×10^{-12} C/V m)
ε_m	Relative permittivity of the medium (−)
ε_p	Relative permittivity of the particle (−)
φ	Friction factor (−)
η_A	Anode overpotential (V)
η_C	Cathode overpotential (V)
λ	Stoichiometric coefficient of the electrolysis reaction (−)
λ_{TW}	Wavelength of travelling electric field (m)
μ	Viscosity of the liquid (Pa/s)
$\vec{v}_{eo}$	Electroosmotic flow velocity (−)
$\vec{v}_{ef}$	Electrophoretic velocity (−)
σ	Electrical conductivity (S/m)
ω	Angular frequency of the A.C. field (rad/s)
ζ	Potential zeta of particles (V)
∇	*del* (gradient) vector operator

Abbreviations

A.C.	Alternative current
BSA	Bovine serum albumin
CFEF	Crossflow electrofiltration
C.C.	Constant electrical current
cr	Critical
C.V.	Constant voltage
DEEF	Dead end electrofiltration
DEP	Dielectrophoresis
D.C.	Direct current
EC	Energy consumption
ED	Electrodewatering
EK	Electrokinetic
f	Final
HGES	High-gradient electric field system
i	Initial
m	Medium
MD	Mechanical dewatering
p	Particle
PEOD	Pressurized electroosmotic dewatering
pk–pk	Peak-to-peak
TMP	Transmembrane pressure

1. INTRODUCTION

The suspensions containing fines (<10 μm) or gelatinous-like particles usually exhibit slow solid–liquid separation rate during a mechanical process such as centrifugation, filtration or sedimentation. The electrical assistance allows improving the separation. It is based on the mobility of ions, charged or neutral particles in an electric field. When an electric field is applied to a material (suspension or semi-solid material) placed between two electrodes, it induces electrokinetic phenomena: electrophoresis, electroosmosis, ion electromigration, and/or dielectrophoresis. To take advantage of these phenomena, several processes were designed such as the electroosmotic dewatering, the electrowashing, the dielectrophoretic separation and the electrofiltration (or other related processes depending on which kind of mechanical separator they are applied: electrosedimentation, electrocentrifugation, etc.). The electrofiltration allows accelerating the filtration and relies on the electrophoresis and electroosmosis (moving in the opposite direction of charged particles and ionized mobile liquid close to the particles surface, respectively, due to the external electric field). The electroosmotic dewatering allows accelerating the consolidation and increasing the dewatering of semi-solid materials and relies on the electroosmosis (moving of ionized mobile liquid close to the particles surface due to the external electric field). The electrowashing allows accelerating the washing of product using at same time less washing liquor and relies on the electroosmosis and ion electromigration (removing of ion due to external electric field). The dielectrophoretic separation allows improving the fine particle removal from liquid and relies on the dielectrophoresis force (moving of polarized particle in a nonuniform electric field).

Currently, those electrotechnologies are in a development stage. Industrialization has been limited to niche areas in soil mechanics, civil engineering, ceramic industry and petroleum refineries. However, there is a growing interest since they show some capabilities to enhance either suspension concentration, liquid or solid purification, particle fractionation, suspension dewatering or filter cake consolidation. Some problems have hampered their diffusion in industry, such as the electrical power consumption, the safety requirements when using high power generators, a lack of knowledge on the right suspension properties and some improperly designed operating systems. This chapter aims at providing a fundamental understanding of these electrotechnologies. It presents bench and emerging commercial equipment and their potential applications. It also presents the specifications for the suspension for these electrotechnologies to be applied efficiently.

2. ELECTROPHORETIC AND ELECTROOSMOTIC TREATMENTS

2.1 Electric Field-Induced Effects

According to standard model (Gouy, 1910; Stern, 1924; Lyklema, 1995) as the fine suspended particles possess a surface charge in aqueous medium, they are surrounded by a layer with a high density of opposite charges, the electric double layer. The existence of the particle's charge and the electric double layer give electrokinetic properties to the particle. When an electric field is applied across two electrodes in the suspension, it interacts with the surface charge of the solid phase and with the dissolved ions, inducing electrokinetic phenomena such as electrophoresis, electroosmosis and ion electromigration. As illustrated in Figure 1, the electromigration is the motion of ions in a pure fluid in an applied electric field. The electrophoresis corresponds to the suspended charged-particle migration relative to the liquid in an applied electric field. The electrophoretic

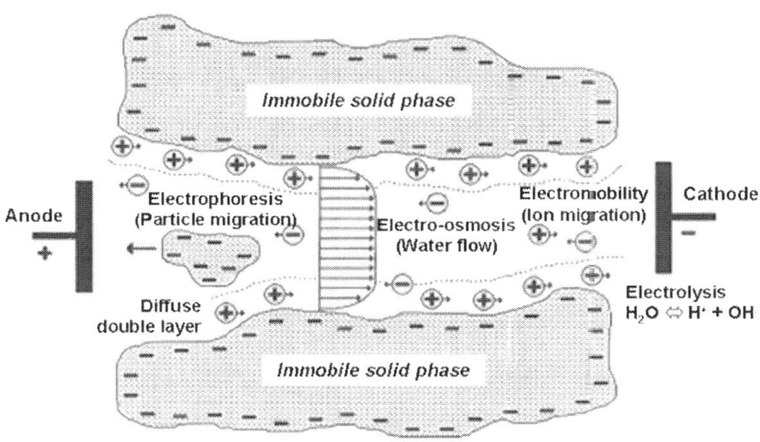

Figure 1 Illustration of the major electrokinetic phenomena. *Jennings and Mansharamani (1999). With permission.*

movement $\overrightarrow{v}_{ef}$ is proportional to the electric field strength E and the electrophoretic mobility of the particle u_e as:

$$\overrightarrow{v}_{ef} = u_e \nabla E \tag{1}$$

The electrophoretic mobility of the particle defined by the Helmholtz-Smoluchowki theory can be written as: $u_e = D\zeta/\mu$, where ζ is the zeta potential of particles, D is the liquid dielectric constant, and μ is the viscosity of the liquid. The electroosmosis refers to the liquid flow relative to the fixed charged layer in a porous matrix such as in filter cake and in filter medium. The electroosmotic velocity profile $\overrightarrow{v}_{eo}$ is analogous to the electrophoresis but in the opposite direction.

Thus, the electrokinetic efficiency depends mainly on the surface electrical charge of the particles reflected by the zeta potential ζ, which is in turn pH and ionic strength dependent. Electrokinetic efficiency is more important for particles which have higher zeta potential. High electrolyte content in liquid phase compresses the electric double layer that decreases zeta potential (in absolute value). And the isoelectric point corresponds to the pH value where zeta potential becomes zero. During the electric treatment, the fluid pH and electrolyte content may change due to the ion migration and electrochemical reactions at the electrodes.

To maintain charge equilibrium, electrons are released from solution into the anode by the oxidation reaction and consumed into the cathode by the reduction reaction. The electrochemical reaction with the most negative potential (U^0) at the anode (or least positive at the cathode) is favoured until the species involved are used up and/or until a high voltage is applied to stimulate competing reactions (Lockhart, 1983a). Thus, the main electrochemical reaction is the water electrolysis at the cathode (C1). The reaction (C2) is favboured over (C1) only for cation (M^{n+}) of Pb, Cu, Ag, Hg, Pt, or Au (Lockhart, 1983a).

$$2H_2O_{(l)} + 2e^- \rightarrow H_{2(g)} + 2OH^-_{(aq)} \quad U^0 = -0.83\ V \tag{C1}$$

$$M^{n+}_{(aq)} + ne^- \rightarrow M_{(s)} \tag{C2}$$

Depending on the electrode material, the main electrochemical reactions at the anode are the metal oxidation (A1) with M_A the metal at anode, or the water electrolysis (A2). The use of an insoluble material (or not consumed coulometrically) favours the reaction (A2).

$$M_A \rightarrow M^{n+}_{A(aq)} + ne^- \tag{A1}$$

$$\begin{gathered} 6H_2O_{(l)} \rightarrow O_{2(g)} + 4H_3O^+_{(aq)} + 4e^- \\ U^0 = 0.82\ V(pH\ 7)\ \text{to}\ 1.23\ V(pH\ 0) \end{gathered} \tag{A2}$$

The ion concentration produced by electrolysis at electrode C, is related to time t and amperage I, according to Faraday states:

$$C = \frac{\lambda \cdot I \cdot t}{n_{e^-} \cdot F} \tag{2}$$

where λ is the stoichiometric coefficient of the electrolysis reaction, n_{e^-} is the number of electron mole, and F is the Faraday number (96,500 C/mol). If the reactions (C1) and (A2) are favoured, gases and hydronium and hydroxide electrolytic ions are produced at the electrodes. Studies show that these hydronium and hydroxide electrolytic ions migrate partly towards the electrode of opposite charge. In a case of unbuffered suspension, it may induce a pH variation from 1 at the anode side to 12 at the cathode side both in the concentrated suspension and in the filtrate (Yuan and Weng, 2003; Citeau et al., 2012a). The electrical effects include also, other electrochemical reactions in the treatment cell such as sorption/desorption, dilution/precipitation, and oxidation/reduction (Kim et al., 2002; Tuan and Sillanpää, 2010). For example, a white layer formation on the cathode surface can be sometimes observed during the electrical treatment. It is mainly composed of calcium hydroxide and other insoluble salts produced by electrochemical reactions at the electrode (Yamaguch et al., 1993; Virkutyte et al., 2002). This can be a great issue since this hydroxide/oxide layer tends to increase the electrical resistance and energy consumption.

The voltage drop between electrodes U_{AC} is divided into the electrochemical potentials at anode and cathode ($U_{A/eq}$; $U_{C/eq}$), the anode and cathode overpotential (η_A;η_C), the Ohmic potential in the cell (IR_{cell}). Over time, a supplementary potential drop may result from the increase of a thick and highly insulating oxide barrier at electrodes and/or a layer of unsaturated cake mainly at anode side due to the electroosmosis (Mahmoud et al., 2010).

$$U_{AC} = U_{A/eq} - U_{C/eq} + \eta_A - \eta_C + IR_{cell} \tag{3}$$

The electrical resistance of the material between the electrodes (R_{cell}) converts a portion of the electric energy into thermal energy, inducing a temperature increase by Ohmic heating. This affects the suspension properties, reducing notably the liquid viscosity (Weber and Stahl, 2002). This in turn could be advantageous since it increases the solid–liquid separation rate.

As the electrotechniques consist of superimposing an electric field to a standard mechanical process, the filtrate flow is then influenced by the combination of the standard mechanical parameters (pressure drop, filter cake specific resistance) and the electrical effects (electrokinetic phenomena, electrochemical reactions at electrodes, Ohmic heating). Among different electric field assisted separation processes, we can define the electrofiltration, the electroosmotic dewatering, the electrowashing, the electrosedimentation and the electrocoagulation. The last-named corresponds to the electrolytic decomposition of corrodible electrode; it is an alternative to chemical conditioning. It would not be investigated in this chapter.

2.2 Dead end Electrofiltration and Electroosmotic Treatment

2.2.1 Mechanisms

2.2.1.1 Dead end Electrofiltration

Figure 2(a)and (b) presents a schematic description of the cake formation steps in a filter-press chamber during a dead end filtration and electrofiltration, respectively. In conventional filtration, the filter cake builds up on the filtering surface (Figure 2(a-I)). As a result, the filtrate flux declines with the cake growth at each filter chamber sides (Figure 2(a-II)). In dead end electrofiltration, the electric field strength is applied in parallel to the filtrate flow direction. It induces an electrophoretic flux of suspended charged particles towards the oppositely charged electrode. In the case of negatively charged particles, the filter cake gradually grows at the anode side. In principle, only a thin film of particles deposits on the cathode side, allowing nearly the whole filtrate to flow through this filter cloth (Figure 2(b-I)–(b-III)). Electroosmotic flow is created in the building filter cake and in the filter cloth, removing a supplementary water amount in comparison to mechanical pressure. As a result, the filtrate flow rate is accelerated at cathode side and the filter cake is compressed further by the electroosmotic pumping force (Moulik, 1971). Over time, the suspension is concentrating in the filter chamber, and the electrophoretic mobility of suspended charged particles becomes negligible compared to the electroosmosis motion

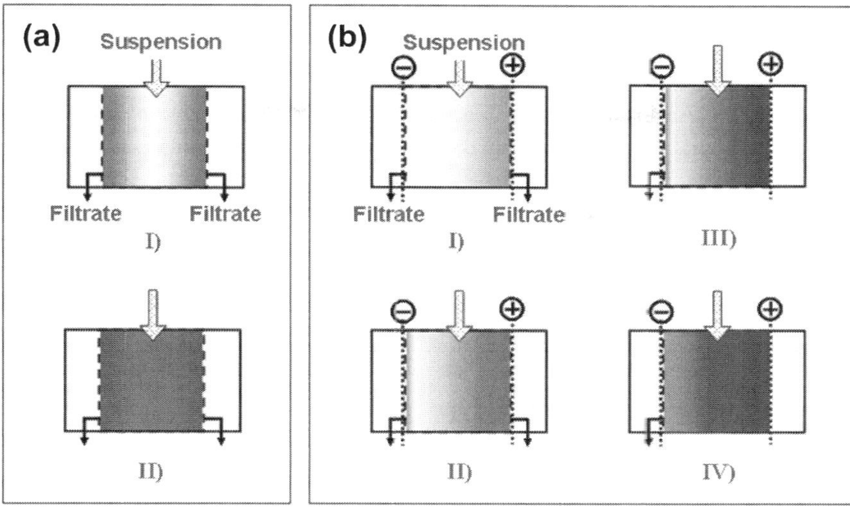

Figure 2 Schematic representation of cake formation steps in a filter press chamber during (a) filtration and (b) electrofiltration where the liquid flow through the perforated electrodes (Vorobiev and Jany, 1999).

(Figure 2(b-IV)). When the filter cake fills up the chamber, the solid–liquid separation mechanism corresponds to the electroosmotic dewatering.

The flow rate in electrofiltration can be written as:

$$q = q_{\mathrm{P}} + q_{\mathrm{E}} \tag{4}$$

where q is the superficial liquid velocity, q_{p} displays the Darcy law, and q_{E} is the electroosmotic contribution described by Helmholtz-Smoluchowski equation (Eqn (1)). Yukawa et al. (1976) modified Darcy's law to describe the filter cake formation during the electrofiltration taking into account the electrokinetic effects. The electrophoretic effect takes part by introducing the electrophoretic coefficient given as $(E_{\mathrm{cr}} - E)/E_{\mathrm{cr}}$, where E_{cr} is the critical electric field strength, describing the equilibrium state between the electrophoretic force and the hydrodynamic resistance force. The electroosmotic effect is considered as an additional pressure Δp_{E} added to the applied hydraulic pressure. Thus, the total flow rate is given as:

$$\frac{\mathrm{d}V_{\mathrm{L}}}{\mathrm{d}t} = \frac{(\Delta p_{\mathrm{H}} + \Delta p_{\mathrm{E}}) \cdot A}{\mu \cdot \left(\alpha_{\mathrm{av}} \cdot w \cdot \left(\frac{E_{\mathrm{cr}} - E}{E_{\mathrm{cr}}}\right) \cdot \frac{V}{A} + R_{\mathrm{m}}\right)} \tag{5}$$

where t is the filtration time, V_L is the filtrate volume, μ is the dynamic viscosity, α_{av} is the average specific filter cake resistance, w is the ratio between the solid content and the filtrate volume, R_m is the filter cloth resistance, A is the filtration surface area, and Δp_H is the hydraulic pressure as in the Darcy's law. Integration of this equation with assumption of constant coefficient, leads to a linear dependence of the inverse volume flow t/V_L on the filtrate volume V_L:

$$\frac{t}{V_L} = \frac{\mu \cdot \alpha_{av} \cdot w \cdot \left(\frac{E_{cr} - E}{E_{cr}}\right)}{2 \cdot (\Delta p_H + \Delta p_E) \cdot A^2} \cdot V_L + \frac{\mu \cdot R_m}{(\Delta p_H + \Delta p_E) \cdot A} \quad (6)$$

2.2.1.2 Electroosmotic Dewatering

The electroosmotic flow is principally based on the interplay between the applied electric field and the electric charge density of ions existing in the liquid close to the particle surface, Figure 1. Thus, under the effect of the external electric field, the water displacement is ensured by the propagation of ionized mobile liquid in the central channel through viscous momentum transfer as shown in Figure 3 (Yang et al., 2009). It allows removing some adsorbed water on the solid surface that cannot be released by mechanical methods (Zhou et al., 2001; Smollen and Kafaar, 1994). As particles are mainly negatively charged, the electroosmotic flow goes from the anode to the cathode. And a moisture gradient increases inside the filter cake, the cake at anode side becoming drier than at the cathode side (Chu et al., 2004; Saveyn et al., 2006a). The resulted flow rate is controlled by

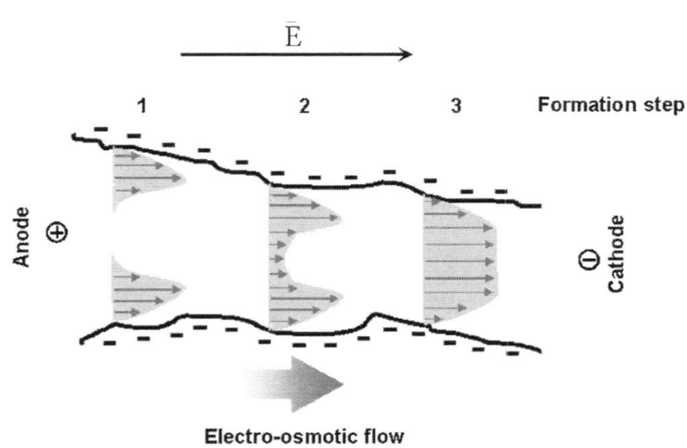

Figure 3 Formation steps of electroosmosis flow in a pore.

the balance between the electric forces and the frictional forces retarding water movement. The process of pressurized electroosmotic dewatering (PEOD) combines a conventional pressure consolidation with an electric field applied through a concentrated suspension between two electrodes.

Some electroosmotic technologies have been also developed for depollution of soils. The process named electrokinetic soil remediation is mainly based on the electromigration, the electroosmosis and the electrochemical reactions similarly to the electroosmotic dewatering. However, the method carried at a lower electric field, does not aim to solid–liquid separation but to remove contaminants such as heavy metals, radionuclides, and hydrophobic organic contaminants (Acar et al., 1995; Virkutyte et al., 2002).

2.2.1.3 Electrowashing

Displacement washing process is used to remove the retained mother liquor in a porous matrix when this liquor is regarded as contaminant or must be recovered as a valuable component. Thus, soluble impurities are flushed out of the voids by the application of fresh wash liquor or solvent that is miscible with the retained mother liquor. An electric field applied simultaneously to the fresh wash liquor flux accelerates the removal of solute and ions previously adhered to particle surfaces, otherwise inaccessible ions towards the bulk flow. These ionic species migrate towards the electrode of opposite charge by electroosmosis, electromigration and by diffusion. As a result, the time of washing could be shortened and much smaller quantities of water could be required to carry out the desired washing by the electrowashing method.

2.2.2 Equipments

Electrical equipments are often adapted of existing industrial technology, such as plate or diaphragm filter-press, pressure-drum device, belt filter-press, rotating screw-press, and vertical drain. The dead end electrofiltration and the electroosmotic dewatering are carried out in similar devices. However, the term of "electroosmotic dewatering" refers generally to the application of the electric field during the solid–liquid expression phase.

2.2.2.1 Filter-Press for Electrofiltration or Electroosmotic Dewatering

Pilot-scale filter-presses are adaptable to both dead end electrofiltration and electroosmotic dewatering by placing two electrodes at both sides of the filter chamber (Bouzrara and Vorobiev, 2000; Saveyn et al., 2006b; Lee et al., 2007; Yang et al., 2011). Figure 4 presents the construction of the filter

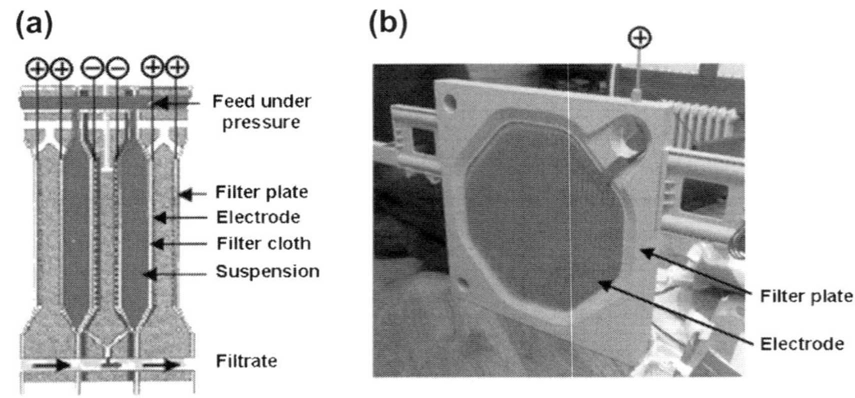

Figure 4 (a) Schematic sectional view of two electrofilter-press chambers (Bouzrara and Vorobiev, 2000). (b) Electrofilter plate.

chamber, including grid-like electrodes installed on filter plates and covered by a filter cloth. It is more indicated to have a two-sided drainage filter chamber compared to one-sided drainage filter chamber. The former presents a larger filtering surface that allows removing more filtrate by mechanical pressure. It allows also evacuating the electrolytic gas generated at the anode and the cathode, avoiding the formation of an insulating layer between the cake and the impermeable electrode (Weber and Stahl, 2002). Furthermore, if the surface charge of the particle is reversed, the electrofiltration may be carried out in spite of the electrophoretic and electroosmotic flow directions reversals.

The cathode material is preferentially in stainless steel or graphite. As the anode is subjected to an oxidation reaction, it is advised to use nonconsumable metal to prevent the electrode corrosion and the metal oxides production that may contaminate the filter cake and clogging the filter cloth. The most suitable electrodes include platinum, graphite or titanium support coated with mixed metal oxide layer, such as iridium oxide, ruthenium oxide or platinum oxide. These latter, named Dimensionally Stable Anode (DSA®), are cheaper than noble metal such as platinum and they combine excellent material properties such as strength and flexibility with a high degree of dimensional stability. They are not yet entirely resistant to chemical attacks. Indeed, when salts are present, then anions such as chloride or sulphate, will enhance the dissolution of the metal anodes, reducing their life time. Nevertheless, they have proven durability in the chore-alkali industry (Trasatti, 2000).

Filter cloth is primarily selected for filtration, strength, and cake discharge properties. The material must stand temperature increasing and pH variation from 1 at the anode side to 12 at cathode side due to water electrolysis. Furthermore, it is advisable to select low thickness filter cloth or manufactured with electrically conductive material to reduce its electrical resistance and then the energy consumption (Saveyn et al., 2006b; Yu et al., 2010; Citeau et al., 2012b). Figure 4 represents a conventional dead end electrofilter suitable to treat filter cakes made of "wastes" but for valuable products, it is necessary to remove the electrolytic products generated at the electrodes otherwise contaminating the filter cake (Larue and Vorobiev, 2004; Citeau et al., 2012a). This can be done in an electrofiltration system with flushing of electrodes.

The first pressurized electroosmotic dehydrator of filter press type (SUPER FILTRON®) has been developed by Shinko-Pantec Co., Ltd and has been adopting in several sewage treatment plants since 1989 (Kondoh and Hiraoka, 1990, 1993; Kondo et al., 199[illegible]). Results showed that D.C. voltage of about 40 V applied to the electrodes increase the final dry solid content of more than 40% w/w DS for an energy consumption of about 0.33–0.52 kWh/kg DS.

2.2.2.2 Electrofilter with Flushing of Electrodes

Figure 5 presents an improved dead end electrofilter using an electrode flushing solution (Hofmann et al., 2006). Compared to the standard electro-filter plate, the compartment between the filter cloth and the electrode is

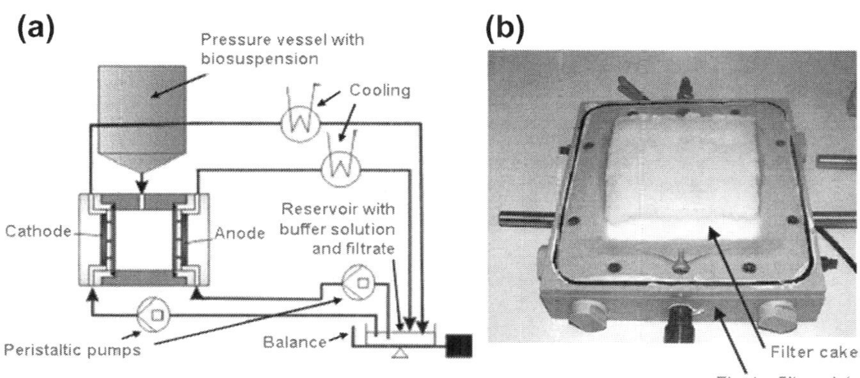

Figure 5 (a) Electrofiltration system with washing of electrodes. *(Käppler and Posten (2007). With permission.)* (b) Photo of the pilot-scale electrofilter plate after electrofiltration of xanthan. *(Hofmann et al. (2006). With permission.)*

flushed by a salt or a buffer solution. Thus, the flushing solution stream rinses out the electrolysis products, preventing their accumulation within the filter chamber. Moreover, it dissipates the Ohmic heat generated by the electrical current out of the filter chamber. The filtrate and the flushing solution from both sides are collected together keeping pH value constant.

Flushing solutions are selected with the same electrolytes present in the filter cake to avoid its contamination. For instance, Hofmann and Posten (2003) and Hofmann et al. (2006) used sodium sulphate solution or phosphate buffer for xanthan polysaccharide suspension and binary biopolymer mixture (xanthan polysaccharide and bovine serum albumin protein). Larue et al. (2006) used as flushing solution, a mix of sodium sulphate salt, sodium hydroxide, and active carbon for bentonite suspension. The active carbon was there to remove the sulphates which could lead to contamination of the filter cake by ion electromigration. Gözke and Posten (2010) used a Tris–HCl buffer for polyester (poly(3-hydroxybutyrate)) and polysaccharide (chitosan and hyaluronic acid) suspensions.

Studies have shown the efficiency of this system for the purification and the isolation of biological particles. Gözke and Posten (2010) suggest that the dead end electrofiltration with flushing of electrodes if it could be used in biotechnology could replace the combination of several steps. Thus, they have compared technological steps carried out for the biopolymer purification in conventional schemes and in the novel approach by electrofiltration. For example in the chitosan production, electrofiltration could be applied instead of precipitation, washing and drying steps, resulting in the processing time reduction and the energy consumption reduction of 64–91% in comparison with precipitation method (Gözke, 2012). For the hyaluronic acid purification, electrofiltration could be applied instead of acid extraction, avoiding furthermore a structural modification of the molecule by organic solvents. And for poly(3-hydroxybutyrate) purification, electrofiltration could be applied instead of centrifugation, freeze drying, extraction and recrystallization steps. This technology has been patented (Patent WO/2002/051874).

2.2.2.3 Dorr-Oliver® Electrofilter

Industrial dead end electrofilter equipment is the Dorr-Oliver Electrofilter® (Freeman, 1979; Klinkowski, 1986). Figure 6(a)and (b) presents the structure of the filter chamber and the anode and cathode assemblies, respectively. The anode side comprises an electrode assembly formed of an ion-exchange membrane shell coated inside with a metal composite layer and

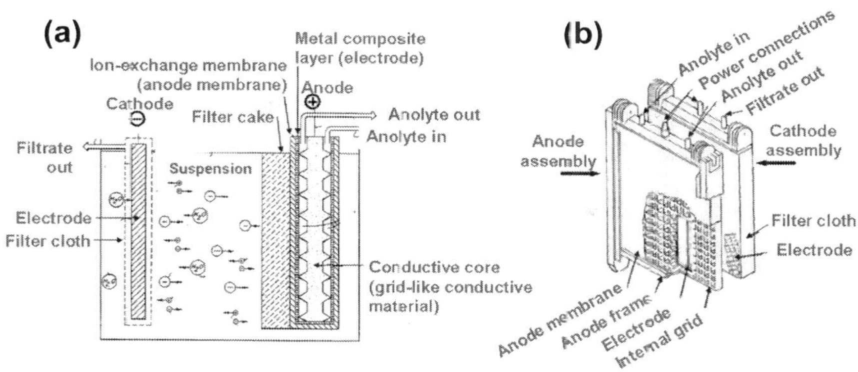

Figure 6 (a) Principle of the Dorr-Oliver Electrofilter (Klinkowski, 1986) and (b) structure of anode and cathode assemblies (Bollinger and Adams 1984).

structurally supported by a conductive core material (or grid-like conductive material). The ion-exchange membrane is substantially impermeable to the liquid flow, it serves to selectively transport cations from the hydrolysis reaction and prevents the entry of small solid particles into the inner compartment of the electrode assembly. It is preferentially made of sulphuric acid or carboxylic cation exchange resin. It is structured into "U" shape on the conductive core. This core serves first to conduct and distribute electrical current to the metal composite layer coated the ion-exchange membrane. As it is principally composed of graphite, the metal composite layer coating is a nonconsumable metal, chosen to catalyze the water hydrolysis and to protect graphite from anodic oxidation reaction. The membrane delimits an electrolyte compartment, where an anolyte solution may be fed-in, flowing through the core. It avoids then concentration polarization, also the build-up of stagnant gas, liquid pockets and solid deposits (such as sodium hydroxide).

This electrode assembly is used for collecting the solid particles. It may be used either as an anode or a cathode, depending on the surface charge of particles. For particles which have a negative charge, the cathode assembly is basically composed of a grid-like electrode covered by a filter cloth which delimits the filtrate chamber.

The electrode assemblies are vertical and extend laterally across the electrofilter cell to define the treatment zone. The entire electrofilter consists of a series of electrofilter cells. The suspension is fed parallel to the filter cloth. The liquid is transported from the filtration chamber through the cathode by the combined forces of vacuum filtration and electric field. Periodically, the anode assemblies are removed from the electrofilter cell. A cake removal device strips the solid from the outside surfaces of the membrane and carries it away from the apparatus. This system was used in particular to concentrate clay suspension such as kaolin and latex polyvinyl chloride. For those products sold as a dry powder or as concentrated slurry, traditional industry leans on the thermal drying or on mechanical dewatering of a flocculated suspension then redispersion which can impair product quality. The electrofiltration provides a better suspension dewatering for energy consumption of 75% and 94% lower than the thermal dewatering for kaolin and latex polyvinyl chloride, respectively (Bollinger and Adams, 1984). Although electrofiltration cannot product a dry powder, it allows reducing the portion of the thermal drying in the process.

2.2.2.4 Belt Filter-Press for Electrofiltration or Electroosmotic Dewatering

Various industrial belt filter-press configurations adapted for electrokinetic dewatering have been proposed (Porta and Kulhanek, 1983; Diaz, 1989; Tije, 1999; Miller et al., 2005; Jones and Lamont-Black, 2012; Smollen and Kafaar, 1994; Raats et al., 2002; Glendening et al., 2007). Figure 7 shows an example of belt electrofilter-press device made up of a gravity-driven thickening belt and a conventional belt filter-press (Raats et al., 2002). This technology allows combining a mechanical compression, an electric field force and a shear stress caused by the drums in the expression zone. Basically, the endless filter belt includes metallic elements to be electrically conductive and acting as a cathode. In the expression zone, Figure 7,

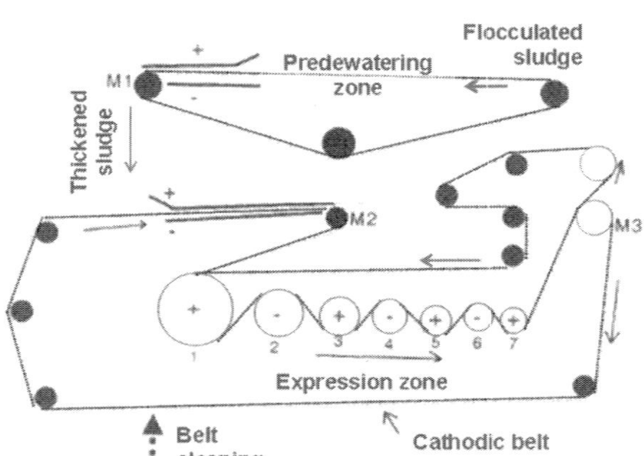

Figure 7 Belt filter-press equipped by electrodes in the thickening zone, before the expression zone and in the expression zone. *Raats et al. (2002). With permission.*

the cathodic belt is in direct contact with the cathodic drums (2, 4, and 6). The other drums (1, 3, 5, and 7) are used as anodes. A second filter belt separates the cathodic belt and the anodic drums. In operation, the suspension is pre-dewatered by gravity through an underlying conveyor. Then, it is compressed between the two moving belts; the water is squeezed from the suspension through the porous belt (cathodic belt) to a drain. Thanks to the solid migration towards the anode side (electrophoresis), the filter fouling is markedly reduced at cathode. In this equipment example, the gravity-driven thickening belt and the belt were also equipped with added electrokinetic facilities thanks to electrodes placed as a static electrokinetic zone before the rollers M1 and M2.

The development of new conductive filter belts (Tije, 1999; Miller et al., 2005; Jones and Lamont-Black, 2012) allows presently using both belts as electrodes without risk of electrical short circuiting when there is no sludge between the belts. The direct contact between the electrodes belt and the dewatering suspension reduces considerably the energy consumption. This equipment composed of 12 drums, is commercialized by the Electrokinetic Limited Company, UK.

An alternative technology, CINETIK® Linear Electro-Dewatering (OVIVO Company, USA), was commercialized and tested in a municipal wastewater treatment plant in 2009 (OVIVO). It consists of a linear filter belt. Perforated cathode is below the filter belt and multiple plates positioned at the topside are used as anodes. The suspension comes from a conventional mechanical dewatering system, then the linear filter belt moves it along the plates that apply pressure and D.C. current. As the cake moves through the belt filter, the gap between anode and cathode narrows, allowing to limit voids in the cake due to the water release. Tests carried out in wastewater treatment plants have demonstrated that the CINETIK® unit reduces the moisture content from 50% to 70% for biologic municipal sludge. The payback period was then estimated at 3–4 years.

2.2.2.5 Pressure-Drum Filter for Electroosmotic Dewatering

The Figure 8 presents a pressure-drum type electroosmotic dewaterer (Yamaguch et al., 1993). It consists of a rotating drum carrying an anode electrode and a metal caterpillar like press conveyor used as cathode electrode which is superposed on an endless filter belt. Pre-dewatered suspension is feed-in between the filter belt and the rotary drum. It is then subjected to an electric field and mechanical compression as the gap between the electrodes gradually decreases. Thus, liquid is removed from the cake

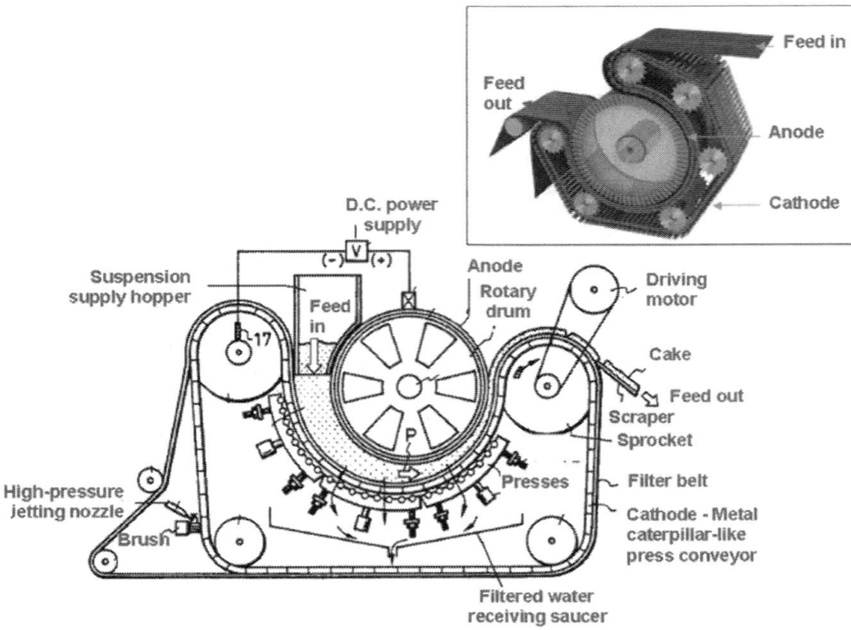

Figure 8 Pressure-drum type electroosmotic dewaterer (Yamaguchi et al., 1993) and photo of the ELODE® system (ACE Korea Incorporation Company, Korea).

and discharged towards the cathode through the filter cloth belt. The final dewatered filter cake is removed from the belt by scraper blade. On this concept, a system named ELODE® (ACE Korea Incorporation Company, Korea) was commercialized in 2005. It is adapted in three models: a model (SELO) installed at the end of an existing dewatering system, such as a press, decanter or gravity dehydrator, or two other models integrating a pre-dewatering system by gravity (GELO) and by filtration in a belt-press built-in machine (BELO). Tests carried out in wastewater treatment plants have demonstrated that the ELODE® unit reduces the moisture content from 80% to 60% for biologic municipal sludge and from 85% to 47% for chemistry paper mill sludge, achieving 48–47% annual cost-saving in comparison to dewatering process by belt-press.

2.2.2.6 Depth Bed Filter for Electrofiltration

A depth bed filtration system assisted by electric field was tested at pilot-scale in a drinking water plant (Li et al., 2009). The device consists of a filtration column filled with sand and gravel layers with an average grain size from 0.51 to 10 mm. This media is similar to that currently utilized in the local drinking water treatment plant. Vertical mesh electrodes are embedded in the column with 2.5 cm spacing and alternatively connected to the positive and negative output of a power supply. The particle removal coefficient was increased from 1.48 without electric field to 1.86 at 480 V/m and 34 A/m^2, with a significantly enhancement of the capture of small particles (<4 μm).

2.2.2.7 Rotating Screw-Press for Electroosmotic Dewatering

The electrically assisted rotating screw-press, presented in Figure 9, consists of a casing and conveyor screw used as anode and cathode, respectively. Only the middle part of the casing is subjected to electrification. It is insulated from the casing ends by means of spacer of an insulating material. The screw blades are made of an insulating material. It drives the solids particles deposited in the casing through the press. The dewatered suspension is collected into a container at the end of the conical portion of the casing. The container is composed of two walls. The inner wall is a flexible corrugated material, resulting in a compaction effect on the continuously transferred dewatered suspension. It is furthermore negatively charged by a connection to the D.C. power supply, repelling negatively charged particles from the wall. The supplementary separated filtrate passes into the annular

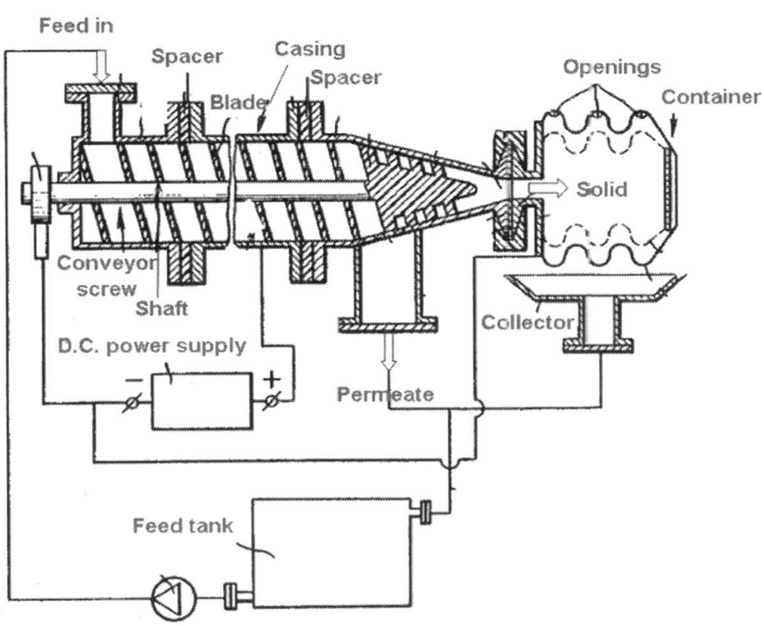

Figure 9 Concept of an electrically assisted rotating screw-press (Alekhin et al., 1982).

space between the inner and outer container wall and through the openings (Alekhin et al., 1982). This device was designed for mineral suspensions such as drilling or mining mud or slurries from the manufacture of construction material.

2.2.2.8 Electroosmotic Drain

Some electroosmotic technologies have been also developed for soil consolidation. Figure 10(a) presents a schematic illustration of electroosmotic vertical drains (or electrokinetic geosynthetic) in the soil. And Figure 10(b) illustrates the reducing of the suspension volume in a container thanks to the electroosmosis. The electrokinetic geosynthetic® (Electrokinetic Limited Company, UK) comprises conducting element coated in a corrosion-resistant material, incorporated into a geosynthetic material (Jones, 2003). The electrodes are connected in parallel to a D.C. supply.

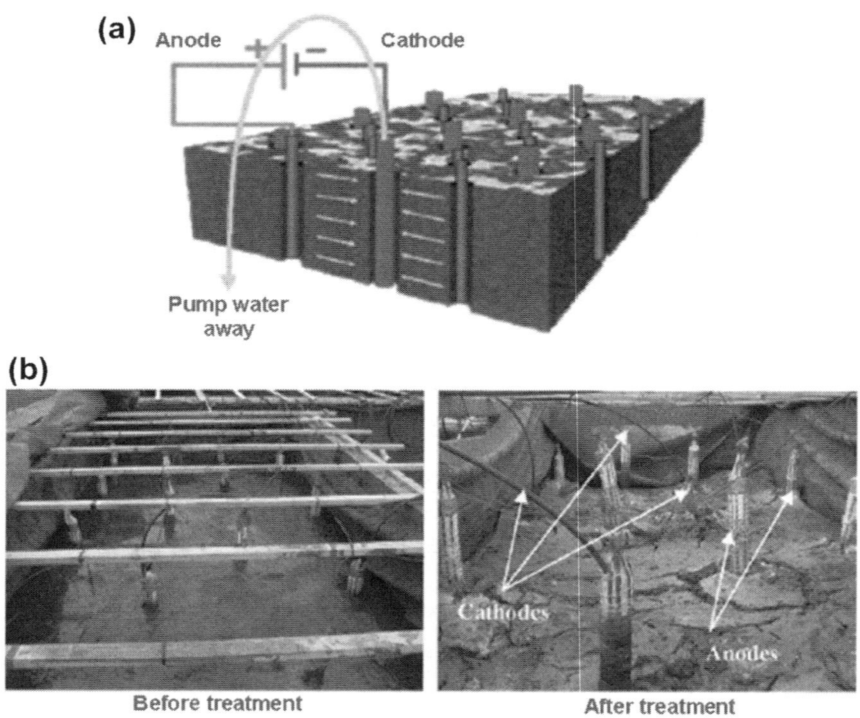

Figure 10 (a) Electroosmotic consolidation *(Jones et al. (2011). With permission.)*; (b) photos of lagooned sewage sludge before and after electroosmotic consolidation using EKG. *(Glendinning et al. (2007). With permission.)*

A pump removes water from the drains. It is estimated that such a system used for fine grained soils (tailings or sludges) allows reducing treatment time by 50–80% thanks to the electroosmosis in comparison to a draining without any electric field. This technology is commercialized under various forms such as EKG filtration bags and electrokinetic soil nails.

2.2.2.9 Electrowashing Device

The device, as shown in Figure 11, comprises two electrodes positioned on each side of the porous matrix separated by a filter cloth. A wash liquor feed reservoir completes the device. The fresh liquor flows either towards the matrix from anode to cathode as the particles initially carry a negative surface charge, Figure 11(a) or tangentially to the matrix surface. Figure 11(b). Few studies have been carried out on the electrowashing method. It was tested on titania filter cake (Tarleton et al., 2003; Kilchherr et al., 2004), on vulcanized fibre, pulp for power cable paper and condenser paper, photographic plate and film (Komagata, 1938). It was found generally that the time of washing is shortened and required quantities of water are reduced.

2.2.3 Typical Ancillaries and Fitment within a Flowsheet

The entire setup consists of a D.C. power supply, the filter module, a suspension tank (and washing liquor tank in case of electrowashing), a filtrate collecting tank and a container for the dewatered suspension. The flowsheet

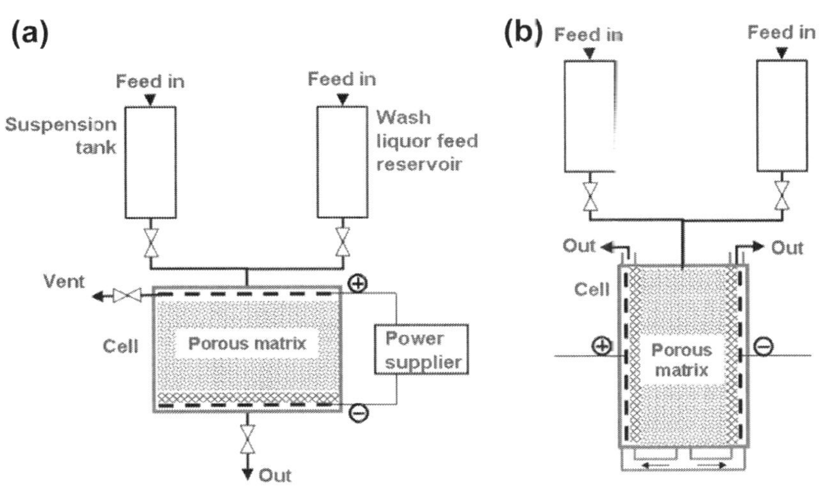

Figure 11 Schematic illustration of electrowashing cells where the fresh liquor flows (a) towards the matrix, or (b) tangentially to the matrix surface.

may previously incorporate a conditioning and thickening equipment to concentrate suspension.

To trigger off significant electrokinetic effects during electrical treatment, it is actually necessary to implement electric field intensity typically below 10 kV/m (Tarleton, 1992). If we consider here the fluid in the suspension is water, a hood and venting system are required to remove O_2 and H_2 generated by water electrolysis at the anode and the cathode, respectively. For security, these gases (combustive and combustible gas, respectively) must be separately evacuated. The temperature evolution at electrodes must be controlled.

In the case of electrically assisted filter, the suspension is fed by a pump. The filtrate from both sides may be collected together to neutralize the pH. This device works in batch fashion. When the filter cake is dewatered, the filter is opened, the cake is discharged and the filter clothes are washed before a new treatment cycle. For the continuous-feed system such as belt filter-press and pressure-drum, the conditioned suspension is first thickened by gravity or by vacuum. Then, it is dewatered in the electrokinetic module, and discharged. After the cake has left the belt, the conveyor may be brushed and cleaned with a high pressure water spray. Precipitates such as calcium hydroxide and other insoluble salts, can be removed by treating the electrodes with an acid such as 10% w/w sulphuric acid (Jones and Lamont-Black, 2012). Then, the washing liquid is sent to a suitable recovery process.

2.2.4 Factors Affecting the Treatment

Table 1 indicates the influence of the suspension properties and process parameters on the electrodewatering, electroosmotic dewatering or draining and electrowashing.

2.2.5 Typical Uses and Applications

Table 2 describes the potential applications of dead end electrofiltration and electroosmotic dewatering or draining for several industrial sectors. Investigations carried out on both dead end electrofiltration and pressurized electroosmotic dewatering show the potential of those techniques (1) to dewater (or concentrate) mineral slurry and biological suspension by accelerating the solid–liquid separation rate in comparison to conventional mechanical method. The dead end electrofiltration seems also suitable (2) to fractionate particles having same size but different surface charge. And the electroosmotic treatment may potentially be used (3) for soil

Table 1 Suspension properties and parameters affecting the electrofiltration, electroosmotic dewatering or draining and electrowashing

Factor	Comment
Suspension type	Aqueous suspension containing fine particles with high surface charge
Feed concentration	All concentrations In case of a low feed concentration, as long as electrophoresis dominates the electrokinetic phenomena, the process is frequently named electrofiltration. When the concentration increases and the particles are embedded in a porous matrix, the electroosmosis becomes the dominate electrokinetic phenomena, the process is more frequently named electroosmotic dewatering or consolidation.
Particles size distribution	Between 3 nm and 50 μm (with at least 30% of the particles below 2 μm for electrofiltration (Shang and Lo, 1997)).
pH	The pH must be away from the isoelectric point: For a zeta potential of particles close to zero, there is no electrokinetic effect.
Zeta potential of particles	The higher the zeta potential in absolute value the better. Typically $\lvert\zeta\rvert > 15-20$ mV.
Salinity	High electrolytes content compresses the electric double layer of particles leading to the zeta potential decrease in absolute value. As a result, the electrophoretic mobility of charged particle and hence the electric field effect are reduced. Furthermore, higher electrolyte concentration promotes higher electrical current, increasing the energy consumption. Initial liquid conductivity in the suspension should not be higher than 6−10 mS/cm. Low electrolyte content can also be prejudicial since it increases the electrical resistance and generated Ohmic heat. Initial liquid conductivity in the suspension should not be lower than 0.5−1 mS/cm.
Polyelectrolytes	Basically, a biological suspension can be previously flocculated using polyelectrolytes to enhance the filtration stage without damaging the electrokinetic performances (Saveyn et al., 2005). On the contrary, the polyelectrolytes use may affect the electrokinetic properties of mineral suspension, by compression of the electric double layer or neutralization of the surface charge (Lockhart, 1983b; Dussour et al., 2000).
Mechanical pressure	Mechanical pressure provides a continuous electrical contact between the electrode and filter cake. The majority of studies has demonstrated that it has a significant contribution to accelerate the flow rate and to reduce the water content in a filter cake.

(*Continued*)

Table 1 Suspension properties and parameters affecting the electrofiltration, electroosmotic dewatering or draining and electrowashing—cont'd

Factor	Comment
Electric field mode	Best current modes are constant voltage (C.V.) or constant electric current (C.C.). Alternative current has not displayed any benefit. Electrical current interruptions and electrode polarity reversals have been reported to be efficient for decreasing the total energy input and for avoiding the increase of a moisture and pH gradient inside the filter cake due to the continuously applied electric field.
Electric field intensity	Electric field intensity (in V/m) governs The electrokinetics and drives the flow rate. Electric field strength: <10 kV/m
Current density	Current density is related to the electric field intensity.
Electrochemical products generated at the electrodes	Due to water electrolysis, over the time, an acid front propagates towards the cathode for negatively charged particles (Yuan and Weng, 2003). The hydronium ions tend to be adsorbed by the particles reducing therefore their zeta potential in absolute value. When the pH narrows the isoelectric point, there is no longer any electrodewatering. But it was observed that the particles coagulate inducing a more opened cake structure which can be favourable for the mechanical dewatering (Larue and Vorobiev, 2004; Saveyn et al., 2005). The impact of the hydronium ions migration is reduced in a buffered suspension (Saveyn et al., 2005). To avoid depression of electrodewatering by the hydronium ions, a flushing solution stream can be used to rinse the electrodes from the electrochemical products. The gas formation, such as oxygen gas can be useful oxidizing impurities from suspension.
Temperature	The temperature rises by Ohmic heating in the filter chamber. It decreases the liquid viscosity which allows an acceleration of the filtrate flow. But, to prevent overheating and damage of the filter and filter cloth, the temperature control is advised. Overheating may be prevented by the additional flushing solution stream at the electrode.
Time of current application	It is not necessarily more effective to apply the current right at the beginning. Indeed, in the earlier stage the mechanical dewatering could predominate on the electrokinetic effects. Then, if the power is applied right from the start of a filtration for instance, a large energy amount is expended needlessly to remove water that can easily be removed by pressure alone (Miller et al., 1998; Lee et al., 2007). This was put in evidence in the electrodewatering of sewage sludge.
Effect of time	The electrical efficiency decreases with the moisture reduction over the time.

Table 2 Examples of mineral and biological suspensions treated by electrofiltration and/or electroosmotic dewatering

Industrial sector	Application field	Suspensions	References
Mining, drilling, and civil engineering	• Dewatering	Swelling clay (bentonite, silica, quartz sand, phosphate, diamond, iron ore, silicate, china clay, coal and coal waste, tailings lagoon)	Moulik (1971), Shang and Lo (1997), Larue et al. (2001), Weber and Stahl (2002), Larue et al. (2006), Saveyn et al. (2006b), Xian-shu et al. (2009), Fourie et al. (2007), Lamont-Black et al. (2005)
	• Refinement of concentrate	Calcium carbonate suspension and white clay slurry	Yukawa et al. (1976)
	• Consolidation	Soil	Jeon et al. (2010)
	• Soil remediation	Soil comprises petroleum hydrocarbons, heavy metals, radioactive species, and/or organic compounds (halogenated hydrocarbons, polynuclear aromatic hydrocarbons)	Acar et al. (1995), Virkutyte et al. (2002), Jeon et al. (2010)
Paper, rubber, ceramic, paint, or plastic industry	• Refinement of concentrate	Kaolin suspension, calcium carbonate suspension, titanium dioxide suspension (anatase, brookite, and rutile particles)	Lockhart (1983a), Krishnaswamy and Bahnii (1986), Vorobiev and Jany (1999), Dussour et al. (2000), Genç and Tosun (2002), Larue and Vorobiev (2004)
Food, pharmaceutical and cosmetic industry	• Concentration	Protein suspension (bovine serum albumin, lysozyme); Polysaccharide suspension (xanthan, chitosan, Hyaluranic acid, and polyester solution)	Iritani et al. (2000), Hofmann and Posten (2003), Hofmann et al. (2006), Gözke and Posten (2010)

(Continued)

Table 2 Examples of mineral and biological suspensions treated by electrofiltration and/or electroosmotic dewatering—cont'd

Industrial sector	Application field	Suspensions	References
	• Fractionation	Protein suspension (BSA, lysozyme, xanthan)	Iritani et al. (1992), Hofmann et al. (2006), Käpler and Posten (2007)
	• Dewatering	Tofu residue, tomato pomace, oily sludge, biscuit sewage sludge	Lite and Xiuqu (1999), Xia et al. (2003), Al-Asheh et al. (2004), Jumah et al. (2005), Yang et al. (2005), Li et al. (2007), Mujumdar and Yoshida (2008), Citeau et al. (2011)
Water industry	• Sewage sludge dewatering	Activated sludge, digested sludge, drinking wastewater, lagooned sewage sludge	Kondoh and Hiraoka (1990), Kondoh and Hiraoka (1993), Miller et al. (1998), Raats et al. (2002), Yuan and Weng (2003), Lamont-Black et al. (2005), Saveyn et al. (2006a,b), Lee et al. (2007), Glendinning et al. (2007), Mahmoud et al. (2011), Citeau et al. (2012b)
	• Liquid purification	Waterborne yeast cells suspended in water	Li et al. (2009)

consolidation or (4) for decontamination purposes, similarly to well-known "electroremediation" used for the depollution of soils. Recently, the cake decontamination was studied on sewage sludges that brought to light notably the removal of large quantity of dissolved organic matter content in liquid part of the sludge (Smollen and Kafaar, 1994; Kim et al., 2002; Tuan and Sillanpää, 2010; Navab Daneshmand et al., 2012). The combination of electricity, heat, and pressure disinfects the biomass, eliminating *Escherichia coli*, *salmonella*, enteric viruses, and parasites (Navab Daneshmand et al., 2012).

Tables 3 and 4 present an efficiency overview in terms of increase of the filtrate flow rate and of the final dry matter content, respectively. Depending on the experimental device, the process parameters and the treated suspension, the filtrate flow rate may increase from 1 to 37 times with the electric field strength application in comparison to the pressure alone. The energy consumption ranged from 8 to 51 kWh/m^3 of water removed (Hofmann et al., 2006; Larue and Vorobiev, 2004). Investigations reveal that for large and compressible biological particles (e.g., particles that compose industrial and domestic sludges), the electrical efficiency is more pronounced in the consolidation stage than in the filtration stage. Indeed, in the earlier stage the mechanical pressure effect may predominate on the electrophoresis. And if power is applied right from the start of the filtration, a large energy amount is expended needlessly to remove water that can easily be removed by pressure alone (Miller et al., 1998; Lee et al., 2007). A final dry solid content of 30–50% w/w may be obtained by superimposing an electric field on a standard consolidation process. It corresponds to a water content reduction of 10–57% in comparison to conventional mechanical process. The energy consumption ranged from 10 to 700 kWh/m^3 of water removed (Zhou et al., 2001; Larue et al., 2001; Yuan and Weng, 2003; Glendinning et al., 2010; Citeau et al., 2012a). As a comparison, thermal drying processes require 1000–1200 kWh/m^3 of water removed (Mahmoud et al., 2010). For sewage sludge, process economic assessments have shown the benefit of this technique in comparison to traditional treatment ways. Taking account of transport, disposal, and supplementary electrical costs, the total cost savings are estimated to 47–64% at industrial scale depending on the sludge type: agro-industrial, chemistry paper mill, biochemical semiconductor, or municipal sludge (ACE Korea Incorporation).

Table 3 Overview of dead end electrofiltration efficiency (electric field applied before the consolidation of the material to increase the filtration rate)

Suspension	Process parameters	Dewatering efficiency	References
Kaolin clay Rutile suspension	Lab. scale in Dorr-Oliver Electrofilter® type; $\Delta P_H = 6.9 \times 10^3$ Pa; 25–100 A/m^2	Filtrate flow rate increased by 7–36.5-fold for kaolin clay and by 1.2–1.9-fold for rutile suspension in comparison to the pressure alone.	Krishnaswamy and Bahnii (1986)
Anatase suspension (TiO_2) (1% w/w DS; 0.3 μm)	Lab. scale; $\Delta P_H = 1.7 \times 10^5$ Pa; 200 V (2.5 kV/m); ~6 A/m^2; ~16 min	Increase of the filtrate volume for 1000 s: 53%; EC: 1.87 kWh/m^3 removed water	Genç and Tosun (2002)
Kaolin clay (13.7 μm; pH 9; ζ: 8 mV)	Lab. scale in vertical filter cell type; $\Delta P_H = 1.2 \times 10^5$ Pa; 50 A/m^2 (~0.58 kV/m) to 100 A/m^2 (~0.85 kV/m)	Filtrate flow rate increased by 1.2–2.2-fold in comparison to the pressure alone; EC: 32 to 51 kWh/m^3 removed water	Larue and Vorobiev (2004)
Pre-sedimentation basin water ($C_i = 42–55 \times 10^3$ particles/mL)	Pilot-sand filter; 0.84 mm/s; $E = 0.32–0.48$ kV m (8–12 V)	Total collected cell factor: 1.79 to 1.86 at 8–12 V compared to 1.48 without electric field; EC: 100 to 273 Wh/m^3 removed water at 8 and 12 V, respectively	Li et al. (2009)
Raw bentonitic drilling sludge (23.4% w/w DS; 3 μm; pH 9.3)	Lab. scale in vertical filter cell type; $\Delta P_H = 5 \times 10^5$ Pa; 80 A/m^2	Filtrate flow rate increase by 5-fold in comparison to the pressure alone; EC: 340 kJ/kg^3 removed water at 64% w/w DS (95 kWh/m^3 removed water)	Loginov et al. (2013)

Xanthan ($C_i = 5$ kg/m^3; 1.2 kDa; pH 5.9; ζ: 10^4 mV)	Lab. scale in filter-press cell type with flushing electrodes; $\Delta P_H = 4 \times 10^5$ Pa; $E = 4$ kV/m	Filtrate flow rate increased by 23-fold in comparison to stirred filter cell; $C_{ED\text{-}f} = 175$ kg/m^3; EC: 44.4 kWh/m^3 removed water	Hofmann and Posten (2003)
Poly(3-Hydroxybutyrate) ($C_i = 2$ kg/m^3; pH 9; ζ: 84 mV)	Lab. scale in filter-press cell type with flushing electrodes; $\Delta P_H = 4 \times 10^5$ Pa; $E = 2–4$ kV/m; 1.5 h	Filtrate flow rate increased by 2.1–3.8-fold in comparison to the pressure alone; $C_{MD\text{-}f} = 9$ kg/m^3; $C_{ED\text{-}f} = 30–41$ kg/m^3	Gözke and Posten (2010)
Chitosan ($C_i = 2$ kg/m^3; pH 3; ζ: +65 mV)	Lab. scale in filter-press cell type with flushing electrodes; $\Delta P_H = 4 \times 10^5$ Pa; $E = 2–4$ kV/m; 3 h	Filtrate flow rate increased by 1.6 to 1.9– to3.1-fold in comparison to the pressure alone; $C_{MD\text{-}f} = 32–56$ kg/m^3; $C_{ED\text{-}f} = 60–100$ to 78–168 kg/m^3	
Hyaluronic acid ($C_i = 2$ kg/m^3; pH 6; ζ: 22 mV)		Filtrate flow rate increased by 2.7–3.6-fold in comparison to the pressure alone; $C_{MD\text{-}f} = 10$ kg/m^3; $C_{ED\text{-}f} = 30–36$ kg/m^3	

$C_{MD\text{-}f}$: final concentration after mechanical dewatering; $C_{ED\text{-}f}$: final concentration after electrodewatering; EC: energy consumption; cells removal factor = cells removal with electric field/cells removal without electric field; flow rate increase factor = filtrate flow rate with EK/filtrate flow rate without EK.

Table 4 Overview of dead end electroosmotic dewatering efficiency (electric field applied during the consolidation of the material to increase the consolidation rate and dewatering)

Suspension	Process parameters	Dewatering efficiency	References
Mineral sand tailings	Electroosmotic drain pilot; gravity pressure; $E = 11$ V/m; 63 days	Water content reduction of 0.38–0.47 times compared with initial suspension amount; EC: 1.25 kWh/m^3 removed water or 0.9 kWh/t DS	Fourie et al. (2007)
Thickened kimberlite slimes	Vertical lab. piston cell; $\Delta P_H = 70 \times 10^3$ Pa; 35 min	Water content reduction of 57% compared with initial condition	Lamont-Black et al. (2005)
Silica suspension (2–2.2 mS/cm)	Horizontal lab. piston cell; $\Delta P_H = 5 \times 10^5$ Pa; 400 A/m^2; <100 V; 140 s	Water content reduction of 20% in comparison to pressure alone; EC: 270 kWh/m^3 removed water (33.4% w/w DS)	Larue et al. (2001)
Bentonite suspension (8.5% w/w DS, 6 mS/cm)	Horizontal lab. piston cell with continuous removal of electrolysis products; $\Delta P_H = 15 \times 10^5$ Pa; 47 A/m^2; 5000 s	Water content reduction of 33% in comparison to pressure alone; EC: 700 kWh/m^3 removed water (40% w/w DS)	Larue et al. (2006)
Activated sludges (3% w/w DS; 1.0–1.7 mS/cm; pH 6–7)	Vertical lab. Piston cell; $\Delta P_H = 3 \times 10^5$ Pa; 100 A/m^2; <100 V; 1 h	Water content reduction of 74–80% in comparison to pressure alone; EC: 2142–2978 kWh/t DS (40% w/w DS)	Miller et al. (1998)

Anaerobically digested sludges (3% w/w DS; 3.6–8.0 mS/cm; pH 7–8)		Water content reduction of 53–70% in comparison to pressure alone; EC: 1900–2538 kWh/t DS (40% w/w DS)	
Sewage sludge (3% w/w DS)	Filter-press pilot; $\Delta P_H = 3 \times 105$ Pa; 100 A/m^2; <100 V	Water content reduction of 82% in comparison to pressure alone; EC: 880 kWh/t DS (40% w/w DS)	
	Belt filter-press pilot; $\Delta P_H = 3 \times 10^5$ Pa; 100 A/m^2; <100 V	Water content reduction of 75% in comparison to pressure alone; EC: 1615 kWh/t DS (40% w/w DS)	
Drinking wastewater	Belt filter-press pilot; 30 V, 35–25 A/m^2; 4 m^3/h	Water content reduction of 8% in comparison to pressure alone; EC: 1 kWh/t sludge or 60 KWh/t DS (24% w/w DS)	Raats et al. (2002)
Flocculated activated sludge	Diaphragm filer-press pilot; $\Delta P_H = 16 \times 10^5$ Pa; 30 min	Water content reduction of 21–30% in comparison to pressure alone; EC: 222–377 kWh/m^3 removed water or 352–569 kWh/t DS (42.0–45.6% w/w DS)	Saveyn et al., 2006a, b
	Diaphragm filer-press pilot; $\Delta P_H = 16 \times 10^5$ Pa; 50 min	Water content reduction of 24–33% in comparison to pressure alone; EC: 313–364 kWh/m^3 removed water or 528–654 kWh/t DS (43.0–52.5% w/w DS)	

(*Continued*)

Table 4 Overview of dead end electroosmotic dewatering efficiency (electric field applied during the consolidation of the material to increase the consolidation rate and dewatering)—cont'd

Suspension	Process parameters	Dewatering efficiency	References
Sewage sludge	Filter-press pilot; $\Delta P_H = 5.88 \times 10^5$ Pa; $E = 70 \times 10^2$ V/m; 150 min	Water content reduction of 31% in comparison to pressure alone; EC: 370–450 kWh/t DS (45% w/w DS)	Lee et al. (2007)
Lagooned sewage sludge (10.6% w/w DS)	Electroosmotic drain pilot; gravity; $E = 33$ V/m; 63 days	A reduction in volume of 23–30% compared with initial suspension volume; EC: 128 kWh/m^3 wet sludge	Glendinning et al. (2007)
Activated sludge and primary sludge mixture (2.5% w/w DS)	Belt filter press pilot; 14 m^3/h; $E = 1700$ V/m; 9 min	Water content reduction of 10% in comparison to pressure alone; EC: 19.50 kWh/t DS (30% w/w DS)	
Flocculated activated sludge (0.3% w/w DS; 0.8 mS/cm; pH 7.7)	Horizontal lab. Piston cell; $\Delta P_H = 5 \times 10^5$ Pa; 40 A/m^2; 1 h	Water content reduction of 16% in comparison to pressure alone; EC: 250 kWh/m^3 supplementary removed water (32% w/w DS)	Citeau et al. (2012b)
Flocculated digested sludge (3.0% w/w DS; 8.2 mS/cm; pH 7.6)	Horizontal lab. Piston cell; $\Delta P_H = 5 \times 10^5$ Pa; 40 A/m^2; 2 h	Water content reduction of 18% in comparison to pressure alone; EC: 720 kWh/m^3 supplementary removed water (47% w/w DS)	

Water content reduction = |(final water content without EK − final water content with EK)|/final water content without EK; EC: energy consumption; DS: dry sludge.

2.3 Crossflow Electrofiltration

2.3.1 Mechanism

2.3.1.1 Crossflow Electrofiltration

Figure 12(a) and (b) presents the crossflow filtration mechanism without and with an external electric field, respectively. In conventional crossflow filtration, the suspension flows parallel to the membrane, and the permeate flows normally through the membrane ought to the transmembrane pressure. Solutes are also transported by convection towards the membrane surface, where they form a concentration polarization layer and gel layer as shown in Figure 12(a). In crossflow electrofiltration (CFEF), the electric field is superimposed orthogonally to the hydrodynamic flow. Critical electric field strength (E_{cr}) is the electric field at which hydrodynamic transport of particle towards the membrane is counter-balanced by the electrokinetic transport of foulants away from the membrane (Huotari et al., 1999a). It depends on the suspension properties (pH, conductivity, particle size, feed concentration). The E_{cr} value can be estimated as the ratio of the permeate flux of the solvent (J) and the electrophoretic mobility of the particle (u_e):

$$E_{cr} = \frac{J}{u_e} \tag{7}$$

Above E_{cr}, the particles velocity due to the electrophoresis is greater than the one due to the bulk flow towards the filtration medium. As a result, the charged particles move away from the membrane surface and migrate towards the anode. The concentration polarization layer and the membrane fouling by deposits are reduced in comparison to the crossflow filtration without electric field, as shown in Figure 12(b). Simultaneously, the

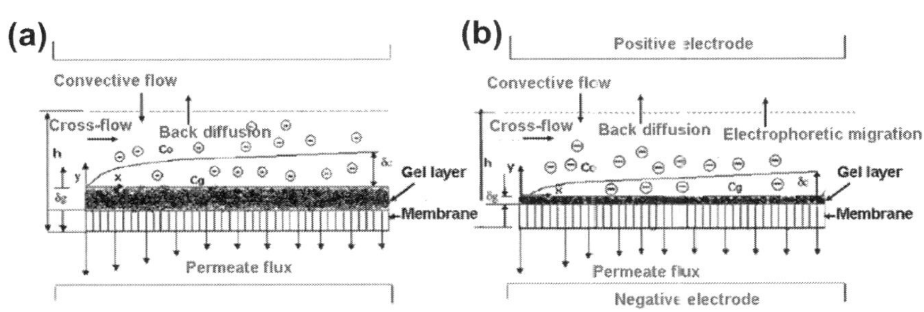

Figure 12 Schematic representation of crossflow filtration mechanism (a) without electric field and (b) in presence of an external electric field. *Sarkar et al. (2008). With permission.*

electroosmosis movement through the membrane increases the permeation flux. The shear force limits the filter deposit layer formation on the anode assembly. The service life of the membrane is thus greatly extended. Nevertheless, periodic backwashings are still required but less frequent than conventional crossflow filtration (Yang et al., 2003; Tsai et al., 2011).

2.3.1.2 Pulsed Electrokinetic Cleaning of Membrane

Electric field can also be used to periodically clean the membrane in conventional crossflow filtration. The electric field is applied for short time intervals after a period of filtration, releasing the fouling layer from the membrane surface (Bowen and Sabuni, 1992).

2.3.2 Equipments

2.3.2.1 Crossflow Filtration Equipped by Electrodes Parallel to the Flow

Figure 13(a) and (b) shows rectangular- or tubular-channel processing modules used for crossflow electrofiltration. Basically, the two electrodes are positioned parallel to the flow behind the filter membranes. As particles are generally negatively charged, cathode is placed on the permeate side. The feed channel height is typically 1–2 mm to limit energy consumption. To prevent anodic oxidation, Dimensionally Stable Anode (DSA®) is preferentially used as anode. And the cathode material is preferentially in stainless steel that is cheaper than coated titanium.

The electric field application allows separating particle sizes smaller than the membrane pore size. Thus, the membrane pore size may be chosen larger than the particle size to improve the permeate flux. Micro- and ultrafiltration membranes are typically used. The membrane must stand the temperature increase and pH variation. To avoid the contamination of suspension or filtrate by the electrolysis products, the anode is generally

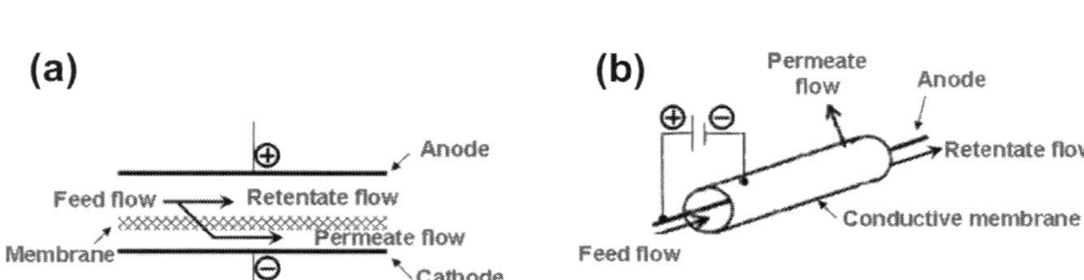

Figure 13 Schematic illustration of crossflow electrofiltration module when (a) an electric field is applied across a flat sheet membrane in a rectangular or tubular- channel processing module and (b) a tubular membrane is used as an electrode. *Huotari et al. (1999a). With permission.*

shielded by an ion exchange membrane or a dialysis membrane and can be rinsed with an electrolyte (Weigert et al., 1999). Filtration membranes made of conductive material, such as carbon fibre-carbon composite membrane (Huotari et al., 1999b), stainless-steel membrane (Bowen et al., 1989) or conductive ceramic membrane (Guizard et al., 1989), were investigated as cathode to replace the conventional nonconductive membranes. It was suggested that the gas bubbles generated at the electrode, when it is used as a membrane, disperses the deposits formed on membrane reducing the fouling effect (Bowen et al., 1989). However, in other case, it was demonstrated that the formation of foam on the membrane surface due to gas bubbles at cathode may cause the filtration flux to decrease (Huotari et al., 1999b). Thus, some studies have shown that the electrode could not be placed closer than 3 mm from the membrane surface in order to carry away the gas bubbles in the filtrate flow (Lin et al., 2007).

Figure 14 is a schematic representation of the Dorr-Oliver® crossflow electrofiltration device (Bollinger and Klinkowski, 1986). It is a rectangular-channel system alternating anode and cathode assemblies spaced by feed-product channels. The anode assembly consists of an anode immersed in an electrolyte stream and covered by an ion exchange

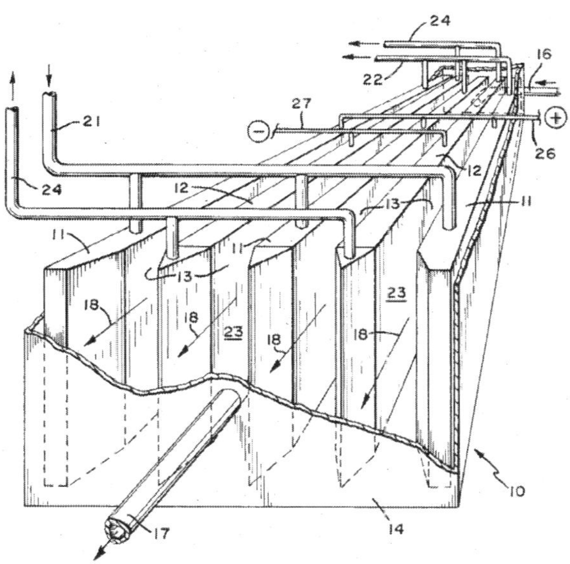

11. Anode assemblies
12. Cathode assemblies
13. Feed-product channels.
14. Housing
16. Feed inlet
17. Retentate outlet
18. Direction of flow
21. Anolyte inlet conduit
22. Anolyte outlet conduit
23. Filter cloth
24. Filtrate withdrawal conduit
26. Electrical conductors
27. Electrical conductors

Figure 14 Schematic representation of the Dorr-Oliver® crossflow electrofiltration device (Bollinger and Klinkowski, 1986).

membrane to prevent back migration or electroosmotic pumping of the electrolyte through the membrane. The anolyte flux (Na_2SO_4 solution) carries away the electrolysis products. The cathode assembly consists of an electrode separated from the feed channel by a membrane made from polymer or ceramic or stainless steel or anisotropic cellulosic type. Results on a kaolin suspension show that the application of a current density of about 50 A/ft^2 (538 A/m^2) in combination with permeate vacuum of 28 inch Hg (9.3×10^4 Pa) produces a permeate rate around 4 times greater than without any current.

2.3.2.2 Crossflow Filtration Equipped by Electrodes Perpendicular to the Flow

Figure 15 shows a schematic diagram of the ceramic ultrafiltration module assisted by electric field. Contrarily to common crossflow electrofiltration devices previously presented, the electrodes are positioned perpendicular to the flow at the inlet and outlet ports of the membrane module. The flow is directed either from anode to cathode for positively charged particle (as illustrated in Figure 15) or from cathode to anode for negatively charged particle The correct combination of applied electric field and pressure allows overcoming the charge repulsion force experienced by the charged particles near the electrode prior to entering the membrane module. Once the particles are in the module, they are attracted by the opposite charged electrode at the outlet side. Deposition into the electrode can be inhibited thanks to a relatively larger velocity drag force. The membrane is made of ceramic with a zirconium dioxide (ZrO_2) active layer. This gives a surface charge to the membrane either positive or negative respectively below or above the isoelectric point.

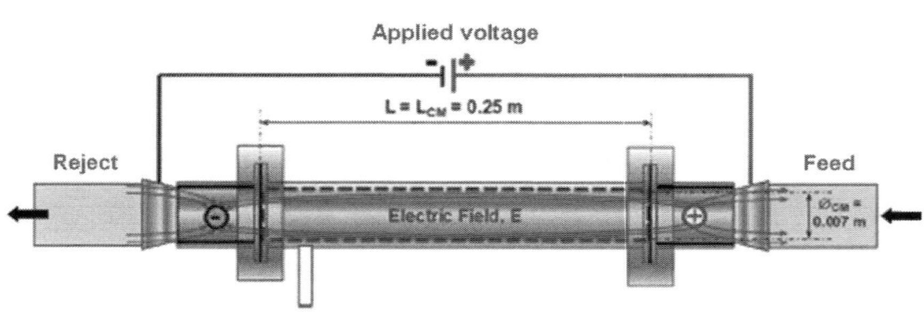

Figure 15 Schematic representation of the ceramic ultrafiltration module with an applied voltage (Agana et al., 2012).

2.3.3 Typical Ancillaries and Fitment within a Flowsheet

The suspension is pumped into the electrofilter. The differential pressure across the membrane to remove the filtrate, is achieved by pressurizing the feed-product channel hydraulically, such as with a feed pump, and/or applying vacuum to the filtrate chamber. The electric field is provided by a power supplier providing 1 to max 30 kV/m. The electrode polarity is selected in order to repel the charged particles away from the membrane surface. The permeate and the retentate are removed to two outlet conduits. The setup includes a tank for the flushing solution storage. This solution is pumped into the anode assemblies. It circulates in a closed loop. The pH and conductivity of the permeate, retentate and flushing solution must be controlled and adjusted at same pH and conductivity than the feed-suspension. Temperature control may also equip the electrodes. Furthermore a hood and venting system are required to remove O_2 and H_2.

Typically washing is made with alkaline solution (e.g.. sodium hydroxide 0.1 N), then followed by an acidic solution (citric acid 0.01 N or sulphuric acid) and finalized by Milli-Q water. The washing liquid is pumped into the feed-channels.

2.3.4 Factors Affecting the Treatment

Table 5 summarizes the influence of the suspension properties and process parameters on the crossflow electrofiltration. In addition to Table 5 information, it is worthwhile noted that particles larger than 6 μm preferentially sediment and are not affected by the electric field (Wakeman and William, 2002). The CFEF is particularly advantageous compared to conventional crossflow filtration for suspensions involving fast concentration polarization or for highly concentrated suspensions. Table 6 displays examples of suspensions which were dewatered by CFEF.

2.3.5 Typical Uses and Applications

Investigations on CFEF reported the potential of the technique: (1) to increase the concentration of suspension by preventing concentration polarization at membrane surface and membrane pore fouling; (2) to purify the permeate by removing nano-sized and colloidal particles, and (3) to fractionate particles by using two molecular weight cut-off membranes or to fractionate particles taking advantage of their different isoelectric points. Table 6 shows potential applications per industrial sectors. Some success was reported in ultra- and microfiltration of microbial cells, proteins, in

Table 5 Suspension properties and parameters affecting the crossflow electrofiltration

Factor	Comments
Suspension type	Aqueous suspension containing fine particles with high surface charge.
Feed concentration	Tested feed concentration range: <50 kg/m^3 Increasing the concentration may increase the viscosity, which in turn would decrease the electrophoretic mobility. However, the filtrate flux is still very high compared with flux without an electric field, thanks to electroosmosis.
Particle size distribution	Range: <8 μm (Wakeman and Williams, 2002)
pH	The pH must be away from the isoelectric point: For a zeta potential of particles close to zero, there is no electrokinetic effect.
Zeta potential of particles	Same as dead end electrofiltration (see Table 1). The sign of the particle surface charge required to be kept away from the membrane must be the same as the electrode in the permeate channel.
Salinity	As explained for dead end electrofiltration (see Table 1), electric field effect is highly influenced by electrolyte concentration.
Crossflow velocity	Range: 0.02–3 m/s; typically: 0.1 m/s (Wakeman and Williams, 2002) The crossflow velocity acts on the shear force that prevents the particle deposition on the anode side. However, because of an increased back diffusion towards the membrane surface, it reduces the filtrate flow (Huotari et al., 1999a). It is not necessary to use a high crossflow velocity in CFEF compared to the conventional crossflow filtration.
Transmembrane pressure	A Critical transmembrane pressure (TMP) may be defined below which the membrane fouling occurs in spite of the electric field application (Huotari et al., 1999a; Song et al., 2010). Typically the critical TMP is below 4×10^5 Pa.
Electric field mode	The use of pulsed electric field mode can minimize power consumption and prevent the changes in stream properties due to electrode reactions or heating. However, contrary results are reported regarding the flux enhancement by pulsed electric field in comparison to continuous one.
Electric field intensity	The electric field strength is preferentially closed to the critical electric field strength (Yang et al., 2003;

Table 5 Suspension properties and parameters affecting the crossflow electrofiltration—cont'd

Factor	Comments
	Tsai et al., 2011). The required electric field strength is increased by the increase of the transmembrane pressure or the membrane pore size. Electric field intensity range: 1–30 kV/m
Current density	Current density is related to the electric field intensity.
Electrochemical products generated at the electrodes	To reduce the impact of hydronium ions produced at anode on the concentrating suspension (for a high value product) a flushing solution stream can be used to rinse the electrode. The gas bubbles generated at the cathode can result either in the filtration flux decrease or in the fouling effect reducing.
Temperature	The temperature of suspension can be maintained constant by the use of a heat exchanger external to the filter. However, if a layer of particles is formed on the membrane, it is subjected to the local heating effect.

the concentration of surfactants and in the separation of molecular components like surfactant and protein. Table 7 presents an overview of the CFEF efficiency. The method is shown to increase permeate flux by factor ranging from 1.5 to 10 during experiments involving different types of solutions. The energy consumption ranges from 0.2 to 6.9 kWh/m^3 with pumping representing less than 13% of the total energy consumption (Hsieh et al., 2008; Bowen et al., 1989; Wakeman and Sabri, 1995; Weigert et al., 1999; Chiu and Garcia Garcia, 2011; Weng et al., 2012).

3. DIELECTROPHORETIC TREATMENT

3.1 Nonuniform Electric Field-Induced Effects

In uniform electric field, Coulomb forces are generated on both sides of a neutral particle. The dipole moment induced in the particle can be represented by two equal and opposite charges (δ^+ and δ^-) at the particle boundary, resulting in zero net force on the particle. However, when the electric field is nonuniform, the unequal fields operating on each side create a force, called dielectrophoretic force (DEP), directed along the electric field gradient, which impels the neutral body towards the region of stronger (positive DEP) or weaker electric field (negative DEP). The dielectrophoretic

Table 6 Examples of mineral and biological suspensions dewatered by crossflow electrofiltration

Industrial sector	Application field	Suspensions	References
Chemical industry	• Liquid purification	Wastewater containing nano and colloidal particles such as metallic compounds (oxide-CMP, alumina, arsenic, TiO_2), mineral compounds (silica dioxide, cristobalite, kaolin) and natural organic matter (humic substances)	Bollinger and Klinkowski (1986), Weigert et al. (1999), Yang et al. (2003), Yang and Li (2007), Lin et al. (2007), Tsai et al. (2011), Hakimhashemi et al. (2012)
Food and beverage industry, Pharmaceutical and cosmetic industry, Drinking water treatment plant	• Liquid purification	Fruit juice (pectin-sucrose solution), fermentation broth (microbial cells, proteins, amino acids), gelatine, surfactant, Oily waste water (oil emulsion)	Akay and Wakeman (1997), Park (2006), Sarkar et al. (2008), Bazinet and Firbaous (2009), Huotari et al. (1999b)
	• Concentration	Protein solution (albumin suspension)	Robinson et al. (1993), Zumbusch et al. (1998), Song et al. (2010), Wakeman (1998)
	• Fractionation (bio-product purification)	Protein mixture (enzyme solution, albumin, suspension, gelatin), peptide mixture	Bargeman et al. (2002), Lentsch et al. (1993), Oussedik et al. (2000), Enevoldsen et al. (2007), Chuang et al. (2008), Sarkar et al. (2009), Brisson et al. (2007)

Table 7 Overview of crossflow electrofiltration efficiency

Suspensions	Process parameters	Electric field efficiency	References
Cristobalite suspension (0.2 vol%; mean size 3 μm; 2 μS/cm; pH 7–8; ζ: 52–66 mV)	TMP: 0.1×10^5 Pa; crossflow velocity: 0.6 m/s; $E = 35 \times 10^3$ V/m	Permeate rate increased by 10-fold in comparison to the single crossflow filtration; EC: ~1.8 kWh/m^3	Weigert et al. (1999)
Groundwater containing solids (0.427 kg/m^3; size range 297–1007 μm; pH 6.6; ζ: 42 mV)	TMP: 1–2 psi (6895–13,790 Pa); Filtration rate: 5×10^{-6} m^3/s; $E = 15.65 \times 10^3$ V/m	Removal efficiency of solids increased to 95% with the electric field in comparison to 7% without electric field	Lin et al. (2007)
Colloidal alumina suspension (0.1 kg/m^3; mean size 209 nm; pH 5.6; ζ: +70 mV)	TMP: 1–2 psi (6895–13,790 Pa); Filtration rate: 5×10^{-6} m^3/s; $E = 9.68 \times 10^3$ V/m	Removal efficiency of colloids increased to 80% with the electric field in comparison to 3% without electric field	
Silica dioxide suspension (0.242 kg/m^3; size range 76–126; pH 5; ζ: 42 mV)	TMP: 1–2 psi (6895–13,790 Pa); Filtration rate: 5×10^{-6} m^3/s; $E = 18.77 \times 10^3$ V/m	Removal efficiency of colloids increased to 98% with the electric field in comparison to 3% without electric field	
Humic acid and kaolin mixtures	TMP: 0.49×10^5 Pa; $E = 6.08 \times 10^3$ V/m	Permeate rate increased by 1.7- to 3.2-fold; DOC rejection increase by 8-fold; DOC aromatic removal increase by 0.3– to0.5-fold	Tsai et al. (2011)
Aluminium oxide suspension (0.5 kg/m^3) +2 mM KNO_3 (pH 6; ζ: +50 mV)	TMP: 0.75×10^5 Pa; Velocity: 0.13 m/s; $E = 2 \times 10^3$ V/m	Permeate rate increased by 52%	Hakimhashemi et al. (2012)

(Continued)

Table 7 Overview of crossflow electrofiltration efficiency—cont'd

Suspensions	Process parameters	Electric field efficiency	References
Surfactant (10 kg/m^3; mean size 0.46 μm)	TMP: 2.04×10^5 Pa; Velocity: 1.5 m/s; $E = 16.7 \times 10^3$ V/m	Permeate rate of colloids increased by 15-fold	Akay and Wakeman (1997)
Fermentation broth	TMP: 10 kg cm^2; Velocity: 0.02 m/s; $E = 8–24$ V	Permeate rate increased by 60–133%	Park (2006)
Mixture of pectin (1–5 kg/m^3) and sucrose (10–14 brix) (57–170 μS/cm; pH 3.1–3.5; ζ: 19.4–20.5 mV)	TMP: 3.6×10^5 Pa; Velocity: 0.12 m/s; $E = 0.6 \times 10^3$ V/m	Permeate rate increased by 2.3- to 3.4-fold	Sarkar et al. (2008)
Treated mosambi fruit juice (1.5 kg/m^3 pectin; 8.60 brix; 3.20 mS/cm; pH 3.84; ζ: 24.2 mV)	TMP: 3.6×10^5 Pa; Velocity: 0.09 m/s; $E = 0.4 \times 10^3$ V/m	Permeate rate increased by 1.4-fold	
Oily waste water (0.21 kg/m^3 oil; 0.1% cleaning agent; 40 μS/cm; pH 8.8; ζ: 67 mV)	TMP: 4×10^5 Pa; $E = 2.4 \times 10^3$ V/m	Permeate rate increased by 4.7-fold	Huotari et al. (1999b)
Mixture of BSA solution (pH 7.4; ζ: 28 mV) and Lysozyme (pH 7.4; ζ: +6.7 mV) (ratio 0.1 kg/m^3: 0.1 kg/m^3; pH 7.4; ζ: 7.1 mV)	TMP: 3.6×10^5 Pa; Velocity: 12×10^{-2} m/s; $E = 1 \times 10^3$ V/m	Permeate rate increased by 1.6-fold (57%); Membrane surface concentration of BSA and lysozyme varied by 0.2 and 4.2-fold in comparison to the single crossflow filtration	Sarkar et al. (2009)

Permeate increase factor = permeate flow rate with EK/permeate flow rate without EK; solids removal increase = (final water content without EK – final water content with EK)/final water content without EK.

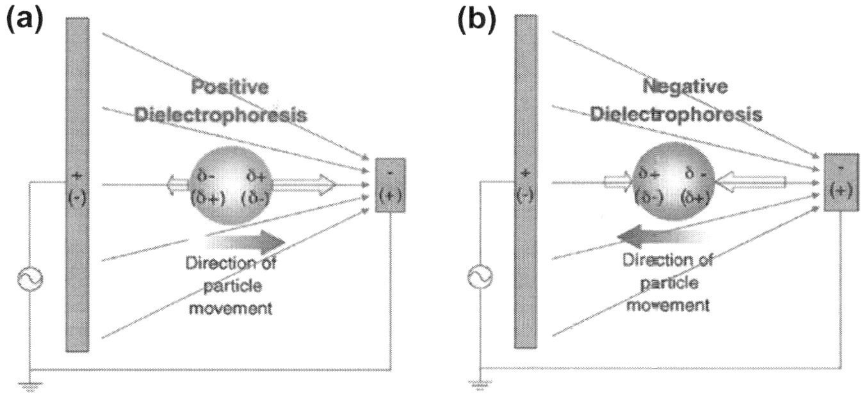

Figure 16 Direction of a neutral particle in nonuniform electric fields depending on dielectrophoresis response: (a) positive DEP and (b) negative DEP. *Doh and Cho (2005). With permission.*

motion of neutral body suspended in medium is illustrated in Figure 16, in which an A.C. signal applied between plate and tip electrodes is used to generate a nonuniform electric field.

In the case of an ideal dielectric sphere suspended in medium (here a fluid) and exposed to a nonuniform electric field E, the dielectrophoretic force $\vec{F}_{\mathrm{DEP}}$ can be written as:

$$\vec{F}_{\mathrm{DEP}} = 2\pi\varepsilon_0\varepsilon_{\mathrm{m}}r^3\mathrm{Re}\left[f_{\mathrm{CM}}\right]\nabla E^2 \tag{8}$$

where r is the radius of the spherical particle, ε_0 is the permittivity of the vacuum (8.854×10^{-12} C/V m), ε_{m} is the relative permittivity of the medium, and f_{CM} is the effective polarizability. The effective polarizability is a complex variable known as the Clausius-Mossotti factor. It depends on the electric properties of the particle and the surrounding medium and on the frequency f of the applied A.C. field. In the case of a spherical particle with permittivity ε_{p} and electrical conductivity σ_{p}, suspended in a medium with permittivity ε_{m} and conductivity σ_{m}, the effective polarizability is given by:

$$f_{\mathrm{CM}} = \frac{\varepsilon_{\mathrm{p}}^* - \varepsilon_{\mathrm{m}}^*}{\varepsilon_{\mathrm{p}}^* + 2\varepsilon_{\mathrm{m}}^*} \tag{9}$$

where ε^* is the complex permittivity, expressed as: $\varepsilon^* = \varepsilon_0\varepsilon + i\frac{\sigma}{\omega}$; with i: the imaginary vector ($i = \sqrt{-1}$) and ω: the angular frequency of the A.C. field ($\omega = 2\pi f$). The polarizability factor determines whether the DEP force on a particle is positive (attracting) or negative (repelling), Figure 16. Its value can theoretically range between −0.5 and +1.0 (Pethig and Markx, 1997; Khoshmanesh et al., 2011). From this equation, it can be seen that for low frequencies the effective polarizability mostly depends on the electrical conductivities, whereas at high frequencies, the influence of permittivity becomes dominating. This means that particles may experience either positive or negative DEP force in the same system depending on the frequency of the applied A.C. signal.

For high frequencies, or with negligible electrical conductivity of the fluid, the dielectrophoretic force increases with the difference between the particle and the fluid permittivity. When the permittivity of the particle is greater than the medium one ($\varepsilon_p > \varepsilon_m$), the particle migrates in the direction of high field region (positive DEP). Indeed, the asymmetric field forces accelerate the particle motion more than the liquid, concentrating it in the highest field (Ballantyne and Holtham, 2010). On the contrary, particles with permittivity lower than the medium one ($\varepsilon_m > \varepsilon_p$) are pushed towards the region of weakest field strength. Permittivity (also named dielectric constant or polarizability) of some liquids, minerals, ceramics and plastics are presented by Lin and Benguigui (1983) and Ballantyne and Holtham (2010).

While the real part of Clausius-Mossotti factor $\mathrm{Re}[f_{CM}]$ determines the dielectrophoretic force, the imaginary part $\mathrm{Im}[f_{CM}]$ determines travelling-wave dielectrophoretic force (TW-DEP). This force exists only when the electric field has phase gradient. The corresponding travelling-wave dielectrophoretic force is given by (Khoshmanesh et al., 2011):

$$\vec{F}_{\mathrm{TW-DEP}} = -\frac{4\pi\varepsilon_0\varepsilon_m r^3 \mathrm{Im}\left[f_{CM}\right] E^2}{\lambda_{TW}} \tag{10}$$

where λ_{TW} is the wavelength of travelling electric field of value equal to the repetitive distance between electrodes of same phase.

3.2 Mechanisms

3.2.1 Dielectrophoresis

The dielectrophoretic method consists to achieve nonuniform electric field with a high electric field gradient. Electrical density and hence electrochemical reactions are preventing by the use of high-frequency fields, nonconductive liquid or insulted electrodes. Nonuniform electric fields may be generated by variations of the A.C. electric field, as well as various microelectrode configurations or insulating structures (designed matrices or variations in the channel geometry) placed between energized electrodes.

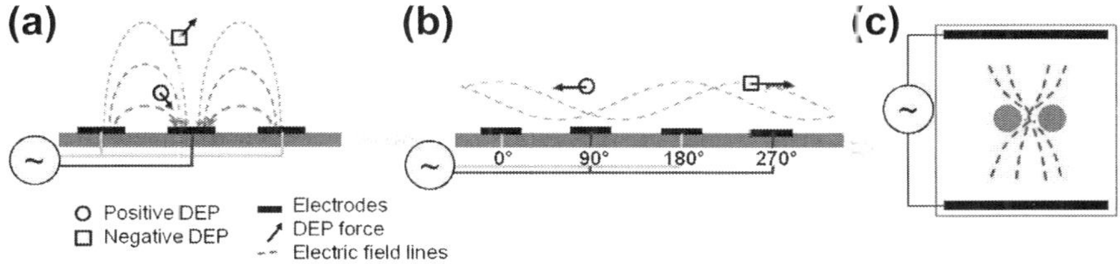

Figure 17 Nonuniform electric field distribution induced (a) by planar parallel microelectrode array for dielectrophoresis, (b) by planar parallel microelectrode array for travelling-wave dielectrophoresis *(adapted from Khoshmanesh et al. (2011). With permission.)*, or (c) by an insulating structure (consisted here of dielectric pillars) placed in an A.C. or D.C. electric field.

Figure 17 present the electric field distribution over three cell configurations. Thus, when a potential is applied across alternating electrodes, in Figure 17(a), this array establishes a nonuniform electric field depending on dimensional and material properties of the electrodes. Particles move in the direction along (positive DEP) or against (negative DEP) the electric field gradient. They are then trapped or levitated, respectively.

A similar arrangement of microelectrodes is now energized by phase-shifted A.C. signals, Figure 17(b) (e.g., a phase-quadrature signals: 0, 90, 180, and 270°). This generates a travelling-wave dielectrophoretic force, which moves the levitating particles parallel to the channel formed by the electrode array.

The dielectrophoretic force decreases approximately exponentially with distance from the electrodes (Pethig and Markx, 1997). To create high electric field gradients over a large volume, insulating structures array can also be employed between electrodes, Figure 17(c). The presence of the insulating structures (or matrix) modifies the electric field distribution, generating sites of high electric field gradients. Thus, when the dielectric constant of insulating structures is larger than the surrounding liquid one, the field is stronger near the structures (especially at their contact points) and weaker in the gaps between them.

Dielectrophoresis is commonly employed in separation or filtration systems where particle is exposed to dielectrophoretic, electric, gravitational, buoyancy and drag forces. The particle velocity induced by dielectrophoresis $\overrightarrow{v}_{\mathrm{DEP}}$ is proportional to the square of the field gradient E and the dielectrophoretic mobility of the particle u_{DEP} as (Aldaeus, 2006):

$$\overrightarrow{v}_{\mathrm{DEP}} = u_{\mathrm{DEP}} \cdot \nabla E^2 \tag{11}$$

In the more general form, the dielectrophoretic mobility may be expressed depending on the particle volume V_P and the friction factor φ, such as:

$$u_{DEP} = \frac{V_P}{4\varphi} \mathrm{Re}\left[f_{CM}\right] \tag{12}$$

The effectiveness of dielectrophoresis decreases as the suspended particle size decreases. For a spherical particle with radius r in a liquid of viscosity μ, the friction factor is given by $\varphi = 6\pi\mu r$ and the dielectrophoretic mobility is:

$$u_{DEP} = \frac{r^2}{18\mu} \mathrm{Re}\left[f_{CM}\right] \tag{13}$$

3.2.2 High-Gradient Dielectrophoretic Separation

The common dielectrophoretic separation mechanisms are based either on the sign or on the magnitude of the DEP force in a laminar flow stream. Figure 18 illustrates the basic principle of three hydrodynamic dielectrophoresis processes. Two kinds of particles are carried by the flow through the chamber, confined within a certain range in the channel. DEP force is generated in the nonuniform electric field provided by intermittent microelectrodes. Sign-based DEP separation illustrated in Figure 18(a), is achieved (1) by applying a frequency where one kind of particle is levitated by negative DEP and flushed out by the flow stream, while the other kind is immobilized on the electrodes by positive DEP. The particles levitate above the electrode surface according to their physical properties and those of the fluid. As the negative DEP force decreases with distance from electrodes and the sedimentation force acts on each particle, repelled particles are driven to an equilibrium height where the sedimentation and levitation forces are balanced. Electric field is periodically switched off (2) to release the trapped particles into a noncontaminated flow stream. Continuous separation system may be achieved using the viscous drag of feed flow.

Magnitude-based DEP separation illustrated in Figure 18(b), allows differentiating particles based on their effective polarizability and density. The frequency and the medium properties are selected in such a way that both kinds of particle are repelled under the nonuniform electric field. Depending on their density and dielectric properties, the particle groups are levitated to different heights. They move along the fluid flow without

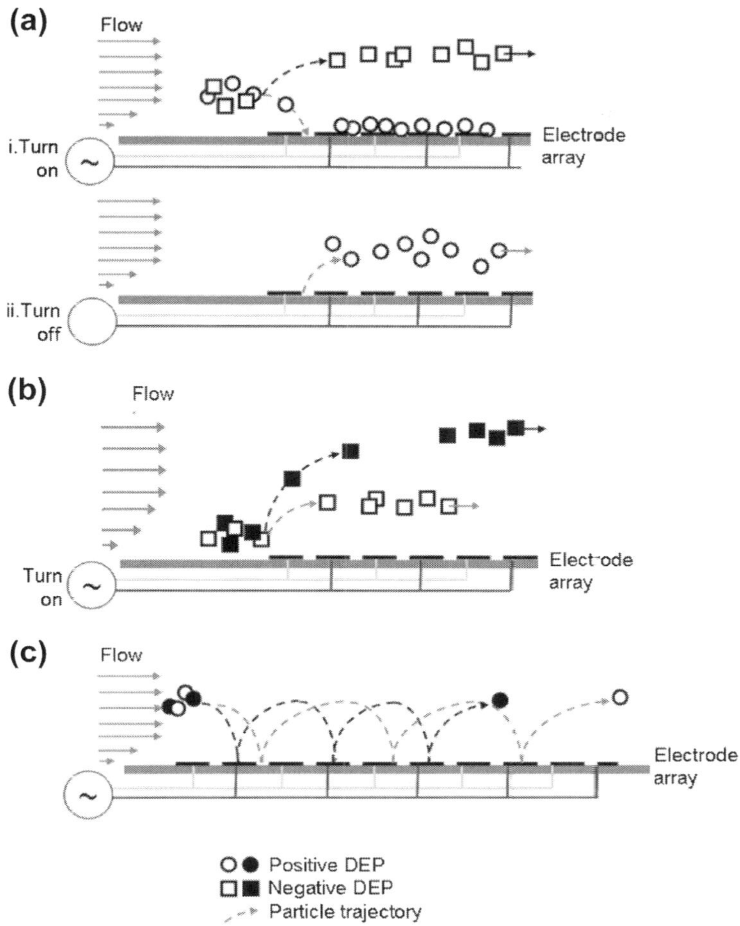

Figure 18 Separation mechanisms based (a) on the dielectrophoretic sign, (b) on the dielectrophoretic magnitude, and (c) on the dielectrophoretic trap-and-release procedure.

change of height. Combined with fluid flow with a parabolic velocity profile, those groups may achieve different velocities in the channel, improving the selective separation along the fluid path.

Figure 18(c) illustrates the trap-and-release procedure. Using first positive dielectrophoresis, two kinds of particles move towards regions of strong electric. According to their dielectrophoretic mobility in nonuniform electric field, they are trapped to different positions. Then, the voltage is switched off; particles are released and picked up by the flow stream. Release of the particles can also be accomplished by changing the A.C. frequency to a value which induces negative dielectrophoresis, bringing out further the separation. The degree of fractionation is gradually enhanced by repetition of the procedure.

As the dielectrophoresis affects particles at a small distance from the electrodes, those particle separation methods are typically used in microfluidics. Other strategies exploiting DEP force are carried out to manipulate particles in liquids by levitation, translation or orientation, and/or to discriminate them (Khoshmanesh et al., 2011; Pethig and Markx, 1997). In this chapter, the investigation subject is restricted to particle separation systems.

3.2.3 High-Gradient Dielectrophoretic Filtration

A standard way to make dielectrophoretic filtration, Figure 19, consists to inject suspension through a porosity dielectric matrix placed between energized electrodes that allow squeezing the electric field. The matrix has a dielectric constant larger than the surrounding liquid one. It may either be finely divided filamentary dielectric material or pieces or foam. It is necessary that the carrier liquid has much lower permittivity than the contaminant particle. Thus, in the nonuniform electric field, the particles are attracted and trapped on the matrix using positive dielectrophoresis. Captured particles form a layer over the matrix surface, creating a new capture site with a

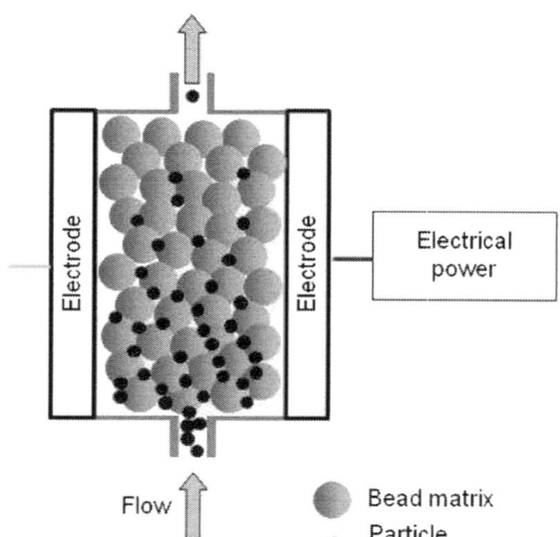

Figure 19 Schematic diagram of a high gradient dielectrophoretic separator.

different permittivity. Particles flowing through a dielectrophoretic filter are attracted and captured in it even though their average size is substantially smaller than the bed pore size. The matrix layers are successively loaded. In addition to the dielectrophoretic particles trapping, particles are also collected between the interstices of the matrix in same way as in a depth bed filtration system. The particle is then subjected to gravitational effects, dielectrophoresis and fluid drag that may release particles from the matrix. The filtration occurs when the ratio of the dielectrophoretic force to the sum of the opposing forces is greater than 1.

Another new way is to combine negative DEP with crossflow filtration to intensify membrane process. Electrodes are positioned at the membrane, separated from the feed stream. Large particles are thus, repelled from the membrane. They are flushed out by the crossflow stream while the fluid flows through the membrane. It results in the membrane fouling prevention. However, DEP is not the dominating force for small particles. Consequently, they can move towards the membrane due to permeate flow and settle on membrane, forming pearl-chain. The use of pulsed electric field allows repelling these agglomerates from the membrane (Du et al., 2009).

3.3 Equipments

3.3.1 Dielectrophoretic Separation Platforms in Laboratory-on-Chip Systems

Commonly, the cell separation chip is composed of the microchannel with an array of microelectrodes fabricated by photolithography on a flat substrate (such as glass microscope slide). Sample volumes are then, limited to microlitres. Various configurations of microelectrodes and electrodeless have been investigated to realize dieletrophoresis. Some examples are schematically shown in Figure 20(a): interdigitated electrodes, sidewall patterned electrodes, insulator-based or electrodeless system, or contactless (Khoshmanesh et al., 2011). The small dimension of the electrodes (typically 10–40 μm wide) allows the generation of high electric fields at lower voltages (Morgan et al., 2001). Usually, the applied electric field is an alternating current signal, created with a common frequency generator, delivered a frequency generally greater than 100 kHz with magnitudes below 20 V peak-to-peaks. To insure visual observation along the channel, transparent conducting materials such as indium tin oxide (ITO) are employed as electrodes (Willams, 2008). Example of fluidic system integrating a dielectrophoretic separation platform is presented in Figure 20(b).

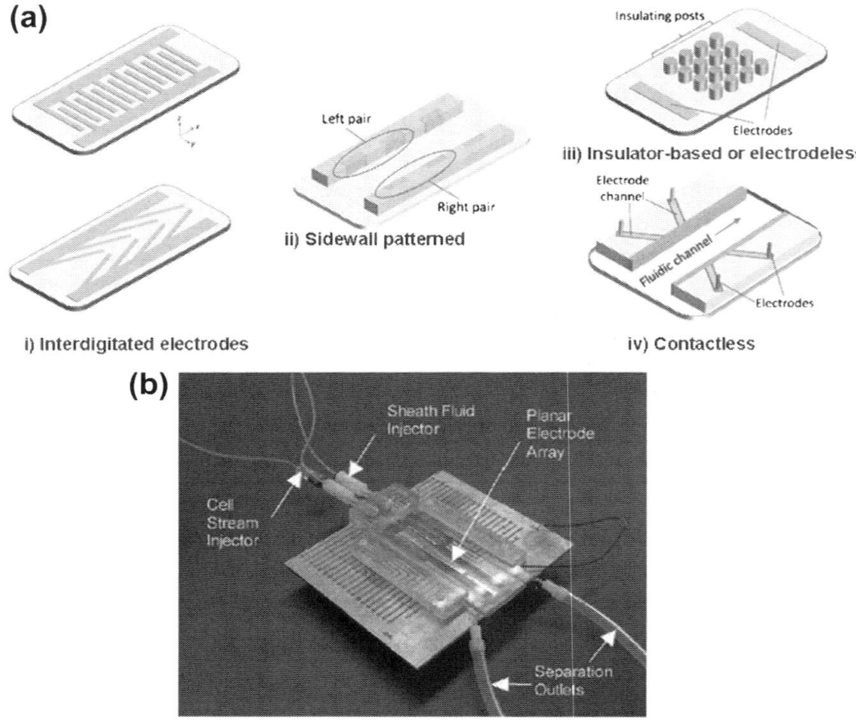

Figure 20 (a) Dieletrophoretic Lab-on-a-Chip systems for separation according to the configuration of microelectrodes (i) parallel or interdigitated electrodes, (ii) sidewall patterned electrodes, (iii) insulator-based or electrodeless system, and (iv) contactless (Khoshmanesh et al., 2011); (b) fluidic system integrating a dielectrophoretic separation platform. *Li and Kaler (2004). With permission.*

3.3.2 Wire Cloth Electrodes for Large-Scale Dielectrophoretic Separation

A cloth, Figure 21(a), weaving together conducting and insulating threads was developed to improve the large-scale dielectrophoretic separation (Abidin et al., 2007). The cloth is made from stainless steel wires (100 μm diameters) oriented at right angles to flexible polyester yarn (75 decitex) with a permittivity around 3.5. The wires act as electrodes and the polyester yarn allows distorting the electric field. Figure 21(b) presents a schematic diagram of dielectrophoretic separation chamber using five wire cloth electrodes. The wire cloth electrodes are sandwiched between sets of Perspex slabs to form the chamber allowing the fluid to pass through the cloth. The device was successfully tested to separate viable from nonviable yeast cells which have significantly different dielectric properties.

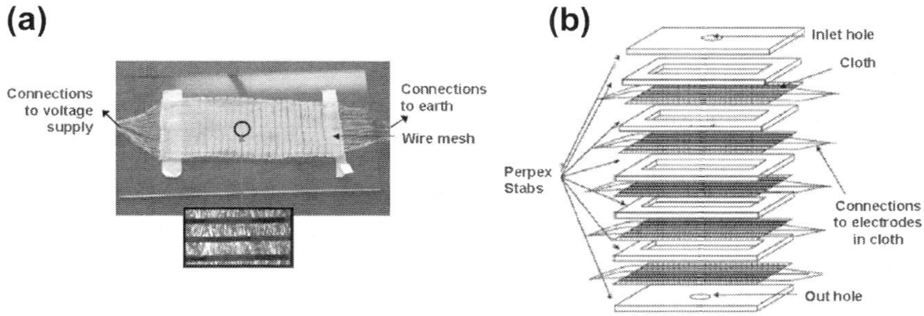

Figure 21 (a) Actual wire cloth electrode, the small insert picture is the wire cloth electrode viewed under microscope at 20× magnification. (b) Schematic diagram of dielectrophoretic separation chamber. *Abidin et al. (2007). With permission.*

3.3.3 Industrial Dielectrophoretic Filter for Nonconducting Liquids

Figure 22 shows two patented dielectrophoretic filters. Basically, the dielectrophoretic module consists of a filtration column filled with a dielectric filtering matrix and equipped of a plurality of parallel electrodes embedded in the column and alternatively connected to the high voltage supply and the earth source (corresponding to energized and grounded electrodes,

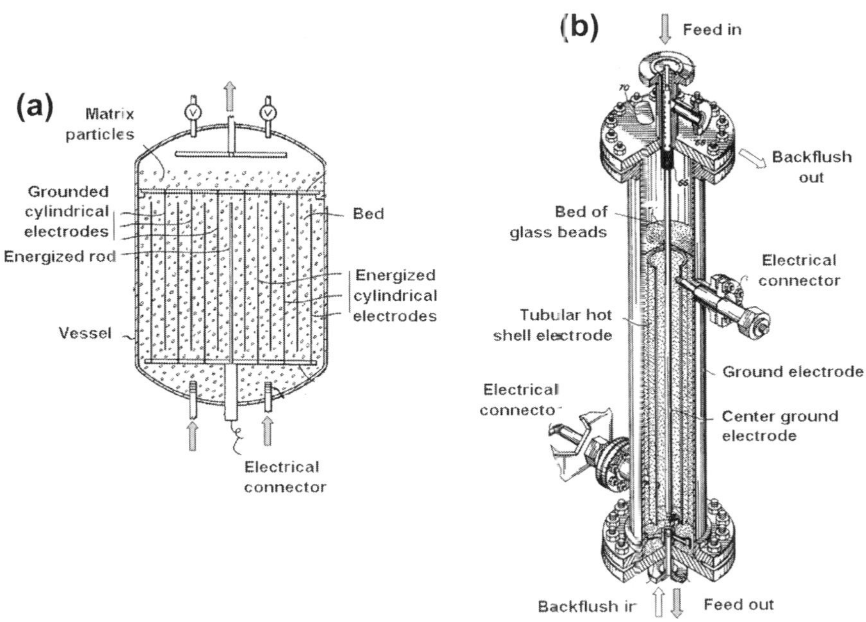

Figure 22 Dielectrophoretic module (a) Oberton (1977); (b) Fritsche et al. (1994).

respectively). Examples of industrial dielectrophoretic filter unit are described by US Patents (Franse, 1974; Fritsche and Haniak, 1975; Oberton, 1977; Watson et al., 1983; Fritsche et al., 1994; Halm and Hayes, 2000).

The electrical current flow has to be limited between electrodes. That is why a nonconductive liquid is preferentially used as carrier. Some time, electrodes are also embedded into filter wall, insulated by a thin layer of glass or plastic. Thus, the anode is not subjected to corrosion, and the electrodes may be in stainless steel or carbon steel. They are either rectangular or concentric cylindrical electrodes, spaced from 25 to 125 mm (Watson et al., 1983).

The matrix has a dielectric constant larger than that of the carrier liquid. It may either be finely divided filamentary dielectric material (fibres, nylon), pieces (glass beads, porcelain beads, ceramic beads, particles of a silicon dioxide containing mineral, furnace slag, sand, gravel, limestone) or foam (polyurethane foam) containing 50–95% void space (Lin and Benguigui, 1983). It is chosen depending on the characteristics of the particles to be separated. It must promote a high field gradient. Thus, smaller beads induce highest electric field strengths around them, increasing the volume in which the particles may be captured. The matrix must also easily release particles during the washing stage. That is why; it is preferentially comprised of smooth-surface collectors. It is made of a chemically inert and nonconductive material. The larger matrix permittivity value helps to enhance the electrical yield. Thus, barium titanate bed possessing a permittivity of 1600 was successfully used by Abidin et al. (2008).

In operation, the suspension passes through the matrix, parallel to the electrodes. It flows either in a tangential direction through the cylindrical porous bed Figure 22(a) (Oberton, 1977) or in a radial direction from the outer periphery to the centre (Watson et al., 1983). A unidirectional electric field is maintained between adjacent electrodes, allowing the migration and the adhesion of finely dispersed solids to the surface of the matrix. The liquid flowing in the column is then cleaned. The dielectrophoretic filtration devices work in a batch mode. As particle accumulation occurs in the matrices, the retentive force and the number of capture sites decrease. When the bed is too loaded, the electric field is interrupted and a cleaning fluid passes through the bed. The porous bed is regenerated by switching off the electric field. Then, a high velocity fluid is flushed into the matrix to purge it from the particles still trapped. Solvents such as kerosene or compatible oils, preferably feed stocks, are effective for backflushing. Special techniques of backflushing are illustrated in US Patents (Shirley, 1968; Franse, 1974; Oberton, 1977). To displace the residual volume of the purified product and of the washing liquid, an inert gas stream such as nitrogen, may be injected through the matrix (Oberton, 1977; France, 1974).

Figure 22b presents one module of commercial equipment: GULFTRONIC® Separator system (GENERAL ATOMICS ELECTRONIC SYSTEMS). The module consists of a cylindrical shell containing three concentric electrodes surrounded by glass beads. The required voltage is usually 30 kV per module. For standard scale, the entire system comprises more than 10 modules, and the energy consumption is around 30 kWh. This technology was used in particular to remove catalyst fines from petroleum or crude oil.

3.3.4 Dielectrophoretic Engine Oil Filter

Shen et al. (2011) developed a dielectrophoretic engine oil filter, presented in Figure 23. It consists of two concentric sheets of woven metal wire-mesh with several sheets of woven insulating wire-mesh sandwiched in between, Figure 23(a). Oil is pumped into the core tube, forced to flow outward through wire-mesh sheets arranged around the core tube, and then is collected downstream, Figure 23(b). Tests were carried out on oil used for lubricating shipboard diesel engines. The oil is contaminated by aluminium oxide, glass beads, or silicon metal particles. The filter demonstrated a high efficiency to remove particles from the oil by applying several hundred volts at a standard frequency of 60 Hz. Shen et al. (2011) reported that the electrofiltration efficiency depends on the combined influence of the particles polarizability and size, the oil viscosity and flow rate, and the electric field gradient.

3.3.5 Multistage Dielectrophoretic Filter for Conducting Liquids

When the carrier liquid has much lower permittivity than the particle once ($\varepsilon_m < \varepsilon_p$), but an electrical conductivity larger than the particle once ($\sigma_m > \sigma_p$), after a short time, the dielectrophoresis is dominated by the conductivity differential between the particle and the liquid ($\sigma_p - \sigma_m$) (Lin and Benguigui, 1982). As a result, effective polarizability becomes negative with increasing of the liquid conductivity and the trapping effect in a matrix decreases with time. The experimental device, Figure 24, was investigated to ensure the positive dielectrophoretic effect at D.C. or low A.C. signal for conducting liquids and to reduce the repulsion probability. It is divided in

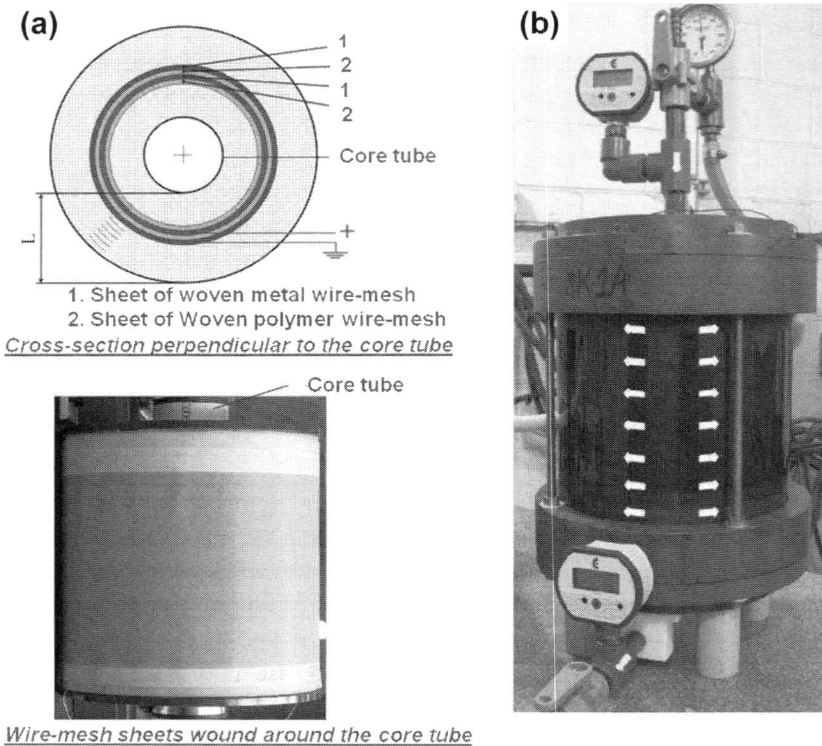

Figure 23 A dielectrophoretic engine oil filter; (a) cross section view of the module; (b) photo of wire-mesh sheets wound around the core tube; and (c) photo of the filter mounted inside a housing. The arrows show the direction in which the oil flows through the filter. The filter mounted inside volume is 8 L. *Shen et al. (2011). With permission.*

three stages (short length cylinder 1, 2 and 3) filled with a matrix (glass balls) and equipped with a cylindrical outer electrode and coaxial inner one, between which a voltage is applied. Those stages alternate with empty zero-field zones A and B. Thus, the flow stream is subjected to repetitively short period of electric field.

3.3.6 Continuous Dielectrophoretic Filter

Figure 25 presents a continuous dielectrophoretic filter (continuous Carousel separator) (Lin and Benguigui, 1983). The matrix which is set in a revolving annular ring passes consecutively in and out of the electric field region. The suspension is fed between the electrodes and flows through the

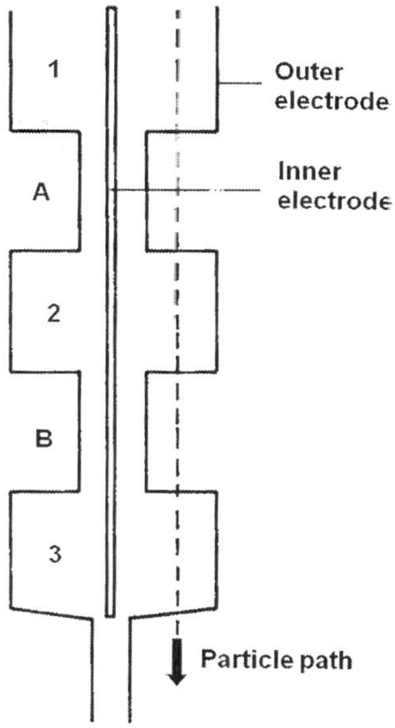

Figure 24 Multistage filtration apparatus for conducting liquids (active field regions) 1, 2, 3 filled with glass balls; zero field regions A and B empty (Lin and Benguigui, 1982).

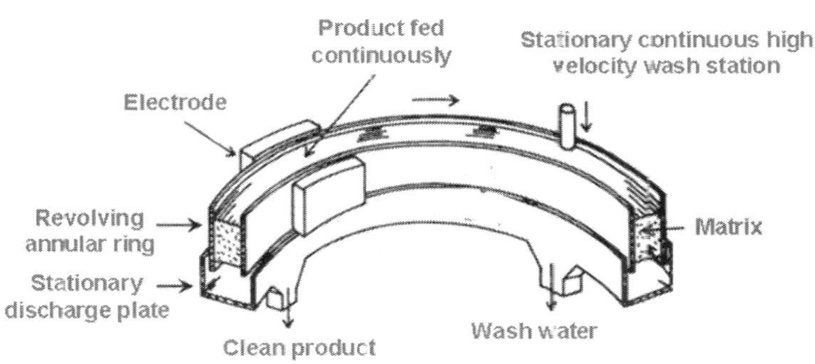

Figure 25 Continuous dielectrophoretic filter or separator. *Lin and Benguigui (1983). With permission.*

matrix where the particles are trapped. Then, thanks to the continuous matrix rotation, the collectors pass through a washing liquid flow out of the electric field region, where trapped particles are removed. The dead time is therefore reduced compared to previously mentioned batch systems. Tests carried out at 5–15 kV allowed a recovery of 80–95% of MgO particles initially suspended in oil.

3.3.7 Dielectrophoretic Membrane Filtration

Recent studies carried out at lab-scale present the opportunity to combine dielectrophoresis with crossflow filtration (Molla and Bhattacharjee, 2005; Du et al., 2009). Figure 26 presents two electrode configurations in experimental setups of a crossflow filtration process assisted by dielectrophoresis. In the filtration module Figure 26(a), a bare grid electrode and an electrically insulated plate electrode are positioned on each side of the rectangular flow channel spaced out of ~1 mm. The filter membrane is placed on the grid electrode allowing the permeate flow. A.C. signal is applied between electrodes to form a nonuniform electric field across the filter membrane. In the filtration module Figure 26(b), the filter membrane constituting the flow channel is embedded with an array of thin parallel bar microelectrodes (30–50 μm wide). Negative DEP is used to repel suspended particles from the membrane. The transmembrane pressure may be applied to enhance the permeate flow. In both case, the electrode surfaces are coated with a layer of low-dielectric material such as plastic film to avoid the occurrence of electrochemical reaction and short-circuiting of the electrodes. As this

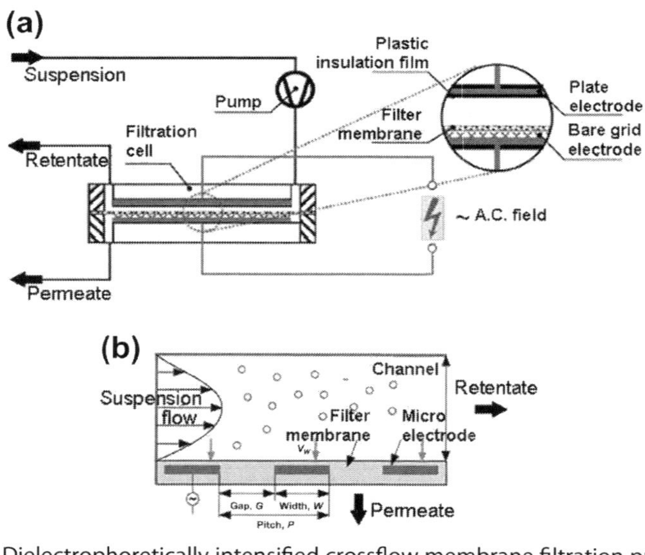

Figure 26 Dielectrophoretically intensified crossflow membrane filtration process consisting (a) with a bare grid electrode and an electrically insulated plate electrode positioned on each side of the rectangular flow channel *(Du et al. (2009). With permission.)*, or (b) with an array of thin parallel bar microelectrodes embedded in the filter membrane. *(Molla and Bhattacharjee (2005). With permission.)*.

coating attenuates the electric field, it must be sufficiently thin to limit high frequency and voltage requirement.

3.4 Typical Ancillaries and Fitment within a Flowsheet for Industrial Dielectrophoretic Filter

The entire plant includes a first circuit for the feed stream treatment. It consists of a heat exchanger to adjust the feed stream temperature, and hence to decrease the fluid viscosity to 10^{-3} Pa s, and a pump to feed the suspension into the dielectrophoretic module. A power supplier energizes the electrodes of the dielectrophoretic module. The flow rate and the temperature are controlled. The clean product is removed to an outlet conduit. Once the matrix becomes saturated with particles, the electric power that energizes the electrodes is switched off and the backflush is set. The blackflush circuit consists of a washing liquid storage tank, a pump and a separator for the washing liquid treatment. The backflushing step may be triggered off either by monitoring in time the increase particle content in the purified liquid stream or by an increase in amperage across the matrix due to the particles accumulation. The washing liquid is pumped into the dielectrophoretic module. In the case of matrix composed of beads, mainly developed at industrial scale, the washing liquid should be introduced at a sufficient rate to fluidize the beads and provided a scrubbing action.

Finally, the setup may be equipped by a circuit for the gas stream. It is injected under pressure in the filter module to displace residual liquid volume. Then, it can be quickly separates from the liquid phase in the separator and it escapes through the gas vent to any suitable disposal or recovery system.

3.5 Factors Affecting the Treatment

Table 8 summarizes the influence of the suspension properties and process parameters on the high-gradient dielectrophoretic separation and filtration. The dielectrophoretic motion is determined by the magnitude and polarity of the charges induced in a particle by the applied field.

3.6 Typical Uses and Applications

Investigations carried out on the dielectrophoretic treatment showed the potential of the technique: (1) to purify the liquid part of a suspension (to recycle liquid, to improve chemical properties, colour and visual appearance of liquid), (2) to separate particles of different dielectric constant, (3) to

Table 8 Suspension properties and parameters affecting the dielectrophoretic treatment

Factor	Comment
Suspension type	The difference between the particle and the fluid permittivity is the dominant factor. The particle permittivity must either be much lower than the liquid's one to ensure repel of particles from high field gradient (negative DEP) or be much higher than the liquid's one to ensure migration and adhesion of particles to the dielectric surface (or the electrode) (positive DEP). The water has a very high relative permittivity: 78.5 against 3.1 and 2.0 for corn oil and kerosene, respectively. In aqueous dispersions, the permittivity of particles, except for pure metal, are significantly smaller than the permittivity of water. It means that such particles suspended in aqueous medium will be repelled from the higher electric field.
Particle size distribution	The dielectrophoresis effect rises with the particle size increase. For small particles, the dielectrophoretic force may be smaller than the fluid drag forces, resulting in an inefficient of the dielectrophoretic separation. Range: 1−1000 μm
Zeta potential of particles	No matter whether particles are charged or uncharged.
Salinity	Dielectrophoretic treatment is preferentially carried out on nonconductive liquids. The use of a nonconductive liquid medium (generally an organic liquid) limits the electrical current. However, the high salinity may be compensated by the use of high frequency electric fields. This allows in particular preventing electrode polarization, electrocheminal reactions and electrical heating.
Viscosity	As the fluid drag forces may release particles from the dielectric surface (or the electrode), the fluid viscosity is preferentially low ($\sim 10^{-3}$ Pa s).
Feed velocity	As the fluid drag forces depend on the feed velocity, the latter is preferentially low. Range: 13−19 m^3/m^2 h (Wakeman and Butt, 2003)
Electric field mode	The fact that the field appears as ∇E^2 in Eqn (8) indicates that reversing the polarity of the applied voltage does not reverse the DEP force. Both electric regimes: Direct (D.C.) or alternating (A.C.) voltage can be applied, since the direction of the dielectrophoretic motion is independent of that of the electric field. However, A.C.

Table 8 Suspension properties and parameters affecting the dielectrophoretic treatment—cont'd

Factor	Comment
	voltage may help particularly in preventing electrical current establishment. It also prevents bridging between electrodes. Indeed, it allows releasing particles partially captured on other particles in the matrix (or electrode) first layers to better trap them in the following layers. The matrix (or electrode array) is them homogeneously loaded and short circuits are avoided. Frequency range: 500 Hz—50 MHz (Pethig and Markx, 1997)
Electric field ntensity	Range: $50-10 \times 10^3$ kV/m The dielectrophoretic treatment requires a high potential gradient. It must be increased with the increase of the liquid carrier permittivity.
Electric density	The electric density is minimized by the high electric resistance between electrodes with nonconductive liquid or insulted electrodes, or by high-frequency field application. However, as the fluid achieves an electrical conductivity, the electrical current increases inducing electrochemical reactions, damage of the electrode metal which is not isolated from the liquid and electrical heating.
Temperature	The fluid permittivity, the electrical conductivity, the density and the viscosity depend on the temperature. Thus, higher temperature increases the permittivity difference between the particle and the liquid and reduces the density and the viscosity. It results in a smaller fluid drag forces and the increase of dielectrophoretic separation efficiency.

upgrade particle concentration, and (4) for particle analysis. Table 9 indicates the potential applications of dielectrophoretic filtration and separation and related industrial sectors. Lin and Benguigui (1983) have described in details some of them. This technology was successfully transferred to the industrial scale for the removal of solid particles from hydrocarbon streams (petroleum or crude oil) or various distillates. Commonly, the solids are catalyst fines or metallic compounds introduced by storage and transfer through the various processing equipment. The demulsification was also performed at the laboratory scale. The coalescence of the droplets in emulsion was obtained with the film disintegration thanks to the strong dielectrophoretic forces. The dielectrophoretic method is also under development for the particle

Table 9 Example of mineral and biological suspensions treated by dielectrophoresis

Industrial sector	Application field	Suspensions	References
Microfluidic, biorefineries, food, pharmatical and cosmetic industry	• Particle separation, sorting and analysis	Biological cell (plant, mammalian, bacterial, yeast, live and dead lung cancer cells, live and dead yeast cells)	Pethig and Markx (1997), Yang et al. (2000), Morgan and Green (2003), Chiou et al. (2005), Tai et al. (2007), Khoshmanesh et al. (2011)
	• Cell concentration to produce biofilm, Liquid purification	Biological cell (yeast)	Abidin et al. (2007), Abidin and Markx (2005)
Petroleum refineries, petrochemical industry, metal and vegetable-oil industries	• Liquid purification	Metals or metal oxides suspended in oil (activated carbon black, aluminium oxide powder, silicon metal powder, nickel, catalyst particles)	Fritsche and Haniak (1975), Oberton (1977), Watson et al. (1983), Fritsche et al. (1994), Wakeman and Butt (2003), Shen et al. (2011)
		• Polymer suspended in oil (PVC, polyethylene particle)	Benguigui and Lin (1984), Wakeman and Butt (2003), Du et al. (2007)
Petrochemical industry, chemical industry (printing, rubber, leather industries) electronics industry, vegetable oil industry	• Demulsification	Water-in-oil emulsions or organic compound-in-water emulsions (*meta*-xylene in wastewater)	Wakeman and Butt (2003), Du et al. (2007), Hosseini et al. (2012)
Lead mining petrochemical industry	• Particle separation	Clay mixture (illite, kaolinite, montmorillonite, prochlorite, halloysite galena, quartz, pyrite, chalcopyrite, plagioclase, sphalerite)	McEuen (1964),[1] Ballantyne and Holtham (2010)

[1]Particle separation using centrifugation assisted by dielectrophoretic force.

Table 10 Overview of dielectrophoretic treatment efficiency

Suspension	Particles	Process parameters	Efficiency	References
Contaminated hydrocarbon oil	Suspended electrically conductive contaminants, $d_{Particle} < 5$ μm; $C_i = 23.7 \times 10^{-3}$ pd/lb (6.75×10^{-2} kg/m^3)	Pilot scale (matrix: glass beads ¼-inch in diameter (6.35 mm)), $Q = 1.4$ gallon/min (8.83×10^{-5} m^3/s) at 265 °F (129 °C) for 46 h; 27 kV DC (~425 kV/m)	>80% of total solids removal in comparison to the initial content; $C_{DEP\text{-}f} = 4.9 \times 10^{-3}$ pd/lb (1.40×10^{-2} kg/m^3)	Fritsche and Haniak (1975)
Contaminated hydrocarbon oils	Catalytic cracking contaminants, $C_i = 2161–3222$ ppm	Pilot scale in GULFTRONIC® separator (matrix: Potassium beads 1/8-inch in diameter (6.2 mm)); $Q =$ N.D; Residence time: 140 s, 250–275 °F (121–135 °C); 30 kV DC at 9.2 mA max	96.9–99.9% Of total solids removal in comparison to the initial content; $C_{DEP\text{-}f} < 100$ ppm	Fritsche et al. (1994)
Contaminated hydraulic oil (ε_f: 2.1;	Activated carbon black; $d_{Particle} > 50$ μm; $C_i = 180$ ppm	Lab. scale (matrix: glass beads 500 μm);	65% of total solids removal in comparison to the initial content.	Wakeman and Butt (2003)

(Continued)

Table 10 Overview of dielectrophoretic treatment efficiency—cont'd

Suspension	Particles	Process parameters	Efficiency	References
$\mu = 23.5 \times 10^{-3}$ Pa s at 50 °C)	PVC; $d_{Particle} > 50$ μm; $C_i = 180$ ppm	$Q = 8.68$ m^3/s at 50 °C; $E = 11.2 \times 10^2$ kV/m^3	~20% of total solids removal in comparison to the initial content.	
Yeast suspension	Yeast cells in deionized water	Lab. scale (matrix: glass beads 150–250 μm); AC voltage of 40 $V_{pk\text{-}pk}$ (~30 kV/m) and 1 MHz	Cells removal factor: 1.4 times higher than in a depth bed filtration; $C_{DEP\text{-}f} = 3.25 \times 10^6$ cells/mL	Abidin and Markx (2005)
Yeast suspension	Yeast cells in deionized water; $C_i = 7 \times 10^6$ cells/mL	Lab. scale (matrix: cloth made from 100-μm diameter stainless steel wires and 75 decitex polyester yarn); $Q = 5 \times 10^{-9}$ m^3/s; AC voltage of 40 $V_{pk\text{-}pk}$ (235–267 kV/m) and 1 MHz	Cells removal factor: 6.1 times higher than in a depth bed filtration. 92% of cells removal compared to 15% without electric field. Separation of dead and live cells	Abidin et al. (2007)

Percentage of solids removal from the suspension = (initial solids content − final solids content after DEP)/initial solids content in the suspension; cells removal factor = cells removal with DEP/Cells removal without DEP; C_i: initial concentration of the suspension; $C_{DEP\text{-}f}$: final concentration of the treated suspension after dielectrophoretic treatment.

separation, sorting and analysis in microfluidic. Table 10 presents an overview of the dielectrophoretic treatment efficiency. Although dielectrophoretic separators are suitable for oils or solvents, recent studies have been carried out with deionized water as media, allowing to collect yeast cells, to separate dead and live cells thanks to the significant difference of their dielectric properties or to purify wastewaters. Usually the dielectrophoretic method requires a low energy input to effect the separation since the application of relatively high voltages is associated with negligible current. However, the use of water as carrier liquid or the presence of water may be problematic as the ions dissolved in the water increase the electrical conductivity and hence the electrical current.

4. COMPARISON OF ELECTRICALLY ASSISTED SEPARATION PROCESSES

Table 11 shows the advantages and disadvantages of the reviewed electrically assisted separations.

Table 11 Advantage/drawbacks of the electrically assisted filters and separators

Assisted separation process	Advantages	Drawbacks
Dead end electrofiltration (electrofiltration and electroosmotic dewatering)	• Good alternative to accelerate the dewatering when it is observed that dead end filtration and consolidation rates are too slow and when particles present high surface charge ($\lvert\zeta\rvert > 15-20$ mV). • Efficient for compressible or gelatinous material. • Efficient for all feed concentrations. • Efficiency depends mainly on the surface electrical charge of the particles. • Typical increase of filtration rate by 1–37 fold; typical increase of final dry content by 10–57% during a consolidation in comparison to mechanical pressure alone.	• Efficiency depends mainly on the surface electrical charge of particles. • Energy consumption by electrofiltration and electroosmotic dewatering lower than 51 and 700 kWh/m^3 of water removed, respectively. • Safety requirements: a Faraday cage around filter for industrial operations, (or limit the access around filter), a purge and venting system to remove electrolysis gases. • The hydronium and hydroxide electrolytic ions generation may affect the filtrate flow and induce also a risk of acidification of the filter cake. For these

(*Continued*)

Table 11 Advantage/drawbacks of the electrically assisted filters and separators—cont'd

Assisted separation process	Advantages	Drawbacks
	• Energy requirement can be significantly lower than thermal drying (1.4–23.5 times less) for an electroosmotic dewatering. • Side-effect of decontamination of the dewatered solid material: Removal of heavy metals, specific ions, soluble COD (especially beneficial for biological sludge and soil decontamination). • Electrically assisted separation can be adapted to existing equipment limiting the investment.	reasons special design of filter cell can be necessary to flush out the electrolysis products: it represents an extra-cost compared to standard electrofilter and a risk to increase significantly the energy consumption. • Electric field induces an Ohmic heating over time; temperature must be controlled at electrodes to avoid damage of filter cloth.
Crossflow electrofiltration	• Good alternative to accelerate the dewatering when crossflow filtration is too slow due to the fouling effect. • Efficiency depends mainly on the surface electrical charge of the particles. • Low feed pressure required. This reduces the shear rate at the membrane surface in comparison to crossflow filtration, improving the possibilities of processing shear sensitive stream. • Typical increase of filtration rate by 1.4–10 folds in comparison to standard crossflow filtration.	• Efficiency depends mainly on the surface electrical charge of particles. • Energy consumption lower than 7 kWh/m^3. • Safety requirements: a Faraday cage around filter for industrial operations, (or limit the access around filter), a purge and venting system to remove electrolysis gases. • Electrochemical reactions induced. • As the filtrate is drained at one side, the electrode polarity has to be adequately fixed and the sign of the particle surface charge must be constant.

Table 11 Advantage/drawbacks of the electrically assisted filters and separators—cont'd

Assisted separation process	Advantages	Drawbacks
Dielectrophoretic treatment	• Suitable for manipulating particle that do not have a net intrinsic charge. • The use of a nonconductive liquid medium (generally an organic liquid) limits the electrical current. Then, contrarily to the previous electric treatments (electrofiltration and pressurized electroosmotic dewatering), the electrochemical reactions are not considered ought to the limited electrical current. • Especially advisable when the liquid medium is an organic solvent. • Typical particle removal from 20% to 99.9% of particles in comparison to the initial content, reducing the suspension concentration to less than 100 ppm. • Low feed pressure required. • Low electrical energy input required. • Electrochemical reactions at the electrodes can be avoided. • Easily cleaning and discharge of solids after operation.	• Efficiency depends mainly on the effective permittivity of particles that fixed in particular the magnitude and the sign of the DEP force. • Energy consumption of 3 KWh per module (GULFTRONIC®). • High potential gradient required. It is generated by a high power supply (typically generator of 10–35 kV) and by the use of small dimension electrodes or insulating porosity dielectric structures placed between electrodes that both increase locally the potential gradient by distorting the electric field. • Low feed concentration required. • Low fluid viscosity required that may involve the feed stream heating (between about 100 and 200 °C for organic liquid). • Presence of water or conductive media is problematic (maximum moisture content of 1000 ppm)

5. CONCLUSIONS

Principle of electrically assisted separators or filters is to superimpose an electric field to a conventional separation process. Electric field then induces electrokinetic phenomena or dielectrophoresis which can have beneficial effects. For instance, it can improve the suspension concentration, the liquid or solid purification, the particle fractionation, the dewatering or the consolidation. But, as presented in this chapter, it is necessary to check the properties of the suspension to make sure that the electric field will be efficient. Thus, the electrophoretic and electroosmotic treatments are suitable to remove colloids or fine particles possessing a high surface charge and more usually dispersed in water. The surface properties of particles are a dominant factor, affected by the pH and the electrolyte content. Examples found in literature show that in comparison to conventional mechanical process, the crossflow electrofiltration increases the filtration rate by 1.4–10 folds with an energy consumption lower than 7 kWh/m^3. The dead end electrofiltration increases the filtration rate by 1.1–37 folds with an energy consumption lower than 51 kWh/m^3. The electroosmotic dewatering carried out on pre-dewatered suspension, allows rising the final dry content by removing 8–50% of supplementary water in comparison to mechanical pressure alone and with an energy consumption lower than 700 kWh/m^3. By exploiting the difference of surface charge between particles, both dead end and crossflow electrofiltration lead to a fractionation step. The electroosmosis treatment may potentially be used for decontamination purposes by drainage of heavy metal ions, desalinization, and organic matter oxidation or by electrowashing.

On the contrary, the dielectrophoretic treatment is more effective to separate neutral finely divided solids from a nonaqueous media. Its efficiency depends mainly of the permittivity difference between the particle and the media. Thus, the purification by dielectrophoresis allows removing 20–99.9% of particles from an organic liquid, reducing the suspension concentration to less than 100 ppm. It proved also efficiency to collect or separate solids from deionized water, or to purify wastewaters. Although the dielectrophoretic treatment requires high electric field gradients, the energy consumption is reduced by the negligible electrical current flow induced in low or nonconductive liquid medium.

Efficiency of those technologies has been mainly demonstrated at lab and bench scale, showing potential benefits for a wide range of sectors. In the future, more industrial implementation could be realized.

REFERENCES

Abidin, Z.Z., Markx, G.H., 2005. High-gradient electric field system for the dielectrophoretic separation of cells. J. Electrost. 63 (6–10), 823–830.

Abidin, Z.Z., LDownes, L., Markx, G.H., 2007. Large scale dielectrophoretic construction of biofilms using textile technology. Biotechnol. Bioeng. 96 (6), 1222–1225.

Abidin, Z.Z., Baik, D.A., Hafifudin, N., Markx, G.H., 2008. Factor affecting dielectrophoretic separation of cells using high-gradient electric field strength system. J. Eng. Sci. Technol. 3 (1), 30–39.

Acar, Y.B., Gale, R.J., Alshawabkeh, A.N., Marks, R.E., Puppala, S., Bricka, M., Parker, R., 1995. Electrokinetic remediation: basics and technology status. J. Hazard. Mater. 40 (2), 117–137.

ACE Korea Incorporation, November 2012. Electro-osmosis Dehydrator ELODE® on Commercial Brochure. http://cms.esi.info/Media/documents/Aquat_ELODE_ML.pdf.

Agana, A.B., Reeve, D., Orbell, J.D., 2012. The influence of an applied electric field during ceramic ultrafiltration of post-electrodeposition rinse wastewater. Water Res. 46 (11), 3574–3584.

Akay, G., Wakeman, R.J., 1997. Electric field enhanced crossflow microfiltration of hydrophobically modified water soluble polymers. J. Membr. Sci. 131 (1–2), 229–236.

Al-Asheh, S., Jumah, R., Banat, F., Al-Zou'Bi, K., 2004. Direct current electroosmosis dewatering of tomato paste suspension. Food Bioprod. Process. 82 (3), 193–200.

Aldaeus, F., 2006. New Concepts for Dielectrophoretic Separations and Dielectric Measurements of Bioparticles (Ph.D. dissertation thesis). KTH Chemical Science and Engineering, Stockholm, Sweden.

Alekhin, S.A., Kuznetsov, E.B., Bakhir, V M., Klimenko, V.I., Zadorozhny, J.G., 1982. US Patent 4323445 (United States Patent, 1982).

Ballantyne, G.R., Holtham, P.N., 2010. Application of dielectrophoresis for the separation of minerals. Miner. Eng. 23 (4), 350–358.

Bargeman, G., Koops, G.-H., Houwing, J , Breebaart, I., van der Horst, H.C., Wessling, M., 2002. The development of electro-membrane filtration for the isolation of bioactive peptides: the effect of membrane selection and operating parameters on the transport rate. Desalination 149 (1–3), 369–374.

Bazinet, L., Firdaous, L., 2009. Membrane processes and devices for separation of bioactive peptides. Recent Patents Biotechnol. 3, 61–72.

Benguigui, L., Lin, I.J., 1984. Phenomenological aspect of particle trapping by dielectrophoresis. J. Appl. Phys. 56, 3294–3297.

Bollinger, J.M., Adams, R.A., 1984. Electrofiltration: an energy efficient alternative in solid-liquid separations. In: Proceedings of the 6th Annual Industrial Energy Technology Conference. Energy Systems Laboratory, Houston (TX), pp. 320–326.

Bollinger, J.M., Klinkowski, P.R., 1986. US Patent 4604174 (United States Patent, 1986).

Bouzrara, H., Vorobiev, E., 2000. Expression of cellular materials intensified by electric field pulses. In: Rushton, A. (Ed.), Proceedings of the 8th World Filtration Congress. Elsevier Science, Brighton (UK), pp. 299–302.

Bowen, W.R., Kingdon, R.S., Sabuni, H.A.M., 1989. Electrically enhanced separation processes: the basis of in situ intermittent electrolytic membrane cleaning and in situ electrolytic membrane restoration. J. Membr. Sci. 40, 219–229.

Bowen, W.R., Sabuni, H.A.M., 1992. Pulsed electrokinetic cleaning of cellulose nitrate microfiltration membranes. Industrial Eng. Chem. Res. 31, 515–523.

Brisson, G., Britten, M., Pouliot, Y., 2007. Electrically-enhanced crossflow microfiltration for separation of lactoferrin from whey protein mixtures. J. Membr. Sci. 297 (1–2), 206–216.

Chiou, P.Y., Ohta, A.T., Wu, M.C., 2005. Massively parallel manipulation of single cells and microparticles using optical images. Nature 436 (7049), 370–372.

Chiu, T.Y., Garcia Garcia, F.J., 2011. Critical flux enhancement in electrically assisted microfiltration. Sep. Purif. Technol. 78 (1), 62–68.

Chu, C.P., Lee, D.J., Liu, Z., Jin, W.H., 2004. Morphology of sludge cake at electroosmosis dewatering. Sep. Sci. Technol. 39 (6), 1331–1346.

Chuang, C.J., Wu, C.Y., Wu, C.C., 2008. Combination of crossflow and electric field for microfiltration of protein/microbial cell suspensions. Desalination 233 (1–3), 295–302.

Citeau, M., Larue, O., Vorobiev, E., 2011. Effect of electrolytes content on the electro-osmotic dewatering of agro-industrial sludge. In: Taoukis, P.S., Stoforos, N.G., Karathanos, V.T., Saravacos, G.D. (Eds.), Proceedings of the 11th International Congress on Engineering and Food. Cosmosware, Athens, pp. 1215–1216.

Citeau, M., Larue, O., Vorobiev, E., 2012a. Influence of filter cell configuration and process parameters on the electro-osmotic dewatering of sewage sludge. Sep. Sci. Technol. 47 (1), 11–21.

Citeau, M., Olivier, J., Mahmoud, A., Vaxelaire, J., Larue, O., Vorobiev, E., 2012b. Pressurised electro-osmotic dewatering of activated and anaerobically digested sludges: electrical variables analysis. Water Res. 46 (14), 4405–4416.

Diaz, M., 1989. US Patent 4861496 (United States Patent, 1989).

Doh, I., Cho, Y.H., 2005. A continuous cell separation chip using hydrodynamic dielectrophoresis (DEP) process. Sens. Actuators A: Phys. 121 (1), 59–65.

Du, F., Baune, M., Thöming, J., 2007. Insulator-based dielectrophoresis in viscous media-simulation of particle and droplet velocity. J. Electrost. 65 (7), 452–458.

Du, F., Hawari, A., Baune, M., Thöming, J., 2009. Dielectrophoretically intensified crossflow membrane filtration. J. Membr. Sci. 336 (1–2), 71–78.

Dussour, C., Favoriti, P., Vorobiev, E., 2000. Influence of chemical additives upon both filtration and electroosmotic dehydration of kaolin suspension. Sep. Sci. Technol. 35 (8), 1179–1193.

Electrokinetic Limited Company, UK. http://www.electrokinetic.co.uk.

Enevoldsen, A.D., Hansen, E.B., Jonsson, G., 2007. Electro-ultrafiltration of industrial enzyme solutions. J. Membr. Sci. 299 (1–2), 28–37.

Fourie, A.B., Johns, D.G., Jones, C.J.F.P., 2007. Dewatering of mine tailings using electrokinetic. Can. Geotechnical J. 44, 160–172.

Franse, A.D., 1974. US Patent 3799857 (United States Patent, 1974).

Freeman, M.P., 1979. US Patent 4168222 (United States Patent, 1979).

Fritsche, G.R., Haniak, L.W., 1975. US Patent 3928158 (United States Patent, 1975).

Fritsche, G.R., Bujas, R.S.V., Caprioglio, G.C., 1994. US Patent 5308586 (United States Patent, 1994).

Genç, A., Tosun, I., 2002. Experimental studies on the effect of electrode configuration in electrofiltration. Sep. Sci. Technol. 37 (13), 3053–3064.

Glendinning, S., Lamont-Black, J., Jones, C.J.F.P., 2007. Treatment of sewage sludge using electrokinetic geosynthetics. J. Hazard. Mater. 139 (3), 491–499.

Glendinning, S., Mok, C., Kulumba, D., Rogers, C., Hunt, D., 2010. Design framework for electrokinetically enhanced dewatering of sludge. J. Environ. Eng. 136 (4), 417–426.

Gouy, A., 1910. Sur la constitution de la charge électrique à la surface d'un électrolyte. J. Phys. Theor. Appl. 9 (1), 457–468.

Gözke, G., Posten, C., 2010. Electrofiltration of biopolymer. Food Eng. Rev. 2 (2), 131–146.

Gözke, G., 2012. Electrofiltration of Biopolymers: Spatially Distributed Process Analysis (Ph.D. dissertation thesis). Karlsruher Institut für Technolgie, Karlsruhe: KIT Scientific Publishing.

Guizard, C., Legault, F., Idrissi, N., Larbot, A., Cot, L., Gavach, C., 1989. Electronically conductive mineral membranes designed for electro-ultrafiltration. J. Membr. Sci. 41, 127–142.

Hakimhashemi, M., Gebreyohannes, A.Y., Saveyn, H., Van der Meeren, P., Verliefde, A., 2012. Combined effects of operational parameters on electro-ultrafiltration process characteristics. J. Membr. Sci. 403–404, 227–235.

Halm, R.L., Hayes, K.Q., 2000. US Patent 6126803 (United States Patent, 2000).

Hofmann, R., Posten, C., 2003. Improvement of dead-end filtration of biopolymers with pressure electrofiltration. Chem. Eng. Sci. 58 (17), 3847–3858.

Hofmann, R., Käppler, T., Posten, C., 2006. Pilot-scale press electrofiltration of biopolymers. Sep. Purif. Technol. 51 (3), 303–309.

Hosseini, M.M., Hosseini, M., Mohebbi-Kalhori, D., 2012. Meta xylene removal from industrial wastewater using dielectrophoretic technique. In: Proceedings of International Conference on Chemical, Environmental and Biological Sciences (ICCEBS'2012), pp. 24–28. Penang (Malaysia).

Hsieh, L.-H.C., Weng, Y.-H., Huang, C.-P., Li, K.-C., 2008. Removal of arsenic from groundwater by electro-ultrafiltration. Desalination 234 (1–3), 402–408.

Huotari, H.M., Trägårdh, G., Huisman, I.H., 1999a. Crossflow membrane filtration enhanced by an external DC electric field: a review. Chem. Eng. Res. Des. 77 (5), 461–468

Huotari, H.M., Huisman, I.H., Trägårdh, G., 1999b. Electrically enhanced crossflow membrane filtration of oily waste water using the membrane as a cathode. J. Membr. Sci. 156 (1), 49–60.

Iritani, E., Ohashi, K., Murase, T., 1992. Analysis of filtration mechanism of dead-end electro-ultrafiltration for proteinaceous solution. J. Chem. Eng. Jpn. 25 (5), 383–387.

Iritani, E., Mukai, Y., Kiyotomo, Y., 2000. Effects of electric field on dynamic behaviors of dead-end inclined and downward ultrafiltration of protein solutions. J. Membr. Sci. 164 (1–2), 51–57.

Jennings, A.A., Mansharamani, P., 1999. Modeling electrokinetically-enhanced aggregate remediation. Environ. Model. Softw. 14 (6), 625–634.

Jeon, C.S., Yang, J.S., Kim, K.J., Baek, K., 2010. Electrokinetic removal of petroleum hydrocarbon from residual clayey soil following a washing process. Clean. Soil. Air Water 38 (2), 189–193.

Jones, C.J.F.P., 2003. US Patent 2003/0173221 (United States Patent, 2003).

Jones, C.J.F.P., Lamont-Black, J., Glendinning, S., 2011. Electrokinetic geosynthetics in hydraulic applications. Geotext. Geomembranes 29 (4), 381–390.

Jones, C.J.F.P., Lamont-Black, J., 2012. US Patent 2012/0097599 A1 (United States Patent, 2012).

Jumah, R., Al-Asheh, S., Banat, F., Al-Zoubi, K., 2005. Electroosmotic dewatering of tomato paste suspension under AC electric field. Dry. Technol. 23, 1465–1475.

Käppler, T., Posten, C., 2007. Fractionation of proteins with two-sided electro-ultrafiltration. J. Biotechnol. 128 (4), 895–907.

Kilchherr, R., Koenders, M.A., Wakeman, R.J., Tarleton, E.S., 2004. Transport processes during electrowashing of filter cakes. Chem. Eng. Sci. 59, 1103–1114.

Kim, S.O., Moon, S.H., Kim, K.W., Yun, S.T., 2002. Pilot scale study on the ex situ electrokinetic removal of heavy metals from municipal wastewater sludges. Water Res. 36 (19), 4765–4774.

Klinkowski, P.R., 1986. US Patent 4569739 (United States Patent, 1986).

Khoshmanesh, K., Nahavandi, S., Baratchi, S., Mitchell, A., Kalantar-zadeh, K., 2011. Dielectrophoretic platforms for bio-microfluidic systems. Biosens. Bioelectron. 26 (5), 1800–1814.

Komagata, S., 1938. An electro-washing process. J. Electrochem. Soc. 73 (1), 511–522.

Kondoh, S., Hiraoka, M., 1990. Commercialization of pressurized electroosmotic dehydrator (PED). Water Sci. Technol. 22 (12), 259–268.

Kondo, S., Suwa, T., Sano, S., Muroi, O., 1991. US Patent 5034111 (United States Patent, 1991).

Kondoh, S., Hiraoka, M., 1993. Studies on the improving dewatering method of sewage sludge by the pressurized electro-osmotic dehydrator with injection of polyaluminum chloride. In: Proceedings of the 6th World Filtration Congress, pp. 765–769. Nagoya (Japan).

Krishnaswamy, P., Bahnii, J.R., 1986. Electrically coupled flow through fine particle cakes. Filtr. Sep. 23 (5), 289–293.

Lamont-Black, J., Glendinning, S., Jones, C., Huntley, D.T., Smith, R., 2005. The development of in-situ dewatering of lagooned sewage sludge using electrokinetic geosynthetic (EKG). In: Lowe, P., Horan, N. (Eds.), Proceedings of the 10th European Biosolids and Biowastes Conference, Paper 21. Aqua Enviro Technology Transfer, Wakefield (UK), pp. 1–8.

Larue, O., Mouroko-Mitoulou, T., Vorobiev, E., 2001. Filtration, cake washing and pressurized electroosmotic dewatering of a highly conductive silica suspension. Trans. Filtr. Soc. 1 (2), 31–37.

Larue, O., Vorobiev, E., 2004. Sedimentation and water electrolysis effects in electrofiltration of kaolin suspension. AIChE J. 50, 3120–3133.

Larue, O., Wakeman, R.J., Tarleton, E.S., Vorobiev, E., 2006. Pressure electroosmotic dewatering with a continuous removal of electrolysis products. Chem. Eng. Sci. 61, 4732–4740.

Lee, J.E., Lee, J.K., Choi, H.K., 2007. Filter press for electrodewatering of waterworks sludge. Dry. Technol. 25, 1985–1993.

Lentsch, S., Aimar, P., Orozco, J.L., 1993. Enhanced separation of albumin-poly(ethylene glycol) by combination of ultrafiltration and electrophoresis. J. Membr. Sci. 80, 221–232.

Li, Y., Kaler, K.V.I.S., 2004. Dielectrophoretic fluidic cell fractionation system. Anal. Chim. Acta 507 (1), 151–161.

Li, F.D., Li, X., Sotome, I., Isobe, S., 2007. Effect of different electric fields on temperature rise, energy efficiency ratio, and solids content during electro-osmotic dewatering of tofu residue (okara). Rep. Natl. Food Res. Inst. 71, 15–26.

Li, Y., Ehrhard, R., Biswas, P., Kulkarni, P., Carns, K., Patterson, C., Krishnan, R., Sinha, R., 2009. Removal of waterborne particles by electrofiltration: pilot-scale testing. Environ. Eng. Sci. 26 (12), 1795–1803.

Lin, I.J., Benguigui, L., 1982. Dielectrophoretic filtration of liquids. II. Conducting liquids. Sep. Sci. Technol. 17 (5), 645–654.

Lin, I.J., Benguigui, L., 1983. High-intensity, high-gradient dielectrophoretic filtration and separation processes. In: Wakeman, R.J. (Ed.), Progress in Filtration and Separation, vol. 3. Elsevier, Oxford (UK), pp. 149–204.

Lin, Y.T., Sung, M., Sanders, P.F., Marinucci, A., Huang, C.P., 2007. Separation of nano-sized colloidal particles using cross-flow electro-filtration. Sep. Purif. Technol. 58 (1), 138–147.

Lite, L., Xiuqu, L., 1999. Electroosmotic dewatering of okara at different electric fields. In: Zeng, Dechao, Wang, Maohua, Zhang, Senwen, Lei, Tingwu, Wang, Baoji (Eds.), Proceedings of the 1999 International Conference on Agricultural Engineering. China Agricultural University Press, Beijing (China), pp. 58–63.

Lockhart, N.C., 1983a. Electroosmotic dewatering of clays, III. Influence of clay type, exchangeable cations, and electrode materials. Colloids Surf. 6 (3), 253–269.

Lockhart, N.C., 1983b. Electroosmotic dewatering of clays, II. influence of salt, acid, and flocculants. Colloids Surf. 6 (3), 239–251.

Loginov, M., Citeau, M., Lebovka, N., Vorobiev, E., 2013. Electro-dewatering of drilling sludge with liming and electrode heating. Sep. Purif. Technol. 104, 89–99.

Lyklema, J., 1995. Fundamentals of Interface and Colloid Science. In: Solid-liquid Interfaces, vol. 2. Academic Press.

Mahmoud, A., Olivier, J., Vaxelaire, J., Hoadley, A.H.A., 2010. Electrical field: a historical review of its application and contributions in wastewater sludge dewatering. Water Res. 44, 2381–2407.

Mahmoud, A., Olivier, J., Vaxelaire, J., Hoadley, A.F.A., 2011. Electro-dewatering of wastewater sludge: influence of the operating conditions and their interactions effects. Water Res. 45 (9), 2795–2810.

McEuen, R.B., 1964. Dielectrophoretic behaviour of clay minerals I. Dielectrophoretic separation of clay mixtures. In: Bradley, W.F. (Ed.), Proceedings of the 12th National Conference on Clays and Clays Minerals. Pergamon press, Atlanta, pp. 549–556.

Miller, S.A., Murphy, A., Veal, C.J., Young, M., 1998. Improved Filtration of Sewage Sludges Using Electrodewatering. End of project report, CSIRO Investigation Report N_ET/IR140.

Miller, S.A., Johnston, B.K., Veal, C., 2005. US Patent 6871744 (United States Patent, 2005).

Molla, S.H., Bhattacharjee, S., 2005. Prevention of colloidal membrane fouling employing dielectrophoretic forces on a parallel electrode array. J. Membr. Sci. 255 (1–2), 187–199.

Morgan, H., Izquierdo, A.G., Bakewell, D.J., Green, N.G., Ramos, A., 2001. The dielectrophoretic and traveling wave forces for interdigitated electrode arrays: analytical solution using Fourier series. J. Phys. Appl. Phys. 34, 1553–1561.

Morgan, H., Green, N.G., 2003. AC Electrokinetics: Colloids and Nanoparticles. Research Studies Press Ltd, Hertfordshire (UK).

Moulik, S.P., 1971. Physical aspects of electrofiltration. Environ. Sci. Technol. 5 (9), 771–776.

Mujumdar, A.S., Yoshida, H., 2008. Electro-osmotic dewatering (EOD) of bio-materials. In: Vorobiev, E., Lebovka, N. (Eds.), Electrotechnologies for Extraction from Food Plants and Biomaterials. Springer, New York, pp. 121–154.

Navab Daneshmand, T., Beton, R., Hill, R.J., Gehr, R., Frigon, D., 2012. Inactivation mechanisms of bacterial pathogen indicators during electro-dewatering of activated sludge biosolids. Water Res. 46 (13), 3999–4008.

Oberton, S.B., 1977. US Patent 4040926 (United States Patent, 1977).

Oussedik, S., Belhocine, D., Grib, H., Lounici, H., Piron, D.L., Mameri, N., 2000. Enhanced ultrafiltration of bovine serum albumin with pulsed electric field and fluidized activated alumina. Desalination 127 (1), 59–68.

OVIVO, November 2012. Advanced Dewatering RAEBL, Montreal, Canada. http://www.ovivowater.com/en/CINETIKLinearElectroDewatering.

Park, Y.G., 2006. Effect of hydraulic pressure and electric field on electro-microfiltration in the wine-brewery industry. Desalination 191, 411–415.

Pethig, R., Markx, G.H., 1997. Applications of dielectrophoretic in biotechnology. Trends Biotechnol. 15, 426–432.

Porta, A., Kulhanek, A., 1983. US Patent 4376022 (United States Patent, 1983).

Raats, M.H.M., Van Diemen, A.J.G., Laven, J., Stein, H.N., 2002. Full scale electrokinetic dewatering of waste sludge. Colloids Surf. A Physicochem. Eng. Asp. 210 (2–3), 231–241.

Robinson, C.W., Siegel, M.H., Condemine, A., Fee, C., Fahidy, T.Z., Glick, B.R., 1993. Pulsed-electric-field crossflow ultrafiltration of bovine serum albumin. J. Membr. Sci. 80, 209–220.

Sarkar, B., DasGupta, S., De, S., 2008. Effect of electric field during gel-layer controlled ultrafiltration of synthetic and fruit juice. J. Membr. Sci. 307 (2), 268–276.

Sarkar, B., DasGupta, S., De, S., 2009. Electric field enhanced fractionation of protein mixture using ultrafiltration. J. Membr. Sci. 341 (1–2), 11–20.

Saveyn, H., Pauwels, G., Timmerman, R., van der Meeren, P., 2005. Effect of polyelectrolyte conditioning on the enhanced dewatering of activated sludge by application of an electric field during the expression phase. Water Res. 39 (13), 3012–3020.

Saveyn, H., Curvers, D., Pel, L., De Bondt, P., Van der Meeren, P., 2006a. In situ determination of solidosity profiles during activated sludge electrodewatering. Water Res. 40 (11), 2135–2142.

Saveyn, H., Meeren, P.V., Pauwels, G., Timmerman, R., 2006b. Bench-and pilot-scale sludge electrodewatering in a diaphragm filter press. Water Sci. Technol. 54 (9), 53–60.

Shang, J.Q., Lo, K.Y., 1997. Electrokinetic dewatering of a phosphate clay. J. Hazard. Mater. 55 (1–3), 117–133.

Shen, Y., Elele, E., Khusid, B., 2011. A novel concept of dielectrophoretic engine oil filter. Electrophoresis 32, 2559–2568.

Shirley, W.L., 1968. US Patent 3394067 (United States Patent, 1968).

Smollen, M., Kafaar, A., 1994. Electroosmotically enhanced sludge dewatering: pilot-plant study. Water Sci. Technol. 30 (8), 159–168.

Song, W., Su, Y., Chen, X., Ding, L., Wan, Y., 2010. Rapid concentration of protein solution by a crossflow electro-ultrafiltration process. Sep. Purif. Technol. 73 (2), 310–318.

Stern, O., 1924. Zur Theorie der Elektrolytischen Doppelschicht. Z. Elektrochem. Angew. Phys. Chem. 30, 508–516.

Tai, C.H., Hsiung, S.K., Chen, C.Y., Tsai, M.L., Lee, G.B., 2007. Biomed. Microdevices 9 (4), 533–543.

Tarleton, E.S., 1992. The role of field assisted techniques in solid/liquid separation. Filtr. Sep. 29 (3), 246–252.

Tarleton, E.S., Wakeman, R.J., Liang, Y., 2003. Electrical enhanced washing of ionic species from fine particle filter cakes. Trans. Inst. Chem. Eng. 81 (A), 201–210.

Tije, A.T., 1999. US Patent 5891342 (United States Patent, 1999).

Trasatti, S., 2000. Electrocatalysis: understanding the success of DSA®. Electrochim. Acta 45 (15–16), 2377–2385.

Tsai, Y.T., Weng, Y.H., Lin, A.Y.C., Li, K.C., 2011. Electro-microfiltration treatment of water containing natural organic matter and inorganic particles. Desalination 267 (2–3), 133–138.

Tuan, P.A., Sillanpää, M., 2010. Migration of ions and organic matter during electro-dewatering of anaerobic sludge. J. Hazard. Mater. 173 (1–3), 54–61.

Virkutyte, J., Sillanpää, M., Latostenmaa, P., 2002. Electrokinetic soil remediation – critical overview. Sci. Total Environ. 289 (1–3), 97–121.

Vorobiev, E., Jany, S., 1999. Etapes de filtration frontale assistée par champ électrique. Entropie 219, 23–28.

Wakeman, R.J., Sabri, M.N., 1995. Utilizing pulsed electric fields in crossflow microfiltration of titania suspensions. Trans. Inst. Chem. Eng. 73 (A), 455–463.

Wakeman, R.J., 1998. Electrically enhanced microfiltration of albumin suspensions. Food Bioprod. Process. 76 (1), 53–59.

Wakeman, R.J., Williams, C.J., 2002. Additional techniques to improve microfiltration. Sep. Purif. Technol. 26 (1), 3–18.
Wakeman, R.J., Butt, G., 2003. An investigation of high gradient dielectrophoretic filtration. Chem. Eng. Res. Des. 81 (8), 924–935.
Watson, F.D., Mayse, W.D., Franse, A.D., 1983. US Patent 4372837 (United States Patent, 1983).
Weber, K., Stahl, W., 2002. Improvement of filtration kinetics by pressure electrofiltration. Sep. Purif. Technol. 26 (1), 69–80.
Weigert, T., Altmann, J., Ripperger, S., 1999. Crossflow electrofiltration in pilot scale. J. Membr. Sci. 159 (1–2), 253–262.
Weng, Y.-H., Wang, Y.-C., Tsai, Y.-T., Chuang, C.-J., Huang, C.P., Li, K.-C., 2012. Effect of hydrophobicity of humic substances on electro-ultrafiltration. Desalination 284, 128–134.
Willams, S., 2008. AC dielectrophoresis lab-on-chip device. In: Dongqing, L. (Ed.), Encyclopedia of Microfluidics and Nanofluidics. Springer, New York pp. 1–8.
Xia, B., Sun, D.-W., Li, L.-T., Li, X.-Q., Tatsumi, E., 2003. Effect of electro-osmotic dewatering on the quality of tofu sheet. Dry. Technol. 21 (1), 129–145.
Xian-shu, D., Xiao-jie, H., Su-ling, Y., Wei-peng, R., Zhi-zhong, W., 2009. Vacuum filter and direct current electro-osmosis dewatering of fine coal slurry. Procedia Earth Planet. Sci. 1 (1), 685–693.
Yamaguchi, M., Nakagawa, M., Ohanamori, H., Yoshida, M., Arai, T., Matsushita, H., 1993. US Patent 5192413 (United States Patent, 1993).
Yang, J., Huang, Y., Wang, X.B., Becker, F.F., Gascoyne, P.R.C., 2000. Differential analysis of human leukocytes by dielectrophoretic field-flow-fractionation. Biophysical J. 78 (5), 2680–2689.
Yang, G.C.C., Yang, T.Y., Tsai, S.H., 2003. Crossflow electro-microfiltration of oxide-CMP wastewater. Water Res. 37 (4), 785–792.
Yang, L., Nakhla, G., Bassi, A., 2005. Electro-kinetic dewatering of oily sludges. J. Hazard. Mater. 125 (1–3), 130–140.
Yang, G.C.C., Li, C.J., 2007. Electrofiltration of silica nanoparticle-containing wastewater using tubular ceramic membranes. Sep. Purif. Technol. 58 (1), 159–165.
Yang, Z., Peng, X.F., Lee, D.J., 2009. Electroosmotic flow in sludge flocs. Int. J. Heat. Mass Transf. 52 (13–14), 2992–2999.
Yang, G.C.C., Chen, M.C., Yeh, C.F., 2011. Dewatering of a biological industrial sludge by electrokinetics-assisted filter press. Sep. Purif. Technol. 79 (2), 177–182.
Yu, X., Zhang, S., Xu, H., Zheng, L., Lü, X., Ma, D., 2010. Influence of filter cloth on the cathode on the electroosmotic dewatering of activated sludge. Chin. J. Chem. Eng. 18 (4), 562–568.
Yuan, C., Weng, C., 2003. Sludge dewatering by electrokinetic technique: effect of processing time and potential gradient. Adv. Environ. Res. 7 (3), 727–732.
Yukawa, H., Kobayashi, K., Tsukui, Y., Yamano, S., Iwata, M., 1976. Analysis of batch electrokinetic filtration. J. Chem. Eng. Jpn. 9 (5), 396–401.
Zhou, J.X., Liu, Z., She, P., 2001. Water removal from sludge in a horizontal electric field. Dry. Technol. 19 (3–4), 627–638.
Zumbusch, P.V., Kulcke, W., Brunner, G., 1998. Use of alternating electrical fields as antifouling strategy in ultrafiltration of biological suspensions—introduction of a new experimental procedure for crossflow filtration. J. Membr. Sci. 142 (1–2), 75–86.

CHAPTER NINE

Ultrasonic

Pentti Pirkonen[1] and B. Ekberg[2]
[1]Technical Research Centre of Finland (VTT), Jyväskylä, Finland
[2]Outotec Oyj, Finland, Turku, Finland

Contents

1. INTRODUCTION

The use of ultrasonic force field to aid separation processes has gained an increasing interest in recent years. The basic phenomena affecting the ultrasonically assisted separation processes are studied intensively during the recent decades.

Properties and small-scale uses of ultrasound have been studied extensively by physicists, chemists, and others. The resulting applications can be found in several areas of industrial process engineering e.g., in extraction processes, cleaning, atomization, emulsification and cell disruption, dispersion of solids, nucleation and growth of crystals, and degassing (Tarleton and Wakeman, 1990; Povey and Mason 1998).

The present text deals with ultrasonically assisted liquid and air separation processes. The advantages expected from using US for separation processes include higher liquid removal rate, higher dry matter (DS) content in product, lower processing temperature, maintenance of product integrity, more selective product, and higher product recovery.

Progress in Filtration and Separation
http://dx.doi.org/10.1016/B978-0-12-384746-1.00009-4

The majority of studies has been performed in lab or pilot scale, but only very few applications have been proceeded to large commercial use. The efforts have been focused on changing suspensions properties to be more favourable for separation and or preventing fouling substance from sticking to the filter surface or cleaning the filter element itself. Combinations of ultrasound and electric field have also been used (Pirkonen, 2001; Pirkonen et al., 2010). Only some US separation applications which are worth mentioning are powder screens, cell separators, electro-acoustic dewatering press (EAD), CERTUS-, Sofi-, Fractor-, Fuji- and Scamsonic-screening filter and Outotec Larox CC capillary action filter (Wakeman and Tarleton, 1991; Mason and Cordemans, 1996; Tarleton and Wakeman, 1998; Pirkonen, 2001; Pirkonen et al., 2010; Tanaka et al., 2012). US assisted screens, Scamsonic filter, cell separators, and ceramic capillary filters have found industrial use. Sofi-polishing filter is in demonstration stage looking for valid commercial applications.

The main reasons hindering breakthrough of US separation techniques are nondevelopment of transducer technology, high energy consumption, problems to scale up the technology, and control of erosion caused by cavitation at high ultrasonic intensities (Pirkonen, 2001; Pirkonen et al., 2010).

2. ORIGIN OF ULTRASONICALLY INDUCED EFFECTS

In general, power US is characterized by an ability to transmit substantial amounts of mechanical power through small mechanical movements. The passing of ultrasonic waves of a suitable high intensity through liquid and gaseous media is accomplished by primary phenomena such as cavitation, radiation pressure, and secondary phenomena of a physicochemical nature such as dispersion, coagulation, and change in liquid properties. In many cases the effect of ultrasound is due to a combination of many effects acting synergistically (Lamminen et al., 2004; Ensminger, 1988; Muralidhara et al., 1986, 1987; Tuori, 1998).

Like with any sound wave, US is propagated via a series of compression and rarefaction waves induced in the molecules of the medium through which it passes (Figure 1). At sufficiently high power, the rarefaction cycle may exceed the attractive forces of the liquid molecules and cavitation bubbles will form. Cavitation occurs at frequencies of roughly 20–1000 kHz. In aqueous systems, each cavitation bubble acts as a localized "hot spot" generating temperatures of about 4000–6000 K and pressures of 100–200 MPa.

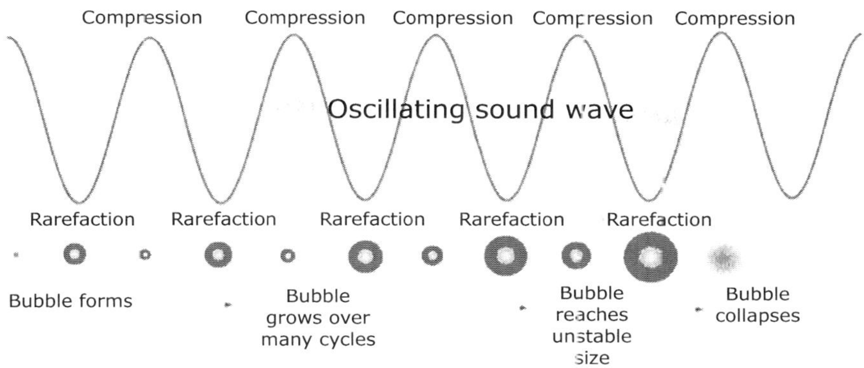

Figure 1 Phenomena in ultrasonic cavitation (Leong et al., 2011).

The implosion occurs with lifetimes of <10 μs and speed of liquid jet can be of the order of 110 m/s (Mason and Cordemans, 1996; Price, 1992). Bubble size can be as much as 100–200 μm before implosion (Price, 1992), but the most effective bubble collapse occurs at a bubble size of several micrometres (Ensminger, 1988). In a heterogeneous solid–liquid situation, liquid jet will have significant mechanical and chemical effects, e.g., on cell destruction and formation of free radicals. Liquid jet is for example used widely used in commercial cleaning processes, where ultrasonic power varies between 0.5 and 6 W/cm^2 and ultrasound frequency in the range of 16–70 kHz (Tarleton and Wakeman, 1998; Ensminger, 1988). Frequency has strong effect on particle scouring from surfaces (Figure 2) The lower the frequency the stronger will be the cavitation-implosion (www.bransoncleaning.com, 2013). General observations on the effects of liquid properties on the transient cavitation threshold are summarized in Table 1.

Ultrasonic radiation without cavitation can cause particle agglomeration due to particle collision; orthokinetic, radiation, Oseen and Stokes forces, and due to attraction of particles; and Bernoulli's effort (Muralidhara et al., 1987). An ultrasonic standing wave is the sum of two oppositely directed travelling waves, which can be originated from two independent transducers or from a transducer and a reflector. In a standing wave, the pressure amplitude has maximum and zero values twice over a distance of one wavelength. Therefore, suspended particles tend to move toward and concentrate at the positions of minimum acoustic potential energy. These localized regions are generally close to pressure nodes, which are separated from each other by distances of half a wavelength (Coakley, 1997).

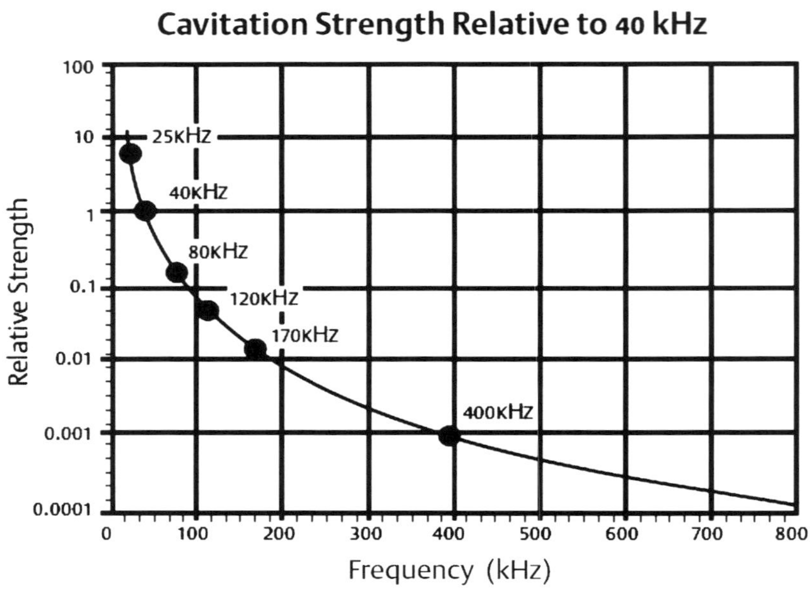

Figure 2 The impact of ultrasonic frequency on particle removal (www.bransoncleaning.com, 2013).

Table 1 Parameters affecting the cavitation threshold (Atchley and Crum, 1986; Wakeman and Tarleton 1991; Tarleton and Wakeman 1998; Tuori 1998)

Parameter	Influence of increasing parameters on threshold
Dissolved gas saturation	Decrease
Hydrostatic pressure	Increase
Surface tension	Decrease
Temperature	Decrease
Solids concentration	Decrease
Particle size	Decrease
Frequency	Increase

Ultrasonic and electric field has some synergetic effects in filtration based on following mechanisms (Ensminger, 1988; Chauhan et al., 1987; Yukawa et al., 1976):

- Greater compaction of the cake through the influence of ultrasonic stresses across the cake. This maintains electrical continuity longer and thus, increases the amount of water removal due to electro osmosis.

- Ultrasound cleans the electrode surface promoting better electrical contact.
- Combined assistance accelerates liquid transfer through the filter for example by lowering liquid viscosity and surface tension.

3. STANDING WAVE SEPARATION

The effect of particle agglomeration in a standing wave is known for a long time. A concept utilizing a standing field is shown schematically in Figure 3 (Petersson et al., 2004).

Due to the primary acoustic radiation force the suspended particles are collected in planes perpendicular to the direction of sound propagation. Subsequently, agglomeration of collected particles occurs, aided by the acoustic interaction force. After switching off the sound field, the larger agglomerates sediment due to the gravity. Method can be applied only to particles forming stable agglomerates like municipal sludge (Muralidhara et al., 1987) and fibre particles (Moreno de Barrera et al., 1995) or in general, for very low solid contents and small particle sizes (Ensminger, 1986). In principle, acoustic separation system can be designed to operate in batch mode or in continuous-flow mode. In the latter case nodal planes move slowly in pseudo-standing wave (Gröschl, 1988). Typical frequency area used for agglomeration/filtering is 2–3.3 MHz (Spengler and Jekel, 1999) but successful application of 10 and 40 kHz is also reported (Gallego-Juárez et al., 1999; Moreno de Barrera et al., 1995; Muralidhara et al., 1987).

All tough, extensive research has been dedicated to acoustic particle agglomeration/separation, most of the works have been restricted to the demonstration of fundamentals and or the development of laboratory scale devices. Particular high potential of applications exist in the field of

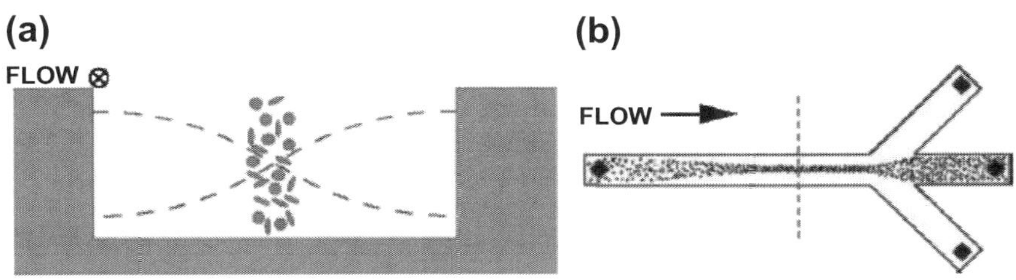

Figure 3 Principle of particle separation in a standing acoustic field. (a) Particles are positioned in the pressure nodal plane of a standing wave. (Cross-section of the channel in (b), dashed line.) (b) Top view of a continuous separation of particles, positioned in the pressure node, from a fraction of their medium (Petersson et al., 2004).

biotechnology. This is due to low flow through volumes and high product values (Gröschl, 1988). Commercial standing wave applications are published under the product marks SonoSep™ and BioSep. SonoSep™ flowrates may range from ml/h of separation for analytical and research applications, up to hl/h of separation of suspensions and emulsions, such as for recycling of fluid streams in closed loop industrial processes. Based on the required separation capacity and other application-specific parameters such as particle concentration, type, and size SonoSep™ optimizes geometry of acoustic separation chambers and electrical parameters such as power output and frequency of ultrasonic controllers and transducers. Several modes of BioSep operation are available making acoustic perfusion applicable for mammalian and animal cell culture. BioSep separation principle is based on gentle acoustically induced loose aggregation followed by sedimentation using frequency of about 2.1 MHz. BioSep operation range is between 1 and 200 L/day. Typical separation efficiency of BioSep (Figure 4) ranges from 90 to 99% (www.sonosep.com, 2013, Applikon Dependable Instruments BV, 2013).

Comprehensive state of art for the standing wave applications is presented by Spengler and Jekel (1999). They have studied standing wave

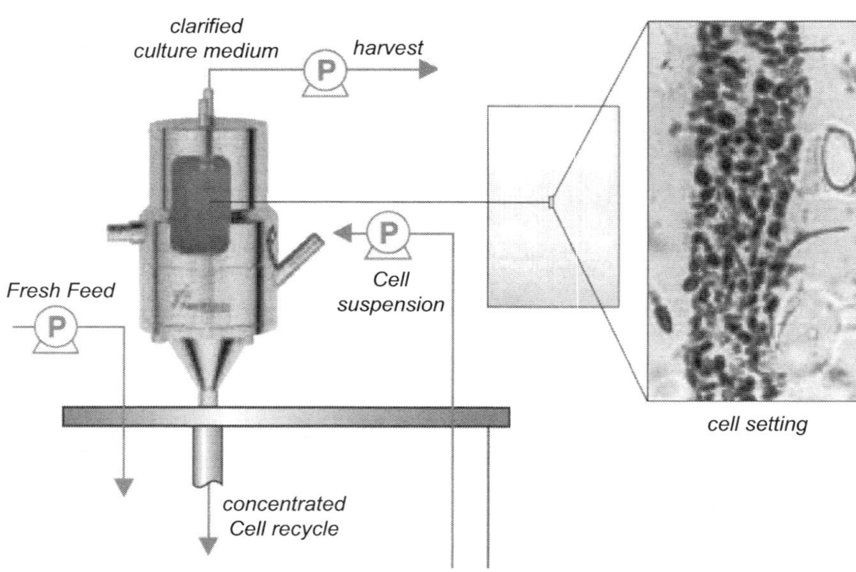

Figure 4 Typical configuration of the acoustic cell retention system (www.applikon.com, 2013).

principle using flowrate of 210 L/h for filter backwash water to separate particles using ultrasonic power of 1000 W equivalent to 0.8 W/cm^2. The separation chamber was divided by acoustically transparent film elements (ATF). Insertion of these barriers perpendicular to the acoustic/convective streaming and sound direction prevents the establishment of large range acoustic and convective streaming (Spengler and Jekel, 1999).

Instituto de Acustica in Spain has developed a semi-industrial pilot plant for reduction of particle emissions in coal combustion fumes. Removal of fine particles (smaller than 2.5 μm) from industrial flue gases is, at present, one of the most important problems in air pollution abatement. These particles penetrate deeply into the lungs and are dificult to remove. High-intensity acoustic field induces agglomeration processes that change the particle size to larger resulting in easier precipitation with a conventional separator. Installation consists of an acoustic agglomeration chamber with a rectangular cross-section, driven by four high-power and highly directional acoustic transducers of 10 and/or 20 kHz, and an electrostatic precipitator (ESP). A fluidized bed coal combustor was used as fume generator, and a sophisticated air sampling station was set up to carry out measurements with fume-flow rates up to about 2000 m^3/h, gas temperatures of about 150 °C, and mass concentrations in the range 1–5 g/m^3. The fine particle reduction was about 40% (Gallego-Juárez et al., 1999).

Large industrial applications utilizing standing wave principle are still missing due to difficulty to scale up this technology.

4. ULTRASOUND ASSISTED SIEVING

Ultrasound assisted particle sieving has been utilized for decades in industrial scale (www.russellfinex.com, 2013). Screen blocking or blinding is a common problem when sieving difficult powders on screens of 300 μm and below 100 μm sieving of powders is nearly impossible. It is particularly common with powders which are sticky or that contain a lot of particles that are similar in size to the mesh apertures (O'Connell, 2002).

Vibrational and wave motion of the sieve is responsible for throughput in conventional sieves. Blinding problems are usually tried to tackle by mechanical cleaning or by using mechanical devices, such as discs or balls which bounce up and down. Unfortunately, these actions may damage the sieve. Instead, sonoscreen transmits an oscillating motion in micron range on the screen surface. This motion reduces friction between the sieve mesh and the bulk material resulting in significant increase in throughput, and at the same time screen blinding and clogging are substantially reduced. Noise level can also be reduced when using US assisted sieving (O'Connell, 2002; Telsonic UK Ltd, 1995).

The system is composed of three parts (Figure 5):

1. The control unit, which houses all of the electronic components driving the system;
2. The acoustically developed transducer, often referred to as the probe; and
3. The mesh screen, which includes a special velocity transfer plate (VTP) to which the probe is connected.

The probe is bolted to the VTP, which, in turn, is bonded to the stainless steel wires of the sieving mesh. When the system is activated, the control box sends signals to drive the piezoelectric element in the probe through a single cable, and the probe is excited at its resonant frequency of 35 kHz. This frequency excites the velocity transfer plate, which, in turn, vibrates each individual wire of the mesh and prevents the powder from sticking to them. Ultrasonic systems have no mechanical or wearing parts, so there is no risk of mesh damage or product contamination (O'Connell, 2002). Vibration can also be arranged with a continuously varying frequency preventing high resonance amplitudes resulting in considerably less mechanical wear on the screens and significantly reducing heating (no hot spots) (Haver and Boecker, 2013).

Ultrasonic vibrating screens are widely used in industry. Ten times increase in throughput has been reported and powders having particle diameter of 20 μm have been successfully screened (O'Connell, 2002). Energy

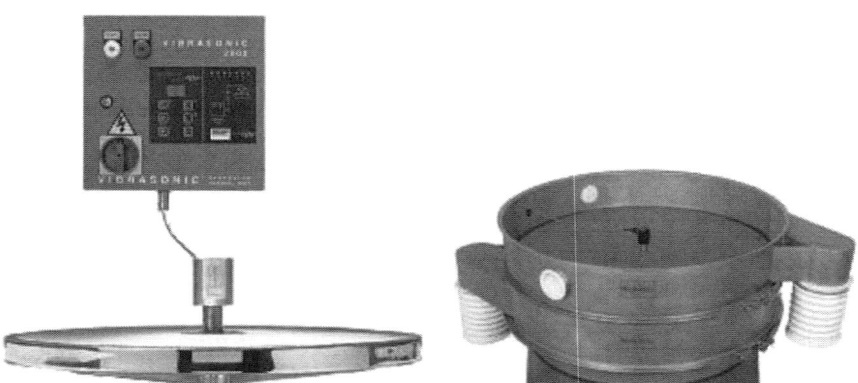

Figure 5 Ultrasonic sieving systems (O'Connell, 2002).

consumption is low; for example an ultrasonic sieve of 2 m diameter has installed power of 4 kW (www.ecvery.com, 2013).

5. POLISHING FILTRATION

US assisted clarification/classification filtration filters called Certus-, Sofi-, Fractor- and Scamsonic-filters have been developed during the recent years. All these equipments utilize ultrasound by keeping tube like filter clean preventing formation of blogging particle layer on the surface of filter element.

Certus MF filter has been built in industrial pilot-scale. Pilot filter consists of one ceramic microfilter tube having filter area of 0.33 m^2. Tube rotates against the direction of suspension to increase the efficiency of cross flow velocity. Ultrasonic transducers are placed in front of the tube and cleaning is carried out periodically to open the blocked pores. Pilot filter has been tested successfully with several process waters using pore sizes of 0.25, 1, and 3 μm. Capacities of 1.9–6.8 $m^3/(m^2h)$ have been achieved in different applications (Rantala and Kuula-Väisänen, 1999; Rantala, 1999). Unfortunately, Certus-filter has not achieved real industrial applications.

The Sofi-Filter (Figure 6) is a continuation of the Certus-filter. Sofi-Filter is based on the same cross flow phenomena as the Certus-filter, but the design is completely different. The innovation of Sofi-Filter enables higher cross flow velocity and thus remarkably higher capacity without

Figure 6 Sofi MF filter (www.sofi-filtration.fi, 2013).

any moving parts inside the filter vessel. The main argument is that the cross flow filtration is a shearing force that is produced by the flow velocity. In the new type of cross flow filter, the velocity is produced very simply and effectively. Ceramic, sintered metal, or silicon carbide can be used as filter elements. Computational fluid dynamics modelling has been made at Aalto University, and calculations show a cross flow velocity of about 20 m/s is achieved in Sofi-Filter. Patent application for the innovation has been left in. The first industrial-scale filtrations show that the shearing forces are significantly higher than, e.g., with Certus-filter. Sofi-Filter is designed for industrial processes for closed water circulation or minimizing waste contamination.

The filtration is done with over pressure, i.e., the bubble-point of the membrane is exceeded. The feed pressures are varying between 0.05 and 0.2 MPa. Sofi-Filter is a compact unit consisting of an acid-proof steel vessel, a sintered metal filtrate element, ultrasonic cleaning unit, and PLC control unit. The modular structure enables high capacity installations by economic combination of units for virtually all capacity needs. Sofi-Filter operates automatically and the filtration data is collected on-line. Feed water is pumped into a steel cylinder at a sufficient speed (e.g., 3 m/s) and the feed design accelerates the flow to about 20 m/s on the element surface to achieve a high permeate flow. To achieve the high cross flow velocity, a part of the feed, typically about 30%, is discharged through two concentrate outlets. The concentrate can be fed to next set of Sofi-Filters to lower the total concentrate stream volume. The filter element requires automated, periodic cleaning so that the surface pores remain clear. The design enables several cleaning options: ultrasonic cleaning, vapour and chemical washing, and backwash. Ultrasonic cleaning causes cavitation on the surface of the metallic membrane. A laboratory and a pilot unit are available for industrial tests. Typical filter element sizes are 1, 5, 10, and 20 μm (www.sofi-filtration.fi, 2013).

VTT in Finland has transformed Certus-filter to mechanical fractionation equipment called Fractor (Figure 7). It consists of feed container, flowmeter, pump, and fractionation modules (rotating fractionation element/cylinder tube unit). Pressure gauge and temperature sensor are also installed, so that these properties can be monitored by a data acquisition system. Ultrasound cleaning system prevents filter cake formation. There are two optional channels in the device; one with rotating fractionation element and another with cylinder tube unit. The rotating fractionation element

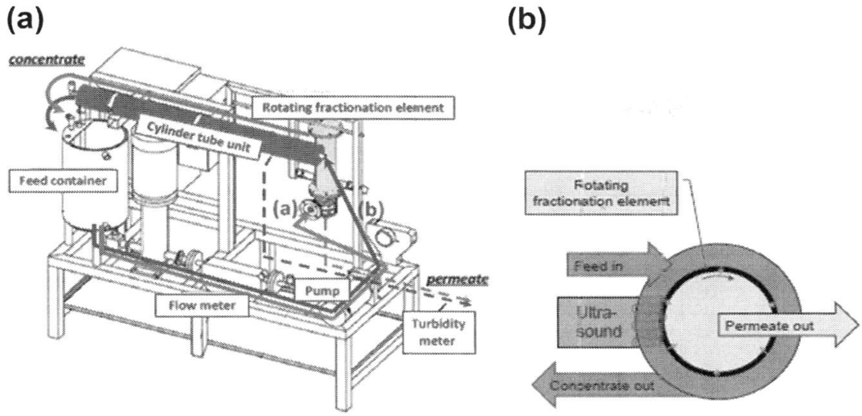

Figure 7 (a) Schematic image of the fractionation device in VTT; (a) channel with rotating fractionation element and (b) channel with membrane concentration tube, (b) Cross-sectional image of Fractor operation (Tanaka et al., 2012).

has an interchangeable screen unit inside. There are alternative screen types and sizes (apertures) as follows:

- Net type of metallic screens; 150, 75, 38, 30, 20, 10, and 1 μm,
- Sintered screens; 0.5 and 0.1 μm, and
- Slot type metallic screens; 50, 40, 30, 20, 10 and 5 μm.

The rotating fractionation element has an ultrasound cleaning system. The centrifugal force together with this ultrasound system prevents formation of a filter cake on the screen. Cleaning by backwashing is also available. The channel with cylinder tube unit is used for final concentration of suspension using polymeric membranes of ~0.1 and ~0.01 μm. Altogether 18 tubes are installed in the cylinder. The device fractionates with one screen (or membrane) at a time, i.e., in a step-wise process. The fraction which does not pass through the screen (concentrate) returns back into the feed tank. As an example of US assisted mechanical fractionation the case of nanocellulose is presented in Figure 8 (Tanaka et al., 2012).

In Japan ultrasonically aided classification filter has been developed. Ultrasound keeps filter element clean and pushes particles through the filter (Miazaki et al., 1990).

Liquid arrives in Scamsonic clarification filter (Figure 9) at the top of the filter and then passes through the filter element from inside to outside. The impurities are stopped at the surface of the cartridge but cannot settle because of intense ultrasonic activity. Sediment falls to the bottom of the

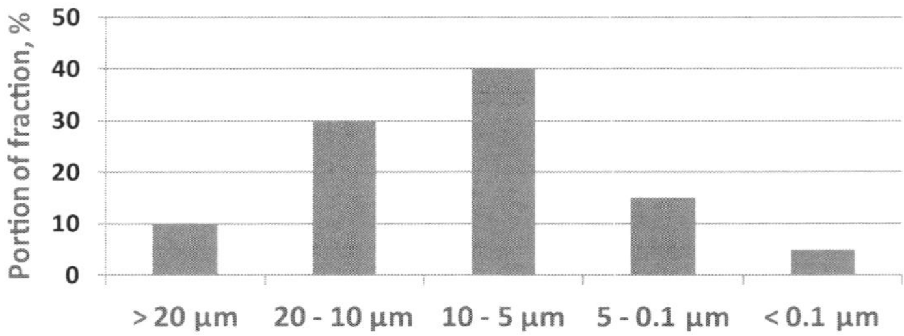

Figure 8 Effects of variables on size profile using three-stage fractionations (Tanaka et al., 2012).

Figure 9 Scamsonic clarification filter (www.scam-filtres.fr, 2013).

filter where it is extracted by cyclical cleaning—manual or automatic. Scamsonic filter can be applied for river water and drilling water filtration. Filtration grades range from 10 to 50 μm. Installation and running costs are very low (www.scam-filtres.fr, 2013).

6. SLUDGE DEWATERING

6.1 Sludge Pretreatment

Typical industrially used pretreatment methods to enhance sludge dewatering are: dosing filter aids, coagulation, flocculation, grinding, freeze/thawing, heating, wet carbonization, and pH regulation. Floc properties in sludge dewatering plays an important role. Flocs must be high in strength in centrifugation. Instead, flocs must be loose aggregates, which release their loose water rapidly in belt filtration (Friedrich et al., 1999). The effects of US in pretreatment can be as follows:

- Breaking down and formation of floc structures (Grönroos, 2010; Friedrich et al., 1999; Spengler and Jekel, 1999; Gröschl, 1988; Kopp and Dichtl 1998; Coakley, 1997; Moreno-Barrera et al., 1995; Muralidhara et al., 1987),
- Cleaning and activation of particle surfaces (Bien, 1988; Singh and Singh, 1997; Fairbanks et al., 1986),
- Disintegration of cell walls and deforming polymer chains (Grönroos, 2010; Bien and Wolny, 1997; Tiehm et al., 1997; Kowalska et al., 1988),
- Inactivation of microorganisms (Singh and Singh, 1997; Bien, 1988; Fairbanks et al., 1986) and
- Even partial breaking of hydrogen bonds and fragmentation of 2-, 4- and 8-hydrogen associates is said to occur (Bien, 1988)

DS-content of ultrasonically treated sludge has been increased by 5-units in laboratory scale filtration while the quality of effluent water remained constant (Friedrich et al., 1999). Even higher enhancements has been reported when time of sonication was 90–120 s using 20 kHz transducer prior to filtration. Sonication has even lowered the dose of polyelectrolyte (Bien, 1998) but, in another case ultrasonic disintegration of sludge has increased the dose of polyelectrolytes even by 200% (Friedrich et al., 1999).

Ultrasonic pretreatment of polyelectrolyte using 5 s 20 kHz transducer prior to laboratory scale vacuum filtration of digested sludge resulted in higher DS-content of cake with lower polyelectrolyte dosage compared to filtration without any ultrasonic treatment (Bien and Wolny, 1997).

Fairbanks has observed that treating of coal coated with sulphur clay bloom in an ultrasonic trough literally exploded the bloom away from the surface of coal particles due to the collapse of cavitation bubbles. Water content was also decreased in subsequent screening and screening rate improved significantly. Fairbanks also treated settling pond water and obtained even 10 times faster settling rates than without ultrasonic treatment and even flocculent dosage was smaller. Advanced Sonic Processing Systems (USA) has commercialized ultrasonic tray treatment process of cleaning for example contaminated earth and the capacity can be 14 t/h/tray.

Ultrasound can be used to destroy partly or totally thread bacteria in floating and bulking sludge resulting in a dramatic increase in sedimentation rate (Friedrich et al., 1999).

In spite of intensive research excluding pretreatment developed by Advanced Sonic Systems no industrial application has been performed.

6.2 Cake Filtration

Power ultrasound has been used for cleaning of cake filter surfaces in slurry filtration (Tuori 1998; Pirkonen 2001; Pirkonen et al., 2010). Electric field combined to power ultrasound has also been used to obtain synergetic effects (Chauhan et al., 1987, 1992; Golla et al., 1992; DOE/ID/13132-T1, 1994; Tuori, 1998; Tuori et al., 2000; Pirkonen, 2001).

Ceramec filter discs membranes with micropores create strong capillary suction action in contact with water. This microporous filter medium allows only liquid to flow through. Filtrate is drawn through the ceramic discs as they are immersed into the slurry bath, and a cake forms on the surface of the discs. Despite an almost absolute vacuum, no air penetrates the filter media. Consequently, Ceramec filters require only a small vacuum pump to transfer filtrate from the discs. Instead, conventional vacuum filters have high air flow through the filter cake, and need a large vacuum pump (Pirkonen et al., 2010).

Ultrasound assisted cleaning of ceramic filter elements is performed together with back flushing with filtrate or with washing chemical solution. Back flushing removes residual cake and cleans the microporous structure. Backflow washing is automatic and adjustable for each application. US transducer boxes are located between all the ceramic filter discs. Cleaning of filter media is carried out typically twice per day. During the cleaning phase slurry basin is filled with water. Alternatively pulsed ultrasonic cleaning is carried out using the slurry in the filter basin as cleaning medium (Pirkonen et al., 2010; www.Outotec.com, 2013).

The capillary action dewatering system is exceptionally efficient for applications in the mining industry. The filter disc material is resistant to most chemicals, and it has a long operational life. Ceramec filters are best suited for dewatering of slurries with high and consistent concentration of solids having the major part of the particles in the size range 30–150 μm, The Ceramec filter technology is used widely for dewatering base metal concentrates, ferrochrome, and iron ore products. Areas of application concentrates: copper, nickel, zinc, lead, and pyrite. The range of machine size covers 6–240 m^2. Industrial installation can also include several Ceramec units (Figure 10). Several hundreds of filter units have been delivered worldwide

Figure 10 Industrial installation of Outotec Larox CC capillary action filters which have ultrasonic cleaning system (www.Outotec.com, 2013).

during the past two decades. The benefits of Ceramec filters are: very low energy consumption; dry filter cake; particle-free filtrate; high filtration rates; simple installation, operation, and maintenance; integrated filter and ancillaries system, continuous operation, and high availability (www.Outotec.com, 2013).

Battelle Memorial Institute has been carrying out comprehensive ultrasonically enhanced electro-acoustic vacuum filtration studies by EAD-method. The basis of the EAD-process is a synergistic effect of combining electric and ultrasonic fields with a conventional driving force; vacuum, or pressure. Battelle Institute has tested the applicability of over 50 different types of sludge for the EAD-process. The key applications for the waste treatment are; sewage sludge, waste water treatment sludge, process effluent sludge, and hazardous waste sludge. Continuously operating vacuum filtration device was designed to be operated along the lines of commercial vacuum filters with two endless belts serving as two electrodes. The press was utilized as secondary dewaterer and could remove up to 50% water from filtered sludge cake at a fraction of the cost compared to conventional thermal drying devices. The dominant mechanism of sludge dewatering by EAD-press was electro osmosis due to the application of direct current field. The ultrasonic field helps electro osmosis by consolidation of the filter cake and release of inaccessible liquid. Later this technology was commercialized together with Ashbrook-Simon-Hartley Inc; but no real applications seem to be in the markets (Chauhan et al., 1987, 1992; DOE/ID/13132-T1, 1994).

Tuori obtained the best filtration results with postfeculent acid when combined ultrasonic and electric fields were applied and on the contrary to the results of Battelle Institute ultrasonic field had the dominant effect. Combined use of these fields increased the filtration capacity 15-fold, use of only ultrasonic field more than 10-fold and of an electric field alone 2-fold. The purity of the filtrate was clearly below the target value (solid matter content <0.5 m%). Two continuously operating transducers of 40 kHz transmitters at input power of 400 W was used in a lab scale device together with electric field using steel anode (DC, 5.5 A, 5 V) that was placed between the transducers. Viscosity of postfeculent acid was 0.13 Pa s which is over 100 times higher than the viscosity of water. Thus, mechanisms of the effect of electro-acoustic treatment consisted of lowering the viscosity, electro osmosis, electrophoresis, and also prevention of filter clogging (Tuori, 1998).

7. MEMBRANE FILTRATION

Severe problems in decline of membrane flux consist of buildup of a boundary concentration layer near the membrane surface, and fouling. Fouling includes blocking of pores by particles, adsorption of substances on the filter medium and within the pores, and formation of a cake layer of particles or colloids on the top of filter medium. Level of membrane fouling depends on feed suspension properties (particle size, particle concentration, pH, and ionic strength), membrane properties (hydrophobicity, charge, and pore size) and hydrodynamics (cross flow velocity and, transmembrane pressure) (Kyllönen, 2005; Huisman, 1998).

Ultrasonic cleaning of MF, UF, NF, and RO membranes has widely been studied, but still commercial applications are missing (Landi et al., 2011; Reuter et al., 2009; Kyllönen, 2005; Chai et al., 1999; Yamaguchi et al., 1998; Sabri et al., 1997; Masuzawa et al., 1997; Sabri 1997; Matsumoto et al., 1996; Li et al., 1995, 1996; Tarleton and Wakeman, 1990). One or more ultrasonic transducers of 27–40 kHz have been mostly used in the studies and typically they have been installed in front of the membrane surface. An example of US arrangement has been presented in Figure 11 (Kyllönen, 2005).

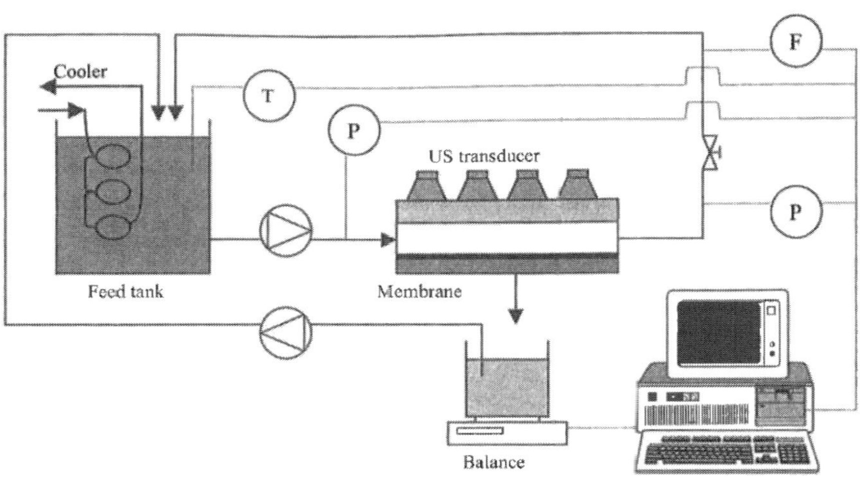

Figure 11 Schematic picture of the cross flow membrane filtration equipment (Kyllönen, 2005).

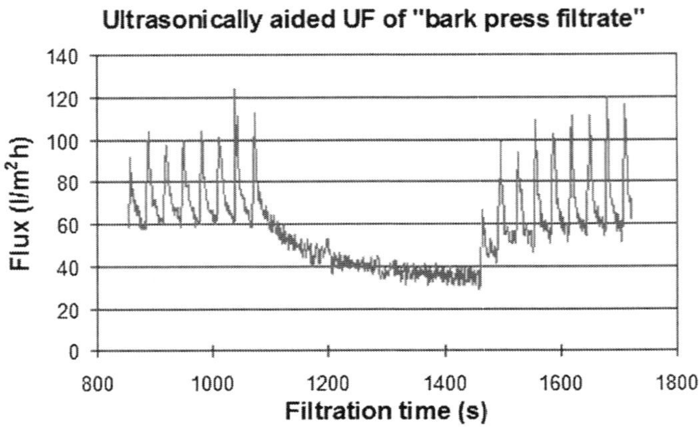

Figure 12 Flux as a function of filtration time when filtering bark press filtrate using a pressure of 3 bar and a cross flow velocity of 0.45 m/s. Intermittent ultrasound (40 kHz, 400 W) was irradiated for 5 s every 30 s (Kyllönen, 2005).

High flux enhancements have been reported. Some examples of the results are presented in Figure 12 (Kyllönen, 2005). Severe wear problem was noticed when ultrasonic radiation was used for long periods in the cavitation conditions which are expected for effective cleaning of membrane surface. Especially polymeric membranes broke down quite fast due to uncontrolled peak intensities of ultrasound (Kyllönen, 2005; Sabri et al., 1997; Sabri, 1997). Ceramic membranes tolerate better ultrasonic radiation than the polymeric ones (Lamminen et al., 2006). Energy consumption may be high in continuous use of ultrasound, but in pulsed use of ultrasound energy consumption can be decreased (Kyllönen, 2005).

MPI (www.MPIconsulting.com, 2013) offers modulated, multimode, multifrequency (MMM) ultrasonic generators, which stimulate efficient wideband acoustic energy in liquid filled chamber.

Other fixed-frequency systems are driving the total acoustic system (converter, filter housing, and liquid) at a frequency optimized for the converter without full consideration of how the filter housing and contents are changing the whole system resonant frequency.

MPI offers new design flexibility and adaptability to any size inline filtration system for most of the micron or submicron materials dispersed in a liquid. High power ultrasonic transducers are strongly (mechanically and acoustically) coupled to the external cylindrical housing of the filter element. The active filter element inside of the housing is radially stimulated (360°) by intensive ultrasonic vibrations transmitted through the liquid or liquid slurry.

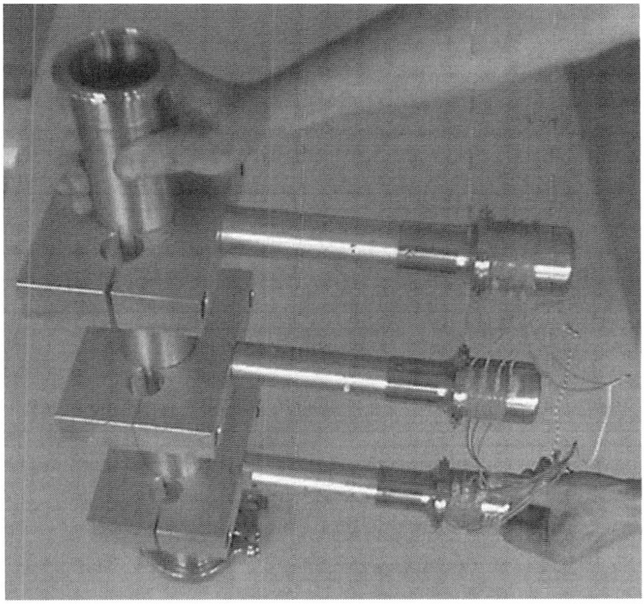

Figure 13 Example of MPI's clamp on ultrasonic (www.MPIconsulting.com, 2013).

With sufficient power and amplitude the ultrasonic vibrations will produce ultrasonic cavitation and streaming effects within the liquid and to the filter element. Oversized particles, powders, impurities, or other materials which usually fill or blind the fine filter structure are forced out of the active filter body. Flowrates are improved by a cleaner filter element plus the ultrasonic acoustic effects are actively working to reduce surface tension, cohesive forces, and or adhesive forces.

Example Applications:

- Micron and submicron materials dispersed in liquids,
- Metal powder slurry,
- Ceramic slurry,
- Biotech and Pharmaceutical products, and
- Water and waste water processing.

Transducer clamp-on technology (Figure 13) allows easy adaptation to existing housing designs and retrofit possibilities. It is suitable for any filter type: sintered metal filters, membrane, ceramic, etc. To achieve a uniform distribution of ultrasonic amplitudes and ultrasonic pressure MMM ultrasonic generator performs frequency sweeping around the optimal resonant frequency of the resonating system (e.g., 35–45 kHz). The speed of applied frequency sweeping is in the range of 50–100 sweep intervals per second. Frequency sweeping also helps particle removal from the filter surface.

To eliminate the creation of standing waves and other vibration stationary (or stable) structures inside the filtering tube, low frequency, and full power On/Off pulsing is applied to the ultrasonic signals. Pulsing shock pressure waves can be many times stronger than the effects of continuous operation.

Although the MMM technology will drive any pipe of most wall thicknesses (e.g., 1–30 mm) there are trade-offs that must be considered. In normal applications with pipe diameters of 25–100 mm the MMM technology delivers the most amplitude and best multifrequency harmonic modes with a thinner wall thickness from 1 to 2.5 mm. Applications requiring a wall thickness greater than 2.5 mm may also be driven with good success. However, more power will be required to drive the system with somewhat less amplitudes and some lesser excitation of multifrequency harmonic modes. The length of pipe effectively activated by one clamp is very dependent on many factors and must be tested for each application. Standard modules in 300, 600, 1200 W, and up to 20 kW on request are provided. Unfortunately MPI does not present any capacity range in microfiltration.

The main hindrance to apply ultrasound cleaning systems to industrial membrane filtration is high surface area needed in membrane filtration. One possibility is to find applications from the US filter cleaning presented in Chapter 5. Perhaps still some small-scale Niche application can be found in membrane filtration of expensive and low capacity suspensions.

REFERENCES

Atchley, A.A., Crum, L.A., 1986. Acoustic cavitation and bubble dynamics. In: Suslick, K.S. (Ed.), Ultrasound. VCH Publishers Inc, New York, pp. 1–63, 1988.

Bien, J., November/December, 1988. Ultrasonic preparation of sludges to improve dewatering. Filtr. Sep., 425–426.

Bien, J., Wolny, L., 1997. Changes of some sewage sludge parameters prepared with an ultrasonic field. Water Sci. Technol. 36 (11), 101–106.

Chai, X., Kobayashi, T., Fuji, N., 1999. Ultrasound associated cleaning of polymeric membranes for water treatment. Sep. Purif. Technol. 15, 139–146.

Chauhan, S.P., Muralidhara, H.S., Kim, B.,C., Senapati, N., Beard, R.E., Jirjis, B.,F., 1987. Electroacoustic dewatering (EAD) – a novel process. In: Proceedings of Summer National AICHE Meeting. Minneapolis, August 16–19, 1987, p. 15.

Chauhan, S.P., Johnson, H., Senthilnathan, P., February 1992. Application of electroacoustic for dewatering of pharmaceutical sludge. Environ. Prog. 11 (1), 74–79.

Coakley, W.,T., December 1997. Ultrasonic separations in analytical biotechnology. Trends Biotechnol. 15, 506–511.

DOE/ID/13132-T1, February 1994. Improved Electroacoustic Dewatering (EAD) Belt Press for Food Products, Phase III. Technical Progress Report, 38 p.

Ensminger, D., 1986. Acoustic dewatering. Chapter 13. In: Muralidhara, H.S. (Ed.), Advances in Solid-liquid Separation. Battelle Press.

Ensminger, D., 1988. Ultrasonic: Fundamentals, Technology and Applications. Revised and Expanded, second ed. Marcel Dekker Inc. 581 p.

Fairbanks, H.V., Morton, W., Wallis, J., 1986. Separation process aided by ultrasound. In: Proceedings of 4th World Filtration Congress 1986.

Friedrich, H., Potthoff, A., Friedrich, E., Hielscher, H., 1999. Improving settling properties and dewaterability of sewage sludges by application of the ultrasound technology. In: TU Hamburg-Harburg Reports on Sanitary Engineering, 25, pp. 189–204.

Gallego-Juárez, J., Riera, J., Rodríguez, C., Hoffmann, T., Gálvez, J., 1999. Application of acoustic agglomeration to reduce fine particle emissions from coal combustion plants. Environ. Sci. Technol. 33, 3843–3849.

Golla, P., Johnson, H., Senthilnathan, P., 1992. Application of Electro Acoustics For Dewatering Pharmaceitical Sludge. Env. Progress 27 (1), 74–77.

Grönroos, A., 2010. Ultrasonically Enhanced Disintegration. VTT Publications 734. Dissertation for the Degree of Doctor in Philosophy. University of Jyväskylä, Department of Chemistry. 2010, 216 p.

Gröschl, M., 1998. Ultrasonic separation of suspended particles-part II: design and operation of separation devices. Acust. Acta Acust. 84, 632–642.

Gröschl, M., Burger, W., Handl, B., 1988. Ultrasonic separation of suspended particles – part III: applications in biotechnology. Acust. Acta Acust. 84, 815–822.

Huisman, I., 1998. Crossflow microfiltration of particle suspensions (Thesis). Department of Food Processing, Lund University 1998.

Kopp, J., Dichtl, N., 1998. Konditierungs- und Entwässerungsverhalten von aufgeschlossenen und gefaulten schlammen. In: Klärschlammdesintegration-Forschung und Anwendung, vol. 61. Veröffentlichung des Instituts fur Siedlungswasserwirtschaft der Tech. Univ. Braunsweig, pp. 215–228.

Kowalska, E., Bien, J., Zielewicz-Madej, E., 1988. Ultrasound in suspension separation methods. Dry. Technol. 6 (3), 447–471.

Kyllönen, H., 2005. Electrically or Ultrasonically Enhanced Membrane Filtration of Wastewater. Dissertation for the degree of Doctor of Science in Technology. VTT publications 576. Espoo 2005. 79 p. + app. 54 p.

Lamminen, M.O., et al., 2004. Mechanisms and factors influencing the ultrasonic cleaning of particle-fouled ceramic membranes. J. Membr. Sci. 237, 213–223.

Lamminen, M.O., et al., 2006. Cleaning of particle fouled membranes during crossflow filtration using an embedded ultrasonic transducer system. J. Membr. Sci. 283, 225–232.

Landi, M., Naddeo, V., Belgorno, V., 2011. Membrane ultrafiltration enhanced by ultrasound: effect of different frequencies on fouling control. In: Proceedings of the 12th International Conference on Environmental Science and Technology. Rhodes, Greece September 8–10, 2011, A-1038-a-1045.

Leong, T., Ashokkumar, M., Kentish, S., August, 2011. No 2. The Fundamentals of Power Ultrasound–A Review, vol. 39, pp. 54–63.

Li, H., Ohdaira, E., Ide, M., 1995. Enhancement in diffusion of electrolyte through membrane using ultrasonic dialysis equipment with plane membrane. Jpn. J. Appl. Phys. 34, 2725–2729.

Li, H., Ohdaira, E., Ide, M., 1996. Effects of ultrasonic irradiation on permeability of dialysis membrane. Jpn. J. Appl. Phys. 35, 3255–3258.

Mason, T., Cordemans, D., July 1996. Ultrasonic intensification of chemical processing and related processes. A review. Trans. IChemE 74 (Part A), 511–516.

Masuzawa, N., Yoshimura, T., Ohdaira, E., Ide, M., 1997. A study on speed-up of integrated separation with ultrasound in semi-permeable tube membrane. Jpn. J. Appl. Phys. 36, 3136–3137.

Matsumoto, Y., Miwa, T., Nakao, S., Kimura, S., 1996. Improvement of membrane permeation performance by ultrasonic microfiltration. J. Chem. Eng. Jpn. 29 (4), 561–567.

Miazaki, K., Takahashi, Y., Shiomi, K., January/February 1990. Classifying of particles by ultrasonic filter. Separation, 28–31.

Moreno de Barrera, L., Gonzales, I., Carbo-Fite, R., Gallego-Juarez, J., 1995. Effects of high power ultrasound on suspensions. In: Proceedings of 1995 World Congress on Ultrasonics, pp. 749–752.

Muralidhara, H.S., Senapati, N., Ensminger, D., Chauhan, S.P., November/December 1986. Electro-acoustic separation process for fine particle suspensions. Filtr. Sep., 351–353.

Muralidhara, H.S., Beard, R.B., Senapati, N., Novemeber/September 1987. Mechanisms of ultrasonic agglomeration for dewatering colloid suspensions. Filtr. Sep., 409–413.

O'Connell, R., 2002. Ultrasonic Deblinding. Reprinted from the November 2002 issue of Ceramic Industry.

Petersson, F., Nilsson, A., Jönsson, C., Laurell, T., 2004. Separation of lids from blood utilizing ultrasonic standing waves in microfluidic channels. Analyst 129, 938–943.

Pirkonen, P., 2001. Ultrasound in filtration and sludge dewatering. In: Mason, T., A.Tiehm (Eds.), Advances in Sonochemistry, Ultrasound in Environmental Protection, vol. 6. Jai Press, Amsterdam, pp. 221–246.

Pirkonen, P., Grönroos, A., Heikkinen, J., Ekberg, B., 2010. Ultrasound assisted cleaning of ceramic capillary filter. Ultrason. Sonochem. 17, 1060–1065.

Povey, J., Mason, T. (Eds.), 1998. Ultrasound in Food Processing. Blackie Academic & Professional, London, 282 p.

Price, G.J., 1992. Current Trends in Sonochemistry. Royal Society of Chemistry, Cambridge, 183 p.

Rantala, P., 1999. CERTUS-Kirkastussuodin-EKY-08. In the Year Book 1999 of Technology Programme: Water Management in Papermaking, pp. 87–88. Abstract in English.

Rantala, P., Kuula-Väisänen, P., 1999. Filtration of paper mill process waters using Certus-polishing filter with ceramic membrane. In: Proceedings of 6th IAWQ Symposium on Forest Industry Wastewaters. June 6–10, 1999 Tampere, pp. 239–244.

Reuter, F., Mettin, F., Lauterborn, W., 2009. The effects of ultrasonic parameters on pressure fields in a membrane cleaning application. In: NAG/DAGA 2009-Rotterdam, pp. 1703–1706.

Sabri, M., 1997. Effective Utilization of Energy by Filter System: Introducing Electric and Ultrasonic Methods to Control Fouling in Membrane Filtration. VTT Energy-European Union; Non-Nuclear Programme. Final Report; Contract No. JO-CT97–5002, 32 p.

Sabri, N., Pirkonen, P., Sekki, H., 1997. Ultrasonically enhanced membrane filtration of paper mill effluent. In: Proceedings of Applications of Power Ultrasound in Physical and Chemical Processing. Toulouse November 18, 1997, pp. 99–104.

Singh, B.P., Singh, R., 1997. Investigation on the effect of ultrasonic pretreatment on selective separation of iron values from iron ore tailings by flocculation. Sep. Sci. Technol. 32 (5), 993–1002.

Spengler, J., Jekel, M., 1999. Ultrasound assisted solid-liquid separation in environmental and water technology-large scale processing performance enhancement and small scale analytical applications. In: TU-Hamburg-Harburg Reports on Sanitary Engineering, vol. 25, pp. 189–204.

Tanaka, A., Seppänen, V., Houni, J., Sneck, A., Pirkonen, P., 2012. Nanocellulose characterization with mechanical fractionation. Nordic Pulp Pap. Res. J. 27 (4), 689–694.

Tarleton, E., Wakeman, R., May/June 1990. Microfiltration enhancement by electrical and ultrasonic force fields. Filtr. Sep., 192–194.

Tarleton, E., Wakeman, R., 1998. Ultrasonically assisted separation processes. In: Povey, J., Mason, T. (Eds.), Ultrasound in Food Processing. Thomas Science, pp. 193–218.

Telsonic UK Ltd., May 1995. Product news. Filtr. Sep., 364.

Tiehm, A., Nickel, K., Neis, U., 1997. The use of ultrasound to accelerate the anaerobic digestion of sewage sludge. Water Sci.Technol. 33 (11), 121–128.

Tuori, T., 1993. Enhancing filtration by electro-acoustic methods (Ph.D. thesis). Loughborough University, UK, 162 p.

Tuori, T., Heikkinen, J., Wakeman, R., Ekberg, B., Gallego-Juarez, J , 2000. Development of deliquoring method enhanced by electric and acoustic force fields. In: Proceedings of 8th World Filtration Congress. April 3–7, 2000.

Wakeman, R., Tarleton, E., September 1991. An experimental study of electro-acoustic cross flow microfiltration. Trans. IChemE 69 (Part A), 387–397.

Yamaguchi, R., Ohdaira, E., Masuzawa, N., Ide, M., 1998. Dependence of permeability of semipermeable membranes on frequency of ultrasonic radiation. Jpn. J. Appl. Phys. 37, 2988–2989.

Yukawa, H., Yoshida, H., Kobayashi, K., Hakoda, M., 1976. Fundamental study on electro-osmotic dewatering of sludge at constant electric current. J. Chem. Eng. Jpn. 9 (5), 402–407.

LIST OF RELEVANT WEB SITES

www.applikon.com, 2013.
www.bransoncleaning.com, 2013.
www.ecvery.com, 2013.
www.haverboecker.com, 2013.
www.MPIconsulting.com, 2013.
www.outotec.com, 2013.
www.russellfinex.com, 2013.
www.scam-filtres.fr, 2013.
www.sofi-filtration.fi, 2013.
www.sonosep.com, 2013.

SECTION THREE

Membranes

CHAPTER TEN

Ion Exchange

Claudia Cobzaru[1] and Vassilis Inglezakis[2]
[1]Technical University "Gh Asachi" Isai, Faculty of Chemical Engineering and Environmental Protection, Romania
[2]Chemical Engineer (MSc, PhD) Environmental Consulting & Project Management Romania

Contents

1. ION EXCHANGE PROCESS

1.1 General

Ion exchange is a process in which mobile ions from a liquid phase are exchanged for ions that are electrostatically bound to the functional groups contained within a solid matrix. When the functional groups are negatively charged the exchange will involve cations and when they are positively charged they involve anions (Braun et al., 2002). By taking advantage of the fact that, under certain conditions, ion exchange media has a greater affinity for certain ionic species than for others, a separation of these species can be made. For example, the hydrogen form of a cation exchanger will

Progress in Filtration and Separation
http://dx.doi.org/10.1016/B978-0-12-384746-1.00010-0

release its hydrogen ion into solution and pick up another ion from the solution according to the following equation (Braun et al., 2002):

$$R{-}H + X^{+} \leftrightarrow R{-}X + H^{+}$$

where R represents the insoluble matrix of the ion exchange resin; X is ion from the solution.

The negative counter ion is not affected by the exchange since every ion removed from solution is replaced by a hydrogen ion and electroneutrality is maintained. For example, Inglezakis et al. (2003a) have studied the removal of Pb^{2+} in a batch system using the chelating resin Lewatit TP-207. The ion exchange process is followed by an irreversible reaction in the resin phase namely:

$$R{-}2Na + Pb^{2+} \leftrightarrow R{-}Pb + 2Na^{+}$$

where (R) is the resin organic part.

To achieve the removal of both positively and negatively charged ions from solution, a mixture of cation and anion resins in a mixed bed system is often used. For a NaCl solution the ion exchange process will be:

$$R{-}H + Na^{+} \leftrightarrow R{-}Na + H^{+}$$

$$R{-}OH + Cl^{-} \leftrightarrow R{-}Cl + OH^{-}$$

$$2H^{+} + OH^{-} = H_2O$$

Since H_2O is only weakly dissociated, the reactions of ion exchange are driven in this case to the right hand side of the equation (Braun et al., 2002).

It was noticed that ion exchange can be seen as a reversible chemical process. Inglezakis and Poulopoulos (2006) have claimed that the characterization of an ion exchange as a "chemical process" is rather misleading. Ion exchange is a redistribution of ions between two phases by diffusion, and chemical factors are less significant or even absent. Only when an ion exchange is accompanied or followed by a reaction such as neutralization can the whole phenomenon be characterized as "chemical". For example, in chelating resins the ion exchange is followed by a chemical reaction and bond formation between the incoming ion and the solid matrix (Inglezakis and Poulopoulos, 2006).

Also, it was accepted that adsorption and ion exchange can be grouped together as sorption for a unified treatment in practical applications. Ion exchange is similar to adsorption process because the mass transfer from a fluid to a solid phase is common in both processes. There are, however, some significant differences namely (Inglezakis and Poulopoulos, 2006):

- The sorbed species are ions in ion exchange, whereas electrically neutral substances are sorbed in adsorption.
- In ion exchange the ions removed from the liquid phase are replaced by ions from the solid phase. So, there occurs actually an exchange of ions and not only a removal in the latter process.
- In ion exchange, the fluxes of at least two different ions are coupled with each other and this process cannot be described by the use of a simple diffusion equation, as in the case of adsorption and the exchange of isotopes (or isotopic exchange).

The electric coupling of the ionic fluxes and the stoichiometric nature of the ion exchange process are the most important features, which distinguish ion exchange from adsorption and isotopic exchange. As a result, the quantitative treatment of ion exchange is much more complicated than adsorption or isotopic exchange (Inglezakis and Poulopoulos, 2006).

The ion exchange is one of the most important method applied for modifying the zeolite characteristics. As already known, due to the (AlO_4) tetrahedrons the zeolite network contains an excess of negative charge that is balanced by the cations distributed in a heterogeneous manner inside the structure. Generally the zeolites contain two types of cations, namely, "located" cations which are attached to certain structure sites (by electrostatic attraction) and "free", hydrated cations, randomly distributed within the large zeolite cavities. The number, size, valency, and location of the interchangeble cations in the crystalline network have significant effects on the size and shape of the pores of access into the zeolite cavities and cause the modification of their adsorptive and catalytical properties. The ion exchange makes possible the systematic size modification of the pores of access into the mono-, di-, or tri-dimensional channel system characteristic of zeolites. The zeolite behavior to the ion exchange depends on the following factors (Cruceanu et al., 1986):

- The cation nature in virtue of its size, electric charge, and the hydrated or anhydrous state;
- The zeolite internal state, meaning its secondary structural units, shape, and size of the cavities as well as the type of the resulting channels that can be mono-, di-, tri-dimensional, and either closed or opened;

- The cation concentration in the exchanging solution;
- The temperature of the ion exchange procedure;
- The nature of the anion associated to the exchange cation;
- The pH value of the exchange solution;
- The solvent nature: aqueous solution, organic solvents, or their mixture.

Also, ion exchange has proven to be one of the most utilized techniques to remove polluting cations from wastewater, because it allows (Colella, 1995; Liguori et al., 2006):

- Good performances.
- Reasonable costs.
- Metal recovery.

Four aspects are important with ion exchange applications (Post, 1991):

- The *ion exchange equilibrium* is the most important aspect with ion exchange.
- *Kinetics of ion exchange* consisting in the making evident the evolution in time of the migration of the exchange ion to the site or the replacement of the initial cations in the structure.
- *Ion exchange capacity* is expressed by the number of cation milliequivalents (meq) per ion exchange material gram. The ion exchange capacity can be affected by the following factors: pH number; temperature; nature and concentration of cations; solvent nature; and presence of complex ions (Barrer and Galabova, 1973).
- *Selectivity for certain cations.* The selectivity is an essential factor since it influences upon the potential technical applications in various activity fields (agriculture, industry, medicine, etc.).

As regards *the ion exchange equilibrium*, its particular importance should be mentioned due to the significant influence on the thermal and chemical stability as well as on the adsorption and catalytical activity of ion exchange materials. Ion exchange equilibrium can be considered to be analogous to chemical equilibrium. Cruceanu et al. (1986) gave detailed theoretical information on the ion exchange as follows:

The ion exchange reaction in an ion exchange material can be expressed by the following general equation:

$$z_A B^{z_B^+}_{(Z)} + z_B A^{z_A^+}_{(S)} \rightleftharpoons z_A B^{z_B^+}_{(S)} + z_B A^{z_A^+}_{(Z)} \quad (1)$$

where $Z^+{}_A$ and $Z^+{}_B$ denote the charges of the exchanging cations, A and B, respectively, and the indexes (Z) and (S) stand for ion exchange material and solution, respectively.

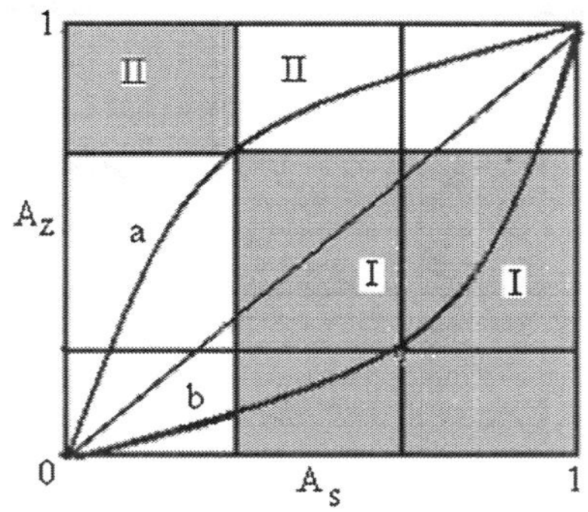

Figure 1 Graphical representation of an exchange isotherm.

The equivalent fractions of the exchanging cation in solution, (A_S), and in material, (A_Z), are given by the relationships (2):

$$A_S = \frac{z_A m_S^A}{z_A m_S^A + z_B m_S^B}; \quad A_Z = \frac{\text{number echiv. exchange cations A}}{\text{total number echiv. cations in material}} \tag{2}$$

where Z_A and Z_B have the same significance as in the relationship (1) and m_S^A and m_S^B stand for the molalities of the ions, A and B, in the equilibrium solution.

It must be taken into account that: $A_z + B_z = 1{,}0$ and $A_S + B_S = 1{,}0$.

An isotherm of ion exchange is obtained by plotting the A_Z versus A_S for a given total concentration of the equilibrium solution and a constant temperature (Figure 1).

The material affinity toward one of two or several cations involved in the ion exchange is expressed quantitatively by means of the separation factor, α_B^A, given by the definition relationship:

$$\alpha_B^A = \frac{A_Z/A_S}{B_Z/B_S} = \frac{A_Z \cdot B_S}{A_S \cdot B_Z} \tag{3}$$

The separation factor depends on the solution total concentration, temperature, and ion exchange time. There are two possible cases:

1. For $\alpha_B^A = 1$, the ionic exchange is ideal and it obeys the law of mass action (the diagonal in Figure 1);
2. For $\alpha_B^A \neq 1$, there is a certain preference:
   a. $\alpha_B^A > 1$ is indicative of the ion exchange material preference for the A cation (Figure 1, curve a),

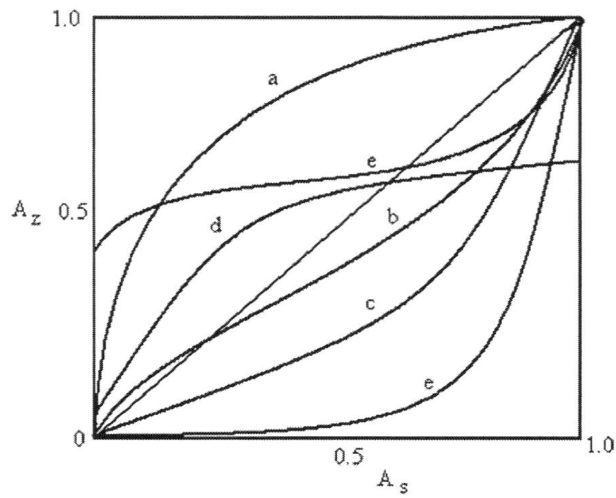

Figure 2 Types of the ion exchange isotherms in zeolite molecular sieves: a – $\alpha_B^A > 1$; b – α_B^A vary with the exchange degree; e – $\alpha_B^A < 1$; d-effect of "ionic sieve"; e-hysteresis effect of the ion exchange.

b. $\alpha_B^A < 1$ is indicative of the A cation "rejection" by material, the cation remaining in the exchanging solution (Figure 1, curve b).

The separation factor (α_B^A) can be estimated from the exchange isotherm graph for any of its points by means of the ratio of the corresponding areas. For example, the separation factor for a certain point on the isotherm (Figure 1) is the following:

$$\alpha_B^A = \frac{A_Z \cdot B_S}{B_Z \cdot A_S} = \frac{\text{Area I}}{\text{Area II}}$$

Unlike the most noncrystalline exchangers some cations in materials are blocked inside the structure while others cannot enter due to their size and hydration degree so that very often a complete ion exchange cannot be achieved. In such cases, the exchange isotherm could come to an end in a point where the exchange degree (x) is below 1. For example, Breck (1974) has settled the main types of ion exchange isotherms characteristic of the zeolite molecular sieves (Figure 2).

According to the data in Figure 2, the following conclusions are drawn:

- Isotherms of the (a) curve shape are indicative of the zeolite molecular sieves preference to the entering cation, A, $(\alpha_B^A > 1)$ and are always situated above the diagonal;
- The isotherms of the (c) curve shape are indicative of the "rejection" of the entering A cation by zeolite molecular sieves $(\alpha_B^A < 1)$, its preference

to solution, being always situated under the diagonal of the perfect equilibrium;

- The isotherms of the (b) curve sigmoidal shape reveal the selectivity dependence on the exchange degree;
- The isotherms characteristic of the (d) curve indicate the zeolite molecular sieves preference to the entering cation, but the maximum exchange degree is not attained due to an "ionic sieve" effect ($x_{max} < 1$);
- The isotherms of the (e) curve shape are rather seldom encountered, in their cases a hysteresis effect arises due to the formation of two zeolite phases.

When the mass action law is applied to the ion exchange equilibrium the selectivity rational coefficient or the concentration constant, K_B^A can be obtained:

$$\text{When } z_A = z_B, \text{ it} \quad K_B^A = \frac{A_z^{zB} \cdot B_z^{zA}}{B_z^{zA} \cdot A_z^{zB}} \tag{4}$$

follows $K_B^A = (\alpha_B^A)^{zA}$

$$\text{When } z_A \neq z_B, \text{ it follows } \left(\alpha_B^A\right)^{zA} = K_B^A \left(\frac{A_z}{A_s}\right)^{z_A - z_B} \tag{5}$$

For the monovalent ions, $K_B^A = \alpha_B^A$

By taking the activity coefficients of the γ_A, γ_B ions in the equilibrium solution into account the corrected selectivity coefficient K_c can be obtained:

$$K_c = \frac{A_z^{zB} \cdot B_s^{zA} \cdot \gamma_z^{zA}}{B_z^{zA} \cdot A_s^{zB} \cdot \gamma_A^{zA}} \tag{6}$$

In case of the electrolyte solutions,

$$K_c = K_B^A \cdot \frac{\left[\gamma_{\pm BY_{zB}}^{(zB+1)}\right]^{zA}}{\left[\gamma_{\pm AY_{zA}}^{(zA+1)}\right]^{zB}} \tag{7}$$

where $\gamma_{\pm AY_{zA}}$ and $\gamma_{\pm BY_{zB}}$ are the activity average molal coefficients of the salts in the mixture phases of the salt solutions (Y = anion in solution) that can be estimated from the Glueckauf equation (Glueckauf, 1949).

The equilibrium thermodynamic constant, K_a is defined by the relationship:

$$K_a = K_c \frac{f_{A(z)}^{zB}}{f_{B(z)}^{zA}} \tag{8}$$

where $f_{A(z)}$ and $f_{B(z)}$ are the activity coefficients of the A and B ions in ion exchange material.

It can be determined by means of the Gaines and Thomas equation (Gaines and Thomas, 1953), which is similar to that advanced by Ekedal et al. (1950):

$$\ln K_a = (z_B - z_A) + \int_0^1 \ln K_c dA_z \tag{9}$$

where $(z_A - z_B)$ denotes the charge difference between the two competing ions.

The terms refering to the modification of the water activity in the solution and material phases have been neglected.

The integral in Eqn (9) can be graphically evaluated from the area of the $\ln K_C - A_Z$ curve (Figure 3).

In this case, the following conditions are mentioned (Cruceanu et al., 1986):

- When the variation $\frac{d\ln K_c}{dA_z} = 0$, $K_a = K_c$ for $\frac{f_{A(Z)}}{f_{B(Z)}} = 1$ (straight line 1);
- When the variation is linear, $\frac{d\ln K_c}{dA_z}$ is constant (straight line 2);
- When the variation $\frac{d\ln K_c}{dA_z}$ is not constant the resulting curves show minima (curve 3), inflexion points (curve 4) or maxima (curve 5).

In case of the ion exchange of the monovalent ions at equilibrium, the isotherms represented by the equivalent fractions $\frac{A_z}{A_s}$ (Figure 3) are almost independent on the solution ionic strength since:

$$K_C = \frac{A_z B_s}{B_z A_s} \frac{\gamma^2_{\pm BY}}{\gamma^2_{\pm AY}} = \text{constant for } A_z = \text{constant} \tag{10}$$

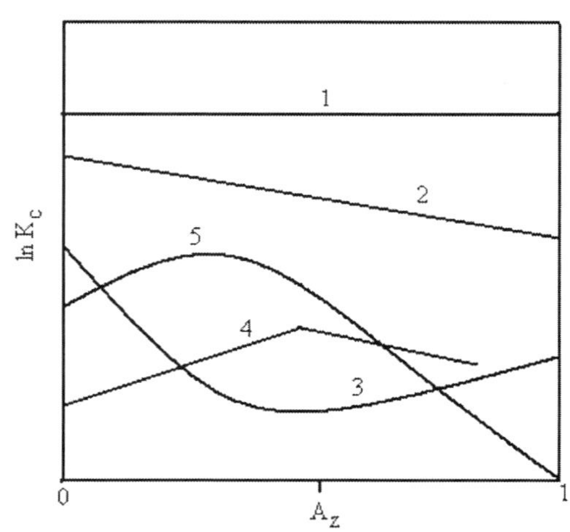

Figure 3 Variation of the corrected selectivity coefficient, ln K_c, with material composition.

where $\gamma_{\pm BY}$ and $\gamma_{\pm AY}$ are the activity coefficients of the AY and BY salts in solution at equilibrium and Y is the anion accompanying the cations.

With diluted solutions the ratio $\frac{\gamma^2_{\pm AY}}{\gamma^2_{\pm BY}} \approx 1$, (although $\gamma_{\pm BY}$ and $\gamma_{\pm AY} \neq 1$) thus:

$$K_c = \frac{A_Z B_S}{B_Z A_S} = \alpha^A_B \tag{11}$$

In case of the mono-divalent ion exchange the corrected selectivity coefficient becomes:

$$K_c = \frac{A_z B_S^2}{B_Z^2 A_S} \cdot \frac{\gamma^4_{\pm BY}}{\gamma^3_{\pm AY_2}} \tag{12}$$

And the equilibrium thermodynamic constant is given by the relationship:

$$K_a = K_c \frac{f_{A(z)}}{f^2_{B(z)}} \tag{13}$$

The K_c must also be of a unique value for a given A_z value so that the ratio $K_c = \frac{B_s^2}{A_s} \cdot \frac{\gamma^4_{\pm BY}}{\gamma^3_{\pm AY_2}}$ must remain constant. This could be possible only if the ratio of the activity coefficients of the ions in solution is close to unity. But the exchange isotherms depend on the entire cation amount in solution and, hence, on the $\frac{B_s}{A_s}$ ratio. Unlike the monovalent exchange, the divalent exchange can vary continuously. In function of the normality of the exchanging solution, the exchange isotherm will suffer a transposition making evident the concentration-valency effect. As a consequence, the more diluted the solution the more selective toward the divalent cation the ion exchange material becomes. As regards the concentration-valency effect, it was determined quantitatively by Barrer and Klinowski (1974). The authors have pointed out that for the mono-divalent exchange the equilibrium condition is given by the relationship:

$$K_a = \frac{2A_Z \cdot B_S^2 \cdot f_A \cdot \gamma_B^2 \cdot N}{B_Z^2 \cdot A_S \cdot f_B^2 \cdot \gamma_a} \tag{14}$$

where N is the total normality of the solution.

Although the above-mentioned information given by Cruceanu et al. (1986) on the ion exchange support its better understanding, the cations do not obey always the conventional rules of the ion exchange characteristic of the organic or inorganic ion exchangers. This is why the assertion has been made that the ion exchange behavior of every ion exchange material to every metallic or organic cation has to be known. Numerous studies are to be found in literature on the subject. For example, those performed by Barrer and Meier (1958, 1959), Barrer et al. (1963, 1968), Barrer and Klinowski (1974), Sherry and Walton (1967), and Sherry (1966) and who laid the foundations of the thermodynamics and thermochemistry of the ion exchange on zeolite molecular sieves are among the first of this type, especially for the A, X, and Y type zeolites.

1.2 Kinetics of Ion Exchange

The kinetics of ion exchange consists in making evident the evolution in time of the migration of the exchanging ion to the center or the replacement of the initial cations in the structure. The migration of the exchanging ion proceeds within the stationary film around the particle whose thickness decreases when stirred. The cation diffuses through this film, through pores and channels to the locations of the internal cations where the exchange is performed and then the replaced cation moves in the opposite direction (Szostak, 1989).

The ion exchange rate is influenced by ion exchange material grain size and temperature. Thus, the small particles assure a higher diffusion rate of the ions through pores than the bigger ones due to the shorter way along the channels. If the fact is taken into account that the cation location inside material is not rigid, the ion exchange can proceed into one or several stages. The overall rate of the process developing into several stages is limited by the slowest stage rate.

The ion exchange is known as a mass transfer process in a heterogeneous system that can be described by a diffusion model. The diffusion is the rate stage for the migration of the mobile cations only and not for the bound ones. The diffusion develops according to the Fick second law and the exchange of the bound and mobile cations is of the first rate order:

$$\frac{\partial C_{\mathrm{i}}}{\partial t} = D_{\mathrm{i}} \frac{\partial^2 C_{\mathrm{i}}}{\partial x^2}$$

The ion exchange materials in fine crystalline powders form showing high self-diffusion coefficients kinetic studies are rather difficult to perform due to the extremely short reaction time. For this reason two working techniques have been advanced, namely (Cruceanu et al., 1986):

1. The method of filtration under pressure, when the exchange solution is quickly passed over an ion exchange material thin disk;
2. The method of temperature rising, when the ion exchange material is introduced quickly into a solution at a high temperature and cooled rapidly after a while at $-110\ ^{\circ}\mathrm{C}$ (ethylic alcohol cooled with liquid air).

In both cases the isotopic exchange is used.

In case of crystalline ion exchange material the ion exchange is controlled by the ion diffusion into the structure and can be described by the Boyd–Adamson relationship:

$$U(t) = 1 - \left(\frac{6}{\pi^2}\right)\sum_{n=1}^{\infty}\left(\frac{1}{n^2}\right)\exp\left(-n^2 Bt\right) \quad (15)$$

where $B = \frac{\pi^2 D_1}{R^2}$ is a diffusion coefficient in material (characteristic frequency); D_1 is the apparent diffusion coefficient in material (diffusibility of mobile cation); R is the radius of the material spherical particle; t is the exchange time; n is the integral (number of particles); and $U(t)$ is the degree of equilibrium attaining at the time t.

The relationship Eqn (15) represents the solution of the equation of the Fick second law for the ion exchange between the mobile ions where the diffusion is the limiting rate stage (Cruceanu et al., 1986):

$$Me^{z+} \xrightarrow{D_1} \text{diffusion}$$

The variation of the mobile cation concentration (C_1), in spherical coordinates is given by the relationship:

$$\frac{\partial C_1}{\partial t} = D_1\left(\frac{1}{R^2}\right)\frac{\partial}{\partial R}\left(R^2\frac{\partial C_1}{\partial R}\right) \quad (16)$$

The degree of equilibrium attaining at the time t is also defined by the equation:

$$U(t) = \frac{[C(0) - C(t)]}{[C(0) - C(\infty)]} \quad (17)$$

where C(0), $C(t)$, and $C(\infty)$ denote the ion concentrations in the ion exchanger at the time zero (initial), time t and at equilibrium, respectively.

The values of the parameter (Bt) corresponding to the experimental values, $U(t)$, have been calculated and put into tables by Reichenberg (1953). When the ion concentration is not constant the exchange degree for the spherical particles in the initial stages is estimated by the relationship Eqn (15) that was simplified by Barrer and Hinds (1953):

$$U(t) = \frac{Q_0}{Q_\infty} = \frac{Q_0}{Q_0 - Q_\infty}\frac{2S}{V}\sqrt{\frac{D_1 t}{\pi}} = \frac{6}{R}\sqrt{\frac{D_1 t}{\pi}} \quad (18)$$

where Q_t and Q_∞ denote the amount of the ions exchanged at the time t and at equilibrium (∞), respectively; Q_0 is the initial concentration of the exchanging ion in solution; and S, V, R are the area, volume, and radius, respectively, of the ion exchange material particles.

The obeying of the linear dependence $U(t) = K\sqrt{t}$ was found to confirm practically the decisive role of the „diffusion in particle" in the kinetics of the ion exchange on ion exchange materials. The methods that can be applied for estimating the diffusion coefficient are the measuring of the material area (S) by gas adsorption and the estimation of the D_1 value from the slope of the initial part of the curve in the coordinates $U(t) - \sqrt{t}$ or $\frac{Q_t}{Q_\infty} - \sqrt{t}$, etc. (Sherry and Walton, 1967).

However, the method mostly applied is that based on the estimation of the B coefficient from the graphical representation $Bt - t$ and the calculation of the diffusion coefficient D_1 from the equation: $B = \frac{\pi^2 D_1}{R^2}$

The effective or apparent diffusion coefficient, D_1, in the equation is referred to the interdiffusion of two exchange cations, A and B, of the same valency for which: $D_1^A = D_1^B = D_1$ where D_1 is independent on composition. For example, for zeolite molecular sieves the coefficients (D_1) show values between 1×10^{-8} and 1×10^{-13} cm^2/s (Cruceanu et al., 1986). The apparent diffusion coefficient varies with temperature according to the Arrhenius equation:

$$D_1 = D_0 \cdot \exp\left(\frac{-E_a}{RT}\right)$$

The activation energy of the ion exchange (E_a), may also be calculated from the plot in the log $D_1 - T^{-1}$ coordinates. When the ratio between the cation radius and the diameter of the access window to the channels is not too low the ion exchange is a simple diffusion process. With the structures little opened or in case of the bulky cations the steric factor prevails and the activation energy increases fast with increasing ionic radius.

Many reported studies are to be found in literature regarding the ion exchange kinetics. The examples to be presented in the following were taken from some of the first studies on the subject. Rao and Rees (1966a, 1966b) have studied the ion exchange kinetic of some monovalent and divalent ions on natural materials, i.e. mordenite ($SiO_2/Al_2O_3 = 10.5$) by applying the radiochemical technique resulting in the data in Table 1.

Table 1 Self-diffusion of some cations in mordenite

Cations	Radius, Å	D_1, cm^2/s	S(BET), cm^2/g	*T*, °C	E_a, kcal/mol
Na^+	0.95	2.5×10^{-13}	2520	24.0	8.74
K^+	1.33	1.22×10^{-13}	6370	20.0	7.14
Rb^+	1.48	1.31×10^{-13}	4980	22.0	20.26
Cs^+	1.69	0.05×10^{-13}	4980	28.75	4.31
Ca^{2+}	0.90	1.13×10^{-17}	9090	27.0	10.45
Sr^{2+}	1.13	0.15×10^{-15}	10,600	20.5	15.97
Ba^{2+}	1.35	0.45×10^{-15}	10,100	18.5	10.14

The data in Table 1 afforded the following authors' conclusions:

- The diffusion coefficients, D_1 vary between 0.05 and 2.5×10^{-13} cm^2 s;
- The cesium diffuses 50 times slower than sodium and E_a is of the lowest value. The cesium ions cannot diffuse through the windows of 2.8 Å (among the adjacent channels) but only through the mordenite big channels (6.95 × 5.81 Å);
- For the rubidium ions the E_a value is higher that would indicate they must enter through the windows of 2.8 Å;
- As for the kalium ions, it is not known if they could be located in the window of 2.8 Å;
- When the $Na^+ \rightarrow K^+$ exchange in the kalium-containing mordenite is made the kalium ions are replaced by the sodium ions in the big channels and when the exchange degree of 50% is attained this replacement is also extended in the smaller side channels.

In comparison with the native mordenite, for sodium-containing zeolites (A, X, Y, and the ZK-4 zeolite which is isostructural with A), the ion exchange with alkali and alkaline-earth cations is very quick and the variation of the exchange degree versus $\sqrt{t}$ leads tu curves with their concavity returning to the base. This corresponds to a very quick start of the exchange followed by an almost constant exchange rate. For example, the values of activation energies at the time to (the time required for attaining an exchange degree of 10%), the D_o constant in the Arrhenius equation and the entropy of activation, ΔS, for the self-diffusion of the Ca^{2+}, Sr^{2+}, and Ba^{2+} cations in the zeolites of the A, X and ZK-4 types have been calculated (Cruceanu et al., 1986). In this case, the entropy of activation, ΔS, was estimated from the value of the D_o coefficient by means of the relationship:

$$D_o = 2.72 \frac{KT}{h} d^2 \exp \frac{\Delta S}{R} \tag{19}$$

where K is the Boltzmann constant; h is the Planck constant; R is the gas constant; and d is the distance or the cation jump in the diffusion process.

The calculated values of the self-diffusion parameters of the divalent cations in zeolites are given in Table 2.

The data in Table 2 afford the following conclusions (Cruceanu et al., 1986):

- For the A zeolite (with smaller pores) E_a varies with the divalent cation size;
- For the zeolite X (with bigger pores) the E_a values are very close, no effect of the hydrated cations is noticed and the ΔS values are positive and high, the ions migrating as a cation-hydrated water complex;
- For the ZK-4 zeolite the E_a values for the Ca^{2+} and Sr^{2+} ions are higher than for the A zeolite. The diffusion of the Ca^{2+} and Sr^{2+} strongly hydrated cations is possible to require the partial diminution of the hydration cloud;
- In ZK-4 zeolite where the H_2O/cation ratio is higher, the hydration is stronger and the diffusion of the above-mentioned cations assumes the elimination of more water molecules;
- The Ba^{2+} cation shows an abnormal behavior in ZK-4, the diffusion activation energy being lower than for Ca^{2+} and Sr^{2+}. An explanation could be advanced by taking the lower hydration energy of the Ba^{2+} ions into account since their hydration cloud during diffusion has not been reduced. Moreover, the electrostatic field of the ZK-4 zeolite (with more silicium) is lower than that of the A zeolite.

Barrer (1964) has presented the correlation between E_a of the self-diffusion of alkaline cations in analcite and their polarizability (Figure 4).

The author has drawn the conclusion that with the zeolites with a too low ratio of the cation radius to the diameter of the access window the steric factor is negligible and the self-diffusion and activation energy depend on the cation charge and the zeolite hydration state (Table 3).

As can be seen in Table 3, for the alkali cations the E_a values are almost identical and about twofold for the anhydrous chabazite. This would indicate a cation–network electrostatic interaction, on one hand, and the role of the hydration water for the cation mobility, on the other hand. Moreover, in chabazite, the diameter of the channel entrance windows is big enough ($\approx$3.9 Å) for an easy access of the cations (Barrer, 1964). Barrer (1964) has also obtained the values of the self-diffusion coefficient (D_1) in the chabazite hydrated at 298.3 K, by plotting Q_t/Q_∞ versus $\sqrt{t}$ and these values are given in Table 4.

Table 2 Values of the self-diffusion parameters of the divalent cations in zeolites (Cruceanu et al., 1986)

Zeolite	Average radius of zeolite particles, μm	Si/Al	Cation	E_a, kcal/ion·g	t_o	T, °C	D_0, cm²/s	ΔS, cal/mol·grad
A	2.64	1.0	Ca^{2+}	16.1	70 ore	30	5.1×10^{-6}	−4.0
A	2.64	1.0	Sr^{2+}	19.6	8 ore	30	1.7×10^{-5}	2.7
A	2.64	1.0	Ba^{2+}	21.6	6 min	35	1.48×10^{-3}	11.6
X	2.04	1.26	Ca^{2+}	19.9	1 min	−18[a]	1×10^{-7}	20.0
X	2.04	1.26	Sr^{2+}	20.0	9 min	−20[a]	1×10^{-7}	20.0
X	2.04	1.26	Ba^{2+}	20.1	20 min	−18[a]	1×10^{-7}	20.0
ZK-4	1.10	1.33	Ca^{2+}	20.1	30 s	28	3.6×10^{-4}	8.6
ZK-4	1.10	1.33	Sr^{2+}	22.6	1 min	30	7.35×10^{-3}	14.8
ZK-4	1.10	1.33	Ba^{2+}	15.8	1 min	29	7.6×10^{-7}	−3.4

[a]In 50% ethylic alcohol.

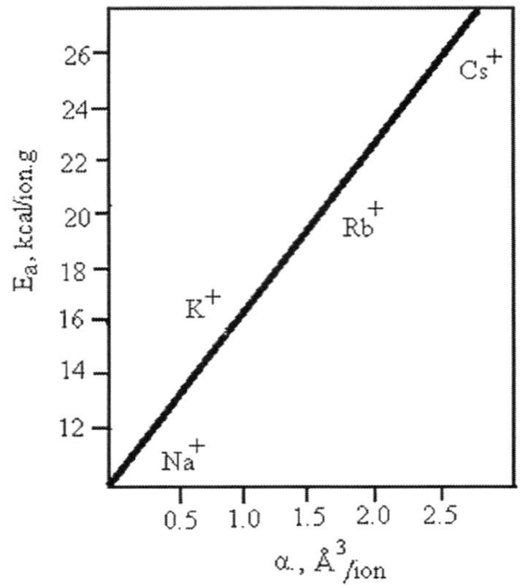

Figure 4 Activation energy of the self-diffusion of alkaline cations versus their polarizability.

Table 3 Effect of the cation charge and zeolite hydration state on the activation energy of the self-diffusion in hydrated and anhydrous chabazite

	Activation energy, E_a, kcal/ion·g	
Cation	**Hydrated chabazite**	**Anhydrous chabazite**
Li^+	—	16.4
Na^+	6.55	15.0
K^+	7.01	16.0
Rb^+	6.75	17.3
Cs^+	7.55	18.3
Ca^{2+}	13.81	—
Sr^{2+}	14.59	—
Ba^{2+}	8.83	—

Table 4 Self-diffusion in chabazit hydrated at 298.3 K

Cations	Radius, Å	D_1, cm^2/s	S, cm^2/g	D_0, $cm^2/s \times 10^{-7}$
Na^+	0.95	6.1×10^{-12}	750	4.0
K^+	1.33	6.9×10^{-12}	1120	9.4
Rb^+	1.48	19.5×10^{-12}	503	17.0
Cs^+	1.69	4.9×10^{-13}	1170	1.7
Ca^{2+}	0.90	4.1×10^{-16}	8100	55.0
Sr^{2+}	1.13	1.3×10^{-16}	7690	101.0
Ba^{2+}	1.35	1.3×10^{-13}	817	4.0

The authors found that the Ba^{2+} which is the bulkiest one among the presented cations is an exception, with E_a and D_1 values lower than for Ca^{2+} and Sr^{2+}. This fact might be explained by the higher polarizability and lower content of hydration water.

Brown et al. (1971) have advanced a new model of the self-diffusion in the hydrated zeolites. They admitted a single process of quick diffusion of the cations in the big cavities and and a slow diffusion (desorption) of the cations in the sodium-containing cavities and the hexagonal prisms in super-cavities. This model was also sustained by the independence of the slow process on the crystallite size.

Another model advanced, by Brown et al. (1971) includes a combination of the diffusion and exchanging processes between the bound and mobile cations in the structure of the A and X zeolites. With this model the assumption is made that all the cations must diffuse alike in the A zeolite since there is only one diffusion way, namely that through the big cavities. The nonlocated mobile cations diffuse from the supercavities according to the Fick second law being characterized by a self-diffusion coefficient (D_1). The bound cations are detached from their sites, shifted to the cavities where they replace the mobile ones becoming alike and diffusing in the same manner to them, having also the same (D_1) value:

$$\text{ions located in cavities} \Leftrightarrow \text{mobile ions} \xrightarrow{D_1} \text{diffusion to surface}$$

The differential equations describing of desorption and adsorption of cations for spherical particles are:

$$\frac{\partial C_1}{\partial t} = \underbrace{D_1\left(\frac{1}{R^2}\right)\frac{\partial}{\partial R}\left(R^2\frac{\partial C_1}{\partial R}\right)}_{\text{The first stage}} + \underbrace{K_2(C_2 - \alpha C_1)}_{\text{The slow stage}} \tag{20}$$

$$\frac{\partial C_2}{\partial t} = K_2(\alpha C_1 - C_2) \tag{21}$$

where t stands for the time; R is the radius; D_1 is the self-diffusion coefficient of mobile cations; K_1, K_2 are the IInd order rate constants characteristic of the cation adsorption and desorption, respectively; $\alpha = \frac{K_1}{K_2}$ C_1 is the concentration of mobile ions; and C_2 is the concentration of the ions located in fixed positions.

The model, advanced by Brown et al. (1971) was also tested on the zeolites A and X of 25–100 μm size by means of the radiometric technique. In the zeolite X the intracrystalline exchange between the bound and the mobile ions in the big cavities is admitted to be faster than the diffusion and the fast stage is an apparently uniform process involving the both cation types. The slow stage is the intracrystalline exchange between the bound cations in the small and big cavities.

Ions strap in small cavities $\Leftrightarrow^{K_4}_{K_3}$ ions strap in super − cavities $\Leftrightarrow^{K_2}_{K_1}$ mobile ions → diffusion to surface. By assuming that the cations shift from the small cavities to the supercavities and the adsorption in S_n and desorption from S_{11} of the cations proceed faster than the diffusion in supercavities the differential equations describing the ion exchange on the X zeolite could be of the form:

$$\frac{\partial C_{1+2}}{\partial t} = D_1 \frac{1}{R^2} \frac{\partial}{\partial R} \left(R^2 \frac{\partial C_1}{\partial R} \right) + K_4 (C_3 - \alpha_{23} C_2) \quad (22)$$

$$\frac{\partial C_3}{\partial t} = K_4 (\alpha_{23} C_2 - C_3) \quad (23)$$

where C_1 is the concentration of the mobile ions in supercavities; C_2 is the ionic concentration in the S_{II} positions; C_3 is the ionic concentration in the small cavities; K_4 is the rate constant of the "jumping" reaction of an ion from the sodalite cell in a S_{II} position; and C_{1+2} is the concentration of the mobile and bound ions (S_{II}) in supercavities.

The experimental runs have also made evident the isotopic exchange of the monovalent and divalent cations in the zeolites of the A, X, and Y types to proceed into two stages apart from some exceptions. The thermodynamic preference for the positions in supercavities with the X zeolite decreases with increasing ionic radius (excepting for Li^+), while in Y zeolite it decreases with increasing the hydration energy of ions. In the small cavities of the X and Y zeolites the exchange positions show preference to the ion of the lowest charge and ionic radius. The exchange of the polyvalent ions by Na^+ ions in the X and Y zeolites shows two stages of much different rates (Cruceanu et al., 1986).

The kinetics of the isotopic exchange on zeolites of the A, X, and Y types was also studied with methanol and ethanol alcoholic solutions by Dyer and Gettins (1970). With the A zeolite the solvent influence on the self-diffusion of Na^+ is not significant while in X and Y zeolites the self-diffusion of the divalent cations is much lowered, the diffusion activation energy increasing with increasing cation radius and with decreasing dielectric constant of the solvent. The ion exchange of Sr^{2+} on X zeolite takes place at a very low rate in methanol while in the case of Ca^{2+} and Ba^{2+} ions the rate is higher

($E_a = 29.1$ kcal/mol, 24.5 kcal/mol, respectively). On the Y zeolite of a Si/Al = 1.86 ratio in methanol the exchange with Ca^{2+}, Sr^{2+}, Ba^{2+} takes place while in ethanol with Ba^{2+} ($E_a = 28.6$ kcal/mol) only. With the Y zeolite of a Si/Al = 2.60 ratio, the ion exchange is possible in both these alcohols (Dyer and Gettins, 1970).

The ion exchange in nonaqueous or in mixed solvents affords the influence of the solvent nature on the ionic interactions with the zeolite network to be studied especially by means of its dielectric constant. In this connection, Dizdar (1972) has studied the ion exchange on the Na-A zeolite in the water-dioxan mixture. At 25 °C, the dioxan dielectric constant is of 2.1 and the critical size of molecule of 6 Å which is larger than the entering pore size of ~4,2 Å in Na-A zeolite. By drawing the exchange isotherm for LiCl 0.01 *n* solution at la 5, 25, 45, and 65 °C the author has noticed the exchange equilibrium does not modify with the temperature till a 0.35% Li concentration in zeolite but above it does. The point corresponding to the concentration $X_{Li(z)} = 0.35$ represents the inflexion point of the isotherm. The temperature influence decreases with increasing dioxan concentration in the aqueous phase following then the same isotherm to be again obtained at a dioxan concentration of 82%. The $Na^+ \rightarrow Li^+$ exchange on the Na-A zeolite becomes increasingly difficult with increasing dioxan concentration and the equilibrium thermodynamic constant decreases. In this case, the ΔG^o_{298} values are of positive and Li^- noticed to have no affinity to the Na-A zeolite.

Dizdar and Popovič (1972) have studied the $Na^+ \rightarrow Li^+$ exchange on the Na-A zeolite in the dimethylsulphoxide-water mixed medium and found the same inflexion at $X_{Li(z)} = 0.35–0.40$ Li^+ concentration in zeolite which just corresponds to the cation exchange in the two positions. The $Na^+ \rightarrow Li^+$ exchange becomes increasingly difficult and incomplete with increasing dimethyl sulphoxide concentration in solution.

Radak and Šušić (1971) have studied the ion exchange of the alkali cations on the A zeolite in the methanol–water media in the function of the mixture dielectric constant. They found that the apparent equilibrium constant increases with increasing alcohol concentration in the mixture (excepting for Li^+ for which the corrected selectivity coefficient K^{Me}_{Na} varies linearly with the cation fraction in zeolite) and the ΔG^o_{298} values are positive as it is with the exchange in aqueous solution. The ion exchange in nonaqueous solvents is more difficult to proceed and the selectivity coefficient is lower in the organic medium (excepting for NH_4^+ in ethanol) than in water.

Huang et al. (1964) have studied the ion exchange of some mono- and divalent cations on Na-X zeolite in alcoholic solutions at 30 °C. The selectivity coefficients for the exchange of mono- and divalent cations on Na-X zeolite in water and alcohols, the influence of the temperature and of the solvent on the exchange enthalpy (ΔHo) as well as some kinetic data are presented in Tables 5 and 6.

The conclusion was drawn by the author that the obtained results sustain the dependence of the selectivity coefficient logarithm on the reversed dielectric constant of the medium that is an almost linear dependence. The exchange rate $K^+ \rightarrow Na^+$ and $Ag^+ \rightarrow Na^+$ in water and ethanol and $\frac{1}{2}Ca^{2+} \rightarrow Na^+$ in water decreases with decreasing dielectric constant and the exchange proceeds slower in nonaqueous solvents than in water.

Table 5 Ion exchange of some mono- and divalent cations on Na-X zeolite in alcoholic media at 30 °C in 0.05 M solutions

Solvent	Dielectric constant	Selectivity coefficient, K_{Na}^{Me}				ΔH_{Na}^{K}, kcal
		K^+	Li^+	Ca^{2+}	Ag^+	
Water	81.70	1.10 0.96	8.4×10^{-2}	0.44	45.9	−993
Methanol	33.66	0.96 0.75	4.2×10^{-3}	18.1	—	−1614
Ethanol	23.56	0.70 0.61	3.9×10^{-3}	6.34	—	−859
n-Propilic alcohol	19.80	0.50	1.5×10^{-2}	—	—	—
Isopropanol	17.90	0.19	1.3×10^{-2}	—	—	—
Isobutyl alcohol	17.10	0.19	1.1×10^{-2}	—	—	—

Table 6 Kinetic data on the exchange of some metallic ions with Na^+ from Na-X in aqueous solution, methanol, and ethanol at 30 °C

Exchange ion	Solvent	$T_{o'5'}$ min	$T_{o'75'}$ min
K^+	Water	18	55
	Methanol	19	60
	Ethanol	48	142[a]
Ag^+	Water	40	120
	Methanol	120	330
	Ethanol	295	850[a]

[a]Extrapolated data.

1.3 Thermodynamics of Ion Exchange

As already mentioned above the ion exchange reaction in an ion exchange material can be described by the following general equation:

$$z_A B^{z_B^+}_{(Z)} + z_B A^{z_A^+}_{(S)} \rightleftharpoons z_A B^{z_B^+}_{(S)} + z_B A^{z_A^+}_{(Z)}$$

where Z_A^+ and Z_B^+ denote the charges of the exchange cations, A and B, respectively, and the indices (Z) and (S) are referred to ion exchange material and solution, respectively.

Cruceanu et al. (1986) have pointed out that the standard free enthalpy (ΔG^o), of the ion exchange process (with $X_{max} = 1$) described by the reaction (1) is given by the relationship:

$$\Delta G^o = -\frac{RT}{z_A z_B} \ln K_a \tag{24}$$

where K_a is the equilibrium thermodynamic constant

The standard entropy (ΔS^o) is given by the relationship :

$$\Delta S^o = \frac{\Delta H^o - \Delta G^o}{T} \tag{25}$$

The standard enthalpy (ΔH^o) is obtained from the variation of K_a with temperature:

$$\frac{\mathrm{dln}\, K_a}{\mathrm{d}T} = \frac{\Delta H^o}{RT^2}, \text{ respectively } \Delta H^o = -\frac{R}{z_A z_B} \frac{\mathrm{dln}\, K}{\mathrm{d}\left(\frac{1}{T}\right)} \tag{26}$$

The enthalpy variations were found to be low in the ion exchange on ion exchange material. For explaining the mechanism of the ion exchange, Eisenmann (1962) has advanced a model where the free enthalpy for the monovalent exchange reaction $A_s^+ + B_z^+ \Leftrightarrow A_z^+ + B_s^+$ is given by the relationship:

$$\Delta G^o = \left(\Delta G_Z^A - \Delta G_Z^B\right) - \left(\Delta G_S^A - \Delta G_S^B\right) \tag{27}$$

The first term represents the difference between the free enthalpies of the cations A^+ and B^+ in material. The second term is the difference between the hydration free enthalpies of the exchange ions.

The author concluded that the electrostatic contribution to the free enthalpy of the exchange depends partially on the strength of the electrostatic field of ion exchange material (density of the fixed negative charge), namely:

- When the electrostatic field is very strong (as it is with the aluminous zeolite molecular sieves with a low Si/Al ratio and high charge density) the electrostatic contribution to the exchange free enthalpy is higher in comparison with the hydration term$(\Delta G_s^A - \Delta G_s^B)$. In such a case, the ions of the smallest ionic radius which can interact in the strongest way with the cations in the exchange positions are preferable in materials;
- When the electrostatic field in ion exchange material is very weak (as it is with the siliceous zeolite molecular sieves of a high Si/Al ratio and low charge density), the electrostatic contribution $(\Delta G_z^A - \Delta G_s^B)$ is low in comparison with the hydration one and the hydration term in the Eqn (19) will prevail. In such cases the materials prefer the ions of bigger radius and less hydrophilic. For instance, in the zeolites A (Si/Al = 1) and X (Si/Al = 1.2) the activity series is $Na^+ > K^+ > Rb^+ > Cs^+ > Li^+$, while for the zeolite Y (Si/Al = 2.8) the order is $Cs^+ > Rb^+ > K^+ > Na^+ > Li^+$ (Cruceanu et al., 1986).

Thus, this model was applied and developed by Sherry and Marinsky (1969) for zeolites.

It was noticed that apart from the ΔG^o variation of the ion exchange, the differences between the entropies of the ions to be exchanged ΔS^o must also be taken into account. For example, Barrer et al. (1963) and Rosseinsky (1965) have calculated the entropies for the complete ion exchange on the Na-A zeolite molecular sieves, the differences between the hydration entropies of the exchange ions as well as the differences between the entropies of the exchange ions in the zeolite phase. The authors made use of the following calculating relationship:

$$\Delta S^o = \left(\Delta S_x^A - \Delta S_z^B\right) - \left(\Delta S_s^A - \Delta S_s^B\right) \tag{28}$$

The obtained results are given in Table 7.

Table 7 Variations of the ion exchange entropies of the alkali metals on the molecular sieve Na-A

Reacţia	ΔS^o	$(\Delta S_x^A - \Delta S_z^B)$	$(\Delta S_s^A - \Delta S_s^B)$
$Li_s^+ + Na_z^+$	3.2	−4.3	−7.5
$K_s^+ + Na_z^+$	−8.5	0	8.5
$Rb_s^+ + Na_z^+$	−11.4	0	−11.4
$Cs_s^+ + Na_z^+$	−19.4	7.3	−12.1

The authors concluded that the difference between the hydration entropies in solution corresponds to the variation of the exchange total entropy only in case of K^+ and Rb^+ ions. In the zeolite phase the water molecules have fewer freedom degrees than in the solution phase so that the water adsorption would result in the entropy decrease. The water content in zeolite decreases continuously with increasing atomic number of the alkali metal ion and consequently the entropy increases. It was found that by calculating similarly the ΔG^o variation for the divalent ions (alkaline-earth) the selectivity series can be settled in function of the anionic field strength of the zeolite. In case of the strong electrostatic field the alkali-earth ion of the smallest radius is preferable while with a weak electrostatic field the ion of the biggest radius is (Cruceanu et al., 1986).

1.4 Processes: Fixed and Fluidized Bed Systems—Basic Principles and Application in Ion Exchange

1.4.1 Ion Exchange Fixed Beds

Ion exchange is of the most common process used for separation purposes and in the most of the cases takes place in fixed beds packed with solid ion exchange materials (Babel and Kurniawan, 2003; Braun et al., 2002; Inglezakis and Poulopoulos, 2006). Although a large number of mathematical models exist for the description of ion exchange fixed bed operation their use in practice is complex and one of the basic reasons is the lack of data for several parameters that normally should be derived by separate experiments. This is one of the major reason explaining why adsorption models are used for ion exchange fixed bed operations. As is well known, the two phenomena share many common features, of the most basic being the transfer and resulting equilibrium distribution of one or more solutes between a fluid phase and particles (Inglezakis, 2010; Perry and Green, 1999). In the related literature, ion exchange and adsorption are grouped together as sorption for a unified treatment. The objective of this chapter is to present the basic fundamentals of the operation (fixed bed) and not the nature of the phenomenon (ion exchange).

In Figures 5 and 6 a typical fixed bed operation and its breakthrough (exit) curve are presented.

As is shown in Figure 5, the operation is upflow, i.e. the liquid phase is introduced in the bottom of the bed. This mode of operation is preferred as it provides better flow quality and complete wetting of the material. In Figure 6 the dimensionless exit concentration C/C_o versus effluent volume V_{eff} is plotted, i.e. the breakthrough curve of the operation. The breakpoint is the moment when the operation is stopped, typically when the exit

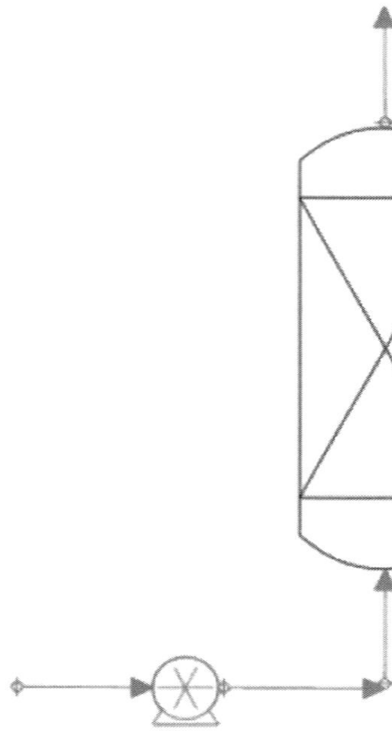

Figure 5 Simplified schematic fixed bed upflow operation.

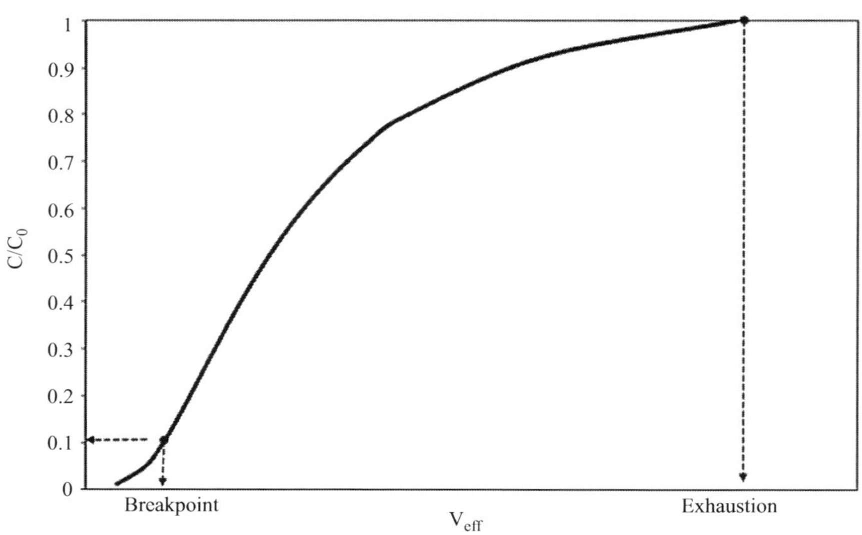

Figure 6 Breakthrough curve.

concentration reaches a certain predefined level according to the needs. For example, in all wastewater treatment operations, this is the point when the maximum permissible exit concentration is reached. In the optimum fixed bed operation the breakthrough curve is a single step, a vertical line, but in the most actual cases is a sigmoidal curve and the steeper the breakthrough curve the higher is the fixed bed utilization. High degree of utilization means that the ratio of the breakthrough capacity to the total capacity is closer to unity (Inglezakis, 2010).

Fixed bed operation is influenced by equilibrium, kinetic, and hydrodynamic factors, constituting the design of the reactor a complicated task and in many occasions the full process modeling is practically impossible (Inglezakis and Poulopoulos, 2006; Inglezakis, 2010). The effectiveness of a fixed bed operation is heavily depended on its hydrodynamic performance, a fact that in many cases is overlooked and inevitably leads many operations to failure. In liquid–solid systems these phenomena are termed as "liquid maldistribution" and they are related to inadequate liquid distribution and low liquid holdup, especially in downflow operation. Other complications occur due to the nature of the contacting system. For example, even in the simple case of the treatment of mono-component solutions in ion exchange fixed beds it is experimentally found that equilibrium parameters and/or solid diffusion coefficients seem are unexpectedly influenced by contact time, making the modeling of the operation a complex task (Inglezakis and Grigoropoulou, 2003).

1.4.1.1 Fixed Bed Mass Balances

In a fixed bed, the continuity equation around a segment of the solid phase between the height x and $x + dx$ and during time dt is (Inglezakis and Poulopoulos, 2006; Perry and Green, 1999; Ruthven, 1984):

$$-\rho_{\mathrm{b}}\frac{\partial \overline{q}}{\partial t} = \varepsilon\frac{\partial C}{\partial t} + u_{\mathrm{s}}\frac{\partial C}{\partial x} - D_{\mathrm{ax}}\varepsilon\frac{\partial^2 C}{\partial x^2} \tag{29}$$

where ε and ρ_b is bed porosity and bulk density, respectively, u_{s} is the superficial liquid velocity, D_{ax} is the axial dispersion coefficient, C is the solution concentration, and $\overline{q}$ is the mean concentration of solute in the solid phase. In dilute aqueous solutions the superficial fluid velocity (u_{s}) is considered to be constant throughout the fixed bed length. However, under high concentration levels u_{s} is variant and this term should be incorporated into the derivative. Such effects are usually significant only for adsorption in gaseous systems (Ruthven, 1984).

Dividing the second and third term of Eqn (29) by the bed height Z we obtain:

$$-\rho_b \frac{\partial \overline{q}}{\partial t} = \varepsilon \frac{\partial C}{\partial t} + \frac{u_s}{Z} \frac{\partial C}{\partial (x/Z)} - \frac{D_{ax}\varepsilon}{Z^2} \frac{\partial^2 C}{\partial (x/Z)^2} \tag{30}$$

and rearranging

$$-\rho_b \frac{Z}{u_s} \frac{\partial \overline{q}}{\partial t} = \varepsilon \frac{Z}{u_s} \frac{\partial C}{\partial t} + \frac{\partial C}{\partial (x/Z)} - \frac{D_{ax}\varepsilon}{u_s Z} \frac{\partial^2 C}{\partial (x/Z)^2} \tag{31}$$

the term $\varepsilon \cdot Z/u_s$ is the liquid residence time (or contact time) while the term $\frac{D_{ax} \cdot \varepsilon}{u_s \cdot Z}$ is the inversed bed Peclet number. Both C and q are dependent on time t and height Z and Eqn (31) is a partial differential equation. As the solute uptake is controlled by the diffusion rates, the rate equations to be used for the liquid and solid phase mass transfer are (Perry and Green, 1999):

$$\frac{\partial q}{\partial t} = \frac{k_f a_u}{\rho_b}(C - C^o) \tag{32}$$

$$\frac{\partial q}{\partial t} = \frac{D_s}{\rho_b}\left(\frac{\partial^2 q}{\partial r^2} + \frac{2}{r}\frac{\partial q}{\partial r}\right) = \frac{D_s}{\rho_b r_p^2}\left(\frac{\partial^2 q}{\partial (r/r_p)^2} + \frac{2}{r/r_p}\frac{\partial q}{\partial (r/r_p)}\right) \tag{33}$$

where C^o is the concentration of the solute in the fluid phase, which is assumed to be in equilibrium with q. The solid diffusion coefficient D_s is theoretically independent of the flow type and rate while liquid mass transfer coefficient is a flow dependent parameter. Furthermore, α_u is the total external solid surface area per unit bed volume:

$$\alpha_u = \frac{6(1 - \varepsilon)}{d_p} \tag{34}$$

The mean solid phase concentration $\overline{q}$ is (Hall et al., 1966; Fleck et al., 1973):

$$\overline{q} = \frac{3}{r_p^3} \int_0^R q r^2 dr \tag{35}$$

where r is the distance from particle center and d_p, r_p is the particle diameter and radius, respectively. Equation (35) is used only for the case of solid diffusion control as for liquid film control $\overline{q} = q$.

As in the case of adsorption, the physical process of ion exchange is considered to be so fast relative to diffusion steps that in and near the solid particles a local equilibrium exists and thus, the equilibrium isotherm $q = f(C^o)$, relates the stationary and mobile phase concentrations at equilibrium (Inglezakis, 2005). For the purposes of simplified fixed bed modeling, Langmuir and Freundlich equilibrium equations are used (Fleck et al., 1973; Hall et al., 1966; Ruthven, 1984). The continuity, rate, and equilibrium equations are solved simultaneously using the appropriate initial and boundary conditions. This system consists of four equations and four unknowns, i.e. C, $\overline{q}$, q, and $C^{\star}$. For combined resistances Eqns (32), (33) and (35) should be used simultaneously.

1.4.1.2 Analytical Solutions to the Fixed Bed Model

Predictive models could be used in order to model the process and determine the controlling step, which for the most cases in ion exchange is the solid phase diffusion. Nevertheless, the controlling step depends on liquid velocity as well because in low velocities the liquid film diffusion step could influence the process. In order to use simplified models the following assumptions should be made (Inglezakis, 2010; Perry and Green, 1999):

1. Plug flow. In this case the last term in the Eqn (31) is neglected. This assumption holds only if the bed Peclet number is greater than 100. Generally in upflow operation the quality of the flow is better, especially in low velocities.
2. Constant pattern condition. With a favorable isotherm and a mass-transfer resistance or axial dispersion, the concentration front approaches a constant pattern, which is an asymptotic shape beyond which the wave will not spread. The wave is said to be "self-sharpening". In contrast, if the concentration front is seen to spread out more and more as it propagates toward the column outlet, is a dispersive or spreading behavior (Figures 7 and 8). Constant pattern condition reduces the continuity Eqn (31) to the simple relation: $C/C_o = \overline{q}/q_{\max}$, or, to put it in words, the dimensionless profiles of concentration and exchanger loadings are identical. This means that all points on the breakthrough curve are moving in the bed under the same velocity and thus a constant shape of this curve is established. Practically, the constant pattern assumption holds if the equilibrium is favorable and under high residence times, i.e. deep bed and low superficial velocity. However, constant pattern assumption is weak if the system exhibits very slow kinetics (Figure 9).

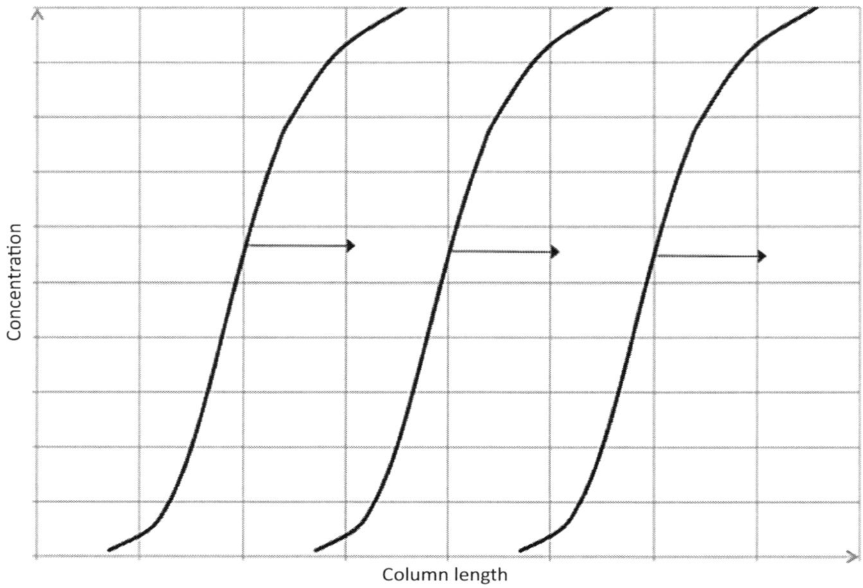

Figure 7 Self-sharpening behaviour.

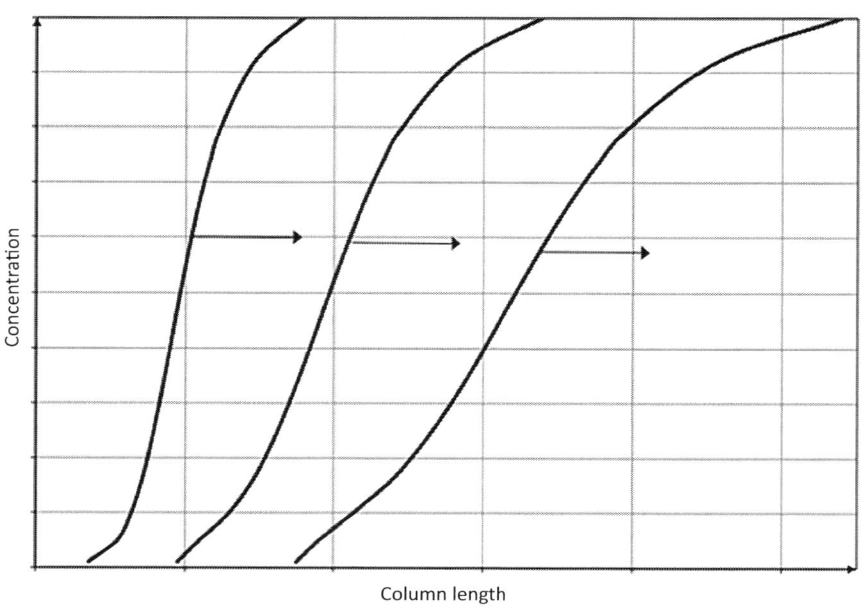

Figure 8 Spreading behaviour.

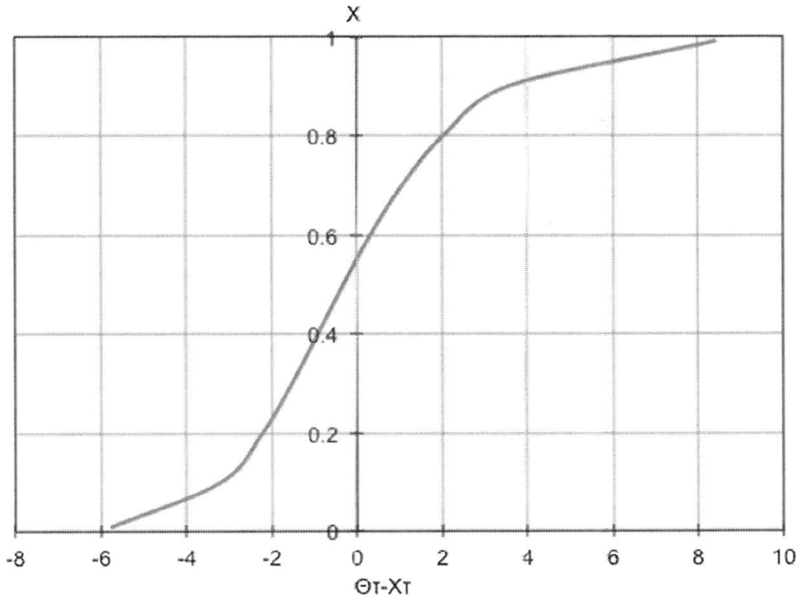

Figure 9 Graphical representation of the model solution for Langmuir equilibrium and combined resistances ($La = 0.5$ and $\zeta = 1000$).

Simplified models under the above assumptions have been proposed and analyzed in the related literature and are in the form of either arithmetic or analytical solutions In the present work we include the case of Miura and Hashimoto model, which covers all simplified equations found in the related literature (Inglezakis and Poulopoulos, 2006; Perry and Green, 1999; Ruthven, 1984). For combined diffusion resistances (fluid film and solid diffusion) the concentration in the bulk liquid phase is different than this on interface due to the effect of the fluid film resistance. The following equations can be used for Langmuir and Freundlich equilibrium equations as published by Miura and Hashimoto (1977):

Langmuir Isotherm

$$\theta_\tau - X_\tau = \frac{1}{1+\zeta}\phi_1 + \frac{\zeta}{1+\zeta}\frac{1}{\eta}\phi_2 \tag{36}$$

$$\begin{aligned} \phi_1 = {} & \frac{1}{1-La}\ln X_i - \frac{La}{1-La}\ln(1-X_i) - \ln[La + (1-La)X_i] \\ & - \frac{La}{1-La}\ln La + 1 \end{aligned} \tag{37}$$

$$\phi_2 = \frac{La}{1 - La}\ln X_i - \frac{1}{1 - La}\ln(1 - X_i) - 1 \tag{38}$$

$$\eta = 1 - 0.192(1 - La)^3 \tag{39}$$

Freundlich Isotherm

$$\theta_\tau - X_\tau = \frac{1}{1+\zeta}\omega_1 + \frac{\zeta}{1+\zeta}\frac{1}{\eta}\omega_2 + \frac{1}{1+\zeta}\frac{\zeta}{\zeta+\eta}\omega_3 \tag{40}$$

$$\omega_1 = \frac{Fr}{Fr-1}\ln\left(X_i^{Fr-1} - 1\right) + 1 + \frac{\eta}{\zeta+\eta}\frac{Fr^2}{Fr-1}I_A \tag{41}$$

$$\omega_2 = \frac{1}{Fr-1}\ln\left(1 - X_i^{1-Fr}\right) + \frac{\zeta}{\zeta+\eta}\frac{1}{Fr-1}I_B \tag{42}$$

$$\omega_3 = Fr - 1 + \frac{Fr}{Fr-1}\left(I_A + I_B\right) \tag{43}$$

$$\eta = 0.808 + 0.192Fr \tag{44}$$

$$I_A = \sum_{n=1}^{\infty}\frac{1}{n[n(1-Fr)] + Fr} \tag{45}$$

$$I_B = \sum_{n=1}^{\infty}\frac{1}{n[n(1-Fr)] + 1} \tag{46}$$

where,

$$\theta_\tau = \frac{k_s a_u}{\rho_b(1 + 1/\zeta)}\left(t - \frac{\varepsilon Z}{u_s}\right) \tag{47}$$

$$X_\tau = \frac{k_s a_u \gamma}{1 + 1/\zeta}\frac{Z}{u_s} \tag{48}$$

$$k_s a_u = \frac{15 D_s \rho_b}{r_p^2} \tag{49}$$

$$\gamma = \frac{q_{max}}{C_o} \tag{50}$$

$$\zeta = \frac{k_f a_u}{k_s a_u \gamma} \tag{51}$$

$$a_u = \frac{3}{r_p}(1 - \varepsilon) \tag{52}$$

where, q_{max} is in mass of solute per unit mass of solid and C_o is in mass of solute per unit volume of fluid. For practical use, the infinite series I_A and I_B are shown in Figure 10.

The interface relative concentration X_i is related to the respective bulk relative concentration X as follows (Inglezakis and Poulopoulos, 2006):

Langmuir

$$X = \frac{(\zeta La + \eta)X_i + \zeta(1 - La)X_i^2}{(\zeta + \eta)[La + (1 - La)X_i]} \tag{53}$$

Freundlich

$$X = \frac{\zeta X_i + \eta X_i^{Fr}}{\zeta + \eta} \tag{54}$$

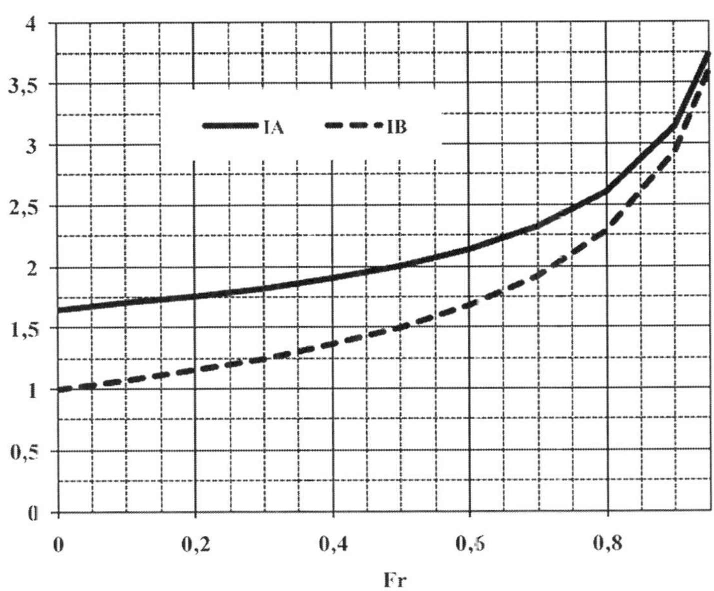

Figure 10 Values of (I_A) and (I_B).

The equations provide only the X_i versus $\theta_\tau - X_\tau$ relation while for a direct relationship between X and $\theta_\tau - X_\tau$, X_i should be explicitly expressed by suing only X. However, there is not an explicit relationship for Freundlich isotherm and in this case an iteration procedure (goal-seek). For Langmuir isotherm:

$$X_i = \frac{-b + \sqrt{b^2 - 4ac}}{2a} \tag{55}$$

where,

$$a = \zeta(1 - La) \tag{56}$$

$$b = (\zeta La + \eta) - (\zeta + \eta)(1 - LaX) \tag{57}$$

$$c = -La(\zeta + \eta)X \tag{58}$$

The same equations can be used as approximations for pore diffusion control, by setting $k_s\, a_u$ equal to $15D_p(1 - \varepsilon)/\gamma r_p^2$.

1.4.1.3 External Mass Transfer

Liquid mass transfer or convection coefficient is a flow dependent parameter, which is evaluated by use of empirical and semiempirical correlations. One of the most well-known correlations for use in water treatment situations is the following (Chern and Chien, 2002):

$$Sh = \left(2 + 0.644Re^{1/2}Sc^{1/3}\right)[1 + 1.5(1 - \varepsilon)] \tag{59}$$

In the adsorption processes for wastewater treatment by use of granular activated carbon the Williamson correlation is proposed (Crittenden et al., 1987):

$$k_f = \frac{2.4u_s}{Sc^{0.58}Re^{0.66}} \tag{60}$$

Kataoka correlation has been used in adsorption systems from liquid phase (Zulfadhly et al., 2001):

$$k_f = 1.85(ReSc)^{-2/3}\frac{u_s}{\varepsilon}\left(\frac{1 - \varepsilon}{\varepsilon}\right)^{-1/3} \tag{61}$$

Finally, for very low Reynolds number (<0.01) the Wilson–Geankoplis correlation has been used in ion exchange and in adsorption from liquid phase (Ben-Shebil et al., 2007; Chen and Wang, 2004; Xiu and Li, 2000):

$$Sh = \frac{1.09}{\varepsilon} Sc^{1/3} Re^{1/3} \tag{62}$$

where

$$Sh = \frac{k_f d_p}{D_f} \tag{63}$$

$$Re = \frac{d_p u_s}{\nu} \tag{64}$$

$$Sc = \frac{\nu}{D_f} \tag{65}$$

where, Re is the particle Reynolds number, Sh is the Sherwood number, Sc is the Schmidt number, D_f is the diffusion coefficient of the solute in the liquid phase, d_p is the particle diameter, and ν is the liquid kinematic viscosity.

1.4.1.4 Fixed bed Hydrodynamics

The analysis of the fixed bed hydrodynamics (hydraulics) involves the evaluation of liquid holdup and the quality of the liquid flow. In this analysis the physical mechanisms that are sensitive to size are investigated separately from chemical (kinetic or equilibrium) studies (Trambouze, 1990). Thus, the fixed bed is examined in respect to its flow patterns, as a vessel, e.g. irrespectively of the specific chemical reaction or physical phenomenon that take place in it.

A fixed bed is an array of voids into which fluid flows at relatively high velocity, accelerating in the ports created by particle–particle intersections and decelerating upon entering the voids, leading to a certain degree of mixing. In the ideal case of plug flow in fixed beds, axial mixing between the several cross-sections of the bed is minimal, whereas radial mixing in each section is maximal (Inglezakis and Poulopoulos, 2006). Ideal flow is studied and represented using the classic dispersion or dispersed plug flow model of Levenspiel (1972). Recall the material balance of a fixed bed reactor (Eqn (30)):

$$-\rho_b \frac{Z}{u_s}\frac{\partial \overline{q}}{\partial t} = \varepsilon \frac{Z}{u_s}\frac{\partial C}{\partial t} + \frac{\partial C}{\partial (x/Z)} - \frac{D_{ax}\varepsilon}{u_s Z}\frac{\partial^2 C}{\partial (x/Z)^2}$$

The last term of this equation is zero in both cases of plug and mixed flow: in plug flow $D_{ax} = 0$ and in mixed flow $\partial^2 C/\partial(x/Z)^2 = 0$. The physical parameter relevant to the flow quality is the axial dispersion coefficient, incorporated into the particle Peclet number:

$$Pe_p = \frac{ud_p}{D_{ax}} \tag{66}$$

where d_p is the particle diameter, D_{ax} the axial dispersion coefficient, and u the interstitial fluid velocity (equal to u_s/ε). The fixed bed (vessel) Peclet number is calculated by multiplying Pe_d with the term Z/d_p, where Z is the fixed bed length:

$$Pe_L = Pe_p \frac{Z}{d_p} \tag{67}$$

The fixed bed (vessel) Peclet number is calculated by multiplying Pe_d with the term Z/d_p, where Z is the fixed bed length. In the case of fixed beds, the higher the vessel Peclet number the better flow quality and if close or higher than about 100 the flow is considered to be ideal, i.e. plug flow (Levenspiel, 1972; Inglezakis, 2010).

For materials that are frequently used in ion exchange wastewater treatment applications, as for zeolites and similar particles of irregular shape, the following equation is proposed (Inglezakis et al., 2001):

$$Pe_p = LRe^k \tag{68}$$

where L is 0.52 for upflow and 0.05 for downflow, k is -0.65 for upflow, and 0.48 for downflow. Peclet correlation (68) holds for $0.6 < Re < 8.5$. Several correlations for the Peclet number can be found elsewhere (Chung and Wen, 1968; Inglezakis, 2010a).

Generally, for spherical and for other irregular-shaped particles (intalox saddles, rasching rings, berl saddles) particle Peclet number is found to be between 0.3 and 0.8 for Reynolds number between 0.01 and 150. For high Reynolds number the Chung correlation holds (Inglezakis and Poulopoulos, 2006):

$$Pe_p = \frac{1}{\varepsilon}\left(0.2 + 0.011Re^{0.48}\right) \tag{69}$$

This expression hold for $25 < Re < 320$. This correlation is derived using glass beads, aluminum beads and steel beads and it has been used for ion exchange systems by use of resins beads (Ben-Shebil et al., 2007). The equation proposed for Kubo can be used for $10 < \mathrm{Rep}/\varepsilon < 2000$ (Inglezakis and Poulopoulos, 2006):

$$Pe_{\mathrm{p}} = 0.243\varepsilon^{-0.27} Re_{\mathrm{p}}^{0.27} \tag{70}$$

Furthermore, according to a literature review presented in Chung and Wen (1968), the particle Peclet number is between 0.06 and 0.3, showing no particular trend, for $0.01 < Re < 10$ and is steadily increased for $Re > 10$. More correlations for Peclet are given in Inglezakis and Poulopoulos, 2006. For the specific case of irregular-shaped particles a review is published by Inglezakis (2010a).

Fixed bed geometric analogies are also essential for a good hydrodynamic behavior in all reactor sizes, especially during scale up. In particular, the following analogies are of the most important for avoiding large-scale flow maldistribution (Inglezakis and Poulopoulos, 2006):

$$\frac{Z}{D} \geq 5 \quad \frac{D}{d_{\mathrm{p}}} \geq 12 - 30 \quad \frac{Z}{d_{\mathrm{p}}} \geq 50 - 150 \tag{71}$$

where D is the bed diameter, Z is the bed height, and d_{p} is the particle diameter. It should be noted that bed porosity is practically constant for low d_{p}/D (<0.1). Finally, in order to avoid extended friction between the packing material in both down and upflow operations a maximum linear velocity is defined (Ruthven, 1984). This velocity is 0.8 times the minimum fluidization velocity for upflow operation and 1.8 times the same velocity for downflow operation.

The second critical parameter in fixed beds is the liquid hold up which represents the part of the bed occupied by the liquid phase. This is important especially in downflow operation as in upflow operation the liquid occupies practically the whole external free void volume of the bed. Total liquid holdup h_{t} consists of two parts: static h_s and dynamic holdup h_{d}. Static holdup is related to the volume of liquid which is adherent to the particles surface, whereas dynamic holdup is related to the flowing part of the liquid (Inglezakis and Poulopoulos, 2006). Liquid holdup (h_{t}, h_{d}, or h_{s}) based on the total volume of the bed occupied by liquid h_e and liquid holdup based the void volume of the bed occupied by liquid h_v are related as follows:

$$h_{\mathrm{v}} = \frac{V_{\mathrm{liquid}}}{\varepsilon V_{\mathrm{totalbed}}} = \frac{1}{\varepsilon}\left(\frac{V_{\mathrm{liquid}}}{V_{\mathrm{totalbed}}}\right) = \frac{h_{\mathrm{e}}}{\varepsilon} \tag{72}$$

As is clear: $0 \leq h_e \leq \varepsilon$ while $0 \leq h_v \leq 1$. Is worthwhile to mention that in the case of partial wetting of the bed ($h_e \leq \varepsilon$) h_e should be used instead of ε in all model equations.

Static holdup is related to the volume of liquid which is adherent to the particles surface, whereas dynamic holdup is related to the flowing part of the liquid (Inglezakis and Poulopoulos, 2006). For zeolites and similar particles of irregular shape the following equation is proposed (Inglezakis et al., 2001):

$$\%h_t = 21 + 99.72u_s^{0.28} \tag{73}$$

In Eqn (69) total liquid holdup (h_t) is expressed as percentage of the total void bed volume, while superficial velocity is in cm/s. The above equation holds for particle size of 1.2–1.3 mm. For different particle size, the dynamic liquid holdup (h_d) is evaluated using the following analogy:

$$\frac{\%h(d_1)}{\%h(d_2)} \cong \left(\frac{d_2}{d_1}\right)^m \tag{74}$$

where m is 0.72. This value is for activated carbon particles, which are of similar shape as zeolites (Colombo and Baldi, 1976; Specchia and Baldi, 1977). The bed voidage is considered to be the same for different particle sizes, which is true for low d_p/D values.

More general is the following equation, derived for two-phase fixed beds operating under downflow operation with liquid distributor at the bed inlet and for particles of several shapes, including irregular-shaped particles of activated carbon of 1 mm diameter and for $0.3 < Re_p < 3000$ (Colombo and Baldi, 1976; Specchia and Baldi, 1977):

$$h_d = 3.86Re_p^{0.545}Ga^{-0.42}\left(\frac{a_u d_p}{\varepsilon}\right)^{0.65} \tag{75}$$

$$Ga = \frac{d_p^3 g \rho^2}{\mu^2} \tag{76}$$

Here, the dynamic liquid holdup is in m^3/m^3 and refers to the portion of void (available) bed volume occupied by liquid. The constant part in liquid holdup correlation (Eqn (69)) is the static liquid holdup (h_s), which is a function of Eotvos number (Eo):

$$Eo = \frac{\rho g d_p^2}{\sigma_L} \tag{77}$$

where (σ_L) is the surface tension in (N/m) and for water at ambient temperature is equal to 71.2×10^{-3} (N/m). The following equation holds for spherical particles:

$$h_s = \frac{0.11}{1 + Eo} \tag{78}$$

where static holdup (h_s) based on the *total* volume of the bed.

1.4.2 Ion Exchange Fluidized Beds

Fluidization is the operation by which fine solids are transformed into a fluid-like state through contact with a gas or liquid (Kunii and Levespiel, 1969). The operation of fluidized beds can be seen as the transition region between continuous-stirred tank and packed-bed modes. For velocities higher than the minimum fluidization velocity, the appearance of fluidized beds is often quite different.

In most liquid–solid systems, as the velocity is increased, the motion of the particles becomes more vigorous, whereas the bed density at a given velocity is the same in all sections of the bed. This is called *particulate fluidization* and its characteristic is the large but uniform expansion of the bed at high velocities. This type of fluidization appears when the fluid and the solids have similar densities. In contrast, the density difference is very high when the fluid is a gas, which results in the so-called *bubbling fluidization*, where the gas moves through the reactor either forming "bubbles" that contain relatively few solid particles or as a continuous "dense" phase where the particle concentration is high. Particulate and bubbling fluidization are the most common operation modes of fluidized beds. Although liquids are associated with particulate fluidization and gases with bubbling, fluidization that is not always the case. The density difference is the decisive parameter and thus bubbling fluidization appears in water systems of heavy solids and particulate fluidization in high-pressure gas systems of fine particles (Inglezakis and Poulopoulos, 2006; McCabe et al., 1993). Finally, another type of fluidization is the *slugging fluidization*. It represents the case where the bubbles form slugs of gas, usually when the size of the bubbles is about one third of the diameter of the bed. In general, slugging is undesirable because is attended by large pressure, which may cause dangerous vibrations to the reactor (Figure 11).

Fluidized beds are used in both catalytic and noncatalytic systems and several applications are found in wastewater treatment and particular in aerobic and anaerobic treatment of municipal and industrial wastewaters

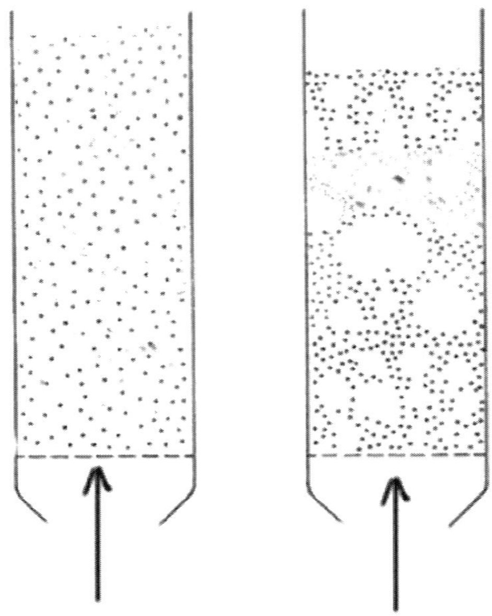

Figure 11 Particulate and bubble fludization regimes.

(Celik et al., 2001; Epstein, 2003; Lee et al., 2004a; Pérez et al., 1999; Saravanane and Murthy, 2000; Turan et al., 2005). Generally, the upward superficial velocity of the gas is usually between 0.15 m/s and 6 m/s and bed heights are not less than 0.3 m or more than 15 m (Perry and Green, 1999). In many applications of fluidization, the particles are in the range of 30–300 μm (Inglezakis and Poulopoulos, 2006).

In lesser extent fluidized beds are used in liquid–solid adsorption and ion exchange processes (Inglezakis and Poulopoulos, 2006; Perry and Green, 1999). In the later applications they are used if the influent contains significant amounts of suspended matter and in combination to small size of the solid phase which could lead in leading in clogging of fixed beds or if problems occur due to the decrease in particles mean radius during the process (Haines, 1978; Inglezakis and Poulopoulos, 2006; Menoud et al., 1998; Yang and Renken, 2000). Another reason for using fludized beds is when yhe distribution of paticle size is widely spreaded, since the mixing of particles is acheived easily and the distribution of temperature is more uniform (Park et al., 1999). Finally, when resins are used, another problem concerns shrinking and swelling of the resin during the adsorption–desorption processes resulting in poor liquid distribution caused by channeling and formation of dead zones in fixed beds (Yang and Renken, 2000).

Among the technologies to carry out ion exchange and adsorption process, the use of batch and fixed bed operation are the most studied. Although, fixed bed process is easy to operate, channeling and formation of dead zones are common phenomena, which reduce the operation performance. Fluidization overcomes operating problems such as bed clogging and high pressure drop, which would be the case if such high-surface area media were used in a packed bed reactor. Fluidized bed is able to minimize or even eliminate this disadvantage because of the intense contact between the solid and the fluid phases that can be achieved. Furthermore, concerning batch-mode operation, the requirement for mechanical agitation can be avoided by using a fluidized bed reactor. In the later case, the agitation and mixing are achieved by means of the moving liquid that carries the solids through the reactor or mixes with the particle phase (Inglezakis and Poulopoulos, 2006).

Fernandez et al. (2008) studied the performance of two laboratory-scale fluidized bed reactors with natural zeolite as support material for treating high-strength distillery wastewater while Valentukeviciene and Rimeika (2007) examined natural powdered zeolite in experiments carried out by fluidized batch process in a laboratory and a pilot-scale unit. Fluidized bed operation in adsorption and ion exchange has been studied for heavy metals removal from liquid phase to a chelating resin and for the determination of mass transfer rates in a liquid magnetically stabilized fluidized bed of magnetic ion exchange particles (Hausmann et al., 2000; Menoud et al., 1998).

Haines (1978) and Giddey (1980) studied the ion exchange process for the recovery of uranium in a fluidized bed filled with resin particles with great success and recognized the usefulness of such operations under certain conditions. Park et al. (1999) studied the mass transfer of cooper ion in semifluidized and fluidized ion exchange beds of resin as solid phase. Yang and Renken (2000) underline the fact that electroplating wastewater frequently often contains solid impurities leading to a plugging of the fixed bed and thus fludized beds are used to avoid column blocking. They studied on heavy metal adsorption to a chelating resin in a binary solid fluidized bed. Fluidized resin ion exchange beds are also studied for drinking water treatment (Li et al., 2009). The authors also mention the need for avoidance of clogging of the ion exchange process due to the scaling which is caused by precipitation of oversaturated salts in water, such as calcium carbonate. Cornelissen et al. (2009) also used resin fluidized beds for treating untreated surface water containing high loads of suspended solids. As they presented, membrane filtration in drinking water treatment is hampered by the occurrence of membrane fouling which leads to operational problems and sometimes to a deterioration of the treated water quality. One of the causes of membrane fouling is the presence of natural organic matter in sources for drinking water and the aim was to reduce membrane fouling by partly removing natural organic matter or eliminating hardness ions by using a fluidized ion exchange process.

There are also few studies dealing with the hydrodynamic analysis of liquid–solid fluidized beds and especially of nonregular shaped solid phase (Siwiec, 2007a,b; Inglezakis, 2010). In another study, Knez et al. (2001) investigated the process parameters for producing granulated zeolite in a semi-industrial scale fluidized bed dryer by utilizing an agglomeration process. Several process parameters investigated, excluding, however, any hydrodynamic parameters as minimum fluidization velocity and fluidized bed voidage. Finally, Khoshtaghaza and Chayjan (2007) investigated the fluidization behavior of several grain samples (millet, barley, paddy, and soya bean) with respect to their physical properties.

More systematic studies have been published on water–solid fluidized beds of nonspherical but regular shaped particles. Escudié et al. (2006) carried out experiments in water-fluidized binary and ternary mixtures of teflon spheres, discs, and rods. The velocity range used (10–30 cm/s), much higher than those normally used in ion exchange systems, where low contact time is required. Liu et al. (2008) studied the sphericity and shape factor in relation to the fluidized-bed dynamics. The experimental results show that nonspherical particles give poor fluidizing quality as compared to spherical particles while with the same volume-equivalent diameter, nonspherical particles have lower minimum fluidization velocity (u_{mf}) (Figure 11).

1.4.2.1 Hydrodynamics of Particulate Fluidized Beds

A typical schematic fluidized bed is shown in Figure 12. The bed normally consists of a flow regulating entry section filled with glass or other inactive beads, covered with stainless steel sieves and a filter, in order to homogenize and evenly distribute the liquid flow before it reaches the active bed section. An identical filter is placed on the top of the column, so as to prevent small particles escaping the column.

The expanded bed height is:

$$Z_f = Z_o \frac{1-\varepsilon}{1-\varepsilon_f} \tag{79}$$

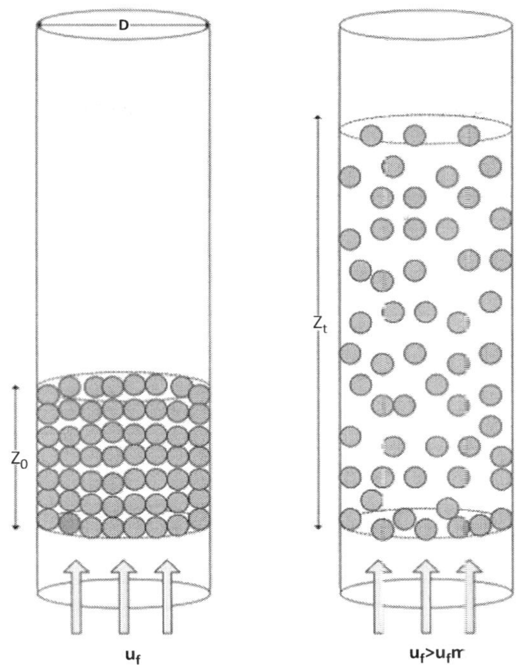

Figure 12 Typical fludized bed scheme (u_f: fluid velocity, u_{fm} minimum fluidization velocity).

The minimum fluidization velocity can be calculated if the pressure drop in a fluidized bed is set equal to the pressure drop in a fixed bed:

$$\frac{150\mu u_{\mathrm{mf}}}{\phi_s^2 d_p^2}\frac{1-\varepsilon_{\mathrm{mf}}}{\varepsilon_{\mathrm{mf}}^3}+\frac{1.75\rho u_{\mathrm{mf}}^2}{\phi_s d_p}\frac{1}{\varepsilon_{\mathrm{mf}}^3}=g(\rho_h-\rho) \tag{80}$$

where

d_p = the diameter for spherical particles or the nominal diameter for irregular-shaped particles; ρ = the density of the fluid; ρ_h = the hudraulic density of the solid; g = the gravity acceleration constant, 9.81 m^2/cm; μ = the dynamic viscosity of the fluid; ε_{mf} = the minimum fluidization voidage; u_{mf} = the minimum superficial fluidization velocity; and ϕ_S = the sphericity of the particle.

It is important to note that when a porous particle is immersed in a fluid, it has an effective density different from the skeletal and particle density. This density is termed "wet" or "effective density" (Inglezakis and Poulopoulos, 2006; Perry and Green, 1999):

$$\rho_h = \varepsilon_p \rho + \rho_p \tag{81}$$

The term *hydraulic density* is used in order to highlight the use of this kind of density in hydrodynamic calculations involving suspension of particles.

The bed voidage at incipient fluidization can be evaluated using the following approximations (Siwiec, 2007; Wen and Yu, 1966):

$$\varepsilon_{mf} = \left(\frac{0.071}{\phi_s}\right)^{1/3} \tag{82}$$

This equation has been proven to be adequate for a wide range of natural materials (Siwiec, 2007). For the determination of the minimum fluidization velocity, the fixed bed porosity (ε) is often used instead of the corresponding voidage a minimum fluidization (ε_{mf}). In particulate fluidization, for $Re_p < 10$, the relationship between the fluidized bed voidage and velocity can be derived from the Ergun equation (Inglezakis and Poulopoulos, 2006; McCabe et al., 1993):

$$\frac{150\mu u_s}{\phi_s^2 d_p^2 g\left(\rho_h - \rho\right)} = \frac{\varepsilon_f^3}{1 - \varepsilon_f} \tag{83}$$

For relatively large particles (of several mm) in water, the equation proposed by Lewis, Gilliland, and Bauer can be used (McCabe et al., 1993):

$$\varepsilon_f = \left(\frac{u_s}{u_{fm}}\right)^{1/m} \varepsilon \tag{84}$$

The exponent (m) is a function of particle Reynolds number based on the minimum fluidization velocity. It can be estimated by the following correlation:

$$m = 4.21\, Re_{fm}^{-0.0804} \tag{85}$$

for $1 < Re_{fm} < 1000$. In this region (m) is approximately between 2.5 and 4.2. The value of (m) is between 4.2 and 4.5 for $0.1 < Re_{fm} < 1$ and an average value of 4.35 can be used as a first approximation.

Another equation that can be used is the empirical Richardson–Zaki equation (Siwiec, 2007a):

$$\frac{u_{\mathrm{s}}}{u_{\mathrm{t}}} = \varepsilon_{\mathrm{f}}^{n} \tag{86}$$

where (n) is an index which depends on the particle size and sedimentation velocity. For the case of zeolites, Siwiec (2007a) verified the Richardson–Zaki equation and proposed the following equation for the exponent (n):

$$n = 8.192\ Re_{\mathrm{t}}^{-0.1598} \tag{87}$$

where (Re_{t}) is the particle Reynolds number based on the superficial sedimentation velocity (u_{t}).

As is proved for the case of zeolite particles, Ergun equation fails to represent the data as the sphericity is a characteristic physical property of the solid material and thus is constant. Similar results and trends have been found elsewhere as well, for different models (Inglezakis et al., 2010; Siwiec, 2007). As it has been proved, the results of Ergun and Richardson–Zaki equations are not satisfactory for zeolite systems and thus there is a need for introduction of a correlation that incorporates the system critical variables and in particular, the fluidized bed voidage, particle size and liquid superficial velocity (Inglezakis et al., 2010). The following equation is proposed:

$$f(\varepsilon_{\mathrm{f}}) = \frac{\varepsilon_{\mathrm{f}}\, d_{p}^{0.29}}{Re_{\mathrm{p}}} = 0.0692\ Re_{\mathrm{p}}^{-0.892} \tag{88}$$

where (d_{p}) is in (m). This equation is applied on the experimental data and the result is shown in Figure 13 (Inglezakis et al., 2010). The validity of the equation is verified for 135 μm $< \overline{d_{\mathrm{p}}} <$ 408 μm and $Re_{\mathrm{p}} < 0.145$ and is safe up to approximately $\varepsilon_{\mathrm{f}} = 0.72$, which corresponds to up to 47% bed expansion (Figure 14).

Furthermore, a correlation for the minimum fluidization velocity was proposed by Inglezakis et al. (2010) for zeolite systems:

$$u_{\mathrm{mf}} = 0.9272\ d_{\mathrm{p}}^{2.6524} \tag{89}$$

where (u_{mf}) and (d_{p}) are in (cm/s) and (mm), respectively. The correlation coefficient of this equation is $R^2 = 0.9738$.

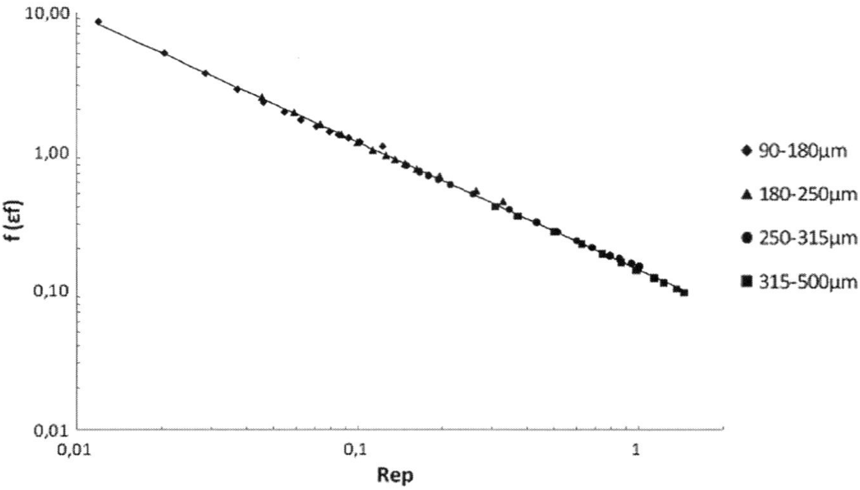

Figure 13 The $f(\varepsilon_f)$ function vs Re_p (Eqn (88)).

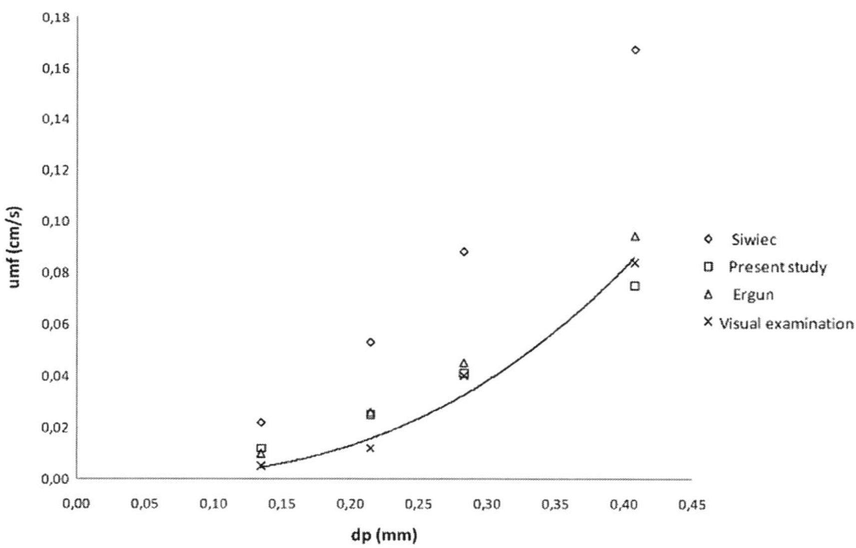

Figure 14 Minimum fluidization velocity vs. particle size.

1.4.2.2 Modeling of Liquid–Solid Particulate Fluidization

Three models of the fluidised bed process can be considered (Inglezakis and Poulopoulos, 2006; Menoud et al., 1998):

- Complete backmixing of the liquid and the solid (CSTR model);

- Nearly plug flow of the liquid and a complete back mixing of the solid, which is the plug flow model with back mixing of the solid;
- Nearly plug flow of the liquid with no back mixing of the solid, which is the plug flow model.

Since particulate flow regime is characterized by relatively uniform expansion of the bed without formation of bubbles, it can be approximated as an expanded fixed bed (Inglezakis and Poulopoulos, 2006; Wen, 1984). Thus, the fixed bed model can be used for particulate flow regime in fluidised beds for ion exchange and the differences are concentrated around the calculation of the mass transfer coefficients.

The main mass transport resistance in liquid fluidized beds of relatively small particles lies in the liquid film. Thus, for ion exchange on small particles the mass transfer limitation provides a simple liquid film diffusion-controlled mass transfer process. It should be noted that in the material balances, the fluidized bed voidage should be used instead of fixed bed voidage and that the appropriate mass transfer correlations should be used for *particulate fluidization* (Inglezakis and Poulopoulos, 2006).

The mass and heat transfer coefficient between the gas or liquid phase and the solid one can be evaluated using the Chu–Kaiil–Wetterath correlation (Smith, 1981):

$$j_{\mathrm{D}} = 1.77\left[\frac{d_{\mathrm{p}}G}{\mu(1-\varepsilon_{\mathrm{f}})}\right]^{-0.44} \tag{90}$$

this equation is valid within the limits:

$$30 < \frac{d_{\mathrm{p}}G}{\mu(1-\varepsilon_{\mathrm{f}})} < 5000 \tag{91}$$

where

$$j_{\mathrm{D}} = \frac{k_{\mathrm{f}}\rho}{G}\left(\frac{a_{\mathrm{m}}}{a_{\mathrm{t}}}\right)\left(\frac{\mu}{\rho D_{\mathrm{f}}}\right)^{2/3} \tag{92}$$

where

ε_{f} = the void fraction of the fluidized bed; $G = u_{\mathrm{s}}\rho$, the fluid mass superficial velocity, kg/m^{2}s; a_{t} is the total mass transfer area, (a_{m}) is the effective mass transfer area; and D_{f} is the solute diffusivity in the fluid phase.

For $5 < Re_p < 100$, the following correlation, obtained by Rahman and Streat for mass transfer, is valid for conventional liquid fluidized beds (Hausmann et al., 2000):

$$Sh = \frac{0.86}{\varepsilon_f} Re_p^{0.5} Sc^{1/3} \tag{93}$$

For lower Reynolds numbers ($0.22 < Re < 1$), the Koloini–Sospic–Zumer correlation is more accurate (Hausmann et al., 2000):

$$Sh = \frac{0.7}{\varepsilon_f} Re_p Sc^{1/3} \tag{94}$$

Here, the (Re_p) is based on superficial velocity.

If a fluidized and a fixed bed are operated at the same Re_p, the mass transfer coefficient is, in general, higher in the later. However, for high flow rates, the fixed bed can be operated only in down flow mode, because the solids would be entrained in up flow, at high Reynolds numbers. Furthermore, for down flow mode, Re_p should be lower than the value of 1.8 times the minimum Re_p for fluidization in order to avoid excessive attrition of the particles in the fluidized bed (Inglezakis and Poulopoulos, 2006). For the typical liquid–solid system shown in Figure 15, the minimum Re_p for fluidization is

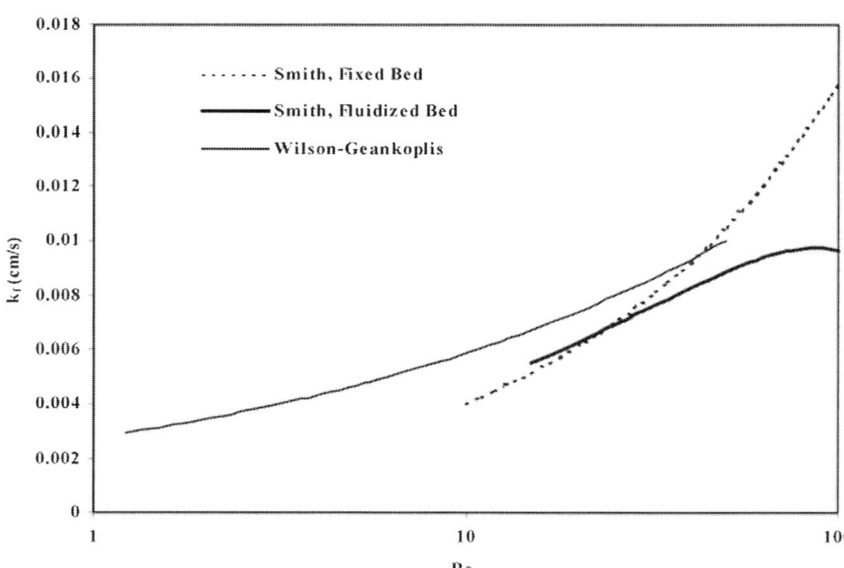

Figure 15 Comparison of mass transfer coefficients for fixed and fluidised beds (System: water at 20 °C, $\Phi_S = 1$, $\rho_h = 2.08$ g/cm^3, $\varepsilon = 0.4$, $D_f = 10^{-5}$ cm^2/s, and $d_p = 1$ mm).

about eight and thus, the upper limit for fixed bed operation in down flow mode is lower than 13. Furthermore, high residence times and thus low superficial velocities and low Re_p are met in fixed beds. For example, Re_p values lower than eight are typical in ion exchange in fixed beds. For the system shown in Figure 15, if the Re_p in the fixed bed is lower than 8, whereas the Re_p in the fluidized bed is higher than this value, the fluidized bed mass transfer coefficient is higher that this of the fixed bed.

Experimental values for the comparison of fixed and fluidized ion exchnage systems were studied by Stylianou (2011). The solid phase is natural zeolite clinoptilolite and the liquid phase aqueous solution of heavy metals (Zn^{2+}, Mn^{2+}, Cr^{3+}). The two systems run under such conditionsin for a direct comparison: the same particle size (90–180 μm), relative flow rate/contact time (12.48 BV/h) and Re_p (0.01). The results were similar for all three metals and in Figure 16 the case of Cr^{3+} is presented.

The results clearly show that the fixed bed exhibits better performance and according to the analysis of mass transfer presented above this is expected due to the higher mass transfer coefficient in the fixed bed. However, another factor that should be taken into account is that fluidized bed operations in natural materials as zeolites frequently suffer from nonidealities in flow, further lowering the performance of the bed. Nevertheless, the advantages of fluidized beds could be evident when working with wastewater

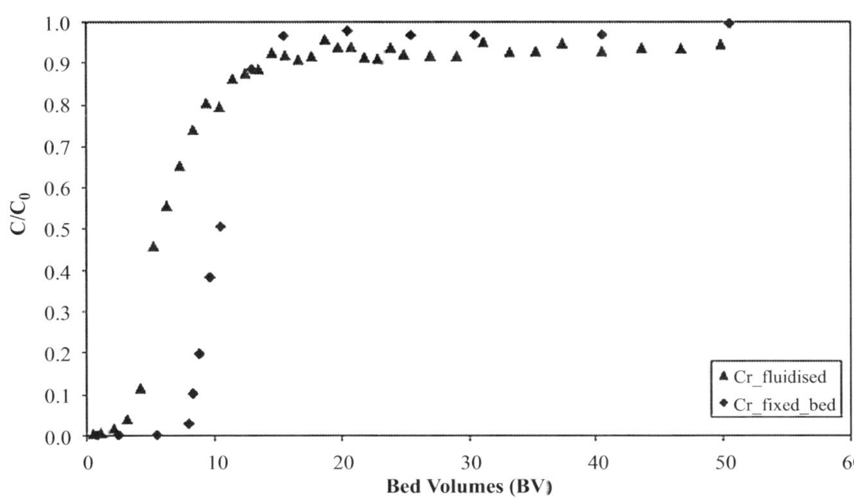

Figure 16 Comparison of ion exchange process in fixed and fluidized beds (Stylianou, 2011).

having high suspended solids concentration, which is very probable to lead in clogging problems in fixed beds with small size of ion exchange particles.

2. ION EXCHANGE MATERIALS

The materials for ion exchange referring to as ion exchangers are solid, water-insoluble substances able to retain certain ions from solutions and to release others. These substances are essentially poly-electrolytes containing an inactive part, namely the support or the matrix with the ion exchanging active groups grafted on. The ion exchange materials are classified according to several criteria. Thus, according to the type of the exchanging ions in their chemical structure the ion exchangers are classified as (Teodosiu, 2001):

- Cation exchangers (cationites);
- Anion exchangers (anionites).

and according to the origin they are grouped into:

1. Inorganic ion exchangers, falling in their turn into two groups:
   a. *Natural ion exchangers,* including: natural zeolites and clays. These materials are aluminum silicates of tri-dimensional macromolecular structures.
   b. *Synthetic ion exchangers* including: synthetic zeolites and permutites.
2. Organic ion exchangers including: *synthetic resins* (where an ionic type is fixed by the insoluble polymer and the counter-ion moves freely and can be exchanged), *polysaccharides* (cellulose); *proteins* (casein, keratin, and collagen); and *carbonic materials*.

2.1 Inorganic Ion Exchangers

2.1.1 Natural Inorganic Ions

A great number of inorganic materials with ion exchanging properties are tobe found in nature. Three types of rocks can be differentiated in function of the main process types involved in the formation of the mineral aggregates, namely: magmatic, sedimentation, and metamorphic. The volcanic tuffs belonging to the magmatic class are the most valuable deposits of natural zeolites among them. This is why in numerous studies the name of the sedimentation host rock is mentioned, namely "volcanic tuffs" or "zeolite tuffs" (Can et al., 2010; Capasso et al., 2005, 2007; Daković et al., 2010; Dávila-Jiménez et al., 2008; Díaz-Nava et al., 2009; Flint and Selker, 2003; García-Mendieta et al., 2009; Gutiérrez-Segura et al., 2009; Krestou et al., 2003; Marañón et al., 2006; Mustafa et al., 2010; Rajic et al., 2009;

Trgo and Perić, 2003; Trgo et al., 2006). The composition of the volcanic tuff includes various minerals such as: clinoptilolite, mordenite, chabazite, phillipsite, quartz, crystobalite, feldspar, and volcanic glass (Gottardi and Galli, 1985). These minerals are classified based on the function of their porus structure into micro-, meso-, and macropores with diameters ranging from a few to 75,000 Å, accessible to big molecules, viruses, and even bacteria (Cruceanu et al., 1986; Kalló, 1992; Papp, 1992).

From structural point of view, the natural zeolites are porous, crystalline materials whose network consists of tetrahedrons containing four oxygen atoms around a silicium (SiO_4) or an aluminum ion (AlO_4). The tetrahedrons are joined together in a crystalline network by means of the oxygen atoms. The fact deserves mention that every silicium atom has a charge of +4 balanced by the charge of the oxygen atoms in the tetrahedron corners so that the silicium tetrahedron is neutral from the electric point of view. The aluminumtetrahedron has a residual charge of −1 due to the aluminum trivalent ion bound to the four oxygen atoms Thus, the crystalline network of zeolites has negative charge and every aluminum tetrahedron requires compensation by a positive charge of a cation to keep the electric neutrality. It follows that all native zeolites contain cations of the elements in the main Ist and IInd groups for balancing the system network (Cruceanu et al., 1986).

From chemical point of view the natural zeolites are different by their cation content and the SiO_2/Al_2O_3 mole ratio. The vacancies in the crystalline network are usually occupied by ions of bigger ionic radius and thus higher coordination number such as Na^+, K^+, Ca^{2+}, Ba^{2+}, and scarcely by cations of small size and hexagonal coordination such as Mg^{2+}, Zn^{2+}, Fe^{3+}. The cations in the network vacancies occupy a rather small part of their volume. Due to the cations a great part of these vacancies is filled with water molecules. In case of natural zeolites the ion exchange was among the first natural phenomena making evident their unique characteristics (Barrer, 1944). The natural zeolites can participate to ion exchange processes when their own cations are replaced by other solution cations (Cruceanu et al., 1986). The cations may be either completely or partily replaced, as can be achieved by the ion exchange process, the zeolites being thus considered as inorganic ion exchangers. Due to this property they are regarded as "proper to anything". The studies made on the separation of gaseous or liquid mixtures resulted in quite remarkable progresses of the ion exchange technique. In this connection, the experiments revealed the substitution of the K^+ for the Na^+ ions in the Na-A synthetic zeolite to determine the decrease of the O_2, CH_3OH, C_2H_6, and C_2H_2 adsorption to practically zero while the substitution of the Ca^{2+} for Na^{+} ions in the same zeolite increases the adsorption from zero to higher values for N_2, n-C_6H_{14}, and C_3H_8 (Breck et al., 1956).

Also, they have a quite large application range: agriculture (Colella, 1999, 2007; Li, 2003; Rehákova et al., 2004; Spedding, 1984), medicine (Bonferoni et al., 2007; Cerri et al., 2004; Colarte, 2006; Grce and Pavelić, 2005; Izmirova et al., 2001; Poljak Blazi et al., 2001), separaration of gases (Ackley et al., 1992, 2003; Jayaraman et al., 2004), catalysis (Cobzaru et al., 2008; Mishima et al., 1998; Moreno-Tost et al., 2004; Onyestyák et al., 2004; Stöcker, 2005; Yilmaz et al., 2005), etc. But the most applications of the native zeolites are directed to the treatments of drinking and wastewaters (Argun, 2008; Armağan, 2003; Doula, 2006; Doula and Dimirkou, 2008; Elizalde-Gonzalez et al., 2001; Elizondo et al., 2000; Erdem et al., 2004; Faghihian et al., 1999; Hodi et al., 1995; Hung and Lin, 2006; Inglezakis et al., 2003; Langella et al., 2000; Zamzow et al., 1990; Zrncevic and Gomzi, 2005). For this reason the commercial interest to them is constantly growing and so does the world production of native zeolites.

The clays are natural colloidal materials showing both ion exchange and adsorption properties. Due to their properties these materials are used in treating drinking and wastewaters. The clays are also, used as an impermeable protective materials for preventing from the contamination of the underground water in the regions with deposits of waste materials. The clay category includes: bentonite, montmorillonite, collinite, beidellite, saponite, and vermiculite (Dumitriu and Lutic, 2002). Many studies are to be found in scientific literature on the potential of these materials in removing organic and inorganic polluting compounds from wastewater (Banat et al., 2002; Gupta and Bhattacharyya, 2008; Harris et al., 2002; Hasgakir and Dolgen, 2005; Koh and Dixon, 2001; Lin and Juang, 2002; Rosic et al., 2000; Sharma et al., 2007; Ugurlu et al., 2003; Vieira dos Santos and Masini, 2007; Yavuz et al., 2003). The ion exchange properties of these materials were found to be very important for determining the physical and economic characteristics crucial for their using in purifying drinking and wastewaters. Their ion exchange capacity depends on the particle size, crystalline structure and nature of the adsorbed ion.

The fact is noticeable that in recent years the native inorganic ion exchangers became very important as supplementary or substitutes of other materials such as the organic resins especially for treatments aimed to remove the radioactive residues. They show better properties than the organic materials due to their increased selectivity toward certain radioactive species: cesium, strontium, etc. (Abusafa and Yücel, 2002; Nilchi et al., 2006). Moreover, the native inorganic materials are much advantageous regarding the immobilization and final removing of contaminants compared to the organic resins (Papelis and Um, 2003).

2.1.2 Synthetic Ion Exchangers

The synthetic ion exchangers are chemical compounds obtained by synthesis to show the desired physical and chemical properties. On the basis of chemical characteristics of synthetic inorganic ion exchangers are classified as follows (Khan et al., 2012):

- Synthetic zeolites (aluminosilicates)
- Hydrous oxides of metals
- Acidic salts of polyvalent metals
- Insoluble salts of heteropolyacids
- Insoluble hydrated metal hexacyanoferrate (II) and (III) (ferro cyanides)
- Other substances with weak exchange properties

Zeolites were the first synthetic inorganic materials to be used for treatment of wastewaters. They are crystalline alumino-silicate based materials and can be prepared as microcrystalline powders, pellets or beads. The main advantages of synthetic zeolites when compared with naturally occurring zeolites are that they can be engineered with a wide variety of chemical properties and pore sizes and that they have a high thermal stability and resistance to radiation (Helfferich, 1962). In addition to that they were found to show an increased selectivity under certain conditions, so that numerous syntheses of inorganic ion exchangers are reported in scientific literature (Brett et al., 2004; Burton et al., 2005; Chareonpanich et al., 2011; Chen et al., 2012; Lee et al., 2004; Mackinnon et al., 2010, 2012; Melo et al., 2012; Oumi et al., 2010; Valuiskaya et al., 1989; Wang and Lin, 2009; Wu et al., 2006).

The main limitations of synthetic zeolites are (Khan et al., 2012):

- They have a relatively high cost compared with natural zeolites;
- They have a limited chemical stability at extreme pH ranges (either high or low);
- Their ion specificity is susceptible to interference from similar sized ions;
- The materials tend to be brittle, which limits their mechanical stability.

Since 1960 the synthetic and some native zeolites have been becoming the most efficient catalysts for the most unexpected chemical reactions. This possibility arouse after replacing the Na^+ ions in the synthetic zeolite structure by other metallic mono- and especially di- or tri-valent ions. The metallic cation in the zeolite structure shows characteristic properties and its nature exerts an influence on the number of the OH groups on the zeolite surface being crucial for the activity and selectivity towards a certain reaction (Breck et al., 1956).

2.2 Organic Ion Exchangers

2.2.1 Synthetic Organic Resins

The synthetic resins contain an organic residue (R) that is a copolymerization reticulated product of the styrene-divinylbenzene type. Active groups either acid (carboxyl-COOH, sulphonic-SO_3H, or phenolic-OH) or basic (amine, $-NH_2$, or N-substituted amines) are grafted on it. The synthetic resins are divided in their turn in two categories: cation exchangers or cationites and anion exchangers or anionites.

The cationites are acid resins in their salts with alkali metals (usually sodium) having the property of exchanging their cations with those in the contact solution. In the exchange reactions they are denoted by R–H, R–Na, (Cat)-Na. The cationites contain acid groups of the –COOH or –COONa; $-SO_3H$ or $-SO_3Na$ types in their molecule. They are able to change Ca, Mg, Na, and bicarbonate ions only. The cationites can easily be regenerated.

The cationites are divided into (Teodosiu, 2001):

- The strongly acid cationites are acrylic polymers branched with divinylbenzene and having sulphonic groups grafted on the aromatic rings. Their commercial names are: C100, C150, C160 (Purolite, SUA), IR-120, IR-252 (Amberlite, SUA), Dowex S 88 (Dow Chemicals, SUA) etc.;
- The weakly acid cationites are acrylic polymers branched with divinylbenzene carrying carboxyl groups. Their commercial names are: C106 (Purolite, SUA), IRC-84 (Amberlite), Wolfatit SBK (Bayer, Germania), Dowex CCR 2 (Dow Chemicals, SUA) etc.

The anionites are basic resins which retain the anions in the contact solution. In the ion exchange reactions they have the symbol R–OH or R–Cl, depending on the anion able to be exchanged (hydroxide or chloride, respectively). The anionites contain basic functional groups of the $-NH_2$ type able to fix the mineral or organic anions and to exchange them with the HO^- ion. According to their basic character the anionites are grouped as follows (Teodosiu, 2001):

- Weakly basic anionites;
  - Contain the $-NH_2$ amino group in the molecule and are denoted as RN^+H_2, HO^-;
  - They are mixtures of amines with aliphatic or aromatic molecular skeleton;
  - They do not exchange the ions of very weak acids;
  - They can easier be regenerated.
- Strongly basic anionites
  - They contain the ammonium group in the molecule and are denoted by $R{-}NR_3^+HO^-$;
  - These resins can exchange the ions of very weak acids;
  - They can be polystyrene or poly-acrylic resins depending on the polymer matrix.

Although the inorganic ion exchangers show several advantages they cannot entirely be replaced by organic resins especially in applications requiring water of higher purity and also in applications involving chemical reactions to be controlled by maintaining some species dissolved. It is about the nuclear stations where the organic resins are still the dominating forms even, the inorganic exchangers play a crucial role in the process of selective elimination of some radioactive species. This fact is in a great extent due to the high efficiency and safety of organic resins in both controlling of chemical reactions developing in the water cooling systems as well as in the processing of the liquid radioactive waste (Braun et al., 2002).

Apart from the organic resins the *polysaccharides* can also act as organic ion exchangers, the *cellulose* being the most important among them. The *proteins (casein, keratin and collagen)* and *carbonic materials* belong also to the category of organic ion exchangers. These organic materials are available in great amounts in nature or in the waste resulting in agriculture and can be used as potential adsorbing agents for purifying the drinking and wastewaters at a low cost. Some waste resulting in the food industry, such as the sunflower seed husk, lentil and rice, nuts peanuts, mandarins, wheaten bran, grounds of coffee, tea residues, various wood residues such as wood dust, wood chips, pine bark and tapers, leaves, ferns, etc.

3. INDUSTRIAL APPLICATIONS OF ION EXCHANGE PROCESSES

3.1 Environmental Processes

At the international level a great number of processes leading to wastewater are developing in all activity domains (agriculture, industry, medicine, nuclear stations etc.). Prior to their capitalization the wastewater require various treatments for removing the contaminants. The natural waters have also to be submitted to some procedures suitable to the purposes to be used. The water quality must be improved especially for the industrial water since the water necessary to certain technological processes has to fulfill certain quality conditions characteristic of every application field. Among the water quality improving operations the following are mentioned: softening (elimination of the Ca^{2+}, Mg^{2+} salts), demineralization (elimination of all mineral components), decantation, treating with precipitating agents, filtration etc.

As for the *water softening* it may be achieved by thermal and chemical methods and by ion exchange on resins.

The softening by ion exchange procedure is performed with a strongly acid cationic exchanger of the R–Na form. In this process the ion exchangers are loaded in columns of sizes corresponding to the flow rate of the water to be treated. Generally, after retaining the Ca^{2+} and Mg^{2+} ions an effluent containing only sodium salts is obtained (Blaga et al., 1983; Dulamita and Stanca, 1999; Parausanu, 1982; Manea et al., 2004):

$$2\text{R}{-}\text{Na} + (\text{Ca},\text{Mg})(\text{HCO}_3)_2 \rightarrow \text{R}_2(\text{Ca},\text{Mg}) + 2\text{NaHCO}_3$$

$$2\text{R}{-}\text{Na} + (\text{Ca},\text{Mg})\text{Cl}_2 \rightarrow \text{R}_2(\text{Ca},\text{Mg}) + 2\text{NaCl}$$

$$2\text{R}{-}\text{Na} + (\text{Ca},\text{Mg})\text{SO}_4 \rightarrow \text{R}_2(\text{Ca},\text{Mg}) + \text{Na}_2\text{SO}_4$$

The following reaction develops by regeneration:

$$\text{R}_2(\text{Ca},\text{Mg}) + 2\text{NaCl} \rightarrow 2\text{R}{-}\text{Na} + (\text{Ca},\text{Mg})\text{Cl}_2$$

The ion exchange is usually performed in two columns containing a cationite and an anionite, respectively, operating simultaneously with other two where the waste exchangers are regenerated. When the water hardness is assumed to be given by $CaCl_2$, the ion exchange resulting in the water softening could be described by the following equations:

$$Z{-}Na + (CaCl_2 \cdot nH_2O) \rightarrow Z_2Ca + (nH_2O + 2NaCl)$$

$$R{-}Na + (CaCl_2 \cdot nH_2O) \rightarrow R_2Ca + (nH_2O + 2NaCl)$$

water softening

The reactions taking place by the passing over a synthetic cationite lead to a softened water of an acid character (Blaga et al., 1983; Dulamita and Stanca, 1999; Parausanu, 1982; Manea et al., 2004):

$$2R{-}COOH_{(s)} + (CaCl_2 \cdot nH_2O)_{(l)} \rightarrow (R{-}COO)_2Ca_{(s)} + (nH_2O + 2HCl)_{(l)}$$

or

$$2R{-}SO_3H_{(s)} + (CaCl_2 \cdot nH_2O)_{(l)} \rightarrow 2RSO_3Ca_{(s)} + (nH_2O + 2HCl)_{(l)}$$

By passing the resulting water over an anionite the demineralized water is obtained with no content of strange ions:

$$R{-}NH_{2(s)} + (nH_2O + HCl)_{(l)} \rightarrow [R{-}NH_3]^+Cl^-_{(s)} + nH_2O_{(l)}$$

With cation exchanger as the Na^+ form, as it is with the filter elements attached to the home taps the reaction is (Blaga et al., 1983; Dulamita and Stanca, 1999; Parausanu, 1982; Manea et al., 2004):

$$2R{-}SO_3Na_{(s)} + (CaCl_2 \cdot nH_2O)_{(l)} \rightarrow 2RSO_3Ca_{(s)} + (nH_2O + 2NaCl)_{(l)}$$

The ion exchangers can be regenerated by using concentrated sodium chloride solutions (brine), concentrated acid solutions (usually 4n HCl) concentrated sodium hydroxide or sodium carbonate solutions of middle concentrations (Blaga et al., 1983; Dulamita and Stanca, 1999; Parausanu, 1982; Manea et al., 2004):

$$R_2Ca + 2NaCl \rightarrow 2\,R{-}Na + CaCl_2$$

$$(R-COO)_2Ca + 2HCl \rightarrow 2R-COOH + CaCl_2$$

$$2[R-NH_3]^+Cl^- + Na_2CO_3 \rightarrow 2R-NH_2 + 2NaCl + CO_2 + H_2O$$

$$[R-NH_3]^+Cl^- + NaOH \rightarrow R-NH_2 + NaCl + H_2O$$

To obtain a high quality of water the demineralization installation is coupled to that for removing the organic compounds. In this case the filtering layer contains both ion exchangers and other media of multifunctional filtration (activated alumina, active coal etc.) which are used for adsorption, neutralization, diminution of the iron, manganese, and hydrogen sulphide amounts.

In the *procedure of water demineralization by ion exchange* the reactions taking place by the passing over a synthetic cationite result in obtaining a softened water of an acid character. By passing the resulting water over an anionite the obtained softened water does not contain any undesired ion. When ion exchange resins are used the hard water is passed firstly over a cationite and then over an anionite. As a rule, the water is softened by treating it successively with (Blaga et al., 1983; Dulamita and Stanca, 1999; Parausanu, 1982; Manea et al., 2004):

- A cation exchanger of the R–H type, when the metallic ions are replaced by hydrogen ions and the neutral salts in water are converted into free acids;
- An anion exchanger (anionite), of the R–OH form retaining the acid anions and releasing the hydroxide ions which with the protons (H_3O) form water.

The demineralization installations consist of five columns successively bound for a consumption of regeneration reagents as low as possible. The following reactions take place in the columns (Blaga et al., 1983; Dulamita and Stanca, 1999; Parausanu, 1982; Manea et al., 2004):

- Column 1:

$$2R-H + (Ca, Mg)(HCO_3)_2 \rightarrow R_2(Ca, Mg) + 2H_2CO_3$$

$$H_2CO_3 \rightarrow H_2O + CO_2$$

- Column 2 contains a strongly acid cationite which eliminates the cations bound to the anions: chloride, sulphuric, siliceous

$$2R-H + (Ca, Mg)SO_4 \rightarrow R_2(Ca, Mg) + H_2SO_4$$

$$2R{-}H + (Ca, Mg)Cl_2 \rightarrow R_2(Ca, Mg) + 2HCl$$

$$2R{-}H + MgSiO_3 \rightarrow R_2Mg + H_2SiO_3$$

- Column 3 contains a weakly basic anionite which eliminates the formed sulphuric acid and hydrogen chloride

$$2R{-}OH + H_2SO_4 \rightarrow R_2SO_4 + 2H_2O$$

$$R{-}OH + HCl \rightarrow RCl + H_2O$$

- The resulting water passes then in the column 4 where CO_2 is removed by air blast till a residual content of about 10 mg CO_2/dm^3
- Column 5 contains a strongly basic anionite for eliminating the silicic acid and remaining CO_2:

$$2R{-}OH + H_2SiO_3 \rightarrow R_2{-}SiO_3 + 2H_2O$$

$$2R{-}OH + H_2CO_3 \rightarrow R_2{-}CO_3 + 2H_2O$$

The anionites are regenerated by treating with NaOH diluted solutions (NaOH 4%).

As regards the *wastewater*, numerous undesired ions can be removed by ion exchange. For instance, for removing the iron and manganese ions the ion exchange is performed similar to the water softening and is efficient with the water of a rather low Fe^{2+} content. The ion exchange is advantageous since an additional stage in water treating is not required and thus it can be made simultaneously with the softening. The main disadvantage is the formation of the ferric hydroxide as a colloidal precipitate depositing on the cationite and blocking the active centers which results in the substratum colmation. To overcome this shortcoming the water is filtered prior to being introduced into the softening column and the cationite is periodically cleaned. The efficient remove of manganese can be achieved with strongly acid cationites as in case of water softening. However, the ionic exchange is not enough with very high concentrations of manganese and its compounds.

As for removing the ammonium ion from the wastewater the common purification procedures (biologic procedures, chemical treatments, chlorination, air stripping etc.) were noticed to be unsatisfactory and thus they were gradually replaced by ion exchange processes (Farkas et al., 2005; Sarioglu, 2005). Numerous reported studies on the ammonium ion retaining from the wastewater by ion exchange were performed with both synthetic ion exchangers (Dowex 50wx8, Purolite MN500) (Jorgensen and Weatherley, 2003) and natural zeolites such as clinoptilolite, heulandite, stilbite, phillipsite, chabazite, and bentonite (Englert and Rubio, 2005; Langella et al., 2000; Levya-Ramos et al., 2004; Weatherley and Miladinovic, 2004).

Many studies are also to be found in scientific literature regarding the removing of heavy metal ions from drinking and wastewater by ion exchange especially with natural zeolites (Argun, 2008; Armağan, 2003; Doula, 2006; Doula and Dimirkou, 2008; Elizalde-Gonzalez et al., 2001; Elizondo et al., 2000; Erdem et al., 2004; Faghihian et al., 1999; Hodi et al., 1995; Hung and Lin, 2006; Inglezakis et al., 2003; Zamzow et al., 1990; Zorpas et al., 2000a,b, 2002; Zrncevic and Gomzi, 2005). These studies made evident the performances of the ion exchange for the purification of drinking and wastewater to depend greatly on both chemical and structural properties of the ion exchangers (ion exchanging capacity, adsorption capacity high microelement content, microporosity, acid-basic, or redox properties of the surface, etc.) as well as the conditions of their treatments.

The studies on the zeolite volcanic tuffs destined to important practical applications revealed their performances to be higher when they are *activated* (Barbat and Marton, 1989). In the main, the activation of the zeolitic volcanic tuffs has in view procedures for modifying, improving and making evident the physical and chemical properties. For instance, for the greatest part of the applications based on the adsorbing properties of zeolites not only the choice of the proper type is important but also the material processing to the necessary granular size followed by the physical (thermal) activation and also by modification of its structure and composition (chemical treatments). Apart from the chemical treatments which are very often enough for improving the zeolite properties a particular attention is paid to the *biological treatments*. Thus, the volcanic tuffs can be "impregnated" by nutritive media and "inoculated" with bacterial population or with mushroom mycelia and used than on various purposes especially for purification of the wastewater. In comparison with the natural process developing more difficult the inoculation with nitrifying bacteria accelerates the transformation of the organic material accumulated within the biofilters with zeolite material. The maintaining of the granular volcanic tuff in pretreated wastewater and especially in the compartments of tertiary biological purification results in the loading of the zeolitic filtering mass with both residual mineral and organic substances and newly formed biogenic material. This material can be periodically collected, dried under sun light and grounded when the resulting flour containing the characteristic mineral part as well as nutrients is suitable for being used as a fodder additive. In this case it is about a double use of the zeolitic material, namely the purification and as a fodder, the first performing also an activation for the second (Barbat and Marton, 1989). In this connection, Hrenović et al. (2003) have studied the adsorption capacity of clinoptilolite after its modification by a biological treatment for the phosphate ion in the wastewater and found that the material thus obtained after adsorption could be a good soil fertilizer.

Apart from the ion exchanging function the native zeolites can also be used as adsorbers for purifying the drinking and wastewaters. The first studies on the subject were reported in 1938 when Barrer has published some information on the properties and characteristics of zeolites (Barrer, 1978). The adsorption technique was then applied at the industrial scale to the purification of wastewater containing noxious organic compounds (McKay, 1996). The adsorption process was also applied to removing the dyes in the wastewater resulting in the textile industry (Armağan et al., 2004; Benkli et al., 2005; Chang et al., 2002). According to the literature data these materials are suitable adsorbers for various polar and nonpolar molecules such as: CO_2, SO_2, NO_2, NO, H_2S, NH_3, H_2O, aliphatic and aromatic hydrocarbons, alcohols, ketones, and other similar molecules (Malherbe et al., 1995a,b; Malherbe, 2000). For making better evident the adsorption capacity of the native zeolites they can be modified by various methods. Thus, several researchers pointed out the significance of using the modified zeolites for removing the undesired ions in the wastewater by means of the adsorption process (Englert and Rubio, 2005; Oliveira and Rubio, 2007a,b).

In recent years a particular attention has been paid to the removing of undesired ions from the *waste and drinking water* by adsorption on organic ion exchangers, especially waste and residues resulting in agriculture and food industry. Numerous studies support these statements. Thus, Ajmal et al. (1998) have studied the adsorption of the copper ions from the wastewater on wood dust as well as the influence of salinity and the influence of the water salinity on the adsorption process. They came to the conclusion that the removal of the copper ion is efficient even when the extraction medium salinity is high. The retaining of the aluminum ion from the waste and drinking water by means of the grounds of coffee as adsorbing material was of a great interest to researchers. The fact was noticed that the grounds of coffee can easily retain high aluminum ion amounts as well as other heavy metals (Adeyiga et al., 1998). Orhan and Buyukgungor (1993) have studied the behavior of various adsorbing materials such as the tea and coffee waste, nut, and peanut shells, for retaining the heavy metals from waste and drinking water. They found these materials to show a high adsorption potential toward the heavy metal ions, the aluminium ions especially.

Ho et al. (2002) have studied the adsorption capacity of fern in water purification and noticed the high efficiency in retaining the Zn^{2+}, Cu^{2+}, and Pb^{2+} ions as well as the influence of temperature and material size on the process yield. Ricordel et al. (2001) have studied the adsorption capacity of the active coal obtained from the peanut in retaining the Pb^{2+}, Zn^{2+}, Ni^{2+}, and Cd^{2+} ions from wastewater by adsorption. They noticed that the Pb^{2+} ion is adsorbed quicker being followed by Zn^{2+}, Ni^{2+}, and Cd^{2+}. Karapinar Kapdan and Kargi (2000) have investigated the removal of some dyes from the wastewater by means of active coal, zeolites, wood chips, and ash resulting by the wood burning. The adsorption capacity of ash was found to be much higher than that of the active coal powder. The adsorption of copper and lead ions from the waste and drinking water on tea waste was studied by Amarasinghe and Williams (2007). The authors found that the adsorption process is influenced by the adsorption rate and the concentration of the adsorbed metallic ion. Al-Asheh and Duvnjak (1997) have studied the adsorption process of the Pb^{2+}, Cd^{2+}, Cu^{2+}, Ni^{2+} ions from the wastewater by means of pine tree bark and have found the following order of the easiness of the ion retaining: $Pb^{2+} > Cd^{2+} > Cu^{2+} > Ni^{2+}$. The removing of the anionic dyes from the wastewater by means of the sunflower seed husk and mandarin rinds was studied by Osma et al. (2007). The first were noticed to retain about 85% of the existing ions after 3 h and the second about 71%. Boonamnuayvitaya et al. (2004) have studied the adsorption process with a mixture of grounds of coffee and clays as adsorbents for retaining the ions of the heavy metals (Cd^{2+}, Cu^{2+}, Pb^{2+}, Zn^{2+}, Ni^{2+}) from the wastewater. This investigation revealed the adsorption process on the material under study to be influenced by temperature, residue–clay amount ratio as well as the diameter of the material particles and the adsorption capacity to increase in the order: $Cd^{2+} > Cu^{2+} > Pb^{2+} > Zn^{2+} > Ni^{2+}$. Aydin et al. (2008) have studied the removal of the copper ion from aqueous solution on various adsorbents (lentil, wheat, rice husks). They found that the copper ion retaining on adsorbent materials is influenced especially by temperature so that the higher temperature is the higher amount of the adsorbed ions is. Other studies were made with egg shells (Tsai et al., 2008), ash obtained by coal burning

(Papandreou et al., 2007; Matheswaran and Karunanithi, 2007), red clay (Namasivayam et al., 2002), grape residues (Farinella et al., 2007), active coal (Dash and Murthy, 2010; Elaigwu et al., 2009; Hasar, 2003; Khalkhali and Omidvari, 2005; Teker and Imamoglu, 1999; Yahaya et al., 2011; Yantasee et al., 2004), cellulose (Greben et al., 2009; Thanh and Nhung, 2009; Yan Yan et al., 2009), maize (Zvinowanda et al., 2010), plants *Typha latifolia* (Mânios et al., 2003), fish scales (Mustafiz et al., 2002), and biomass obtained from mushrooms (Sag, 2001).

3.2 Production Processes

As we have already mentioned the most applications of the ion exchange are directed to purification of waste and drinking water. Apart from this, there are many chemical processes developing with materials modified by ion exchange as catalysts, the native zeolites prevailing among them. Many studies on the applications to chemical processes of native zeolites modified by ion exchange are to be found in literature and the most significant ones are to be discussed in the following:

The hydrocracking is one of the catalytic processes having been performed with native zeolites. The erionite as its Ni form obtained by ion exchange or by impregnation as well as an eronite–clinoptilolite–mordenite mixture in the Pt form can be cited among the zeolite catalysts (Oudejans, 1984). In this process, the C_5–C_9 paraffins are selectively adsorbed and submitted to hydrocracking under the temperature of 525–700 K and pressure of 1.36–13.6 MPa.

The *catalytic reduction of NO* is another process where natural zeolites, mordenite and clinoptilolite were tested after ion exchange as Zn and Ag forms (Moreno-Tost et al., 2005). In this case the two transition metals were found to improve slightly the catalytic performances of the materials used. The zeolites as their Zn form show satisfactory catalytic activities, with conversions of 58% to NO and a high selectivity to N_2 at high temperatures. The mordenite and clinoptilolite exchanged in their Cu forms were also used in catalytic reduction of NO (Moreno-Tost et al., 2004). The authors noticed that the zeolites exchanged as Cu form are very active attaining conversions to NO of 95%, a high selectivity to N_2 and a good tolerance to water.

The catalytic activity and selectivity of clinoptilolite exchanged as the Ni(II) form were tested in the process of *ethylene dimerization* (Choo and Kevan, 2001). The authors have noticed the performances of the natural catalyst to depend on the reaction temperature, the type of the cocations formed and on the amount of nickel ions incorporated into the extra-network sites. A higher amount also the Ni(II) ions incorporated into the extra-network sites of clinoptilolite results in the increase in the amount of resulting n-butenes and, on the other hand, in an easier deactivation due to the reduction of the active Ni(I) ion to the inactive Ni(0) ion.

In the process of *1-butene isomerization to isobutene* the activity and selectivity of the clinoptilolite exchanged in its cobalt and barium forms were tested (Lee et al., 2002). The authors have proved the cobalt form of the natural zeolite to show a higher selectivity to isobutene than in case of the raw material while the barium form resulted in a high activity to isobutene and a quite low selectivity.

Onyestyák and Kalló (2002) have obtained a cadmium form of clinoptilolite by means of both an ion exchange process in liquid phase and the contact of zeolite in its NH_4 form with a solid cadmium salt at the temperature of about 300 C close to the meltin point of the metal. These materials were tested in the process of *hydration of acetylene*. The authors have noticed that the activity of the natural zeolite prepared by ion exchange in the liquid phase was essentially lower than that of the zeolite prepared by contacting with the solid cadmium salt.

Aykaç and Yilmaz (2008) have studied the *citral hydrogenation in liquid phase* on mono- (Ni) and bimetallic Ni–Sn catalysts on mordenite as Na form and clinoptilolite as supports. The zeolite support type was noticed to affect the catalyst activity and selectivity. Yilmaz et al. (2005) have investigated the process of citral hydrogenation in liquid phase with palladium on clinoptilolite support under various reaction conditions. They found that the zeolite support does not affect the properties of the active metal and the catalytic activity is influenced by the reaction temperature. Moreover, the selectivity to citronelal increases from 78% to 90% with increasing catalyst amount.

Mačala and Pandová (2007) have noticed that the clinoptilolite modified by ion exchange and thermally could be used for *reducing the nitrogen monoxide in the noxious gases* evolved by the combustion engines, by adsorption. In addition to that, the adsorption capacity increases significantly by chemical treatment since the clinoptilolite exchanged with ammonium chloride and cobalt chloride show o high ability in decreasing the nitrogen monoxide content in the noxious gases. Faghihian and Pirouzi (2009) have investigated the catalytic properties of raw clinoptilolite and of that modified in the

Na, K, Ca, and Mg forms in the process of adsorption of *cis-trans-2 butene* and noticed the high potential for separating the two hydrocarbons.

Faghihian and Mousazadeh (2007) have investigated the catalytic properties of clinoptilolite exchanged with hexadecyl–trimethylammonium in the process of *elimination of polycyclic aromatic hydrocarbons from n-paraffins*. The fact was noticed that this natural material modified by a surfactant can remove the polycyclic aromatic hydrocarbons from paraffins about as much as 50%.

Hutchings et al. (1988) have studied the deactivation and reactivation of the modified clinoptilolite in *conversion of methanol to hydrocarbons*. In this case they came to the conclusion that the clinoptilolite modified by either ammonium ion exchange or hydrochloric acid treatment, exhibit a short useful catalyst lifetime (ca. 2–3 h) due to a high rate of coke deposition ($3–5 \times 10^{-3}$ g carbon/g catalyst/h).

Dumitriu et al. (2003) have studied the catalytic properties of the natural clinoptilolite in the process of *condensation of lower aldehydes*. The authors have demonstrated that the clinoptilolite volcanic tuff wasted by using to water purification can successfully be used as a catalyst in the aldol condensation of the lower aldehydes (CH_2O, CH_3CHO). Although the natural tuff was found to be active in these reactions its modification by ion exchange with metallic cations (Ni^{2+}, Cd^{2+}, Pb^{2+}, Fe^{3+}, Cr^{3+}) has a positive effect improving the selectivity to acrolein as well as the resistance to deactivation by coking.

Arenamnart and Trakarnprukb (2006) have investigated the catalytic properties of the clinoptilolite modified by ion exchange in the process of *ethylene conversion to ethylene*. The authors pointed out that the modification of dealuminated mordenite by various metals gives a more efficient catalyst and the amount of the coke deposited on the catalyst is quite low.

Yeniova et al. (2007) have studied the activity and selectivity of clinoptilolite exchanged as the ammonium form and dealuminated in the process of *isomerization, disproportionation*, and *hydrocracking* of 1,2,3-trimethyilbenzene. The authors have drawn the conclusion that the catalyst type and temperature have a crucial influence on the process.

REFERENCES

Abusafa, A., Yücel, H., 2002. Removal of ^{137}Cs from aqueous solutions using different cationic forms of a natural zeolite: clinoptilolite. Sep. Purif. Technol. 28, 103–116.

Ackley, W.M., Giese, F.R., Yang, T.R., 1992. Clinoptilolite: untapped potential for kinetics gas separations. Zeolites 12 (7), 780–788.

Ackley, W.M., Rege, S.U., Saxena, H., 2003. Application of natural zeolites in the purification and separation of gases. Microp. Mezop. Mater. 61 (1–3), 25–42.

Adeyiga, A.A., Hu, L., Greer, T., 1998. Removal of metal ions from wastewater with natural wastes. In: Sixth Annual Historically Black Colleges and Universities and Other Minority Institutions Symposium Conference Proceedings: Technology Transfer Session.

Ajmal, M., Khan, H.A., Ahmad, S., Ahmad, A., 1998. Role of sawdust in the removal of copper(II) from industrial wastes. Water Res. 32, 3085–3091.

Al-Asheh, S., Duvnjak, Z., 1997. Sorption of cadmium and other heavy metals by pine bark. J. Hazard. Mater. 56, 35–51.

Amarasinghe, B.M.W.P.K., Williams, R.A., 2007. Tea waste as a low cost adsorbent for the removal of Cu and Pb from wastewater. Chem. Eng. J. 132, 299–309.

Arenamnart, S., Trakarnprukb, W., 2006. Ethanol conversion to ethylene using metal-mordenite catalysts. Int. J. Appl. Sci. Eng. 4 (1), 21–32.

Argun, E.M., 2008. Use of clinoptilolite for the removal of nickel ions from water: kinetics and thermodynamics. J. Hazard. Mater. 150, 587–595.

Armağan, B., 2003. Factors affecting the performances of sepiolite and zeolite for the treatment of textile wastewater. J. Environ. Sci. Health A38, 883–896.

Armağan, B., Turan, M., Çelik, S.M., 2004. Equilibrium studies on the adsorption of reactive azo dyes into zeolite. Desalination 170, 33–39.

Aydin, H., Bulut, Y., Yerlikaya, C., 2008. Removal of copper (II) from aqueous solution by adsorption onto low-cost adsorbents. J. Environ. Manag. 87, 37–45.

Aykaç, H., Yilmaz, S., 2008. Hydrogenation of citral over Ni and Ni–Sn catalysts. Turk. J. Chem. 32 (5), 623–633.

Babel, S., Kurniawan, T., 2003. Low-cost adsorbents for heavy metals uptake from contaminated water: a review. J. Hazard. Mater. B97, 219–243.

Banat, A.F., Al-Bashir, B., Al-Asheh, S., Hayajneh, O., 2002. Adsorption of Phenol by bentonite. Environ. Pollut. 107, 391–398.

Barbat, A., Marton, A., 1989. The Zeolitic Volcanic Tuffs. Dacia Publishing House, Cluj Napoca, Romania.

Barrer, M.R., 1944. Sorption by gmelinite and mordenite. Trans. Faraday Soc. 40, 555.

Barrer, M.R., Hinds, L., 1953. Ion-exchange in crystals of analcite and leucite. J. Chem. Soc. 27, 879–1888.

Barrer, M.R., Meier, M.W., 1958. Structural and ion sieve properties of a synthetic crystalline exchanger. Trans. Faraday Soc. 54, 1074–1085.

Barrer, M.R., Meier, M.W., 1959. Exchange equilibria in a synthetic crystalline exchanger. Trans. Faraday Soc. 55, 130.

Barrer, M.R., Rees, L.V.C., Ward, D.J., 1963. Thermochemistry and thermodynamics of ion exchange in a crystalline exchange medium. Proc. Royal Soc. A 273, 180–197.

Barrer, R.M., 1964. Surface flow of adsorbed gases and Vapours. Proc. Br. Ceram. Soc. I, 145–166.

Barrer, M.R., Davies, J.A., Rees, L.V.C., 1968. Thermodynamics and thermochemistry of cation exchange in zeolite Y. J. Inorg. Nucl. Chem. 30 (12), 3333–3349.

Barrer, M.R., Galabova, M.I., 1973. Advances in chemistry. Am. Chem. Soc. 121, 356.

Barrer, M.R., Klinowski, J., 1974. Ion-exchange selectivity and electrolyte concentration. J. Chem. Soc. Faraday Trans. 1 (70), 2080–2091.

Barrer, M.R., 1978. Zeolites and Clay Minerals as Sorbents and Molecular Sieves. Academic Publishing House, London.

Benkli, Y.E., Can, M.F., Turan, M., Çelik, M.S., 2005. Modification of organo-zeolite surface for the removal of reactive azo dyes in fixed-bed reactors. Water Res. 39, 487–493.

Ben-Shebil, S., Alkan-Sungur, A., Özdural, A., 2007. Fixed-bed ion exchange columns operating under non-equilibrium conditions: estimation of mass transfer properties via non-equilibrium modeling. React. Funct. Polym. 67, 1540–1547.

Blaga, A., Popescu, M., Stoescu, M., 1983. General Chemical Technology and Processes Types. Didactic and Pedagogic Publishing House, Bucharest, Romania.

Bonferoni, M.C., Cerri, G., De'Gennaro, M., Juliano, C., Caramella, C., 2007. Zn^{2+}-exchanged clinoptilolite-rich rock as active carrier for antibiotics in anti-acne topical therapy: in-vitro characterization and preliminary formulation studies. Appl. Clay Sci. 36 (1–3), 95–102.

Boonamnuayvitaya, V., Chaiya, C., Tanthapanichakoon, W., Jarudilokkul, S., 2004. Removal of heavy metals by adsorbent prepared from pyrolyzed coffee residues and clay. Sep. Purif. Technol. 35, 11–22.

Braun, A., Efremenkov, V.M., Evans, D., Gandhi, P.M., Garamszeghy, M., Geiser, H., Hooper, E.W., Kaufmann, F., Luyex, P., Martynov, B., Mezga, L.J., Neubauer, J., Norden, M., Pecha, M., Piacek, P., Sinha, P.K., Stevens, T., Tusa, E., Zika, H., 2002. Application of Ion Exchange Processes for the Treatment of Radioactive Waste and Management of Spent Ion Exchangers. Technical Reports Series No. 408. IAEA, Vienna.

Breck, D.W., 1974. Zeolite Molecular Sieve (New York).

Breck, D.W., Evesole, W.G., Milton, R.M., Reed, T.B., Thomas, T.L., 1956. Crystalline Zeolites. I. The Properties of a New Synthetic Zeolite, Type A. J. Am. Chem. Soc. 78, 5963–5972.

Brett, A., Wang, H., Yan, Y., 2004. High silica zeolite Y nanocrystals by dealumination and direct synthesis. Microp. Mezop. Mater. 74 (1–3), 189–198.

Brown, L.M., Sherry, H.S., Krambeck, F.J., 1971. Mechanism and kinetics of isotopic exchange in zeolites. I. Theory. J. Phys. Chem. 75 (25), 3846–3855.

Burton, W.A., Zones, I.S., Elomari, S., 2005. The chemistry of phase selectivity in the synthesis of high-silica zeolites. Curr. Opin. Colloid & Interface Sci. 10 (5–6), 211–219.

Can, Ö., Balköse, D., Ülkü, S., 2010. Batch and column studies on heavy metal removal using a local zeolitic tuff. Desalination 259 (1–3), 17–21.

Capasso, S., Salvestrini, S., Coppola, E., Buondonno, A., Colella, C., 2005. Sorption of humic acid on zeolitic tuff: a preliminary investigation. Appl. Clay Sci. 28 (1–4), 159–165.

Capasso, S., Coppola, E., Iovino, P., Salvestrini, S., Colella, C., 2007. Sorption of humic acids on zeolitic tuffs. Microp. Mezop. Mater. 105 (3), 324–328

Cerri, G., De'Gennaro, M., Bonferoni, M.C., Caramella, C., 2004. Zeolites in biomedical application: Zn-exchanged clinoptilolite-rich rock as active carrier for antibiotics in anti-acne topical therapy. Appl. Clay Sci. 27 (3–4), 141–150.

Chareonpanich, M., Jullaphan, O., Tang, C., 2011. Bench-scale synthesis of zeolite A from subbituminous coal ashes with high crystalline silica content. J. Clean. Prod. 19 (1), 58–63.

Chang, W.S., Hong, S.W., Park, J., 2002. Effect of zeolite media for the treatment of textile wastewater in a biological aerated filter. Process Biochem. 27, 693–698.

Chen, D., Hu, X., Shi, L., Cui, Q., Wang, H., Yao, H., 2012. Synthesis and characterization of zeolite X from lithium slag. Appl. Clay Sci. 59 (60), 148–151.

Choo, H., Kevan, L., 2001. Catalytic study of ethylene dimerization on Ni (II)-exchanged clinoptilolite. J. Phys. Chem. 105 (27), 6353–6360.

Cobzaru, C., Oprea, S., Dumitriu, E., Hulea, V., 2008. Gas phase aldol condensation of lower aldehydes over clinoptilolite rich natural zeolites. Appl. Catal. General. A 351 (2), 253–258.

Colella, C., 1995. Use of Italian chabazite and phillipsite for the removal of heavy metals from wastewaters: a review. In: Ming, D.W., Mumpton, F.A. (Eds.), Natural Zeolite'93. International Community of Natural Zeolites, Brockport, New York, pp. 363–375.

Colella, C., 1999. Natural zeolites in environmentally friendly processes and applications. Stud. Surf. Sci. Catal. 125, 641–655.

Colella, C., 2007. Natural zeolites. Stud. Surf. Sci. Catal. 133, 20–57.
Colarte, I., 2006. Antacid drug based on purified natural clinoptilolite. Microp. Mezop. Mater. 94 (1–3), 200–207.
Cruceanu, M., Popovici, E., Balba, N., Naum, N., Vladescu, L., Russu, R., Vasile, A., 1986. Zeolite Molecular Sieves. Scientific and Encyclopedic Publishing House, Bucharest, Romania.
Celik, M.S., Ozdemir, B., Turan, M., Koyuncu, I., Atesok, G., Sarikaya, H.Z., 2001. Removal of ammonia by natural clay minerals using fixed and fluidized bed column reactors. Water Sci. Technol. Water Supply 1, 81–88.
Chen, J., Wang, L., 2004. Characterization of metal adsorption kinetic properties in batch and fixed-bed reactors. Chemosphere 54, 397–404.
Chern, J., Chien, Y., 2002. Adsorption of nitrophenol onto activated carbon: isotherms and breakthrough curves. Water Res. 36, 647–655.
Chung, S., Wen, C., 1968. Longitudinal dispersion of liquid flowing through fixed and fluidized beds. Am. Inst. Chem. Eng. J. 14, 857–866.
Colombo, A., Baldi, G., 1976. Solid-Liquid contacting effectiveness in trickle bed reactors. Chem. Eng. Sci. 31, 1101–1108.
Cornelissen, E.R., Beerendonk, E.F., Nederlof, M.N., Van der Hoek, J.P., Wessels, L.P., 2009. Fluidized ion exchange (FIX) to control NOM fouling in ultrafiltration. Desalination 236, 334–341.
Crittenden, J., Hand, D., Arora, H., Lykins Jr., B., Jr., January 1987. Design considerations for GAC treatment of organic chemicals. J. Am. Water Works Assoc., 74–82. Jan.
Daković, A., Kragović, M., Rottinghaus, G.E., Sekulić, Ž., Milićević, S., Milonjić, S.K., Zarić, S., 2010. Influence of natural zeolitic tuff and organozeolites surface charge on sorption of ionizable fumonisin B1. Colloids Surfance B: Biointerface 76 (1), 272–278.
Dash, S.N., Murthy, R., 2010. Preparation of carbonaceous heavy metal adsorbsnt from *Shorea Robusta* Lef Litter using phosphoric acid impregnation. Int. J. Environ. Sci. 1 (3), 296–313.
Dávila-Jiménez, M.M., Elizalde-González, M.P., Mattusch, J., Morgenstern, P., Pérez-Cruz, M.A., Reyes-Ortega, Y., Wennrich, R., Yee-Madeira, H., 2008. In situ and ex situ study of the enhanced modification with iron of clinoptilolite-rich zeolitic tuff for arsenic sorption from aqueous solutions. J. Colloid Interface Sci. 322 (2), 527–536.
Díaz-Nava, C., Olguín, M.T., Solache-Ríos, M., Alarcón-Herrera, M.T., Aguilar-Elguezabal, A., 2009. Phenol sorption on surfactant-modified Mexican zeolitic-rich tuff in batch and continuous systems. J. Hazard. Mater. 167 (1–3), 1063–1069.
Dizdar, Z., 1972. Sodium-lithium exchange on synthetic zeolite 4A in water-dioxane mixtures. J. Inorg. Nucl. Chem. 34 (3), 1069–1081.
Dizdar, Z., Popovič, P., 1972. Sodium-lithium exchange on synthetic zeolite 4A in water-dimethyl sulphoxide (DMSO) mixtures. J. Inorg. Nucl. Chem. 34 (8), 2633–2642.
Doula, M.K., 2006. Removal of Mn^{2+} ions from drinking water by using clinoptilolite and a clinoptilolite-Fe oxide system. Water Res. 40, 3167–3176.
Doula, M.K., Dimirkou, A., 2008. Use of an iron-overexchanged clinoptilolite for the removal of Cu^{2+} ions from heavily contaminated drinking water samples. J. Hazard. Mater. 151, 738–745.
Dulamita, N., Stanca, M., 1999. Chemical Technology, vol. I. Cluj University Publishing House, Cluj-Napoca, Romania.
Dumitriu, E., Lutic, D., 2002. The Catalisys: A General Approach. VIE Press, Iasi.
Dumitriu, E., Hulea, V., Cobzaru, C., Oprea, S., 2003. Aplication of natural zeolites in removal metals from wastewaters and in catalysis of chemical reactions. II. The aldol condensation catalized by natural clinoptilolite modified with metallic cations by ion exchange process. Chem. Mag. Buchar. 54 (2), 142–145.

Dyer, A., Gettins, R.B., 1970. The mobility of cations in synthetic zeolites with the faujasite framework. III. Self-diffusion of cations into X and Y zeolites from non-aqueous solutions. J. Inorg. Nucl. Chem. 32 (7), 2401–2410.

Eisenmann, G., 1962. Cation selective glass electrodes and their mode of operation. J. Biophysics 2, 259–323.

Ekedal, E., Högfeldt, E., Sillen, L.G., 1950. Activities of the components in ion exchangers. Acta Chem. Scand. 4, 556–558.

Elaigwu, S.E., Usman, L.A., Awolola, G.V., Adebayo, G.B., Ajayi, R.M.K., 2009. Adsorption of Pb (II)from aqueous solution by activated carbon prepared from cow dung. Adv. Nat. Appl. Sci. 3 (3), 442–446.

Elizalde-Gonzalez, M.P., Mattusch, J., Wennrich, R., 2001. Application of natural zeolites for preconcentration of arsenic species in water samples. J. Environ. Monit. 3, 22–26.

Elizondo, N.V., Ballesteros, E., Kharisov, B.I., 2000. Cleaning of Liquid Radioactive Wastes Using Natural Zeolites. Appl. Radiat. Isotopes 52, 27–30.

Englert, H.A., Rubio, J., 2005. Characterization and environmental application of a Chilean natural zeolite. Int. J. Mineral Process. 75 (1–2), 21–29.

Epstein, N., 2003. Applications of liquid-solid fluidization. Int. J. Chem. React. Eng. 1, 1–6.

Erdem, E., Karapinar, N., Donat, R., 2004. The removal of heavy metal cations by natural zeolites. J. Colloid Interface Sci. 280, 309–314.

Escudié, R., Epstein, N., Grace, J.R., Bi, H.T., 2006. Effect of particle shape on liquid-fluidized beds of binary (and ternary) solids mixtures: segregation vs. mixing. Chem. Eng. Sci. 61, 1528–1539.

Faghihian, H., Ghannadi Marageh, M., Kazemian, H., 1999. The use of clinoptilolite and its sodium form for removal of radioactive cesium, and strontium from nuclear wastewater and Pb^{2+}, Ni^{2+}, Cd^{2+}, Ba^{2+} from municipal wastewater. Appl. Rad. Isotop. 50 (4), 655–660.

Faghihian, H., Mousazadeh, M.H., 2007. Removal of PAHs from n-paraffin by modified clinoptilolite. Iran. J. Chem. Eng. 26 (3), 121–127.

Faghihian, H., Pirouzi, M., 2009. Cis/trans-but-2-ene adsorption on natural and modified clinoptilolite. Clay Miner. 44 (3), 405–409.

Farkas, A., Rozic, M., Barbaric-Mikocevic, Z., 2005. Ammonium exchange in leakage waters of waste dumps using natural zeolite from the Krapina region, Croatia. J. Hazard. Mater. B117, 25–33.

Farinella, N.V., Matos, G.D., Arruda, M.A.Z., 2007. Grape bagasse as a potential biosorbent of metals in effluent treatments. Bioresource Technol. 98 (10), 1940–1946.

Fernández, N., Montalvo, S., Borja, R., Guerrero, L., Sánchez, E., Cortés, I., Colmenarejo, M.F., Travieso, L., Raposo, F., 2008. Performance evaluation of an anaerobic fluidized bed reactor with natural zeolite as support material when treating high-strength distillery wastewater. Renew. Energy 33, 2458–2466.

Fleck, R., Kirwan, D., Hall, K., 1973. Mixed-resistance diffusion kinetics in fixed bed adsorption under constant pattern conditions. Ind. Eng. Chem. Fundam. 12, 95–99.

Flint, L.E., Selker, J.S., 2003. Use of porosity to estimate hydraulic properties of volcanic tuffs. Adv. Water Res. 26 (5), 561–571.

Gaines Jr., G.L., Jr., Thomas, H.C., 1953. Adsorption studies on clay minerals. II. A formulation of the thermodynamicas of exchange adsorption. J. Chem. Phys. 21, 714–718.

García-Mendieta, A., Solache-Ríos, M., Olguín, M.T., 2009. Evaluation of the sorption properties of a Mexican clinoptilolite-rich tuff for iron, manganese and iron–manganese systems. Microp. Mezop. Mater. 118 (1–3), 489–495.

Giddey, T.B.S., 1980. The benefits of heavy resins in fluidized-bed ion-exchange columns. J. South Afr. Inst. Min. Metallurgy 12, 431–435.

Glueckauf, E., 1949. Activity coefficients in concentrated solutions containing several electrolytes. Nature 163, 414–415.
Gottardi, G., Galli, E., 1985. Natural Zeolites. Springer, Berlin.
Grce, M., Pavelić, K., 2005. Antiviral properties of clinoptilolite. Microp. Mezop. Mater. 79 (1–3), 165–169.
Greben, H., Sigama, J., Burke, L., Venter, S.N., 2009. Cellulose Fermentation Products as an Energy Source for Biological Sulphate Reduction of Acid Mine Drainage Type Wastewaters. Report NO.1728/1/08, WRC.
Gutiérrez-Segura, E., Solache-Ríos, M., Colín-Cruz, A., 2009. Sorption of indigo carmine by a Fe-zeolitic tuff and carbonaceous material from pyrolyzed sewage sludge. J. Hazard. Mater. 170 (2–3), 1227–1235.
Gupta, S.S., Bhattacharyya, K.G., 2008. Immobilization of Pb(II), Cd(II) and Ni(II) ions on kaolinite and montmorillonite surfaces from aqueous medium. J. Environ. Manag. 87 (1), 46–58.
Haines, A.K., 1978. The development of continuous fluidized-bedion exchange in South Africa and its use in the recovery of uranium. J. South Afr. Inst. Min. Metallurgy 7, 303–315.
Hall, K., Eagleton, L., Acrivos, A., Vermeulen, T., 1966. Pore- and solid-diffusion kinetics in fixed bed adsorption under constant-pattern conditions. Ind. Eng. Chem. Fundam. 5, 212–223.
Harris, R.G., Wells, J.D., Johnson, B.B., 2002. Selective adsorption of dyes and other organic molecules to kaolinite and oxide surfaces. Colloids Surfaces A: Physicochem. Eng. Aspects 180, 131–140.
Hasar, H., 2003. Adsorption of nickel from aqueous solution onto activated carbon prepared from almond husk. J. Hazard. Mater. B97, 49–57.
Hasgakir, B., Dolgen, D., 2005. Kil minerallerinin atiksu aritiminda kullanilabilirligi: Kaolinit ile organik madde giderimi. In: XII Ulusal Kil Sempozyumu, September 5–9, Van, Turkey.
Hausmann, R., Hoffmann, C., Franzreb, M.G., Holl, W.H., 2000. Mass transfer rates in a liquid magnetically stabilized fluidized bed of magnetic ion-exchange particles. Chem. Eng. Sci. 55, 1477.
Helfferich, F., 1962. Ion Exchange. McGraw-Hill, New York.
Hrenović, J., Büyükgüngör, H., Orhan, Y., 2003. Use of zeolite to Upgrade activated sludge. Food Technol. Biotechnol. 41 (2), 157–165.
Ho, Y.S., Huang, C.T., Huang, H.W., 2002. Equilibrium sorption isotherm for metal ions on tree fern. Process Biochem. 37, 1421–1430.
Hodi, M., Polyak, K., Htavay, J., 1995. Removal of pollutants from drinking water by combined ion-exchange and adsorption methods. Environ. Int. 21, 325–329.
Huang, P.C., Mizany, A., Pauly, J.L., 1964. Ion exchange of synthetic zeolites in various alcohols. J. Phys. Chem. 68, 2575.
Hung, H.W., Lin, T.F., 2006. Adsorption of MTBE from contaminated water by carbonaceous resins and mordenite zeolite. J. Hazard. Mater. B135, 210–217.
Hutchings, G.J., Themistocleous, T., Copperthwaite, R.G., 1988. Methanol conversion to hydrocarbons using modified clinoptilolite catalysts: investigation of catalyst lifetime and reactivation. Appl. Catal. 43 (1), 133–140.
Inglezakis, V.J., Lemonidou, M., Grigoropoulou, H., 2001. Liquid holdup and dispersion in zeolite packed beds. Chem. Eng. Sci. 56, 5049–5057.
Inglezakis, V.J., Loizidou, M.D., Grigoropoulou, H.P., 2003a. Pb^{2+} removal from liquid phase using chelating resin lewatit TP 207: the effect of operating variables and application of shrinking Core model. In: 8th International Conference on Environmental Science and Technology, Lemnos, Greece.
Inglezakis, V.J., Zorpas, A.A., Loizidou, M.D., Grigoropoulou, H.P., 2003. Simultaneous removal of metals Cu^{2+}, Fe^{3+} and Cr^{3+} with anions SO_4^{2-} and HPO_4^{2-} using clinoptilolite. Microp. Mesop. Mater. 61 (1–3), 167–171.

Inglezakis, V.J., Grigoropoulou, H., 2003. Modeling of ion exchange of Pb^{2+} in fixed-beds of clinoptilolite. Microp. Mezop. Mater. 61 (1–3), 273–282.

Inglezakis, V.J., 2005. The concept of "capacity" in zeolite ion exchange systems. J. Colloid Interface Sci. 281, 68–79.

Inglezakis, V.J., Poulopoulos, S., 2006. Adsorption Ion Exchange and Catalysis: Design of Operations and Environmental Applications. Elsevier, International Edition.

Inglezakis, V.J., 2010a. Non ideal flow in liquid-solid fixed beds of irregular-shaped particles: a critical review. Int. J. Chem. React. Eng. 8. R3.

Inglezakis, V.J., 2010. Ion exchange and adsorption fixed bed operations for wastewater treatment – part I: modeling fundamentals and hydraulics analysis. J. Eng. Stud. Res. 16, 29–41.

Inglezakis, V.J., Stylianou, M.A., Loizidou, M.D., 2010. Hydrodynamic studies on zeolite fluidized beds. Int. J. Chem. React. Eng. 8. A126.

Izmirova, N., Aleksiev, B., Djourova, E., Blagoeva, P., Gendjev, Z., Mircheva, Tz, Pressiyanov, D., Minev, L., Bozhkova, T., Uzunov, P., Tomova, I., Baeva, M., Boyanova, A., Todorov, T., Petrova, R., 2001. Clinoptilolite and the possibilities for its application in medicine. Stud. Surf. Sci. Catal. 135, 375.

Jayaraman, A., Hernandez-Maldonado, A.J., Yang, R.T., Chinn, D., Munson, C.L., Donald, D.H., Mohr, H., 2004. Clinoptilolites for nitrogen/methane separation. Chem. Eng. Sci. 59 (12), 2407–2417.

Jorgensen, T.C., Weatherley, L.R., 2003. Ammonia removal from wastewater by ion exchange in the presence of organic contaminants. Water Res. 37, 1723–1728.

Kalló, D., 1992. Natürliche Zeolithe-Herkunft und Wirkungsmechanismen. Abwassertechnik 43 (2), 40–43.

Karapinar Kapdan, I., Kargi, F., 2000. Atiksulardan tekstil boyar maddelerinin adsorpsiyonlu biyolojik aritim ile giderimi. Turk. J. Eng. Environ. Sci 24, 161–169.

Khalkhali, R.A., Omidvari, R., 2005. Adsorption of mercuric ion from aqueous solutions using activated carbon. Pol. J. Environ. Stud. 14 (2), 185–188.

Khan, A., Asiri, A.M., Rub, M.A., Azum, N., Khan, A.A.P., Khan, I., Mondal, P.K., 2012. Review on composite cation exchanger as interdicipilinary materials in analytical chemistry. Int. J. Electrochem. Sci. 7, 3854–3902.

Khoshtaghaza, M.H., Chayjan, R.A., 2007. Effect of some physical properties on fluidisation stability of grain products. Biosyst. Eng. 98, 192–197.

Knez, S., Strazisar, J., Golob, J., Horvat, A., 2001. Agglomeration of zeolite in the fluidized bed. Acta Chim. Slov. 48, 487–504.

Koh, S.M., Dixon, J.B., 2001. Preparation and application of organo-minerals as sorbent of phenol, benzene and toluene. Appl. Clay Sci. 18, 111–122.

Krestou, A., Xenidis, A., Panias, D., 2003. Mechanism of aqueous uranium (VI) uptake by natural zeolitic tuff. Miner. Eng. 16 (12), 1363–1370.

Kunii, D., Levespiel, O., 1969. Fluidization Engineering (New York).

Langella, A., Pansini, M., Cappelletti, P., De Gennaro, B., de 'Gennaro, M., Colella, C., 2000. NH_4^+ Cu^{2+}. Zn^{2+}, Cd^{2+} and Pb^{2+} exchange for Na^+ in a sedimentary clinoptilolite, North Sardinia. Italy. Microp. Mezop. Mater. 37 (3), 337–343.

Lee, H.C., Woo, H.C., Chung, S.H., Kim, H.J., Lee, K.H., Lee, J.S., 2002. Effects of metal cation on the skeletal isomerization of 1-butene over clinoptilolite. J. Catal. 211 (1), 216–225.

Lee, C.I., Yang, W.F., Hsieh, C.I., 2004a. Removal of copper(II) by manganese-coated sand in a liquid fluidized-bed reactor. J. Hazard. Mater. 114, 45–51.

Lee, H., Zones, S.I., Davis, M.E., 2004. A new synthesis method for zeolites. Stud. Surf. Sci. Catal. 154 (Part A), 102–109.

Levenspiel, O., 1972. Chemical Reaction Engineering. Wiley Eastern Limited, International Edition.

Levya-Ramos, R., Aguilar-Armenta, G., Gonzales-Gutierrez, L.V., Guerrero-Coronado, R.M., Mendoza-Barron, J., 2004. Ammonia exchange on clinoptilolite from mineral deposits located in Mexico. J. Chem. Technol. Biotechnol. 79, 651–657.

Li, Z., 2003. Use of surfactant-modified zeolite as fertilizer carriers to control nitrate release. Microporous Mezoporous Mater. 61 (1–3), 181–188.

Li, S., Heijman, S.G.J., Verberk, J.Q.J.C., Van Dijk, J.C., 2009. An innovative treatment concept for future drinking water production: fluidized ion exchange – ultrafiltration – nanofiltration – granular activated carbon filtration. Drink Water Eng. Sci. 1, 41–47.

Liguori, B., Cassese, A., Colella, C., 2006. Safe immobilization of Cr(III) in heat-treated zeolite tuff compacts. J. Hazard. Mater. 137 (2), 1206–1210.

Lin, S., Juang, R., 2002. Heavy metal removal from water by sorption using surfactant-modified montmorillonite. J. Hazard. Mater. B92, 315–326.

Liu, B., Zhang, X., Wang, L., Hong, H., 2008. Fluidization of non-spherical particles: sphericity, Zingg factor and other fluidization parameters. Particuology 6, 125–129.

Mackinnon, I.D.R., Millar, G.J., Stolz, W., 2010. Low temperature synthesis of zeolite N from kaolinites and montmorillonites. Appl. Clay Sci. 48 (4), 622–630.

Mackinnon, I.D.R., Millar, G.J., Stolz, W., 2012. Hydrothermal syntheses of zeolite N from kaolin. Appl. Clay Sci. 58, 1–7.

Malherbe, R., Fernandes, L., Colado, L., 1995a. Phsico-chemical properties of natural zeolites used for adsorption of water. In: Ming, D.W., Munpton, F.A. (Eds.), Natural Zeolites 1993. International Community of Natural Zeolites, Brockport, New York, USA, pp. 299–308.

Malherbe, R., Fernandez, L., Lopez, L., Pozas, C., Montes, A., 1995b. Natural zeolites. In: 1993 conference Volume of International Community of Natural Zeolites, Brockport, New York, pp. 299–308.

Malherbe, R., 2000. Complementary approach to the volume filling theory ofadsorption in zeolites. Microp. Mezop. Mater. 41 (1), 227–240.

Marañón, E., Ulmanu, M., Fernández, Y., Anger, I., Castrillón, L., 2006. Removal of ammonium from aqueous solutions with volcanic tuff. J. Hazard. Mater. 137 (3), 1402–1409.

Mačala, J., Pandová, I., 2007. Natural zeolite-clinoptilolite-raw material serviceable in the reduction of toxical components at combustion engines noxious gases. Gospod. Surowcami Miner. 23 (4), 19–26.

Mânios, T., Stentiford, E.I., Millner, P., 2003. Removal of heavy metals from a metaliferous water sohitia by Typha latifolia plants and sewage sludge compost. Chemosphere 53, 487–494.

Manea, F., Marsavina, D., Ursoiu, I., 2004. Principles, Methods and Applications in Water Analysis. Polytechnica Publishing House, Timisoara, Romania.

Matheswaran, M., Karunanithi, T., 2007. Adsorption of chrysoidine R by using fly ash in batch process. J. Hazard. Mater. 145, 154–161.

McCabe, W.L., Smith, J.C., Harriott, P., 1993. Unit Operations of Chemical Engineering, fifth ed. McGraw-Hill, New York.

McKay, G., 1996. Use of Adsorbents for the Removal of Polutants from Wastewaters. CRC Press, Boca Raton.

Melo, C.R., Riella, H.G., Kuhnen, N.C., Angioletto, E., Resmini, Melo A., Bernardin, A.M., Roberto da Rocha, M., Da Silva, L., 2012. Synthesis of 4A zeolites from kaolin for obtaining 5A zeolites through ionic exchange for adsorption of arsenic. Mater. Sci. Eng. 177 (4), 345–349, 15.

Menoud, P., Cavin, L., Renken, A., 1998. Modelling of heavy metals adsorption to a chelating resin in a fluidized bed reactor. Chem. Eng. Process 37, 89.

Mishima, H., Hashmoto, K., Ono, T., Anpo, M., 1998. Selective catalytic reduction of NO with NH_3 over natural zeolites and its application to stationary diesel engine exhaust. Appl. Catal. B: Environ. 19 (2), 119–126.

Miura, K., Hashimoto, K., 1977. Analytical solutions for the breakthrough curves of fixed-bed adsorbers under constant pattern and linear driving force approximations. J. Chem. Eng. Jpn. 10, 490–493.

Moreno-Tost, R., Santamaría-González, J., Rodríguez-Castellón, E., Jiménez-López, A., Autié, M.A., González, E., Glacial, M.C , Pozas, C.D.L., 2004. Selective catalytic reduction of nitric oxide by ammonia over Cu-exchanged Cuban natural zeolites. Appl. Catal. B: Environ. 50 (4), 279–288.

Moreno-Tost, R., Santamaría-González, J. Rodríguez-Castellón, E.. Jiménez-López, A., Autié, M.A., Glacial, M.C., Castro, G.A., Guerra, M., 2005. Selective catalytic reduction of nitric oxide by ammonia over Ag and Zn-exchanged Cuban natural zeolites. Z. für Anorg. Allg. Chem. 631 (11), 2253–2257.

Mustafa, F., Al-Ghouti, M.A., Khalili, F.I., Al-Degs, Y.S., 2010. Characteristics of organosulphur compounds adsorption onto Jordanian zeolitic tuff from diesel fuel. J. Hazard. Mater. 182 (1–3), 97–107.

Mustafiz, S., Basu, A., Islam, M.R., Dewaidar, A., Chaalal, O., 2002 A novel method for heavy removal. Energy Sources 24, 1043–1051.

Namasivayam, C., Yamuna, R.T., Arasi. D.J.S.E., 2002. Removal of procion orange from wastewater by adsorption on waste red mud. Sep. Sci. Technol. 37 (10), 2421–2431.

Nilchi, A., Maalek, B., Khanchi, A., Ghanadi Maragheh, M., Bagheri, A., Savoji, K., 2006. Ion exchangers in radioactive waste management. Natural Iranian zeolites. Appl. Radiat. Isotopes 64, 138–143.

Oliveira, C.R., Rubio, J., 2007a. Adsorption of ions onto treated natural zeolite. Mater. Res. 10 (4), 407–412.

Oliveira, C.R., Rubio, J., 2007b. New basis for adsorption of ionic pollutants onto modified zeolites. Miner. Eng. 20 (6), 552–558.

Onyestyák, G., Pál-Borbély, G., Kalló, D., 2004. Acetylene hydration on diff erent Cd-zeolites. Stud. Surf. Sci. Catal. 154 (3), 2831–2838.

Onyestyák, G., Kalló, D., 2002. Catalytic behavior of Cd-clinoptilolite prepared by introduction of cadmium metal onto cationic sites. Stud. Surf. Sci. Catal. 142B, 1047–1054.

Orhan, Y., Buyukgungor, H., 1993. Removal of heavy metals by using agricultural wastes. Water Sci. Technol. 28 (2), 247–255

Oudejans, J.C, 1984. Zeolite catalysts in some organic reactions, TH-diss 1394.

Oumi, Y., Manabe, T., Sasaki, H., Inuzuka, T., Sano, T., 2010. Preparation of Ti incorporated Y zeolites by a post-synthesis method under acidic conditions and their catalytic properties. Appl. Catal. General. 388 (1–2), 256–261.

Osma, J.F., Saravia, V., Toca-Herrera, J.L., Couto, S.R., 2007. Sunflower seed shells: a novel and effective low-cost adsorbent for the removal of the diazo dye Reactive Black 5 from aqueous solutions. J. Hazard. Mater. 147, 900–905.

Papandreou, A., Stournaras, C.J., Panias, D., 2007. Copper and cadmium adsorption on pellets made from fired coal fly ash. J. Hazard. Mater. 148, 538–547.

Papelis, C., Um, W., 2003. Evaluation of Cesium, Strontium and Lead Sorption, Desorption and Diffusion in Volcanic Tuffs from Frenchman Flat, Nevada Test Site: Macroscopic and Spectroscopic Investigations. U.S.Department of Energy, Las Vegas, Nevada. Publication no. 45189.

Papp, J., 1992. Einsatzmöglichkeiten von Zeolith in der Abwassertechnik. Abwassertechnik 43 (2), 44–47.

Parausanu, V., 1982. Chemical Technologies. Scientific and Encyclopedic Publishing House, Bucharest, Romania.

Park, Y.G., Cho, S.Y., Kim, S.J., Lee, B.G., Kim, B.H., Park, S.J., 1999. Mass transfer in semi-fluidized and fluidized ion-exchange beds. Environ. Eng. Resour. 4 (2), 71–80.

Perry, R.H., Green, D., 1999. Perry's Chemical Engineer's Handbook, seventh ed. McGraw-Hill, International Edition.

Pérez, M., Romero, L.I., Sales, D., 1999. Anaerobic thermophilic fluidized bed treatment of industrial wastewater: effect of F: M relationship. Chemosphere 38, 3443–3461.

Poljak Blazi, M., Katic, M., Kralj, M., Zarkovic, N., Marotti, T., Bosnjak, B., Sverko, V., Balog, T., Pavelic, K., 2001. In vitro and in vivo effect of natural clinoptilolite on malignant tumors. Stud. Surf. Sci. Catal., 135–374. 32-P-09.

Post, M.F.M., 1991. Diffusion in zeolite molecular sieves. In: Van Bekkum, H., Flanigen, E.M., Jansen, C.J. (Eds.), Studies in Surface Science and Catalysis, 58. Elsevier, Amsterdam, p. 391.

Radak, V.M., Šušić, M.V., 1971. Ion exchange equilibria in synthetic zeolite 4A in a mixed solvent system. J. Inorg. Nucl. Chem. 33 (6), 1927–1931.

Rajic, N., Stojakovic, D., Jevtic, S., Logar, N.Z., Kovac, J., Kaucic, V., 2009. Removal of aqueous manganese using the natural zeolitic tuff from the Vranjska Banja deposit in Serbia. J. Hazard. Mater. 172 (2–3), 1450–1457.

Rao, A., Rees, L.V.C., 1966a. Kinetics of ion exchange in mordenite. Trans. Faraday Soc. 62, 2505.

Rees, L.V.C., Rao, A., 1966b. Self-diffusion of various cations in natural mordenite. Trans. Faraday Soc. 62, 2103.

Rehákova, M., Čuvanová, S., Dzivák, M., Rimár, J., Gavalová, Z., 2004. Agricultural and agrochemical uses of natural zeolites of the clinoptilolite type. Curr. Opin. Solid State & Mater. Sci. 8 (6), 397–404.

Reichenberg, D., 1953. Properties of ion-exchange resins in relation to their structure. III. Kinetics of exchange. J. Am. Chem. Soc. 75, 589.

Ricordel, S., Taha, S., Cisse, I., Dorange, G., 2001. Heavy metals removal by adsorption onto peanut husks carbon: characterization, kinetic study and modeling. Sep. Purif. Technol. 24, 389–401.

Rosic, M., Cerjan-Stefanovic, S., Kurajica, S., Vancina, V., Hodzic, E., 2000. Ammoniacal nitrogen removal from dswater by treatment with clays and zeolites. Water Res. 34 (14), 3675–3681.

Rosseinsky, D.R., 1965. Electron potentials and hydration energies. Theories and correlations. Chem. Rev. 65, 467–490.

Ruthven, D., 1984. Principles of Adsorption and Adsorption Processes. John Wiley and Sons, USA.

Sag, Y., 2001. Bioadsorption of heavy metals by fungal biomass and modelling of fungal biosorption: a review. Sep. Purif. Technol. 30, 1–48.

Saravanane, R., Murthy, D.V., 2000. Application of anaerobic fluidized bed reactors in wastewater treatment: a review. Environ. Manag. Health 11, 97–117.

Sarioglu, M., 2005. Removal of ammonium from municipal wastewater using natural Turkish zeolite. Sep. Purif. Technol. 41, 1–11.

Siwiec, T., 2007a. The experimental verification of Richardson-Zaki law on example of selected beds used in water treatment. Electron. J. Pol. Agric. Univ. 10 (2), 14–23.

Siwiec, T., 2007b. The sphericity of grains of filtration beds applied for water treatment on examples of selected minerals. Electron. J. Pol. Agric. Univ. 10 (1), 23–30.

Sherry, H.S., 1966. The ion-exchange properties of zeolites. I. Univalent ion exchange in synthetic faujasite. J. Phys. Chem. 70, 1158–1168.

Sherry, H.S., Walton, H.F., 1967. The ion-exchange properties of zeolites. II. Ion-exchange in the synthetic zeolite Linde 4A. J. Phys. Chem. 71, 1457.

Sherry, H.S., 1969. The ion-exchange properties of zeolites. In: Marinsky, J.A. (Ed.), Ion Exchange: A Series of Advances, Vol. 2. Marcel Dekker, New York.

Smith, J.M., 1981. Chemical Engineering Kinetics, third ed. McGraw-Hill, International Edition.

Specchia, V., Baldi, G., 1977. Pressure drop and liquid holdup for two-phase concurrent flow in packed beds. Chem. Eng. Sci. 32, 515–523.

Spedding, C.R.W., 1984. In: Pond, W.G., Mumpton, F.A. (Eds.), Zeo-agriculture: Use of Natural Zeolites in Agriculture and Aquaculture. Westview Press, Boulder, Colorado, p. 285.

Stylianou, M., 2011. Removal of Heavy Metals from Aqueous Solutions by Using Natural minerals (Ph.D.thesis). National Technical University of Athens.

Stöcker, M., 2005. Gas phase catalysis by zeolites. Microp. Mezop. Mater. 82 (3), 257–292.

Sharma, Y.C., Srivastava, V., Srivastava, J., Mahto, M., 2007. Reclamation of Cr(VI) rich water and wastewater by wollastonite. Chem. Eng. J. 127, 151–156.

Szostak, R., 1989. Molecular Sieves-priciples of Synthesis and Identification. Van Nostrand Reinhold, NewYork.

Teker, M., Imamoglu, M., 1999. Adsorption of copper and cadmium ions by activated carbon from Rice Hulls. Turk. J. Chem. 23, 185–191.

Teodosiu, C., 2001. The Technology of Drink and Industrial Water. Matrix Rom Press, Bucharest.

Thanh, N.D., Nhung, H.L., 2009. Cellulose modified with citric acid and its absorption of Pb^{2+} and Cd^{2+} ions. In: 13rd International Electronic Conference on Synthetic Organic Chemistry 1–3 November.

Trambouze, F., 1990. Reactor scaleup methodology. Chem. Eng. Prog. 23, 31.

Trgo, M., Perić, J., 2003. Interaction of the zeolitic tuff with Zn-containing simulated pollutant solutions. J. Colloid Interface Sci. 260 (1), 166–175.

Trgo, M., Perić, J., Vukojević Medvidović, N., 2006. Investigations of different kinetic models for zinc ions uptake by a natural zeolitic tuff. J. Environ. Manag. 79 (3), 298–304.

Turan, M., Gulsen, H., Celik, M.S., 2005. Treatment of Landfill Leachate by a combined anaerobic fluidized bed and zeolite column system. J. Environ. Eng. 131, 815–819.

Tsai, W.T., Hsien, K.J., Hsu, H.C., Chiu, C.H., Lin, K.Y., Lin, C.M., 2008. Utilization of ground eggshell waste as an adsorbent for the removal of dyes from aqueous solution. Bioresource Technol. 99 (6), 1623–1629.

Ugurlu, M., Girses, A., Yalgin, M., Dogar, C., 2003. Kagit endistrisi atik sularindan lignin ve fenolin sepiyolit minerali ile giderimi. In: XI Ulusal Kil Sempozyumu, September 3–6, Izmir, Turkey.

Valentukeviciene, M., Rimeika, M., 2007. Development of a fluidized batch process using natural powdered zeolite (Clinoptilolite). Pol. J. Environ. Stud. 16, 283–288.

Valuiskaya, O.M., Nefedov, B.K., Kanakova, O.A., Zubkov, A.M., 1989. Industrial synthesis of types X and Y zeolites. Chem. Technol. Fuels Oils 25 (7), 340–342.

Vieira dos Santos, A.C., Masini, J.C., 2007. Evaluating the removal of Cd(II), Pb(II) and Cu(II) from a wastewater sample of a coating industry by adsorption onto vermiculite. Appl. Clay Sci. 37, 167–174.

Wang, Y., Lin, F., 2009. Synthesis of high capacity cation exchangers from a low-grade Chinese natural zeolite. J. Hazard. Mater. 166 (2–3), 1014–1019.

Weatherley, L.R., Miladinovic, N.D., 2004. Comparison of the ion exchange uptake of ammonium ion onto New Zealand clinoptilolite and mordenite. Water Res. 38, 4305–4312.

Wen, C.Y., Yu, Y.H., 1966. Mechanics of fluidization. Chem. Eng. Prog. Symp. Ser. 62, 100–111.

Wen, C.Y., 1984. Flow regimes and flow models for fluidised bed reactors. In: Doraiswamy, L.K. (Ed.), Recent Advances in the Engineering Analysis of Chemically Reacting Systems. Wiley Eastern Limited, India.

Wu, D., Zhan, B., Yan, L., Kong, H., Wang, X., 2006. Effect of some additives on synthesis of zeolite from coal fly ash. Int. J. Mineral Process. 80 (2–4), 266–272.

Xiu, G., Li, P., 2000. Prediction of breakhrough curves for adsorption of lead (II) on activated carbon fibers in a fixed bed. Carbon 38, 975–981.

Yahaya, N.K.E.M., Latiff, M.F.P.M., Abustan, I., Bellob, O.S., Ahmadb, M.A., 2011. Adsorptive removal of Cu (II) using activated carbon prepared from rice husk by $ZnCl_2$ activation and subsequent gasification with CO_2. Inter. J. Eng. & Technol. 11 (1), 207–211.

Yan Yan, F., Krishniah, D., Rajin, M., Bono, A., 2009. Cellulose extraction from palm kernel cake using liquid phase oxidation. J. Eng. Sci. Technol. 4 (1), 57–68.

Yang, J., Renken, A., 2000. Heavy metal adsorption to a chelating resin in a binary solid fluidized bed. Chem. Eng. Technol. 23, 1007–1012.

Yantasee, W., Lin, Y., Fryxell, G.E., Alford, K.L., Busche, B.J., Johnson, C.D., 2004. Selective removal of copper(II) from aqueous solutions using fine-grained activated carbon functionalized with amine. Ind. Eng. Chem. Res. 43, 2759–2764.

Yavuz, O., Altunkaynak, Y., Giizel, F., 2003. Removal of copper, nickel, cobalt and manganese from aqueous solution by kaolinite. Water Res. 37, 948–952.

Yeniova, H., Karaduman, A., Alibeyli, R., 2007. Isomerization, disproportionation and hydrocracking of 1,3,5- trimethylbenzene and n-decane mixture over natural clinoptilolite zeolites. Petroleum Sci. Technol. 25 (3), 387–398.

Yilmaz, S., Ucar, S., Artok, L., Gulec, H., 2005. The kinetics of citral hydrogenation over Pd supported on clinoptilolite rich natural zeolite. Appl. Catal. General. 287 (2), 261–266.

Zamzow, M.J., Eichbaum, B.R., Sandgren, K.R., Shanks, D.E., 1990. Removal of heavy metals and other cations from waste water using zeolites. Sep. Sci. Technol. 25 (13–15), 1555–1569.

Zorpas, A.A., Constantinides, T., Vlyssides, A.G., Haralambous, I., Loizidou, M., 2000a. Heavy metal uptake by zeolite and metais partitioning in sewage sludge compost. Bioresour. Technol. 72, 113–119.

Zorpas, A.A., Kapetanios, E., Zorpas, G.A., Karlis, S.P., Vlyssides, A.G., Haralambous, I., Loizidou, M., 2000. Compost produced from organic fraction of municipal solid waste, primary stabilized sewage sludge and natural zeolite. J. Hazard. Mater. B77, 149–159.

Zorpas, A.A., Inglezakis, V.J., Loizidou, M., Grigoropoulou, H., 2002. Partide size effects on uptake of heavy metais from sewage sludge compost using natural zeolite clinoptilolite. J. Colloid Interface Sci. 250, 1–4.

Zrnčević, S., Gomzi, Z., 2005. CWPO: An environmental solution for pollutant removal from wastewater. Ind. Eng. Chem. Res. 44, 6110–6114.

Zulfadhly, Z., Mashitah, M., Bhatia, S., 2001. Heavy metals removal in fixed-bed column by the macro fungus *Pycnoporus sanguineus*. Environ. Pollut. 112, 463–470.

Zvinowanda, C.M., Okonkwo, J.O., Agyei, N.M., Staden, M.V., Jordaan, W., Kharebe, B.V., 2010. Recovery of lead(II) from aqueous solutions by *Zea mays* tassel biosorption. Am. J. Biochem. Biotechnol. 6 (1), 1–10.

CHAPTER ELEVEN

Hot Gas Filters

Steffen Heidenreich
Pall Filtersystems GmbH, Werk Schumacher Crailsheim, Crailsheim, Germany

Contents

1. INTRODUCTION

The use of hot gas filters has gained significant interest over the last years. This is mainly driven by more stringent dust emission regulations concerning total dust concentration and fine particle fractions, such as PM 10 and PM 2.5. Moreover, energy efficiency, process intensification, water shortage and water quality as well as overall process costs are topics which raise an increasing interest in hot gas filtration.

Hot gas filtration offers also some significant advantages concerning the product quality as well as the protection of downstream equipment. Depending on the operating temperature, contamination of products or

Progress in Filtration and Separation
http://dx.doi.org/10.1016/B978-0-12-384746-1.00011-2

blocking of downstream equipment by condensation or desublimation can be prevented.

Hot gas filters are applied in a broad range of industrial processes, for example in coal and biomass gasification, in chemical industry, in fluid catalytic cracking units, in oil refineries, in waste incineration and pyrolysis, in glass industry, and in metal refining and metal recycling.

Filtration at temperatures above 260 °C is called hot gas filtration according to the VDI (Association of German Engineers) guideline 3677-3 (VDI, 2012). Hot gas filtration is performed at pressures up to 80 bar, in both oxidizing and reducing atmospheres, and often with chemically aggressive compounds. This places high demands on the mechanical, thermal, and chemical stability of hot gas filter elements. Only rigid ceramic and metal filter media are applied in hot gas filters due to their high mechanical resistance. For a stable long-term operation of a filter, the filter elements need to be regenerated frequently when the pressure drop increases by the dust cake buildup. In most cases hot gas filter elements are back pulsed by using conventional jet pulse blowback systems. These jet pulse systems originally applied for flexible filter media provide a limited cleaning performance for rigid filter elements. Due to this reason some blowback systems have been specially developed for hot gas filters in order to provide high cleaning intensities for rigid filter media.

Hot gas filters were technically firstly applied for the incineration of low-level contaminated radioactive waste from nuclear power generation at the beginning of the 1970s. For example, hot gas filters were used at the former nuclear research centre of Karlsruhe in Germany (Leibold et al., 1989) and in some incineration plants in France (Perkins, 1976). Since 1978 hot gas filtration has been broadly used in nuclear power plants in Japan (Torii, 2005).

The interest in advanced coal-based power generation techniques such as the pressurized fluidized bed combustion (PFBC) and the pressurized integrated gasification combined cycles (IGCC) had a significant influence on the development of hot gas filters from the end of the 1980s until the end of the 1990s. Different hot gas filter media and systems were developed and tested in a number of research and demonstration projects during this decade.

The coal gasification IGCC demonstration power plant (250 MW_{el}) in Buggenum in the Netherlands (Nuon, 2013), the demonstration IGCC power plant (300 MW_{el}) in Puertollano in Spain (NETL, 2013a; Elcogas, 2013), the PFBC test facility (10 MW_{th}) in Karhula in Finnland (Chalupnick et al., 1999), the PFBC demonstration plant in Wakamatsu in Japan

(71 MW_{el}) (Yamada et al., 2003), the IGCC demonstration plant in Wabash River (262 MW_{el}) (NETL, 2013b), and the Power System Development Facility (15 MW_{th}) in Wilsonville (USA) (Guan et al., 2008) are some examples of demonstration projects mainly performed in Europe, the USA, and in Japan.

While the PFBC technology has not been transferred from demonstration phase to commercial application, coal gasification has got high interest in the last 10 years. The IGCC demonstration plants in Buggenum, Puertollano, and Wabash River have been successfully transferred into commercial running plants.

Meanwhile more than 25 large hot gas filter units are in operation or in commissioning in coal gasification plants worldwide. About 20 of them were installed in China in the last 8 years. The syngas is here mainly used for production of chemicals and less for power production. Several new hot gas filter installations in coal gasification are under construction, e.g., in the 582 MW IGCC power plant in the Kemper County project in Mississippi (USA) (Mississippipower, 2013), or will be installed in the near future, e.g., in the POSCO's Gwangyang coal to substitute natural gas (SNG) project (Energy Business Review, 2013; Engineerlive, 2013).

2. HOT GAS FILTRATION—ADVANTAGES/ DISADVANTAGES

2.1 Advantages

Hot gas filters offer advantages in improving processes as well as products. The use of hot gas filters in high temperature processes for the production of metal oxide powders, catalysts, pigments, and metals has shown that product quality and process efficiency can be improved. Sometimes, high filtration temperatures are required to avoid undesirable condensation or desublimation reactions. These reactions would result in fouling and blocking of filters or in pollution of products. The filtration of hot process gases also generates the opportunity to utilize particle free, high temperature gas in coupled or subsequent process steps. Furthermore, the economic and process-technological advantages of hot gas filtration are obvious in high temperature processes that involve the circular movement of substances or process gases as well as the recovery of products or expensive materials, such as catalysts or noble metals. Hot gas filters are used to protect downstream equipment, such as heat exchangers, catalyst units, scrubbers, or gas turbines. Besides the general protection of these equipments, the efficiency and the performance of these units can be increased.

2.2 Disadvantages

Higher temperatures place higher demands on the properties and the mechanical, thermal, and chemical stability of the materials which are used. Due to the increased demands on the materials the investment costs for hot gas filters increase with the operating temperature. Moreover, the gas volume increases with the operating temperature and thus the size of the filter and correspondingly the investment costs. With an increase of the temperature, the pressure drop is higher caused by the increased viscosity of the gas.

Depending on the dust properties, softening or sintering of the dust can occur at higher temperatures. This can lead to a sticking dust layer on the filter element surface and correspondingly to an instable filtration. In the worst case, breakage of filter elements can result from bridging of the dust. At higher temperatures, also the risk of chemical solid phase reactions in the dust cake or chemical reactions of the dust cake with gas components is higher which can result in changes of the properties of the dust cake.

However, in most cases the advantages of hot gas filtration dominate and higher investments for the filter system are compensated by a reduction of the overall process costs.

3. FILTER MEDIA FOR HOT GAS FILTRATION

A hot gas filter medium has to be resistant against the temperature, requires a high thermal shock resistance and a sufficiently high mechanical stability. Furthermore, the filter medium has to have a good chemical stability against gas components, such as H_2S, HCl, NH_3, Cl_2, H_2O, etc., and against alkalis and aggressive dust components. When selecting the right material for the application and gas conditions it has to be taken into account that the properties of the porous filter material with the high surface area will be different compared to the compact material.

Only rigid self-supporting filter elements made of ceramic or metal are applied for hot gas filtration. Flexible filter media would not be resistant enough to withstand the high mechanical stress caused by the movement during back pulsing at high temperatures.

Filter candles closed at one end (see Figures 1 and 2) are the preferred geometry of hot gas filter elements. This geometry enables a very reliable

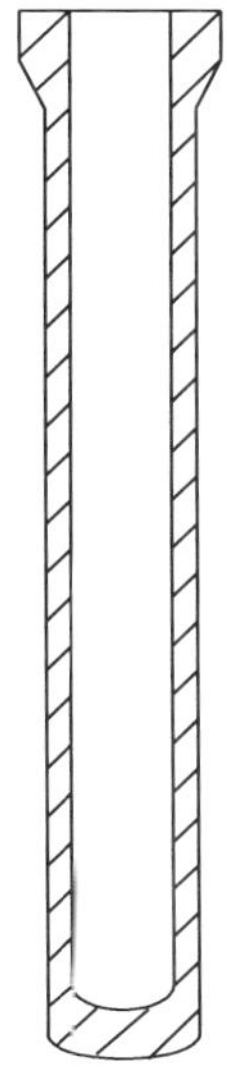

Figure 1 Scheme of a filter candle closed at one end.

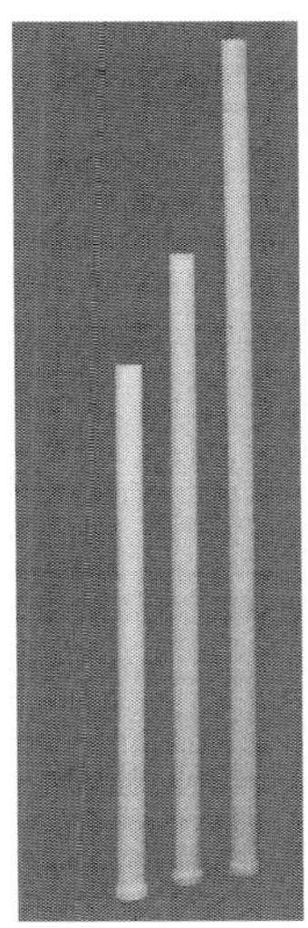

Figure 2 Examples of high-density ceramic filter candles.

operation. The cake is typically built up at the outside of the candles and can be easily and reliably detached from the outside surface of the candles by back pulsing. The length of hot gas filter candles can vary between 1 and 3 m depending on the material. Typical outer candle diameters are between 60 and 150 mm.

3.1 Ceramic Media

Ceramic filter media are very important for hot gas filtration due to their high temperature and chemical resistance. Ceramic filter elements may be made of ceramic grains, ceramic fibres, or a compound ceramic—a fibre compound embedded in a grain matrix. Fibre media, also called low density ceramic, have higher porosities compared to granular media, also called high-density ceramic. This results in lower initial differential pressures of fibre ceramic filter media. The porosity ranges from approximately 40% with pure grain ceramic up to approximately 90% with pure fibre ceramic. Due to the fibre structure, fibre elements have a very large inner surface besides their high porosity as well as their low flow resistance. Figure 3 shows an example of the structure of a fibre ceramic consisting of aluminium silicate fibres. Fibre ceramic filter media provide a high dirt holding capacity inside their structure. They have high fracture toughness as a result of their loose structure, but their mechanical strength is relatively low and the risk of candle breakages is correspondingly high. The intensity of the back pulse for regeneration needs to be controlled so that no fibres are released from the filter structure. Low density ceramic filter media are typically vacuum-formed from fibre slurry and the fibres are bonded together at the contact points by inorganic and/or organic binders.

Figure 3 Structure of a fibre ceramic.

Grain ceramic has a mechanically stable and strong structure. They are manufactured by pressing or extrusion followed by sintering at high temperature. High-density ceramic filter elements made of silicon carbide, e.g., can have bursting pressures of higher than 5 MPa and O-ring pressure strength of more than 20 MPa (Pall, 2013). The size and size distribution of the pores of the filter structure is defined by the selection of the used grain sizes. Moreover, an option to adjust the pore size and porosity is to add pore-forming materials which burn out during the sintering process, such as polymers, sawdust, or graphite. By the amount and particle size of the pore-forming material, the pore size and the porosity can be controlled.

There are high-density ceramic filter elements available with symmetric or asymmetric structures. With a symmetric structure, the pore size is distributed evenly throughout the entire material. Contrary to this, a membrane layer with small pore size is applied on the upstream side of a coarser support material providing an asymmetric structure (see Figures 4 and 5). With their finer pore size, the membrane separate smaller particles. The differential pressure of the filter element is limited by making the membrane as thin as possible. A membrane layer which is so thin that the support is just covered by the layer without defects would be optimal. By the membrane, surface filtration is achieved even for very fine dusts. A penetration of particle in the support structure of the filter element is prevented and the element can be effectively regenerated by back pulsing. This is an advantage

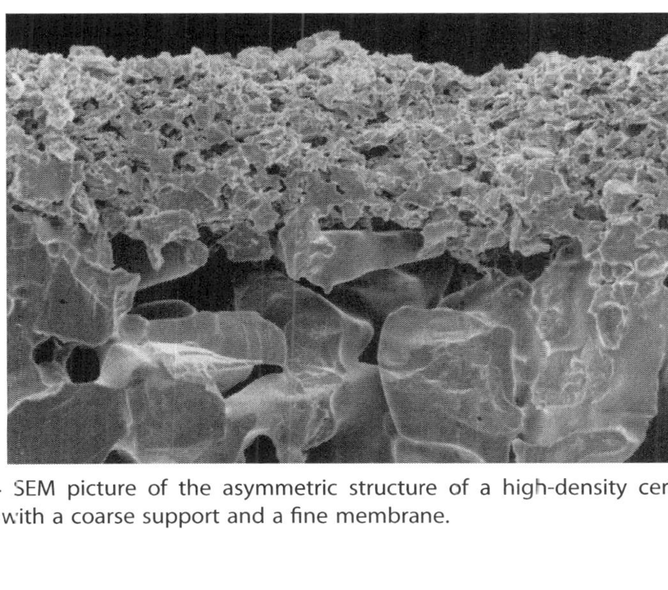

Figure 4 SEM picture of the asymmetric structure of a high-density ceramic filter element with a coarse support and a fine membrane.

Figure 5 High-density ceramic filter element with a coarse support and a fine membrane.

with regard to the long-term behaviour of the filter elements and their life time. The initial rigid ceramic filters used depth filter technology, but the advantages of minimum pressure drop and high efficiency back-pulse cleaning quickly led to the development of thin surface membrane filter elements. The collection efficiency of high-density ceramic filter elements is very high and reaches nearly 100% (Cummer, 2002), and even for submicron particles this performance can be achieved (Heidenreich and Scheibner, 2002). Asymmetric structures allow higher throughputs at low pressure drop and higher filtration grade and high mechanical stability at the same time.

High-density ceramic hot gas filter elements mainly consist of silicon carbide, aluminium oxide, and mullite. Due to the high mechanical strength of granular ceramic filter media those are the best choice for high pressure applications. Thermal stability of ceramic filter media depends on the material used. Temperatures up to 1000 °C and higher are in general possible (Pall, 2013; Filtraguide, 2013).

3.2 Metal Media

Sintered porous metal filter media are also manufactured of either metal fibres or powders. Additionally, metal fabrics are available as filter media.

Sintered powder metal media are typically manufactured by pressing and subsequent sintering under vacuum or inert gas atmosphere. Metal fibre media are made starting from wires which are drawn down in several steps to a certain diameter, typically in the range from about 2 to about 40 μm. The wires are chopped to a length of typically about 20–25 μm and a web is formed and sintered. By rolling or pressing the web to a defined thickness, the permeability and filter grade of the media is adjusted. The fine fibre web is sintered on a coarser wire support layer to get higher mechanical stability. High porosities of up to 85% can be achieved.

Metal filter media made of different steel grades and various metal alloys are available. The material applied needs to be selected by its thermal and chemical stability. Oxidation and corrosion can lead to irreversible plugging of the pores due to the fact that the volume of the reaction products is higher than that of the pure metal. Stainless steel can be typically applied for temperatures up to 420 °C. High temperature steels can be applied up to 650 °C (Pall, 2013). If sulphur or chlorides are present in the gas, which is the case in most industrial applications, special alloys, such as Inconel[1] 600, Monel[1], Hastelloy[1] X, or HR 160[1], need to be used. For applications with high levels of hydrogen sulphide, special metal filter media have been developed. These filter media are made of special metal alloys which generate a protective surface of alumina during heat treatment under oxidizing atmosphere (Tortorelli et al., 1998; June and Sawyer, 1998). Examples for such metal alloys are FeCralloy[1] and iron aluminide alloys. Filter elements made of iron aluminide can be applied to a maximal operating temperature of about 780 °C (Ondrey, 2001). The maximal operating temperature of FeCralloy filter elements is reported to about 1000 °C (Mott, 2013).

3.3 Catalytically Active Filter Media

Multifunctional hot gas filter elements have been recently developed. These filter elements combine removal of dust by filtration and removal of gaseous impurities from the gas phase by a catalytic reaction. The advantage of catalytic filter elements is that two functions can be combined in one unit instead of having two. This simplifies the process setup and reduces costs and footprint. Furthermore, the combination allows the use of the thermal

[1] Dia-Schumalith is a trademark of Pall Corporation, Inconel and Monel are trademarks of Special Metals Corporation Hastelloy and HR 160 are trademarks of Haynes International Inc. Fecralloy is a trademark of UKEA, UK.

energy of the gas for the catalytic reaction and prevents plugging of the catalyst by particle deposition.

Catalytically activated ceramic hot gas filter elements have been developed for the combined removal of particles and the selective catalytic reduction (SCR) of nitrogen oxides from exhaust gas streams (Nacken et al., 2007a; Heidenreich et al., 2008a; Hackel et al., 2008). The inner structure of the filter element is impregnated with a catalyst which is based on TiO_2-V_2O_5-WO_3. A fine filtering membrane on the outer side of these filter elements protects the catalyst of any particle deposition. Thus, deactivation of the catalyst by blocking of the active sites of the catalyst by dust particles is avoided. A scheme of the structure of such a catalytically activated filter element is shown in Figure 6. Filtration as well as reduction of the nitrogen oxides is performed at temperatures typically between 280 and 320 °C. If a low temperature SCR catalyst is used, the reaction temperature can significantly be decreased to temperatures down to 180 °C (Long et al., 2002). By using noble metals, such as platinum, the reaction temperature can be further decreased down to 150 °C (Kim et al., 2008). However, a stable operation can only be achieved at these lower temperatures when the gas is sulphur free. The higher the sulphur content in the gas is, the higher the risk of the formation of ammonium sulphate is with decreasing temperature. The formation of ammonium sulphate leads to the deactivation of the catalyst by deposition on the surface of the catalyst.

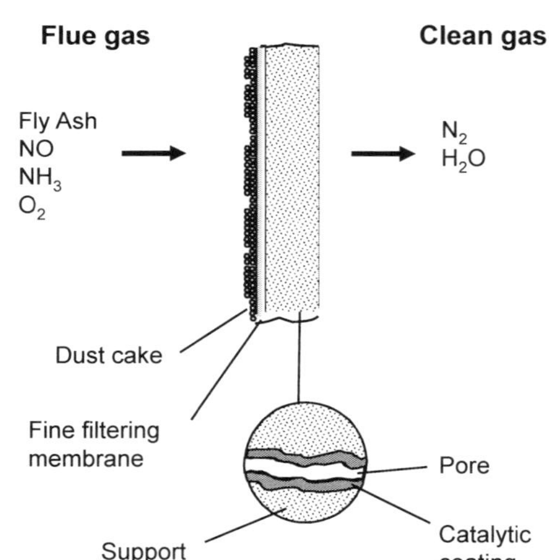

Figure 6 Schematic structure of a catalytically activated hot gas filter element for combined removal of particles and selective catalytic reduction of nitrogen oxides.

A second application where catalytic filter elements have been developed is the combined removal of particles and tars from the syngas in biomass gasification. Catalytic filter elements with nickel-based tar reforming catalysts integrated as fixed bed, as catalytic coating of the porous inner structure of the filter elements or with catalytically coated ceramic foam inlets have been developed (Heidenreich and Nacken, 2004; Nacken et al., 2007b, 2009). In order to achieve a high tar reforming performance reaction temperatures between 800 and 850 °C are required. First measurements with real biomass gasification syngas showed that tar conversion rates of up to 94% could be achieved.

4. SURFACE FILTRATION AND HOT GAS FILTER ELEMENT CLEANING

Most industrial applications have high dust concentrations. For this reason surface filters are the favoured solution for long-term operation. In contrast to depth filters where the dust penetrates into the filter material and a continuous increase of the pressure drop results, the use of surface filters leads to the buildup of a dust cake on the surface of the filter elements. After the formation of a first dust layer on the surface of the filter element, this dust or filter cake takes on the actual filter effect. The filter element is regenerated when a certain pressure drop of the dust cake is reached. State-of-the-art technology for the regeneration of filter elements is a back pulse with a particle-free gas countercurrent to the flow direction of the filtration. Depending on the process, air, nitrogen, carbon dioxide, or recycled and cleaned process gas are used as blowback gas. Conventional back-pulse systems which operate according to the so-called jet pulse method or back pulse systems specially developed for the cleaning of rigid hot gas filter elements, such as the Pall Schumacher venturi ejector jet pulse system (Heidenreich et al., 2013) or the coupled pressure pulse cleaning system (Mai et al., 2002; Heidenreich et al., 2010), are applied.

4.1 Conventional Jet Pulse Cleaning

In conventional jet pulse systems, the back pulse is achieved by short and quick blowing a free jet from the clean gas side into the filter element.

Jet pulse systems which were at first developed for flexible bag filter media have been used for the regeneration of rigid filter media too. When a flexible filter medium is used, the removal of the dust layer is mechanically supported by a fast inflation and a subsequent deflation of the filter surface. Rigid filter media are regenerated only by pressure and fluid forces.

In a jet pulse system, a blowpipe is located at a defined distance above the head of the filter candle, see Figure 7. The blowpipe is connected to a pressurized back-pulse gas reservoir. Normal back-pulse gas pressures are between 0.5 and 1 MPa for a filter system operated at atmospheric pressure. For high-pressure applications, such as pressurized coal gasification, the cleaning gas pressure for jet pulse regeneration should be approximately twice as high as the system pressure. For regeneration of a rigid filter candle, the back-pulse gas streams into the filter candle and the kinetic energy of the free jet is converted into static pressure. The over pressure inside the candle results in the detachment of the dust cake and regeneration of the candle.

The cleaning intensity depends on the flow velocity of the jet at the outlet of the blow pipe, the mass flow rate of the jet, and the geometry of the blowback system. The flow velocity of the jet is limited to sonic speed. In order to increase the amount of a secondary flow and correspondingly to reduce the gas amount of the primary jet often a small venturi ejector is installed on top of each filter element.

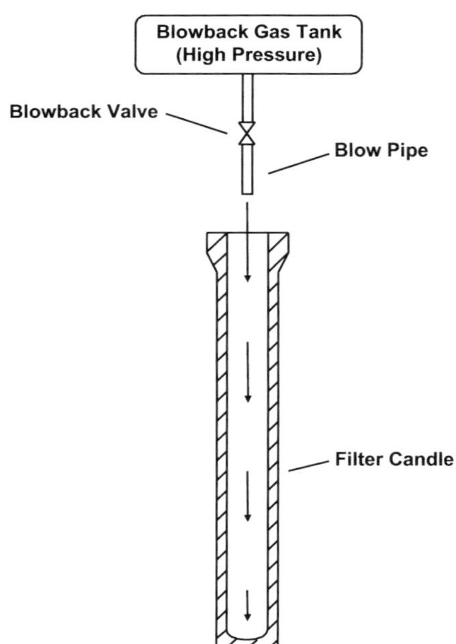

Figure 7 General principle of jet pulse cleaning.

Problems which can occur when a conventional jet pulse system is used for the regeneration of rigid hot gas filter elements are that the filtration of very fine dusts and of sticking dusts can result in an incomplete detachment of the dust layer from the rigid surface of the filter elements and consequently in an unstable filtration behaviour with a continuous increase of the pressure drop of the filter elements.

4.2 Pall Schumacher Venturi Ejector Jet Pulse System

In the Pall Schumacher venturi ejector jet pulse system, the filter elements are installed in a tube sheet and arranged in circular groups, as shown in Figure 8. Above each group of filter elements a venturi ejector is installed on the upper clean gas side of the tube sheet (see Figure 9). The regeneration of the filter elements is performed by a reverse jet pulse generated from the clean gas side into the venturi ejector. By this primary jet, a secondary flow is sucked into the venturi ejector from the surrounding clean gas cavity. Thus, the blowback flow for the regeneration of the filter elements is increased. The venturi ejector enables a homogeneous distribution of the reverse

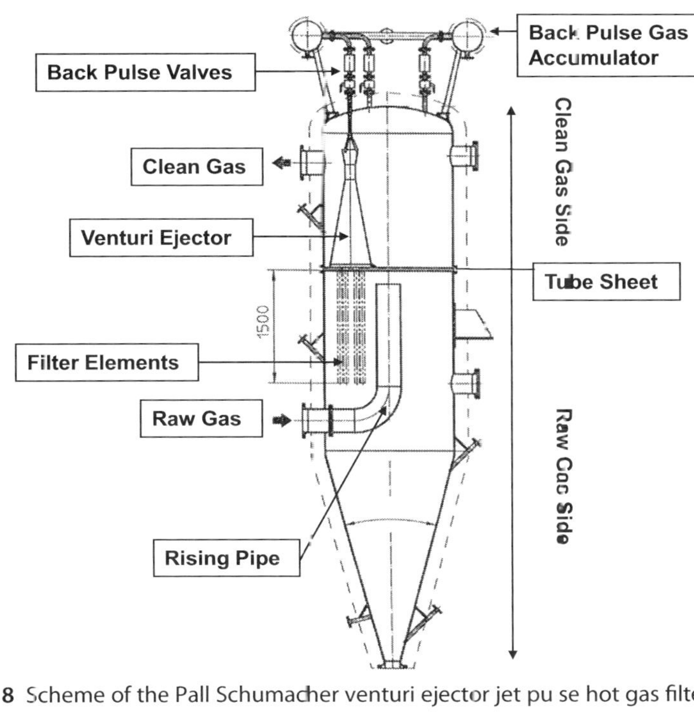

Figure 8 Scheme of the Pall Schumacher venturi ejector jet pulse hot gas filter system.

Figure 9 Venturi ejectors installed on the upper clean gas side of the tube sheet.

flow across all filter elements of a group and a uniform pressure development along the filter elements. By mixing the primary jet, which in most cases is colder than the clean gas, with the secondary flow of clean gas and the clean gas contained in the venturi, the thermal shock and stress on the filter elements during the blowback is reduced compared to the direct back pulse into single filter elements with colder gas. The different groups are sequentially cleaned on-line. This allows filtration to continue in the remaining groups and hence the continuous operation of the filter.

4.3 Coupled Pressure Pulse Cleaning

In order to improve the cleaning efficiency of rigid filter media, a new cleaning method called Coupled Pressure Pulse (CPP) cleaning has been jointly developed by the Karlsruhe Institute of Technology and by Pall Corporation (Mai et al., 2002; Heidenreich et al., 2010). In contrast to a conventional jet pulse cleaning, the CPP cleaning method is based on the direct coupling of the cleaning system and the clean gas side of the filter elements. The general principle of the direct coupling of the CPP cleaning is shown in Figure 10. A scheme of a hot gas filter with CPP cleaning is shown in Figure 11.

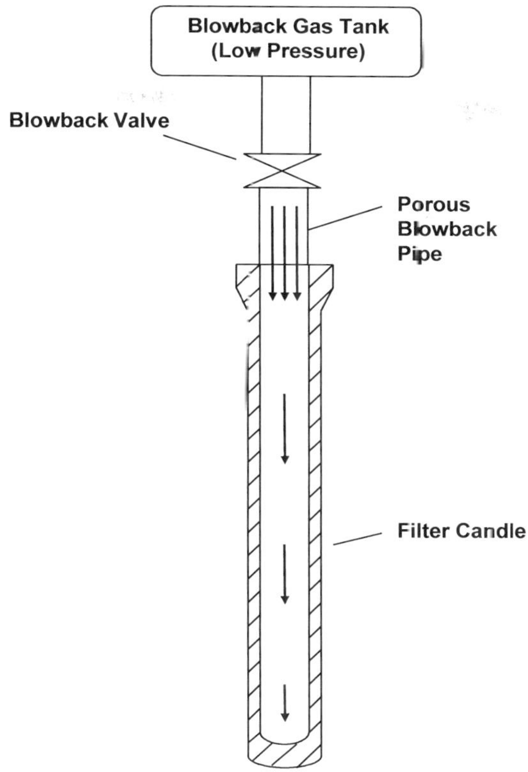

Figure 10 General principle of the direct coupling of the CPP cleaning method.

During filtration the cleaning system is separated from the clean gas room by the blowback valve. By opening this valve the pressure inside the clean gas plenum and the filter candles is fast increased by the direct coupling with the higher pressure inside the cleaning gas reservoir for a short time. This fast increase of the pressure inside the filter elements results in a high cleaning intensity.

The cleaning intensities which can be achieved with the CPP cleaning method are considerably higher compared to conventional jet pulse cleaning. Besides this the CPP process requires only a low cleaning gas pressure. The cleaning gas pressure has typically to be only 0.1–0.2 MPa higher than the pressure of the system. In case of a conventional jet pulse cleaning system, the cleaning gas pressure is typically in the order of 0.5–1 MPa for a filter system operating at atmospheric pressure. For high pressure applications, for example pressurized coal gasification, the cleaning

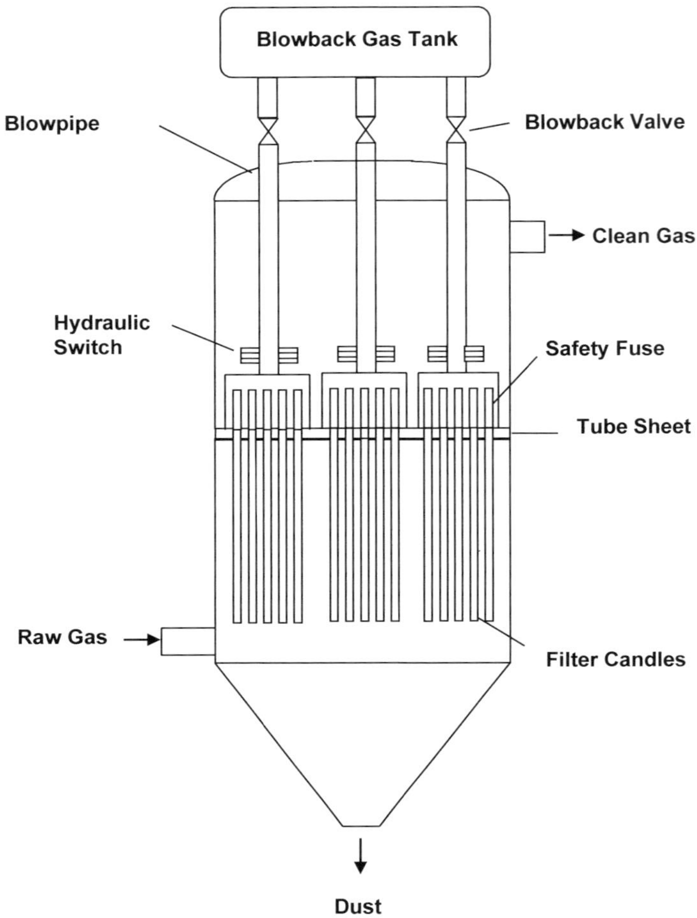

Figure 11 Schematic of a hot gas filter system with CPP cleaning.

gas pressure for pulse jet cleaning should be at least twice as high as the system pressure.

In order to build up the pressure in the clean gas plenum and in the filter candles without losing cleaning gas through the clean gas outlet, a special system component, the so-called hydraulic switch, is installed in the clean gas outlet. The function of the hydraulic switch is characterized by its flow resistance. For low flow velocities as in case of filtration the flow resistance of the switch is very low. During the cleaning the flow is highly turbulent and the resistance of the switch is high, effectively stopping flow from that section of elements.

The CPP cleaning system offers the possibility to add an individual safety filter element for each filter candle, as shown in Figure 11.

4.4 Pulse-less Crossflow Filtration Concepts

Recently developed new hot gas filter concepts try to make use of the crossflow filtration principle. By using the method of crossflow filtration, shear forces on the surface of the filter elements are generated by turbulence of the axial gas flow along the filter elements. These shear forces cause a removal of dust from the surface of the filter elements and correspondingly reduce the increase of the pressure drop by the buildup of the dust cake. Thus, the need for regeneration by back pulsing is reduced which results in a lower number of regeneration pulses and in longer cycle times.

Experimental studies of such hot gas filter concepts working according to the crossflow filtration principle have been performed on laboratory scale so far. Sibanda et al. (2001, 2010) for example investigated a concept (see Figure 12) in which the raw gas flow enters the inside of a filter tube. Part of the gas flow passes as crossflow radially through the wall of the filter tube and the remaining part flows axially through the tube and is directed to a downstream cyclone. The particles need to agglomerate at the surface of the filter tube and the shear forces resulting from the axial flow need to remove the agglomerates from the surface and carry them to the downstream cyclone so that the concept works. Agglomeration and detachment

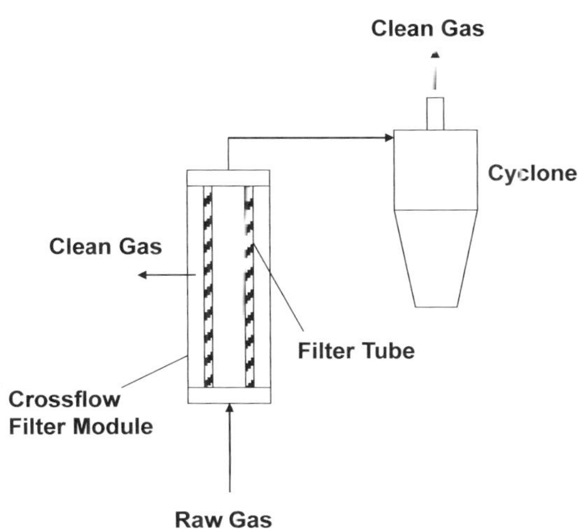

Figure 12 Schematic drawing showing the crossflow concept investigated by Sibanda et al. (2001).

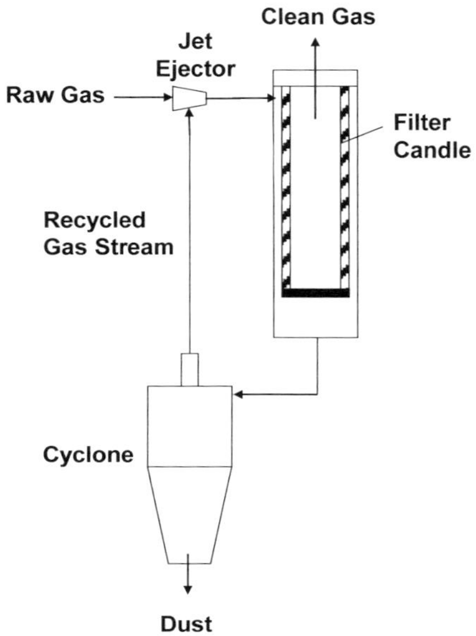

Figure 13 Schematic drawing showing the crossflow concept investigated by Sharma et al. (2010).

of the particles depend strongly on their sticking properties. Sibanda et al. (2001, 2010) could achieve collection efficiencies of 99% for optimal operating conditions. The collection efficiency of the cyclone alone was 90% in this case (Sibanda et al., 2001).

A quite similar concept has been tested by Sharma et al. (2010, 2011) which they called pulse-less filtration. In this concept, a part of the gas flow which enters the filter vessel passes through a filter candle from the outside to the inside of the candle and exits as cleaned gas. The remaining part of the gas flow is recycled and mixed with the inlet flow (see Figure 13). The gap between the housing and the filter candle is kept small in order to have high enough shear forces resulting from the downward flow along the candle. Agglomerates are removed from the surface of the filter candle by turbulent shear forces of the recycled gas stream flowing along the candle surface. The agglomerates released from the filter surface are carried to a downstream cyclone where they are separated. Sharma et al. (2010, 2011) reported successful testing of their concept on laboratory scale.

If fine particles are in the gas which cannot be agglomerated, then particles smaller than the cut size of the cyclone will be concentrated in the recycled gas stream. As long as agglomeration can be achieved or just coarse dusts have to be removed then the crossflow concepts will work. However, scale-up of these interesting concepts have to be evaluated.

5. HOT GAS FILTER DESIGN

5.1 Filter Element Arrangements

Mostly applied for hot gas filtration are filter candles which are closed at one end (see Figure 1). These filter candles are typically installed vertically hanging in a tube sheet which separates the filter vessel in a raw and a clean side part (see, e.g., Figure 8). The raw gas flows from the outside of the filter candles through the porous wall of the elements to the clean gas part of the filter. The dust is collected on the outer surface of the candles building a dust cake. The filter candles are typically arranged into groups (clusters). By arranging the filter elements into groups, the different groups can be sequentially cleaned on-line by a back pulse with clean gas. During one group is cleaned the others are operating and the process runs without any interruption. The filter elements can be clustered, e.g., into circular, trapezoidal, in cake piece like forms or in lines depending on the design, geometry, and dimension of the filter vessel. For high operating pressures or high operating temperatures typically circular vessels are used. For applications at atmospheric pressure and lower temperatures rectangular vessels are preferably applied. Installation and maintenance of the filter candles can be easily performed from the clean gas side of the filter vessel.

Sometimes, filter tubes which are open at both ends were used. In one hot gas filter design developed by Asahi Glass Co. Ltd. in Japan the raw gas flows into the inside of vertically arranged filter tubes downward from the top. The dust cake which is built up at the inside surface of the tubes is detached by a back pulse from the outside to the inside and discharged at the bottom of the tubes. The tubes are supported by water cooled tube sheets which divide the filter vessel into several clean gas chambers. Ceramic filter tubes with a length of 3 m and an outer and inner diameter of 168 mm and 140 mm, respectively, were used (Oda and Hanada, 1996). The main advantages of this filter design have been seen in the fact that dust bridging between the filter tubes and thus the risk of candle breakage is avoided. The filter system was tested in the 71 MW PFBC demonstration plant at Wakamatsu in Japan at temperatures between 650 and 850 °C

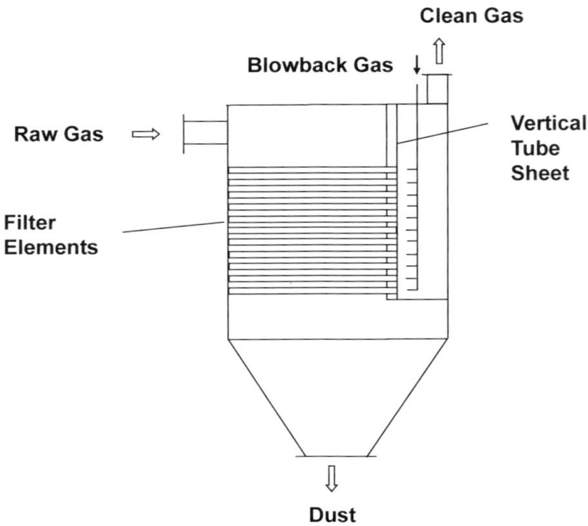

Figure 14 Scheme of a hot gas filter system with horizontally installed filter tubes.

(Sasatsu et al., 1999, 2001). However, several problems, such as breakage of the tubes and dust leakages at the tube sealings, occurred (Sasatsu et al., 2002).

In other hot gas filter designs, filter tubes are horizontally arranged in vertically installed tube sheets (see, e.g., Smolders and Baeyens (2000); Hajek (2003); Tomonori (2005)). In these designs, the raw gas flows from the outside to the inside of the filter tubes and the dust cake is built up on the outside surface of the tubes. A principle scheme of such a filter system is shown in Figure 14. The back pulse for the regeneration of the filter elements needs to have a sufficiently high intensity so that the dust cake on the top side of the filter tubes can be detached. Furthermore, the arrangement of the tubes needs to enable the settlement of the dust through the gaps between the filter tubes down to the bottom of the filter vessel without the formation of dust bridges.

In order to increase the filtration area in circular vessels and to reduce the vessel costs by decreasing the vessel diameter, hot gas filter systems have been developed in the past, where the filter candles are arranged in multistages inside the filter vessel. Examples of such filter systems are the systems developed by Lurgi Lentjes Babcock (LLB) (Dehn et al., 1998) and by Westinghouse (Haldipur and Dilmore, 1992). Main feature of the LLB filter system is that ceramic filter candles are mounted standing instead of hanging.

LLB filter systems were installed and operated in the past at the High Temperature Winkler gasification demonstration plant in Berrenrath (Germany) (Schumacher and Renzenbrink, 1996) and at the 300 MW_{el} IGCC coal gasification plant in Puertollano (Spain) (Krein, 1999). In the Westinghouse filter system, filter candles are installed hanging in single holders (see Figure 8). A group of holders, about 30–60, are connected to a common plenum. Two or more plenums are stacked vertically and supported by a common pipe which is connected to a main tube sheet. The Westinghouse filter system has been mainly installed and tested in different pilot and demonstration projects in the United States (see, e.g., Ahmadi and Smith (1998, 2002), Guan et al. (2008)).

5.2 Flow Distribution

Dust movement by gravity is mainly influenced by the particle size and geometry as well as by the particle density. The settlement velocity of larger and heavier particles is higher than that of smaller and lighter particles. For fine dusts and dusts with a low bulk density which tend to form a weakly agglomerated dust cake, a downdraft flow of the raw gas in the filter vessel is preferably used. The downdraft flow supports the movement of the dust to the hopper after the detachment of the dust cake and prevents a re-entrainment of the dust to the surface of the filter elements. In order to have a downdraft flow, the raw gas is guided by an internal rising pipe (see Figure 8) upward to the lower side of the tube sheet where its flow direction is changed downward. In larger filter vessels more than one rising pipe are often used to achieve a homogeneous distribution of the raw gas in the filter vessel (see, e.g., Durst et al. (1996), Dries (2012)). One example of such a special flow distribution system developed by Durst et al. (1996) is shown in Figure 15.

The design of a filter system operated in updraft flow is simpler than a downdraft system. The raw gas enters the filter vessel below the filter candles. No internals for the gas distribution are needed in this case. If it is expected that the dust cake will break into larger pieces by the backpulse regeneration and falls easily down to the hopper or dusts with a high bulk density are filtered, then the updraft flow is the preferred design.

5.3 Special Filter Design

A special filter system design recently developed (Heidenreich et al., 2008b) is the integration of a hot gas filter directly into the freeboard of a biomass

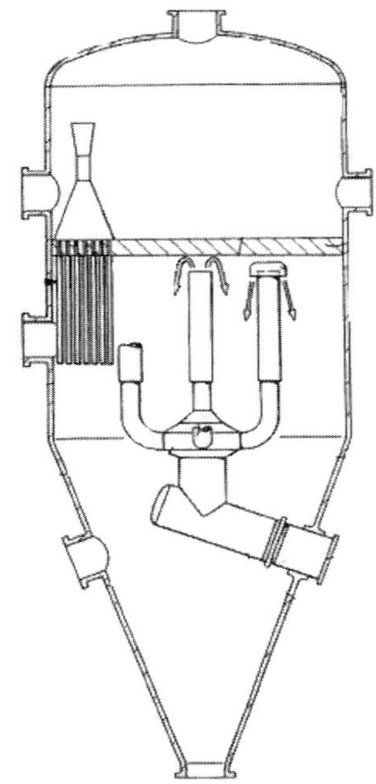

Figure 15 Schematic drawing showing a filter housing with special raw gas distribution invented by Durst et al. (1996).

gasifier. Catalytically active filter elements for combined particle and tar removal are used. This new concept offers a compact and cost-effective gasification and hot gas cleaning and conditioning system for biomass gasification in one unit. Filter candles are installed hanging in a tube sheet which is integrated in the freeboard of the gasifier. The concept offers additionally the option to add sorbents into the gasifier to remove gaseous components, such as H_2S, as well as to remove alkali metals. By this cost-effective approach, the investment costs of the gas cleaning equipment for a biomass gasification plant can be reduced as well as the needed space for the installation. Figure 16 shows schematically the principle of this new compact cleaning concept. Remarkable system simplification and process intensification can be achieved by this concept.

Investigations at a bench-scaled fluidized bed biomass gasifier, operated at atmospheric pressure and temperatures between 800 and 820 °C with a catalytic filter candle integrated in the freeboard of the gasifier have demonstrated the general feasibility of the concept (Heidenreich et al., 2008c; Rapagna et al., 2010).

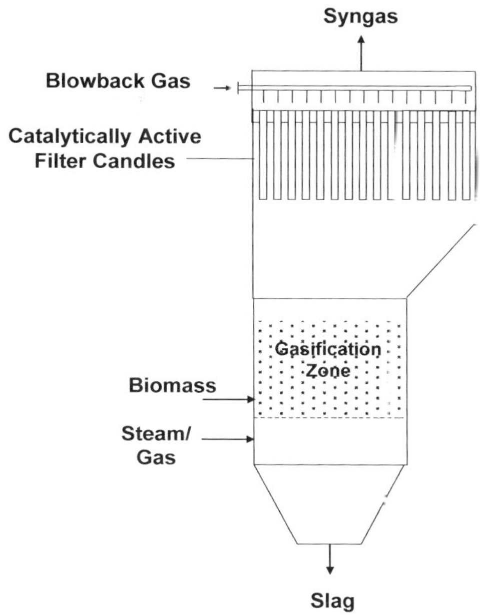

Figure 16 Schematic principle of the integration of catalytically activated filter candles into the freeboard of a fluidized bed gasifier.

6. APPLICATIONS

The total number of hot gas filters worldwide used is in the order of some hundreds. Hot gas filters have already been applied in many different processes, such as for the incineration and pyrolysis of radioactive waste, in coal combustion and gasification, in gasification and pyrolysis of biomass, in refineries, in cement and glass industry, in recycling of metals, in production of chemicals and catalysts, in production of pigments and nanoparticles, in smelting processes and metal production, as well as in waste incineration and pyrolysis.

First technical application where hot gas filters have been used was the incineration of low-level contaminated radioactive wastes from nuclear power generation at the beginning of the 1970s. A two stage hot gas filter system was operated at temperatures between 650 and 900 °C at the former nuclear research centre of Karlsruhe in Germany. Each filter stage comprised a refractory lined vessel containing 85 ceramic filter elements of silicon carbide (Leibold et al., 1989). Hot gas filters were also used in some incineration plants for low contaminated radioactive wastes in France (Perkins, 1976).

Since 1978 hot gas filters have been broadly used for the incineration of low contaminated radioactive wastes in nuclear power plants in Japan. The hot gas filter systems used have typically two stages each one comprising a refractory lined vessel with about 100 ceramic filter candles. The first stage is operated at 600–800 °C, the second one at 500–600 °C (Torii, 2005).

Hot gas filters are also applied for the pyrolysis of low contaminated radioactive waste (Hesbol and Mason, 1999; Mason et al., 1999; Studsvik, 2013; Nukem, 2013).

Coal gasification which was one of the main drivers for the development of hot gas filters has got an increasing interest in the last 10 years. Meanwhile, more than 25 large hot gas filter units are in operation or in commissioning in coal gasification plants. The largest filter vessels have a diameter of about 6.5 m, a height of more than 25 m, and a weight of about 300 t containing about 1200 filter candles. About 90% of the filter units are equipped with ceramic filter candles of silicon carbide. The remaining 10% are operated with iron aluminide or FeCralloy metal filter candles.

Interesting applications for hot gas filters are gasification and pyrolysis of biomass. Many hot gas filters containing 1 to 100 candles have been operated in lab and demonstration plants around the world since the beginning of the 1990s (Heidenreich, 2013).

Hot gas filters are also used in refineries for flue gas cleaning of fluid catalytic cracking units (FCCU) to accomplish increased particulate emission regulations. For many years hot gas filters of smaller size containing up to about 100 candles have been used as 4th stage underflow filters of the 3rd stage separator. About 3–5% of the gas flow of the third separator stage is directed as underflow to the 4th stage filter. In recent years, some refineries have replaced their third stage cyclone by a third stage hot gas filter to fulfill more stringent emission levels.

Furthermore, many hot gas filters are applied in waste incineration and pyrolysis. Mostly low density ceramic filter candles are used in these applications. Filtration temperature in waste incineration is usually between 200 and 350 °C. Filtration surface areas of up to 1000 m^2 are installed. Filtration temperature in waste pyrolysis is typically between 350 and 500 °C. It is reported that more than 30 hot gas filters are installed in pyrolysis plants where most of these applications are in Japan (Startin, 2007).

Hot gas filters are also often applied in the production of metal oxide powders, catalysts, pigments, and metals. In these applications, hot gas filters can achieve significant process and economical advantages.

Moreover, hot gas filters are also used in metal recycling processes, for example in the recycling of aluminium (see, e.g., Beattie and Withers (1993), Startin and Elliott (2001)).

7. CONCLUSIONS

Hot gas filtration is a reliable and well proven technology which was firstly technically applied for flue gas cleaning at the incineration of low-level contaminated radioactive wastes from nuclear power generation at the beginning of the 1970s.

The development of advanced coal-based power generation techniques in the 1980s and the 1990s has significantly influenced the development of hot gas filter systems and filter media.

Nowadays, hundreds of hot gas filters are already applied in different processes for many years. Hot gas filter media have shown an excellent filtration performance by achieving highly efficient separation of particles with sizes down to the submicron range and by achieving clean gas concentrations down to below 1 mg/m^3.

Due to the high requirements on the mechanical stability of the filter elements at higher temperature, only rigid self-supporting ceramic or metal filter elements are applied for hot gas filtration.

Hot gas filters can be advantageously used in many industrial processes. Downstream equipment, such as catalyst units, heat exchangers, turbines, and scrubbers can be protected from erosion, fouling, and pollution. Furthermore, undesired condensation or desublimation reactions can be avoided by keeping the process temperature high.

In general, hot gas filtration attracts more and more interest due to the possibility to intensify and simplify processes, to increase the process and energy efficiency, to improve the product quality, as well as to achieve lower emission levels.

REFERENCES

Ahmadi, G., Smith, D.H., 1998. Aerosol Sci. Technol. 29, 206–223.

Ahmadi, G., Smith, D.H., 2002. Powder Technol. 128, 1–10.

Beattie, C.J C., Withers, C.J., 1993. Applications of low density ceramic filters for gas cleaning at high temperatures. In: Clift, R., Seville, J.P.K. (Eds.), Gas Cleaning at High Temperatures, pp. 173–189. London.

Chalupnick, R., Isaksson, J., Jones, B., Krein, J., Köbernick, O., Schulz, K., 1999. Demonstration of hot gas filtration systems. In: Dittler, A., Hemmer, G., Kasper, G. (Eds.), High Temperature Gas Cleaning II, pp. 276–290. Karlsruhe.

Cummer, K.R., Brown, R.C., 2002. Biomass Bioenergy 23, 113–128.

Dehn, G., Möllenhoff, H., Wegelin, R., Krein, J., 1998. US Patent 5769915.
Dries, H., 2012. US Patent Application 2012/012003.
Durst, M., Müller, M., Schnell, M., Schulze-Dieckhoff, R., Minzenbach, G., 1996. US Patent 5514195.
Elcogas, 2013. http://www.elcogas.es (accessed 15.04.13.).
Energy Business review, 2013. http://fossilfuel.energy-business-submitted-for-publication.com/suppliers/egas-technology-energy-business-submitted-for-publication/news/e-gas-technology-selected-for-posco-gasification-project-press-release-21422 (accessed 15.04.13.).
Engineerlive, 2013. http://www.engineerlive.com/content/23842 (accessed 15.04.13.).
Filtraguide, 2013. http://www.filtraguide.com (accessed 15.04.14.).
Guan, X., Gardner, B., Martin, R.A., Spain, J., 2008. Powder Technol. 180, 122–128.
Hackel, M., Schaub, G., Nacken, M., Heidenreich, S., 2008. Powder Technol. 180, 239–244.
Hajek, S., 2003. Cit. Plus 5, 90–92.
Haldipur, G.B., Dilmore, W.J., 1992. US Patent 5143530.
Heidenreich, S., Scheibner, B., May 2002. Filtration + Separation, 22–25.
Heidenreich, S., Nacken, M., 2004. Chem. Ing. Technol. 76, 1508–1512.
Heidenreich, S., Nacken, M., Hackel, M., Schaub, G., 2008a. Powder Technol. 180, 86–90.
Heidenreich, S., Nacken, M., Foscolo, P.U., Rapagna, S., 2008b. International Patent Application WO 2008/135226.
Heidenreich, S., Nacken, M., Salinger, M., Foscolo, P.U., Rapagnà, S., 2008c. Integration of a catalytic filter into a gasifier for combined particle separation and tar removal from biomass gasification gas. In: Proceeedings of the 7th International Symposium on Gas Cleaning at High Temperature (Newcastle, Australia).
Heidenreich, S., Haag, W., Mai, R., Leibold, H., 2010. Gefahrst. Reinhalt. Luft 70, 237–241.
Heidenreich, S., 2013. Fuel 104, 83–94.
Heidenreich, S., Haag, W., Salinger, M., 2013. Fuel 108, 19–23.
Hesbol, R., Mason, J.B., 1999. US Patent 5909654.
June, M., Sawyer, J.W., 1998. Iron aluminide hot gas filter development. In: Proceedings of Advanced Coal-Based Power Environment System 1998, US Department of Energy Paper PB.16.
Kim, Y.A., Choi, J.H., Scott, J., Chiang, K., Amal, R., 2008. Powder Technol. 180, 86–90.
Krein, J., 1999. LLB candle filter for the elcogas IGCC power plant puertollano. In: Dittler, A., Hemmer, G., Kasper, G. (Eds.), High Temperature Gas Cleaning II, pp. 253–260. Karlsruhe.
Leibold, H., Dirks, F., Rüdinger, V., 1989. Waste Manag. 9, 87–94.
Long, R.Q., Yang, R.T., Chang, R., 2002. Chem. Commun., 452–453.
Mai, R., Leibold, H., Seifert, H., Heidenreich, S., Haag, W., Walch, A., 2002. Chem. Ing. Technol. 74, 1438–1441.
Mason, J.B., Oliver, T.W., Carson, M.P., Hill, G.M., 1999. Studsvik processing facility pyrolysis/steam reforming technology for volume and weight reduction and stabilization of LLRW and mixed wastes. In: Proceedings of the Waste Management Conference 99 (Tucson).
Mississippipower, 2013. http://www.mississippipower.com/kemper/home.asp (accessed 15.04.13.).
Mott, 2013. http://www.mottcorp.com (accessed 15.04.13.).
Nacken, M., Heidenreich, S., Hackel, M., Schaub, G., 2007a. Appl. Catal. B: Environ. 70, 370–376.
Nacken, M., Ma, L., Engelen, K., Heidenreich, S., Baron, G.V., 2007b. Ind. Eng. Chem. Res. 46, 1945–1951.
Nacken, M., Ma, L., Heidenreich, S., Baron, G.V., 2009. Appl. Catal. B: Environ. 88, 292–298.

NETL, 2013a. http://www.netl.doe.gov/technologies/coalpower/gasification/gasifipedia/6-apps/6-2-6-3_elcogas.html (accessed 15.04.13.).
NETL, 2013b. http://www.netl.doe.gov/technologies/coalpower/gasification/gasifipedia/6-apps/6-2-6-2_Wabash.html (accessed 15.04.13.).
Nukem, 2013. Pyrolysis of Radioactive Organic Waste. http://www.nukemgroup.com/fileadmin/pdf/Brochure_Pyrolyse_April_2007.pdf (accessed 15.04.13.).
Nuon, 2013. http://www.nuon.com/company/core-business/energy-generation/power-stations/buggenum/ (accessed 15.04.13.).
Oda, N., Hanada, T., 1996. Performance of the advanced ceramic tube filter (ACTF) for the wakamatsu 71 MW PFBC and further improvements for commercial plants. In: Schmidt, E., Gäng, P., Pilz, T., Dittler, A. (Eds.), High Temperature Gas Cleaning, pp. 794–805. Karlsruhe.
Ondrey, G., November 2001. Chem. Eng., 29–39.
Pall, 2013. http://www.pall.com (accessed 15.04.13.).
Perkins, B.L., 1976. Incineration Facilities for Treatment of Radioactive Wastes – a Review. US Department of Commerce. Report LA-6252.
Rapagna, S., Gallucci, K., Di Marcello, M., Matt, M., Nacken, M., Heidenreich, S., Foscolo, P.U., 2010. Bioresour. Technol. 101, 7134–7141.
Sasatsu, H., Misawa, N., Abe, R., Mochida, I., 1999. Prediction for pressure drop across CTF at wakamatsu 71 MW_e PFBC combined cycle power plant. In: High Temperature Gas Cleaning II, Dittler, A., Hemmer, G., Kasper, G. (Eds.), pp. 261–275. Karlsruhe.
Sasatsu, H., Misawa, N., Shimizu, M., Abe, R., 2001. Powder Technol. 118, 58–67.
Sasatsu, H., Misawa, N., Kobori, K., Iritani, J., 2002. Hot gas particulate cleaning technology applied for PFBC/IGCC. In: Proceedings of the 5th International Symposium on Gas Cleaning at High Temperature. Morgantown, Paper 1.6.
Schumacher, H.J., Renzenbrink, W., 1996. Gas cleaning with the ceramic candle filter of the HTW demonstration plant at berrenrath. In: Schmidt, E., Gäng, P., Pilz, T., Dittler, A. (Eds.), High Temperature Gas Cleaning, pp. 794–805. Karlsruhe.
Sharma, S.D., Dolan, M., Ilyushechkin, A.Y., McLennan, K.G., Nguyen, T., Chase, D., 2010. Fuel 89, 817–826.
Sharma, S.D., McLennan, K., Dolan, M., Nguyen, T., Chase, D., 2013. Fuel 108, 42–53.
Sibanda, V., Greenwood, R.W., Seville, J.P.K., 2001. Powder Technol. 118, 193–202.
Sibanda, V., Greenwood, R.W., Seville, J.P.K., Ding, Y., Iyuke, S., 2010. Powder Technol. 203, 419–427.
Smolders, K., Baeyens, J., 2000. Powder Technol. 111, 240–244.
Startin, A., Elliott, G., November 2001. Filtration + Separation, 38–40.
Startin, A., 2007. The Application of Ceramic Filters to Pyrolysis Gas Cleaning. ThermalNet Newsletter. Issue 5. Available from: http://www.thermalnet.co.uk (accessed 15.04.13.).
Studsvik, 2013. http://www.studsvik.com (accessed 15.04.13.).
Tomonori, A., 2005. The field test of hot Gas filter at MSW incineration plant. In: Kanaoka, C., Makino, H., Kamiya, H. (Eds.), Advanced Gas Cleaning Technology. Jugei Shobo, Tokyo, pp. 340–347.
Torii, A., 2005. Ceramic filter for radioactive waste treatment system. In: Kanaoka, C., Makino, H., Kamiya, H. (Eds.), Advanced Gas Cleaning Technology. Jugei Shobo, Tokyo, pp. 58–64.
Tortorelli, P.F., Lara-Curzio, E., McKamey, C.G., Pint, B.A., Wright, I.G., Judkins, R.R., 1998. Evaluation of iron aluminides for hot gas filter applications. In: Procedings of Advanced Coal-Based Power Environment System 1998, US Department of Energy, Paper PB.7.
VDI, 2012. VDI Guideline 3677. Part 3. Beuth, Berlin.
Yamada, et al., 2003. Coal combustion power generation technology. Bull. Jpn. Inst. Energy 82 (11), 822–829.

CHAPTER TWELVE

Air Tabling—A Dry Gravity Solid–Solid Separation Technique

Gjergj Dodbiba and Toyohisa Fujita
Department of Systems Innovation, The University of Tokyo, Tokyo, Japan

Contents

GLOSSARY

Grade and recovery of a product Grade (%) and recovery (%) are defined as follows:

$$\text{Grade}_A = \frac{\text{Mass of material A in the collected fraction}}{\text{Total mass of the collected fraction}} \times 100\%$$

$$\text{Recovery}_A = \frac{\text{Mass of material A in the collected fraction}}{\text{Total mass of material A in feed}} \times 100\%$$

Where "A" denotes one of the components of the mixture.

Total efficiency of the separation "Total efficiency of separation" incorporates the "grade", and the "recovery" of both fractions, satisfying the requirement that in ideal separation 100% efficiency would be achieved only when 100% of each constituent of

Progress in Filtration and Separation
http://dx.doi.org/10.1016/B978-0-12-384746-1.00012-4

the mixture is recovered at 100% grade. Likewise, if no separation occurs, then the efficiency is zero. Total efficiency of separation is defined as follows:

$$\text{Efficiency} = \left[\left(\frac{M_{a(1)}}{M_{a(i)}}\right) \times \left(\frac{M_{a(1)}}{M_1}\right)\right] \times \left[\left(\frac{M_{b(2)}}{M_{b(i)}}\right) \times \left(\frac{M_{b(2)}}{M_2}\right)\right] \times 100\%$$

Where $M_{a(1)}$ is the mass of material "a" in output stream 1; $M_{a(i)}$ is the mass of materials "a" in feed; M_1 is the mass of the output stream 1; $M_{b(2)}$ is the mass of material "b" in output stream 2; $M_{b(i)}$ is the mass of materials "b" in feed; M_2 is the mass of the output stream 2.

NOMENCLATURE

a	Displacement amplitude of vibration	m
a_c	Constant required to secure capacity C in terms of tonnes of solid per hour, ($a_c = 0.278$)	—
a_x	Acceleration of the solid body (placed on the deck) in direction x, at equilibrium conditions	m/s^2
a_x^p	Acceleration of the particle placed on the deck in direction x	m/s^2
a_y	Acceleration of the solid body (placed on the deck) in direction y, at equilibrium conditions	m/s^2
a_x^D	Acceleration of the deck in direction X	m/s^2
a_y^D	Acceleration of the deck in direction Y	m/s^2
A_d	Cross-sectional area of the separating device (i.e., surface of the porous deck)	m^2
A	Particle's area projected onto the horizontal pane	m^2
D	Diameter of particle, i.e., particle size	m
D^*	A Dimensionless measure of particle size,	—
D_a^{Min}	The minimum size of fraction of material "a"	m
D_b^{Max}	The maximum size of fraction of material "b"	m
C	Capacity of classifier in tonnes of solid per hour	t/h
C_L	Coefficient of lift	—
f	Longitudinal vibrating frequency of the deck	1/s
F_f	Frictional force	N
g	Acceleration due to gravity, i.e., $g = 9.81$ m/s^2	m/s^2
h_p	Suspension height	m
h_r	Height of riffles	m
m	Mass	kg

Q_v	Reaction force	N
R	Lift force	N
r_p	Radius of particle	m
t	Time	s
u	Superficial velocity of air	m/s
u_{mf}	Superficial velocity of air at minimum fluidizing conditions	m/s
u_t	Terminal velocity	m/s
u_t^*	A Dimensionless measure of terminal velocity	—
V_x	Velocity of the deck in direction X	m/s
V_y	Velocity of the deck in direction Y	m/s
w	Feed flow rate of air tabling	$(\text{Kg/h})/\text{m}^2$
x	Horizontal location of the solid body on the porous deck	m
X	Horizontal location of the porous deck	m
y	Vertical location of the solid body on the porous deck	m
Y	Vertical location of the porous deck	m

Greek Letters

α	End slope of the porous deck	degrees
β	Side slope of the porous deck	degrees
η	The percentage of solid by volume in the classifier intake	%
μ	Absolute viscosity of air, i.e., $\mu m = 1.8 \times 10^{-5}$ Pa s	Pa s
μ_f	Coefficient of friction	—
π	Archimedes number (Pi); rounded value: 3.14159	—
θ	Angle of attack, i.e., angle between the air flow and the body	degrees
ρ_a	Density of material "a"	kg/m^3
ρ_b	Density of material "b"	kg/m^3
ρ_g	Density of air ($\rho_g = 1.2$ kg/m^3)	kg/m^3
ρ_s	Density of solid material	kg/m^3
ϕ_s	Sphericity of particle	—
φ	Angle of vibration	degrees
ω	Angular frequency of vibration or natural frequency	rad/s

Superscripts

D	Porous deck of air table
Min	Minimum
Max	Maximum

1. INTRODUCTION

Gravity separation is the simplest of all methods of separation, which has been and remains one of the most important means of separating solid species. It may be defined as the separation of two or more solid species, usually of different densities, by their relative movement in response to the force of gravity and one or more other forces, one of which is generally the resistance of motion by a medium such as air, water, etc.

During the process, particles are primarily separated by virtue of the difference in density (Clifford, 1999). In wet separation methods, the medium (usually a water-base solution) in which the separation takes place is a large factor in securing the differential movement between particles, which eventually result in separation (Dodbiba et al., 2002). Similarly, when the medium in which, and largely by which, separation is affected is air, such separation is described as *dry gravity separation* or better known as air or pneumatic separation (Truscott, 1923).

Generally speaking, dry gravity separation has the attraction of low capital and operating cost which together with the lack of water, chemicals and drying requirements means it is environmentally friendly (Falconer, 2003). It should be noted that dry gravity separation is not a recent discovery, since many patents can be found dating back as far as 1850. They cover early attempts to separate materials of various densities or shapes by means of air (Arms, 1924). The majority of dry separation processes, which have been developed or are commercially available, are similar in principle to their wet gravity separation counterparts. These processes include devices such as: (1) pneumatic jig, (2) dry pinched sluice, and (3) air table (Burt, 1984). In addition, other dry gravity separation devices have been developed, which do not have a direct counterpart in wet separation processes, include: (1) fluidized bed separator and (2) air classifier, which separate by differential deflection of falling particles. Moreover, air classifiers, according to their shape, are divided into three main categories: (a) rotary air classifiers,

(b) horizontal air classifiers (such as air knife), or (c) vertical air classifiers (such as zigzag air classifier, pneumatic jig, air table, etc.)

All dry gravity separation devices, except the pneumatic jig, utilize a constant controlled upflow of fluidizing air. The pneumatic jig (also know as active pulsed-flow air classifier), however, utilizes a pulsating airflow to affect the stratification of particles to be separated (Vesilind et al., 1981).

Vertical air classifiers in general and zigzag air classifiers in particular are not considered to have a simple design (Stessel, 1996), since they need to be connected with a device (commonly a cyclone) in order to subsequently separate the low-density fraction entrapped in the air stream from air. Moreover, these air classifiers, which generally do not allow for separation into more than two fractions (known as extract and reject), require a large working space (because of their considerable height) and are not an answer where the space is at premium. Consequently, the need for a more compact dry gravity separation device with a simpler geometry leads to the study on air tabling.

The air table, to be discussed in this chapter, is more compact device with a simpler geometry. Unlike most of the vertical air classifiers, air table does not require separating the low-density fraction from the air stream and is capable of effectively treating materials of different densities.

2. APPLICATIONS OF AIR TABLING

Air tables or pneumatic tables (Knapp, 1953), more than any other dry gravity separation devices, have found their major applications particularly in food industry as they are originally developed for seed separation (Burt, 1984; Jaman, 1985). However, air tables have also an important use in treatment of heavy minerals sand deposit (Hudson, 1962; Canning, 1980), in cleaning of coal (Appleyard, 1931; McCulloch et al., 1950; Llewellyn, 1977), in upgrading of tungsten (Osborn, 1927), and in other applications where the water is at premium (Sivamohan et al., 1985; Wills, 1997). More recently, air tables have found favour in separating a wide variety of secondary materials such as abrasive grains, scrap glass, scrap wire from its insulator, metal from crushed crucibles, lead from plastics in old batteries, etc. (Jaman, 1985).

Being a dry process, separation by means of air table has a clear advantage over wet separation techniques because it does not require chemical pretreatment (such as in the case of sink-float separation, flotation, etc.).

Furthermore, some potential problems associated with wet separating methods in general, such as treatment of water from the process for reuse or discharge, the requirement of expensive wetting reagents, and most importantly, dewatering or drying the mixture after separation can be avoided. Another point in favour of air tabling is that air keeps the bed of particles loose, permitting the particles to move readily to their respective position under density; wet tabling by means of shaking table, on the other hand, tends to compact the bed, cramping the necessary freedom of movement.

3. APPARATUS

Because of its external appearance, the device has been called air table, although it functions essentially as a pneumatic jig with a constant upward airflow of fluidizing air. However, the longitudinal vibration conveying from a standard shaking table is also performed by the eccentric drive incorporated into this device.

The air table is mainly comprised of a hopper, a vibrating feeder, a porous deck powered by an eccentric drive to impart the longitudinal vibration, and an electric fan located below the porous deck to generate the upward airflow at a controlled value of superficial velocity. Sketch of a laboratory-scale air table are shown in Figure 1. A collecting bin is arranged alongside the discharge end of the separator. It generally consists of two compartments separated by a splitter. The left-hand compartment collects the so-called low-density fraction, whereas the right-hand compartment collects high-density fraction. Not shown in Figure 1 is a truncated pyramid-shaped hood of trapezoidal base, suspended above the porous deck. The hood is sometimes employed to prevent the initial decrease of airflow close to the sides of the deck.

The flat trapezoidal deck of the air table is constructed of porous material, which is fitted with a perforated plate with circular openings (opening area ca. 11.8%) and with a series of riffles arranged in parallel to the direction of longitudinal vibration (Figure 2). The size of the openings is always smaller than the size of the smallest particle in the mixture, preventing particle from falling through the deck. The deck can be adjusted in transverse and longitudinal inclination. In other words, deck can be tilted from inlet end to discharge end and from side to side, creating the "side slope" β and the "end slope" α respectively (Figure 2). Controllers, mounted near

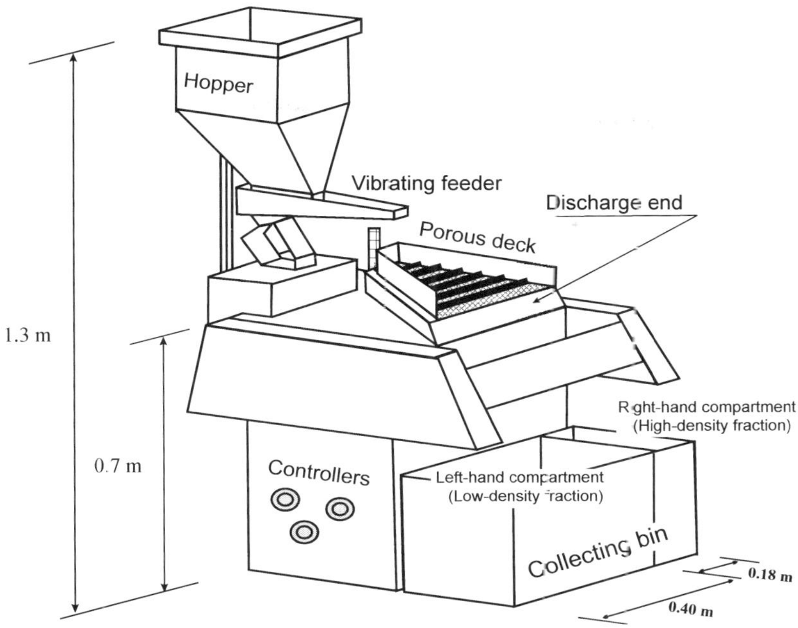

Figure 1 Schematic design of the laboratory-scale air table.

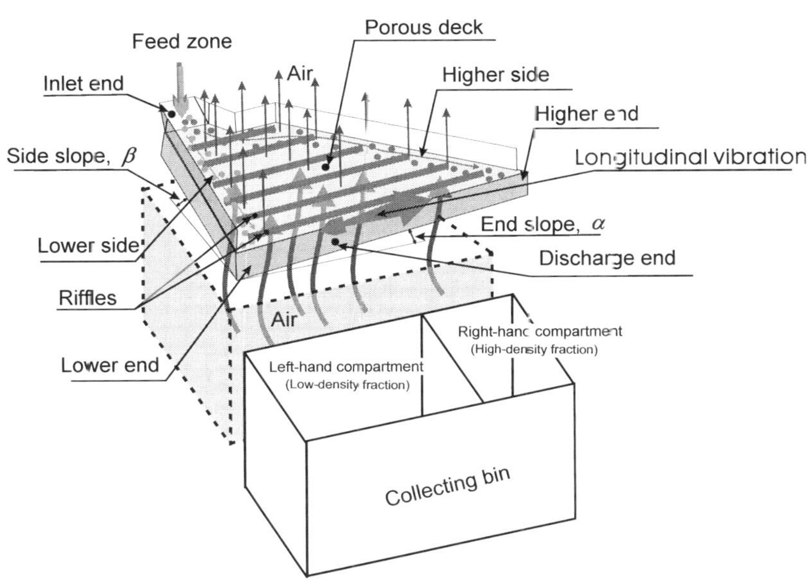

Figure 2 Schematic diagram illustrating the principle of separation by air table.

the deck (Figure 1), allow adjusting of the end slope, the side slope, the longitudinal vibration frequency, and the superficial velocity of the air.

4. PRINCIPLES OF AIR TABLING

The basic principles of air tabling hardly differ from those of wet tabling. In operation, initially particles of the same size are discharged from the hopper, and then are fed by the vibrating feeder onto the deck of air table (Figure 1), creating a uniform bed over its surface. The eccentric drive vibrates the deck in a side-to-side motion, along the direction of riffles, at a frequency f (f can range from 0.09 to 13.33/s) with corresponding stroke length of about 4.5×10^{-3} m. Simultaneously, the electric fan blows air upward through the porous deck at a superficial velocity u (u can vary from one to 4 m/s), (Figure 2).

The longitudinal vibration and airflow spread and lift the bed of particles on the surface of the deck; then, as the bed falls, it is expanded and fluidized. This stratifies the material according to density, causing the high-density particles to settle on the deck and contact its surface, while the low-density particles to float on top of the high-density ones (Figure 3). As the eccentric drive vibrates the deck using a slow forward stroke and a rapid return, the high-density particles move along the deck between the riffles, uphill the end slope α towards the higher side (Figures 2 and 4).

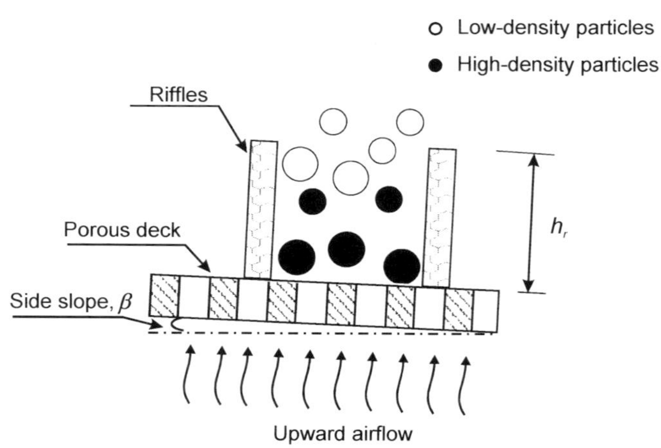

Figure 3 Vertical stratification of particles between rifles of air table (cross-section of the side view of the deck).

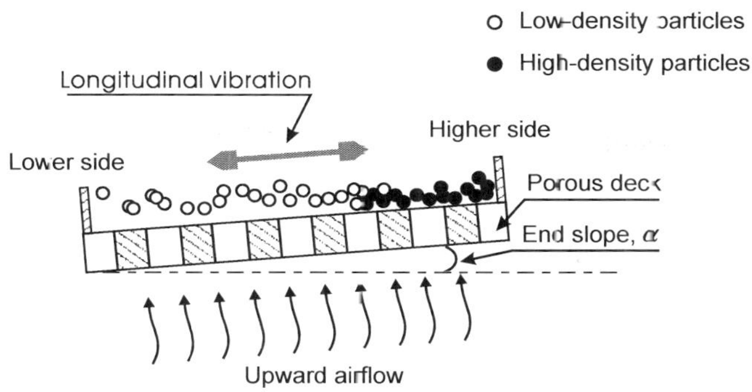

Figure 4 Arrangement of particles of different densities in porous deck due to longitudinal vibration and upward airflow (cross-section of the front side of the deck).

Subsequently, the high-density particles are flowed off the deck through the higher side, which channels them downward to the discharge end, and then drop into the right-hand compartment of collecting bin (Figure 2). On the other hand, the low-density particles, which remain fluidized, drift downhill in the direction of the deck's inclination due to gravitational pull and are discharged from the deck at its lower end. The low-density fraction is then collected in the left-hand compartment of the collecting bin (Figure 2).

The motion of a particle resting on the porous deck is primarily depended upon the acceleration of the deck. However, the analysis of the movement of solid materials along vibrating deck is not simple (Colijn, 1985), as it is influenced by many factors related to the deck movement and material property. Figure 5 depicts a particle placed on an inclined

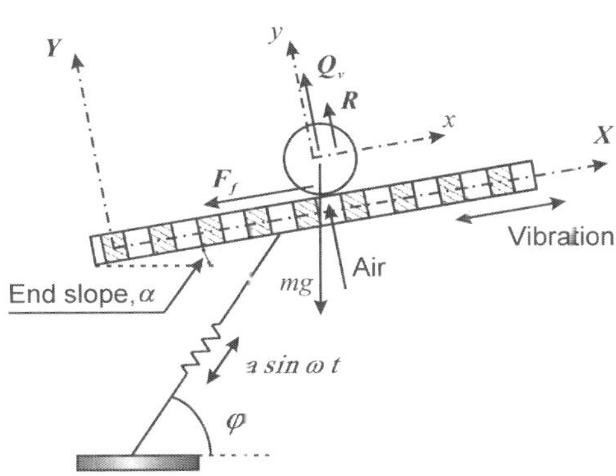

Figure 5 Forces acting on a particle placed on a porous deck. *Modified version of Figure 1 of Taniguchi et al. (1962).*

porous deck, which is vibrated in a side-to-side motion. Suppose that the inclined deck, which creates an angle α to the horizontal line, is moving with a sinusoidal motion of $a \sin \omega t$ at an angle φ to horizontal line (Figure 5). Thus, the horizontal location X and vertical location Y of the porous deck may be described mathematically by the following equations (Taniguchi at al., 1962):

$$X = a \cos(\varphi - \alpha)\sin \omega t \tag{1}$$

$$Y = a \sin(\varphi - \alpha)\sin \omega t \tag{2}$$

Considering the first and the second derivation of Eqns (1) and (2) with respect to time t, the velocity and acceleration of the deck in direction X and Y respectively (Figure 5) are obtained and given below:

$$V_x = \frac{dX}{dt} = a\,\omega \cos(\varphi - \alpha)\cos \omega\, t \tag{3}$$

$$V_y = \frac{dY}{dt} = a\,\omega \sin(\varphi - \alpha)\cos \omega t \tag{4}$$

$$a_x^{D} = \frac{d^2X}{dt^2} = -a\,\omega^2 \cos(\varphi - \alpha)\sin \omega t \tag{5}$$

$$a_y^{D} = \frac{d^2Y}{dt^2} = -a\,\omega^2 \sin(\varphi - \alpha)\sin \omega t \tag{6}$$

where a is the displacement amplitude of vibration and ω is the angular velocity, which is related with the vibration frequency f of the deck by the following formula:

$$\omega = 2\pi f \tag{7}$$

Hence, with the expressions of ω (Eqn (7)), Eqns (5) and (6) become:

$$a_x^{D} = -4\, a(\pi f)^2 \cos\left(\varphi - \alpha\right)\sin\left(2\pi ft\right) \tag{8}$$

$$a_y^{D} = -4\, a(\pi f)^2 \sin(\varphi - \alpha)\sin(2\pi ft) \tag{9}$$

On the other hand, the equations of motion of the particle on the porous deck, in direction x and y, may be described as follows:

$$ma_x = m\frac{dx^2}{dt^2} = F_f - mg \sin \alpha \tag{10}$$

$$ma_y = m\left(\frac{\mathrm{d}y^2}{\mathrm{d}t^2}\right) = Q_\mathrm{v} + R - mg\cos\alpha \tag{11}$$

Where F_f is the frictional force, Q_v is the reaction force, R is the lift force, and m is the mass of particle. Suppose that the particle has a spherical shape, then m is given as follows:

$$m = \rho_s \cdot \frac{\pi D^3}{6} \tag{12}$$

Moreover, the reaction force is expressed by the following equation (Taniguchi at al., 1962):

$$Q_\mathrm{v} = mg\cos\alpha - ma_y^\mathrm{D} \tag{13}$$

With the expressions of a_y^D (Eqn (6)), Eqn (13) becomes:

$$Q_\mathrm{v} = m\left[g\cos\alpha - a\,\omega^2\sin(\varphi - \alpha)\sin(\omega t)\right] \tag{14}$$

The lift force R is the upward force that is exerted on a particle as it passes through air. The lift produced from air flowing can be calculated from Eqn (15):

$$R = C_\mathrm{L} A \rho_\mathrm{g}\left(\frac{u^2}{2}\right) \tag{15}$$

where ρ_g is the density of air, u is the velocity of air, and A is the particle's area projected onto the horizontal plane that in case of a spherical particle can be given as:

$$A = \pi\left(\frac{D}{2}\right)^2 \tag{16}$$

The coefficient C_L in the Eqn (15) is the coefficient of lift, which depends on the shape of the particle and the Reynolds number. No simple relationship can be given for calculating the lift for a solid body, but the theoretical coefficient of lift for a thin plate in two-dimensional flow at a low angle of attack θ, is given by Eqn (17):

$$C_\mathrm{L} = 2\pi\sin\theta \tag{17}$$

In addition, the frictional force can be given as follows:

$$F_\mathrm{f} = \mu_\mathrm{f}(Q_\mathrm{v} + R),\ \text{if } a_x < a_x^\mathrm{D} \quad \text{(i.e., the particle will move with the deck)} \tag{18}$$

or

$$F_{\rm f} = -\mu_{\rm f}\left(Q_{\rm v} + R\right), \text{if } a_x > a_x^{\rm D} \quad \left(\text{i.e., the deck will slip away from the particle}\right) \tag{19}$$

With the expressions of $F_{\rm f}$ (Eqn (18)), $Q_{\rm v}$ (Eqn 14), R (Eqn (15)), and $C_{\rm L}$ (Eqn (17)), Eqn (10) becomes:

$$a_x = \frac{\mu_{\rm f}\left[m\left[g\cos\alpha - a(2\pi f)^2\sin(\varphi-\alpha)\sin(2\pi f\, t)\right] + 2\pi\sin(\theta)A\,\rho_{\rm g}\left(u^2/2\right)\right]}{m} - (g\sin\alpha) \tag{20}$$

Equation (20) mathematically describes the acceleration a_x of a particle on the porous deck at equilibrium conditions (i.e., the particle does not move, as the balance of forces acting on it is zero), as a function of vibrating frequency f, end slope α, mass of particle m, etc. Hence, since the motion of the deck is reciprocating and asymmetrical, the particle may move in one direction with the deck if its acceleration $a_x^{\rm P}$ is higher than the one at equilibrium condition (Eqn (20)), i.e., $a_x^{\rm P} > a_x$.

The eccentric drive vibrates the porous deck in a side-to-side motion creating an angle φ (usually about 35°) with the horizontal line and displacement amplitude of the vibration a, (Figure 5). The velocity u of the upward airflow blown through the porous deck is supposed to be 1.8 m/s. The angle θ between the airflow and the particle is considered to be 90°.

In an air table as well as in a fluidized bed or pneumatic jig, the bed of particles is lifted by the airflow. Lemaitre et al. (1990) studied theoretically and experimentally the suspension height of a disc as a function of the airflow through the table. Considering that the air distribution out from the table is continues, the ejection speed from the porous deck is constant and vertical as well as assuming that fluctuations are weak enough so that the disc remains horizontal, he calculated the suspension height by the following formula (Eqn (21)):

$$h_{\rm p} = \sqrt{\frac{\pi\,\rho_{\rm g}\,u^2\,r_{\rm p}^2}{4\,m\,g}} \tag{21}$$

Equation (21) indicated that the suspension height is primarily dependent on the superficial air velocity u and the radius of particle $r_{\rm p}$.

Moreover, the stratification of particle in an air table is similar to the stratification of particle in a fluidized bed or pneumatic jig. When the superficial air velocity, u, is zero (i.e., air is not passing thought the porous deck), the particle weight is entirely supported by the deck. However, when the air is forced through the porous deck of air table and the bed of particles, the bed expansion eventually occurs when the superficial velocity of the airflow reaches a so-called minimum fluidizing value. At this point the weight of the bed of particles is fully supported by the aerodynamic drag force produced by the air (Galvin et al., 1999). The superficial velocity of air at minimum fluidizing conditions, u_{mf}, for coarse particles is expressed as follows (Kunii et al., 1991):

$$u_{\mathrm{mf}} = \left(\frac{\mu}{D \cdot \rho_{\mathrm{g}}}\right) \times \left\{ \left[823.69 + 0.0494 \left(\frac{D^3 \rho_{\mathrm{g}} \left(\rho_{\mathrm{s}} - \rho_{\mathrm{g}}\right) g}{\mu^2} \right) \right]^{1/2} - 28.7 \right\} \tag{22}$$

Equation (22) points out that the minimum fluidizing velocity (u_{mf}) depends on the size (D) and the density of particle (ρ_{s}), density of air (ρ_{g}), and air absolute viscosity (μ). Furthermore, increasing the superficial velocity of airflow more than u_{mf} value, the basic premise is that the low-density particles will be caught in the upward current of air and carried with the air, while the high-density ones will drop down, unable to be supported by the air current. However, to avoid the carryover of particles from the bed, the superficial velocity of air should be kept between the minimum fluidizing velocity u_{mf}, and the terminal velocity of particles u_{t} (Kunii et al., 1991). Haider and Levenspiel, (1989), presented the following equation (Eqn (23)) for direct evaluation of the terminal velocity u_{t} of coarse particles:

$$u_{\mathrm{t}} = u_{\mathrm{t}}^{*} \left[\frac{u \left(\rho_{\mathrm{s}} - \rho_{\mathrm{g}}\right) g}{\rho_{\mathrm{g}}^2} \right]^{1/3} \tag{23}$$

Where u_{t}^{*} was defined as a dimensionless air velocity and calculated as follows:

$$u_{\mathrm{t}}^{*} = \left[\frac{18}{\left(D^{*}\right)^2} + \frac{2.335 - 1.744\, \phi_{\mathrm{s}}}{\left(D^{*}\right)^{0.5}} \right]^{-1} \quad 0.5 < \phi_{\mathrm{s}} < 1 \tag{24}$$

Accordingly, u_t^* (Eqn (24)) depended on D^*, a dimensionless particle size that was calculated by the following equation (Eqn (25)):

$$D^* = D\left[\frac{\rho_g\left(\rho_s - \rho_g\right)g}{\mu^2}\right]^{1/3} \tag{25}$$

Note that u_t^* (Eqn (24)) is also related to the one-parameter measure called the sphericity ϕ_s. Sphericity was defined as the ratio of the surface area of a sphere of volume equal to that of the particle, to the surface area of the particle (Kunii et al., 1991; Foust et al., 1980), and was given as follows (Eqn (26)):

$$\phi_s = \left(\frac{\text{surface area of sphere}}{\text{surface area of particle}}\right)_{\text{of same volume}} \tag{26}$$

Figure 6 shows a three-dimensional plot of u_{mf} (Eqn (22)) as a function of particle size (D ranged from 0 to 8 mm), and density (ρ_s varied from 900 to 1400 kg/m^3). The minimum fluidizing velocity increases sharply with increasing particle size and, to a lesser extent, with increasing density, (Figure 6). Therefore, processing particles of a narrow range of density and/or of a wide range of size will affect the separator's performance, since the minimum fluidizing velocities of the particles do not differ much. Hence, in order to obtain a satisfactory separation of a mixture by means

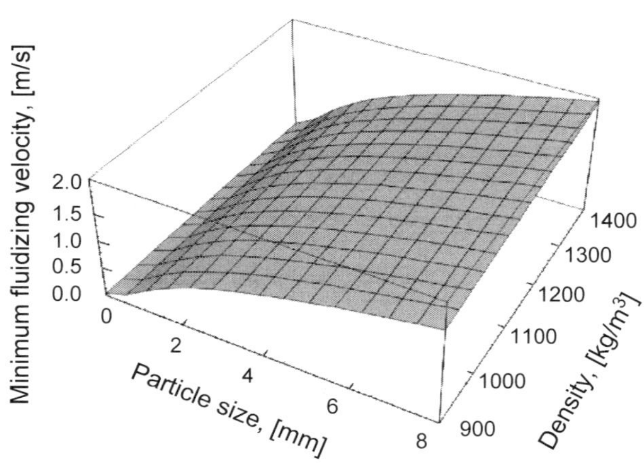

Figure 6 A three-dimensional plot of minimum fluidizing velocity, u_{mf}, as a function of particle size D, and density ρ_s as air tabling; ($g = 9.81$ m/s^2; $\rho_g = 1.2$ kg/m^3; $\mu = 1.8 \times 10^{-5}$ Pa s).

of air table, the size range of materials to be processed and the superficial air velocity should be carefully controlled. This is important because the upward airflow should lifts up only the low-density particles, whereas the high-density ones should remain settled on the deck.

In order to further clarify the phenomenon, the range of particle size, which can be separated by air table, has been calculated. Suppose, for instance, that two particles "a" and "b" with "a" being denser than "b", having different settling velocities, are placed in an upward airflow. If the velocity of the air is adjusted between the terminal velocities of the two particles, stratification will occur. However, if the particle size range of two particles is wide, the terminal velocity of the largest particle "b" may be greater than the one of the smallest particle "a", and hence no complete stratification and eventually no effective separation is likely to be completed. Assuming turbulent flow, the particle size range of particles, which can be effectively separated, was calculated by recognizing the occurrence of the equality of the terminal velocities of particles "a" and "b" (Foust et al., 1980). Hence:

$$\frac{D_{\mathrm{a}}^{\mathrm{Min}}}{D_{\mathrm{b}}^{\mathrm{Max}}} > \left[\frac{\rho_{\mathrm{b}} - \rho_{\mathrm{g}}}{\rho_{\mathrm{a}} - \rho_{\mathrm{g}}}\right] \tag{27}$$

Consequently, in order to achieve a satisfactory separation of a mixture by means of air table it is important to carefully control the size range of particles to be processed.

5. CASE STUDY: AIR TABLING OF PVC/PP MIXTURE

Although the air tabling seems simple, the study on performance and optimization of air table is not an easy task, as there are five main operating variables that simultaneously affect the performance of air table. Thus, the set of experiments is concerned with developing an understanding of the effects of varying superficial velocities of the upward airflow u, longitudinal vibrating frequency f, end slope α, side slope β, height of riffles h_{r}, and feed flow rate w respectively. Subsequently, the effects of these variables upon separation efficiency are demonstrated experimentally.

The experiments are carried out on 50/50% artificial mixtures of polyvinyl chloride (PVC), (density of PVC: 1400 kg/m^3) and polypropylene (PP), (density of PP: 900 kg/m^3). Initially, the mixture is shredded, and classified by size using a series of screens. For the reasons that stem from the analysis presented above (Eqn (27)), only $-2.38 + 1.63$ mm ($-6 + 8$ mesh) size fraction is subject to air tabling.

The experiments are performed in a laboratory-scale operation with the air table. Except where stated otherwise, during each test a sample of approximately 0.20 kg is processed creating a feed flow rate of 41.4 (kg/h)/m^2. A splitter cuts the product stream at the discharge end producing only two fractions. Thus, part of product stream, i.e., low-density fraction, is entered the left-hand compartment of collecting bin, while the remaining part (i.e., high-density fraction) is channelled into the right-hand compartment (Figures 1 and 2). At the end of each experiment, the individual separated fractions are collected and weighed. Then, the evaluations of the separation for the two-component mixture are made in terms of grade (%), recovery (%), and total efficiency of separation (%).

5.1 Variation in Superficial Air Velocity

In this series of tests the superficial air velocity is the variable and ranges from 1.1 to 2.8 m/s. The other variables of the operation are kept constant at the following values: longitudinal vibrating frequency of 11.95/s; end slope of 4.5°; side slope of 2.5°; height of riffles of 7.0×10^{-3} m. Table 1 gives the numerical results of the tests.

Results presented in Table 1 and Figure 7 make clear that the variation in superficial velocity of air greatly affects the total efficiency of separation. The higher is the velocity of air, the higher becomes the recovery of the PP particles collected in the left-hand compartment and the lower becomes the

Table 1 Results for separation of PVC/PP mixture. Superficial air velocity is the variable; (experimental conditions: PVC/PP mixture (50/50%); $D = -3.36 + 2.38$ mm; $f = 11.95$/s; $\alpha = 4.5°$; $\beta = 2.5°$; $h_r = 7.0 \times 10^{-3}$ m; $w = 41.4$ (kg/h)/m^2)

Test No.	Superficial air velocity (m/s)	PP fraction left-hand compartment: Grade (%)	PP fraction left-hand compartment: Recovery (%)	PVC fraction right-hand compartment: Grade (%)	PVC fraction right-hand compartment: Recovery (%)	Total efficiency (%)
83	1.1	96.91	7.81	45.83	94.21	7.36
112	1.4	99.05	56.33	88.09	94.43	53.19
110	1.8	95.84	96.70	99.90	97.70	94.46
79	2.2	95.25	96.57	94.01	69.46	67.08
80	2.5	79.72	97.97	94.60	31.28	30.65
81	2.8	68.21	98.82	91.84	12.00	11.86

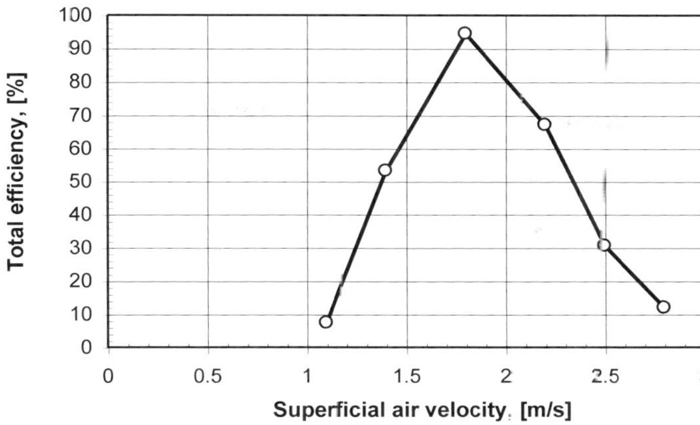

Figure 7 Total separation efficiency of PVC/PP mixture (50/50%), showing the effect of superficial air velocity; (experimental conditions: particle size $= -3.36 + 2.38$ mm; $\alpha = 4.5°$; $\beta = 2.5°$; $f = 11.95$/s; $h_r = 7.0 \times 10^{-3}$ m; $w = 41.4$ (kg/h)/m^2).

recovery of the PVC ones collected in the right-hand compartment (Table 1). This behaviour is due to the fact that with increasing superficial air velocity more particles (especially the low-density ones) are suspended above the deck and collected as low-density fraction into the left-hand compartment of collecting bin. Table 1 shows that the recovery of collected PP fraction increased from 7.81% to 98.82% in going from low (1.1 m/s) to high (2.8 m/s) superficial air velocity. It is important to note that the upward airflow with a velocity higher than 2.8 m/s tends to raise all the particles off the deck, destroying any stratification effect and making the separation of the mixture an extremely difficult task. On the contrary, the lower is the velocity of air, the greater becomes the mass of the high-density fraction of PVC collected into the right-hand compartment (Table 1). In turn, the grade of PP, collected in the left-hand compartment, is increased (Table 1). Table 1 also shows that the recovery of collected PVC fraction increases from 12.0% to 94.2% in going from high (2.8 m/s) to low (1.1 m/s) superficial air velocity, reaching its peak of 97.7% at 1.8 m/s. Nevertheless, superficial air velocity lower than 1.1 m/s, caused the bed of particles to be sluggish and accumulate at the higher end of the deck (Figure 2).

Referring to data given in Figure 7, the total efficiency of separation rises from 7.36% and reaches its peak at 94.46%, in going from low (1.1 m/s) to high (1.8 m/s) superficial air velocity. Thus, the experimental results demonstrate that the optimal value of superficial velocity of air (i.e., 1.8 m/s) is higher than the velocity of air required for minimum fluidizing condition of low-density particles of PP (i.e., $u_{mf(PP)} = 1.26$ m/s) but similar to the one required for minimum fluidizing condition of high-density particles of PVC (i.e., $u_{mf(PVC)} = 1.60$ m/s). It is important to note that the values of superficial air velocity at minimum fluidizing condition are calculated by using Eqn (22). Indeed, this optimal value of the velocity of the upward airflow creates the basic premise for low-density materials to be caught in an upward current of air and carried with it, while the high-density fraction to remain in close contact with the surface of the deck, as are unable to be supported by the air current.

5.2 Variation in Longitudinal Vibrating Frequency of the Deck

The effects of changes in longitudinal vibrating frequency of the deck f on the total efficiency of separation is studied by ranging f from 6.21 to 12.55/s with corresponding stroke length of 4.5×10^{-3} m. The other operating variables are maintained at the following constant values: superficial air velocity of 1.8 m/s; end slope of 4.5°; side slope of 2.5°; height of riffles of 7.0×10^{-3} m. Values of the total efficiency corresponding to the separation at various vibrating frequencies shown in Figure 8 are listed in Table 2 together with other results of separation.

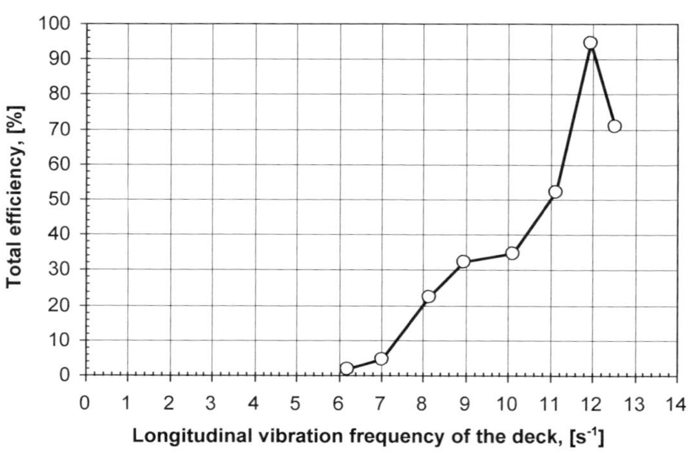

Figure 8 Total separation efficiency of PVC/PP mixture (50/50%), showing the effect of longitudinal vibration frequency of deck; (experimental conditions: particle size $= -3.36 + 2.38$ mm; $u = 1.8$ m/s; $\alpha = 4.5°$; $\beta = 2.5°$; $h_r = 7.0 \times 10^{-3}$ m; $w = 41.4$ (kg/h)/m^2).

Table 2 Results for separation of PVC/PP mixture. Longitudinal vibrating frequency is the variable; (experimental conditions: PVC/PP mixture (50/50%); $D = -3.36 + 2.38$ mm; $u = 1.8$ m/s; $\alpha = 4.5°$; $\beta = 2.5°$; $h_r = 7.0 \times 10^{-3}$ m; $w = 41.4$ (kg/h)/m^2)

Test No.	Vibrating frequency (1/s)	PP fraction left-hand compartment Grade (%)	PP fraction left-hand compartment Recovery (%)	PVC fraction right-hand compartment Grade (%)	PVC fraction right-hand compartment Recovery (%)	Total efficiency (%)
91	6.21	68.69	88.12	32.29	1.86	1.64
90	7.05	93.40	86.87	50.42	5.15	4.47
89	8.15	94.84	88.97	74.95	25.04	22.28
88	8.97	96.83	91.97	77.01	35.11	32.29
87	10.12	96.82	92.16	84.15	37.67	34.72
86	11.16	96.99	94.13	92.22	55.07	51.84
110	11.95	95.84	96.70	99.90	97.70	94.46
108	12.55	87.03	97.51	98.80	72.53	70.77

Considering the results, no separation is observed at the longitudinal vibrating frequency of 6.21/s or lower, as the PVC particles in contact with the deck shift towards the lower side and fall into the left-hand compartment (Figure 2) contaminating the low-density fraction (Table 2, Figure 8). Indeed, at $f = 6.21$/s, the recovery of PVC fraction in right-hand compartment is 1.86%, while the grade of PP fraction in left-hand compartment is 68.69%. Moreover, when the vibrating frequency increases, more PVC particles (i.e., particles closest to the deck) are directed towards the higher side of the deck. In turn, the recovery of PVC fraction increases, reaching its peak of 97.70% at 11.95/s (Table 2).

Data tabulated in Table 2 demonstrate that the effect of vibrating frequency on the recovery of PP particles is irrelevant, because those particles are not in direct contact with the surface of the deck due to the fluidizing effect of the upward airflow. Indeed, the recovery of PP slightly changes from 86.87% to 96.70% in going from 7.05 to 11.95/s. Moreover, data plotted in Figure 8 show that the total efficiency of separation rises from 7.36%, reaching its peak at 94.46% in going from low (6.21/s) to high (11.95/s) longitudinal vibration frequency. However, when the vibrating frequency is higher than 11.95/s no further improvements on total efficiency of separation are observed (Figure 8). This is due to the fact that at higher frequencies, the jump height of the particles in direct contact with the surface of deck is increased. As a result, the recovery of PVC and the grade of PP decrease, as more PVC particles overflow the riffles and contaminate the collected low-density fraction of PP (Table 2). It can be concluded that the longitudinal vibrating frequency of the deck has a pronounced influence on the total efficiency of separation.

5.3 Variation in Slope of the Deck (End Slope and Side Slope)

Initially, the effect of the end slope α (Figure 2) is studied over a range of 0.0–8.0°, with other variables held constant at the following values: superficial air velocity of 1.8 m/s; longitudinal vibrating frequency of 11.95/s; side slope of 2.5°; height of riffles of 7.0×10^{-3} m.

Figure 9 depicts a plot of total efficiency against the end slope of the deck. It shows that the highest efficiency (i.e., 94.46%) is achieved while the end slope is set at 4.5°. Moreover, Table 3 shows a summary of experimental results of the tests. Indeed, an end slope higher than 4.5° decreases the total efficiency of separation (Figure 9) and the grade of the collected PP fraction, as more PVC particles are directed towards the lower side of the deck and collected into the left-hand compartment of collecting bin contaminating the PP fraction (Table 3). On the contrary, the lower is the end slope, the higher becomes the recovery of PVC particles (Table 3), as more particles are directed towards the higher side of the deck (Figure 2).

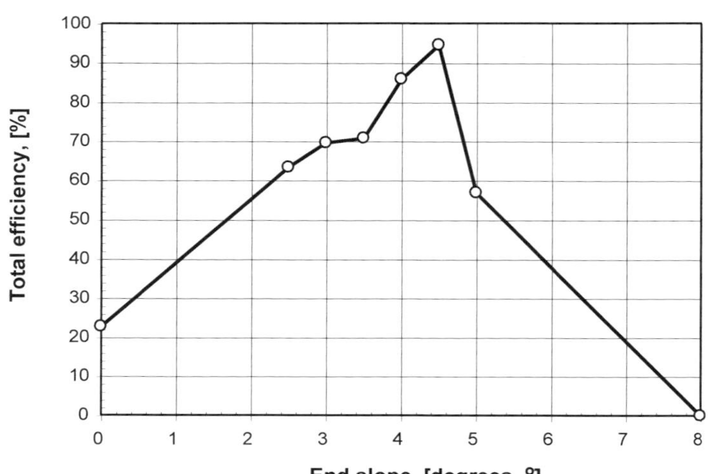

Figure 9 Total separation efficiency of PVC/PP mixture (50/50%), showing the effect of end slope; (experimental conditions: particle size $= -3.36 + 2.38$ mm; $u = 1.8$ m/s; $f = 11.95$/s; $\beta = 2.5°$; $h_r = 7.0 \times 10^{-3}$ m; $w = 41.4$ (kg/h)/m^2).

Table 3 Results for separation of PVC/PP mixture. End slope is the variable; (experimental conditions: PVC/PP mixture (50/50%); $D = -3.36 + 2.38$ mm; $u = 1.8$ m/s; $f = 11.95$/s; $\beta = 2.5°$; $h_r = 7.0 \times 10^{-3}$ m; $w = 41.4$ (kg/h)/m^2)

Test No.	End slope (°)	PP fraction left-hand compartment		PVC fraction right-hand compartment		Total efficiency (%)
		Grade (%)	Recovery (%)	Grade (%)	Recovery (%)	
119	0.0	97.37	22.84	66.67	99.50	22.73
112	2.5	96.95	64.22	80.59	98.59	63.32
113	3.0	97.13	70.92	93.61	98.04	69.53
114	3.5	96.41	72.17	93.51	98.00	70.72
111	4.0	96.70	87.77	98.11	97.91	85.94
110	4.5	95.84	96.70	99.90	97.70	94.46
115	5.0	87.48	99.63	99.74	57.16	56.95
118	8.0	49.96	99.98	0.00	0.00	0.00

Thus, an increase in end slope results in a shift of materials towards the lower end of the deck, whereas a decrease in end slope results in a shift towards the higher end.

Experimental data indicate that the recovery of PVC fraction increases from 97.70% to 98.59% in going from 4.5 to 2.5° end slope (Table 3). However, an end slope lower than 3.5° also caused the accumulation of the low-density particles at the higher end of the deck. This is due to the fact that at a low end slope the direction of gravitational pull of the particles, lifted by the upward airflow (i.e., PP particles), changes in such a direction that they fall to the right-hand compartment of collecting bin contaminating the high-density fraction. In turn, the efficiency of separation process sharply decreases (Figure 9). Nevertheless, this set of tests indicates that variation in end slope affects, to a large extent, the direction of travel of the high-density particles suggesting that an efficient separation is accomplished at an end slope of 4.5°.

Table 4 and Figure 10, on the other hand, show the experimental results in a series of tests in which the side slope β varies from 0.0 to 8.0°, while the other variables are kept constant at the following values: superficial air velocity of 1.8 m/s; longitudinal vibration frequency of 11.95/s; end slope of 4.5°; height of riffles of 7×10^{-3} m.

Considering the results presented in Table 4, the recovery of PP fraction slightly increases with increasing side slope. Indeed, the recovery of PP increases from 92.86% to 99.73% in going from 0.0 to 8.0° side slope. However, a very steep side slope (let us say more than 5°) resulted in a

Table 4 Results for separation of PVC/PP mixture. Side slope is the variable; (experimental conditions: PVC/PP mixture (50/50%); $D = -3.36 + 2.38$ mm; $u = 1.8$ m/s; $f = 11.95$/s; $\alpha = 4.5°$; $h_r = 7.0 \times 10^{-3}$ m; $w = 41.4$ (kg/h)/m^2)

Test No.	Slide slope (°)	PP fraction left-hand compartment Grade (%)	PP fraction left-hand compartment Recovery (%)	PVC fraction right-hand compartment Grade (%)	PVC fraction right-hand compartment Recover (%)	Total efficiency (%)
120	0.0	88.07	58.61	90.91	47.23	27.68
54	1.9	88.68	95.15	93.60	68.55	65.23
98	2.3	90.46	95.22	94.69	71.61	68.18
110	2.5	95.84	96.70	99.90	97.70	94.46
99	3.5	80.18	97.98	95.50	44.63	43.73
100	4.5	72.94	98.87	96.18	20.28	20.05
101	5.0	71.40	98.36	96.91	8.31	8.17
102	6.0	71.71	98.21	97.14	8.23	8.09
103	7.0	68.50	99.65	97.44	8.12	8.09
104	8.0	66.74	98.73	97.88	7.92	7.82

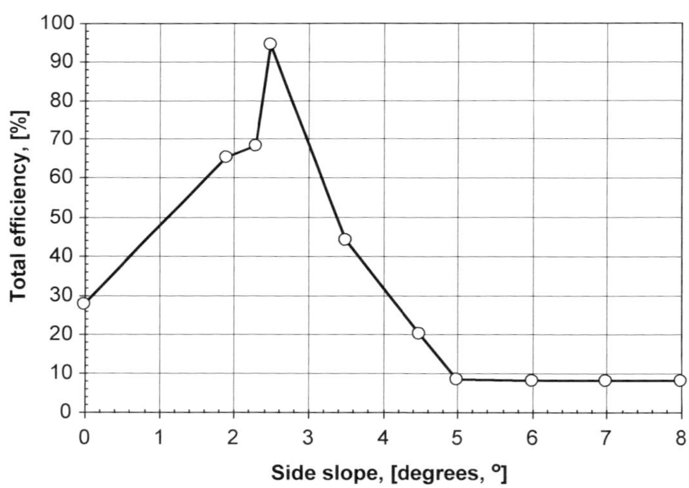

Figure 10 Total separation efficiency of PVC/PP mixture (50/50%), showing the effect of side slope; (experimental conditions: particle size $= -3.36 + 2.38$ mm; $u = 1.8$ m/s; $f = 11.95$/s; $\alpha = 4.5°$; $h_r = 7.0 \times 10^{-3}$ m; $w = 41.4$ (kg/h)/m^2).

decrease of the total efficiency of separation to less than 8.17% (Figure 10). This is due to the fact that an increase in the side slope tends to decrease the residence time of particles on the deck and thus directs more high-density particles of PVC to the left-hand compartment of the collecting bin contaminating the low-density fraction of PP (Table 4). Reduction of the side slope, conversely, enhances the separation efficiency as the residence time of particles on the deck increases. Experimental data indicate that the recovery of PVC fraction increases from 8.31% to 97.70% in going from 5.0 to 2.5° side slope (Table 4). However, a side slope lower than 2.5° makes the sliding or rolling motion of particles difficult. In turn, the recovery of PVC and PP (Table 4), as well as the total efficiency of separation decrease (Figure 9), since more particles are retained on the deck. Therefore, side slope should be set to the minimum at which it is possible to achieve good distribution of materials on the deck of the air table. The experimental results suggest that a side slope of 2.5° is suitable for achieving the highest separation efficiency (Figure 9).

Subsequently, the experimental data, depicted in Figures 9 and 10, are combined together with other experimental results of air tabling of more than 25 samples processed separately at varying end slope and side slope. The results are then plotted in Figure 11. Figure 11 shows the surface plot and the distribution of contour lines of the variation in "total separation efficiency" as a function of the "end slope" and the "side slope". These plots are three-dimensional representations of the experimental data that are displayed with combination of "end slope," "side slope" and "total separation efficiency" values. Contours define lines of constant "total separation efficiency";

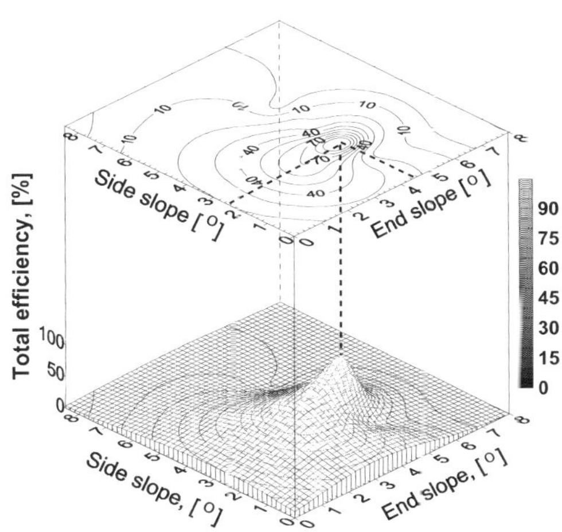

Figure 11 Surface plot (at the bottom) and contour lines (at the top) of variation in total efficiency as a function of end slope and side slope of the deck; (experimental conditions: PVC/PP mixture (50/50%); particle size $= -3.36 + 2.38$ mm; $u = 1.8$ m/s; $f = 11.95$/s; $h_r = 7.0 \times 10^{-3}$ m; $w = 41.4$ (kg/h)/m^2).

or in other words, lines of constant values of total efficiency, across the extent of the "map" of the end slope and the side slope values.

Figure 11 shows that variation in the side slope and particularly in the end slope have a pronounced effect on the total separation efficiency of PVC/PP mixture. It also shows that the process for separating this specific mixture is a very difficult task if performed with an end or side slope higher than 6°. Moreover, it can clearly be seen that the total efficiency reaches its peak, while the end slope and side slope are set up at 4.5 and 2.5° respectively (Figure 11).

5.4 Variation in Height of Riffles

Riffles are added to the deck for three purposes: (1) to build a deeper bed of particles, (2) to prevent the high-density particles from contaminating the low-density fraction, and (3) to increase the residence time of materials on the deck. With this in mind, the other studied variable is a constructive parameter of the design, named as height of riffles h_r. Hence, the porous deck is fitted with riffles of 4.5×10^{-3} (i.e., original design), 7.0×10^{-3} or 10.0×10^{-3} m high respectively, and the experiments are carried out at superficial air velocity of 1.8 m/s; longitudinal vibrating frequency of 11.95/s; end slope of 4.5°; and side slope of 2.5°.

The way in which the separation results of PVC/PP mixture are affected, as the height of riffles varies, is recorded in Table 5. Moreover, the variation in the total separation efficiency of PVC/PP mixture with changes in the height of riffles is shown in Figure 12. It was found that the deck fitted with riffles of 7.0×10^{-3} m high ensures higher total efficiency than the one fitted with riffles of 4.5×10^{-3} m high. This is due to the fact that riffles of 7.0×10^{-3} m high are suitable to retain more

Table 5 Results for separation of PVC/PP mixture. Height of riffles is the variable; (experimental conditions: PVC/PP mixture (50/50%); $D = -3.36 + 2.38$ mm; $u = 1.8$ m/s; $f = 11.95$/s; $\alpha = 4.5°$; $\beta = 2.5°$; $w = 41.4$ (kg/h)/m^2)

Test No.	Height of riffles ($\times 10^{-3}$ m)	PP fraction left-hand compartment		PVC fraction right-hand compartment		Total efficiency (%)
		Grade (%)	Recovery (%)	Grade (%)	Recovery (%)	
106	4.5	83.85	85.12	73.47	53.75	45.75
110	7.0	95.84	96.70	99.90	97.70	94.46
107	10.0	97.49	91.42	97.09	97.06	88.74

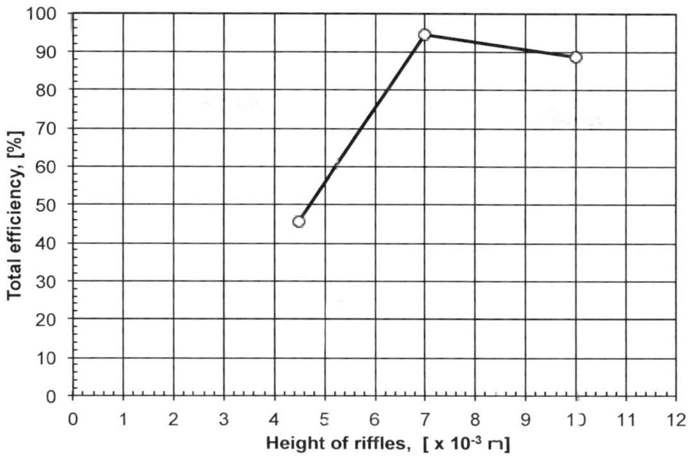

Figure 12 Total separation efficiency of PVC/PP mixture (50/50%), showing the effect of height of riffles; (experimental conditions: particle size = −3.36 + 2.38 mm; $u = 1.8$ m/s; $f = 11.95$/s; $\alpha = 4.5°$; $\beta = 2.5°$; $w = 41.4$ (kg/h)/m^2).

high-density particles (i.e., PVC particles) on the deck, thus preventing them from being reported to the left-hand compartment of the collecting bin, and at the same time insuring their movement towards the higher side (Figure 2).

Experimental results indicate that the recovery of PVC and grade of PP increase from 53.75% to 97.70% and 83.85–95.84% respectively in going from 4.5 to 7.0×10^{-3} m high riffles (Table 5). However, at the given experimental conditions, riffles higher than 7.0×10^{-3} m (i.e., 10.0×10^{-3} m) hold back also the low-density particles (i.e., PP particles). As a result, the total efficiency decreases, since the recovery of PP and grade of PVC also decrease (Table 5, Figure 12). Considering the results, riffles of 7.0×10^{-3} m high are suitable to be employed for the separation of this specific mixture (Table 5, Figure 12).

5.5 Variation in Feed Flow Rate

The capacity of a separator (such as air table) is directly proportional to (1) the cross-sectional area of the separating device, A_d; (2) the rising velocity of fluid; and (3) the solid contend in the separator intake (Gaudin, 1939). Hence,

$$C = a_c A_d u \eta \rho_s \tag{28}$$

in which C is the tonnes of solid per hour, A_d the cross-sectional area (such as surface area of the porous deck), u velocity of fluid, η the percentage of solid by volume, ρ the density of solids, and a_c the constant (usually $a_c = 0.278$) required to secure C in terms of tonnes of solid per hour (Gaudin, 1939). Nevertheless, the actual capacity for any separation can only be fully determined by test work (Burt, 1984).

Following, the effect of changes in feed flow rate w on the total efficiency of separation is studied by ranging w from 41.4 to 68.2 (kg/h)/m^2. This is accomplished by varying the mass of sample introduced into the porous deck (surface area of 0.1216 m^2) from 0.25 to 0.41 kg, while the operation time of the air table remains constant at 180 s (i.e., 0.05 h). The other operating variables of air table are kept at following constant values: superficial air velocity of 1.8 m/s; longitudinal vibrating frequency of 11.95/s; end slope of 4.5°; side slope of 2.5°; height of riffles of 7.0×10^{-3}.

The way in which the separation results of PVC/PP mixture are affected, as the feed flow rate varies, is recorded in Table 6. Moreover, the variation in the total separation efficiency of PVC/PP mixture with changes in the feed flow rate is shown in Figure 13. Considering the results, the separation efficiency decreases with rising feed flow rate. This behaviour is due to the fact that with increasing feed flow rate, the volume of bed of particles also increases, leading to worsening the stratification of the particles. As a result, more high-density particles of PVC overflow the riffles and less low-density particles of PP flow on the top of the bed of materials, as they are "trapped" inside the bed. A look at Table 6 shows that the recovery of PVC fraction (collected in the right-hand compartment) drops from 97.70% to 63.08%, and recovery of PP fraction (collected in the left-hand compartment) decreases from 96.70% to 80.93%, in going from 41.4 to 68.2 (kg/h)/m^2. Consequently, the experimental results indicate that the feed flow rate should be controlled carefully.

Table 6 Results for separation of PVC/PP mixture. Feed flow rate is the variable (experimental conditions: PVC/PP mixture (50/50%); $D = -3.36 + 2.38$ mm; $u = 1.8$ m/s; $f = 11.95$/s; $\alpha = 4.5°$; $\beta = 2.5°$; $h_r = 7 \times 10^{-3}$ m)

		PP fraction left-hand compartment		PVC fraction right-hand compartment		
Test No.	Feed rate ((kg/h)/m^2)	Grade (%)	Recovery (%)	Grade (%)	Recovery (%)	Total efficiency (%)
110	41.4	95.84	96.70	99.90	97.70	94.46
116	60.0	70.15	85.55	86.96	69.97	59.85
117	68.2	68.98	80.93	77.84	63.08	51.05

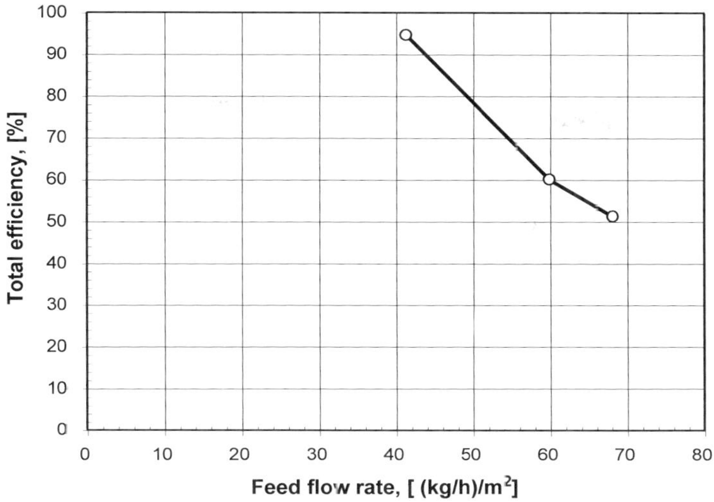

Figure 13 Total separation efficiency of PVC/PP mixture (50/50%), showing the effect of feed flow rate; (experimental conditions: particle size $= -3.36 + 2.38$ mm; $u = 1.8$ m/s; $f = 11.95$/s; $\alpha = 4.5°$; $\beta = 2.5°$; $h_r = 7.0 \times 10^{-3}$ m).

The best separation occurs when the feed flow rate is as low as possible, while maintaining a full cover of particles on the deck.

6. PERFORMANCE CURVE OF AIR TABLING

The air tabling is a dry gravity separation technique whereby particles of mixed sizes, or different densities are separated from each other due to the differential settling in an upward airflow and under the effect of a vibrating action. The performance curve is, therefore, used to quantify the separation by air table. It relates the total efficiency of separation to the respective density differential between components of the mixture. Hence, various 50/50% mixtures, composed of materials of known density (Table 7), are sorted in only two density fractions by means of air table. During each test, samples of similar shape and 2.38–3.36 mm in size are processed. Next, each separated fraction discharged from the air table is collected to measure the mass and calculate the total efficiency. Thus, the points for the performance curve are then determined, knowing the difference in density between the components of the binary mixtures taken under investigation. For density differential ranges from 10 to 7560 kg/m^3, Figure 14 shows the performance curve of air

Table 7 Composition of various binary mixtures (size fraction: 3.36 + 2.38 mm)

Mixtures (50/50%)	Density, ρ_s (kg/m^3)			
Component "a"/ component "b"	Component "a"	Component "b"	Differences in density (kg/m^3)	Total efficiency (%)
Copper/PVC	8960	1400	7560	99.9
Aluminum/ABS	2702	1060	1642	99.0
PVC/PP	1400	900	500	96.1
PET/PP	1350	900	450	92.5
PVC/ABS	1400	1060	340	46.0
PET/PS	1350	1050	300	36.0
ABS/PP	1060	900	160	28.1
PE/PP	960	900	60	25.7
PVC/PET	1400	1350	50	25.8
ABS/PS	1060	1050	10	25.0

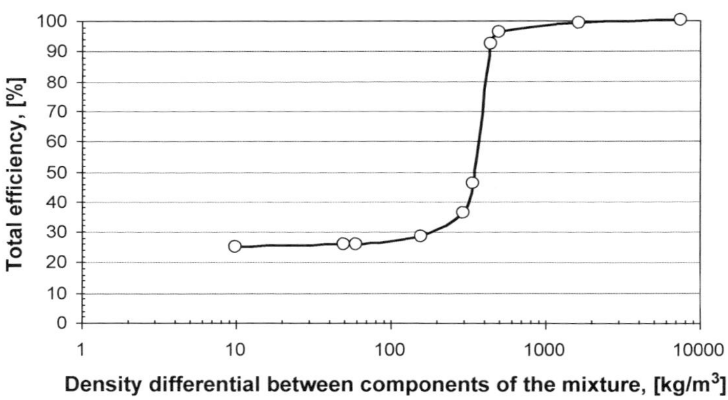

Figure 14 Performance curve of air tabling.

tabling. The performance curve expresses the separation efficiency as a function of the density differential on logarithmic scale.

Figure 14 clearly shows that the efficiency of separation increases with increasing density differential between components of the mixture being processed. As a general rule, the performance curve indicates that sharp separations (i.e., efficiency higher than 90%, which corresponds to a grade higher than 90%) are achieved, if there is a difference in density of at least 450 kg/m^3. Obviously, with a composition of particles of a wider density range the results improve, while they get worse if the composition is more restricted.

REFERENCES

Appleyard, K.C., 1931. Dry cleaning of coal in England. Trans. Am. Inst. Min. Metall. Eng. 94, 235–265.

Arms, R.W., 1924. Dry cleaning of coal. Trans. Am. Inst. Min. Metall. Eng. 70, 758–774.

Burt, R.O., 1984. Gravity Concentration Technology. Elsevier Science Publishers B.V. pp. 80–107 and 352–372.

Canning, E., 1980. Dry mineral sand separation plants of associated mineral consolidated Ltd., new South Wales and Queensland. Min. Metall. Pract. Aust., 758–761. No. 10.

Clifford, D., 1999. Gravity concentration. Miner. Mag. 180 (3), 136.

Colijn, H., 1985. Mechanical Conveyors for Bulk Solids. Elsevier Science Publishers B.V., pp. 236–329.

Dodbiba, G., Haruki, N., Shibayama, A., Miyazaki, T., Fujita, T., 2002. Combination of sink–float separation and flotation technique for purification of shredded PET-bottle from PE or PP flakes. Int. J. Miner. Process. 65, 11–29.

Falconer, A., 2003. Gravity separation – old technique/new methods. Phys. Sep. Sci. Eng. 12 (1), 31–48.

Foust, A.S., Waenzel, L.A., Clump, C.W., Maus, L., Andersen, L.B., 1980. Principles of Unit Operations, second ed. John Wiley & Sons Inc. pp. 611–617 and 711.

Galvin, K.P., Pratten, S.J., 1999. Application of fluidization to obtain washability data. Miner. Eng. 12 (9), 1051–1058.

Gaudin, A.M., 1939. The Principles of Mineral Dressing. Mc-Graw Hill Book Company, Inc, pp. 217–219.

Haider, A., Levenspiel, O., 1989. Drag coefficient and terminal velocity of spherical and nonspherical particles. Powder Technol. 58 (1), 63–70.

Hudson, S.B., 1962. Air tabling of beach sand non-conductors for zircon recovery. Proc. Aust. Inst. Min. Metall. 204, 81–108.

Jaman, W., 1985. Pneumatic concentration. In: Weiss, N.L. (Ed.), SME Mineral Processing Handbook. Published Society of Mining Engineers. Section 4, (Chapter 7), pp. 4-51 to 4-52.

Knapp, E.A., 1953. Concentrating Ores by Pneumatic Tables. Recent Development in Mineral Dressing. A symposium arranged by IMM, held on September 23–25, 1952 at the Huxley building, Imperial college of science and technology, London, pp. 299–311.

Kunii, D., Levenspiel, O., 1991. Fluidization Engineering, second ed. Butterworth-Heinemann, Oxford. pp. 68–83.

Lemaitre, J., Gervois, A., Peerhossaini, H., Bideau, D., Troadec, J.P., 1990. An air table designed to study two-dimensional disc packings – preliminary test and first results. J. Phys. Appl. Phys. 23 (11), 1396–1404.

Llewellyn, R.L., 1977. Pneumatic Cleaning of Fine Coal for Sulfur Reduction, 3rd Symposium on Coal Preparation, NCA/BCR conference, October 1977, pp. 123–136.

McCulloch, W.C., Llewellyn, R.L., Humphereys, K.H., Leonard, J.W., 1950. Dry concentration at coal preparation. In: Mitchell, D.R., Seeley, W. (Eds.), Mudd Series pp. 11.3–11.20.

Osborn, W.X., 1927. Tabling tungsten ore without water. Eng. Min. J. 123 (7), 287–289.

Sivamohan, R., Forssberg, E., 1985. Principle of tabling. Int. J. Miner. Process. 15, 281–295.

Stessel, R.I., 1996. Recycling and Resource Recovery Engineering – Principles of Waste Processing. Springer, pp. 155–180.

Taniguchi, O., Sakata, M., Suzuki, Y., Osanai, Y., 1962. Studies on vibratory feeder. Kikai Gakkai Ronbunshyu 28, 485–492 (in Japanese with English abstract).

Truscott, S J., 1923. A Text-book of Ore Dressing. Macmillan and Co. Ltd., pp. 570–590.

Vesilind, P.A., Rimer, A.E., 1981. Unit Operations in Resource Recovery Engineering. Prentice-Hall, Inc., 154–177.

Wills, B.A., 1997. Mineral processing Technology, an Introduction to the Practical Aspects of Ore Treatment and Mineral Recovery, sixth ed. Butterworth-Heinemann, Oxford. pp. 207–214, 228–232 and 241.

CHAPTER THIRTEEN

Gas–Gas Separation by Membranes

Thomas Harlacher and Matthias Wessling
Aachener Verfahrenstechnik, Chemical Process Engineering, RWTH Aachen University, Aachen, Germany

Contents

Progress in Filtration and Separation
http://dx.doi.org/10.1016/B978-0-12-384746-1.00013-6

GLOSSARY

Selectivity Characteristics of a membrane to differentiate among various components.
Stage cut Parameter which defines the fraction of the feed stream which permeates through the membrane.
Pressure ratio Dimensionless parameter which describes the ratio of the total feed pressure to the total permeate pressure.
Retentate The part of the feed stream, which is rejected by the membrane, and leaves the module without passing the membrane on the high pressure side.
Permeate The part of the feed stream, which passes the membrane and leaves the module on the low pressure side.
Recovery The ratio of the amount of a product, enriched in a useful product stream, to the total amount of this component in the feed stream.
Permeability Parameter for the trans-membrane flux per unit driving force per unit membrane thickness.
Permeance The trans-membrane flux per unit driving force.

NOMENCLATURE

D	Diffusivity	m^2/s
E_a	Activation energy	J/kmol
$\dot{n}''$	Specific mole flow	$kmol/m^2/s$
P	Permeability	Barrer
S	Solubility	$kmol/m^3/bar$
T	Temperature	K
x	Mole fraction	–
y	Mole fraction	–
δ	Membrane thickness	μm
ΔH_L	Heat of solution	J/kmol
Π	Permeance	GPU
Φ	Pressure ratio	–
Θ	Stage cut	–

INDICES

i, j, k	Component i, j, k
0	Standard conditions
F	Feed
P	Permeate

1. INTRODUCTION

In recent decades, next to conventional technologies, such as physical adsorption, chemical or physical absorption and cryogenic fractioning, and membrane processes have become an industrial alternative for gas separation. For separation tasks with moderate flow rates and moderate purities, gas permeation is the state-of-the-art technology in many cases. Compared to conventional techniques, gas permeation processes are characterized by a simple, flexible, and compact design. Table 1 shows a detailed collection of advantages and drawbacks of gas permeation processes.

This chapter gives a condensed overview of the basics of gas permeation processes. Both the design and operation of membrane modules as well as approaches for process design are described. Examples of industrial membrane gas separation processes illustrate the application. Detailed information

Table 1 Advantages and drawbacks of membrane gas separation processes in comparison to conventional techniques (Favre 2011)

Advantages	Drawbacks
+ Operating flexibility causing membranes to tolerate fluctuating feed conditions without losing quality	- No economies of scale due to modular design, penalty for large-scale applications
+ Low energy requirement and/or high energy efficiency leading to low operating costs	- Pretreatment can be heavy, and, in some cases, as expensive as the costs of the modules
+ Absence of chemicals, make-up and solvent emissions	- Sensitivity to chemical compounds
+ Easy start-up and shut-down	- Requires a high-quality energy source (electrical energy for compression), whereas absorption processes can use low-quality energy such as heat (often heat excess in chemical processes)
+ No moving parts and minimal maintenance and operator attention	
+ Modular design, easy to scale up when increased capacity is required	
+ Small footprint and compact system (advantage for offshore applications)	
+ Minimal utilities required, easy to control, simple sensors required (pressure, temperature, and flow rate)	

about gas separation processes can be found in Baker (2004), Drioli (2010, 2011), Favre (2011), Melin 2007), Merkel 2010), Mulder (1996), Nunes (2006), Ohlrogge (2006), and Yampolskii (2010). Yampolskii reports particularly on membrane materials for gas permeation Yampolskii (2006).

2. MEMBRANE MODULES FOR GAS SEPARATION

2.1 Module Construction

To use membranes in technical processes, the membrane material has to be assembled into a manageable, custom-designed membrane configuration. This membrane configuration is called the membrane module which is the main building block of a membrane plant. In a membrane module, the feed stream is separated into a concentrate stream (retentate) rejected by the membrane and a filtrate stream (permeate) passing through the membrane.

Membrane modules for gas separation applications are generally designed as 3-end modules with one inlet-stream and two outlet streams; 4-end modules are used for applications with an additional sweep stream in the permeate.

For gas separation module development, the following requirements have to be taken into account:

- uniform fluid flow without dead zones
- mechanical, chemical, and thermal stability
- high packing density (ratio of membrane area and module volume)
- cost-efficient production
- low pressure losses.

Based on these requirements and on the availability of the membranes in terms of flat sheet and hollow fibre membranes, three types of membrane modules are used for gas separation processes. Hollow fibre membranes are attached to hollow fibre modules and flat sheet membranes to spiral wound or envelope type modules. Table 2 compares the module properties of these three module types. Table 3 gives information about industrial gas separation membrane suppliers.

2.1.1 Hollow Fibre Modules

Hollow fibres with an outer diameter between 80 and 500 μm are usually arranged in parallel in a pressure pipe. Some module suppliers apply a chaotic fibre arrangement to introduce turbulent flow on the shell side of the module. Hence, the boundary layers on the shell side are reduced.

Table 2 Key characteristics of the three major types of modules used for industrial applications of gas separation processes (Drioli 2010)

	Hollow fibre	Spiral wound	Envelope
Packing density (m^2/m^3)	500–10,000	200–1000	30–500
Approximate are per module (m^2)	300–600	20–40	5–20
Pretreatment requirement	High	Fair	Minimal
Resistance to fouling	Poor	Fair	Good
Pressure drop	High	Fair	Low
Flow distribution	Good	Moderate	Fair
Manufacturing costs	Low	Medium	High

Table 3 Membrane suppliers and their membrane module characteristics for gas permeation processes (Scholz 2013)

Supplier	Module type	Polymer
Air liquide medal	Hollow fibre	Polyimide, polyaramide
Air products	Hollow fibre	Polysulfone
Borsig	Envelope	Polyethylene oxide
Evonik	Hollow fibre	Polyimide
IGS generon membrane technology	Hollow fibre	Tetrabromide polycarbonate
Kvaerner membrane systems	Spiral wound	Cellulose acetate
MTR Inc.	Spiral wound	Perfluoro polymer, silicone rubber
Parker	Hollow fibre	Polyphenylene oxide
Praxair	Hollow fibre	Polyimide
UBE membranes	Hollow fibre	Polyimide
UOP	Spiral wound	Cellulose acetate

The capillaries are bedded in resin for sealing against the inlet and outlet sides (Figure 1).

The most important advantage of hollow fibre modules is their high packing density which is up to 10,000 m^2/m. However, the modules are prone to impurities and pressure losses, due to the small fibre diameter. Feed pressures up to 15 bar lumen-sided and 70 bar shell-sided can be applied.

2.1.2 Spiral Wound Modules

For a spiral wound module, one or several membrane envelopes combined with a spacer are spirally wrapped around a permeate collecting pipe. The membrane envelope is built up of two membrane sheets with a permeate

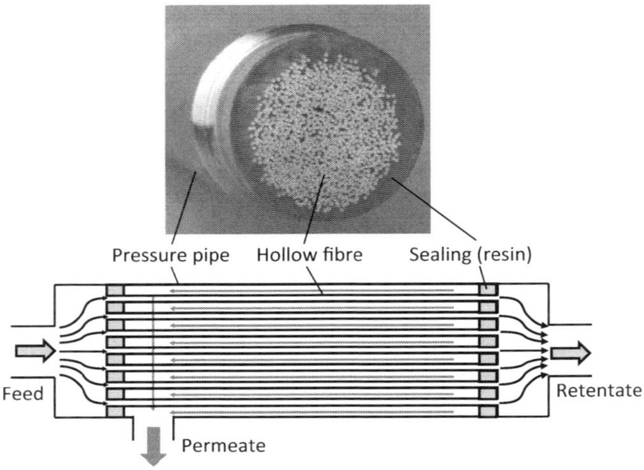

Figure 1 Hollow fibre module.

spacer in-between. The envelope is sealed on three sides and the fourth, open side is connected to the collecting pipe. The module is placed inside a tubular pressure vessel. The feed gas enters the module at the end wall and passes axially through the module. Meanwhile, the permeate streams spirally through the spacer to the collecting pipe (Figure 2).

The permeate flow path to the collecting pipe might be very long which causes significant pressure losses. These pressure losses can be reduced by using multiple short envelopes.

2.1.3 Plate and Frame Modules

For gas separation applications, the envelope type module, as a further development of a classical plate and frame module, is used. Two circular membrane sheets are glued onto the brink with a spacer in-between to form the envelopes. A centered bore hole is inserted for a permeate collection from the inner envelope area. In alternation with seals, the membrane

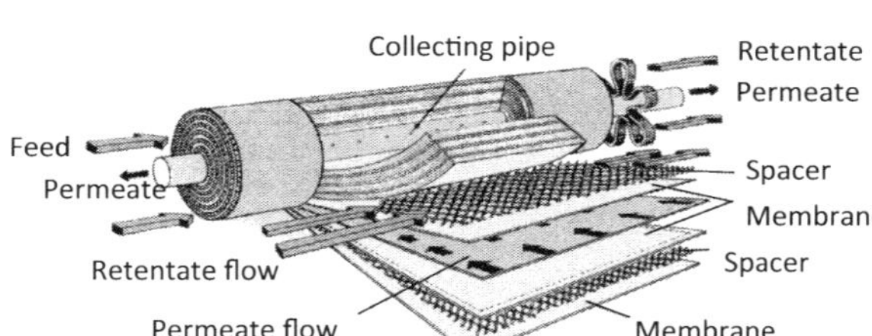

Figure 2 Spiral wound module.

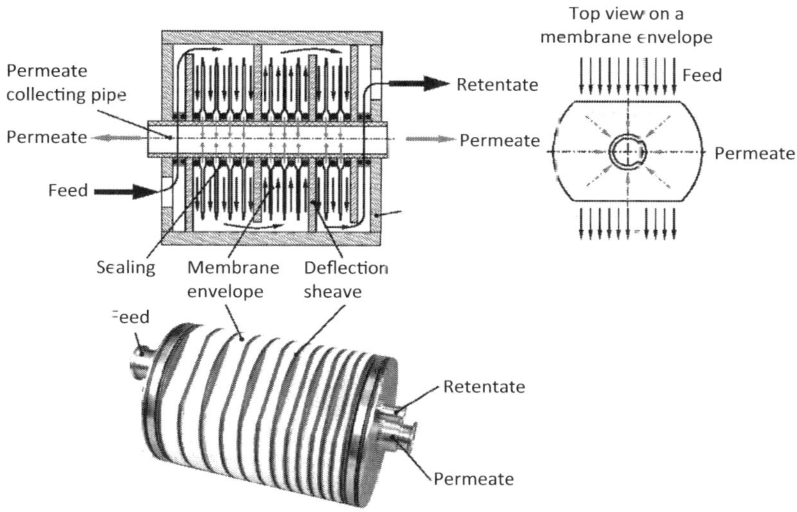

Figure 3 Plate and frame module (Helmholtz–Zentrum Geesthacht) (Melin 2007).

envelopes are pushed onto a collecting pipe. For flow conduction, deflection sheaves are added, whereby the feed flow velocity can be kept in a defined range by the selection of various envelope numbers per compartment (Figure 3). The modular placing of the single membrane envelopes enables the exchange of individual envelopes, but in contrast to the hollow fibre and spiral wound modules, the envelope modules have lower packing densities and more complex sealing is required.

2.2 Membrane Material

Several materials have been developed for membranes in gas permeation processes. Actually, polymeric membranes are mostly used for industrial applications. Metal and inorganic membranes show advantages, especially in high temperature applications, but these membranes are still under investigation and have not yet reached the industrial scale. Therefore, this work just outlines polymeric membranes.

A variety of polymers and polymer blends are available for the manufacture of membranes. The selection of the polymers determines the chemical, mechanical and thermal stability, but also the mass transfer of gases through the membrane. These properties can be adjusted to the gas separation task. However, the number of potential membranes for industrial applications is restricted, because only a few polymer gas permeation membranes are

Table 4 Permeabilities of common gas separation polymers in Barrer (Nunes 2006)

Polymer	Permeability at 30 °C (Barrer)				
	H_2	N_2	O_2	CH_4	CO_2
Cellulose acetate (CA)	2.63	0.21	0.59	0.21	6.3
Ethyl cellulose	87	3.2	11	19	26.5
Polycarbonate, brominated (PC)		0.18	1.36	0.13	4.23
Polydimethylsiloxane (PDMS)	550	250	500	800	2700
Polyimide (Matrimid) (PI)	28.1	0.32	2.13	0.25	10.7
Polymethylpentene (PMP)	125	6.7	27	14.9	84.6
Polyphenylenoxide (PPO)	113	3.81	16.8	11	75.8
Polysulfone (PSF)	14	0.25	1.4	0.25	5.6
Polyetherimide (PEI)	7.8	0.047	0.4	0.035	1.32

1 Barrer = 10^{-10} cm^3 cm/cm^2 s cmHg.

commercially available. An overview of the used polymers and their pure gas permeabilities is given in Table 4.

Membranes for gas separation generally feature an asymmetric structure. The active layer, responsible for the separation of gas mixtures, is manufactured as thinly as possible to minimize the mass transfer resistance. To ensure mechanical stability, the active layer is combined with a porous structure. The thickness generally is around 0.05–0.5 μm for the active layer and 30–200 μm for the porous layer, respectively.

Asymmetric membranes can be differentiated as integrally asymmetric and composite asymmetric. Integrally asymmetric membranes are produced by phase inversion and consist of one material, so that the active layer and the porous support structure only differ by the pore size. During the manufacturing of composite membranes, a thin polymer layer is applied on a porous structure, where different polymers for the active layer and the porous layer can be selected, although a separate optimization of the two layers with regard to stability and separation performance is possible.

2.3 Mass Transfer through Polymer Membranes

The separation of gases through membranes is a partial pressure driven separation process. The partial pressure difference between the feed and the permeate generates the driving force for the respective gas components.

During the permeation process, the gas molecules absorb into the membrane polymer on the feed side, diffuse through the membrane, and desorb on the permeate side. According to this solution-diffusion model, the gas transfer through the membrane can be calculated:

$$\dot{n}''_k = \Pi_k \cdot \left(x_k p_F - y_k p_P\right) \tag{1}$$

The flux of a component k through the membrane is proportional to the partial pressure difference. The permeance Π_k is the permeability P_k related to the membrane thickness δ. The permeance is a gas specific and module specific parameter and has to be determined experimentally. The permeance is often indicated in GPUs (gas permeation units) whereby one GPU is equal to $2.7 \cdot 10^{-3} m_N^3 (m^2 h\ bar)^{-1}$.

The permeance for component k is proportional to the diffusivity D_k and the solubility S_k of component k in the membrane and inversely proportional to the membrane thickness δ:

$$\Pi_k \sim \frac{D_k \cdot S_k}{\delta} \tag{2}$$

This correlation illustrates the importance of a preferably thin active layer. Diffusivity and solubility depend on the respective gas but also on the polymer.

The diffusion coefficients of gas molecules characterize their mobility in the polymer and diffusion coefficients generally increase as gas molecule sizes decrease.

The solubility describes the number of gas molecules sorbed in a polymer and solubility depends on the boiling point of the respective gas. The solubility rises with a rising boiling point, so that condensable gases generally have higher solubility than permanent gases.

In view of the influence of the polymer on diffusivity and solubility, the polymers can be divided into two groups characterized by their glass transition temperature: Polymers with a glass transition temperature lower than the operating temperature are in a rubbery state, ones with a glass transition temperature higher than the operating temperature are glassy. Diffusion effects dominate mass transfer through glassy polymers. Therefore, smaller molecules preferentially pass the membrane even if their solubility is relatively low. By contrast, for rubbery polymer membranes the mass transfer mainly depends on solubility.

Membranes in a glassy state generally have high permeabilities for water vapor, helium and hydrogen, but are less permeable for nitrogen, methane and higher hydrocarbons. In contrast, membranes of rubbery polymers have high permeabilities for organic solvents/vapors compared to the permeabilities of permanent gases and are suited for the separation of solvents from exhaust gases. Rubbery membranes also show high permeabilities for water vapor.

Selectivity is the characteristic of a membrane to differentiate among various components. The ideal selectivity, the maximum achievable selectivity of a membrane, for any two gases (*i* and *j*) is given by the pure gas permeance coefficient:

$$\alpha_{ij} = \frac{\Pi_i}{\Pi_j} \tag{3}$$

Where *i* is the faster permeating gas so that $\alpha_{ij} > 1$. Statistically the selectivity of a polymer rises with decreasing permeances. This effect was introduced by Robeson (1991) and theoretically described by Freeman (1999).

2.3.1 Influence of Gas Phase Pressure

The permeances are often assumed to be pressure independent for the first calculations with the solution-diffusion model. This assumption applies for ideal permanent gases in both glassy and rubbery polymers. In contrast, for gases which show real gas behavior, a considerable influence of the pressure on the permeances is visible.

Influenced by pressure-dependent solubility, the permeances increase progressively in rubbery membranes with increasing pressures, while the permeances in glassy membranes are reduced. This effect can be superposed by a second effect in glassy membranes. Plasticization increases the chain spacing in the polymer and increases the chain mobility so that the permeances increase. The changed permeances can also result in changed selectivities, which influence not only the required membrane area but also the separation performance (Yampolskii 2006).

2.3.2 Influence of Gas Phase Temperature

Since solubility and diffusivity depend on the temperature, the permeabilities of gases through polymers also depend on the temperature. The temperature dependency can be described by an Arrhenius type approach:

$$\Pi_k(T) = \Pi_{k0} \exp\left[-\frac{E}{R}\cdot\left(\frac{1}{T} - \frac{1}{T_0}\right)\right]; E = E_a + \Delta H_L \tag{4}$$

with E_a as the activation energy for diffusion and ΔH_L as the heat of the solution. Therewith, the permeabilities increase exponentially with rising

Table 5 Joule-Thomson coefficients at 30 °C and 10 bar (Scholz 2013)

	J-T coefficient (K/bar)
N_2	0.20
O_2	0.26
CH_4	0.42
CO_2	1.05
C_3H_6	1.95
C_3H_8	2.03

temperatures. For both glassy and rubbery membranes, selectivity decreases with increasing temperature.

For high pressure applications, the Joule-Thomson effect, influencing the operating temperature, has to be attended. Components, passing the membrane at high pressure difference and with a high Joule-Thomson coefficient, cause temperature reduction, which results in lower permeances and a changed module separation performance. Example for the pressure- and temperature-dependent Joule-Thomson coefficients of different gases is given in Table 5.

2.3.3 Concentration Polarization Effects

The membrane preferentially retains certain components on the feed side, which have to be transported back to the bulk stream. This diffusive transport results from a concentration gradient between the membrane surface and the bulk stream. Therefore, the concentration of the retained components at the membrane surface is higher than the concentration in the bulk stream and accordingly, the concentrations of the components preferentially passing through the membrane are lower at the membrane surface than in the bulk stream. This effect results in a reduced driving force for the favored permeating components and enhances the mass transfer of the retained gas components. Concentration polarization arises with higher fluxes through the membrane and higher selectivities. The effects of concentration polarization significantly influence the separation performance, when the permeability exceeds 1000 GPU with a selectivity higher than 100 (Mourgues 2005).

2.4 Concepts of Flow Control

The form of module structure allows different flow control for certain technical membrane modules. The feed stream can pass the module inside the hollow fibres or on the shell side in hollow fibre modules. Furthermore, the feed and permeate can flow in co-current, counter-current, cross flow, and ideal-mixed compartments.

For pressures higher than 15 bar, the feed has to flow on the shell side because of the pressure stability of the hollow fibres. Otherwise, the feed flow through the hollow fibre lumen is advantageous. In the lumen a defined flow is realized, whereas channeling often occurs on the shell side, which reduces the module separation performance. Furthermore, the influence of pressure losses is generally lower for applications with a lumen-side feed.

Counter-current flow is to predominate flow conduction in hollow fibre modules. Since not only the concentrations but also the pressure influences the driving force, counter current is not generally the superior flow conduction. If the influence of pressure losses on the driving force is higher than the influence of concentration changes, the parallel flow leads to a favorable driving force along the membrane.

In spiral wound modules the feed and permeate stream are led in cross flow whereupon the channels for feed and permeate are constructional predetermined. In envelope modules (plate and frame modules) the flow construction is a combination of co-current and counter-current flows.

3. PROCESS DESIGN

3.1 Pressure Ratio and Stage Cut

Next to selectivity as a membrane property, two operating parameters determine the separation performance of a gas separation system: the partial pressure ratio and the stage cut.

According to the mass transfer equation, a flux of a component i through the membrane only occurs if the partial pressure of component i on the feed side is higher than its partial pressure on the permeate side:

$$x_i p_F > y_i p_P. \tag{5}$$

Hence, the maximum separation which can be achieved by the membrane is:

$$\frac{y_i}{x_i} < \frac{p_F}{p_P} = \varphi \tag{6}$$

The enrichment in the permeate relative to the feed is always smaller than the feed-to-permeate pressure ratio, even at high membrane selectivities. Even though the permeate purity rises with increasing pressure ratios, maximum industrial applied pressure ratios are limited and depend on the individual process. Since large pressure ratios require expensive compressors or vacuum equipment and significant energy demands are required, the applied pressure ratio is usually between five and 20.

The stage cut is the fraction of the feed stream which permeates through the membrane:

$$\theta = \frac{\text{permeate flow rate}}{\text{feed flow rate}} \tag{7}$$

The stage cut has an influence on the trade-off between a high purified retentate stream and a high concentrated permeate stream. With a relatively low membrane area for a defined feed stream, low stage cuts are realized and the permeate stream is highly concentrated. However, only a small fraction of the feed stream is separated and the resulting retentate stream resembles the feed stream in flow rate and composition. In contrast, high retentate purity is achieved at high stage cuts. In this case, a high fraction of the feed stream permeates through a relatively high membrane area and the enrichment of fast permeating components in the permeate is insignificant.

These parameters illustrate the limit of a single module separation performance. A retentate-sided product can be concentrated to any purity by a single membrane module with any pressure ratio higher than one and selectivity. Yet, the achievable product recovery declines with increasing product purity. In contrast, the achievable purity of a permeate product in a single membrane module is limited by the membrane selectivity and the pressure ratio.

3.2 Driving Force Generation

The driving force for the mass transfer in gas permeation processes results from the partial pressure difference between the feed and permeate of the respective components. The partial pressure difference can be achieved by increasing the feed pressure by compression or decreasing the partial pressure on the permeate side by applying subambient pressure or by using a sweep gas (Figure 4).

A high pressure difference is adjustable by feed compression, whereby the required membrane area is reduced, while the energy demand increases. By expansion of the retentate stream, mechanical energy can be recovered. Furthermore, the capital cost for compression equipment is two-thirds to

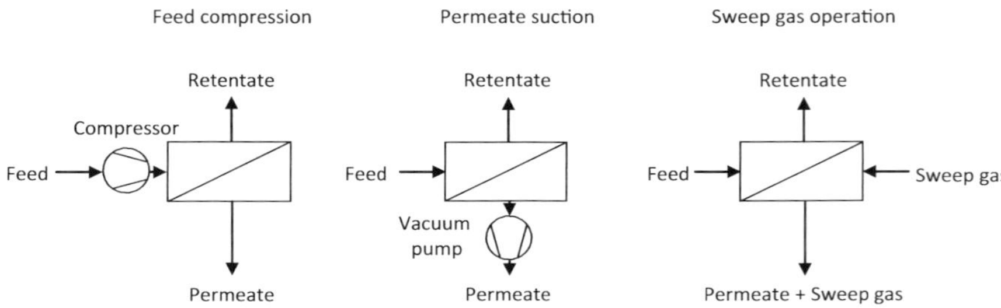

Figure 4 Different concepts of driving force generation (Follmann 2010).

half of the cost of vacuum equipment of the same power requirement (Merkel 2010).

For subambient pressure, not the whole feed stream but just the permeate stream has to be treated. Therefore, suction becomes more and more advantageous to compression with decreasing stage cut. Furthermore, the pressure ratio for a certain pressure difference and therewith the maximum separation applying suction is higher than using compression, but the maximum pressure difference is limited. Actually, for a vacuum in industrial gas separation applications a minimum permeate pressure of 200 mbar is realizable. Because of the limited pressure difference, higher membrane areas are required. Furthermore the efficiency of vacuum pumps is much lower than for compressors.

To generate the driving force by sweeping, the permeate partial pressure for the membrane passing component is reduced by diluting the permeate by an inert gas stream. Because membranes do not separate ideally, the intended sweep should be present in the feed mixture, too. If not, the sweep gas diffuses from the permeate side through the membrane to the feed and contaminates the retentate stream. The sweep potentially has to be separated from the permeate easily, which has to be taken into account for sweep selection. Condensable vapors are potential sweep streams.

Especially for sweeping, the achievable partial pressure difference does not provide sufficient driving force. Therefore, a combination of sweeping with feed gas compression, suction, or a combination of both is applied. The combination of feed compression and permeate suction can also be advantageous to profit from the benefits of both concepts (Follmann 2010).

3.3 Feed Pretreatment

The chemical, mechanical, and thermal membrane stabilities have to be ensured for long-term applications of industrial membrane processes.

Acidic sulfur-containing gases (H_2S, SO_2) easily react with the membrane components and cause destruction of the membrane. Water vapor can destroy the selective active layer as well. Moreover, water vapor can induce plasticization which significantly reduces the membrane selectivity. Contamination of the membrane surface and support layer with liquids, such as water or condensed hydrocarbons, reduces the mass transfer rate through the membrane by inducing additional resistance. Particles can physically damage the active layer or block the feed or permeate channels, in particular the thin hollow fibres have to be protected from particles. The operating temperature significantly influences the module separation performance and high temperatures even cause a thermal degradation of the membrane material.

To reduce the impact of the treated gas on the membrane module, a feed pretreatment has to be implemented. Applicable units in a pretreatment process are as follows:

- a coalescing filter, knockout drum or mist eliminator vessel for liquid elimination;
- an adsorbent guard bed for the removal of trace contaminants, such as hydrocarbons, H_2S, and SO_2;
- a particle filter for dust removal after the adsorbent bed; and
- a heat exchanger to set the desired feed temperature.

3.4 Module Interconnection

The membrane selectivity and the industrially applicable pressure ratio are limited. Therefore, a one-stage system may not provide the desired separation performance in view of purity and recovery so that several interconnected modules are applied. For module interconnection two different general concepts are used: In multistep processes the retentate is further treated in an additional module, while the permeate is further purified in multistage processes (Figure 5).

In a two-step process, the retentate is led into a second module. The permeate of this module is recycled in front of the compressor. The permeate of the first module and the retentate of the second module are the outlet streams of the process. A higher product recovery for a desired purity can be achieved, though the energy consumption for compression and the required membrane area increase.

In two-stage processes the permeate of the first module is recompressed which is fed into a second module. The permeate stream of the second stage module has a high quality as it is enriched in the first and in the second

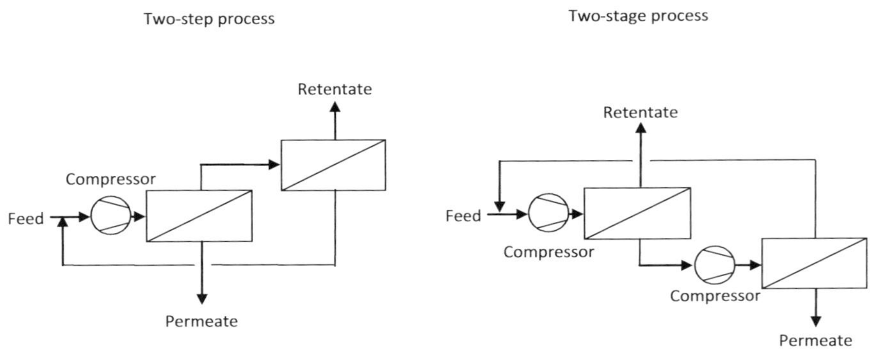

Figure 5 Membrane process design concepts.

modules. The retentate of the second stage is recycled to the first module inlet. A higher permeate purity compared to the one-stage process is realized, but an additional compressor and higher membrane area are required. Even several membrane stages can be realized theoretically, economically the number of stages is limited to two, exceptionally to three in industrial applications.

Combinations of a multistep and multistage design are possible, but they are only rarely used due to the higher complexity of the process and higher costs. Generally, several processes with module interconnection can be designed. The optimum process design and economically acceptable process complexity depend on the separation task.

Within a certain module interconnection, there are several parameters which influence the module separation performance and have to be taken into account. Next to specifying pressures and temperatures, the membrane material and membrane area of each module have to be determined. The use of various membrane materials in the different steps or stages might improve process separation performance. The partition of the membrane area in several identical constructed parallel lines potentially allows a dynamic adapting of the membrane area to a change of the feed flow by switching on and off the parallel lines.

Combining a membrane process with a conventional separation technology in a hybrid process may be attractive in using the advantages of the different technologies. The membrane process may outperform conventional technologies for bulk removal and enrichment, but for further separation to high purities (parts per million range) the conventional technologies may perform better.

3.5 Economic Parameters for Process Design

For the cost calculation of membrane processes, several full cost models are described in the literature, but the data collection is quite extensive. Since the main costs are caused by only a few components, a first rough appraisal gives a significant predication about the economics of the process. In the detailed engineering of the respective processes, the cost calculation can be further refined.

The main process costs for a gas separation process result from the driving force generation and from the membranes. For driving force generation both the investment costs for the compressor or the vacuum pump and also the operating costs for the required energy have to be included. For the calculation of the depreciation of the investment costs, the limited life of the membranes compared to that of the driving force equipment, has to be taken into account. While a maximum compressor life of up to 25 years can be assumed, membranes usually have to be replaced after 5 years at the latest (Calabrò 2011).

Generally, the costs for driving force generation increase with increasing pressure ratios, while membrane areas decrease. Calculating the total process costs, a minimum for one particular pressure ratio can be determined (Figure 6). Depending on the product value, it might be necessary to include the costs for product losses because of the incomplete recovery in determining the most economic operating point.

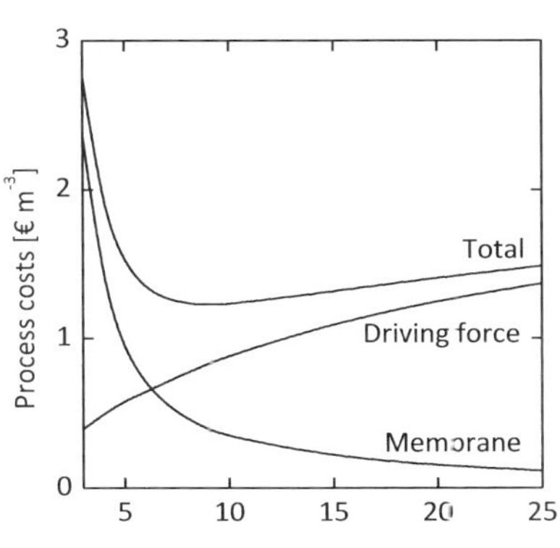

Figure 6 Example of the main costs for gas permeation processes as a function of the pressure difference.

4. APPLICATIONS OF GAS PERMEATION PROCESSES

Gas permeation processes with polymeric membranes are state-of-the-art processes for the separation of several gas mixtures. An overview of the industrial areas, where membrane processes are applied, is given in Table 6. Next to the challenge of further improving these membrane processes, membrane processes for H_2/CO_2 separation during hydrogen production and CO_2 removal from power plant flue gas are major research areas in which gas permeation processes are applied.

By means of nitrogen/oxygen separation and the treatment of biogas, the applications of membrane gas permeation processes are discussed in detail with regard to different membrane materials and process designs. The further applications, listed in the table, are described in the literature referred to in the introduction.

4.1 Oxygen/Nitrogen Separation

4.1.1 Nitrogen Enrichment

Meanwhile, gas permeation processes are industrially applied for inert gas production from air next to cryogenic processes and pressure swing adsorption (PSA). Membrane processes are the most economic process for nitrogen purification with low to medium flow rates up to 5000 m^3/h and nitrogen purities up to 99.5 Mol-%.

Since the product N_2 accumulates on the retentate side of the membrane, the desired purity can always be adjusted even with a single membrane module. However, with increasing N_2 purity, the achievable product recovery in a single-stage process is reduced. In N_2 production, N_2 recovery is less important as the raw material is air. The important parameters for process design and optimization are the required membrane area and the energy demand; though energy demand strongly depends on recovery, because a high fraction of the compressed feed stream permeates is useless to the permeate for lower recoveries.

In this chapter, the exemplary process design is discussed for three different membrane materials. PI is a membrane material with a high O_2/N_2 selectivity but low permeances, whereas PDMS is a high flux membrane with lower selectivities. Both the permeances and selectivities of PPO for O_2 and N_2 are in-between the values of PDMS and PI (Table 7). Since a high fraction of the feed stream has to permeate through the membrane to achieve the required purities, driving force generation by compression is favored. Sometimes, the compression is supplemented by a sweep stream

Table 6 Industrial application areas for gas permeation processes (Melin 2007)

Gas components	Field of application	Remarks and technical challenges
H_2/N_2	H_2-recovery during ammonia synthesis	Industrially applied, but condensable vapor has to be removed
H_2/CH_4	H_2-recovery during refinement	Industrially applied, but VOC vapors disturb (fouling, plasticization)
H_2/CO	Methanol synthesis gas adjustment	Industrially applied, but methanol has to be removed if necessary
O_2/N_2	Inert gas production	Industrially applied, for moderate amounts and purities (up to 5000 m^3/h and 99% N_2)
O_2/N_2	O_2-enriched air for oxidation processes or medical attendance	Up to 60% O_2 with polymer membranes achievable, (issue of economics)
CO_2/CH_4	Natural gas/bio gas treatment, heat value adjustment	Industrially applied, but precleaning necessary, higher separation factors for higher CH_4 recovery preferable
CO_2/CH_4	CH_4-recovery for enhanced oil recovery	Industrially applied, but precleaning necessary, higher separation factors for higher CH_4 recovery preferable
H_2O/CH_4 VOC/CH_4	Natural gas drying and/or separation of VOC	Applicable, but CH_4 losses are mostly too high
H_2O/air	Compressed air drying	Industrially applied, air losses caused by the internal sweep stream reduce economic efficiency
VOC/air	VOC/petrol vapor recovery	Industrially applied, concentration polarization and explosion control in permeate are problems
CH_4/N_2	Natural gas treatment at low gas qualities	Current available membranes do not show sufficient selectivity for acceptable CH_4 losses
He/VOC	He-recovery from natural gas	Applicable, but low feed concentrations require multistage processes
He/N_2	He-recovery from diving air mixtures	Applicable, but only a small market

Table 7 Membrane properties (Melin 2007)

Membrane material	Permeance O_2 (GPU)	Permeance N_2 (GPU)	Ideal selectivity (–)
Poyimide (Matrimid) (PI)	20	3	6.7
Polyphenylenoxide (PPO)	110	25	4.4
Polydimethylsiloxane (PDMS)	1185	590	2.0

by using a partial stream of the retentate to increase the driving force for oxygen. In this example, only compression as the driving force generator is investigated.

4.1.1.1 Single-step Process

First, the separation performance for the single-stage processes is discussed. The recoveries depend on the membrane selectivity. Low selective membranes, such as PDMS, achieve only very low recoveries, especially when high purities are required. Generally, the recovery decreases with increasing purities (Figure 7 left) and increases with an increasing pressure ratio between the feed and the permeate (Figure 7 right). The achievable product recovery is only depending on the membrane selectivity and the pressure ratio for single-stage processes.

Figure 8 shows the influence of the membrane material and the pressure ratio on the required membrane area as well as the energy demand. The required membrane area depends on the respective permeances. Implementing the high flux membranes, the lowest membrane areas are

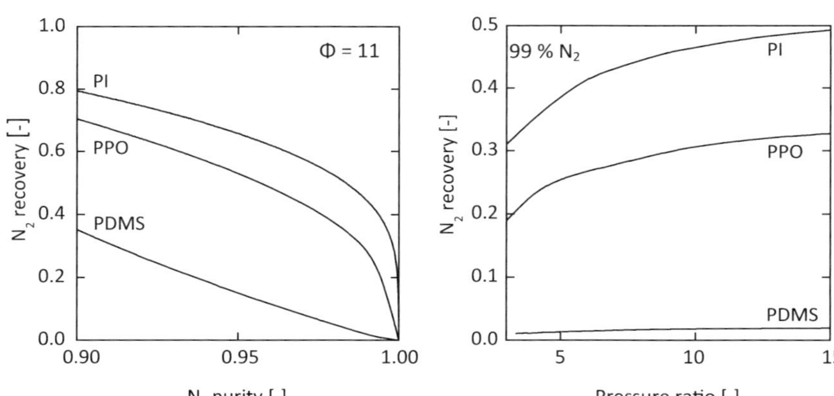

Figure 7 Recoveries with different polymer membranes for nitrogen enrichment.

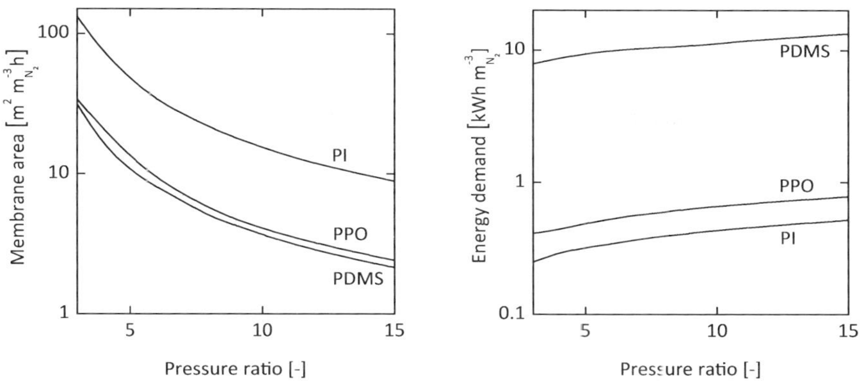

Figure 8 Specific energy demand and required membrane area for a fixed product purity of 99% N_2.

required, whereas low permeances demand greater membrane areas (Figure 8 left). With increasing pressure ratios, the required membrane is reduced. In contrast, the energy demand increases for higher pressure ratios. Thereby, the highest energy demand occurs for the material with the highest fluxes.

4.1.1.2 Multistep Processes

Since the achievable recovery for a defined retentate purity is limited in a one-step process, multistep processes may significantly improve the process performance in terms of energy demand and required membrane area. For the N_2-enrichment of air, two- and three-step processes were developed as given in Figure 9. The data represents the operating point of 11 bar in the feed. The same membrane area was implemented in all the process steps. The nitrogen fraction of the permeate streams is higher than that of the feed air for the modules in the second and third steps. By recycling these streams back to the feed, the oxygen fraction of the feed streams is reduced. This process modification results in a reduced membrane area and reduced energy demand. Compared to single-step processes, the membrane area and energy demand can be reduced by about 5% for the investigated PI membrane. This process improvement can be realized without additional costs because only a piping change is necessary. In contrast, for a further improvement of 2% by implementing a three-step process, a second compressor is necessary. Therefore, three-step processes are generally implemented economically in large systems, where the energy and membrane savings compensate for the higher process complexity and the costs of the second compressor (Baker 2004). Figure 10

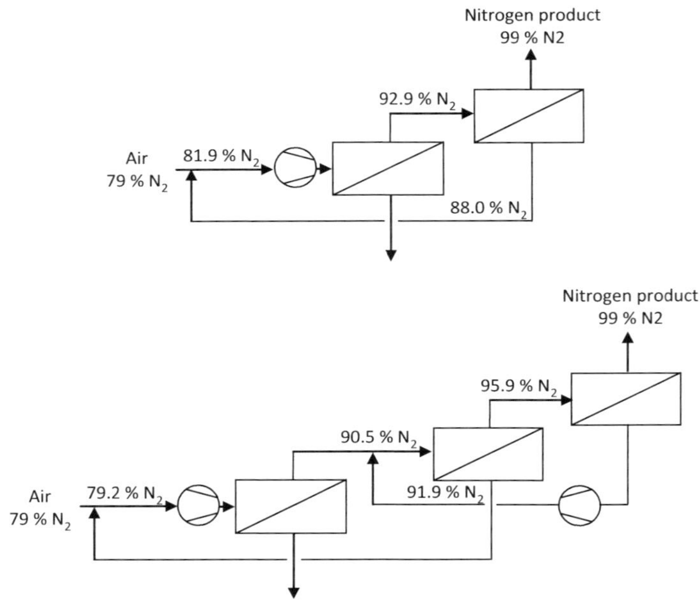

Design	Relative membrane area	Relative compression energy
One-Step	1	1
Two-Step	0.95	0.95
Three-Step	0.93	0.93

Figure 9 Multistep processes for N_2 enrichment (values for PI membrane with $p_{Feed} = 11$ bar). *(Based on Baker (2004)).*

shows the product recovery for the different process designs as a function of the product purity and the pressure ratio. For multistep processes, the product recovery is significantly increased compared to the one-step process. This effect is particularly important, when the feed is a valuable product and is not gratis, as the air in this example. Next to the membrane material and pressure ratio selection, the membrane area ratio between the respective process steps can be adapted for economic process optimization.

4.1.2 Oxygen Enrichment

In contrast to N_2 enrichment, with the application of membrane processes for the production of an O_2-enriched gas stream only limited O_2 purities can be achieved. Here, the permeate stream is the product stream. Since the O_2 fraction in the feed is only 21 Mol-%, the partial pressure difference as the driving force is always higher for N_2 than for O_2. In combination

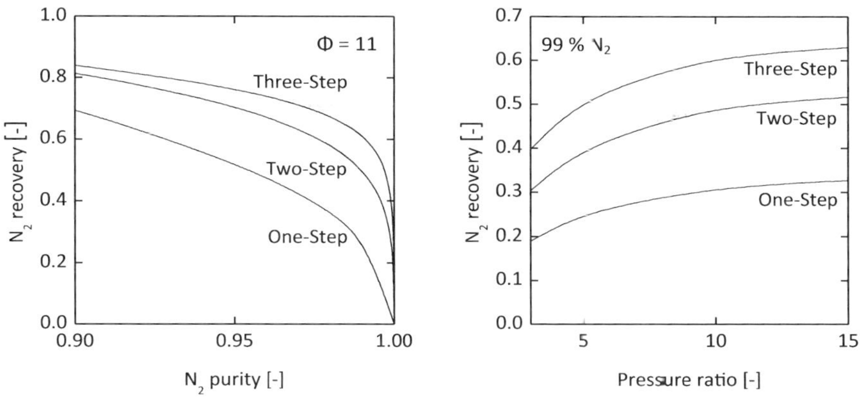

Figure 10 Product recovery of multistep processes compared to the single-step process.

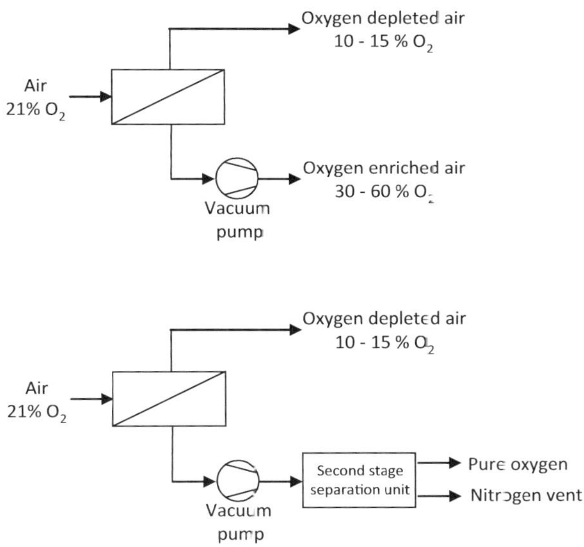

Figure 11 Membrane processes for the production of oxygen enriched air and pure oxygen. *(Baker 2002).*

with limited membrane selectivities, significant quantities of N_2 permeate through the membrane. Therefore, only O_2-enrichment and no O_2-purification is economically feasible in applying membrane processes (Figure 11).

For O_2-enrichment, a vacuum on the permeate side of the membrane is more favorable than compression of the feed, since only a small fraction of

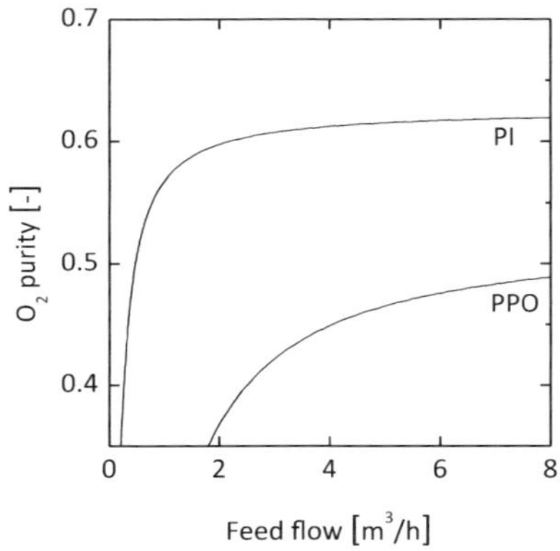

Figure 12 O_2 purity in a one-step membrane process.

the feed permeates through the membrane. The maximum pressure difference is limited, so that the required membrane area is comparatively high (Baker 2002).

The separation performance of a single-stage membrane process using PI and PPO membranes is shown in Figure 12. By increasing the feed flow, the O_2-depletion in the feed stream of the membrane module is reduced. Thus, the driving force for O_2 is kept higher resulting in higher O_2 purity in the permeate. For a minimum O_2 depletion a maximum O_2 purity is achieved. For the PPO membrane with a N_2/O_2 selectivity of 4.4, the maximum O_2 purity is around 50%, for the PI membrane with a N_2/O_2 selectivity of 6.3 the maximum purity is around 60%.

For higher O_2 purities, a hybrid process combining a membrane step with an additional purification step can be applied. For small units in which flow rates of up to 200 tons a day are treated, vacuum swing adsorption is used; cryogenic processes are the most economic processes for product flow rates of more than 200 tons a day. By air pretreatment using membrane units, the second stage unit is smaller and cheaper compared to the adsorption or cryogenic stand-alone units fed with air (Baker 2002).

4.2 Biogas Treatment

Raw biogas is mostly polluted with carbon dioxide and hydrogen sulfide. In order to use these energy sources, the gas has to be delivered to the central gas grid. Here, defined gas specifications in terms of accepted carbon dioxide and hydrogen sulfide as well as gas humidity have to be maintained.

Predominantly, the established processes, such as amine scrubbing, water scrubbing or PSA, are used for CO_2 removal. Nevertheless, the conventional process equipment shows following drawbacks:

- he processes are energy intensive.
- he CH_4 rich product gas is at low pressure.
- omplex systems involving several process steps are required to remove CO_2, H_2S as well as H_2O, hence, reducing the robustness of the system.
- Large equipment sizes.
- Associated materials such as water, amines, or activated carbon are required.

Next to the afore mentioned advantages of membrane processes in contrast to the conventional technologies, treatment additional significant advantages occur in the case of biogas. First, the product gas is already at natural gas grid pressure. Therefore, no additional compression unit is necessary for gas supply into the gas grid. Furthermore, not only can the CO_2 be separated by the membrane, but, since the trace components in the raw biogas, like hydrogen sulfide or water vapor, permeate faster through the membrane than CO_2, the membrane unit can separate CO_2, hydrogen sulfide and water in one step. Instead of three conventional units, only one membrane unit could be required which depends on the driving force for the trace components. However, the membrane stability for H_2S has to be ensured (Figure 13).

Especially for small to intermediate systems, the membrane process is economically superior to the conventional technologies. Therefore, the application of membrane gas separation is very promising for local biogas treatment. Instead of an extensive transport of biomass to a central biogas plant, the biomass can be used at source, for example at a farm. The resulting biogas can be purified on-site and directly delivered to the gas grid. The

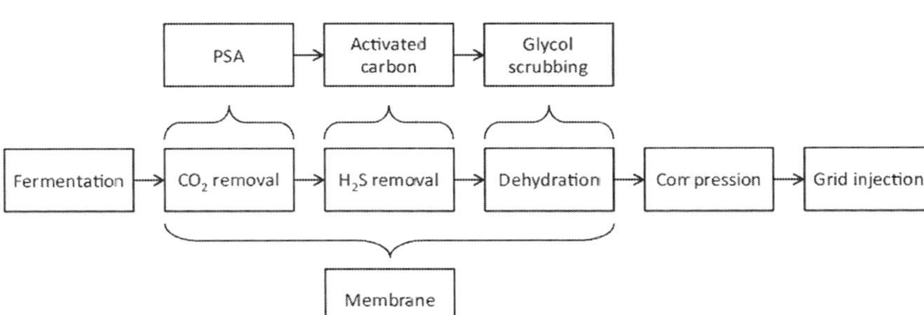

Figure 13 Process simplification by membrane unit implementation (Scholz 2013).

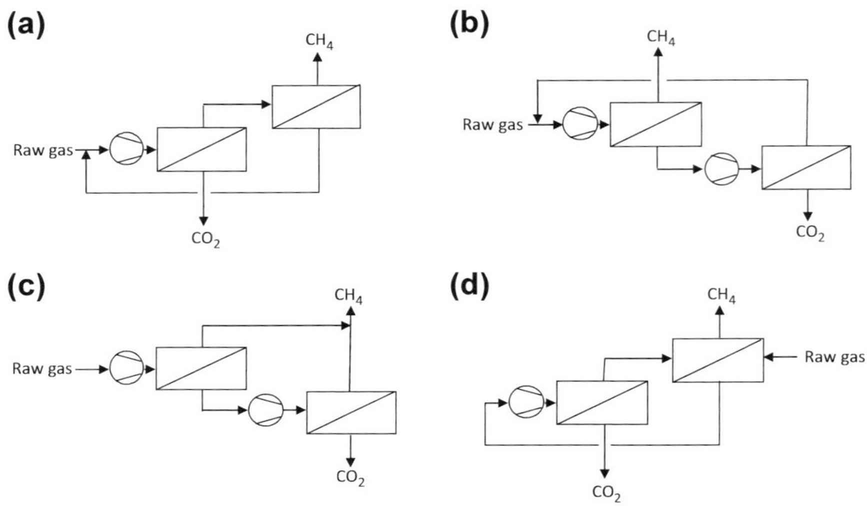

Figure 14 Membrane processes for biogas upgrading (Scholz 2013).

minimal maintenance and operator attention allows the operating of the unit without detailed technical know-how.

Several process designs are discussed in the literature for membrane supported biogas treatment. The selection of the most suitable process strongly depends on the feed composition and the required methane purities. Figure 14 shows a selection of the discussed process designs.

Since single-stage gas permeation processes are inefficient due to significant CH_4 losses, multistage gas permeation processes are applied to obtain high product purity and simultaneously increase the CH_4 recovery of the upgrading system.

Figure 14 shows different membrane stand-alone processes. In process (a), the first module is in charge of the CO_2 bulk removal. A relatively high CO_2 fraction is received in the permeate, but the product stream leaving this module does not comply with the gas grid requirements. In order to purify the CH_4 stream, a second membrane module is applied which controls the product purity. In this module, the CH_4 losses into the permeate are quite high. By recycling this stream in front of the compressor, the CH_4 recovery of the process is increased.

In process (b) the required product composition is obtained in one step. Next to the CO_2 and the trace components, a huge amount of CH_4 permeates through the membrane. Here, a second stage is added to reduce the CH_4 losses. The driving force for this module is generated by a second compressor. The CH_4-enriched retentate stream is recycled in front of the first compressor, the CO_2-rich permeate stream is led out of the process.

Process (c) is similar to process (b), but the retentate of the second stage is not recycled. Instead, the stream is also purified up to gas grid conditions and mixed with the retentate of the first module.

Process (d) has the same module connection as process (a). In addition to the driving force generated by the compressor, the raw biogas is used as a sweep stream in the second module. This sweep stream can be applied to increase the driving force for CO_2, when the CO_2 fraction in the permeate of the second module is higher than the CO_2 fraction in the raw gas.

Since the highest CH_4 losses occur in a membrane process when high product purity is required, hybrid processes can also be economic for biogas treatment. In general, two different arrangements for a hybrid process are possible: the retentate configuration and the permeate configuration.

In the retentate configuration (Figure 15(a)), the membrane unit is used for bulk removal. In the following absorption or adsorption unit, the required product purity is obtained. In contrast, the gas grid specification is obtained in the membrane module of the permeate configuration (Figure 15(b)). In a second unit, using a conventional separation technique, the permeate stream of the membrane unit is treated. A huge fraction of the CH_4 is recovered and mixed with the CH_4 rich retentate stream of the membrane unit (Scholz 2013).

The selection of a suitable process design should be carried out based on the given boundary conditions with regard to feed composition, feed volume and product specification.

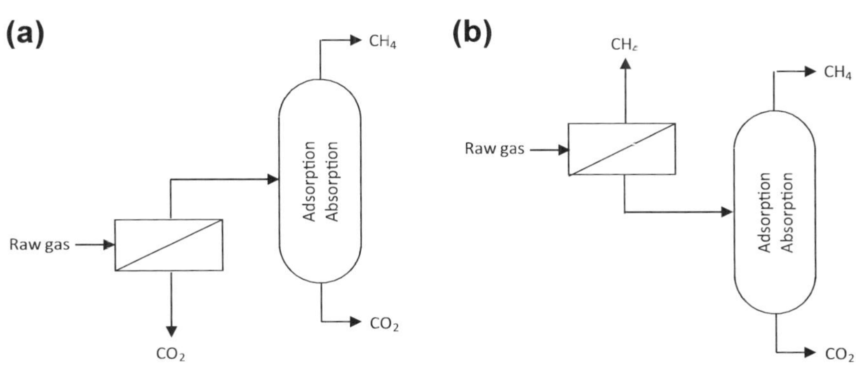

Figure 15 Hybrid process for biogas upgrading (Scholz 2013).

REFERENCES

Baker, R.W., 2002. Future directions of membrane gas separation technology. Ind. Eng. Chem. Res. 41 (6), 1393–1411.

Baker, R.W., 2004. Membrane Technology and Applications, second ed. John Wiley and Sons.

Calabrò, V., Basile, A., 2011. Economic analysis of membrane use in industrial application. In: Basile, A., Nunes, S.P. (Eds.), Advanced Membrane Science and Technology for Sustainable Energy and Enviromental Applications. Woodhead Publishing.

Drioli, E., Giorno, L., 2010. Comprehensive Membrane Science and Engineering. Elsevier.

Drioli, E., Barbieri, G., 2011. Membrane Engineering for the Treatment of Gases. Royal Society of Chemistry.

Favre, E., 2011. Polymeric membranes for gas separation. In: Drioli, E., Barbieri, G. (Eds.), Membrane Engineering for the Treatment of Gases. Royal Society of Chemistry.

Follmann, P.M., 2010. Membrane gas separation processes for post combustion CO_2 capture from coal fired power plants, dissertation.

Freeman, B.D., 1999. Basis of permeability/selectivity tradeoff relations in polymeric gas separation membranes. Macromolecules 32, 375–380.

Melin, T., Rautenbach, R., 2007. Membranverfahren – Grundlagen der Modul- und Anlagenauslegung. aktualisierte und erweiterte Auflage, third ed,. Springer.

Merkel, T., Lin, H., Wei, X., Baker, R.W., 2010. Power plant post-combustion carbon dioxide capture: an opportunity for membranes. J. Membr. Sci. 359, 126–139.

Mourgues, A., Sanchez, J., 2005. Theoretical analysis of concentration polarisation in membrane modules for gas separation with feed inside hollow fibres. J. Membr. Sci. 252, 133–144.

Mulder, J., 1996. Basic Principles of Membrane Technology, second ed. Springer.

Nunes, S.P., Peinemann, K.V., 2006. Membrane Technology in the Chemical Industry. revised and extended edition, second ed. WILEY-VCH.

Ohlrogge, K., Ebert, K., 2006. Membranen. WILEY-VCH.

Robeson, L.M., 1991. Correlation of separation factor versus permeability for polymeric membranes. J. Membr. Sci. 62, 165–185.

Scholz, M., Wessling, M., 2013. Transforming biogas into biomethane using membrane technology. Renew. Sustain. Energy Rev. 17, 199–212.

Yampolskii, Y., Pinnau, I., Freeman, B., 2006. Materials Science of Membranes for Gas and Vapor Separation. WILEY-VCH.

Yampolskii, Y., Freeman, B., 2010. Membrane Gas Separation. WILEY-VCH.

CHAPTER FOURTEEN

Surface Area: Brunauer–Emmett–Teller (BET)

Majid Naderi
Surface Measurement Systems, Ltd., Alperton, London, UK

Contents

1. INTRODUCTION

The measurement of the surface properties is of increasing importance to a wide range of materials. One of the most fundamental of these properties is the surface area available for adsorption of gas molecules. Surface area is the means through which a solid interacts with its surroundings, especially liquids and gases. Surface area can be created by particle size reduction, e.g., grinding and milling as well as making materials porous. Surface area

Progress in Filtration and Separation
http://dx.doi.org/10.1016/B978-0-12-384746-1.00014-8

may also be destroyed by high temperatures, e.g., melting. Surface area of a solid material is typically determined by physical adsorption of a gas on the surface of the solid and by calculating the amount of adsorbate gas corresponding to a monomolecular layer on the surface.

2. GAS–SOLID INTERFACE

Gas-solid sorption phenomena (i.e., surface adsorption, condensation in pores/tunnels and bulk sorption) often overlap and complicate the interpretation of the gas sorption studies. Moreover, the surface of a real solid material is often heterogeneous/anisotropic; therefore, one might encounter several sorption mechanisms in the same material concurrently. Surface adsorption is the accumulation of an adsorbate or a solute at an interface. At the vapour solid interface, the amount of adsorbate adsorbed on a surface (including irregularities and pore interiors) is described by the adsorption isotherm which is a function of the amount adsorbed versus the partial pressure at constant temperature. Hence, it is known that the tendency for an adsorption to occur is strongly dependent on the vapour pressure, temperature, and strength of the interfacial intermolecular interactions (Zografi, 1988).

2.1 Gas–Solid Interfacial Thermodynamic

In the thermodynamic description of interfaces, the interfacial excess volume V^σ (or surface phase thickness = 0) is zero, the Gibbs dividing plane is located at the solid surface, and the surface excess, Γ, can now be defined by:

$$\Gamma = \frac{N^\sigma}{A}$$

where N^σ is the number of moles adsorbed and A is the total surface area of the solid. The driving forces for adsorption are influenced by the various quantities (enthalpies, energies, and entropies) of adsorption. When the adsorption reaches equilibrium at a certain temperature and pressure, the chemical potential of the molecules in the gas phase is equal to that of the adsorbed molecules, therefore the Gibbs free energy of adsorption, ΔG_{ad} becomes zero.

$$\Delta G_{\text{ad}} = \mu^\sigma - \mu^g = 0$$

3. SURFACE ADSORPTION PHENOMENA

Surface adsorption can occur via two mechanisms: physisorption, and chemisorption (Adamson and Gast, 1997). They are distinct by the nature of the intermolecular attractions between the molecule and the surface, with the former being a physical interaction while the latter interacts via chemical bonding. The difference between these two types of adsorption is fundamentally arbitrary, and for cases exist, there are no absolutely clear distinction can be made between the two.

Physisorption involves relatively weak van der Waals forces and therefore it takes very little energy (1–2 kcal/mol) to remove physisorbed species from a solid surface. This type of adsorption exhibits relatively fast kinetics in the gas phase and all surfaces in high vacuum ($\sim 10^{-8}$ Pa or $\sim 10^{-10}$) can be considered to be free of physisorbed species. There is no significant redistribution of electron density in either the molecule or at the substrate surface, resulting in the association of water molecules in a quasiliquid layer on the substrate surface. Physisorption is therefore reversible and desorption can be induced by simply increasing the temperature and/or reducing the vapour pressure. Physisorption is also a nondissociative adsorption mode, capable of accommodating multilayer molecules. However, physisorption information is important, as it provides information about the material, from which surface area, pore size and pore size distribution can be derived.

On the other hand, chemisorption is associated with chemical valence forces – the stronger chemical interactions (10–100 kcal/mol) formed between the adsorbate and substrate. Hence it requires a great deal of energy comparable to those associated with chemical bond formation to remove the adsorbed species. The energy being a function of the solid surface to which, the adsorbing species attaches itself and the character of the adsorbing species as well (Trapnell, 1955). Chemisorbed molecules are linked to reactive parts of the surface and the adsorption process. There is substantial rearrangement of electron density, thus it is often dissociative and possibly an irreversible phenomena. Chemisorption is necessarily confined to merely a monolayer (Rouquerol et al., 1999), therefore it is useful for the characterization of catalyst surfaces, and is also the key mechanism of heterogeneous catalysis of chemical reactions.

3.1 Adsorption Isotherms

Several experimental adsorption isotherms can be distinguished depending on the physicochemical conditions of the interactions. Figure 1 depicts

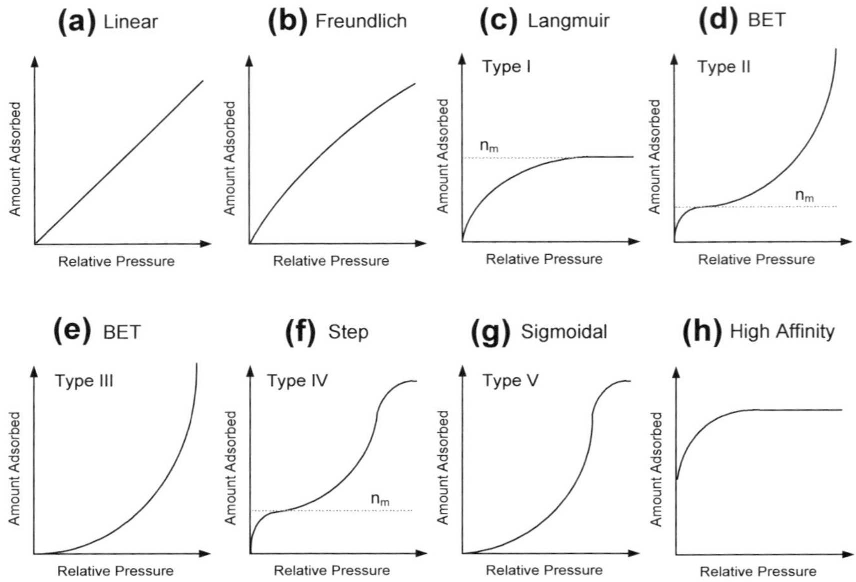

Figure 1 Schematic diagrams of eight commonly observed adsorption isotherms.

the eight commonly observed adsorption isotherms which can be distinguished based on the physicochemical conditions of the interactions (Sing et al., 1985; Brunauer et al., 1938). These include the five major classifications of adsorption isotherms (Type I–Type V) by Brunauer, Deming, Deming and Teller (BDDT) in 1940 (Brunauer et al., 1940). The BDDT classification has since become the core of the International Union of Pure and Applied Chemistry (IUPAC) classification of gas adsorption isotherms.

Adsorption isotherm in Figure 1(a) is described by Henry's Law, where Γ increases linearly with the partial pressure and is the ideal limiting case for low Γ:

$$\Gamma = K_H P$$

where K_H is the Henry constant.

The Freundlich adsorption isotherm (Freundlich, 1923) in Figure 1(b) is for heterogeneous surfaces with high and low adsorption affinity regions. The high affinity regions are first occupied, accounting for a steep rise at the front of the isotherm, followed by reduced adsorption affinity due to lateral repulsion between adsorbed molecules.

The Langmuir isotherm (Figure 1(c)) is a characteristic Type I isotherm in the BDDT classification. The isotherm portrays a rapid rise which approaches a maximum value asymptotically as the vapour pressure increases, indicating the completion of monolayer adsorption (Langmuir, 1918). The monolayer capacity, from an energetic point of view (Adolphs and Setzer, 1996), is defined as the amount of adsorbed molecules at the maximum of surface area interaction with the adsorbent. The equation shown below assumes that at equilibrium, the rate of adsorption and desorption are equal, regardless the occupancy of the neighbouring position. Every molecule coming from the gas phase to the first layer of molecule is elastically reflected. Therefore due to the negligible forces of attraction of the adsorbed molecules, multilayered molecule adsorption is not possible.

$$\theta_c = \frac{\mathrm{K_L} P}{1 + \mathrm{K_L} P}$$

With

$$\theta_c = \frac{\Gamma}{\Gamma_{\mathrm{mono}}}$$

where $\mathrm{K_L}$ is the Langmuir constant, θ_c is the surface coverage, and Γ_{mono} is the surface excess corresponding to a monolayer of molecules.

Though Langmuir isotherm behaviour is rarely observed for gases, porous materials can observe this type of adsorption as their pores become saturated. The limit of the uptake is governed by the accessible micropore volume instead of the internal surface area. It is therefore primarily for adsorption on microporous solids with relatively small external surfaces, for instance, activated carbon, salts of heteropoly acids, molecular sieve zeolites, and several porous oxides (Sing et al., 1985; Gregg et al., 1982).

Type II and III isotherms (Figure 1(d) and (e)) can both be described by the BET adsorption model 6, given as the equation:

$$\theta_c = \frac{n}{n_m} = \frac{c_{\mathrm{BET}} \cdot x}{[1 - x] \cdot [1 + x(c_{\mathrm{BET}} - 1)]}$$

With

$$x = \frac{P}{P_0}$$

where n is the amount adsorbed, n_m is the monolayer capacity, c_{BET} is the BET constant, and P/P_0 is the relative pressure.

Type II BET isotherm refers to reversible and unrestricted multilayer physical adsorption on nonporous (e.g., metal; alumina) or macroporous solids, displaying a point of inflexion which is attributed to the formation of a monolayer. After the monolayer adsorption, further increase in partial pressure result in extensive adsorption and followed by multilayer coverage.

Type III BET isotherm is characterized by the convexity towards the partial pressure axis, starting at the origin. It occurs in situations where interaction between the adsorbate molecules approaches that between adsorbate and adsorbent, i.e., heat of adsorption is similar to the heat of condensation. Therefore it is necessary to have significant partial pressure of adsorbate before the adsorption process commences. Having the surface covered with adsorbate, the favourable adsorbated–adsorbate interaction would then lead to a very rapid adsorption process, as the partial pressure increases. Clustering of the adsorbate onto the surface also results in this type of isotherm.

Type IV (Figure 1(f)) is analogous to Type II isotherm, having an inflection (or knee) as the monolayer formation. This isotherm is commonly exhibited by many mesoporous industrial adsorbents and inorganic oxide xerogels (Gregg et al., 1982), of which the monolayered surface coverage of the pore walls is followed by capillary condensation or pore filling. The adsorption ceases, once all the pores are completely filled.

Type V isotherm (Figure 1(g)) is commonly observed for flat, homogeneous adsorbents. The initial pathway of this isotherm is similar as in Type III. In this instance, the adsorbate preferentially interacts with the monolayer than the adsorbent surface, due to the lower heat of adsorption compared to the heat of liquefaction. A high affinity isotherm (Figure 1(h)) meanwhile is typical for very strong adsorption interactions.

Isotherms are useful for predicting the gas sorption properties, however, the BDDT and IUPAC classifications have a few deficiencies, as they give the inaccurate impression that adsorption isotherms are always monotonic functions of pressure, and they only consider adsorption at subcritical temperatures.

4. BET SURFACE AREA MEASUREMENTS

Surface area is best described as the external surface area of a solid object including surface attributable to pores. Gas adsorption provides a distinct advantage as many classical models for particle measurement and characterization fail to consider porosity. As mentioned earlier, physisorption is generally weak and reversible, the solid must be cooled and a method used to estimate the monolayer coverage from which surface area can be calculated. The area covered may be calculated by considering the amount of gas/vapour used to form the monolayer as well as the dimensions and the number of molecules.

4.1 BET Equation

Though there are a number of methods in use, which take into account the full range or part of the isotherm, by far the most prevalent and successful methods are based on the BET method for gas adsorption onto a solid surface. The BET equation, as shown below, is the most commonly used method to determine the monolayer and specific area values in various physicochemical areas.

$$\frac{x}{V(1-x)} = \frac{1}{V_m \cdot c_{\text{BET}}} + \frac{x \cdot (c_{\text{BET}} - 1)}{V_m \cdot c_{\text{BET}}}$$

where V is the volume of adsorbed molecules, V_m is the monolayer volume, c_{BET} is the BET constant, and x is the relative pressure (P/P_0). The BET constant c_{BET} arises from the algebraic rearrangement of the series approximation for the determination of subsequent ith layer adsorption volumes. c_{BET} is related to the adsorbate–adsorbent interaction strength, and so to the heat of adsorption. The higher the value of c_{BET}, the higher the interaction. In any case, BET equation is generally only used to give an apparent surface area related to the adsorption capacity of the solid.

A plot of $x/V(1-x)$ as the ordinate and x as the abscissa should give a straight line over a certain x (P/P_0) range. The data are considered acceptable if the correlation coefficient, r, of the linear regression is not less than 0.9975; that is, r^2 is not less than 0.995 (Specific Surface Area by Gas Adsorption). By determining the slope and the intercept of the resulting linear plot, V_m the amount of gas adsorbed if a monolayer was to form can be determined.

Slope

$$\text{Slope} = \frac{c_{\text{BET}} - 1}{V_m \cdot c_{\text{BET}}}$$

Intercept

$$\text{Intercept} = \frac{1}{V_m \cdot c_{\text{BET}}}$$

V_m (volume of monolayer)

$$V_m = \frac{1}{\text{Slope} + \text{Intercept}}$$

Specific surface area (ssa) can then be derived

$$\text{ssa} = \frac{V_m \cdot N_A \cdot a_m}{v_m \cdot m_s}$$

where N_A is the Avogadro's number (6.022×10^{23} mol^{-1}), a_m is the effective cross-section area of one adsorbed molecule, v_m is the molar volume of one adsorbed molecule (22,400 mL of volume occupied by 1 mol of adsorbate gas at standard condition), and m_s is the mass of substrate/adsorbent.

4.2 Multipoint Measurement

BET equation is a standard evaluation of monolayer values for adsorbate activity of between 5% and 35% P/P_0. This is mainly due to its simplicity and also the approval of IUPAC (Timmermann, 2003). A minimum of three data points is typically required along the linear regression, in order to determine the ssa of the sample. However, more data points may be conducted in cases where nonlinearity is observed at region of relative pressure close to 30% P/P_0. For the higher end of the relative pressure, the BET equation is only best fit for data obtained from conditions up to 50% P/P_0, as the prediction reaches infinity rapidly above 50% P/P_0.

4.3 Single Point Measurement

It may be acceptable to determine the ssa of a sample using a single value of V (volume of adsorbed molecules) on the isotherm. The BET single-point model is derived by assuming the BET intercept ($1/c_{\text{BET}}$) equals zero. This approximation becomes valid under certain circumstances of which the c_{BET} approaches infinity, or if c_{BET} is not infinite but may be invariant (Specific Surface Area by Gas Adsorption). The single-point assumption produces only small errors for materials with high c_{BET}, but causes large errors for materials with low c_{BET} for example, a common pharmaceutical lubricant – magnesium stearate (Andrés et al., 2001).

5. SAMPLE PREPARATION

Before the ssa of the sample can be determined, it is important to remove gases and vapours that may have become physically or chemically adsorbed onto the surface after manufacture and during treatment, handling and storage. If outgassing is not achieved, the ssa values measured can be low and may be variable because an indeterminate area of the surface might be covered with molecules of the previously adsorbed gases or vapours.

The outgassing conditions, defined by the temperature, pressure, and time, are critical for obtaining the reproducible and accurate ssa measurements.

Outgassing of many substances can be achieved by applying a vacuum or by purging the sample in a flowing stream of an inert, dry gas. In either case, elevated temperatures are sometimes applied to increase the rate at which the contaminants (preadsorbed molecules) leave the surface and remove chemisorbed species. Nonetheless, this has to be conducted without affecting the nature of the surface and the integrity of the sample, i.e., melting, dehydration, sintering, and decomposition.

To illustrate of the importance of pretreatment conditions, Magnesium stearate (MgSt) is an excellent case study. MgSt, which is a waxy, lamellar (platey) solid, is the most widely used solid excipient in pharmaceutical industry. One of the least understood physical properties, and exactly how to measure it, is surface area as moisture content and method of drying has huge effect on the BET surface area. The removal of tightly bound water at 110 °C significantly increases surface area. Table 1 shows the surface area values for MgSt at different background humidities. The surface area drops as percentage relative humidity (% RH) increases.

The adsorption method of BET assumes a simple physical adsorption mechanism where the surface sites are populated homogeneously and there are no adsorbate–adsorbate interactions. Therefore in principle the equation can be applied to any adsorption system where this holds true. Historically, this work has been typically undertaken using inert gases such as nitrogen, argon, and krypton. For the samples to adsorb at least a monolayer of these inert species, the experimental temperatures must be close to their boiling point (i.e., 77 K for nitrogen adsorption).

In practice this is most likely to be applicable to nonpolar, alkane molecules with boiling points between 350 and 400 K. Highly polar adsorbates such as water tend to have very strong adsorbate–adsorbate interactions and may adsorb preferentially at specific adsorption sites, therefore in this case the BET equation should be applied with caution, as the physical significance of the calculated surface areas will be highly

Table 1 Surface area values for magnesium stearate at different background % (relative humidity) RH

Background RH (%)	BET S.A. (m^2/g)	R-squared
0	4.08	0.999
35	3.31	0.999
70	3.16	0.999
0 (110 °C drying)	5.95	0.999

suspect. There can be no doubt in the universality or importance of the nitrogen BET method at low temperatures, however, in recent years, the rapid development of newer dynamic gravimetric methods for BET analysis offers interesting options compared with the traditional approach.

6. VOLUMETRIC GAS ADSORPTION TECHNIQUE

In the volumetric method, the recommended adsorbate gas is nitrogen, which is admitted into the evacuated space above the previously powder sample to give a defined equilibrium pressure, *P* of the gas. The use of a diluent gas, such as helium, is therefore unnecessary, although helium may be employed for other purposes, such as to measure the dead volume. Since only pure adsorbate gas is used, instead of a gas mixture, interfering effects of thermal diffusion are avoided in this method (Figure 2).

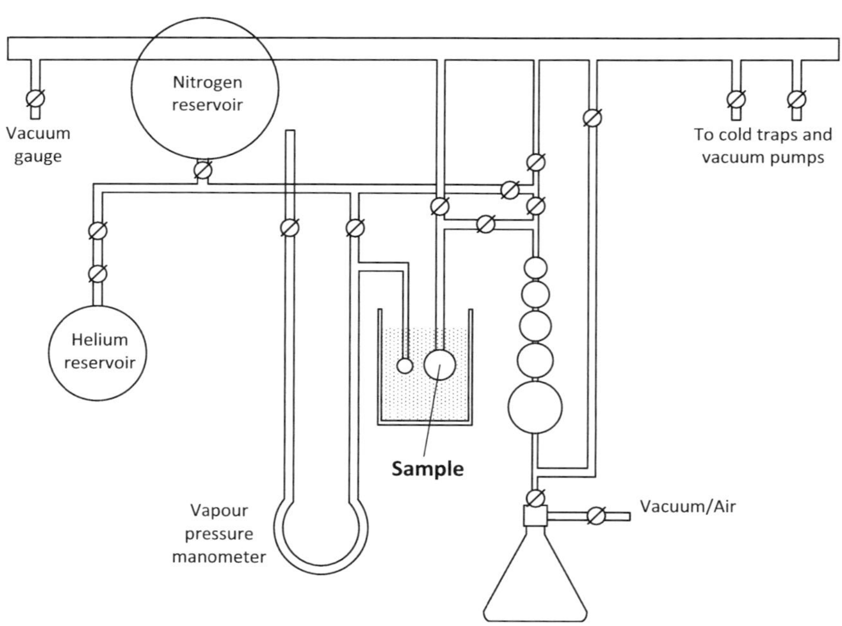

Figure 2 Schematic diagram of the volumetric method apparatus. European Pharmacopeia for measuring the BET surface area by nitrogen gas adsorption. *(Specific Surface Area by Gas Adsorption)*

6.1 Adsorbates

The most common adsorbate used is nitrogen; however, other adsorbates are used in some circumstances. A list and properties of common adsorbates used in volumetric BET experiments are given in Table 2.

For materials of low ssa ($<0.2\ m^2/g$), the proportion adsorbed is low. In such cases, krypton at liquid nitrogen temperature is preferred because the low vapour pressure exerted by this gas greatly reduces error. The use of larger sample quantities, where feasible (equivalent to $1\ m^2$ or greater total surface area using nitrogen), may compensate for the errors in determining low surface areas. All gases used must be free from moisture.

Total surface of the sample is at least $1\ m^2$ when the adsorbate is nitrogen and $0.5\ m^2$ when the adsorbate is krypton. Lower quantities of sample may be used after appropriate validation.

6.2 Advantages and Limitations

The measured surface area values may be influenced by not allowing sufficient drying equilibrium time at different points. The method would not accurately predict the multilayer adsorption behaviour above $P/P_0 = 0.5$ (*capillary condensation*/pore filling with liquid adsorbate). The best straight line through the data should be used by discarding under-equilibriated data points, which are too low or too high in P/P_0. A suitable range of P/P_0 values should be selected for linearity. Figure 3 shows the

Table 2 Adsorbates used in volumetric methods for determining BET surface area

Gas	Temp (°C)	α factor $\times 10^5$ (1/mm Hg)	Cross-sectiona area (Å^2/mol)	Molecular weight (g/mol)
Ar	−195.8	11.4	14.2	39.948
	−183	3.94		
CO_2	−78	2.75	19.5	44.01
	0	1.75		
	25	1.55		
CO	−183	3.42	16.3	28.01
N_2	−195.8	6.58	16.2	28.0134
	−183	3.78		
O_2	−183	4.17		
C_4H_{10}	0	14.2	46.9	58.12
	25	4.21		

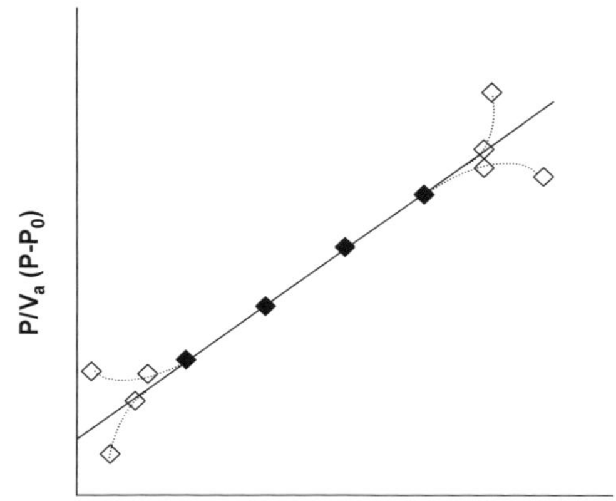

Figure 3 Gradual increase or decrease in slope in the (Brunauer–Emmett–Teller) BET plot.

gradual decrease or increase in slope in the BET plot, at too low or too high pressures.

Samples must be exposed to very low pressures and temperatures, which would completely dry most pharmaceutical hydrates and cause fragile crystals to collapse. Krypton adsorption would be used for surface areas below $0.5 \, m^2/g$. The surface area values calculated from nitrogen and krypton adsorption may differ by a factor of six (Andrés et al., 2001).

7. GRAVIMETRIC DYNAMIC VAPOUR SORPTION TECHNIQUE

A comparison of volumetric and gravimetric techniques is given below:

Gravimetric dynamic vapour sorption (DVS) is a well-established method for the determination of vapour sorption isotherms. It is based on a very accurate gravimetric system, which allows one to follow the adsorption and desorption of extremely small amounts of probe molecule. Sorption experiments can be undertaken with water and organic vapours. The benefit of using nonpolar vapours for the determination of surface areas at ambient temperatures is the pure, dispersive interaction, whereas water can interact in different ways, which makes it less reliable as a probe molecule. DVS experiments on a series of model compounds were reported by Marshall and Cook, 1994. Although vacuum techniques have been well established for studies of porous and inorganic materials (O'Hanlon, 1989), they are not ideal for materials with low surface areas. In addition, the ambient pressure dynamic flow technique would offer a relatively smaller sample size to minimize the adsorbate diffusion path length and speed up vapour sorption equilibrium, accurate control of vapour concentration, dynamic flow of vapour minimized vapour concentration gradients within the sample chamber and optimal heat transfer into and out of the sample compared to vacuum methods. Table 3 gives a comparison between static volumetric and dynamic gravimetric sorption techniques for BET surface area determination.

7.1 Principle

The classical BET Eqn (1) is given below where x is the partial vapour pressure of vapour above the surface and V is the amount of vapour adsorbed.

$$\frac{1}{V}\frac{x}{1-x} = \frac{c-1}{cV_m}x + \frac{1}{cV_m} \qquad (1)$$

A plot of $(1/V)[x/1-x]$ as the ordinate and x as the abscissa should give a straight line. By determining the slope and the intercept of this line, V_{m} the amount of gas adsorbed if a monolayer was to form and c a constant related to the adsorbate–adsorbent interaction strength can be determined.

Table 3 Comparison of volumetric and gravimetric (Brunauer–Emmett–Teller) BET surface area determination

	Traditional volumetric BET	**Gravimetric BET**
Adsorbing species	Nitrogen, argon, krypton	Liquid vapours at 300 K
Temperature	Always at very low T, i.e., 77 K	Can be undertaken at ambient T
Sample size	Typically 1 g	Typically 100 mg
Surface area	Gives the surface area seen by small molecules, i.e., N_2 (limited access to materials with surface areas smaller than 0.5 m^2/g)	Gives the surface area as seen by a "real world" molecule
Experimental conditions	Low temperatures and vacuum	Ambient temperature and pressures

The BET equation assumes a simple physical adsorption mechanism where the surface sites are populated homogeneously and there are no adsorbate–adsorbate interactions. Therefore, in principle the equation can be applied to any adsorption system where this holds true. In practice this is most likely to be applicable to alkane molecules with boiling points between 350 and 400 K. Highly polar adsorbates such as water tend to have very strong adsorbate–adsorbate interactions and may adsorb preferentially at specific adsorption sites, therefore, in this case the BET equation should be applied with caution, as the physical significance of the calculated surface areas will be highly suspect.

7.2 Sample Preparation and Experimental Procedures

The samples were analysed on a DVS automated moisture sorption instrument at 25 °C with sample sizes of 50–90 mg for the analysis. The samples were initially dried for 300 min under a continuous flow of air to establish the dry mass. The samples were then exposed to the following typical partial pressure profile: 0–95% RH in 5% steps.

At 25 °C, the DVS-Advantage instrument has a working concentration range of 0–98% P/P_0; a temperature stability of 0.1 °C; and up to 10 vapour cycles on the same sample can be programmed per experiment. Additionally, multiple experiments on the same sample can be preprogrammed to run in sequence. For instance, experiments could be run over several temperatures. Or, experiments switching between two preloaded solvents can be run without any further user interface. The DVS Advantage can measure and actively control both water and organic vapour concentrations in real time. Further, the temperature range for the DVS Advantage is 5–60 °C. Also, the DVS-Advantage has a sample preheater which allows heating the sample in situ up to 200 °C. In addition to water, a wide range of organic solvents can be used. The exhaust organic vapours from the system were vented to a fume extraction system for safety considerations.

Table 4 below shows a list of typical DVS adsorbates with their molecular size, dispersive component of their surface energy, and their chemical character.

Isotherms were measured on α-Lactose monohydrate in the partial pressure range 0–1.0 using *n-octane* at 25.0 °C.

Figure 4 shows isotherm data for the adsorption of octane on α-Lactose monohydrate. The data is plotted as percent change in mass referenced to the mass after drying in “dry” air (<0.1%RH). The isotherm shows typical type II/IV adsorption shape and hysteresis between sorption and desorption

Table 4 Properties of (dynamic vapour sorption) DVS adsorbates

	Molecular size Å^2	Dispersive surface energy mN/m (at 25 °C)	Specific characteristic
Hexane	51.5	18.4	Neutral
Heptane	57.0	20.3	Neutral
Octane	62.8	21.3	Neutral
Nonane	68.9	22.7	Neutral
Chloroform	44.0	25.0	Acidic
Carbon tetrachloride	46	26.8	Acidic
Acetone	42.5	16.5	Basic
Ethyl acetate	48.0	19.6	Basic
Isopropyl alcohol	44[b]	19.8	Amphoteric
Water	10.4–12.5[a]	21.8	Amphoteric
Nitrogen	16.2		Neutral

[a]Depends on strength of interaction and thus surface packing dependent.
[b]Calculated from bulk properties.

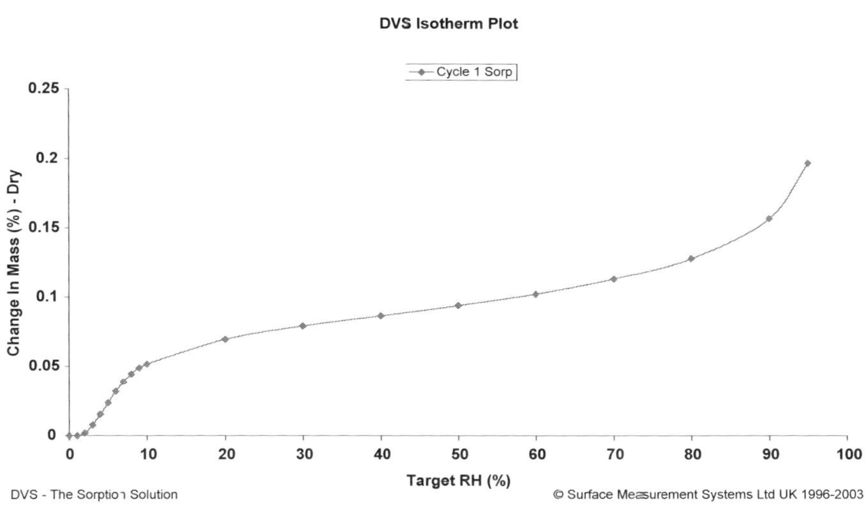

Figure 4 Sorption isotherm plot of octane on lactose.

is minimal. The fact that the uptake is relatively low (<0.1% at saturation partial pressure) together with fast sorption kinetics indicates a surface-only sorption mechanism. Therefore, the BET model may be appropriately applied in this case.

Analysis of the experimental results using the BET equation gave a good straight line data fit shown in Figure 5, over the partial pressure range

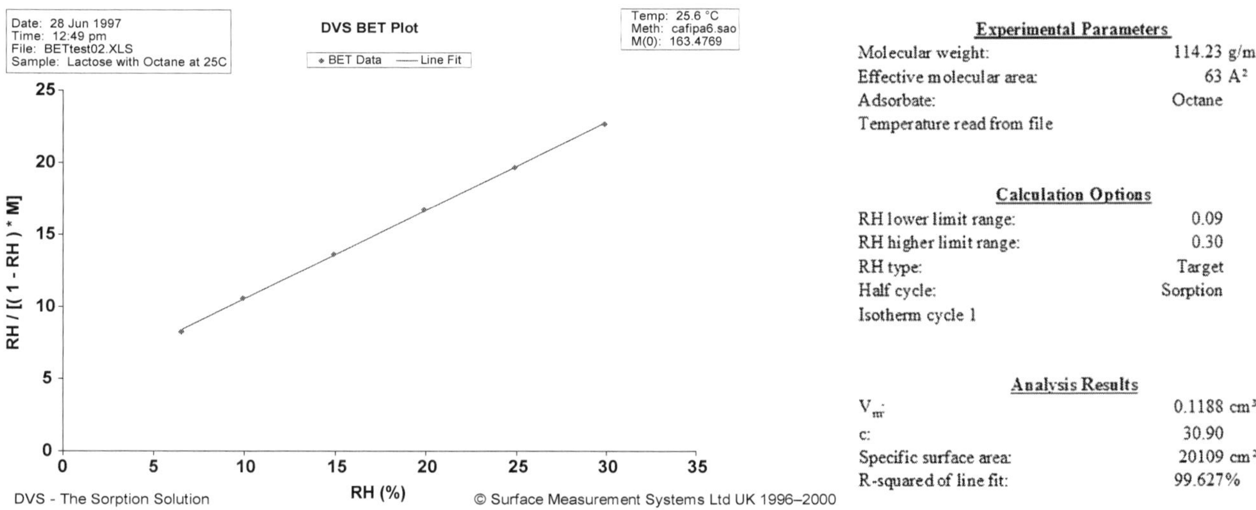

Figure 5 (Brunauer–Emmett–Teller) BET analysis for lactose sample.

0.05–0.3. The surface area obtained of 0.25 m^2/g using octane is consistent with published nitrogen surface area values of between 0.2 and 0.8 m^2/g for this material.

The results were in good coincidence with other methods such as inverse gas chromatography (IGC) and volumetric nitrogen measurements. The dynamic gravimetric BET methods offer new opportunities for studying powder and particle surface area properties when combined with traditional nitrogen BET methods. This includes the possibility of studying small sample sizes at ambient temperature and pressure with a wide range of adsorbing vapour species. In addition, the sensitivity of the ultra-microbalance gravimetric technique also allows relatively low surface areas to be measured.

7.3 Advantages and Limitations

Although the experiments can be performed at ambient temperature and ambient pressure and background humidity may be applied while measuring second isotherm, the most reliable results are obtained by using P/P_0 values between 0.10 and 0.35. Also, the calculations assume only surface adsorption. If a probe that has significant bulk absorption is used, then the results may not be reliable. Therefore, the use of this method for polymeric type materials where bulk absorption can be significant is limited. The method works best with highly crystalline materials or "rigid" inorganic materials (i.e., alumina, silica, minerals, glass fibres, etc.).

Although the use of polar probe molecules such as water is not recommended they may still give the user some valuable information. The values would not be indicative of total surface area. The values may give some information on the water sorption capacity which is not the same as BET surface area. With food materials, water will most likely give a Type III isotherm and be dominated by bulk absorption.

Other limitations include small mass changes for low surface area materials and some probes may not access all surface due to steric effects, e.g., microporous materials.

8. CHROMATOGRAPHIC ADSORPTION TECHNIQUE

For practical purposes an investigation at ambient temperatures or above is of greater interest since elevated temperatures are more relevant in industrial processes and quite often, material behaviour varies with temperature. Testing at ambient temperatures also permits the use of various gases and vapours for the measurement, whereas experiments at 77 K are restricted to just a few probe molecules.

A convenient method for the study of sorption under the desired conditions is IGC. IGC was demonstrated in various papers as a quick method to determine isotherms at finite concentration using organic probe molecules at ambient temperatures. From these isotherms BET surface areas and pore size distribution functions can be derived (Baumgarten and Weinstrauch, 1977; Roginskii, 1960). Due to its sensitivity IGC proves especially beneficial in the determination of small surface areas.

8.1 Principle

IGC has been demonstrated in various papers as a quick method to determine isotherms at finite concentration and ambient temperatures, using organic probe molecules (Condor and Young, 1979). A unique injection mechanism and variable injection pulse sizes provide major improvement in allowing the BET region, within the partial pressure range of 5–35% P/P_0 of the isotherm, to be obtained more accurately, especially in the case of materials with small surface areas.

Generally IGC measurements can be configured in two ways: as an elution or a frontal experiment. In a frontal experiment the probe molecule is added continuously to the carrier gas, whereas a pulse measurement involves a single injection of a certain amount of vapour. The vapour is transported by the carrier gas through the IGC to the column. The amount adsorbed from the sample in the column is eluted by the carrier gas and in the ideal case an equilibrium state is reached. In the case of a frontal experiment a breakthrough curve is observed. A pulse experiment results in a peak, the shape of which depends strongly on the shape of the solid-vapour sorption isotherm (Figure 6).

In the case of infinite dilution a symmetrical (Gaussian) peak is observed representing a linear (Henry) isotherm. At high concentration (finite dilution) tailing or leading will occur. In the case of a type I, II or IV isotherm there is a tailing because adsorbent/adsorbate interactions are much stronger than adsorbate/adsorbate interactions. This is usually the case when nonpolar probe molecules adsorb on solid surfaces. Thus, sorption isotherms in this paper were calculated based on the method of Cremer and Huber (Cremer and Huber, 1962).

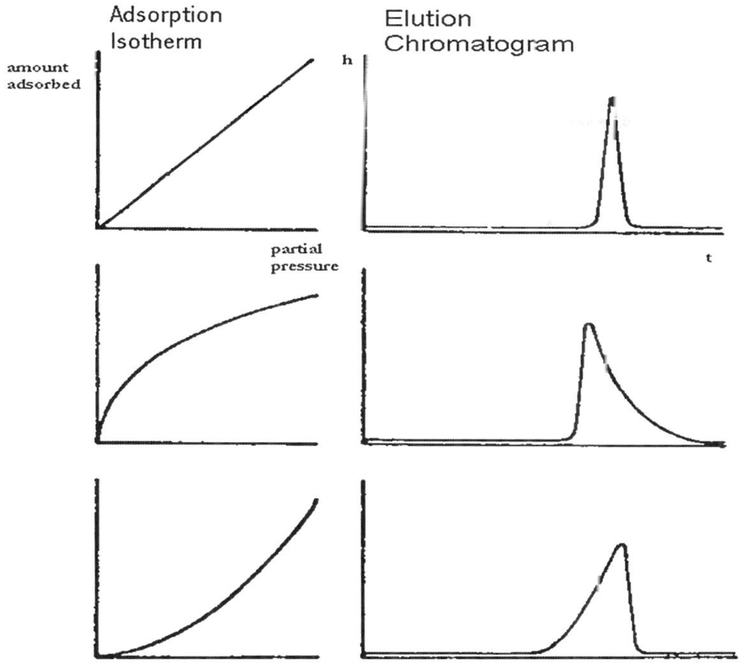

Figure 6 Correlation of peak form and sorption isotherm for finite and infinite dilution.

In this calculation partial pressures are obtained from the peak height h and the net retention volume V_N from the retention time.

$$V_N = j \cdot w/m \cdot (t_c - t_0) \cdot T_C/273 \tag{1}$$

In this equation w is the carrier gas flow rate, m the sample mass, T_C the column temperature, t_c the gross retention time, t_0 the dead time and j the James-Martin correction for the pressure drop across the column. The net retention volume is divided by the gas constant and the column temperature to obtain the pressure related retention volume.

The partial pressure is obtained from the peak height by equation

$$p = h/E \tag{2}$$

where E is the conversion factor calculated according to Equation 3.

$$E = F \cdot w \cdot T_{\text{loop}} / \left(V_{\text{loop}} \cdot p_{\text{sat}} \cdot F/P_0 \cdot 273\right) \tag{3}$$

In Equation 3 F is the area under the peak, p_{sat} the saturation pressure of the gas probe molecule, and V_{loop} the volume of the gas probe loop.

A plot of the pressure related retention volume versus the partial pressure represents the first derivation of the isotherm (with the pulse method only desorption data is available for these types of isotherms). The retention volumes and partial pressures can either be obtained from the maxima of peaks at different concentrations (Peak Maximum method) or from the tailing of a high concentration peak (Elution by a Characteristic Point, ECP method). Graphical or numerical integration gives the desorption isotherm in both cases.

8.2 Sample Preparation and Experimental Procedures

The ssas of two nonporous alumina Reference BET Standards Certified Reference Material 170 and 171 (CRM 170 and CRM 171) as well as hydrophobic glass beads (Supelco) were determined by measuring the octane adsorption isotherms at 30 °C and 0% RH. Alumina CRM 170 and CRM 171 are standard BET reference materials with surface areas of $1.05\ m^2/g$ and $2.95\ m^2/g$, respectively. The mean particle sizes of 9–13 μm for glass beads and $\sim$7 μm for alumina were obtained by laser diffraction. The BET ssas of the samples were subsequently calculated from the corresponding octane isotherms, within the partial pressure range of 5–35% P/P_0.

For the isotherm determination, presilanised glass columns with 30 cm length and a 4 mm ID were filled with 1 g of glass beads, 500 mg of CRM 170 and 200 mg of CRM 171 by gentle tapping. Each column was preconditioned for 2 h at 30 °C and 0% RH with helium carrier gas to remove any physisorbed water. All experiments were carried out at 30 °C with 10 sccm total flow rate of helium. All IGC Surface Energy Analyzer (SEA) analyses were carried out using IGC SEA, and the data were analysed using both standard and advanced SEA analysis software. Table 5 summarizes the BET ssas of the samples obtained by IGC SEA and their reference surface area values.

Table 5 The BET specific surface areas of the samples obtained by IGC SEA and their reference surface area values

	IGC SEA surface area (m^2/g) ($R^2 = 0.9999$)	**Reference surface area values (m^2/g)**
Glass beads	0.55	0.50
Reference CRM 170	1.01	1.05 ± 0.05
Reference CRM 171	2.95	2.95 ± 0.13

IGC, inverse gas chromatography; SEA, Surface Energy Analyzer; BET, Brunauer-Emmett-Teller.

Figure 7 shows isotherm data for the adsorption of octane on of non-porous alumina Reference BET Standards CRM 170 and CRM 171 (Commission of the European Communities Community Bureau of Reference Material Alumina No 170 and 171). The isotherms show a significant uptake at low partial pressures followed by relatively smaller adsorption at intermediate vapour concentrations, which is typical type II/IV isotherm.

The fact that the uptake is relatively low (<0.1% at saturation partial pressure) together with fast sorption kinetics indicates a surface-only sorption mechanism. Therefore the BET model may be appropriately applied in this case.

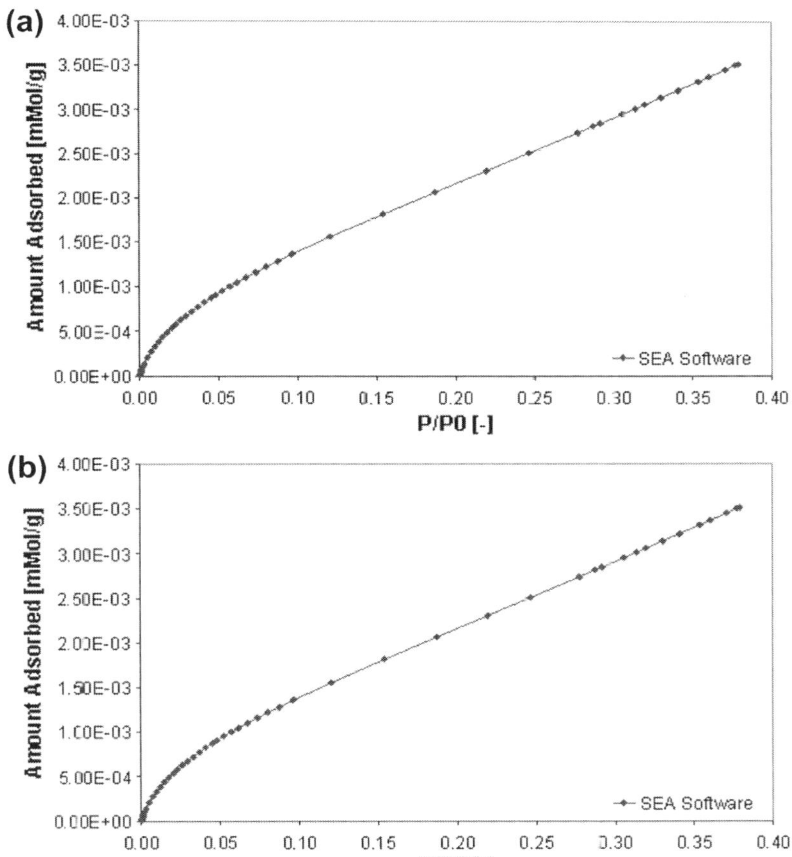

Figure 7 Sorption isotherm plots of octane on (a) CRM 170 and (b) CRM 171 at 30 °C.

Analysis of the experimental results using the BET equation gave a good straight line data fit shown in Fig. 8, over the partial pressure range 0.05–0.3. The BET linearized plots over the partial pressure range 5–35% for the standard CRM 170 and CRM 171 reference materials are shown in Figure 8(a) and (b), respectively. The corresponding surface area values for the standard CRM 170 and CRM 171 reference materials were found to be 1.0125 and 2.9524 m^2/g, which are in good agreement with the published nitrogen surface area values of 1.05 and 2.95 m^2/g, respectively.

Porosity information is often obtained based on BET isotherm of nitrogen. For some materials it can be difficult to differentiate between surface adsorption and micropore filling, although pore size distributions can be

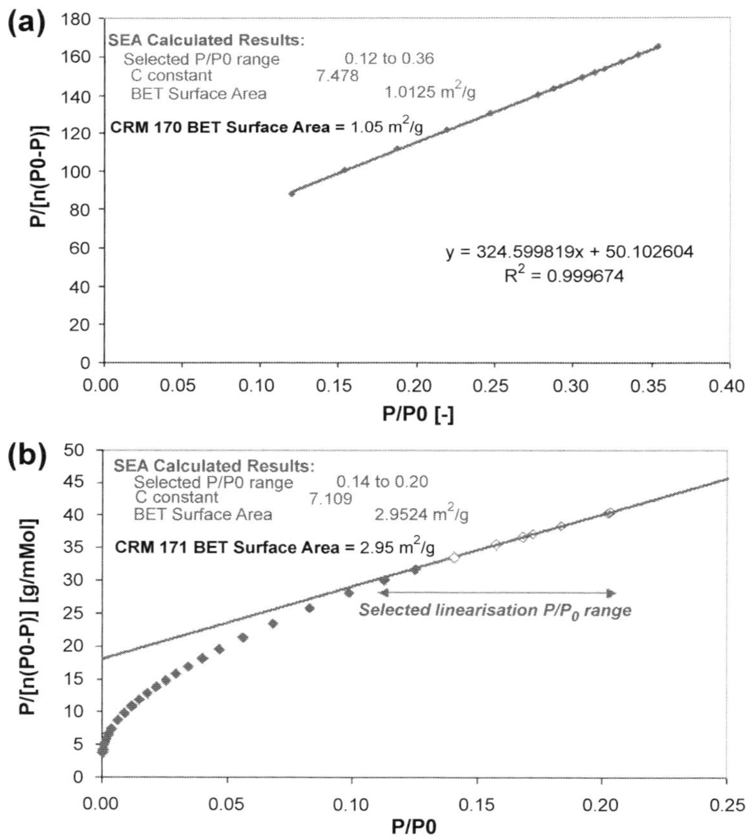

Figure 8 *BET analysis for* the standard (a) CRM 170 and (b) CRM 171 reference materials *at 30 °C.*

estimated from BJH type equations (Brunauer et al., 1940). However, IGC can overcome this problem. After equilibrium adsorption of cyclohexane under conditions of high p/p_0 and low temperatures would be attained, the samples would then undergo a rapid thermal desorption process. Since surface adsorption and pore filling mechanisms are driven by different heats of interactions, the difference in time for cyclohexanes molecules desorbing from the sample bed is indicative of the strength of their interaction.

8.3 Advantages and Limitations

As for the Dynamic Gravimetric Sorption Technique although the experiments can be performed at ambient temperature and ambient pressure and background humidity may be applied while measuring second isotherm, some probes may not access all surface due to steric effects, e.g., microporous materials. The use of polar probe molecules such as water is not recommended due to cluster formation rather than monolayer formation. In situ conditioning allows for direct measure of thermal and RH sample history.

9. SUMMARY

The dynamic gravimetric BET methods offer new opportunities for studying powder and particle surface area properties when combined with traditional nitrogen BET methods. This includes the possibility of studying small sample sizes at ambient temperature and pressure with a wide range of adsorbing vapour species. In addition, the sensitivity of the ultra-microbalance gravimetric technique also allows relatively low surface areas to be measured.

IGC SEA has shown to be an accurate technique for the determination of isotherms and derived parameters such as BET surface areas for alumina standards, allowing low surface areas to be measured. Results were in good agreement with nitrogen sorption experiments demonstrating excellent reproducibility of the data as well as the possibility of studying small sample sizes at ambient temperature and pressure with a wide range of adsorbing vapour species.

REFERENCES

Adamson, A.W., Gast, A.P., 1997. Physical Chemistry of Surfaces. Wiley-Interscience, New York.

Adolphs, J., Setzer, M.J., 1996. A model to describe adsorption isotherms. J. Colloid Interface Sci. 180, 70–76.

Andres, C., Bracconi, P., Pourcelot, Y., 2001. On the difficulty of assessing the specific surface area of magnesium stearate. Int. J. Pharm. 218, 153–163.

Baumgarten, E., Weinstrauch, 1977. J. Chromatogr. 138, 347.

Brunauer, S., Emmett, P.H., Teller, E., 1938. Adsorption of gases in multimolecular layers. J. Am. Chem. Soc. 60 (2), 309–319.

Brunauer, S., Deming, L.S., Deming, W.E., Teller, E., 1940. On a theory of the van der Waals adsorption of gases. J. Am. Chem. Soc. 62 (7), 1723–1732.

Condor, J.R., Young, C.L., 1979. Physicochemical Measurement by Gas Chromatography. John Wiley and Sons, Chichester.

Cremer, E., Huber, H., 1962. Gas. Chromatogr. Intern. Symp. 3, 169.

Freundlich, H., 1923. Kapillarchemie. Akademische Verlagsgesellschaft, Leipzig, Germany.

Gregg, S.J., Sing, K.S.W., 1982. Adsorption, Surface Area and Porosity, second ed. Academic Press Inc, London.

Langmuir, I., 1918. The adsorption of gases on plane surfaces of glass, mica and platinum. J. Am. Chem. Soc. 40 (9), 1361–1403.

Marshall, P.V., Cook, P.A., 1994. A New Analytical Technique for Characterising the Water Vapour Sorption Properties of Powders. International Symposium on Solid Oral Dosage forms. The Swedish Academy of Pharmaceutical Sciences.

O'Hanlon, J.F., 1989. A User's Guide to Vacuum Technology. John Wiley and Sons, New York.

Roginskii, S., 1960. Dokl. Phys. Chem. 133, 717.

Rouquerol, F., Rouquerol, J., Sing, K.S.W., 1999. Adsorption by Powders and Porous Solids. Academic Press, London.

Sing, K.S.W., Everett, D.H., Haul, R.A.W., Moscou, L., Pierotti, R.A., Rouquerol, J., Siemieniewska, T., 1985. Reporting physisorption data for gas/solid systems with special reference to the determination of surface area and porosity (recommendations). Pure Appl. Chem. 57, 603–619.

Specific surface area by Gas adsorption. The European Pharmacopoeia.

Specific Surface Area by Gas Adsorption. In: European Pharmacopoeia, fifth ed. pp. 2811–2814.

Timmermann, E.O., 2003. Multilayer sorption parameters: BET or GAB values? Colloids Surf. A: Physicochem. Eng. Asp. 220, 235–260.

Trapnell, B.M.W., 1955. Chemisorption. Academic Press Inc, New York.

Zografi, G., 1988. States of water associated with solids. Drug Dev. Ind. Pharm. 14 (14), 1905–1926.

CHAPTER FIFTEEN

Particle Shape Characterization by Image Analysis

Marie-Noëlle Pons[1] and John Dodds[2]

[1]Laboratoire Réactions et Génie des Procédés, Université de Lorraine, Nancy cedex France
[2]Professeur honoraire, Ecole des Mines d'Albi, France

Contents

NOMENCLATURE

a	straight line slope
A	projected area (pixel)
A_c	area of the convex bounding polygon (pixel)
b	straight line origin
C	circularity
D_{eq}	equivalent diameter (pixel, length, and unit)
D_f	fractal perimetric dimension
d_{max}	maximal distance (pixel)
E, E'	aspect ratios
$\widehat{e}$	estimated breadth (pixel)
F_{max}	maximal Feret diameter (pixel)

Progress in Filtration and Separation
http://dx.doi.org/10.1016/B978-0-12-384746-1.00015-X

F_{min}	minimum Feret diameter (pixel, length, and unit)
IC_s	surfacic concacity index
k	prefractal factor
l	box side length (pixel)
$\widehat{L}$	estimated length (pixel)
LCI	largest concavity index
N	number of boxes
P	perimeter (pixel)
P_c	Crofton perimeter (pixel)
r	rectangularity
R	radius (pixel)
Rob	robustness
x_i	pixel i position according to image rows
y_i	pixel i position according to image lines
θ	angle (radian)
ω_i	order of morphological erosions
ρ	straight line parameter in polar coordinates
σ	standard deviation

1. INTRODUCTION

It is now well established that particle size analysis is important for understanding the behaviour of powders and suspensions. But it is also realized that many of the physical properties of finely divided systems do not just depend on the size of particles but also on their shape. Clearly a suspension of spherical particles will behave differently from a suspension of needle-like particles. In this case simple methods, usually based on ratios of characteristic sizes often measured manually, have until recently been the only practical means of quantifying particle shape. However, more subtle differences in behaviour due to particle shape may not be revealed by such methods. Even though observation by electronic and optical microscopy or video can distinguish differences visually it is can be difficult to quantify such observations.

This state of affairs has changed radically in the last few years due to the rapid development of image analysis, itself benefitting from recent advances in computation power, software and imaging methods. This has lead to establishing new concepts and methods for describing the shape of particles and new investigations into the practical application for describing the properties of particle systems based on quantification of shape and the structure of particle agglomerates.

Particle shape and size are two concepts which are intrinsically connected. Simple objects such as a sphere or a cube are the only objects whose size can be characterized by a single parameter: their diameter or side length.

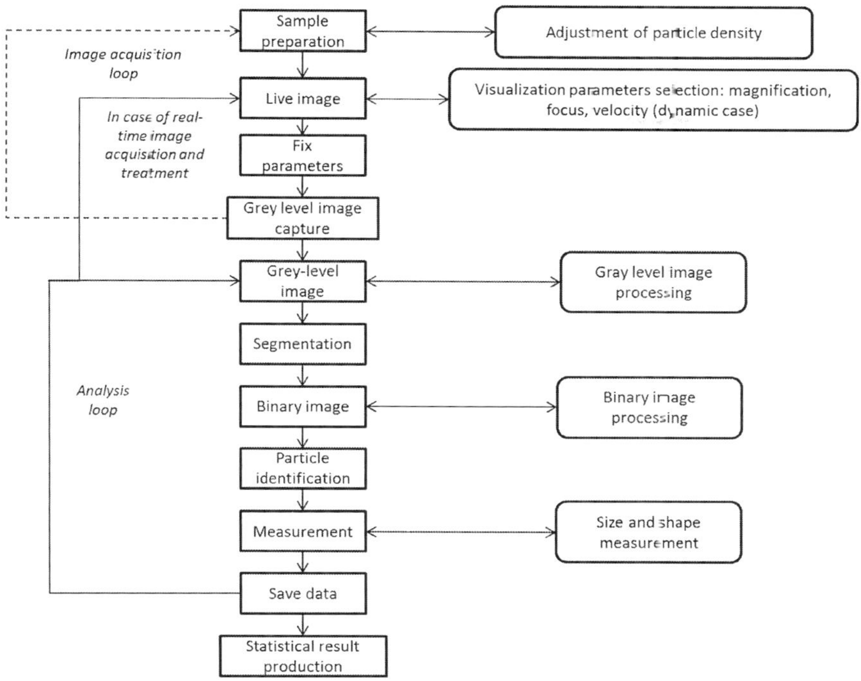

Figure 1 Image acquisition and analysis general flowchart.

For any other object some assumptions are necessary about its shape to provide information on its size. To make these assumptions the best way is to look at the object. Depending upon the size range, various visualization tools can be used. The development of digital image capture and its subsequent automated analysis has been favoured by the progress in computer science (hardware and software). It has however to be recognized that an image is a two-dimensional space, which will introduce a bias for the determination of the three-dimensional shape of particles. For deformable particles such as paper fibres, flocs, or fungal hyphae, the apparent shape depends even more upon the preparation of sample for visualization.

Figure 1 summarizes the different steps to be accomplished whatever the image acquisition might be, whether static or dynamic

2. IMAGE ACQUISITION

Both static and dynamic image acquisition can be used. In static visualization, usually preferred for very small particles, the particles will be disposed on a surface such as a glass slide. Visualization devices range from the transmission electron microscope (TEM) for the very smallest particles to the flatbed scanner for quite large objects. Sample preparation and the subsequent imaging is a key step, which should be performed with the highest care to ensure the best final results in terms of shape and size quantification. Contacts between particles should be avoided, although they are not always easy to detect, especially for particles with irregular shapes such as twinned crystals or agglomerates. Dry dispersion in a chamber under controlled conditions is an efficient technique, but fragile particles could be broken. Dispersion in a viscous liquid (gelatine, sucrose, etc.) in which the particles do not dissolve is another option: a drop of the suspension is deposited on a glass slide and a coverslip is placed on the top to spread the drop. With such preparation methods the particles will be displayed in their most stable mechanical position and it is the shape of their silhouette which will be assessed, rather than their real three-dimensional shape. The use of a sticky surface, such as double-side carbon tape for scanning electron microscopy preparation, may lead to other positions. For optically transparent particles, the addition of a liquid with an adequate refractive index can improve the visualization of their contours (Belaroui et al., 2002), which will be very helpful during the image analysis step.

Although automated image capture has been developed with fully controlled microscope stages, static image acquisition is often judged cumbersome and time-limiting by users. Each slide should be scanned without frame overlapping and without selecting the "best" particles. Several slides may be necessary to get a meaningful number of particles.

Several dynamic image acquisition systems, for particles in motion, have been proposed (ISO 13322-2, 2006). Such systems are convenient for analyzing a large number of particles (Figure 2). Sheath-flow cells can ensure that the particles are well in focus: elongated particles and platelets tend to align along the direction of flow. An image treatment should be used to remove out of focus particles in agitation or circulation cells. In a free-falling system the particles are fed on the focus plane. For the three latter systems, it may be assumed that there is a random positioning with respect to the shape. The speed of motion should be properly selected with respect to the exposure time to avoid blurring. Particles on a pneumatic belt lie in their most stable mechanical position as in the static acquisition mode. Gao et al. (2012) developed an illumination system based on three solid-state lasers to increase the size of the field of view on a conveyor belt.

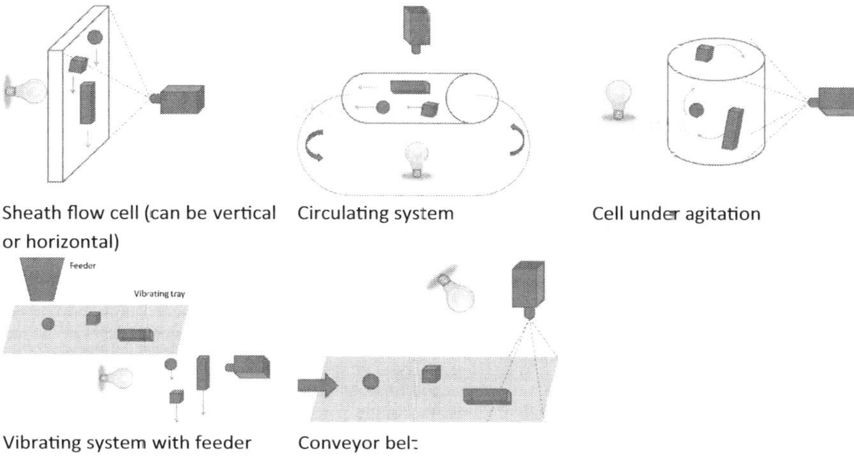

Figure 2 Dynamic image acquisition systems.

In situ microscopes have been developed for real-time visualization in crystallizers (Calderon de Anda et al., 2005; Bluma et al., 2009; Wang et al., 2008; Zhou et al., 2009), liquid/liquid contactors (Maaβ et al., 2011) as well as in bioreactors. In the latter case applications are mainly for yeast cells (Bittner et al., 1998; Akin et al., 2011; Belini et al., 2013) and mammalian cells, either in suspension (Joeris et al., 2002; Guez et al., 2010) or attached to microcarriers (Rudolph et al., 2007). Low-cost in situ imaging using a medical endoscope has been tested for crystallization monitoring (Simon et al., 2012). In all these cases, the main difficulty is to avoid the superposition in the image plane of objects of interest. It is also recommended that the magnification be chosen so that the maximum size parameter (diameter, length) of the largest particle does not exceed one-third of the image height (ISO 13322-2, 2006). Unlike manual image acquisition it is more likely that particles are not fully in view when using dynamic image acquisition, or automated static image acquisition. To avoid missing particles, Liao et al. (2010) have developed an on-line full scan inspection system based on a line scan CCD (Charge-coupled device) camera: the comparison of two successive stored frames enables to retrieve the particles in contact with the frame border.

Technically a digital image is a large matrix. The value taken by each of its elements corresponds to the intensity of the light signal captured by an electronic device on a small surface picture element or pixel of the field of view. For size and shape characterization monochrome 8-bit images (i.e. with 256 grey levels) are sufficient. Except when real time is a necessity (reactor control, for example) image acquisition and image analysis can be decoupled.

3. IMAGE TREATMENT

All the examples of image analysis and treatment used here have been obtained using the image analysis software Visilog version 6.9 (FEI Visualization Sciences Group, Mérignac, France).

The first step of image treatment for size and shape characterization is segmentation, which consists in discriminating the silhouette of the objects of interest from the background. After this binarization, the image grey levels will be reduced to two, with value of 0 for the background and one for the objects. The quality of the image background (presence of a halo or uneven illumination) can complicate the procedure which is not unique and may also depend on the operations offered by the image analysis software. The case of opaque particles is usually simpler, as their contour is better defined. Segmentation should be automated as far as possible so as to eliminate interaction with the operator and to allow for analysis of large series of images. Image cleaning is then performed to eliminate small artifacts (presence of dust or debris, particles of interest but in contact with the image borders or too small to be correctly analyzed later on) and to fill the silhouette of transparent objects. The particles finally retained after this process are then individually identified. Figure 3 shows the critical steps of such a treatment applied to NaCl crystals:

- Grey-level image (Figure 3(a))
- Binary image (Figure 3(b)): an automatic segmentation algorithm has been applied, the selection of the threshold being based on a variance criterion. The idea is to divide the grey-level histogram into two regions, one for the objects of interest and one for the background, the limit between them (i.e. the threshold) being selected so the sum of the variances of the two regions is minimized (Figure 4). In the present case the threshold is equal to 24.
- Hole-filling (Figure 3(c)): to fill any hole within the particle outline. This step is particularly useful for transparent particles such as the sugar crystals depicted in Figure 5.

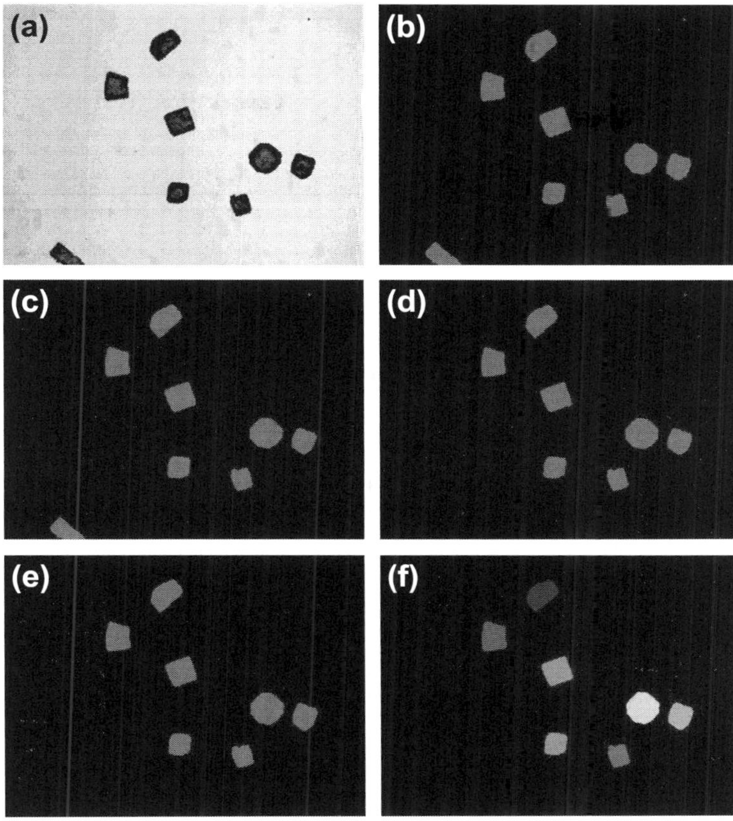

Figure 3 Basic image analysis treatment (a) Grey-level image, (b) Binary image, (c) Hole-filling, (d) Border-killing, (e) Noise elimination, and (f) Particle labelling.

- Border-killing (Figure 3(c)): to remove particles which are in contact with the image frame and which may, therefore, not be fully in view.
- Noise elimination (i.e. dust particles or debris the user does not want to consider further) (Figure 3(c)): a series of morphological erosions is applied first, the number of erosions depending upon the size of the object to be removed. Then the image is reconstructed using as seed particles what remain after the series of erosions.
- Particle labelling: each particle silhouette is identified and given a label. From this step, image analysis (i.e. size and shape characterization) can start.

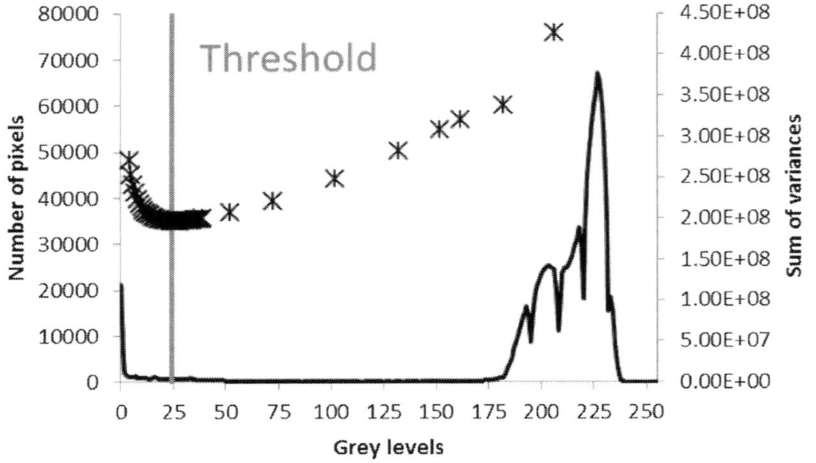

Figure 4 Grey-level histogram of Figure 3(a) image (black line), sum of variances (*), and threshold (red line) (For interpretation of the references to colour in this figure legend, the reader is referred to the online version of this book).

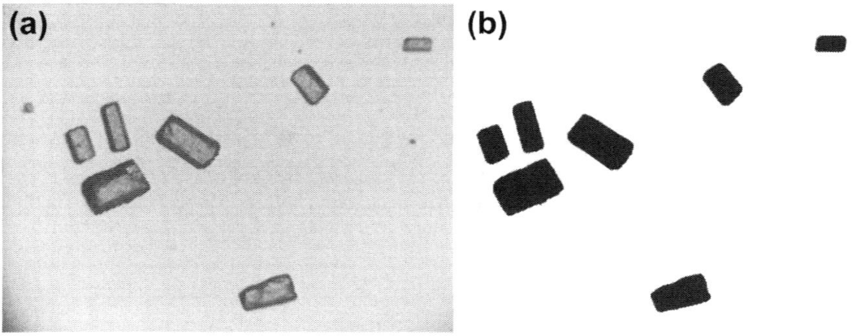

Figure 5 (a) Grey-level image of transparent particles (sugar crystals) and (b) Final binary image.

4. BASIC SIZE DESCRIPTORS

With image analysis size (and shape) characterization is always done in terms of pixels, without considering the real size of a pixel. This means that it can be applied to many different fields of activity, irrespective of the object size range. The conversion of pixels into real-world dimensions should be made prior to reporting results.

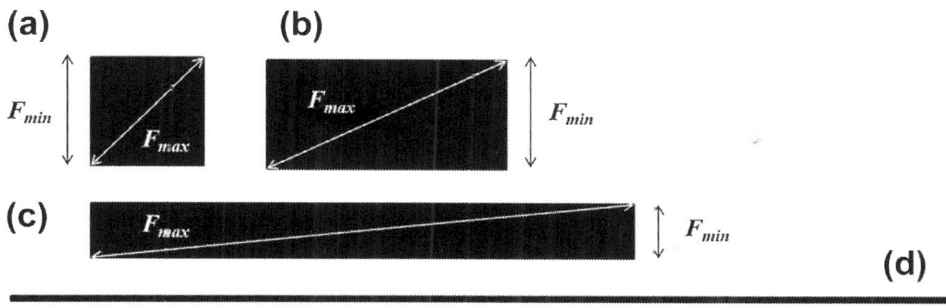

Figure 6 Meaning of maximum (F_{max}) and minimum (F_{min}) Feret diameter for a square (a), rectangles (b) and (c), and straight fibres (d).

The primary parameter measured on a particle silhouette is its projected area (A) (i.e. the area occupied by its silhouette in the binary image), which is the number of pixels with a value of 1. The equivalent diameter (i.e. the diameter of the disc of surface A) is given by:

$$D_{eq} = 2\sqrt{A/\pi}$$

The equivalent diameter is not well suited to elongated particles such as rods, needles, or straight fibres. For these types of particles a first approach is to determine the Feret diameters in several directions: for each of these directions the Feret diameter, defined as the distance between the two parallel planes perpendicular to that direction and bounding the object (i.e. its convex hull). The largest and the smallest of these Feret diameters (respectively F_{max} and F_{min}) are commonly used as estimates of the length ($\widehat{L}$) and the breadth ($\widehat{e}$) of the particle (Figure 6). This is appropriate for rod-like particles (Guo et al., 2012), short fibres (Pabst et al., 2006; Bekker et al., 2012), or straight needles but not so much for more rectangular shapes, as the largest Feret diameter is the diagonal of the bounding box. For them the length can be estimated by:

$$\widehat{L} = \sqrt{F_{max}^2 - F_{min}^2} \quad \text{or} \quad \widehat{L} = A/F_{min}$$

For curved fibres, their length cannot be deduced as simply as their breadth based on the maximum value of their Euclidian Distance Map (EDM), built by applying a distance transform: each pixel taking the value of its distance (in terms of number of pixels) to the nearest boundary point. If d_{max} is the maximum distance (in pixels),

$$\widehat{e} \approx 2.d_{max} - 1 \quad \text{and} \quad \widehat{L} = A/\widehat{e}$$

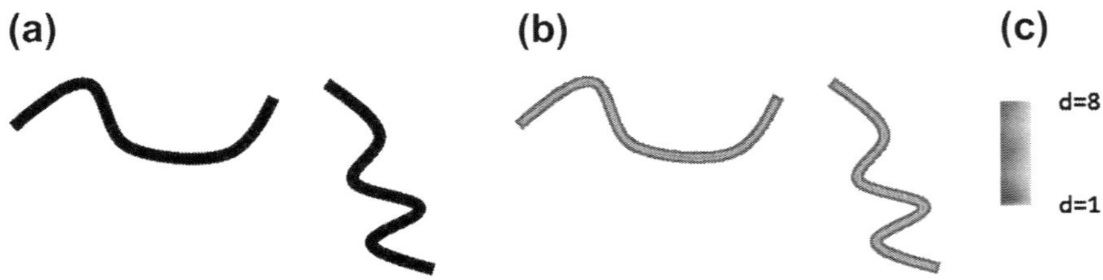

Figure 7 a) Binary image of two curved particles; (b) Corresponding Euclidian Distance Map: the red zones represent the pixels the farthest from the particle's boundary (For interpretation of the references to colour in this figure legend, the reader is referred to the online version of this book).

In the example shown in Figure 7, $d_{\max} = 8$ pixels, $\widehat{e} = 15$ pixels, $A = 6500$ pixels and $\widehat{L} = 435$ pixels.

5. SHAPE DESCRIPTORS

Basic shape descriptors are easily deduced from the equivalent diameters and the estimated length and breath. They are invariant with respect to the size or the particle and its position in the image. These can be given different names depending upon which software and which textbook are used. Thus the formula should be checked to avoid confusion. Aspect ratios are classically evaluated by:

$$E = \widehat{L}/\widehat{e} \quad \text{or} \quad E' = \widehat{L}/D_{\text{eq}}$$

The ratio between the major and minor axes of the equivalent ellipse, obtained from the centroid and the moments of inertia of the particle silhouette can also be used as an aspect ratio (Mulchrone and Choudhury, 2004). A best-fit rectangle approach has been proposed by Wang (2006), which combines the determination of the orientation based on the moments of inertia and of the size descriptors using the Feret diameters.

Other shape factors are based on the comparison between the surface and the perimeter (P) of the particle silhouette. The determination of a particle perimeter is not obvious in terms of image analysis as a simple count of the border pixels, without taking into account their neighbourhood, is far from the real perimeter value. A better approximation is obtained via the Crofton formula (often called Crofton perimeter, P_c) (Russ, 1994). Circularity (C), also called roundness, is based on the comparison with a disc:

$$C = \frac{P_c^2}{4\pi A}$$

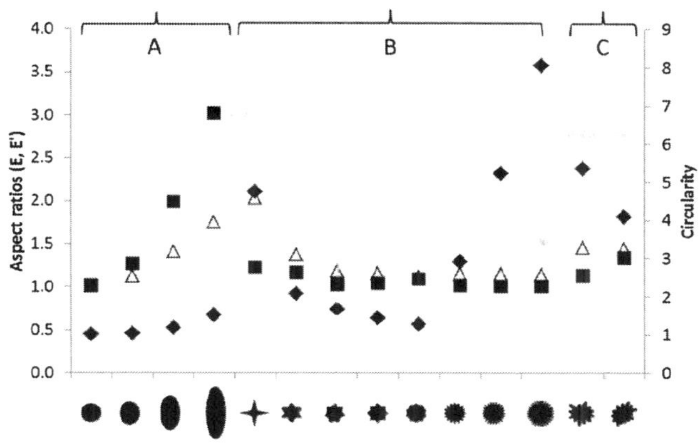

Figure 8 Comparison of values taken by circularity (♦) and aspect ratios (E (■) and E′ (Δ)) descriptors for some shapes.

For a disc, $C = 1$ when calculated with P_c and 1.6 without the Crofton correction. Circularity increases when the silhouette departs from this reference shape, either because it gets elongated or because its roughness increases.

Figure 8 gives examples of circularity and aspect ratios for different shapes. Circularity and aspect ratios increase regularly when going from a disc to an elongated 2D-ellipsoid in sub-set A. In subset B, although the number of branches increase, both aspect ratios remain close to one and circularity is the only discriminating descriptor. For subset C, the three descriptors discriminate the objects one from the other and also from the objects of subsets A and B. For the curved fibres of Figure 7, their true aspect ratio (in this case the term elongation is often used) is about 30 when the ratio F_{max}/F_{min} is close to three.

A rectangularity factor has been proposed by Bekker et al. (2012):

$$r = 2\left(\widehat{L} + A/\widehat{L}\right)/P$$

r takes values close to one for convex shapes (rectangles but also discs) and decreases for nonconvex shapes. For the elongated crystal of Figure 9(a) $r = 1$, whereas for the three particles of Figure 9(b), r varies between 0.66 and 0.72.

More complex shape descriptors have been proposed (Mora and Kwan, 2000; Al-Rousan et al., 2007), based on:

- length measurements as for circularity or aspect ratio,
- analysis of the $R(\theta)$ function: R is the distance between the centroid of the particle and its boundary and θ the directional angle (Figure 10).

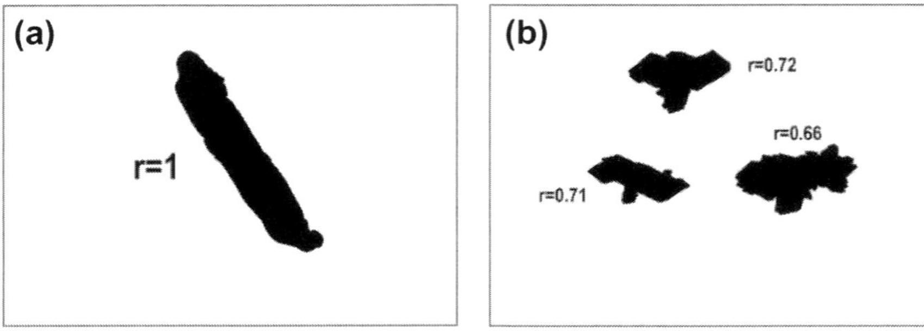

Figure 9 Calculation of rectangularity for different crystalline particles (a) elongated particle (b) twinned crystals.

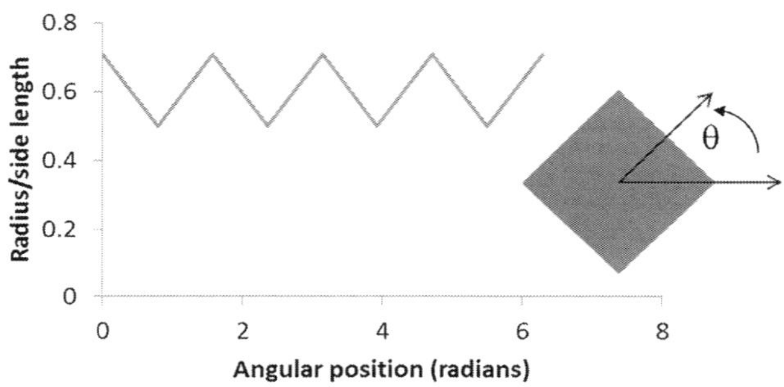

Figure 10 Relative radius versus angular position for a square.

Fourier series and wavelets have been proposed to analyze this periodic function (Gonzalez and Woods, 1992).

- mathematical morphology (i.e. series of erosions and dilations)
- fractals

6. TWINNED CRYSTALS AND AGGLOMERATES

In crystallization twinned crystals and agglomerates are frequently observed. In comparison to their convex bounding polygon their projected silhouette presents many small or large concavities. Snow crystals as those depicted in Figure 11 are good examples of complex crystalline structures. Specific descriptors have been proposed (Pons et al., 1997):

the robustness

$$\mathrm{Rob} = 2.\omega_1/\sqrt{A},$$

the largest concavity index

$$\mathrm{LCI} = 2.\omega_2/\sqrt{A}$$

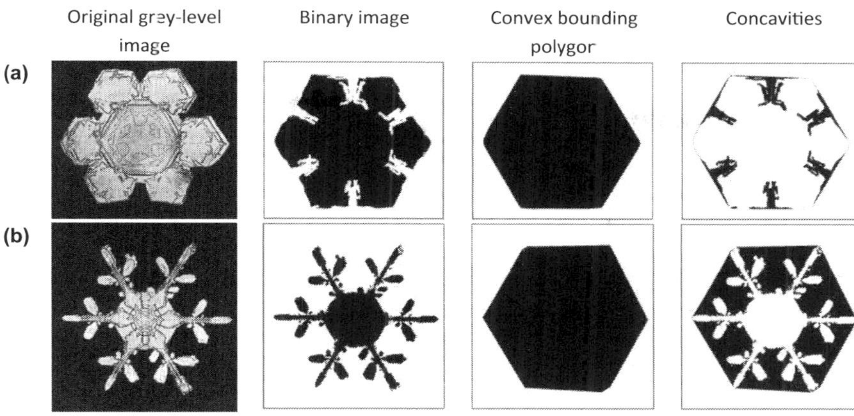

Figure 11 Snow crystals shape analysis. (a) Stellar plate (b) Stellar dendrite. *Snow crystals images adapted from Bentley and Humphreys (1931).*

and the surfacic concavity index

$$IC_s = A/A_c$$

where ω_1 and ω_2 are the maxima of the Euclidian Distance Map of the particle silhouette and of the concavity image, respectively, and A_c the area of the convex bounding polygon (Figure 11). In fact ω_1 and ω_2 are the orders of the morphological erosions necessary to make the silhouette and its residual set (i.e. the concavities) disappear completely. Table 1 gives examples of application for selected crystals.

7. FRACTAL-LIKE PARTICLES

Self-similarity and iterative patterns are characteristics of theoretical fractal objects. Nanoparticles aggregates (Lee et al., 2013; Romanello and de Cortalezzi, 2013) and mineral flocs (Jarvis et al., 2005; Li et al., 2007; Maggi, 2007) are often considered as fractal objects. The structured walk (Kaye et al., 1994) and the covering box method are two of the algorithms used to evaluate the fractal dimension (Wettimuny and Penumadu, 2003). In the latter, assuming $N(l)$ is the number of boxes of side length l (in pixels) necessary to cover the contour of the particle, the fractal perimetric dimension D_f is defined through the equation:

$$N(l) = k \cdot l^{-D_f}$$

Table 1 Crystal Characterization Based on Mathematical Morphology

	Rob	LCI	IC_s
	0.55	0.12	0.85
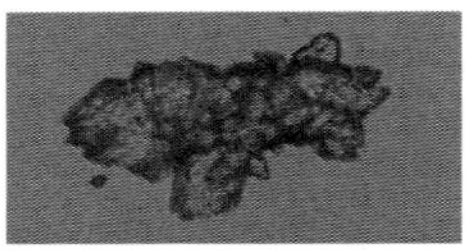	0.59	0.16	0.82
	0.52	0.18	0.81
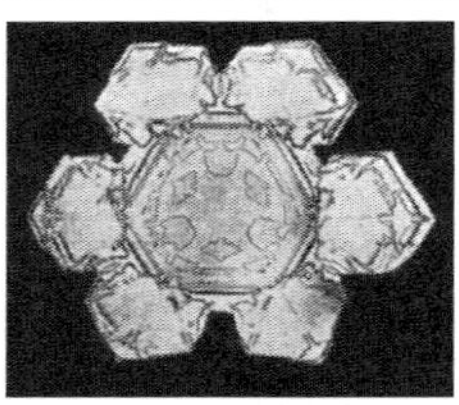	0.55	0.07	0.84
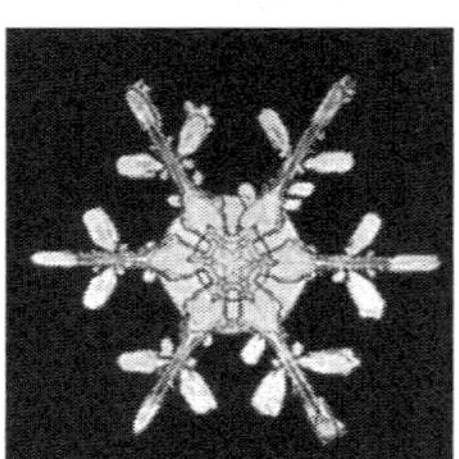	0.50	0.26	0.38

In the example shown in Figure 12, 29 boxes are required to cover the contour of the particle. The fractal perimetric dimension of the particle is 1.09, as calculated from the slope of the plot (Figure 12(c)). In Figure 13, the fractal perimetric dimension of some artificial particles have been plotted. Many parameters such as the image resolution and quality (Wozniak et al., 2012), the roughness and orientation of the aggregates or the box size range, affect the results (Lottin et al., 2013). An example is shown in Figure 14: the spatial resolution of the mineral floc on the left side is not sufficient: the

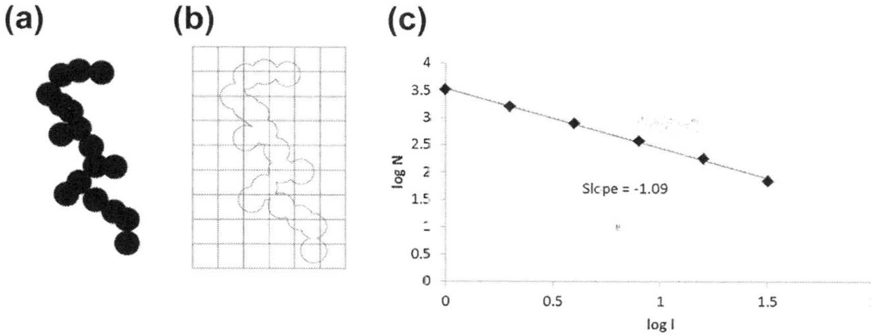

Figure 12 Box covering method (b) for the determination of the fractal perimetric dimension (c) of a virtual particle (a).

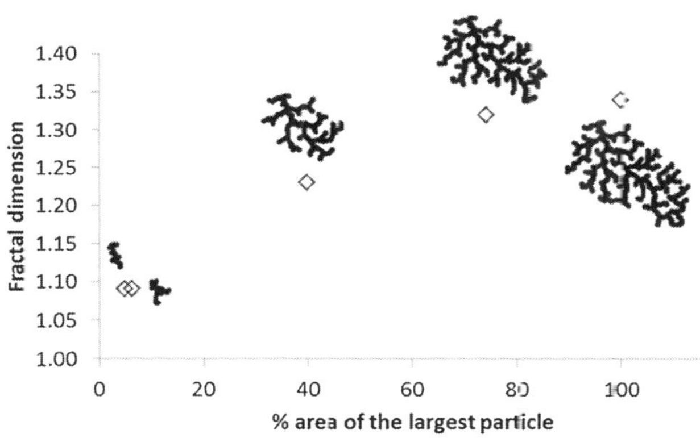

Figure 13 Examples of synthetic complex particles and their fractal perimetric dimension.

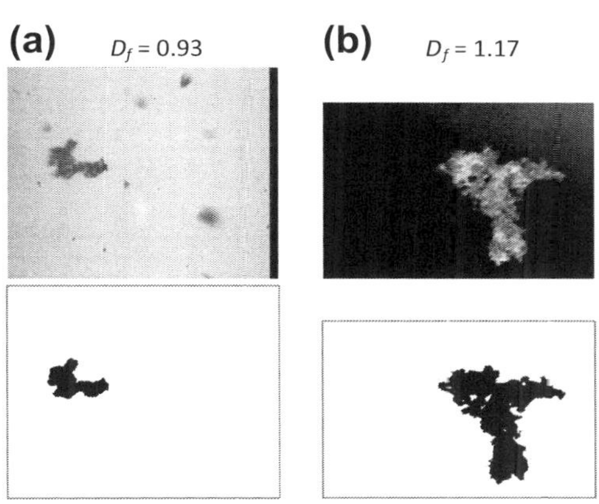

Figure 14 Grey level and binary images of two mineral flocs (a) dynamic image acquisition in a water column (b) static image acquisition with an optical microscope.

calculated D_f is 0.93, which is wrong as it should be between 1 and 2. The floc on the left side has a fractal perimetric dimension of 1.17. Nevertheless, the image segmentation step is critical for that type of shape characterization.

8. BIOLOGICAL PARTICLES

There is an abundant literature on the use of image analysis techniques to assess the size and the shape of microorganisms, either single or developing complex architectures, such as hyphae, filamentous pellets, or flocs. The difficulty is that, in many cases, they are highly deformable. So the apparent shape in a microscope image depends very much upon the way the slide has been prepared.

Budding is an important biological phenomenon for yeast cells such as *Saccharomyces cerevisiae*. Individual *S. cerevisiae* cells are usually modelled as prolate ellipsoids thus, by fitting to an ellipse, their size characteristics can be estimated (Pons et al., 1993; Tibayrenc et al., 2010). It is also possible to count the number of buds and their size. Extracting information on cells in colonies is more difficult: Doncic et al. (2013) have developed a method to detect budding cells based on a set of thresholds instead of using just a single one for segmentation.

Filamentous species (yeast, bacteria, and fungi) are probably the microorganisms which have received the most attention for size and shape quantification by image analysis. For well defined hyphae features such as the number and the size of hyphal branches are often calculated by detecting the end points and the triple points (i.e. branching points) of their skeleton and removing them to obtain individual branches as shown in Figure 15. The skeleton is the result of a morphological

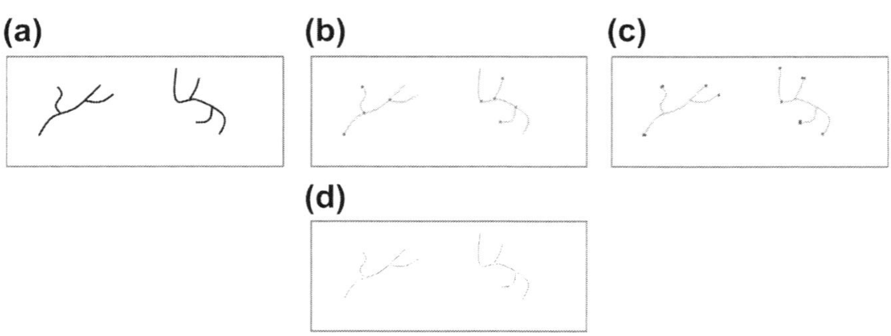

Figure 15 Analysis of individual hyphae: (a) binary image, (b) triple points, (c) end points, and (d) separated branches.

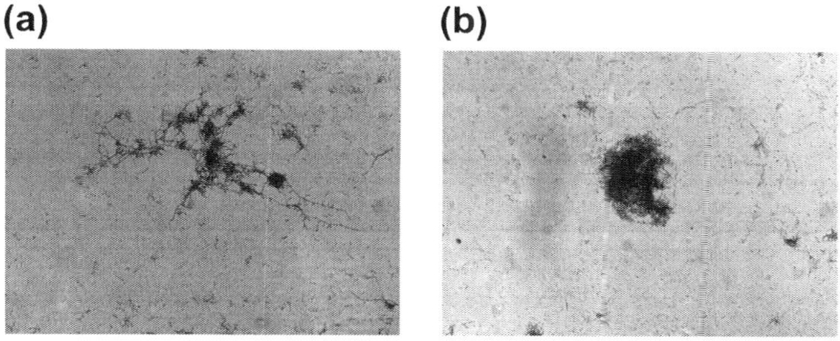

Figure 16 Entanglements of filaments (a) and filamentous pellet (b).

transformation based on the thinning of connected components until a line is achieved (Russ, 1994). The thickness of the hyphae can be estimated using the Euclidian Distance Map, as for fibres (Figure 7). However, this approach is not possible in the case where filaments are entangled. For pellets (Figure 16) several parameters should be combined, based on projected area, concavity index, etc. (Wucherpfennig et al., 2010; Posch et al., 2012).

Activated sludge flocs are assemblies of filamentous and nonfilamentous bacteria involved in biological wastewater treatment. The filamentous bacteria constitute the backbone of the flocs and are the key features for the floc settling in the final clarifier of wastewater treatment plants. Filamentous bacteria overgrowth (Figure 17) is associated to filamentous bulking and severely decreases the floc settling velocity, leading to biomass loss into

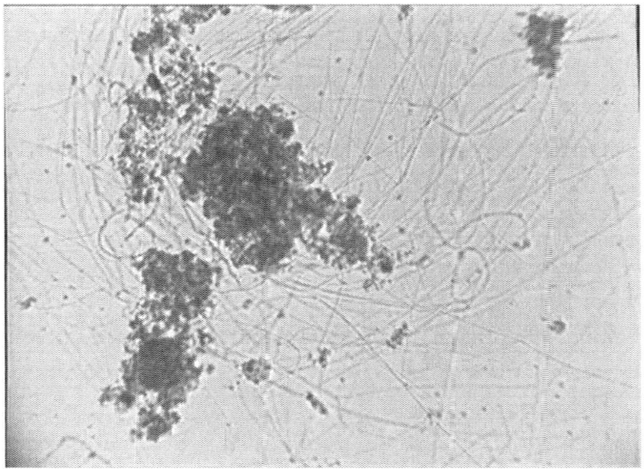

Figure 17 Activated sludge flocs with protruding filamentous bacteria.

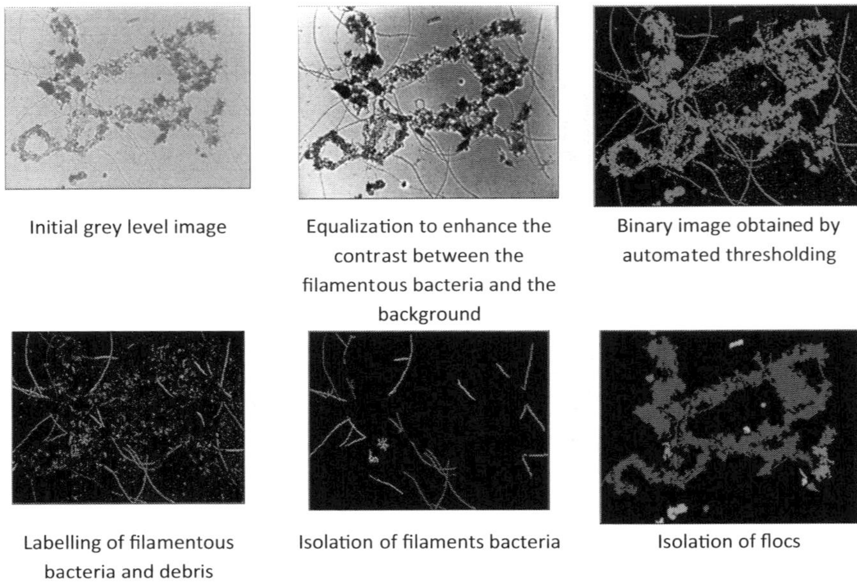

Figure 18 Flow sheet of the treatment of activated sludge image.

the aquatic environment. Automatically quantifying the abundance of protruding filaments out of flocs (da Motta et al., 2001) is a way of improving the control of wastewater treatment plants by being able to quickly detect any onset of filamentous bulking (Figure 18) (see also Costa et al., 2013 for a review).

9. CASE OF IN SITU IMAGES

The quality of images captured by an in situ microscope is generally poorer than that of static microscope images (Zhou et al., 2009): the in situ images require some grey-level enhancement procedure to help the segmentation step. Edge detection algorithms are useful: they compute the magnitude of the gradient of pixel intensity values. Depending upon the threshold selection the edges, which correspond to rapid spatial changes of grey levels, are detected. Figure 19 shows an example of caffeine crystals: the edges are better detected with a Sobel detector (Russ, 1994) than by a simple segmentation based of the original grey-level image.

In the case of elongated crystals (needles, rods), another option could be to detect lines in the image by applying a Hough transform (Gonzalez and

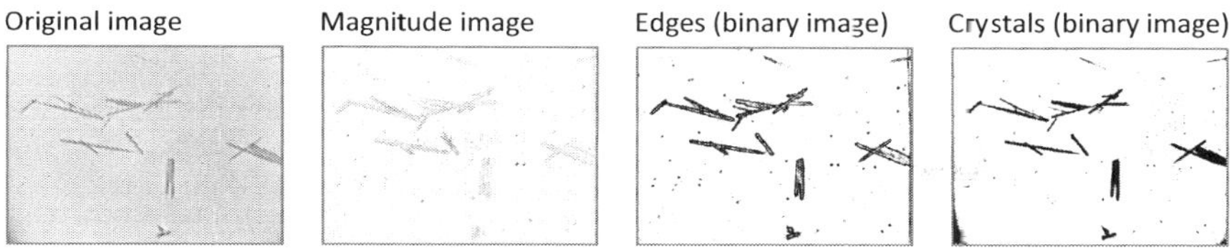

Figure 19 Detection of elongated crystals using a Sobel edge detector or by simple thresholding.

Woods, 1992). This type of transform allows detecting shapes defined by analytical functions (lines, circles, etc.).

If $M_1(x_i,y_i)$ and $M_2(x_i,y_i)$ are two points in an image, all the straight lines passing through M_1 and M_2 respectively are described by the following equations in the slope-intercept form (Figure 20(a)):

$$(\Delta_A) : y_i = a \cdot x_i + b \quad \text{and} \quad (\Delta_B) y_j = a \cdot x_j + b$$

or in the parameter space (a, b) (Figure 20(b)):

$$(\Delta_A) : b = -a \cdot x_i + y_i \quad \text{and} \quad (\Delta_B) : b = -a \cdot x_j + y_j.$$

The only common point $(a';b')$ of Δ_A and Δ_B is the straight line between M_1 and M_2. This means that all the points belonging to a straight line in the image are transformed into a unique point in the parameter space. In general it is easier to work with polar equations:

$$x \cdot \cos(\theta) + y \cdot \sin(\theta) = \rho$$

The (ρ, θ) plane is discretized (Figure 20(c)) and transformed into an image (Hough image). In practice, a unique point per line in the

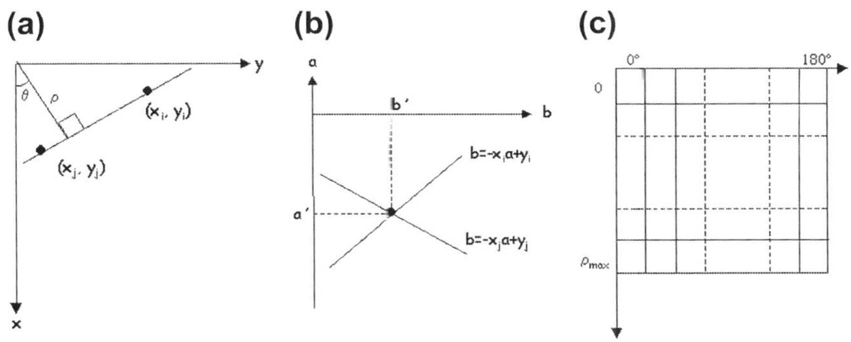

Figure 20 Hough transform for straight lines detection.

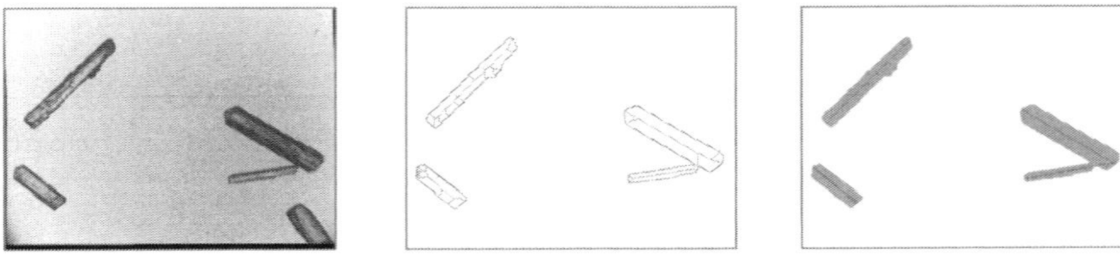

Figure 21 Detection of crystal axes using the Hough transform.

parameter space is not found and only small accumulation zones are found, which should be examined one by one. This is related to the discrete nature of images and to the nonperfect linearity of the axis for many real "straight" features. In Fig. 21 the Hough transform has been applied to the silhouette skeleton to find the main axis of the needles. The Hough transform is efficient but very memory and time consuming as speed is related to the discretization step of the parameter space. Larsen et al. (2006, 2007) prefer to use the Burns line detector (Burns et al., 1986) not only for this reason but also because it can detect short and long straight features in the image.

10. SELECTION OF MAGNIFICATION

The selection of the optimal resolution (i.e., magnification) results from a compromise:

- Each particle should be represented by a number of pixels which allows a minimization of the inaccuracy of its boundary, keeping in mind that pixelization introduces jaggedness along this boundary (Figure 22). This

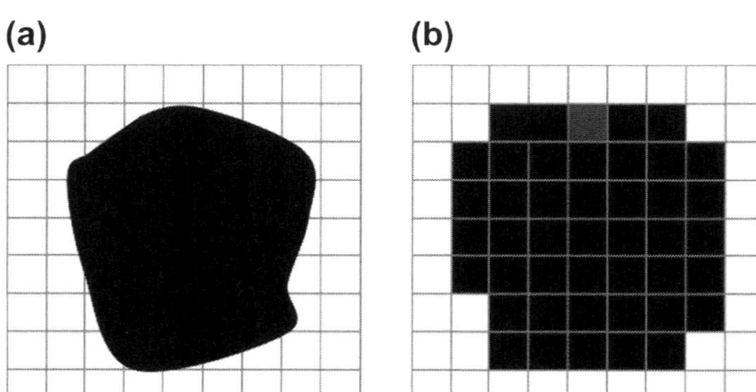

Figure 22 Effect of pixelization (b) on the boundary of a particle (a).

jaggedness affects the definition of the perimeter, which is critical for those shape parameters based on this measure (circularity, perimetric fractal dimension). The accuracy is a function of the shape of particle and increases with the ratio of the real particle boundary length to its real projected area (Zeidan et al., 2007). This means that the accuracy is less for an elongated rod or a needle than for a square. In the case where the set of particles to be analyzed have a distribution of shape, the accuracy depends upon their shape.

- On the other hand, high resolution leads to longer computation times, either because of the larger number of pixels in the image or because a larger number of images are necessary to analyze a sufficient number of particles. One has to also remember that particles truncated by the image boundary cannot be taken into account.

11. DISTRIBUTIONS

In image analysis, the size and shape distributions are always given in numbers. The first question which arises is how many particles should be analyzed. The stability of the means and the standard deviations of the size and shape descriptors can be used for this analysis. Figure 23 gives an example for a sample of sodium chloride particles visualized by optical microscopy on glass slides. Whilst a smooth variation of the equivalent diameter with respect to the number of particles can be seen, this is not so for the aspect ratio, the irregularity around 250 particles being linked to a change of slide.

According to the International Standards Organization for static image analysis (ISO 13322-1, 2004), it is necessary to count about 6100 particles to get the mean volume diameter within 5% error with 95% probability

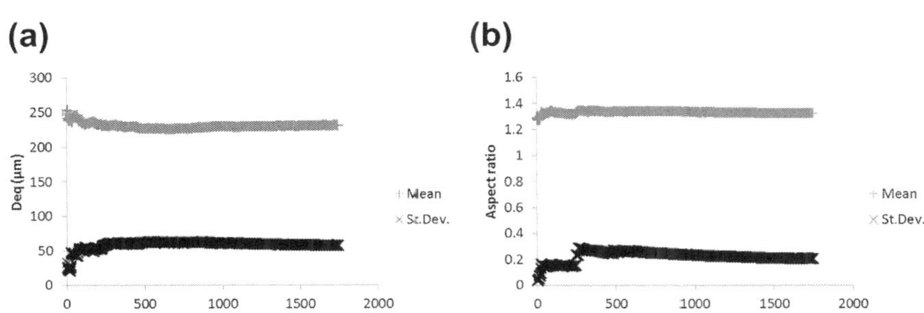

Figure 23 Stabilization of the mean equivalent diameter and its standard deviation.

for a powder sample with a standard deviation of 1.60. This number increases to 61,000 if a mass median diameter is needed.

In any case results obtained by image analysis cannot be directly compared to those provided by other sizing techniques as shape should be taken into account. For poly-disperse short fibres, with average aspect ratios varying between 5 and 16, laser diffraction measurement of medians from volume-weighted size distributions were found to be close to the medians of the minimum Feret diameter distributions, after transformation of the later into volume-weighted distributions (Pabst et al., 2006). Igathinathane et al. (2009) have discussed how difficult it is to compare size distribution of elongated particles obtained by mechanical sieving with particle size distributions obtained by image analysis, due to the way these particles pass through the sieve openings.

It is not easy to visualize simultaneously size and shape distributions. This is illustrated by the following example, for which two sodium chloride samples (salts "B" and "D") have been visualized by static optical microscopy, with the same equipment and the same magnification. Typical images are shown in Figure 24. The image analysis treatment applied is that described in Figure 3. Table 2 summarizes some overall statistical parameters obtained for both of these salts, namely means, standard deviations, minimum and maximum values obtained on counts of more than 1700 particles. The values related to their size (i.e. D_{eq} and F_{min}) differ between the two salts: salt "B" has smaller mean values of D_{eq} and F_{min} than salt "B" although the standard deviations for the latter are smaller. The overall statistical parameters related to shape (aspect ratio, circularity and robustness) are very similar for both salts. Going into more detail with

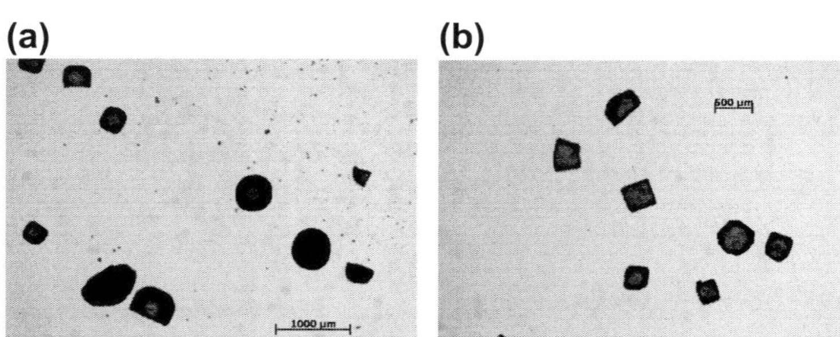

Figure 24 Typical optical microscopy images of the two sodium chloride samples tested (a) Salt "B" (b) Salt "D".

Table 2 Statistical Parameters Obtained for Salts "B" and "D": Mean, Standard Deviation (σ), and (min–max)

	Salt "B" (n = 1855)	Salt "D" (n = 1749)
D_{eq} (μm)	187 (σ = 129) (30–684)	230 (σ = 57) (30–423)
F_{min} (μm)	171 (σ = 121) (24–554)	208 (σ = 52) (27–426)
Aspect ratio	1.34 (σ = 0.33) (1.02–7.91)	1.32 (σ = 0.20) (1.05–4.07)
Circularity	1.16 (σ = 0.28) (1.00–9.56)	1.19 (σ = 0.32) (1.01–9.09)
Robustness	0.76 (σ = 0.07) [0.32–0.96]	0.75 (σ = 0.07) (0.34–0.96)

boxplots of the distributions of single size and shape descriptors leads to similar results (Figure 25).

Although some authors manage to discuss their results based on the statistics of each size and shape descriptor taken individually (Ulusoy, 2008), it is often necessary to combine size and shape to see the difference between the samples, as proposed in Fig. 26:

- salt "B" has a bimodal particle population with, on the one hand, a first subpopulation of small and elongated particles (crystal fragments) and on the other hand a second subpopulation of larger particles (in the range 150–400 μm) with a 2D-circular shape (circularity between 1 and 1.1). It is the larger circularity of the smallest particles which impacts the average circularity of the sample.

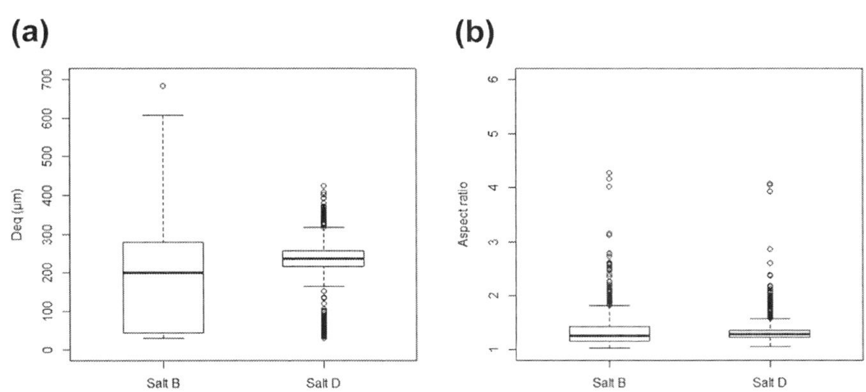

Figure 25 D_{eq} (a) and aspect ratio (b) boxplots for salts "B" and "D".

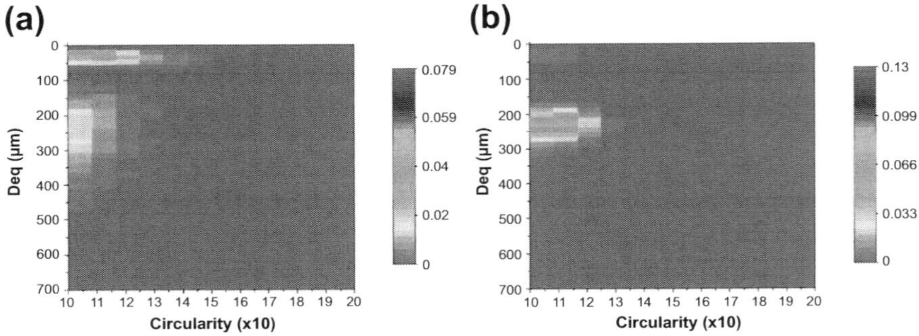

Figure 26 Example of size-shape plots for salts "B" (a) and "D" (b).

- salt "D" has a unimodal particle population, with a narrower size distribution than salt "B" and a larger circularity than the second subpopulation of salt "D".

Another option would be to plot length versus width, especially for elongated particles as proposed by Eggers et al. (2008).

When more shape parameters need to be included, visualization gets to be more difficult. In such cases, data-mining techniques such as clustering or principal components analysis are used (Belaroui et al., 2002; Pons et al., 2002; Duchesne et al., 2012).

12. 3D SHAPE

Shape is essentially a 3D feature of an object and assessing it through 2D images is of course a limitation. Efforts have been made for 3D shape characterization. Yamamoto et al. (2002) have developed a tri-axial viewer to measure the 3D shape from three orthogonal directions, a single 2D image being captured for each of these directions. A particle is then characterized by three 2D shape factors and located in a shape space. However, the particles have to be visualized one by one, which is a strong limitation for statistics. Lee et al. (2007) have proposed a laser triangulation technique to characterize shape in terms of angularity for particles on a belt conveyor. However, the test particles (coarse agglomerates between 8 and 32 mm) were rather large. Atomic Force Microscopy (AFM) has been tested on platy materials such as clay or pigments, the grey level being associated to the thickness of the platelets (Gélinas and Vidal, 2010). Kempkes et al. (2010) have introduced two mirrors in a vertical flow cell setup (Figure 2) for the stereoscopic imaging of particles. 3D information on the shape of the particles can be obtained from the two images captured at a 90° angle.

13. CONCLUSIONS

It has long been recognized that the behaviour of finely divided systems of particles can be strongly influenced by the shape of particles involved and that, whilst observation by electronic and optical microscopy or video can easily distinguish particles by shape, it is necessary to translate observations into quantitative measurements for better characterization of the physical properties of particle systems. Here we show how developments in digital image capture and subsequent automated analysis, favoured by the progress in computer science (hardware and software), can be used for detailed quantification of particle shape.

Common to all methods is the correct acquisition of digital images whether from classical static microscopy or from images of moving systems, necessarily followed by basic digital image treatment leading to binary images. It is then shown how these treated images can be used to give basic particle size descriptors, which can be combined in different ways to give shape descriptors such as circularity and aspect ratios. Such methods are extensions of traditional shape characterization by making ratios of two different particle sizes but can be more precisely targeted.

Furthermore we show that the procedures used in digital image treatment: erosions, dilations, determining bounding perimeters, etc., can be used to give more detailed shape descriptors based on the concepts of mathematical morphology. Examples of these methods are presented to characterize the shape of twinned crystals and particle agglomerates. Other methods are described to characterize fractal structures and biological particles.

In conclusion, the methods described here may seem rather fastidious and over detailed but it should be noted that they can be directly automated to give rapid measurements for process control. Properly used they can quantify very pertinent differences in the behaviour of particle systems. An example is the control of wastewater treatment by clear identification of the onset of filamentous bulking. The examples given here are not exclusive and in general when the critical factors for a given particle system are identified it is usually possible to develop very specific shape descriptors for rapid characterization of any system. Measuring particle shape by image analysis is, therefore, a general method for particle systems characterization applicable to any system where valid images can be obtained.

REFERENCES

Akin, M., Prediger, A., Yuksel, M., Höpfner, T., Demirkol, D.O., Beutel, S., Timur, S., Scheper, T., 2011. A new set up for multi-analyte sensing: at-line bio-process monitoring. Biosens. Bioelectron. 26, 4532–4537.

Al-Rousan, T., Masad, E., Tutumluer, E., Pan, T., 2007. Evaluation of image analysis techniques for quantifying aggregate shape characteristics. Constr. Build. Mater. 21, 978–990.

Bekker, A.V., McShane, J., Vassallo, D., Bedell, D., Li, T.S., Livk, I., 2012. 2-D sizing of sodium oxalate crystals by automated optical image analysis. Adv. Powder Technol. 23 (4), 509–516.

Belaroui, K., Pons, M.N., Vivier, H., 2002. Morphological characterization of gibbsite and alumina. Powder Technol. 127, 246–256.

Belini, V.L., Wiedermann, P., Suhr, H., 2013. In situ microscopy: a perspective for industrial bioethanol production monitoring. J. Microbiol. Methods 93, 224–232.

Bentley, W.A., Humphreys, W.J., 1931. Snow Crystals. McGraw-Hill Book Company, Inc., NY.

Bittner, C., Wehnert, G., Scheper, T., 1998. In situ microscopy for on-line determination of biomass. Biotechnol. Bioeng. 60, 24–35.

Bluma, A., Höpfner, T., Rudolph, G., Lindner, P., Beutel, S., Scheper, T., 2009. Adaptation of in-situ microscopy for crystallization processes. J. Cryst. Growth 311, 4193–4198.

Burns, J.B., Hanson, A.R., Riseman, E.M., 1986. Extracting straight lines. IEEE Trans. Pattern Anal. Mach. Intell. 8 (4), 425–455.

Calderon de Anda, J., Wang, X.Z., Roberts, K.J., 2005. Multi-scale segmentation image analysis for the in-process monitoring of particle shape in crystallisers. Chem. Eng. Sci. 60, 1053–1065.

Costa, J.C., Mesquita, D.P., Amaral, A.L., Alves, M.M., Ferreira, E.C., 2013. Quantitative image analysis for the characterization of microbial aggregates in biological wastewater treatment: a review. Environ. Sci. Pollut. Res. 20 (9), 5887–5912.

Doncic, A., Eser, U., Atay, O., Skotheim, J.M., 2013. An algorithm to automate yeast segmentation and tracking. PLoS ONE 8 (3), e57970.

Duchesne, C., Liu, J.J., MacGregor, J.F., 2012. Multivariate image analysis in the process industries: a review. Chemom. Intell. Lab. Syst. 117, 116–128.

Eggers, J., Kempkes, M., Mazotti, M., 2008. Measurement of size and shape distributions of particles through image analysis. Chem. Eng. Sci. 63, 5513–5521.

Gao, L., Yan, Y., Lu, G., Carter, R.M., 2012. On-line measurement of particle size and shape distributions of pneumatically conveyed particles through a multi-wavelength based digital imaging. Flow. Meas. Instrum. 27, 20–28.

Gélinas, V., Vidal, D., 2010. determination of particle shape distribution of clay using an automated AFM image analysis method. Powder Technol. 203, 254–264.

Gonzalez, R.C., Woods, R.E., 1992. Digital Image Processing. Addison-Wesley, Reading, MA.

Guez, J.S., Cassar, J.P., Wartelle, F., Dhulster, P., Suhr, H., 2010. The viability of animal cell cultures in bioreactors: can it be estimated online by using in situ microscopy? Process Biochem. 45, 288–291.

Guo, Q., Chen, X., Liu, H., 2012. Experimental research on shape and size of biomass particle. Fuel 94, 551–555.

Igathinathane, C., Pordesimo, L.O., Columbus, E.P., Batchelor, W.D., Sokhansanj, S., 2009. Sieveless particle size distribution analysis of particulate materials through computer vision. Comput. Electron. Agric. 66, 147–158.

ISO 13322-1, 2004. Particle Size Analysis – Image Analysis Methods – Part 1: Static Image Analysis Methods. ISO Office, Geneva, Switzerland.

ISO 13322-2, 2006. Particle Size Analysis – Image Analysis Methods – Part 2: Dynamic Image Analysis Methods. ISO Office, Geneva, Switzerland.

Jarvis, P., Jefferson, B., Parsons, S.A., 2005. Measuring floc structural characteristics. Rev. Environ. Sci. Bio/Technol. 4 (1–2), 1–18.

Joeris, K., Frerichs, J.G., Konstantinov, K., Scheper, T., 2002. Insitu microscopy: online process monitoring of mammalian cell cultures. Cytotechnology 38, 129–134.

Kaye, B.H., Clark, G.G., Kydar, Y., 1994. Strategies for evaluating boundary fractal dimensions by computer aided image analysis. Part. Part. Syst. Charact. 11, 411–417.

Kempkes, M., Vetter, T., Mazotti, M., 2010. Measurement of 3D particle size distributions by stereospoci imaging. Chem. Eng. Sci. 65, 1362–1373.

Larsen, P.A., Rawlings, J.B., Ferrier, N.J., 2006. An algorithm for analyzing noisy, in situ images of high-aspect-ratio crystals to monitor particle size distribution. Chem. Eng. Sci. 61, 5236–5248.

Larsen, P.A., Rawlings, J.B., Ferrier, N.J., 2007. Model-based object recognition to measure crystal size and shape distributions from in situ video images. Chem. Eng. Sci. 62, 1430–1441.

Lee, J.R.J., Smith, M.L., Smith, L.N., 2007. A new approach to the three-dimensional quantification of angularity using image analysis of the size and form of coarse aggregates. Eng. Geol. 91, 254–264.

Lee, D., Choi, S.C., Lee, C.S., 2013. Impact of some blended fuel combustion on soot morphological characteristics in a diesel engine. Int. J. Automot. Technol. 14 (5), 757–762.

Li, T., Zhu, Z., Wang, D., Yao, C., Yang, H., 2007. The strength and fractal dimension characteristics of alum–kaolin flocs. Int. J. Miner. Process. 82 (1), 23–29.

Liao, C.W., Yu, J.H., Tarng, Y.S., 2010. On-line full scan inspection of particle size and shape using digital image processing. Particuology 8, 286–292.

Lottin, D., Ferry, D.,, Gay, J.M., Delhaye, D., Ouf, F.X., 2013. On methods determining the fractal dimension of combustion aerosols and particle clusters. J. Aerosol. Sci. 58, 41–49.

Maaß, S., Wollny, S., Voigt, A., Kraume, M., 2011. Experimental comparison of measurement techniques for drop size distributions in liquid/liquid dispersions. Exp. Fluids 50, 259–269.

Maggi, F., 2007. Variable fractal dimension: a major control for floc structure and flocculation kinematics of suspended cohesive sediment. J. Geophys. Res. 112, C07012.

Mora, C.F., Kwan, A.K.H., 2000. Sphericity, shape factor and convexicity measurement of coarse aggregate for concrete using digital image processing. Cem. Concr. Res. 30, 351–358.

da Motta, M., Pon, M.N., Roche, N., Vivier, H., 2001. Characterisation of activated sludge by automated image analysis. Biochem. Eng. J. 9, 165–173.

Mulchrone, K.F., Choudhury, K.R., 2004. Fitting an ellipse to an arbitrary shape: implications for strain analysis. J. Struct. Geol. 26 (1), 143–153.

Pabst, W., Berthold, C., Gregorová, E., 2006. Size and shape characterization of polydisperse short-fibers systems. J. Eur. Ceram. Soc. 26, 1121–1130.

Pons, M.N., Vivier, H., Rémy, J.F., Dodds, J., 1993. Morphology characterization of yeast by image analysis during alcoholic fermentation. Biotechnol. Bioeng. 42, 1352–1359.

Pons, M.N., Vivier, H., Dodds, J.A., 1997. Particle shape characterization using morphological descriptors. Part. Part. Syst. Charact. 14, 272–277.

Pons, M.N., Vivier, H., Delcour, V., Authelin, J.R., Paillères-Hubert, L., 2002. Morphological analysis of pharmaceutical powders. Powder Technol. 128, 276–286.

Posch, A.E., Spadiut, O., Herwig, C., 2012. A novel method for fast and statistically verified morphological characterization of filamentous fungi. Fungal Genet. Biol. 49 (7), 499–510.

Romanello, M.B., de Cortalezzi, M.M.F., 2013. An experimental study on the aggregation of TiO_2 nanoparticles under environmentally relevant conditions. Water Res. 47, 3887–3898.

Rudolph, G., Lindner, P., Gierse, A., Bluma, A., Martinez, G., Hitzmann, B., Scheper, T., 2007. Online monitoring of microcarrier based fibroblast cultivations with in-situ microscopy. Biotechnol. Bioeng. 99, 136–145.

Russ, J.C., 1994. The Image Processing Handbook, second ed. CRC Press, Boca Raton, FL.

Simon, L.L., Merz, T., Dubuis, S., Lieb, A., Hungerbuhler, K., 2012. In-situ monitoring of pharmaceutical an specialty chemicals crystallization processes using endoscopy-stroboscopy and multivariate image analysis. Chem. Eng. Res. Des. 90 (11), 1847–1855.

Tibayrenc, P., Preziosi-Belloy, L., Roger, J.M., Ghommidh, C., 2010. Assessing yeast viability from cell size measurements? J. Biotechnol. 149 (1–2), 74–80.

Ulusoy, U., 2008. Application of ANOVA to image analysis results of talc particles produced by different milling. Powder Technol. 188, 133–138.

Wang, W., 2006. Image analysis of particles by modified Ferret method – best-fit rectangle. Powder Technol. 165, 1–10.

Wang, X.Z., Roberts, K., Ma, C., 2008. Crystal growth measurement using 2D and 3D imaging and the perspectives for shape control. Chem. Eng. Sci. 63, 1173–1184.

Wettimuny, R., Penumadu, D., 2003. Automated digital image based measurement of boundary fractal dimension for complex nanoparticles. Part. Part. Syst. Charact. 20, 18–24.

Wozniak, M., Onofri, F.R.A., Barbosa, S., Yon, J., Mroczka, J., 2012. Comparison of methods to derive morphological parameters of multi-fractal samples of particle aggregates from TEM images. J. Aerosol Sci. 47, 12–26.

Wucherpfennig, T., Kiep, K.A., Driouch, H., Wittmann, C., Krull, R., 2010. Morphology and rheology in filamentous cultivations. Adv. Appl. Microbiol. 72, 89–136.

Yamamoto, K.I., Inoue, T., Miyajima, T., Doyama, T., Sugimoto, M., 2002. Measurement and evaluation of three-dimensional particle shape under constant particle orientation with a tri-axial viewer. Adv. Powder Technol. 13 (2), 181–200.

Zeidan, M., Jia, X., Williams, R.A., 2007. Errors implic in digital particle characterisation. Chem. Eng. Sci. 62, 1905–1914.

Zhou, Y., Srinivasan, R., Lakshminarayanan, S., 2009. Critical evaluation of image processing approaches for real-time crystal meaasurements. Comput. Chem. Eng. 33, 1022–1035.

CHAPTER SIXTEEN

Turbidity: Measurement of Filtrate and Supernatant Quality?

John Bratby
Brown and Caldwell, Golden, CO, USA

Contents

1. IMPORTANCE OF PARTICULATES IN PROCESS AND MUNICIPAL WATERS

In many industries the concentration of particulate matter is important for quality control and process optimization. Particulate matter can adhere to heat transfer surfaces such as those found in heat exchangers, cooling ducts and radiators and interfere with the rate of heat exchange. In cooling towers and many municipal applications, suspended solids can shield microorganisms or create a disinfection demand that interferes with effective disinfection (Allhands, 2007).

Particulate contaminants may vary from several nanometres to several millimetres in size. Consequently, cost-effective removal of these particles from the large volumes of water required by many communities represents a significant challenge for water treatment engineers. In the process of removing contaminants from water, certain particles may also be added to the water. Water treatment facilities must take into account the need to remove particles that are added or created during treatment as well as those present in the raw water. Natural weathering of minerals produces a variety of particles in water. Inorganic particles may consist of iron oxides such as goethite, haematite, and amorphous iron hydroxide; silica dioxide; calcites;

Progress in Filtration and Separation
http://dx.doi.org/10.1016/B978-0-12-384746-1.00016-1

clays such as kaolinite, montmorillonite, muscovite, or bentonite; pure and mixed oxides of aluminium; and many other minerals. Alone, these particles are objectionable in water because they affect the water's aesthetic qualities. However, in the form of small colloids, even small mass concentrations of these particles may present substantial surface areas for the adsorption of natural organic matter (NOM), pesticides and other synthetic organic chemicals (SOCs), metals, and other toxic substances. Bacteria and viruses may also attach to inorganic particles, and there is some concern that particulate materials (organic and inorganic alike) may "shield" pathogens from chemical disinfectants. In addition, asbestos fibres, which may come from natural or anthropogenic sources, are possible carcinogens when introduced into the gastrointestinal tract.

Clays, metal hydroxides, and other particles originating from mineral sources typically vary from several nanometres to several microns in diameter. The distribution of materials over this range is likely to be continuous. However, in surface waters most of these particles appear to be approximately 0.1–1 μm in diameter. Their small size renders them relatively stable in suspension. Moreover, particles in this size range scatter visible light efficiently and may create a cloudy appearance in water at very low concentrations.

Particulate matter is typically expressed as total suspended solids (TSS) measured by gravimetric analysis whereby particles are removed from a volume of water by dead-end filtration using a rinsed membrane of known dry weight. Membranes with a nominal pore size of 0.45–2 μm are often used. The membrane is then dried and weighed to determine the mass of particles deposited on the membrane. If the mass concentration of the suspension is low, relatively high volumes of water need to be filtered to deposit measurable amounts of solid material and minimize inherent errors in measurement. Measurements of the mass concentration of particles in the water by this method will be affected by the specific gravity of the particles, the effective pore size of the membrane, and the mass of material lost during drying. Some particles, such as algae or bacteria, may represent a relatively large particle volume concentration in the raw water. However, because of their high water content they may account for only a small mass of solids after drying. Similarly, organic compounds that volatilize during the drying process will reduce the estimate for the solids concentration in the sample.

The operational definition of particulate and dissolved materials is linked to the effective pore size of the membrane used to collect suspended materials. In addition, a cake of material that builds up on the membrane during filtration may become the effective membrane. In either case, significant concentrations of colloidal material may pass through the membrane and be excluded from the measurement of suspended material. In some cases this error may be important, particularly in estimating the quantities of colloid-bound pollutants.

2. ADVANTAGES OF TURBIDITY MEASUREMENTS

If rigorous laboratory procedures are adhered to, gravimetric measurements have the potential for accurate measurements. However, they are notorious for providing inconsistent results, particularly for more dilute suspensions. They are also time-consuming and are not convenient for real-time process-control work.

Turbidimetry on the other hand, is a more straightforward test and, although still not without its own challenges, is a reliable measure of source-water clarity and the effectiveness of subsequent treatment. In water treatment practice, turbidity is the most widely used indicator of the concentration of particles in water.

Turbidity is a measure of the scattering of light by particles suspended in a medium such as water. The deflection or scattering of light is related to a number of optical phenomena that in turn depend on the concentration, size, and shape of the particles as well as the wavelength of the incident light, the angle of observation, the optical properties of the particles, and the refractive index of the suspending medium (AWWARF/IWSA, 1997).

Originally, turbidity was a parameter used in limnological work to study the depth at which light can penetrate a body of water. Field measurements still use the Secchi disk, shown in Figure 1. The disk is lowered in the water until it becomes obscure. Figure 2 shows a relationship between Secchi disk readings and standard turbidity readings (see also Davies-Colley, 1988).

In the past, the turbidity of water was measured based on the attenuation of light passing through a sample, as viewed through a Jackson candle turbidimeter, shown schematically in Figure 3. This instrument could measure the clarity of water down to 25 Jackson Turbidity Units (JTU). The Jackson candle turbidimeter used a graded glass tube that was gradually filled with water until a person looking down through the tube could no longer discern the candle. This method took into account the effects of both light scattering and absorbance, and the reported value for turbidity was an apparent value.

The appearance of a water sample is the product of the combined effects of scattering and absorption of light by dissolved and suspended material. It is a reasonable approximation that dissolved material absorbs light

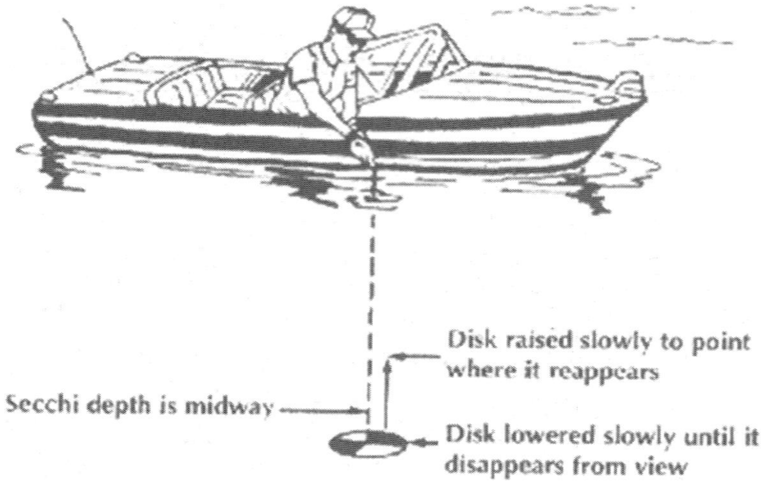

Figure 1 Use of the Secchi disk.

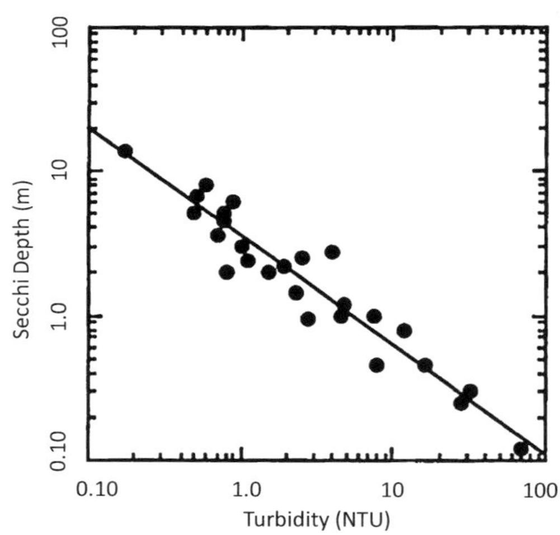

Figure 2 Secchi disk depth versus turbidity in 27 New Zealand Lakes. *From Vant and Davies-Colley 1984.*

but particulate matter absorbs light and reemits part of it as scattered light. Scattering of light is the cause of a hazy or milky appearance which is described by the turbidity, while colour is generally regarded as being imparted by dissolved material. This separation of the two effects is not rigid because substances in solution do scatter light to some degree and particles in suspension, such as iron compounds, may be coloured. At

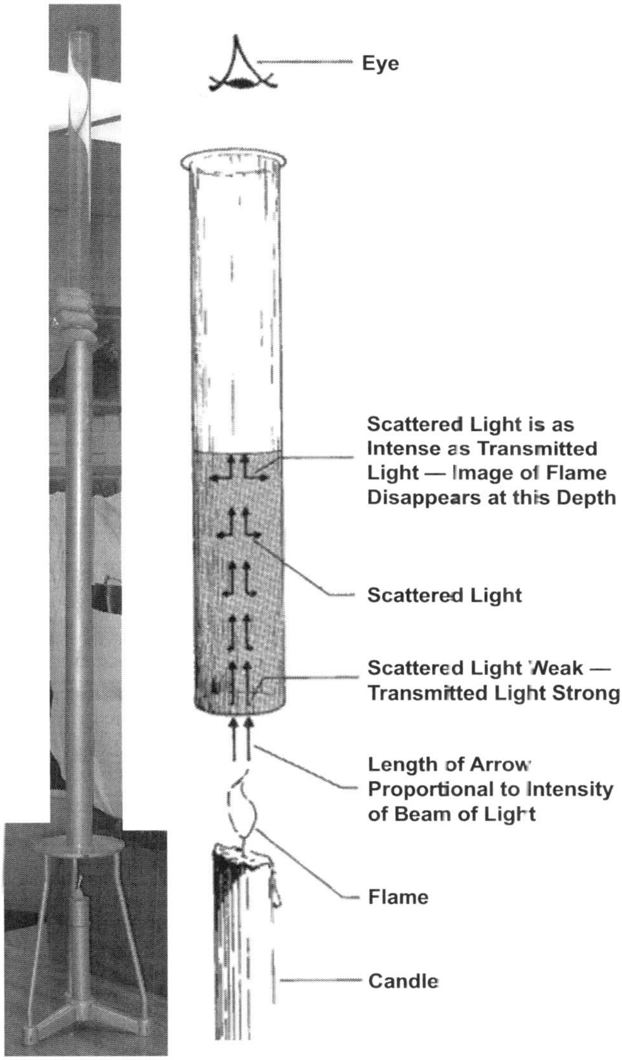

Figure 3 The Jackson candle turbidimeter.

high suspended solids concentrations the appearance of a sample may be due as much to attenuation of light by absorption as to the scattering of light. For practical purposes the contribution of particulate matter to appearance can be measured as turbidity directly on the water sample, but measurement of colour generally requires the removal of suspended material by filtration.

Table 1 Example turbidity ranges at water treatment plants

Location	**Turbidity range (NTU)**
Surface water	1 to >100
Settled water applied to filters	0.5–1.5
Filtered water under steady state operation	0.05–0.15
Filtered water during ripening	0.05–0.50
Filter backwash water	10–500

There was a gradual shift from expressing turbidity as an aesthetic criterion, to stipulating turbidity goals for health reasons. This required accurate measurements of low levels of turbidity, far lower than was possible with the Jackson candle method. Typical ranges of required turbidity measurements at water treatment plants are presented in Table 1 (Burlingame et al., 1998).

As an example of the health-based role of low-level turbidity measurements, the Pennsylvania Department of Environmental Protection in 1996 pointed out the relationship between turbidity spikes and Giardia breakthrough in filtered water: a 10- to 50-fold increase in cyst breakthrough from filter media disturbance could occur with only a corresponding increase in turbidity of 0.1 NTU (Burlingame et al., 1998).

Monitoring turbidity at each step of the treatment process is performed at many water treatment plants. The Long Term two Enhanced Surface Water Treatment Rule (LT2ESWTR, USEPA, 2003) proposed a 0.5 log Cryptosporidium removal credit for a combined filter effluent turbidity less than 0.15 NTU in 95% of samples each month and (with certain conditions) a 1.0 log credit for filtered water turbidities of less than 0.10 NTU in 95% of daily maximum values from individual filters.

As more sensitive instruments developed, they adopted a tungsten filament light source at a colour temperature of 2200–3000 K that impinged on a water sample held in a clear glass sample cell. A spectral peak response detector tuned between 300 and 400 nm was located at an angle to the incident light beam to measure the amount of light reflected towards the detector. Eventually this angle was fixed at 90°. This instrument is called a nephelometer and gives values in nephelometric turbidity units or NTU. However, in the past, a number of other turbidity units were used, based on the synthetic dispersions used to calibrate the instruments. The units are not interchangeable and NTU is currently the preferred unit. Some of the different turbidity units that have been used are listed in Table 2.

Table 2 Different turbidity units

Nephelometric Turbidity Units	NTU
Formazin turbidity units (various)	FTU
Jackson turbidity units	JTU
Silica (Kieselgur) units	$mgSiO_2/L$
Fullers earth (Fulbent 570) units	mgF.E./L
Absolute units (Zeiss-Pulfrich turbidity unit)	A.E.
Mastic units	Drops of mastic
Langrohr units (reciprocal length)	C
Coleman Nephelos units (TiO_2)	CNU
Helm units ($BaSO_4$)	

From Water Research Centre (1979).

3. TURBIDITY AS SURROGATE FOR PARTICLE CONCENTRATIONS

Turbidity is often assumed to be a surrogate for TSS. It is generally true that the higher the TSS then the more particles are expected in suspension, and the higher should be the turbidity. However, suspended particles, particularly in natural waters contain a large variety of material that could include clays, silts, inorganic matter, organic matter, vegetation, and living organisms that produce a large variety of particle sizes and optical characteristics. Therefore, there is no universal means of turbidity calibration that can define turbidity for all water sources.

Bratby (2006) compiled the relationships found between turbidity (NTU) and TSS (mg/L) for a number of raw water sources. The values presented were: 0.74; 0.80; 1.0; 1.25; 1.40; and 1.60 NTU/TSS. Results using the overflow from water treatment sludge gravity thickeners were reported as ranging from 2.0 to 2.27 NTU/TSS. These results demonstrate that turbidity is not a universal or inherent measure of suspended material.

Light scattering and light absorption depends on the range of optical properties of a suspension, which is dependent on the shape, size and refractive index of the particles. For example, particles less than 0.1 μm in diameter do not produce significant turbidity responses; particles less than 1 μm produce the highest turbidity responses; and particles greater than 1 μm can produce progressively lower turbidity readings (Burlingame et al., 1998).

The above is not to say that turbidity is not a good indicator of the presence of suspended particles, but that there is not a clear conversion factor between TSS and turbidity. However, for a given water or application, under a given set of circumstances, empirical correlations between TSS measurements and turbidity readings can be useful, although the relationship between the two parameters may vary over time and particularly seasonally for raw water sources, requiring periodic checks on the empirical relationship.

The true value of turbidity in modern water treatment plants is that it allows on-line monitoring of treated water quality, at extremely low levels of particle concentrations. In this case the conversion between such low levels of turbidity and TSS becomes almost meaningless. Concerns with cyst breakthroughs from water treatment filters are of concern to most water treatment utilities. The advent of modern high precision instruments, particularly those relying on laser technology have been shown to provide reliable indicators of particle breakthroughs from filters. Figure 4 demonstrates one case where an online particle counter was unable to adequately detect some of the particle breakthroughs, whereas an online laser turbidimeter did detect the breakthroughs (Sadar, 2005). The major difference between the two instruments is that the particle counter detected particles larger than 2 μm, whereas the turbidimeter was able to detect submicron particles.

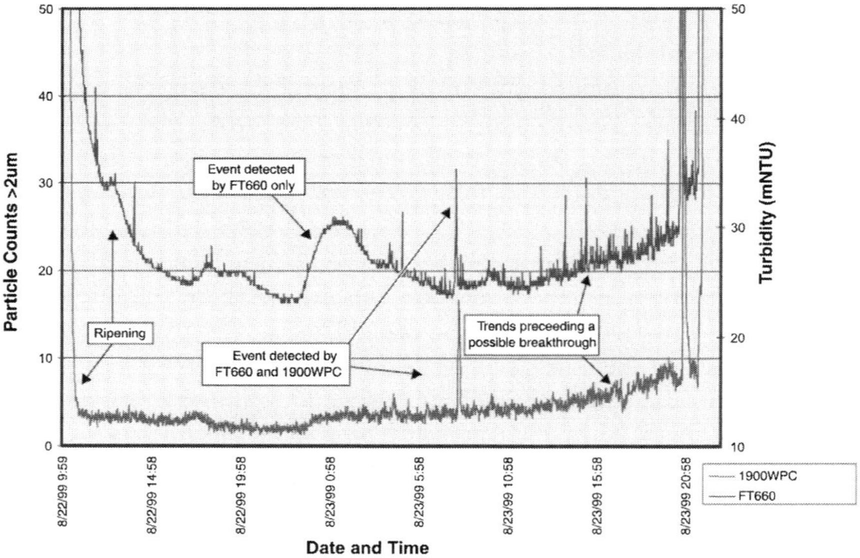

Figure 4 Side-by-side filtered water quality using a laser turbidimeter (FT600; 1 mNTU = 0.001 NTU) and a particle counter. *From Sadar (2005).*

4. PRINCIPLES OF TURBIDITY MEASUREMENT

When light is directed towards a suspension of particles, the intensity of light is reduced due to absorption and scattering of light. Absorption results in loss of energy from the light which is transferred to heat. Absorption of specific light wavelengths manifests as a characteristic colour.

Unlike absorption, light scattering results in no net loss of energy from the light beam and results in radiation of energy in all directions at the same frequency as the incident light radiation. The intensity of the scattered light depends on the refractive index of the particles, and the difference between the refractive indices of the particles and the water, or other suspending medium. The refractive index of pure water is approximately 1.33, although the value changes with temperature and the concentration of dissolved organic and inorganic compounds. Temperature effects are relatively minor, within the ranges commonly encountered in water treatment. The effect of dissolved compounds is generally to increase the refractive index of water.

For a given particle size and wavelength of incident light, turbidity decreases with increasing refractive index of the suspending medium. Turbidity also increases as the particle size increases from zero up to a diameter equal to the wavelength of the incident light. As particle size increases further, turbidity then progressively decreases. These phenomena are mathematically complex and are described sby the theory attributed to Gustav Mie, using Maxwellss equations to solve for regions inside and outside the particles (AWWARF/IWSA, 1997). Figure 5 schematically shows the light scattering phenomenon expected with a small particle, of diameter smaller than 10% the wavelength of the incident light. Such small particles exhibit a relatively symmetrical scattering distribution in both the forward and backward directions. The small particle absorbs the light energy and reradiates the light energy in all directions as if it were a light source itself (Hach, 1985).

Figure 5 Scattered light pattern for very small particles (smaller than 10% the wavelength of the incident light) (Hach, 1985).

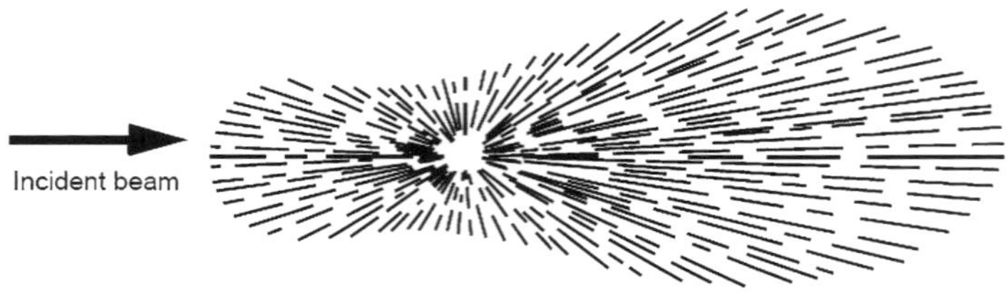

Figure 6 Scattered light pattern for large particles (approximately 25% of the wavelength of the incident light) (Hach, 1985).

Figure 6 shows the scattered light pattern for a relatively large particle, with an assumed diameter approximately one-quarter the wavelength. Light scattered from different positions around the particle create interference patterns that are additive in the forward direction, resulting in forward-scattered light that is a higher intensity than light scattered in other directions.

Figure 7 shows the scattered light pattern for particles with diameters that are larger than the wavelength of the incident light. In this case the scattering in the forward direction is increased even further, with scattering also at other angles.

Since light scatters in all directions, there are essentially two methods of measuring turbidity: by measuring the reduction in intensity of the transmitted light, and by measuring the scattered light intensity, usually at 90° to the transmitted light beam (Gregory, 2006). Although the 90° angle is not the most sensitive to concentration it is probably least sensitive to variations in particle size.

Turbidity is calculated from the ratio of the light scattered by the sample to that produced by a suspension selected as a turbidity standard. However, different instruments can give different readings for the same sample, even when they have been similarly calibrated. Characteristics of the light source and detector determine the effective spectral distribution of the light source which could give rise to differences in the results between different nephelometer instruments.

Figure 7 Scattered light pattern for larger particles (larger than the wavelength of the incident light) (Hach, 1985).

5. TURBIDITY INSTRUMENTS

Light scattered by suspended particles is measured using a nephelometer, a schematic of which is shown in Figure 8. Basically, a light source (tungsten filament or LED) directs light through a sample chamber and a detector measures scattered light, usually at 90° to the incident beam.

There has been an attempt at standardization of turbidimeters so that measurements are relatively reproducible. There are two principal standards, summarized in Table 3. United States Environmental Protection Agency (US EPA) 180.1 is widely used throughout the world. It calls for a tungsten-filament lamp and a spectral peak response for the detector between 400 and 600 nm corresponding to the wavelengths of visible light. The objective is that turbidity measurements should be most sensitive to particles that are 400–600 nm (0.4–0.6 μm) in diameter. This range of sensitivity should not be strongly dependent on the presence of dissolved compounds or the angle at which turbidity measurements are made.

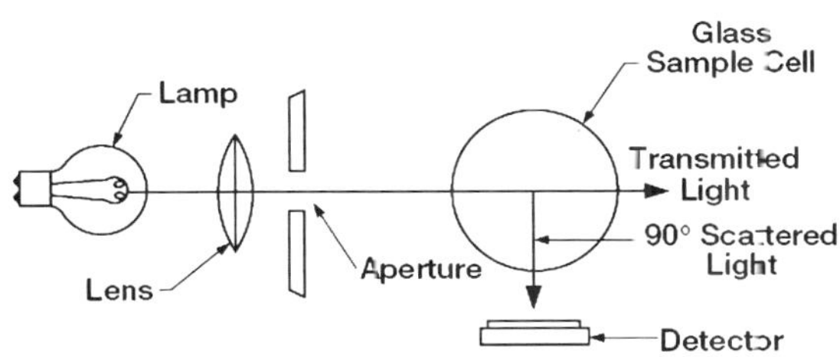

Figure 8 Schematic of a nephelometric turbidimeter (Sadar, 1999).

Table 3 Summary of design standards

Requirement	USEPA 180.1	ISO 7027
Primary detector	90° ± 30°	90° ± 1.5°
Light source	Tungsten filament lamp with colour temperature between 2200 and 3000 K	Wavelength of 860 nm (LED or a combination of tungsten filament lamps with filters can be used)
Spectral response peak for the detector	400–600 nm (primary wavelengths of light)	Spectral bandwidth of the light must be within 860 nm ± 30 nm
Measurement range	0–40 NTU (any sample above 40 NTU must be diluted)	0–4 NTU (any sample above 40 NTU must be diluted)

NTU, nephelometric turbidity units

Method USEPA 180.1 is particularly applicable to clean water samples, below 1.0 NTU, which is the critical range for drinking water treatment. However, the method is sensitive to colour that absorbs light in the 400–600 nm range of wavelength, and the tungsten filament lamps require lengthy warm-up times to achieve stability, and should be calibrated frequently. It is for this reason that laboratory instruments using this technology are often recommended to be permanently switched on for regular use.

Method ISO 7027 originated in the brewing industry and is commonly used in Europe. The method calls for a near monochromatic light source that is stable, has low absorbance interference with samples, and results in low stray light. However, the disadvantage of the long wavelength source adopted in this standard is a reduced sensitivity to small particle sizes. If the reduced sensitivity is electronically amplified this can result in increased background noise at low turbidity levels. Instruments using this method can have a disadvantage when measuring very low turbidities required in water treatment plants.

There are a number of turbidimeter configurations that have been designed to achieve increased accuracy or to reduce interferences. Figure 9 shows a schematic of a ratio turbidimeter. In addition to a primary light detector positioned at 90° to the incident beam, it also includes secondary detectors situated (for example) at 0° and possibly at 15° to the transmitted light beam. The signals from the multiple light detectors combine electronically to give the turbidity reading. Such ratio-type instruments are useful when measuring the turbidity of coloured liquid suspensions, such as beer, by minimizing the effect of light-absorbing soluble substances on the turbidity measurement. Some instruments are provided with a switch to activate or deactivate the secondary detectors thus changing the instrument from a ratio to standard design.

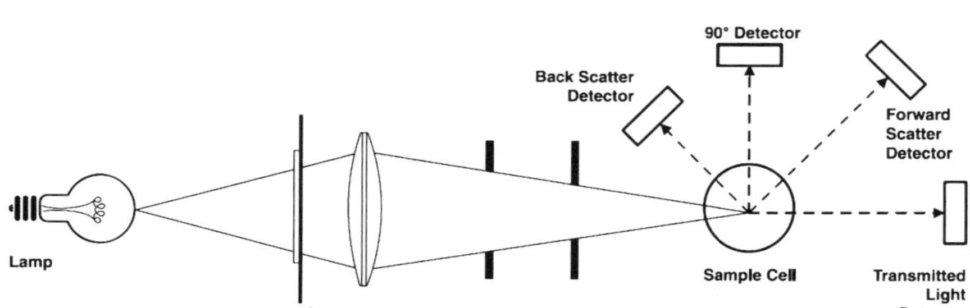

Figure 9 Schematic of a ratio turbidimeter (Sadar, 1999).

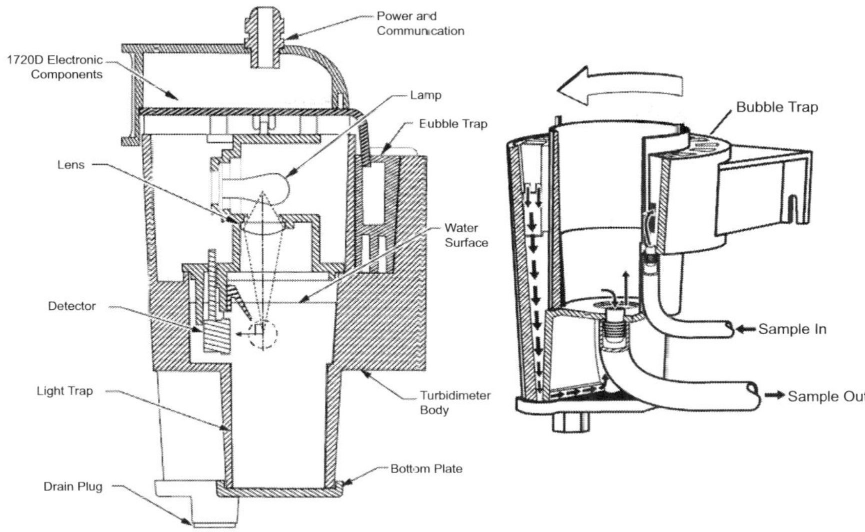

Figure 10 Schematic of an inline process turbidimeter (Sadar, 1998).

A major benefit of turbidity measurements is that they are well suited to on-line process control. Figure 10 shows a schematic of one such instrument which includes a network of baffled chambers exposed to atmosphere thereby minimizing the chance of false readings due to the presence of bubbles. This design also includes a "keyhole" design to reduce stray light and maximize the accuracy of the readings.

An alternative design of process instrumentation is shown in Figure 11. This is an example of a surface scatter turbidimeter in which the light source and detector are mounted above the turbidimeter body thereby isolating optical components from the process water. The sample flow is introduced into the centre of the body and overflows a weir at the top forming a near perfect optical flat surface. The light beam is focused on the sample surface at an acute angle of about 15° and light scattered by the particles in the illuminated area is detected by the photodetector. The stray light in this instrument can be kept to a minimum by providing an entrance slit for the light beam, shielding baffles over the photocell, and providing an optical black surface to the sample container body. Stray light has been measured at an equivalent of less than 0.01 NTU. The instrument is particularly suited to wide-range measurements.

A turbidimeter design suited to very low in-line turbidity measurements is based on laser technology, as shown in Figure 12. The light source

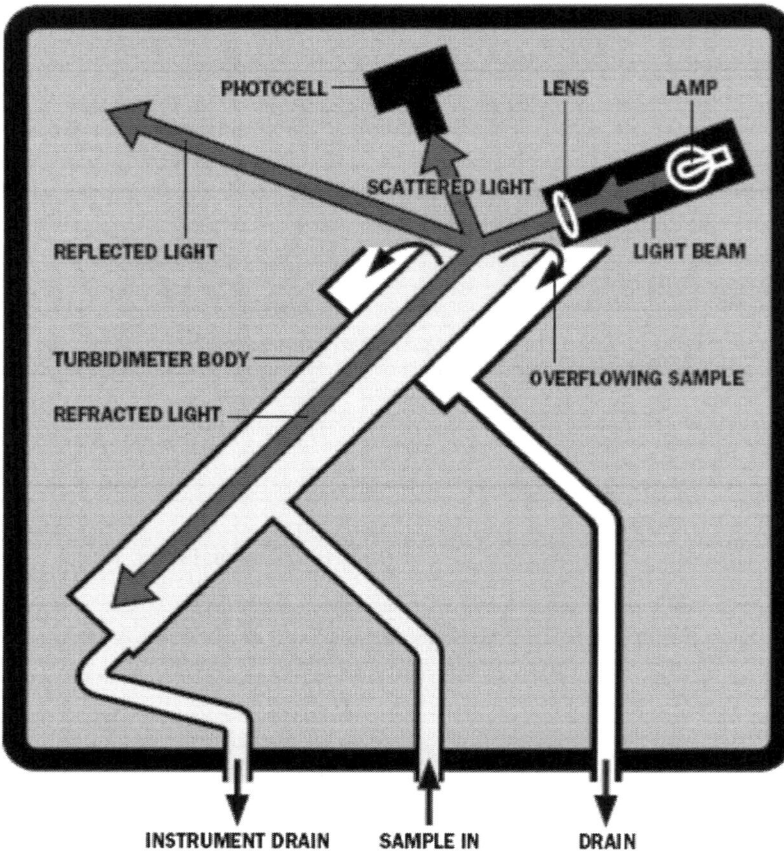

Figure 11 Schematic of an inline process surface scatter turbidimeter (Sadar, 1998).

in the instrument shown is a laser diode with a wavelength of 660 nm. The response system is optimized to detect this specific wavelength. The design produces a high beam power intensity that maximizes scattered light intensity from smaller particles. A light trap at the bottom of the nephelometer body reduces the stray light by about 50% compared with traditional instruments (Sadar, 2005). The instrument includes optical fibres to maximize the signal conveyed to the detector. The instrument is capable of detecting changes in turbidity down to 0.3 mNTU (0.0003 NTU). Some instruments include a Relative Standard Deviation (RSD) feature that quantifies the fluctuations in the laser turbidity measurements. This allows detection of precursors to turbidity events and impending turbidity spikes.

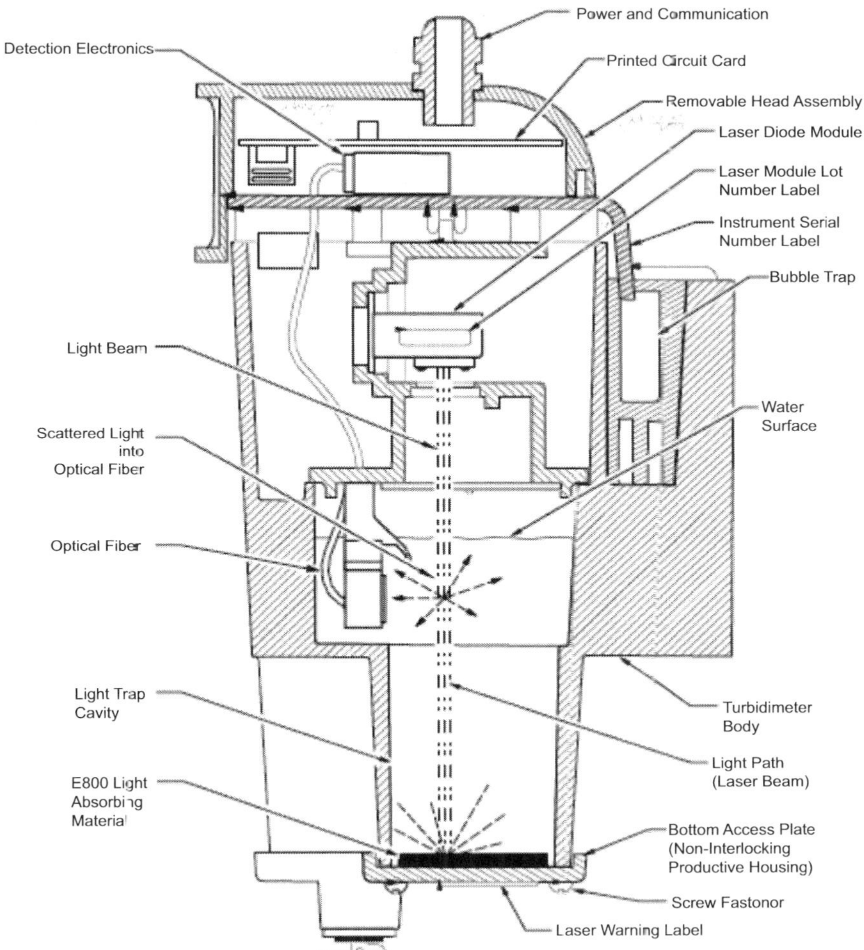

Figure 12 Schematic of an inline process laser turbidimeter (Sadar, 2005).

Figure 13 shows a turbidimeter suited to high solids concentrations. This type of instrument is used in filter backwash troughs or in activated sludge basins. The instrument shown includes an LED light source that transmits a beam of infrared light into the sample stream at an angle of 45° to the sensor face. A pair of photoreceptors detect scattered light at 90° to the transmitted beam. A backscatter photoreceptor can be mounted at 140° to the transmitted beam to detect light scattered in high solids sample streams. The sensor face can also include a self cleaning wiper device to minimize biological growth on the face of the device.

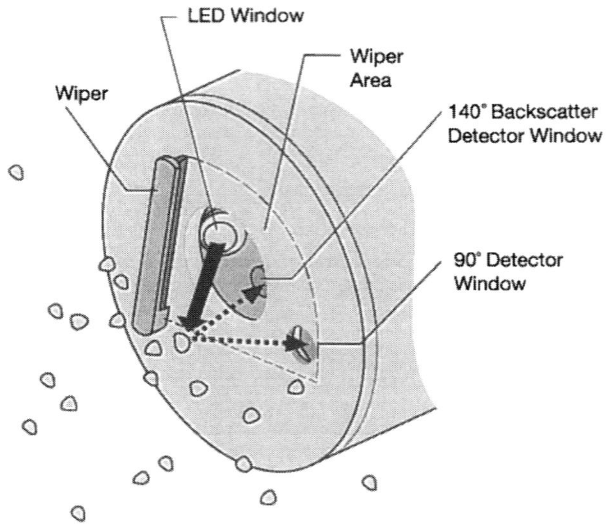

Figure 13 Schematic of a high solids turbidimeter (Hach 1985).

6. INSTRUMENT CALIBRATION

Turbidimeters should be regularly calibrated to ensure consistency between readings. Calibration includes adjusting the instrument to known turbidity standards. It is usually recommended that the instrument be calibrated to a standard suspension that is the closest to the expected range of turbidities of the samples to be measured. The primary standard for calibrating nephelometric turbidimeters has long been considered to be that based on formazin. It is prepared by accurately weighing 10 g of hydrazine sulphate and 100 g of hexamethylenetetramine in 1 L of distilled water (method 180.1, USEPA, 1979). The solution develops a white turbidity after standing at 25 °C for 48 h. The polymer formed consists of random shapes and sizes, ranging from 0.01 to 10 μm and the suspension is highly reproducible. Care is required in handling and storing the chemicals due to their toxicity.

The turbidity of this suspension is 4000 NTU and working suspensions of lower turbidity are prepared by diluting the 4000 NTU suspension using particle-free water, typically less than 0.03 NTU.

Since formazin is considered a primary standard, it can be used in any instrument to calibrate against known turbidity values.

The 4000 NTU stock suspension is generally stable for up to 1 year. However, diluted suspensions have a shelf life that ranges from minutes to weeks, depending on the dilution. A diluted suspension of 2 NTU is generally considered the minimum practicable. Studies have shown that a 40 NTU solution decayed to 28 NTU in 18 h and a 1 NTU solution decayed to 0.4 NTU in 19 h, even with refrigeration (Burlingame et al., 1998).

Because of issues with handling formazin suspensions, one manufacturer provides more stable formazin suspensions prepared in sealed cuvettes down to 0.10 NTU. According to the manufacturer, these preparations are stable for at least 2 years. The suspensions are prepared using an excess of hexamethylenetetramine. The characteristics of the stable suspensions are the same as the original formazin standard suspension (Sadar, 1999).

An alternative standard for calibration provided by one manufacturer uses suspensions of microspheres made of styrene divinylbenzene (SDVB) copolymer. Suspensions of SDVB have been shown to be as stable as concentrated formazin and much more stable than diluted formazin. Similar to the preprepared stable formazin suspensions, the SDVB suspensions are provided prediluted in sealed cuvettes, ready to use. SDVB suspensions are prepared to agree with formazin standards and are considered secondary standards.

Due to the monodispersed nature of the SDVB size distribution, incident light may over-scatter in the forward direction resulting in inaccurate calibration with some machines. This is the case with ratio turbidimeters, for example, where the manufacturer provides a special formulation to obtain acceptable readings when the instrument is tested against formazin standards.

Letterman et al. (2004) conducted comparative tests on a range of filtered, settled or disinfected only waters from different treatment technologies and water treatment plants from 10 geographically different water supply systems. They evaluated analyst techniques; 10 different instruments and 12 different instrument mode categories; sample turbidities ranging from 0 to 0.6 NTU; and four different calibration materials. The calibration materials evaluated were as follows:

- User-prepared formazin (following USEPA method 180.1). Subsequent dilutions used laboratory reverse osmosis (RO) water;
- Commercially supplied formazin stock solutions, subsequently diluted with RO water;
- Stabilized commercial formazin, prepared and sealed in a range of usable dilutions by the manufacturer;
- SDVB microspheres supplied by the manufacturer in a range of prediluted suspensions.

Principal findings from this study are summarized as follows:

- The light source of the various instruments (tungsten filament, green LED, or infrared LED) cause minimal differences in readings. Instruments with LED light sources gave readings approximately 3% higher than tungsten filament instruments. For ratio instruments the discrepancy between the two light sources dropped to less than 0.5%.
- The calibration material used had minimal effect on the results of all instrument modes and water sources.
- A significant difference was noted between two groups of instruments when measuring low turbidity levels. The instruments that automatically set a predetermined reading (for example, 0.02 NTU) when a deionized low particle water sample (for example, the laboratory RO water used for dilution of the calibration material) is placed in the instrument during calibration, subsequently gave significantly lower readings than instruments that do not operate this way. It was suggested that the differences between the two groups is related to the different electronic processes used in "zeroing" and calibrating the instruments (Letterman et al., 2004). Instruments that do not operate this way measure the turbidity of the dilution water and store the result in the instrument software. This value is then used to subtract the turbidity of the dilution water from the measured values of the other turbidity standards used to perform a calibration (Sadar, 1999).
- Analyst technique when calibrating and measuring low turbidities is a significant factor in the accuracy of the results.

7. TECHNIQUES FOR ACCURATE TURBIDITY MEASUREMENTS

A number of workers have pointed out techniques that are important for accurate and reproducible turbidity measurements. It is also important to follow the specific instrument operations manual. A number of the principle techniques are summarized below:

- It takes only a small amount of suspended matter, or a slightly dirty sensor, or a slightly dirty cell to cause significant inaccuracies when measuring low turbidities, for example at the 0.1 NTU range (Burlingame et al., 1998).
- The cleanliness of sample cells is an important issue for low-level turbidity measurements. One manufacturer recommends the following for ultra-low turbidity measurements:

- Wash sample cells with soap and deionized water;
- Soak the washed sample cells in 1:1 hydrochloric acid for at least 1 h. Sample cells can also be placed in a sonic bath to release particles from the glass surfaces;
- Rinse the cells with ultra-filtered deionized water (RO or 0.2 μm filter) at least 15 times;
- Immediately cap the cells to prevent contamination and to prevent drying out of the inner cell walls;
- Polish the outside of the cell with silicone oil to prevent particle attachment and to fill in small imperfertions in the outer glass.

- At low-level turbidity readings, it becomes more important that sample cells be indexed. Cells vary in optical quality; one side reflects more light, whereas another side absorbs more light. Burlingame et al. (1998) describe their practice whereby new cells are indexed and tested against the requirement that they not vary by more than 0.01 NTU. As many as 25% of the new cells are discarded. The recommended procedure is as follows:
  - Once the sample cells have been cleaned, fill them with ultra-filtered low turbidity water;
  - Let samples stand to allow any bubbles to rise;
  - Then polish the cells with silicone oil;
  - Measure the turbidity at several points of rotation of the sample cell in the instrument;
  - Find the orientation where the turbidity reading is the lowest and mark the cell on the side in such a way that it can always be lined up the same way in the instrument's chamber for future measurements.
- If possible, use one sample cell for the low turbidity readings, particularly one that has been demonstrated to be of high optical quality and has been indexed.
- Dust inside the instrument can increase stray light, especially when it is stirred up when closing the instrument lid. Stray light can cause errors from less than 0.01 to greater than 0.04 NTU. It is important to follow the instrument operations manual in this regard.
- Condensation on the exterior glass window of the light source can interfere with low turbidity readings by causing noise and drift of the readings. Follow the instrument operations manual.
- Decay of the incandescent lamp or its variation over time can be significant at low-level measurements and can be a significant source of noise and drift in readings. Follow the instrument operations manual.

- It is important to remove bubbles from the sample before taking readings. Gas bubbles reflect light, produce light scatter and cause false turbidity responses and erratic instrument behaviour. The first measure should be to fill the sample cell slowly to avoid entraining bubbles. Then let the sample stand for several minutes to allow bubbles to vacate the sample. If the sample needs to be mixed, gently invert the sample several times to avoid introducing bubbles.

If bubbles are still observed in the cell, for example attached to the cell wall, the capped cell can be rotated in the horizontal plane to dislodge the bubbles and allow them to rise to the headspace when returned to the upright position.

A vacuum can be applied to the sample, although care must be taken to avoid contaminating the sample with the vacuum device. One method inserts a 50 mL syringe through a single hole in the cap of the sample cell. The plunger on the syringe is retracted to cause a partial vacuum. The vacuum is held until the formation of gas bubbles is no longer evident.

Sonication can also be applied for 1–2 s to samples that are warming up to room temperature. Care should be taken with sonication since it could cause particles to fracture and change size, and cause particles attached to the sample cell wall to detach and cause higher turbidity of the sample.

- The use of in-line instruments is expected to reduce some of the errors caused by sampling. Grab sample collection can be affected by the cleanliness of the sample bottles; the sample holding time; sample storage conditions; particle breakup, clumping or settling; or particles sticking to the sample bottle. Some in-line process models include bubble traps or dissipaters that reduce the effects of bubbles on the turbidity readings.

REFERENCES

Allhands, M., 2007. For Good Measure. Breaking Down Turbidity and Total Suspended Solids. WE&T. Sep., 112–114.

AWWARF/IWSA (American Water Works Association Research Foundation and International Water Supply Association), 1997. Treatment Process Selection for Particle Removal. Cooperative Research Report, USA.

Bratby, J., 2006. Coagulation and Flocculsation in Water and Wastewater Treatment. IWA Publishing, London.

Burlingame, G., Pickel, M., Roman, J., 1998. Practical applications of turbidity monitoring. J. AWWA 90 (8, Aug), 57–69.

Davies-Colley, R., 1988. Discussion of Secchi Disc Transparency and turbidity by Effler S.W., J. Environ. Eng. Div. ASCE, 114(6).

Gregory, J., 2006. Particles in Water. Properties and Processes. IWA Publishing/CRC Press, Taylor and Francis, London/Boca Raton.

Hach, C., 1985. Turbidity Standards. Technical Booklet 12 (Also Updated Version by Sadar M.). Hach Company, U.S.A.

Letterman, R., Johnson, C., Viswanathan, S., 2004. Low-level turbidity measurements: a comparison of instruments. Jour. AWWA 96 (8, Aug), 125–137.

Sadar, M., 1998. Turbidity Science. Technical Information Series-11. Hach Company, U.S.A.

Sadar, M., 1999. Turbidimeter Instrument Comparison: Low Level Sample Measurements. Technical Information Series. Hach Company, U.S.A.

Sadar, M., 2005. Introduction to Laser Nephelometry: An Alternative to Conventional Particulate Analysis Methods. Hach Company, U.S.A.

USEPA, 1979. Methods for Chemical Analysis of Water and Wastes, 600/4/79/020, 180.1-1, Washington.

USEPA, 2003. Long term 2 enhanced surface water treatment rule; proposed rule. Fed. Reg. 68 (154), 47640 (Washington).

Vant, W.N., Davies-Colley, R.J., 1984. Factors affecting clarity of New Zealand lakes. New Zealand Journal of Marine and Freshwater Research 18, 367–377.

Water Research Centre, 1979. Measurement of Color and Turbidity. Notes on Water Research, No. 22, August, U.K.

CHAPTER SEVENTEEN

Capillary Suction Time (CST)

N.F. Gray
Water Technology Research Group, Trinity Centre for the Environment, Trinity College, University of Dublin, Dublin, Ireland

Contents

1. INTRODUCTION

Capillary suction time (CST) was originally developed as a simple method of determining both sludge dewaterability (i.e., the release of water from the sludge matrix) and for evaluating the effects of pretreatment chemicals (i.e., conditioners, flocculants, and coagulants) on sludge filterability. Modern measurements still employ the original method and apparatus developed at the Water Pollution Research Laboratory in Stevenage, England (now the Water Research Centre(WRc)) (Gale and Baskerville, 1967; Baskerville and Gale, 1968).

When a solids suspension is retained in a column a layer or bed of concentrated solids will form at the base. A thin porous layer, such as a filter paper, exposed to the base of this bed will allow the free water not chemically bound to the solids to drain either from the solids or through the solids bed from the water above. This water travels radially from the base of the column due to capillary suction pressure of the porous layer. The rate of movement through the porous layer is primarily dependent on the permeability of the bed of concentrated solids, and the water holding capacity of the solids both settled and in suspension. Where the settled solids form a dense compacted bed with small interstices, water will pass through slowly (i.e., low CST). Where solids form large aggregates or are less compacted with larger

Progress in Filtration and Separation
http://dx.doi.org/10.1016/B978-0-12-384746-1.00017-3

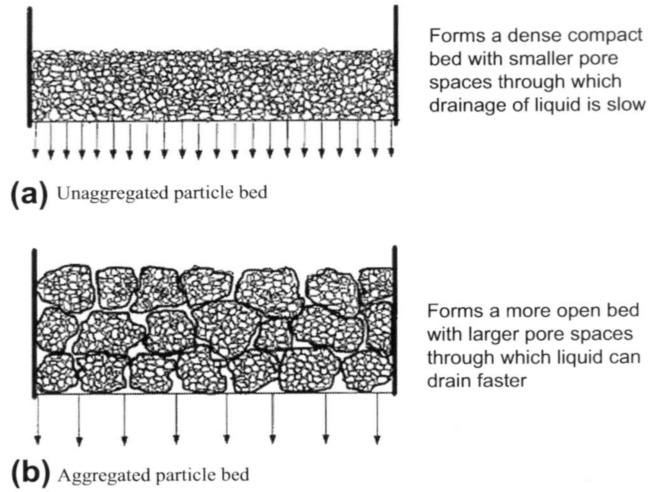

Figure 1 Schematic illustration of the nature of sediment bed formed by (a) dispersed suspension and (b) aggregated suspension. *Reproduced from Besra et al. (2005) with permission from Wiley Inc. New York.*

interstices then water will pass through the bed at a higher rate (i.e., high CST) (Figure 1). In certain circumstances no settlement occurs with the solids equally distributed within the water phase. However, the solids can still bind water chemically by hydration reactions, which can reduce the rate of water movement through the porous layer (i.e., low CST) (Besra et al., 2005). Capillary suction creates a much larger force than that created by the hydrostatic head alone. So as long as there is sufficient sample of test material to give a CST value, the test can be considered independent of the volume of test material when using the standard test apparatus. Outside the standard test environment, the height of the sample in the test column will increasingly become a factor. The CST test is a widely used alternative to the more complex specific resistance to filtration (SRF) test developed by Coakley and Jones (1956). Although the SRF test has been modified it remains a quite difficult and expensive alternative to CST (Christensen and Dick, 1985a,b; Novak and Knocke, 1987; Sawalha and Scholz, 2009).

2. METHODS

The most widely used CST apparatus is produced by Triton Electronics Ltd (www.tritonel.com/) (Figure 2). A similar unit is manufactured in the USA by OFI Testing equipment Inc. (www.ofite.com/products/294-50.asp).

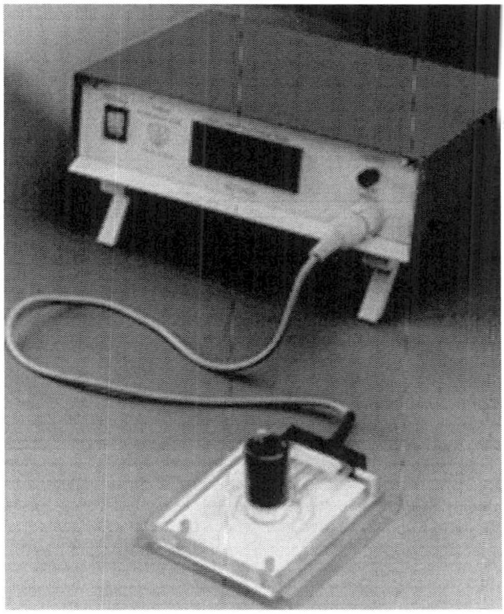

Figure 2 The standard type 304 CST apparatus manufactured by Triton Electronics Ltd, Dunmow, Essex, UK. CST, capillary suction time. *Reproduced with permission.*

The standard test apparatus consists of filtration block and timing unit. The filtration unit comprises two Perspex blocks between which the filter paper is placed (Figure 3). The lower block is recessed to allow the insertion of a standard size piece of absorbent chromatography paper (Whatman No. 17). An upper Perspex plate in which three electrical contacts are embedded is placed on top of the filter paper. A stainless steel funnel (either 10 or 18 mm internal diameter) for holding the test suspension is inserted into a central hole in the upper plate The larger funnel is used for heavier sludges to reduce the time taken for the test to be completed and to reduce variability (Sawalha, 2010). The CST is measured by recording the time in seconds for the water draining from the sludge poured into the central funnel to travel a set distance. This is achieved by two electrodes that are embedded in the upper Perspex plate at a standard distance from the funnel that opens and closes a circuit as the water in the filter passes the electrodes.

The test starts when the funnel is filled with a standard volume of liquid or sludge. Water is pulled from the sludge by capillary action into the filter paper creating a wetted area that slowly expands outwards. When the wetted area reaches 18.6 mm from the centre of the funnel it

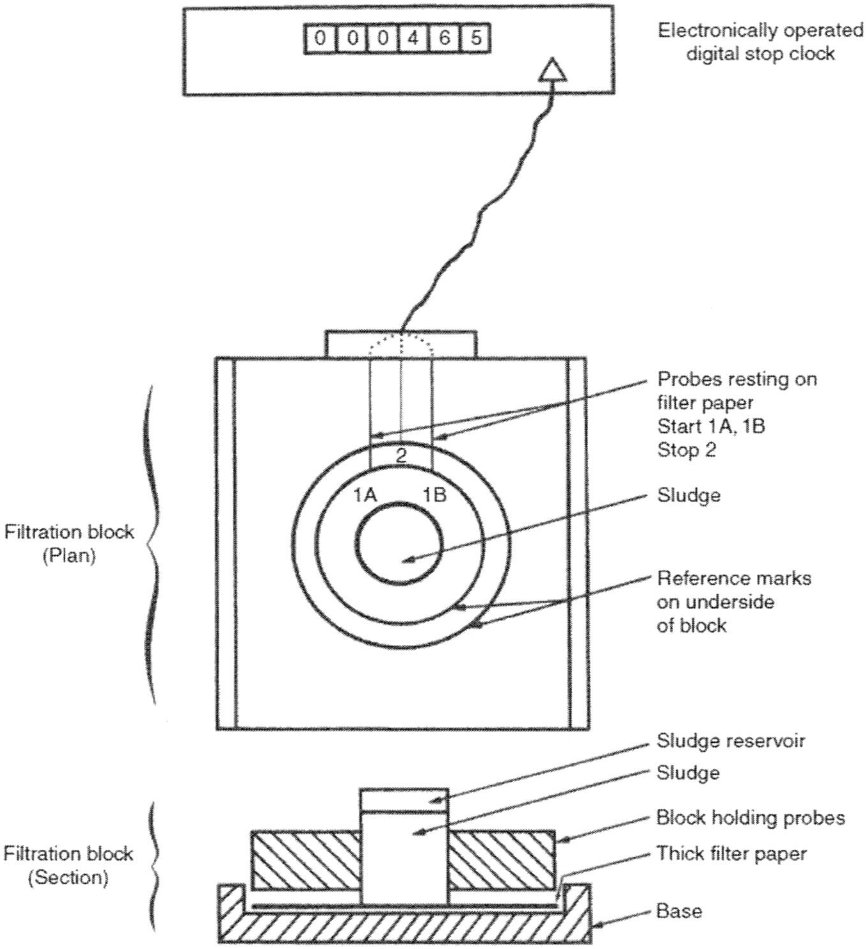

Figure 3 The capillary suction time (CST) test apparatus. *Reproduced from the Department of the Environment (1985) with permission of her Majesty's stationery office, London.*

reaches the first two electrical (Figure 3-1A or 1B) contacts that form a circuit that turns on the timer to start the test. When the wetted area of the filter paper reaches the third contact position located 22.3 mm from the central position (Figure 3–2) the timer stops and displays the CST result in seconds. So in practice, sludges that release their water readily have a low CST and vice versa (Figure 4).

Standard versions of the CST method have been published in the United Kingdom (Department of the Environment, 1985) and the United States as Standard Test Method 2710G CST (Standard Methods, 2012).

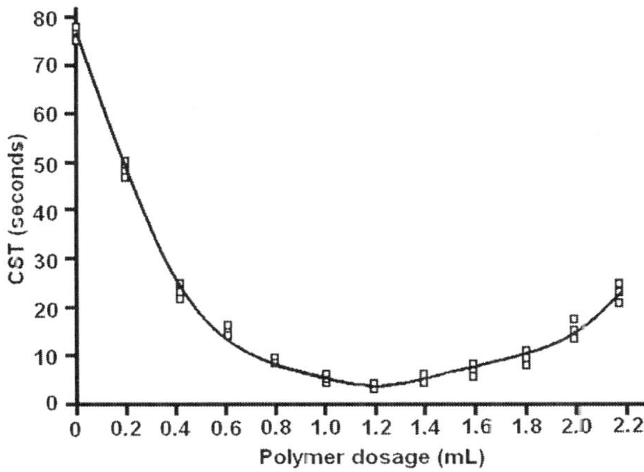

Figure 4 The effect of chemical conditioning on the capillary suction time(s) (CSTs) of a wastewater sludge.

A number of different modifications to the standard apparatus have been proposed. Scholz (2006) reported improved precision by replacing the circular funnel of the standard CST test apparatus with a rectangular funnel. He also recommended using a cheaper filter paper with similar or improved characteristics to reduce consumable costs. Sawalha and Scholz (2007) also studied the funnel geometry of the standard CST apparatus and were able to improve test repeatability by replacing the standard circular funnel with a novel rectangular design. They also proposed the use of a funnel sealant, to reduce leakage between the funnel and the paper, which reduced their test variability by 63% of the coefficient of variation (Sawalha and Scholz, 2009). Schozl (2005, 2006) found that stirring the sludge within the sludge chamber improved CST precision by preventing sedimentation, although this has not been confirmed in subsequent studies (Sawalha, 2010). The test bed has also been modified, most notably into a linear CST unit (Unno et al., 1983; Tiller et al., 1990), although the radial flow unit remains the standard. Herwijn et al. (1995) have replaced the filter paper with a ceramic plate that permits the wetted area to be monitored continuously over time as it advances.

Modelling the CST has proven difficult due to the variability of the equipment, test conditions, and the sludge characteristics (Unno et al., 1983). Vesilind (1988) has explored this in detail and has proposed a model for the test based on a filterability constant. Since then filterability as measured by the CST test has been widely examined and discussed by a number of different authors using the piston-like (Unno et al., 1983; Meeten and

Smeulders, 1995) and diffusion-like models (Lee, 1994; Lee and Hsu, 1992) which are compared by Lee and Lin (1996).

More recently a simple model to estimate filterability has been proposed by Sawalha (2010) as:

$$\log Y = \beta_0 + \beta_1\sqrt{X}$$

where Y is the predicted mean CST value (s), β_0 is the intercept (the predicted mean CST (s) when the distance[4] between the starting and stopping electrodes of the CST device is zero), β_1 is the filterability which is the slope of the regression line (s/m^2), and X is the distance[4] (m) between the electrodes.

Sawalha and Scholz (2010) were able to predict the SRF from CST tests using the model:

$$\log \mathrm{SRF} = 46.128 - 1.346\ T + 0.035\ T^2 + 13.760\ F/\mathrm{TSS}$$

where SRF is the specific resistance to filtration (m/kg), T is the temperature (°C), F is the filterability (log s/m^2), and TSS is the total suspended solids concentration (g/l). This model must be used with caution as it is only valid for the range of solids concentration tested.

3. FACTORS AFFECTING CST MEASUREMENTS

Many papers have been published on the factors affecting the CST test. These have dealt with the characteristics of the filter papers used, the effect of temperature, and solids concentration on the test and harmonization of the test.

Filtration paper: Using four different sludges, Sawalha and Scholz (2007) compared 12 different filter papers to the standard Whatman No. 17 chromatographic paper and found that several cheaper alternative papers could be used to estimate CST. They found that the anisotropic Whatman No. 17 filter paper did not produce the most consistent results of the papers tested or in the shortest time. They observed better repeatability with certain sludges using the alternative isotropic filter papers which were cheaper and quicker. Sawalha and Scholz recommend filter paper SS1107 as most suitable for testing heavy sludges. Problems have occurred when testing very small particles, such as bentonite suspensions, which can become trapped in the large pores of Whatman No. 17 filter paper (pore size 8 μm) giving erroneous results. This can be overcome by placing a filter paper of a suitable pore size between the funnel and Whatman No. 17 paper (Meeten and Smeulders, 1995).

Temperature: Vesilind (1988) showed that CST should vary in a linear manner with viscosity, which can be altered by temperature. This was

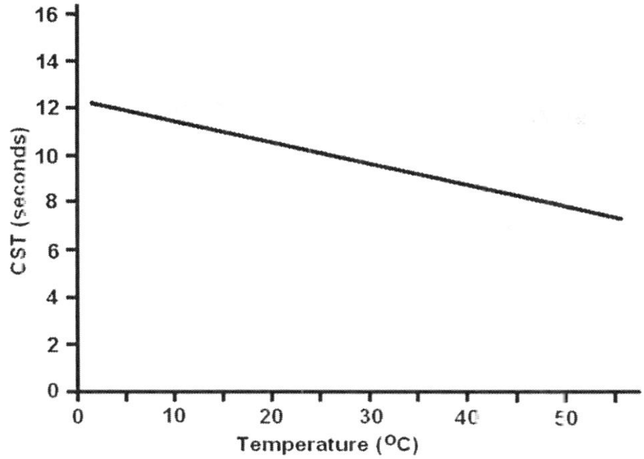

Figure 5 Linear decrease of CST of an inert calcium carbonate slurry as temperature increases. CST, capillary suction time. *After Vesilind (1988).*

shown to be true for inert sludges (Figure 5) and also organic sludges but only up to 60–65 °C, beyond which the cells increasingly lysed which altered the particle distribution within the sludge causing an increase in CST (Figure 6). However, Sawalha (2010) recorded a nonlinear relationship between CST and temperature. She explains this is most likely due to the complex effects that temperature has on sludge viscosity, filterability, settleability, desorptivity, and flocculation behaviour. For this reason she recommends replicate tests should be done at a constant temperature, which should always be recorded.

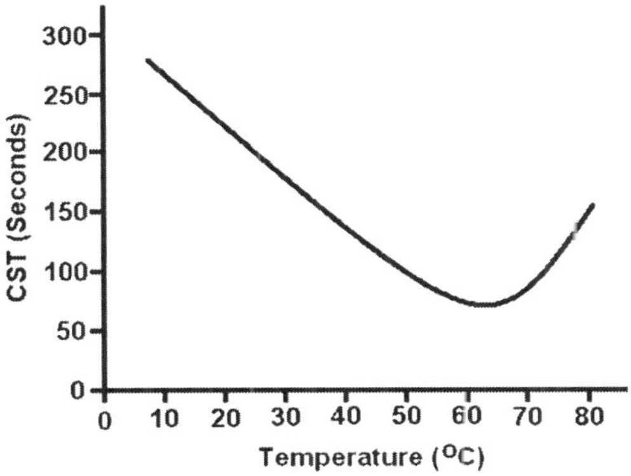

Figure 6 CST of digested sludge falls with increasing temperature until the cells lyse at >65 °C which changes the nature of sludge causing a rapid increase of CST. CST, capillary suction time. *After Vesilind (1988).*

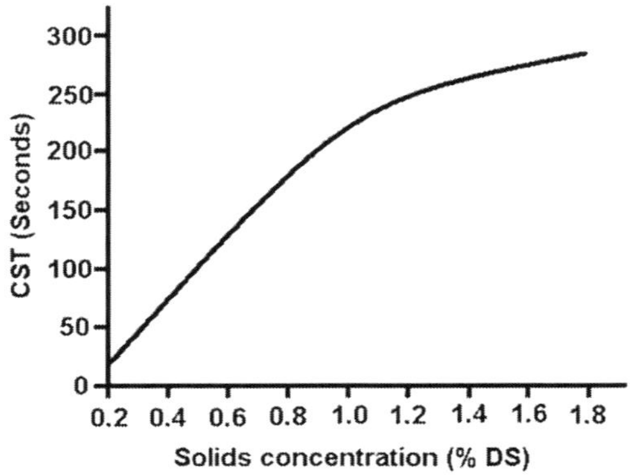

Figure 7 The CST increases with solids concentration of wastewater sludges, although not always linearly as reported by Vesilind (1988). CST, capillary suction time.

Solids concentration: There is a linear relationship between solids concentration of inert solids and CST. Vesilind (1988) also reported a linear relationship using mixed digested sludge, although in practice the rate of increase slows at higher solids concentration (Figure 7).

Harmonization: The complex and diverse nature of wastewater sludges makes it difficult to compare CST test units and to make modifications that can be universally adopted. Sawalha and Scholz (2007) have developed a standardized sludge that can be used to standardize results and harmonize between test units. The application of synthetic sludge can support the testing of experimental methodologies and should be used for subsequent benchmarking purposes (Sawalha and Scholz, 2007). CST varies between the instruments used, the temperature at which the test is conducted (normally ambient air temperature), and the specific solids concentration. Therefore CST values cannot be compared between different water and wastewater treatment plants. For that reason, unlike SRF which is independent of solids concentration, CST is not generally considered to be a fundamental sludge dewatering parameter, (Vesilind, 1988).

4. EXAMPLES OF CST USE

CST has been used for over 40 years as a simple measure of primarily wastewater sludge filterability and conditionability, but also of water treatment sludges. It is also used to calculate the optimum dosage of conditioner required by comparing filterability before and after conditioner is added to the sludge. Sludge filterability, as measured by the CST, is used to optimize the performance and operation of most sludge dewatering processes including belt and filter presses, vacuum filters, centrifuges, and drying beds. CST has also been used to evaluate the characteristics of activated sludge and also to measure the fouling potential by extra cellular polymers produced by the microbial biomass in submerged membrane reactors (Jin et al., 2004). Although other methods are available to measure the dosage of conditioner required, such as SRF, CST remains the most widely used due to its simplicity, speed, and reliability (Department of the Environment, 1985). Typical CST values for conditioned sludges with a 5% dry solids concentration using the 18 mm funnel are 10 s for belt presses, 15 s for filter presses, 40 s for both vacuum filters and centrifuges, and 300 s for drying beds.

While used primarily to assess the filterability of domestic and industrial wastewater sludges (Gale and Chem, 1977; Sengupta et al., 1994, 1997), the CST method has also successfully been used in a number of other applications. These include the study of the colloidal properties of clay suspensions, measuring the dispersion characteristics of powder suspensions such as ceramic powders and the calculation of dispersant or flocculent dosages (Besra et al., 2005) (Figure 8), and the precipitation of hydrous oxides (Soehnel et al., 1980). It is used in for a variety of applications in the mining and quarrying industry, as well as in the production of sugar and paper pulp and by the metal extraction and potash industries. Another major application of the test is in the oil and gas industry to

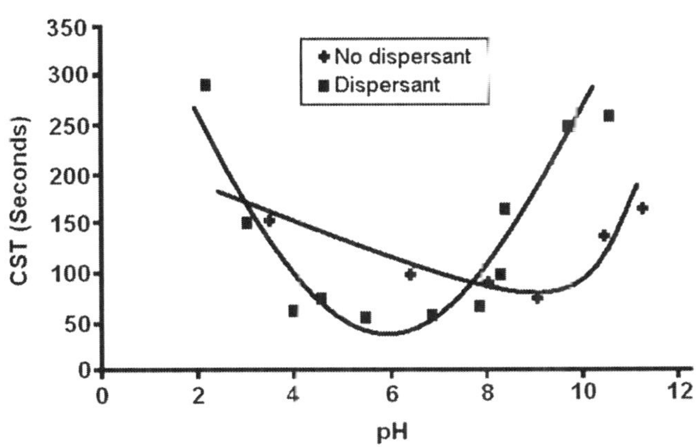

Figure 8 The effect of pH on the CST of a 1.28 vol.% alumina suspension with and without the addition of a dispersant (Darvan-C at 2.29 mg/g). CST, capillary suction time. *Besra et al. (2005).*

characterize shales and to optimize the electrolyte concentration in drilling fluids in order to minimize their effect on shale formations (Wilcox and Fisk, 1983), although there is a general acknowledgement of the limitations of the test for many of these purposes and a need for alternative test methods to be developed (Pagels et al., 2012).

5. CONCLUSIONS

The CST test is a simple and precise measure of the rate at which water is released from a sludge matrix and can be used as a substitute for SRF to assess sludge dewaterability. However, the CST is not a universal parameter in a strict sense but a comparative tool for use with specific sludges and test apparatus. Where standard dose–response curves for sludge conditioners are created care must be taken to minimize any differences between the experimental method and the temperature at which the test is conducted. Consideration should be given to using a standard rather than ambient temperature for CST measurements. The use of a synthetic sludge as suggested by Sawalha and Scholz (2007) for standardization of results and harmonization (i.e., repeatability and reproducibility) between test units will expand the applicability of the CST test.

REFERENCES

Baskerville, R.C., Gale, R.S., 1968. A simple automatic instrument for determining the filterability of sewage sludge. J. Instit. Water Pollut. Control 67, 233–241.

Besra, L., Sengupta, D.K., Singh, B.P., Bhattacharjee, S., 2005. A novel method based on capillary suction time (CST) for assessment of dispersion characteristics of suspensions. J. Am. Ceram. Soc. 88 (1), 109–113.

Christensen, G.L., Dick, R.I., 1985a. Specific resistance measurements: methods and procedures. J. Environ. Eng. Div. Am. Soc. Civ. Eng. 111, 258–271.

Christensen, G.L., Dick, R.I., 1985b. Specific resistance measurements: non-parabolic data. J. Environ. Eng. Div. Am. Soc. Civ. Eng. 111, 243–257.

Coakley, P., Jones, B.R.S., 1956. Vacuum sludge filtration. Sew. Ind. Wastes 28, 963–976.

Department of the Environment, 1985. The Conditionability, Filterability, Settleability and Solids Content of Sludges 1984. A Compendium of Methods and Tests, Methods for the Examination of Waters and Associated Materials. Department of the Environment, Her Majesty's Stationery Office, London.

Gale, R.S., Baskerville, R.C., 1967. Capillary suction method for determination of the filtration properties of a solid/liquid suspension. Chem. Ind. 9, 336–355.

Gale, R.S., Chem, F.I., 1977. The use of pretreatment chemicals. In: Purchas, D.B. (Ed.), Solid/ Liquid Separation Equipment Scale-Up. Upland Press Ltd, Croydon, UK, pp. 39–80.

Herwijn, A.J.M., La Heij, E.J., Jzermans, J.I., Jan Coumans, W., Kerkhof, P.J.A.M., 1995. Determination of specific cake resistance with new capillary suction time apparatus. Ind. Eng. Chem. Res. 34, 1310–1319.

Jin, B., Wilén, B.-M., Lant, P., 2004. Impacts of morphological, physical and chemical properties of sludge flocs on dewaterability of activated sludge. Chem. Eng. J. 98, 115–126.

Lee, D.J., 1994. A dynamic model of capillary suction apparatus. J. Chem. Eng. Jpn. 27, 216–221.

Lee, D.J., Hsu, Y.H., 1992. Fluid flow in capillary suction apparatus. Ind. Eng. Chem. Res. 31, 2379–2385.

Lee, D.J., Lin, W.W., 1996. Some comments of 'Filterability measured by the capillary suction time method.'. Chem. Eng. Sci. 51 (8), 1353–1354.

Meeten, G.H., Smeulders, J.B.A.F., 1995. Interpretation of filterability measured by the capillary suction time method. Chem. Eng. Sci. 50, 1273–1279.

Novak, J.T., Knocke, W.R., 1987. Discussion of specific resistance measurements: non-parabolic data. G.L. Christensen and R.I. Dick J. Environ. Eng. Div. Am. Soc. Civ. Eng. 113, 659–661.

Pagels, M., Hinkel, J.J., Willberg, D., 2012. Moving beyond the Capillary Suction Time Test, 15–17 February 2012. In: Proceedings of the SPE International Symposium and Exhibition on Formation Damage Control. Society of Petroleum Engineers, Lafayette, Louisiana (Allen, Texas, USA).

Sawalha, O., 2010. Capillary Suction Time (CST) Test: Developments in Testing Methodology and Reliability of Results. (Ph.D. thesis). University of Edinburgh, Edinburgh, Scotland. (http://www.era.lib.ed.ac.uk/bitstream/1842/4887/1/Sawalha 2011.pdf)

Sawalha, O., Scholz, M., 2007. Assessment of capillary suction time (CST) test methodologies. Environ. Technol. 28 (12), 1377–1386.

Sawalha, O., Scholz, M., 2009. Innovative enhancement of the design and precision of the capillary suction time testing device. Water Environ. Res. 81 (11), 2344–2352.

Sawalha, O., Scholz, M., 2010. Modelling the relationship between capillary suction time and specific resistance to filtration. J. Environ. Eng.-ASCE 136, 983–991.

Scholz, M., 2005. Review of recent trends in capillary suction time CST dewaterability testing research. Ind. Eng. Chem. Res. 44, 8157–8163.

Scholz, M., 2006. Revised capillary suction time (CST) test to reduce consumable costs and improve dewaterability interpretation. J. Chem. Technol. Biotechnol. 81 (3), 336–344.

Sengupta, D.K., Hamza, H.A., Hashmi, K.A., 1994. Role of cationic and anionic flocculants in the filtration of an industrial sludge. In: Mohant, P., Pickles, C., Lu, W.K. (Eds.), Proceedings of the International Symposium on Resources, Conservation and Environmental Technologies in Metallurgical Industries. CIM, Toronto pp. 357–365.

Sengupta, D.K., Kan, J., Al Taweel, A.M., Hamza, H.A., 1997. Dependence of separation properties on flocculation dynamics of kaolinite suspension. Int. J. Miner. Process. 49, 73–85.

Soehnel, O., Sura, J.M., Nemeckova, L., 1980. Study of the properties of precipitated suspensions by the CST (capillary suction time) method. Chem. Prumsyl 30 (11), 578–580.

Standard Methods, 2012. 2710G the Method for Capillary Suction Time. Twenty-second ed., Standard Methods for the Examination of Water and Wastewater. American Public Health Association, Washington DC the American Water Works Association, Denver, CO, and the Water Environment Federation, Alexandria, VA, USA. http://www.standardmethods.org/.

Tiller, F.M., Shen, Y.L., Adin, A., 1990. Capillary suction theory for rectangular cells. J. Water Pollut. Control Fed. 60, 215–220.

Unno, H., Muraiso, H., Akehata, T., 1983. Theorectical and experimental study of the factors affecting capillary suction time (CST). Water Res. 17, 149–156.

Vesilind, P.A., 1988. Capillary suction time as a fundamental measure of sludge dewaterability. J. Water Pollut. Control Fed. 60 (2), 215–220.

Wilcox, R.D., Fisk, J.V., 1983. Tests show shale behaviour, aid well planning. Oil Gas. J. 9, 106–108.

INDEX

Note: Page numbers followed by "f" and "t" indicate figures and tables respectively.

D

H

N

O

P

R

S

T

U

V

W

X